Programme book of the joint conference of ECPA – ECPLF

Centre for Agricultural Landscape and Land Use Research (ZALF)
Eberswalder Strasse 84
D-15374 Muencheberg
Germany

Institute of Agricultural Engineering Bornim (ATB)
Max-Eyth-Allee 100
D-14469 Potsdam
Germany

Programme book of the joint conference of ECPA – ECPLF

Edited by
A. Werner
A. Jarfe

Wageningen Academic
P u b l i s h e r s

CIP-data Koninklijke Bibliotheek
Den Haag

ISBN 9076998345

paperback

Subject headings:
precision agriculture
precision livestock farming
information driven agriculture

First published, 2003

Photo cover:
Andreas Jarfe

Wageningen Academic Publishers
The Netherlands, 2003

Printed in The Netherlands

Contents

Full non reviewed papers for ECPA 45

Abstracts from oral presentations for ECPA 151

11

Full poster papers for ECPA 343

17

18

20

21

Poster abstracts for ECPA 609

23

Full non reviewed papers for ECPLF

Abstracts from oral presentations for ECPLF 705

Full poster papers for ECPLF 743

Poster abstracts for ECPLF 787

Preagro Section 799

Sponsors of the 4ECPA and 1ECPLF

German Federal
Environmental Foundation

German Agricultural Society

German Research Foundation

German Engineering Federation

Society of Chemical Industry

COMITEK Bürosysteme

Walter Stauß-Stiftung

Co-organizing Institutions of the 4ECPA and 1ECPLF

International Commission
Agricultural Engineering

European Association
for Animal Production

European Society
for Agronomy

European Society of
Agricultural Engineers

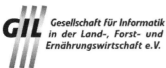

Association for Informatics in Agriculture,
Forestry and Nutrition

Society of Crop Science,
Germany

German Association of
Engineers

International Program Committee of the 4th ECPA and 1st ECPLF:

Organizers:
Prof. J. Zaske, (ATB, Germany)
Dr. A. Werner, (ZALF, Germany)

 **ATB**
Agrartechnik Bornim

Editors of the ECPA Proceedings:
Dr. John Stafford (UK); Dr. Armin Werner (Germany)

Editors of the ECPLF Proceedings:
Sidney Cox (UK)

Country representatives for ECPA:
Dr. S. Christensen, secretary (Denmark)
John Stafford, Editor (UK)
Gilles Rabatel, (France)
Dr. N. Bjugstad, (Norway)
Dr. D. Goense, (Netherlands)
Lars Thylen, (Sweden)
Dr. H. Haapala, (Finland)
Peter Juerschik, (Germany)
Prof. A. Fekete, (Hungary)
Dr. Margaret Oliver, (UK)
Prof. J. de Baerdemaeker, (Belgium)
Dr. E. Molto, (Spain)
Jürgen Zaske, chairman (Germany)

Country representatives for ECPLF:
A. R. Frost, (UK)
R. Kaufmann, (Switzerland)
K. Sallvik, (Sweden)
Prof. Dr. Ir. Daniel Berckmans, (Belgium)
Dr. Ir. Cees van't Klooster, (Netherlands)
Dr. Soeren Pedersen, (Denmark)

Program Committee

Prof. Juergen Zaske,	Institute of Agricultural Engineering Bornim (ATB), Potsdam-Bornim
Dr. Armin Werner,	Centre for Agricultural Landscape and Land Use Research, (ZALF), Müncheberg
Prof. Herman Van den Weghe,	Georg-August-University Göttingen, Vechta
Prof. Hermann Auernhammer,	Weihenstephan Center of Life and Food Sciences, Freising-Weihenstephan
Prof. Eberhard von Borell,	Martin Luther University Halle, Halle/Saale
Dr. Jens-Peter Ratschow,	Association of Engineers, Münster
Dr. Detlef Ehlert	Institute of Agricultural Engineering Bornim (ATB), Potsdam-Bornim
Dr. Reiner Bruinsch	Institute of Agricultural Engineering Bornim (ATB), Potsdam-Bornim
Dr. Andreas Fischer	Centre for Agricultural Landscape and Land Use Research, (ZALF), Müncheberg

Conference Office of the 4ECPA and 1ECPLF:

Andreas Jarfe	Centre for Agricultural Landscape and Land Use Research, (ZALF), Müncheberg

Scientific Panel of the 4th European Conference on Precision Agriculture

Louis Assemat	France	Martin Heide Jørgensen	Denmark
Herman Auernhammer	Germany	Peter Juerschik	Germany
John Bailey	UK	Christian Kersebaum	Germany
José Benlloch	Spain	Newell Kitchen	USA
Jaap van Bergeijk	Netherlands	Herman Kutzbach	Germany
Bill Batchelor	USA	Jürgen Lamp	Germany
Nils Bjugstad	Norway	Murray Lark	UK
Simon Blackmore	Denmark	Kees Lokhorst	Netherlands
Ian Bradley	UK	Dan Long	USA
Rosie Bryson	UK	Jess Lowenberg-Deboe	USA
Don Bullock	USA	John Marchant	UK
Svend Christensen	Denmark	Alex McBratney	Australia
Tom Colvin	USA	Paul Miller	UK
John Conway	UK	Enrique Molto	Spain
Simon Cook	Australia	David Mulla	USA
Rob Corner	Australia	Mike O'Dogherty	UK
Trevor Cumby	UK	Svend Elsnab Olesen	Denmark
Stan Daberkow	USA	Margaret Oliver	UK
Peter Dampney	UK	Bernard Panneton	Canada
Markus Demmel	Germany	Fran Pierce	USA
Anita Dille	USA	Lisa Rew	USA
Horst Domsch	Germany	Pierre Robert	USA
Detlef Ehlert	Germany	John Sadler	USA
Mike Ellsbury	USA	John Schueller	USA
Andras Fekete	Hungary	Thomas Selige	Germany
Richard Ferguson	USA	Martin Smith	UK
César Fernández-Quintanilla	Spain	Henning Sogaard	Denmark
Zoe Frogbrook	UK	Mike Steven	UK
Roland Gerhards	Germany	Marvin Stone	USA
Gilbert Grenier	France	Ken Sudduth	USA
Ken Giles	USA	Lars Thylén	Sweden
Richard Godwin	UK	Bruno Tisseyre	France
Daan Goense	Netherlands	Raphael Viscarra Rossel	Australia
Gillian Goodlass	UK	Peter Wagner	Germany
Hannu Haapala	Finland	Richard Webster	UK
Dale Heerman	USA	Ole Wendroth	Germany
John Hummel	USA	Brett Whelan	Australia
Chris Johannsen	USA	Gavin Wood	UK

Scientific Panel of the 1st European Conference on Precision Livestock Farming

Herman Auernhammer	Germany
John Bailey	UK
Franz Josef Bokisch	Germany
Reiner Brunch	Germany
Sidney Cox	UK
Richard Dewhurst	UK
David Filmer	UK
Andy Frost	UK
Rony Geers	Belgium
David Givens	UK
Henk Hogeveen	Netherlands
Pieter Hogewerf	Netherlands
Anders Jonsson	Sweden
Thomas Jungbluth	Germany
Peter Kettlewell	UK
Joachim Krieter	Germany
Jos Metz	Netherlands
Toby Mottram	UK
Dieter W. Ordolff	Germany
David Parsons	UK
Wim Rossing	Netherlands
Lars Schrader	Germany
Bart Sonck	Belgium
Jack Whittier	USA
Gavin Wood	UK

Reviewers of the 4th European Conference on Precision Agriculture

Ulisses Antuniassi
Hermann Auernhammer
John S. Bailey
Bent S Benedsen
José Benlloch
Michel Berducat
Bachelor Bill
Ralf Bill
T.F.A. Bishop
Nils Bjugstad
Simon Blackmore
Greg Blumhoff
Daniel Boffety
S.C. Borgelt
Stefan Böttinger
Rob Bramley
Basso Bruno
Rosie Bryson
D.G. Bullock
Svend Christensen
TS Colvin
Robert Corner
Trevor Cumby
Michel Dabas
Peter Dampney
Josse De Baerdemaeker
Markus Demmel
Bernd Dohmen
Detlef Ehlert
Mickael Ellsbury
Andras Fekete
Brian Finney
Zoe Frogbrook
Roland Gerhards
Grenier Gilbert
Daan Goense
Simon Griffin
Martine Guerif
Hannu Haapala
Pat Haydock
Dale F. Heermann
Bernd Hohnermeier
John Hummel
Edmund E. Isensee
Keith Jaggard
C. Johannsen

Peter Jürschik
F Juste
Kurt-Christian Kersebaum
Florian Kloepfer
Tadeusz T. Kuczynski
Jürgen Lamp
O. Lavialle
D. Lepoutre
J.A. Marchant
Clouaire Roger Martin
Gaines Miles
Enrique Molto
David Morris
Thomas Mueller
Thomas Muhr
David J. Mulla
Hervé Nicolas
Anna Nyberg
Margaret Oliver
F.J. Pierce
Gilles Rabatel
Pierre C. Robert
E.J. Sadler
Antonio Saraiva
Urs Schmidhalter
John K. Schueller
Reinhart Schwaiberger
Thomas Selige
Keith Smettem
Claus Sommer
John Stafford
Marvin.L. Stone
K.A. Sudduth
Lars Thylen
Nick Tillett
Bruno Tisseyre
Raphael Viscarra Rossel
Els Vrindts
Peter Wagner
Karl-Otto Wenkel
Dwayne Westfall
B. Whelan
Wilhelm Windhorst
Gavin Wood
Douglas Young

Reviewers of the 1st European Conference on Precision Livestock Farming

Thomas T. Amon
Daniel Berckmans
Franz-Josef Bockisch
Reiner Brunsch
Wolfgang Büscher
Andreas Fischer
Gates
Jacobson
Anders Jonsson
Thomas Jungbluth
Otto O. Kaufmann

Joachim Krieter
Tadeusz T. Kuczynski
F. Madec
Jos J. Metz
Soren Marcus Pedersen
Bart Sonck
Herman Van den Weghe
C.E. Van´t Klooster
J.A.M. Voermans
Eberhard von Borell
Georg Wendl

Welcome

in Berlin to the two joint conferences:
4th European Conference on Precision Agriculture (4ECPA)
and the
1st European Conference on Precision Livestock Farming (1ECPLF)

In the tradition of the conference-series that started in Warwick in 1997, an international committee prepared this pair of scientific conference in Berlin for you. We are convinced, that the conferences will offer a unique insight into the actual research and development in precision agriculture and precision livestock farming. Our hopes are, that the conference will offer unrestricted impulses, places and time for vivid discussions and exchange of interesting results, new hypotheses or ideas and visions for a growing field of agricultural sciences.

With this 'booklet' we provide a collection of the abstracts for the conference participants to have access to all oral papers or posters that should be presented during the conferences. The booklet therefore has not the character of conference proceedings. These can be found in two separate books[1]. This booklet is meant to have primarily the character of a program-book to the conference, similar to the program-booklet of a great and long concert. The participants of the conference can use this booklet to have some information at hand to find enough orientation in the full score of the scientific parts of the conference.

The booklet also contains longer papers of presentations, that were not included in the conference proceedings.

Many helpful minds and hands were necessary to prepare this conference and to make it to a success for science, research and the transfer of new knowledge. We owe our greatest thank to all those people who were convinced of the necessity to have this meeting and who kept their fate in the time of almost two years preparation of the conference. Many of them stayed behind the scenes, but were important with every help they gave.

Extensive financial support was supplied by many public and private organisations, companies and people. Without them a conference like this is not possible. We are very grateful that all supporters did not hesitate at all to offer their help when we addressed them. Thank you very much. The list of these contributors is provided below.

The conferences were prepared by a program committee and jointly organized by the Centre for Agricultural Landscape and Land Use Research (ZALF), Müncheberg, and the Institute for Agricultural Engineering Bornim (ATB), Potsdam, both in Germany. Chief coordinator of the conferences was Andreas Jarfe from ZALF, who perfectly managed the very complex preparation of the conferences and finally their realization.

Müncheberg and Potsdam-Bornim, May 2003
The organizers of the 4ECPA, 1ECPLF

---

[1] a) Precision Agriculture. Proceedings of the 4th European Conference on Precision Agriculture (4ECPA)
   b) Precision Livestock Farming. Proceedings of the 1st European Conference on Precision Livestock Farming (1ECPLF)

4th European Conference on Precision Agriculture (4ECPA)
1st European Conference on Precision Livestock Farming (1ECPLF)

An integrated approach in research and development to design the agriculture of the future: *"Precision agriculture meets precision livestock farming"*

Editorial

The European Conference on Precision Agriculture 2003 in Berlin is the fourth in a sequel of biennial conferences. The last one took place in Montpellier in June 2001.In the conference from June 15-19 in 2003 we expect 400 to 500 participants from all over the world. The participants will be scientists, representatives from agricultural industry, from public administrative organisations in agriculture, and those from service companies. Also personnel from extension services as well as farmers registered.

Up to the last one, the conference was solely oriented towards arable cropping (precision agriculture: ECPA). From this year on the first European Precision Livestock Farming Conference (ECPLF) will start in addition. In Berlin the two conferences will take part in parallel and will have joint plenary sessions with topics relevant for both precision productions in agriculture.

By combining the topics of precision agriculture and precision livestock farming in parallel conferences, the links between both technologies are drawn and the possible interactions between them will be shown for the first time. With this first step, the sciences of precision farming (PA), which is the more field-oriented plus the sciences of precision livestock farming (PLF), which is more oriented on in-house husbandry will evolve to a common science of a full *precision agriculture*, which defines the whole farm as its system borders. Many aspects of economic or ecological-impact analysis from agricultural activities on fields or with the livestock can only be determined when defining a joint system-border. Thus limiting the scientific research to one of these strands is mostly not feasible. At least it would result in too limited answers to actual scientific and societal questions.

The two conferences therefore will follow congruent goals:

4th ECPA   to present and discuss the state of the art in research and development for precision agriculture

1st ECPLF   to analyse the state of precision livestock farming components and their integration from the perspectives of science and application

The common objectives of the two conferences are:

to identify relevant scientific problems, present latest methods and results of research and development, define research demands and requirements for standardisation as well as proposing and studying suitable systems for technology transfer into practical farming.

Müncheberg and Potsdam-Bornim, May 2003
Armin Werner
Jürgen Zaske
(Chairpersons of the 4ECPA, 1ECPLF)

Preface

Scientific work on all aspects of *precision agriculture* (PA) is a research field that currently shows a growing awareness in agricultural and related sciences. This is a not only a result of the recent technical developments in cheap systems for identifying positions in the fields (Global Positioning Systems, GPS), powerful PC-based geographic information systems (GIS), bus-systems for communications in tractors or equipment and sensors to identify states in crops and soils. Much interest on precision agriculture research is related to the actual demands of modern crop management. These demands are dominated by the economic pressure to increase production efficiency and are additionally enhanced by the societal requests to control the environmental impacts of agricultural land use and to increase the transparency in the agricultural production processes. To enhance the decision making processes in crop production by having almost unlimited access in the future to information on states and processes in soils and crops becomes the most important option to achieve the described demands. Thus the arable crop production and presumably also the grassland management actually are in the transition towards INFORMATION DRIVEN PRODUCTION systems

Research in the area of *precision livestock farming* (PLF) is not that much well developed yet as a clearly perceptible field of similar interests and goals. This contrasts to the situation, that in the livestock sciences the research and development on information based production systems is already much longer an issue than in crop production sciences. However to have the notion or a closed research field of a precision livestock farming seems less relevant to the livestock sciences. But all efforts that are related to an information-driven livestock production should get the chance to concentrate and find synergistic moments in transparent and sometimes co-ordinated research activities.

It is obvious, that both precision technologies and their development share same goals, as well as similar scientific methods and scientific principles in studying the relevant processes. Many aspects in scientific research of PA and PLF are related almost identically to (see also Cox 2003[2]):

i.   systems research and modelling of complex systems,
ii.   explaining process performance by complex sets of interacting driving forces,
iii.   variability in space and time as a basic feature of data and information,
iv.   developing sensors to assess states or processes within the production systems and designing tools to derive potentials and decisions from the sensor signals and historical information about the system,
v.   designing concise data models and minimum datasets in information processing,
vi.   enhancing the traditional basic research data of process understanding with evidence based information or from on-farm research,
vii. decision making with complex but limited information under uncertainties,
viii. information management on different complexity levels in the farm: from the separately distinguishable object in the process-understanding and thus production management (sub-units in the field, single animal in the flock) over object-aggregates (fields, crop rotations, flocks, herds) to whole farms.

Relevant scientific problems in the area of crop production and livestock farming can be found in the understanding of the states, of the development and of the control for heterogeneous crop stands or herds and thus yield or product-quality. Substantial interest in research is related to the assessment of the impact of different production management options and frame conditions onto relevant states and impacts of crop stands or herds on heterogeneous fields or living conditions.

---

[2] Precision Livestock Farming '01. Proceedings of the 1st European Conference on Precision Livestock Farming (1ECPLF): Editorial

Agriculturally used fields are structures that are composed from highly interacting geophysical, geochemical and biological systems. These systems seldom show spatial homogeneity in their characteristics relevant to crop production and its impacts. The systems can either physically be distinguished into elements (objects) or into virtual structures that are defined by non-physical system boundaries like management structures, administrative borders etc. The overall characteristics of fields are the results of the endogenous processes of the single systems and of the lateral processes between these systems and their elements. Similar aspects can be addressed with herds or flocks in livestock production.

Such interacting systems show very high complexities. Understanding the relevant impacts of specific management measures in agricultural production is only possible, when the complex interactions between different driving forces are known and can be explained or quantified. In addition, these interactions vary in space and time. To deal with these complexities on a scientific way is only feasible and can only be done, when the analysing scientific disciplines try to interact with other disciplines, relevant to the system understanding. Research for precision agriculture and precision livestock farming has primarily interdisciplinary characteristics and can be distinguished by this from research fields that are covered by single disciplines. An integrative shell for research on precision agriculture and on precision livestock farming is the common intention to support the decision making processes of farmers. To achieve this, new information systems are necessary and thus the modelling part. Process oriented simulation models, state-models or meta-models and general information processing models will be developed to understand, analyse and assess the states and dynamics of heterogeneous sites, crop stands or herds. Such information related systems and their implementation in practical farming should be the final goal for all research activities in the fields of information driven agricultural production.

Thus the research for precision agriculture and for precision livestock farming has some specificities. The focus of research for precision agriculture and for precision livestock farming is oriented towards (i) understanding the relevant systems, (ii) their interactions that are influenced by management measures, (iii) the impact of this production systems onto relevant states and functions within the farm or within the landscape and (iv) describing the relevant processes in the form of models of various kind. In addition, the research for precision agriculture and for precision livestock farming is looking into actual and developing production systems as options and feasible solutions that help to manage the agricultural production.

These alternatives and their impact assessment are necessary in the process of finding those production systems in agriculture that fit best to the actual and probably also future demands of the producers and the society. In this context the principles of precision agriculture and precision livestock farming are the key-technologies for a sustainable development of agriculturally land use. It is based on new information-, sensor- and control-technologies. They allow for the first time to implement economic as well as ecological goals in the production systems in an efficient and transparent way. The technologies of ´precise agriculture´ are based on the joint, but specialized knowledge of various agricultural disciplines. They have to collect and combine data and information from multiple sources in an intelligent way to carry on decisions associated with agricultural production, logistics and also with marketing.

For the involved agricultural scientific disciplines the growing field of research in precision agriculture and precision livestock farming offers an unprecedented step towards a new quality of science. The related scientific, technical and practical problems can only be solved, when several disciplines work together or at least share their experiences on common grounds or objects. Very complex systems and their processes have to be analysed and explained, new scientific methods have to be developed or adopted and data have to be gained as base for methods, algorithms or models for decision support and impact assessment.

In this way agricultural research will come to its best: a strongly interdisciplinary field of academic activities that work with excellent science on real world problems. Thus agricultural sciences returns to its roots: being an extraordinary kind of art. It is an interdisciplinary science that steadily expands with new scientific fields, that continues to be dynamic, inspiring, demanding and that deals with almost unlimited scientific and real world problems. An interesting and challenging perspective for all participating scientists, researcher, developers and extension people. The farmers, the environment and the society will benefit from taking up that call.

Armin Werner

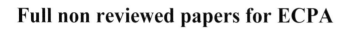

Full non reviewed papers for ECPA

Analysis and theoretical estimation of the errors of the agriculture crop mapping

P. Cârdei, V. Muraru, V. Gângu and I. Pirna
INMA - Bucharest, Bd. Ion Ionescu dela Brad No. 6, Romania

Abstract

The authors try to find relationships between the final error to the combine harvester performing the production mapping and the errors of the process parameters: displacement speed, working width, flow, and transfer through the harvester, supposed to be known. The experimental estimation of the mapping errors are practically impossible, on one hand due to the necessity of destroying the real information if we wish to have knowledge on the specific crop production before mapping and on the other hand, due to the operation unrepeatability because after a mapping the harvest is processed, the repeatability in identical crop conditions being excluded. For these reasons our estimation are pure theoretical and are not based on the experimental data.

Keywords: errors, estimation, crop, mapping

Introduction

The mathematical model is based on the model which is given in [1]. The following functions which describe the model:

$YI=YI(x,y)$, is the real crop production, kg/m^2;
$YI_{rep}=YI_{rep}(x,y)$, is the reported crop production, kg/m^2;
$SP=SP(x,y)$, is the advancing speed , m/s;
$SPY=SPY(x,y)$, is the lateral speed, m/s;
$AW=Aw(x,y)$, is the current cutting width, m;
l_0 is the clean cutting, m;
dr, distance between two consecutive reports or reporting distance, m;
$FR_{in}=FR_{in}(x,y)$, is the inlet flow rate of grains, kg/s;
$FR_{out}=FR_{out}(x,y)$, is outlet flow rate of grains, kg/s;
t, is the time, s.

The functions which describe the model depend on the coordinates x, y, but the coordinates x, y are variable in time, t. Therefore, the model functions depend on the time. The relations between the location coordinates and the time variable are:

$$x = x(t) = x_0 + \int_0^t SP(s)ds, \quad y = y(t) = y_0 + \int_0^t SPY(s)ds,$$

(1)

where s are the current time, $s \in [t_0, t]$. The basic relations given by [1] are natural one:

$$FR_{in}(t) = YI(t) \cdot SP(t) \cdot AW(t),$$

(2)

$$FR_{out}(t) = G(FR_{in}(t)),$$

(3)

where G is a function which can be experimentally obtained.

In the next consideration we using the decomposition of a function in mean value and deviation on an interval $[a, b]$ of real numbers. If $f:[a, b] \rightarrow \mathbf{R}$ is a function, then the mean value, f_{med} and the deviation function, ε_f, for the function f, are :

$$f_{med} = \frac{1}{b-a} \int_a^b f(u)du, \quad \varepsilon_f(u) = f(u) - f_{med}, u \in [a,b]$$

(4)

Then the function f, have the next form:

$$f(u) = f_{med} + \varepsilon_f(u) . \tag{5}$$

This decomposition is used for the estimation of the local error of the crop production. We accept the decomposition in mean value and deviation of the advancing speed, the lateral speed, the current cutting width and the crop production :

$$SP = SP_{med} + \varepsilon_{SP}, SPY = SPY_{med} + \varepsilon_{SPY}, \quad AW = AW_{med} + \varepsilon_{AW}, \quad YI = YI_{med} + \varepsilon_{YI}. \tag{6}$$

Now we introduce the hypothesis that the function G is linear:

$$FR_{out}(t) = G(FR_{int}(t)) = aFR_{int}(t) + \varepsilon(t) \tag{7}$$

In this conditions, from (7) and (2) we have:

$$YI(t) = \frac{FR_{out}(t) - \varepsilon(t)}{a \cdot SP(t) \cdot AW(t)} . \tag{8}$$

Finally, we accept that the reported crop production is give by the outlet flow rate of grains divided by the product between the mean value of the current cutting width and the mean of the advancing speed:

$$YI_{rep}(t) = \frac{FR_{out}(t)}{AW_{med} \cdot SP_{med}} = \frac{a \cdot YI(t) \cdot SP(t) \cdot AW(t) + \varepsilon(t)}{AW_{med} \cdot SP_{med}} . \tag{9}$$

The estimation of the dependence of the reported crop production error on the crop production error, advancing speed error and the current width error

The reported crop production error is defined by the equation:

$$\Delta YI(x, y) = YI(x, y) - YI_{rep}(x, y) . \tag{10}$$

If in the equation (10) we introduce the reported crop production from the relation (9), and we use the relations (6) for the advancing speed and the current width, we obtain the relation (11) for the reported crop production error.

$$\Delta YI(x, y) = \left[1 - a \left(1 + \frac{\varepsilon_{SP}(x, y)}{SP_{med}} \right) \cdot \left(1 + \frac{\varepsilon_{AW}(x, y)}{AW_{med}} \right) - \frac{\varepsilon(x, y)}{SP_{med} \cdot AW_{med}} \right]. \tag{11}$$

On (11), increasing we obtain the inequality.

$$|\Delta YI(x, y)| \le |YI(x, y)| \cdot$$

$$\left\{ |1 - a| + |a| \cdot \left(\frac{|\varepsilon_{SP}(x, y)|}{SP_{med}} + \frac{|\varepsilon_{AW}(x, y)|}{AW_{med}} + \frac{|\varepsilon_{SP}(x, y)| \cdot |\varepsilon_{AW}(x, y)|}{SP_{med} \cdot AW_{med}} \right) + \frac{|\varepsilon(x, y)|}{SP_{med} \cdot AW_{med}} \right\}. \tag{12}$$

If the advancing speed error and the current width error satisfy the condition :

$$|\varepsilon_{SP}(x, y)| \le k_{SP} \cdot SP_{med}, |\varepsilon_{AW}(x, y)| \le k_{AW} \cdot AW_{med}, \tag{13}$$

then we have the approximation (14) for the reported crop production error.

$$|\Delta YI(x, y)| \le |YI(x, y)| \cdot \left[|1 - a| + |a| \cdot (k_{SP} + k_{AW} + k_{SP} \cdot k_{AW}) + \frac{|\varepsilon(x, y)|}{SP_{med} \cdot AW_{med}} \right]. \tag{14}$$

If $a = 1$ and $\varepsilon(x, y) = 0$, then the reported crop production error satisfy the inequality:

$$|\Delta YI(x, y)| \le (k_{SP} + k_{AW} + k_{SP} \cdot k_{AW}) \cdot \max_{(x, y) \in D} (YI(x, y)) , \tag{15}$$

where D is the domain of the location variable (x, y) of the reference point of the combine and the last factor in the right term is the maximum of the real crop production.

Another result assume that if the advancing speed error and the current width error are small comparison with their mean values, then the report between the influence of the current width error and the influence of the advancing speed error is approximately equal with the report between the mean value of the advancing speed and the mean value of the current width.

The estimation of the map crop production error through numerical experiment (simulation)

On the analytical way we show that the mapping error of the crop production can be superior limited. We consider in this paper only the errors of the current width generated by the error of the combine displacement, the advancing and lateral speed. The relation between the coordinates of the combine and the error of the advancing speed are related through the relations (1). The deviation in direction of the combine produces an error of the current width, given by the formula (16).

$$\varepsilon_{AW}(t) = -l_0 \frac{SPY(t)}{\sqrt{SP^2(t) + SPY^2(t)}} \tag{16}$$

Obviously, if the lateral speed, SPY, is null, then the current width is so null.

Another factor which introduces errors in the mapping of the crop production is the distance between two reported values running for the reported crop production, short namely report distance. The report distance may keep constant through measuring here on the combine wheel, but on this way are keeping the direction errors.

Finally, the list of the errors which we consider is: the error of the advancing speed, ε_{SP}, the error of the lateral speed, ε_{SPY}, the error of the cutting width, ε_{AW}, the error between the inlet and outlet flow grains, ε, and the error introduced by the distance between two consecutive reports. The analytical estimation of the global error caused by these errors is very difficult because their random character. For this reason we try a numerical way in order to estimation the effect of these errors in the final report of the crop production. The pure experimental way for the estimation of the errors in the crop production mapping can be use only for a limited control through the harvest the crop before the mapping operation. On the numerical experiment we can known exactly the crop production, and so this crop production is reproducible. Thus, for the same crop production we can apply different crop production mapping processing. On this way we vary the process parameters (the mean values and the errors) and simulate the mapping process. At the final is calculate the statistic deviation and others parameters which give the mapping error. Generally, the crop production mapping can be seeing like a geometrical transform of the real crop production (the density of the crop production, YI). The result of this transform is the mapping crop production (the density o the crop production YI_{rep}). In other words, the mapping crop production is a deformation of the real crop production into the reported crop production. The main problem is to choose the work process parameters which produce a minimum deformation: null or small displacement of the extreme of the real crop production, a minimum deviation of the mapping crop production like the real crop production.

Result of the numerical experiments

We simulate a mapping process on a rectangular field, 120 x 60 m. The mean advancing speed have the value 3 m/s and the clean cutting length, l_0= 6 m. In figure 1 is show the real crop production (in the left) and the mapping crop production (in the right), obtained for a reporting distance, dr=1 m.

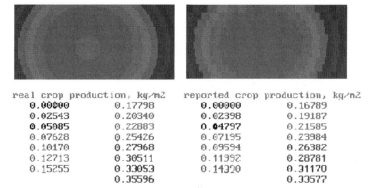

```
real crop production, kg/m2       reported crop production, kg/m2
   0.00000      0.17798              0.00000       0.16789
   0.02543      0.20340              0.02398       0.19187
   0.05085      0.22883              0.04797       0.21585
   0.07628      0.25426              0.07195       0.23984
   0.10170      0.27968              0.09594       0.26382
   0.12713      0.30511              0.11992       0.28781
   0.15255      0.33053              0.14390       0.31170
                0.35596                            0.33577
```

Figure 1. The real crop production (left) and the mapping crop production (right), obtained for $dr= 1$ m .

In the figure 2 are give the maps of the crop production, obtained for values 5, 10 and 15 m of the distance between two consecutive reports, dr, for the same real crop production.

Figure 2. Maps of the mapping crop production for $dr= 5$m (left), $dr= 10$ m (middle), $dr=15$ m (right).

Conclusions

C1) The map of the crop production, obtain through the mapping of the crop production, is different on the real map of the crop production. The resolution of the map obtain through the mapping of the crop production is limited by the distance between two consecutive reports, dr. The mapping crop production, more exactly, the density of the crop production, YI_{rep}, is calculate like mean value of the crop production on the rectangle which have length dr and width the length of the clean cutting, l_0;

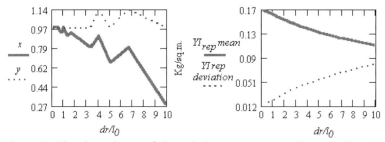

Figure 3. The dependence of the relative extreme coordinate (left) and of the mean and deviation crop production (right), on the report between the reporting distance and the clean cutting length.

C2) The map obtained by mapping crop production will be better as the reporting distance and the working width are less (see figure 3);

C3) The reporting distance must be more little than maximum diameter of the deficient zone which we wait to find;

C4) If the reporting distance is sufficiently small (under 10% on the length of the mapping field) and the advancing speed error under 2 % on the mean advancing speed, then the coordinates of the minim of the mapping crop production, practically coincide with the real crop production (see figure 3);

C5) If the reporting distance is great than 15% of the length of the crop field, or the advancing speed error exceeds 2% on the mean value of the advancing speed is possible to appear displacement of the extreme of the crop production with values of the order of one or two reporting distance. If there is so error of the displacement direction, then the deformation of the real map of the crop production become inadmissible. Thus result a wrong map of the crop production, i.e. the deficit zone of the field is moved at distance on the order of many reporting distances.

References

[1] R. Vansichen, J. De Baerdemaeker, Processing of Signals and Dynamic System when Countinously Measuring the Production Performed on Harvesting Machines, Ag Eng '92, June 1-4, Uppsala, Sweden;
[2] Obtained Data in the Process of Grain Combining, 1998.

Estimation of nitrogen content and prediction of yield using remote sensing technology for rice paddy

Ryu Chanseok, Hisaya Hamada, Suguri Masahiko, Mikio Umeda and Tatsuya Inamura
Graduate School of Agriculture, Kyoto University, Sakyo-ku, Kyoto, 606-8502 Japan

Abstract

In order to decide the amount of fertilizer for variable rate application, it is necessary to detect the nitrogen contents of rice plant at panicle initiation stage. The field B was used for making calibration equation to detect the nitrogen content based on remote sensing technology. The difference between real and predicted nitrogen content of the field B using vegetation indices is less than 0.1mg. With this result, it is possible to predict the nitrogen contents of rice plant at the field A. The difference between the real and predicted nitrogen of the field A was about 2mg. It was, however, difficult to predict yield of the field A using the vegetation indices of the field B.

Keywords: remote sensing, vegetation indices, nitrogen content.

Introduction

For the purpose of deciding the amount of fertilizer for variable rate application, it is necessary to detect the nitrogen contents of rice plant at panicle initiation stage. Normally, the nitrogen contents of rice plant at panicle initiation stage were investigated by chemical analysis. Recently, remote sensing technology is widely used for predicting crop condition and yield. Remotely sensed reflectance data, obtained either by satellite or aircraft, can provide a set of detailed, spatially distributed data on plant growth and development (Plant et al., 1999). The major limitation of the chlorophyll meter in determining crop nitrogen status is that readings are generally taken at a limited number of locations in the field, and the results often provide little information concerning the nitrogen status across the entire field. As such, chlorophyll meters, by themselves, are not practical tolls to characterize crop nitrogen status on a whole-field basis. The SPAD data (by chlorophyll meter) is compared with each of vegetation indices calculated by SPOT satellite images, Emerge and ADAR serial images (Han et al., 2002). NDVI integrated over time showed a significant correlation with lint yield in those experiments in which there was a significant stress effect on yield. The spatiotemporal pattern of NDVI reflected stress factors and was approximately coincident with the onset of measurable water stress (Plant et al., 2002). Unfortunately, the data gathered in this field of research using remote sensing technology for paddy field is insufficient. The coefficient of determinant (COD) between the nitrogen content and GNDVI, and COD between LAI and vegetation coverage ratio were investigated by machine vision system with extremely nitrogen changed depending on plots (Suguri et al., 2001).

In this research, the standard plots were used for making calibration equation for the field A (0.5ha). The objects of this research are 1) to predict nitrogen content and 2) to predict yield using vegetation indices of the field B (0.04ha).

Materials and methods

Experimental field

The experimental farm of Kyoto University located in Takatsuki, Osaka prefecture (135°38'E, 34°51'N, 10m above sea level). The soil of the test field is classified as Typic Fluvaquent or Gary Lowland soil. The tested crop was Oryza sativa L., cv. MINAMI-HIKARI, a high-yield variety. The field A examined is almost flat (0.07%) and has an area of 0.5 ha with 100 sampling plots. The field B is located at northern part of the field A. Fig. 1 shows virtually divided plots at field A and B. The field B, to make a model of nitrogen contents and yield, has 10 plots and each plot is 8.5m×4.5m.

Plant management

Plots of the field A and B were investigated every sampling term with SPAD, plant length and the number of tillers. The samples of the field B except roots were dried, weighted and ground into powder in order to measure the nitrogen content by chemical analysis. For the purpose of getting a wide range of data, plots were applied three different amounts of fertilizer as basal dressing and topdressing applications (0, 30, 60 kgN/ha). Table 1 shows investigating data, items and field management of the field A and B. Field A was applied 30kgN/ha and 30kgP/ha fertilizer as basal dressing and 0~50kgN/ha as topdressing depending on nitrogen content of rice plant. Otherwise, field B was applied amount of fertilizer as shown Figure 1. For the example, b3 plot was 30kgN/ha as basal dressing and 60kgN/kg as topdressing.

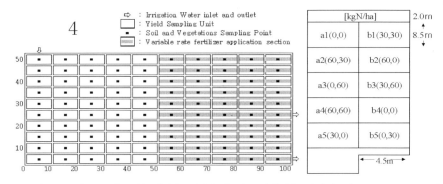

Figure 1 The experimental paddy field A (100 plots) and field B (10plots).

Images

In order to apply remote sensing technology to the rice paddy, satellite images were taken by SPOT5 and aerial images were taken by Airborne Digital Sensor (ADS40, LH system) at panicle initiation stage, heading stage, and ripening stage. Resolution of SPOT5 is 10m/pixel with 3 bands (R, G, NIR) and ADS40 is 0.2m/pixel with 4 bands (R, G, B, NIR). The reflectance panel was set up to calculate the reflectance of bands.

Table 1. Investigation data, items and field management at field A and B.

Item and management	Field A	Field B	Image
Seedling	May 11	May 15	
Basel dressing	May 30	Jun 12	
Transplanting	June 9	June 19	
Maximum tiller number stage	July 25	July 20	
Panicle initiation stage	Aug. 7	Aug. 6	Aug 5 (SPOT5, ADS40)
Top dressing	Aug. 10	Aug. 9	
Meiosis stage	Aug. 25		
Heading stage	Sept. 8	Sept. 6	Aug 26 (SPOT5) Sep 2 (ADS40)
Ripening stage	Oct. 17		Oct 11 (SPOT5) Oct 24 (ADS40)
Yield	Oct 30	Oct. 23	

Results and discussion

Satellite images were not useful to detect nitrogen content of plant at 0.5ha of paddy field. The reason is that our experimental field is too small compared with resolution of SPOT5. However, it is useful to compare spatial variability of wider region of field or more than one field simultaneously. The vegetation indices were compared to assume the nitrogen contents of the field A. The other properties, for the examples, SPAD, plant length, the number of tillers, dry weight, LAI were also compared with the vegetation indices of images and the nitrogen contents of field A. Normally, vegetation indices in late season has high relationship with SPAD, plant length, the number of tillers, dry weight, LAI and the nitrogen content of plant. Figure 2 shows the relationship between the nitrogen content and Green NDVI $\{(R_{NIR}-R_{Green})/(R_{NIR}+R_{Green})\}$ at panicle initiation stage. Figure 3 shows the relationship between the nitrogen content and Green NDVI at heading stage.

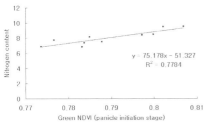
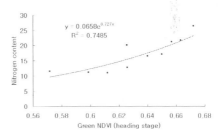

Figure. 2. Relationship between the nitrogen contents and Green NDVI of the field B.

Figure 3. Relationship between the nitrogen contents and Green NDVI of the field B.

The relationships between the yield and vegetation indices, however, have a tendency to depend on the season. At the panicle initiation stage, the relationship between the yield and Green NDVI is less than $R^2=0.1$ as shown in figure 4. The reason for low relationship is the panicle initiation stage is just before topdressing. Figure 5 shows the relationship between the yield and Green NDVI at the heading stage.

It is possible to improve the relationship between the nitrogen contents and vegetation indices multiple by LAI (leaf area index) of the field B. In the previous research, LAI had high correlation with the vegetation coverage ratio (VCR). Figure 6 shows the relationship between the nitrogen contents and LAI multiplied by Green NDVI at the panicle initiation stage of the field B. The NG (R_{NIR}/R_{Green}) value is high correlation with LAI (R^2=0.918), so the nitrogen content of field B was estimated as shown in Figure 7.

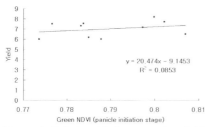

Figure 4. Relationship between the yield and Green NDVI of the field B.

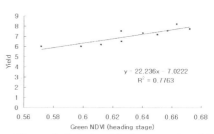

Figure 5. Relationship between the yield and Green NDVI of the field B.

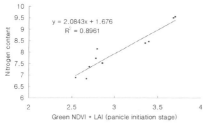

Figure 6. Relationship between the nitrogen contents and LAI multiplied by Green of the field B.

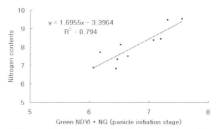

Figure 7. Relationship between the nitrogen contents and NG multiplied by Green NDVI NDVI of the field B.

As the result of this, it is possible to predict the nitrogen contents of the field A using the data of field B. The difference of the real nitrogen contents of the field B is less than 0.67mg. Fig. 8 shows the relationship between real and predicted nitrogen contents of the field A at panicle initiation stage.

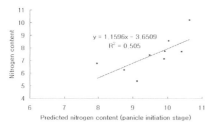

Figure 8. Relationship between of real and predicted nitrogen contents of the field A.

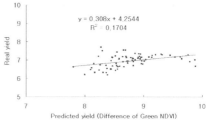

Figure 9. Relationship between real and predicted the yield of the field A.

The difference between the real and predicted nitrogen contents of the field A is about 2mg. The reason for the difference of the nitrogen content might be one week of difference between samples of the field A and field B. On the other hand, it is difficult to predict the yield of the field A based on the equation calculated using vegetation indices of field B. One of the reasons is that the variation of yield of the field A is less than 5%. Fig. 9 shows the relationship between real and predicted yield of the field A. Unfortunately, it was not useful method to predict the yield of the field A based on the data of the field B. It is also difficult to detect a good correlation between the yield and vegetation indices using only the data of field A. The correlation is less than r=0.59.

Conclusions

In order to estimate the nitrogen content of plant and predict yield, the field B was used for making calibration equation. It is possible to estimate nitrogen contents of rice plant at the field A using the vegetation indices of the field B at panicle initiation stage. However, it is difficult to predict yield of the field A based on the vegetation indices of the field B. The reasons for low relationship with yield prediction were low variation yield of field A and vegetation indices that was influenced by canopy coverage ratio, background soil and water reflectance and so on. To make confirm cause and effect, experiment also has been continued.

Acknowledgements

The authors are very grateful to Dr. Tsutomu Matsui, Kyoto University, for his information and discussion of field information of the rice.

References

Haboudane, D., Miller, J., Tremblay, N., Zarco-Tejada, P., Dextraze, L. Integrated narrow-band vegetation indices for prediction of crop chlorophyll content for application to precision agriculture. 2002. Remote Sensing of Environment 81: 416-426.

Han, S., Hendrickson, L. L., Ni, B. 2002. Comparison of satellite and aerial imagery for detecting leaf chlorophyll content in corn. ASAE. 45(4): 1229-1236.

Iida, K., M. Suguri, T.Matsui and M. Umeda. 2000. Estimation of the nitrogen content using machine vision in a paddy field, ASAE paper No. 003021.

Plant, R. E. and M. Keeley. 1999. Relationship among plant growth indices in Acala cotton. J. Prod. Agric. 12(1): 61-68.

Plant, R. E., Munk, D. S., Roberts, B. R., Vargas, R. L., Rains, R. L., Travis, R. L., Hutmacher, R. B., 2000. Relationships between remotely sensed reflectance data and cotton growth and yield. ASAE. 43(3): 535-546.

Yang, C., Bradford, J. M., Wiegand, C. L., Airborne multispectral imagery for mapping variable growing condition and yields of cotton, grain sorghum, and corn. 2001. ASAE 44(6): 1983-1994.

Characterization of stone densities of soils by image processing

M. Chapron[1] and P. Huet[2]
[1]ETIS-ENSEA, 6 avenue du ponceau, 95014 CERGY, France
[2]UMR INRA-INA P.G., Environnement et Grandes Cultures, BP 01, 78850 Thiverval-Grignon, France
chapron@ensea.fr

Abstract

This work is concerned with the stoniness of soils. This consists in computing the density of stones on the soil and extracting some features of stones. The properties of soils are the following : the texture (stony, argillaceous, sandy, ...), the structure (fine, coarse, medium, ...) and the consistency (friable, sticky, damp, ...). Herein, we describe a method which computes the size distribution of stones and a characterization of stone contours. This permits to predict the ability of the soil to stock water and determine the well adapted crops on the observed soil with other agronomical parameters.

Introduction

The image treatments are carried out on grey-level images because the colour provides almost nothing. Usual techniques applied for classifying different textures on images can not be easily utilized for separating stones from soil because the stones are too different to separate them from one another. The proposed method is the following. The first step of the method is to filter the image in order to better distinguish the stones from the soil. From the filtered image, it is easy to calculate the segmentation of the image which permits to compute geometrical parameters such as area, perimeter, compactness factor and invariant moments (translation, rotation and scaling) of the stones, these parameters characterize them. Then, the stone distribution is computed and finally the characterization of significant stone contours is carried out.

Image filtering and multiresolution

For reaching this aim, different processings have been tried : median filter, filterings based upon the coefficient thresholding of the different types of wavelets and wavelet packets , filter founded on the total variation principle. The best results are obtained from the last filter which is applied several times. This process well preserves the stone contours. During the last few years, there have been more and more applications of partial differential equation based techniques (PDE) in image processing (see the works of Morel and Alvarez). These new methods are motivated by a more systematic approach to restore images with sharp edges, this can also be the previous step of the segmentation stage. The soil and stones are denoised according to a non-linear anisotropic diffusion PDE, designed to smooth less near the edges. The main drawback of using quadratic regularization functional is the inability to recover sharp discontinuities (edges). To improve the Tikhonov's regularization with quadratic regularization functionals, Rudin, Osher and Fatemi [1] proposed to use the total variation as a regularization functional. The total variation principle has been used for smoothing images while preserving the contours. The algorithm used is based on the works of Chan, Golub , Mulet [2], Carter [3 and Chambolle [4]. The last author provides a proof of convergence of the iterative algorithm based on a fixed point formulation of the total variation. With the same notations as in [4], the images are NxN matrices, the gradient $(\nabla u)_{i,j} = (((\nabla u^1)_{i,j} , (\nabla u^2)_{i,j})$ is a vector at every pixel located in (i,j) with : $(\nabla u^1)_{i,j} = \begin{cases} u_{i+1,j} - u_{i,j} & \text{if } i < N \\ 0 & \text{if } i = N \end{cases}$

$(\nabla u^2)_{i,j} = \begin{cases} u_{i,j+1} - u_{i,j} & \text{if } j < N \\ 0 & \text{if } j = N \end{cases}$ and u an image .

The total variation of u is defined by the following criteria : $J(u) = \sum_{(i,j)=1..N} |(\nabla u)i,j|$, $|(\nabla u^1)_{i,j}|$ is

the Euclidean norm. The Euclidean scalar product on vectors $\underline{p} = (p^1, p^2)$ and $\underline{q} = (q^1, q^2)$ is defined as follows :

$$<\underline{p},\underline{q}> = \sum_{1 < i,j < N} p^1_{i,j} q^1_{i,j} + p^2_{i,j} q^2_{i,j}$$

For each criteria, the criteria can be written as:

$J(u) = \sup_{\underline{p}} <\underline{p}, \nabla u>$ $\underline{p}$ is such that $|p_{i,j}| \leq 1$

The discrete divergence div is defined from the adjoint of ∇ by : $\text{div} = -\nabla^*$. For every pair (p,u), we get $<-\text{div } p,u>_X = <\underline{p}, \nabla u>_Y$. X is the set of images and Y is the set of gradients of images. For each $\underline{p} = (p^1, p^2)$ of Y, we have :

$(\text{div } p)ij = \begin{cases} p^1_{i,j} - p^1_{i-1,j} & \text{if } 1 < i < N \\ p^1_{i,j} & \text{if } i = 1 \\ -p^1_{i-1,j} & \text{if } i = N \end{cases}$ $+ \begin{cases} p^2_{i,j} - p^2_{i,j-1} & \text{if } 1 < j < N \\ p^2_{i,j} & \text{if } j = 1 \\ -p^2_{i,j-1} & \text{if } j = N \end{cases}$

The algorithm solves the minimization of

Min $\dfrac{|u-g|^2}{2\lambda} + J(u)$ with g the input image and

$u \in X$

$\lambda > 0$. The Euler-Lagrange equation of the above criteria gives : $u = (I + \lambda \partial J)^{-1}(g)$ with ∂J the sub-differential of J. After several equations, one can demonstrate the image solution u corresponds to the non-linear projection :

$\min\{|\lambda \text{ div } p - g|^2 : p \in Y, |pi,j| \leq 1 \ \forall(i,j) \in \{1,2,...N\}^2\}$

The Lagrange multipliers are associated to each constraint $|pi,j| \leq 1$. After some calculations, the fixed point algorithm is given :

$$p\, p^{n+1}_{i,j} = \frac{p^n_{i,j} + \tau(\nabla(\text{div } p^n - g/\lambda)_{i,j}}{1 + \tau |(\nabla(\text{div } p^n - g/\lambda)_{i,j}|}$$

with $p^0 = 0$ and $0 < \tau < \dfrac{1}{4}$ for the convergence of the algorithm.

Below, we can see the input image and the filtered image on the first image line. Before filtering the image, it is approximated by a wavelet transform one time (four pixels become one pixel) in order to get rid of very small or small stones. On the second image line, the distributions of grey-levels of a image row before and after filtering are presented. On this filtered image, the iterative process has been stopped when the norm of $p^{n+1} - p^n$ is less than 0.005.

The input image

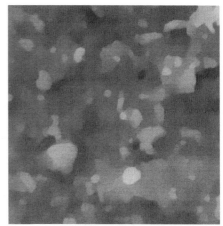
The filtered image

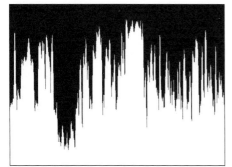
The profile of row 405 of input image

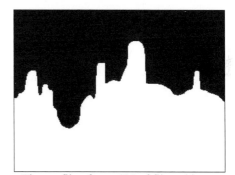
The profile of row 405 of filtered image

Model of contours

In mathematics, singularities are generally characterised by their Lipschitz exponents. In digital images, these singularities can be computed through the different coefficients of wavelet modulus maxima at different scales. Multi-scale edge detectors smooth the rows and columns of images at various scale and detect sharp variation pixels from the first or second-order derivative. The maxima of the first derivative absolute value correspond to the zero crossings of the second derivative and to the inflection points of the smoothed signal.

The stone contours are characterized by using a multi-scale technique founded on the wavelet based multi-resolution representation of the image. The estimation of the coefficients of Lipschitz regularity on significant pixels of stone contours is performed by following the modulus maxima obtained from Canny's edge detector [5] through different resolutions. The Lipschitz regularity α at x_0 is defined as follows :

$|f(x_0+h)-P_n(h)| \leq A|h|^{\alpha}$ where $P_n(h)$ is a polynomial of n order. This can be represented in a scale space plane and the estimation of α use the regression technique in this plane [6,7]. The gradient definitions on different resolutions use tempered distributions. When following the modulus maxima through the different resolutions the orientation of the contour in the neighbourhood of these singular points is taken into account. Short sequences of modulus maxima do not permit to compute the coefficient α with high reliability. This technique can implemented in a real time way.

$f(x)$ is uniformly Lipschitz-α on $]a,b[$ if f is Lipschitz-α for any x in the interval $[a,b]$. f is a tempered distribution if f is Lipschitz-α on $]a,b[$ and its primitive is Lipschitz-$(\alpha+1)$ on $[a,b]$. For the 2D signal of an image $f(x,y)$ is Lipschitz-α at $(x0,y0)$ if it exists a constant A such that :

$\forall (h,k) \in Z^2 \quad |f(x0+h,y0+k)-f(x0,y0)| \le A(h^2+k^2)^{\alpha/2}$. The wavelet transform applied on f is defined as follows :

$$Wf(u,a) = \int_{R^2} f(t) \frac{1}{a} \psi^* (\frac{t-u}{a}) dt$$

with ψ wavelet, u translation parameter, a scale parameter. Wavelets are efficient for singularities detection at coarse scales and localization at fine scales. ψ has n vanishing moments if for $0 \le k \le n$, $\int_{-\infty}^{+\infty} t^k \psi(t) dt = 0$

A smoothing function any function $\theta(x,y)$ whose integral is equal to 1 and that converges to 0 at infinity. One example of this function is a gaussian function : $\theta(x,y) = \frac{1}{2\pi\sigma^2} \exp(-\frac{x^2+x^2}{2\sigma^2})$.

$\psi^a(x,y) = \frac{\partial \theta(x,y)}{\partial x}$ and $\psi^b(x,y) = \frac{\partial^2 \theta(x,y)}{\partial x^2}$ can be considered as wavelets because their integral

is equal to 0. Let us denote the dilation $\xi_s(x,y) = \frac{1}{s^2} \xi(\frac{x}{s}, \frac{y}{s})$ by s the scale factor of any function

$\xi(x,y)$. The image $f(x,y)$ is smoothed at different scales s by a convolution with $\theta_s(x,y)$. $f(x,y)$

is supposed to belong to $L^2(R^2)$. Let us define two wavelet functions $\psi^1(x,y) = \frac{\partial \theta(x,y)}{\partial x}$ and

$\psi^2(x,y) = \frac{\partial \theta(x,y)}{\partial y}$.

The two corresponding scale wavelets $\psi_s^1(x,y)$ and $\psi_s^2(x,y)$ are as follows :

$\psi_s^1(x,y) = \frac{1}{s^2} \psi^1(\frac{x}{s}, \frac{y}{s})$ and $\psi_s^2(x,y) = \frac{1}{s^2} \psi^2(\frac{x}{s}, \frac{y}{s})$. The two components of the wavelet transform of $f(x,y)$ at scale s are defined by the two formulas :

$W_s^1 f(x,y) = f(x,y) * \psi_s^1(x,y)$ and $W_s^2 f(x,y) = f(x,y) * \psi_s^2(x,y)$.

Then, the gradient vector $\vec{V}(f*\theta_s)(x,y)$ is equal to :

$$\begin{pmatrix} W_s^1 f(x,y) \\ W_s^2 f(x,y) \end{pmatrix} = s \begin{pmatrix} \frac{\partial}{\partial x}(f*\theta_s)(x,y) \\ \frac{\partial}{\partial y}(f*\theta_s)(x,y) \end{pmatrix} = s\vec{V}(f*\theta_s)(x,y)$$

The wavelet transform is proportional to the gradient. $W_s^1 f(x,y)$ detects vertical singularities and $W_s^2 f(x,y)$ horizontal singularities at scale s.

We will use the modulus and angle of the gradient vector from the wavelet representation as folows :

$M_s f(x,y) = \sqrt{|W_s^1 f(x,y)|^2 + |W_s^2 f(x,y)|^2}$ and

$A_s f(x,y) = \text{atan}(\frac{W_s^2 f(x,y)}{W_s^1 f(x,y)})$ if $W_s^1 f(x,y) \ge 0$

$= \pi - \text{atan}(\frac{W_s^2 f(u,a)}{W_s^1 f(u,a)})$ if $W_s^1 f(x,y) < 0$

The link between wavelets and Lipschitz regularity α is related below. ψ has n vanishing moments with $\alpha \leq n$. If $f(x,y)$ is Lipschitz-α, then it can be demonstrated that for any (x_1,y_1) in the neighbourhood of (x_0,y_0), we have $|f(x_0,x_0)-f(x_1,y_1)| \leq K((x_0-x_1)^2+(y_0-y_1)^2)^{\alpha/2}$. It can be demonstrated that a function $f(x,y)$ uniformly Lipschitz α over an open set of R^2 if and only if there exists a constant K such that in the open set $M_s f(x,y) \leq Ks^{\alpha}$ and conversely if α is non integer. α is estimated asymptotically from the calculation of the following slope through the different scales :

$\log|M_s f(x,s)|=\log K + \alpha \log(s)$ and for the dyadic wavelet used $\log_2(M_{2j}f(x,y)) \leq \log_2(K) +\alpha_j$

The maxima lines is defined by the modulus maximum : a point (x_0,y_0) where $|M_s f(x_0,y_0)|$ is locally maximum and a maxima line is any connected curve $c(u)$ of modulus maxima in the scale-space plane for the estimation of α. Thus, if $f(x,y)$ is singular at (x,y), there exists a sequence of maxima points (x_s,y_s) converging towards (x,y).

The thresholdings of $M_s f(x,y)$ are performed in order to take into account significant contour pixels. Then, the computation of the wavelet modulus maxima on 3x3 neighbourhood $(x\pm1,y\pm1)$.

The chaining of modulus maxima through the different scales also use the eight point neighbourhood. The significant edges are composed by only the pixel chains which are long enough (above a given length). The modulus maximum reached at the top (fine scale) of the chain is included in a result image. Along a contour, the Lipschitz regularity of a pixel is estimated by the linear regression using the different scales. The scales are numbered between 1 and N, the finest one is 1 and the coarsest one is N. Contours are not always closed. Pieces of contours, defined a number of points whose pixels have similar Lipschitz regularity and modulus maxima belong to a given interval. Below, the Lipschitz regularity parameters on the stone contours of the input image are displayed on the image. The Lipschitz coefficients on the contours characterize the texture in the neighbourhood of stone contours and soil. The parameter α varies in the interval $[\alpha_{min}, \alpha_{max}]$. The Lipschitz coefficient values are represented by colors, the ground is in white.

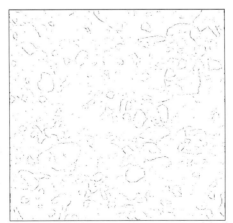

The Lipschitz coefficient image on the contours of the input image

Distribution of stones

After filtering the images by the algorithm described above, we extract the contours of stones and labelled each stone for counting the number of pixels and compute the geometrical parameters of each stone. One of the features used in agronomy is the distribution of stones according to the size.

Below, the segmentation results and the size distribution are shown.

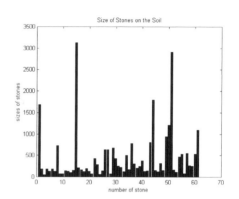

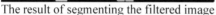
The result of segmenting the filtered image

Conclusion

These image processings give good results on different types of input images. Other filtering methods have been tried, they were not successful. This filter algorithm is quite fast. The characterization of stone contours is relevant in order to classify the stones and soil. This work is important for finding the spatial distribution and movement of rock fragments on soil, the effects of rock fragments on some hydrological processes, the thermal properties of top soils, the physical soil degradation and soil productivity according to fertilizers.

References

[1] L. Rudin, S. Osher, E. Fatemi, "Nonlinear total variation based noise removal algorithms", *Physica D* 60, 259-268, 1992.

[2] T. Chan, G. Golub, P. Mulet, "a non-linear primal-dual method for total variation-based image restoration", *SIAM J. Sci. Comput.* 20(6), p 1964-1977, 1999.

[3] J. Carter, "Dual methods for total variation-based image restoration", *Ph.D. thesis*, UCLA, 2001.

[4] A. Chambolle, "An algorithm for total variation minimization and applications", *preprint CEREMADE*, 2002.

[5] J. Canny, "A computional approach to edge detection", IEEE Trans. Pattern Analysis Machine Intelligence, vol 8, pp 679-698, 1986.

[6] S. Mallat, *A wavelet tour of signal processing*, Academic Press, Second edition.

[7] S. Mallat, S. Zhong, Characterization of signals from multiscale edges, IEEE Transactions on Pattern Analysis and Machine Intelligence, vol. 14, no. 7, july 1992.

Sensor-controlled variable rate real-time application of herbicides and fungicides

K.-H. Dammer, G. Wartenberg, H. Böttger and D. Ehlert
Institute of Agricultural Engineering, Potsdam-Bornim, Germany

Abstract

An important step towards variable rate application of plant protection agents is the development and operating of online sensors for the detection of weeds and plant mass or leaf area index respectively. Field trials were conducted with a sensor operated field sprayer in the last 3 years for quantifying the influence of a site-specific application of herbicides and fungicides in real-time in small row crops (mainly cereals).
In 18 field trials herbicide savings on the average of 24% and fungicide savings on the average of 19% were achieved. There was no yield reduction due to the variable-rate application.

Keywords: variable rate plant protection, herbicides, fungicides, sensors, economic thresholds

Introduction

Weeds and also plant diseases (mainly in there initial phase of epidemy) often occur in patches within the crop stand. The common practise is to apply herbicides and fungicides uniformly over the hole field. But the application is not necessary in the weed- or disease-free areas. A variable rate application of herbicides and fungicides according to the weed occurrence and disease incidence would help optimise the use of production inputs and would reduce the input of biocides into the environment. A prerequisite is the knowledge of the small-scale distribution of weeds and diseases. But an assessment in the field by walking is time and labour consuming. The use of sensors for automatic weed and disease detection helps to get small-scale information about their distribution very fast. At the Institute of Agricultural Engineering Bornim (ATB) sensors were developed which are used for a variable rate application of herbicides and fungicides in real-time.
At present to systems are developed for automatic weed detection. The image analysis system uses CCD cameras and image analysis software to detect the weed species composition and to discriminate weed from crop plants based on colour, shape and texture features (Chapron et al, 1999, Gerhards et al, 2002). On the other hand optoelectronic sensors measure the reflectance of light in an certain range of wave bands. A spot spraying system for the application of non-selective herbicides within culture free areas were developed by Felton and McCloy (1992). Green leaves reflect the light in the near-infrared wave band and absorb the light in the red wave band. The reflectance curve of soil is nearly constant. This features are used for discrimination of green weeds from the soil. In spectral analysis systems the classification between crops, weeds and soil is done by using a certain number of wave bands (Vrinds et al, 1999, Biller and Schicke, 2000). The optoelectronic weed sensor, developed at the ATB, allows a variable rate herbicide application in real-time with treatment speeds, which are common practice.
Similar to weed mapping, the estimation of disease incidence within the field by walking is very time consuming. Therefore reports on the use of disease maps, based on visual assessment of diseases, come from experimental sites (Secher et al, 1995, Bjerre, 1999). If weather conditions are favourable, diseases spread out very fast over the hole field. Under

practical condition, this method causes problems, because disease maps are not available fast enough to make decisions on disease control. Automatic disease detection before their incidence reaches thresholds would help to provide information about parts of the fields, in which diseases occur. Since there are no sensor-based technologies on the market for the automatic detection, an alternative method has been developed to optimise fungicide application in real-time. The strategy behind is to apply the same concentration of active fungicidal substance per unit of plant surface area (leaf area index, LAI). Small-scale information about the present plant surface of the cultivated crop can be obtained indirectly by the pendulum-meter, a mechanical sensor for scanning biomass (Ehlert, 2001). The deviation angle from this sensor is correlated with the LAI (Dammer et al, 2001), which can be measured by hand-held optical sensors like LAI2000® or SunScan®. With this online sensor small-scale information about vegetation differences within the field can be obtained automatically with high data density. Other autors (Bjerre, 1999, Ewaltz, 2000) adapted the fungicide rate to the ratio vegetation index (infrared reflectance/red reflectance) after collecting the data manually by hand held Cropscan radiometer.

In this paper three year results from sensor-controlled real-time spraying are shown. An example, each for real-time application of herbicides and fungicides resp. is presented.

Materials and methods

The aim of field strip trials was to analyse the effect on disease occurrence and yield of a variable-rate real-time fungicide application by comparison with a uniform application.

An optoelectronic sensor detects the weeds within the tram lines of narrow row crops, e.g. cereals and legumes. Based on the found relationship between the weed counts in total and the sum of the calculated yield loss due to certain weed species, the method makes no discrimination between weed species. From this function an economic threshold under consideration of the costs of the herbicide treatment can be calculated. If the sensor measurement values are equal or above the threshold while spraying, the farmer's common dose is applied. If the sensor value is below the threshold the dosage is reduced to 50%. Details are given in Wartenberg and Dammer (2001).

The mode of action of the pendulum-meter is described in detail from Ehlert and Domsch (2002). If the highest expected deviation angle is reached (maximum LAI), the maximum fungicide rate is applied. In areas with LAI below the maximum, the application amount of fungicides is reduced. The sensor-based spatial surveying of the plant surface helps to quantify the treatment area within the whole field, and could be used therefore to optimise the amount of fungicide spraying liquid.

The variable-rate real-time application of herbicides and fungicides was performed using a commercial 4000 litres sprayer (trials in 2000 and 2001: Air Matic® system, 18 meter boom width, 2002 trials: VarioSelect® system, 24 meter boom width). The weed sensor and the pendulum-sensor resp. were mounted in front of the tractor. Based on the sensor signals, the application rate was adjusted by a sprayer control system. The sensor signal, the position and the rate of flow were recorded and processed by a data processing system. The field experiments were conducted in commercially cultivated fields. The variable-rate application was performed in one tramline and the uniform application with the farmers common dose in the neighbouring tramline. Therefore the treatment plots were located next to each other to minimise the soil influence on the experimental results.

Weed and disease assessment was performed within the strips of the experimental area before applying the herbicides and the fungicides. In the herbicide example 0.8 l ha[-1] Basagran® and 2.0 l ha[-1] Stomp® in mixture were sprayed with 200 l ha[-1] water on 29.04.2002 in peas at one runner stage (BBCH 11). In the fungicide example 1.2 l*ha[-1] Caramba® in 300 l*ha[-1] water was applied on 06.06.2001 in winter wheat at the end of ear emergence (BBCH 59). Weed assessment was done before harvesting the peas and disease assessment was performed at growth stage milk ripeness (BBCH 75) of winter wheat.

Two strips per plot were harvested by a combine harvester with a yield monitoring system (herbicide trial: 30.07.2002, fungicide trial: 23.08.2001). From the local yield measurements relative yield levels between the variable rate treatments and the adjacent uniform treatments (quotient: yields of variable rate treatment / yields of uniform treatment) were calculated for 9 meter distances. Because of the resulting high sample size, for yield comparison frequency plots of the relative yields were used.

To justify the comparability of the plots with the uniform and variable rate application, the difference (variable rate – uniform treatment) of the sensor values (sum of "green signals" from the weed sensor and deviation angle from pendulum sensor resp.) of the same 9 meter distances as for yield comparison were calculated. From this differences a frequency plot shows, if the experimental conditions within the experimental design were more or less equal.

Results and discussion

Variable rate application of herbicides

In the example of the herbicide trial in peas the weed counting was done at 29.04.2002 (growth stage: BBCH 11) with help of a counting frame (0.5 m^2) along an 24 m by 50 m sampling grid. The total weed number per m^2 was in the range of 0 to 66. The sampling points with a high weed number were located at the field edges. According to the yield loss function (Wartenberg and Dammer, 2001) the economic threshold was set to 85 weeds m^{-2}. According to the results of visual assessment there were no sampling point equal or higher as the economic threshold But this threshold was reached or exceeded on 4% of the 631 weed sensor values. The sensor offers the possibility to evaluate the weed distribution more accurate within the field

The differences of the sensor values (variable rate-uniform treatment) were in the range of -333 to 122, indicating, that in adjacent treatment plots were a different weed density. But the frequency plot shows that the majority of the differences is found around zero. At mean there were no differences between the two treatments (calculated median = 0) regarding to the weed density (Figure 1, left).

Yield reductions up to 41% (minimum of the relative yield level: 0.59) as well as a yield increases up to 63% (maximum: 1.63) were obtained (Figure 1 right). The majority of the relative yield values is found around one. At mean there was a negligible 2% lower yield in the variable rate treatment plots in comparison with the uniform treatment plots (calculated median = 0,98).

The weed counting before harvesting along the same 25 m by 50 m sampling grid brought only up to 2 weeds per m^2 in a few samples, which are emerge after herbicide spraying. There were found in the variable rate plots as well as in the uniform plots, indicating the equal efficacy against on the weeds of the two variants.

Table 1 shows the results from the variable rate herbicide trials in the three years. The values of herbicide savings reached from around 12 to 43%.

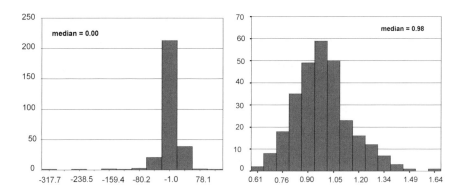

Figure 1. Frequency plots, left: differences of the adjacent weed sensor values (variable rate-uniform treatment, correspond to weeds m^{-2}), right: relative yield values (variable rate/uniform treatment) of the herbicide trial

Table 1. Crop, year, acreage, growth stage, range of application rate and herbicide savings in the herbicide trials.

crop / year	acreage	growth stage	range of application rate	herbicide savings
	(ha)	(BBCH)	(l ha^{-1})	(%)
winter rye, 2000	22	24 - 26	150 - 300	30.5
winter wheat, 2000	32	22 - 23	210 - 300	19.0
triticale, 2000	37	23 - 24	210 - 300	24.5
summer barley, 2000	6	12 - 14	125 - 250	29.5
peas, 2000	8	12 - 13	170 - 280	22.0
winter rye, 2000	28	11 - 12	100 - 200	20.0
winter wheat, 2001	26	11 - 12	100 - 200	12.7
peas, 2002	14	11 - 12	100 - 200	30.0
peas, 2002 *	8	10 - 11	100 - 200	40.9
summer barley, 2002	12	13 - 21	100 - 200	12.8

* example field

Variable rate application of fungicides

The differences of the values of the deviation angle (variable rate - uniform treatment) were in the range of -61° and 43° indicating differences in the vegetation density between adjacent treatment plots. But the frequency plot (Figure 2 left) shows that the majority of the differences is found around zero. At mean there were a little higher deviation angle of 3° (calculated median = 3°) in the variable rate treatment plots. This indicates that in the variable rate application plot the LAI was higher in comparison to the uniform plot.

As in the herbicide field trial, yield reductions as well as a yield increases were obtained (Figure 2 left). At mean there was a 12% higher yield in the variable rate treatment plots by comparison with the uniform treatment area (calculated median of the relative yield levels = 1.12). This was expected because the experimental conditions within the experimental design were no equal. The 3° higher deviation angle two month ago indicated a higher vegetation, which is responsible for the formation of the yield.

The visual assessment of diseases was performed in adjacent areas of uniform and variable rate fungicide treatment mainly at milk ripeness of cereals. There were no differences in disease incidence in the areas surveyed in all three years.

Table 2 shows the results from the variable rate fungicide trials. The values of fungicide savings reached from 7 to 38%.

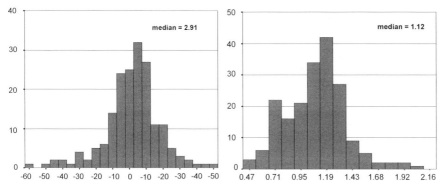

Figure 1. Frequency plots, left: differences of the adjacent deviation angles of the pendel sensor (variable rate - uniform treatment), right: relative yield values (variable rate / uniform treatment) of the fungicide trial

Table 2. Crop, year, acreage, growth stage, range of application rate and fungicide savings in the fungicide trials.

crop / year	acreage	growth stage	range of application rate	fungicide savings
	(ha)	(BBCH)	$(l\ ha^{-1})$	(%)
winter wheat, 2000	44	47 - 51	100 - 250	16.1
winter wheat, 2000	5	47 - 51	119 - 250	12.8
winter wheat, 2000	5	47 - 51	175 - 300	7.0
summer barley, 2000	6	61 - 65	104 - 300	27.4
winter wheat, 2001*	21	55 - 59	120 - 300	25.0
summer barley, 2002	19	69 - 71	40 - 200	37.5
winter wheat, 2002	44	59 - 61	55 - 200	8.7
winter wheat, 2002	5	59 - 61	90 - 200	15.0

* example field

Conclusions

In 10 field trials (total area: 193 hectares) herbicide savings on the average of 24% (12 to 43%) were reached. The level of the savings depends on the sensor detected weed distribution within the field and the chosen economic threshold while spraying. Savings can not reach levels above 50% because this is the minimum application rate in the method to guarantee no weed occurrence when harvesting. The range of application rate is therefore in the range from 50 to 100%.

Fungicide savings on the average of 19% (7 to 38%) were obtained in 8 field trials (total area: 149 hectares). In the case of variable rate fungicide application the level of the savings depends on the extension of the vegetation heterogeneity within the field and the chosen maximum and minimum application amount. By contrast with the variable rate herbicide spraying the minimum application rate depends on the minimum LAI obtained and can be below 50%. In the trials in the year 2002 for example the application rates were in the range of 40 to 200 l ha^{-1}, 55 to 200 l ha^{-1} and 90 to 200 l ha^{-1}.

Beside the variable rate application of herbicides and fungicides in real-time the weed sensor and the pendel sensor offer the possibility to evaluate, if the experimental conditions (weed and crop density resp.) within the experimental design were comparable.

Acknowledgements

This study was funded as part of the BMBF-project "preagro" (No. 0339740).

References

Bjerre, K. D. 1999. Disease maps and site-specific fungicide application in winter wheat. In: Stafford (ed.) Precision Agriculture 1999, Sheffield Academic Press, p. 495-504

Biller, R., Schicke, R. 2000. Multi-frequency optical identification of different weeds and crops for herbicide reduction in precision agriculture. Proceedings of the 5th International Conference on Precision Agriculture, Minnesota, CD-Rom

Chapron, M., Requena-Esteso, M., Boissard, L., Assemat, L. 1999. A method for recognizing vegetal species from multispectral images. In: Stafford (ed.) Precision Agriculture 1999, Sheffield Academic Press, p. 239-284

Dammer, K.-H.; Reh, A.; Wartenberg, G.; Ehlert, D.; Hammen, V.; Dohmen, B.; Wagner, U. (2001): Recording of present plant parameters by pendulum sensor, remote sensing, and ground measurements, as fundamentals for site-specific fungicide application in winter wheat. In: Grenier, G. and Blackmore, S. (eds.) Third European Conference on Precision Agriculture, Vol 2: 647-652

Ehlert, D., Domsch, H. 2002. Sensor pendulum-meter in field tests. International Conference on Agricultural Engineering, Budapest, CD-Rom

Ewaldz, N. A. T. 2000. Radiometric readings as a tool for predicting optimal fungicide dose in winter wheat. Journal of Plant Diseases and Protection 107, p. 594-604

Felton, W.L.; McCloy, K. R. 1992. Spot spraying. Agricultural Engineering, 11, p. 9-12

Gerhards, R., Sökefeld, M., Nabout, A, Therburg, R.-D., Kühbauch, W. 2002. Online weed control using digital image analysis. Zeitschrift für Pflanzenkrankheiten und Pflanzenschutz Sonderheft XVIII, p. 421-425

Secher, B.J.M. 1997. Site specific control of diseases in winter wheat. Aspects of Applied Biology 48, p.57-65

Vrindts, E., De Baerdemaeker, J., Ramon, H. 1999 Weed Detection using canopy reflectance. In: Stafford (ed.) Precision Agriculture 1999, Sheffield Academic Press, p. 257-264

An artificial intelligent approach to identify complex associations among soil properties and weed density

B. Díaz[1], A. Ribeiro[1], D. Ruiz[2], J. Barroso[2] and C. Fernandez-Quintanilla[2]
[1]IAI-CSIC, 28500 Arganda del Rey, Madrid, Spain.
bdiaz@iai.csic.es, angela@iai.csic.es
[2]CCMA-CSIC, Serrano 115 B, 28006 Madrid, Spain.
david.ruiz@ccma.csic.es, judit@ccma.csic.es, cesar@ccma.csic.es ,

Abstract

Weed abundance varies spatially across a field in non-random spatial patterns. Since the site properties also vary across crop fields (i.e. spatially) they could be related to the weed heterogeneous distribution. The proposed approach assumes a complex relationship among site properties and weed density that can be expressed as a rule set and then it searches these rules by a learning process based on Genetic Algorithm (GA); a search procedure based on the mechanics of natural selection and natural genetics. The GA searches the best rules that accurately explain weed density in terms of soil properties.

Keywords: Machine Learning, Soil Properties, Weed Control, Rules Induction, Genetic Algorithms

Introduction

Weed infestations in crops are still a problem to solve in agriculture, since certain sites where weed density is higher than another places of the field. In fact, crop fields weeds tend to appear aggregately while the rest of the site is weed-free (Cardina et al. 1996). This spatial heterogeneity provokes, first, an inefficient use of agrochemical products because generally the management is realised homogeneously over the all field and second, a less crop production in percentages of this specific places due to, for example, the weed-crop competition. These spatial variations and the own stability of weed population have a strong interest in order to find the relations among weed density and the crop yield loss (Cousens et al. 1987). In addition, the persistency of high density areas over time in fields determinate a non-random distribution that can probably depend on physic-chemical irregularities in the field. Since soil characteristics, as well as own properties of plant species have a strong influence on the growth and reproduction of both crop and weed, many scientists are lead their researches toward finding relations or dependencies among soil factors and weed existence. As well, due to the site properties as edaphic factors have spatial variability in agricultural fields the most attempts have been addressed to the search of soil or site relations rather than spatial heterogeneity of weed population. Usually statistical methods are used to define these relations. For example the *Canonical Correlation Analysis* is used to interpret association among several plant species and site properties by covariance-based coefficients (Dieleman et al. 2000) or via combination of semivariograms and a geostatistic interpolation method as kriging. Former method models variability based on spatial dependence in a data set (Cardina et al. 1995).

Although some of these authors have obtained sure dependencies for certain weeds, there is no a unique or generic method to study associations among the site properties for all weeds type. In addition, these statistical approaches required an extreme control of studied input data, for instance in a Canonical Correlation Analysis the data should be analysed and modified until the variables are completely independent. Statistic presents some other restrictions, for example it is based on assumptions as stationary one (Fortin et al. 2002), which assumes a data normal distribution. As well, some times the interpretation of results, as for example numerical indexes of correlation and significance, do not allow to determinate clearly if the analysed variables are or not related. Furthermore the statistic methods are based on centralisation measures (i.e. average or variance) as consequence the obtained results are affected notably by noise in data input. Consequently, obtaining a precise analytical model for weed dynamics is a very complex, probably unreachable, task. A solution could be to find a qualitative and approximate model; such as, a set of rules that explain input data in the same way that an expert farmer:

The objective of this study is to assess how an artificial intelligence approach -a rule induction method- could be used to identify complex association among site properties and the abundance of weeds within agricultural fields. The proposed approach has been tested finding out a rule set that explains the abundance of winter wild oat (*Avena sterilis* L.) in regard to eight soil parameters.

Materials and methods

Data description and data pre-processing

Data have been obtained from a quadrangular grid sampling carried out in barley fields in two different locations in South-East Madrid (Spain). The range of the field sizes was from 0.5 ha. to 1.6 ha. In each grid point, soil samples and wild oat abundance data, were gathered.

Table 1. Site property and wild oat abundance data collected on five fields.

Field	Size (ha)	Topography	Grid size (m)	n	Wild Oat data	Sampling date
1	0.5	Flat	10x10	50	Seedlings	Feb 01
2	1.6	Flat	12x6	38	Seed rain	Jul 00
3	0.9	Hilly	10x10	96	Seed rain	Jul 00
4	1.2	Hilly	10x10	124	Seedlings	Feb 01
5	1.6	Flat	12x6	228	Seed rain	Jul 01

The soil parameters analysed for each sampled point were *pH*, Organic Matter (*OM*), total Nitrogen (*N*), Phosphor (*P*), Potassium (*K*), *sand*, *silt* and *clay* contents. Wild oat counting was realised by a square of $0.1 m^2$ representing the *weed density*. The 536 sampled points were stored in a relational database where records gathered soil properties and weed density at each sample point.

Descriptive statistical analysis of data shows that attributes do not have the same range of values in the studied fields, however all these fields present spatial variations of the weed density. In other hand to reduce other factor effects such as the field history and the landscape characteristics in the weed evolution in each field the inter-fields data are homogenised, making them comparable, by a linear scaling technique. Then all input data will take their values in the range 0-1. As a result, data are analysed under a relative perspective, in other words, in terms of relative higher or lower weed density. Accordingly, data of each field are pre-processed; that is they are transformed into common domain range. After normalisation, all numeric variables are categorised into different intervals that are named by labels: *high*, *middle* or *low*, following the criteria shown in table 2. Each label represents an interval of values of a variable, therefore the use of this technique reduces the data uncertainty. In addition reasoning with symbolic data has an immediately advantage inasmuch as the obtained patterns will be expressed close to natural language. On other hand, the categorisation thresholds have been established according to expert criteria; building regular intervals of values. While for most variables there are three regular intervals, for pH experts estimated two intervals due to the little domain size. In the case of Wild Oat variable the value 0.2 was selected as threshold that determinates a similar percentage of affected points by weeds for every studied field, around 50% of data. Therefore data involved in the experiment have been divided in two classes, *High-density* class and *Low-density* class, using the quantity of wild oat observed at each point. The High *density class* contains 271 (50.56%) points, registers with a weed density higher than 0.2. The rest of 265 data (49.44%) belong to the *Low density* class.

Table 2. Intervals and labels for the studied variables.

Variables	Normalised ranges	Linguistic labels
pH	[0, 0.5)	*low*
	(0, 0.5-1]	*high*
N,K,OM,P	[0, 0.333]	*low*
Sand,Clay,	(0.333, 0.666)	*middle*
Silt	[0.666, 1]	*high*
Wild Oat density	[0, 0.200]	*low*
	(0.200, 1]	*high*

Now then, the supervised learning process to be applied next involves inducing concept descriptions from a set of positive and negative examples of a target concept. Examples are represented as points in a n-dimensional feature space which is a priori defined and for which all the legal values of the features are known. Concepts are therefore represented as subsets of points in the given n-dimensional space.

The supervised learning process

In this point search process conducted by a Genetic Algorithm (GA) is performed to find the set of IF-THEN rules that best explain the two training examples. Genetic algorithms (GAs) [Goldberg 1989] are search and optimisation techniques based on a formalisation of natural genetic processes. The basic idea underlying genetic process is to start with a population of randomly generated solutions, namely chromosomes, that define the first generation *(G(0))*, from which evolution begins. While a specific ending condition is not met, each chromosome is evaluated with respect to its ability to solve the target problem. This evaluation is performed by means of a fitness function. Then a new population is created *(G(t+1))*, by applying a set of genetic operators to the individuals of the generation *G(t)*. The more common operators are the selection, the crossover and the mutation. In our context the chromosomes will be sets of rules that try to explain the data contained in each of the training example sets (*High-density* class and *Low-density* class). The exactitude in the training set classification will allow to calculate the fitness to each rule set (chromosome). Therefore to apply a GAs to a particular problem, we need to select an internal representation of the search space (space of all rule sets) and define a fitness function, which evaluate, to each candidate solution, how well it explain input training examples.

Although there is many ways to represent problems in GAs, the most used and traditional representation for chromosomes are binary strings, which in this case must be represent the rules. In the current application, three linguistic labels are defined for each physical variable {low, medium and high}, except for the pH and Wild_Oat variables, which have only two linguistic labels {low, high}. The definition of an uniform structure facilitates both codification and decodification of the chromosomes. For this reason two bits are used to code each label, so that codification is as follow: (low, 01), (medium, 10), (high, 11) and, for pH and Wild_Oat variables, (low, 10) and (high, 11). The configuration 00 for the three-label variables, and 01 and 00 for the two-label ones are used to represent the absence of rule antecedents or rule consequent respectively. Using this representation the antecedents and the consequent of each rule is internally represented by means of a binary string, such as:

pH	OM	N	P	K	Sand	Silt	Clay	Wild_Oat
-	-	-	Medium	High	Low	-	-	High
00	00	00	10	11	01	00	00	11

Specifically previous binary-string characterises the following rule:

IF *P* is *Medium* and *K* is *High* and *Sand* is *Low* THEN *Wild_Oat* is *High*

As we have said before in addition to selecting a good representation it is important to define a good payoff function (fitness). In our case the selected fitness function evaluates the quality of the rule set codified in the chromosome (their accuracy to explain the training sets) and it is defined as follows:

$$fitness = \frac{L_T + H_T}{L_T + L_F + H_T + H_F}$$ (1)

Where L_T, L_F, H_T and H_F are the number of *low true*, *low false*, *high true* and *high false* respectively (see Table 3). The low and high trues are well-classified examples (i.e. L_T and H_T), while low and high falses are faults (i.e. L_F and H_F). Then fitness represents the number of well-classified examples respect to the total number data. Therefore, when the fitness value is maximum (i.e. 1.00) then the all data are well classified, since the sum of L_T, L_F, H_T and H_F represents the total number of examples.

		Classified by RULE	
		LOW	*HIGH*
REAL values (Instances)	*LOW*	Low True L_T	High False H_F
	HIGH	Low False L_F	High True H_T

Table 3. Parameters of fitness function.

AG generates continuously set of rules, which are hypothesis to model the weed density in terms of soil factors. The complex relations between high weed density and soil factors come towards the evolution to the best rule as consequence of the different selection, mutation and crossing operations applied over the best hypotheses/chromosomes of each population.

Results and discussion

In order to test the proposed approach, the 90% of the input data is chosen randomly forming the *training* set of examples over which is accomplished the search process of the best rule set. The remaining 10% is useful in the validation process that is performed in order to know the predictive ability of the best rule set that has been obtained by the AG.

Taking into account a chromosome is internally represented by a binary string of fixed length, in this experiments and as first step in the research, each solution was restricted to appear only once each variable in resulted rules. Next, applying the explained process, the following set of two rules has been found:

(IF *OM* ≠ *low* and *P* ≠ *high* and *clay* ≠ *low* THEN *Wild_Oat* = *high*) (R1)
(IF *silt* = *middle* and *sand* = *low* THEN *Wild_Oat* = *high*) (R2)

The rule set presents a maximal fitness value of 0.772 that corresponds to approximately 70% of good estimation for the instances of high weed density. In others words, samples that have a non-low content of OM, non-high content of P and a non-low value of clay have a high density of Wild_Oat. Also if the silt content is middle and the sand value is low then the Wild_Oat density is high. Else the samples will have a low content of weed in the tested barley crop fields. Moreover the cover percentage of the first rule (R1) over the input examples is a 10% while the rule (R2) covers about a 60% of examples. This suggests that a better solution that explains or covers more examples will be achieved with a higher number of rules as well as a rule was able to contain the variables repeated.

The result of the validation process is shown in figure 1 where two maps for each field are shown: real and estimated weed density value at each point. Summarising, field (a) has 50 samples and the estimation fails in 12 points (24%), in field (b) 14 estimation fails from 38 sample points (36%), in field (c) 33 faults are computed from 96 samples (34%) in field (d) the number of faults was 49 of 124 points (39.5%), and finally 50 mistakes of 228 sample points (21.9%) in field (e). Consequently results show a similar behaviour of the discovered rules when they cover the samples of the different tested fields.

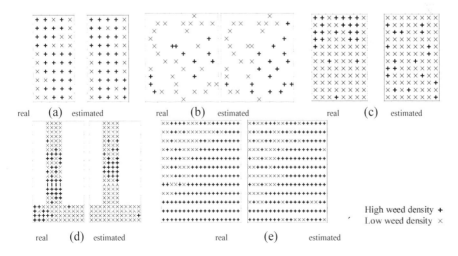

Figure 1. Maps of high (+) and low (×) weed real and estimated density of the tested fields: (a) 1 with 50 samples, (b) 2 with 38 , (c) 3 with 96 , (d) 4 with 124 , and (e) 5 with 228 points.

Conclusion

This paper presents a generic methodology that can be applied to describe weed density in terms of soil factors. The proposed approach consists in three stages. First, the inter-fields data are homogenised, making them comparable, by a linear scaling technique, that normalise all input values. Second, the method used transforms the input data into labels with an associated semantic: a) *Low*, *Medium* and *High* labels to the values of site properties; b) *Low* and *High*, to values of both pH and weed density values. The two labels of the weed abundance divide the input data into two sets, namely training sets. The third step is a learning process, where a Genetic Algorithm (GA) searches a set of rules, which explain the two training examples. The validity of the rule set produced by the GA is measured as the percentage of input data that are correctly classified. The firsts results are very promising. The proposed method explains, with two rules, about a 70% of the input data.

More research must be performed to improve results. For example the proposed method should be extended to allow a limitless number of rules. This requires the definition of a new fitness function that incorporates a specific term to evaluate the number of rules in each chromosome and to reward the smaller rule sets.

Acknowledgement

The authors wish to thank the Spanish Science and Technology Commission for funding this research through research Projects AGF1999-1125-C03-03 and AGL2002-04468-C03-01, and the Ministry of Science and Culture through a pre-doctoral grant. Authors thank Professor Jose Antonio Malpica Velasco and Dr. Maria C. Garcia-Alegre for their valuable comments.

References

Cardina, J., Johnson G. A., Mc. Coy, E.L. (1995). Analysis of spatial distribution of common lambsquarters in notill soybean. Weed Science 43: 258-269.

Cardina, J., Johnson G. A., Sparrow, D.H. (1996). The nature and consequence of weed spatial distribution. Symposium: Importance of weed biology to weed management. Weed Science, 45: 364-373

Cousens, R, Brain, P., O'Donovan, J.T., O'Sullivan, P.A. (1987) The use of biologically realistic equations to describe the effect of weed density and relative time of emergence on crop yield. Weed Science, 35: 720-725.

Dieleman, J.A., D.A. Mortensen, D.D. Buhler, C.A. Cambardella and T.B. Moorman. (2000). Identifying associations among site properties and weed species abundance. I. Multivariate analysis. Weed Science 48:567-575

Fortin, M., Mark, R.T., Dale & Jay ver Hoef. (2002). Spatial analysis in ecology. Encycopedia of Envirometrics, 4: 2051-2058. ISBN 0471 899976. Eds. Abdel H. El-Shaarawi and Walter W. Piegorsch. Wiley & sons.

Goldberg, D. E. (1989). Genetic Algorithms in Search, Optimization, and Machine Learning. Addison Wesley, Reading MA.

Oliver, M.A. (1999) Exploring soil spatial variation geostatisyically. In Proceedings of Precision Agriculture, Ed. John V. Stanfford, 1, 3-17.

Quality of agricultural positioning and communication systems

M. Ehrl[1], W. Stempfhuber[2], H. Auernhammer[1] and M. Demmel[1]
[1]Technische Universitaet Muenchen, Department of Bio Resources and Land Use Technology, Crop Production Engineering, Am Staudengarten 2, 85354 Freising-Weihenstephan, Germany
[2]Technische Universitaet Muenchen, Institute of Geodesy, GIS and Land Management, Chair of Geodesy, Arcisstrasse 21, 80290 Muenchen, Germany

Abstract

The successful use of precision farming technology for automated data acquisition, site specific farming, fleet management and field robots makes it necessary to have a detailed knowledge on the quality of all utilised equipment. To get knowledge on this quality is the aim of the presented investigation. Key technologies as the positioning system or the standardised bus communication are the main subjects. The delays and errors of these systems and their variance allow to point out limiting factors and formulate general design rules for future precision farming developments.

Keywords: precision agriculture, electronic controlled unit, GPS, LBS, ISOBUS

Introduction

Precision farming becomes more and more an accepted way of crop production and helps to achieve a sustainable environmental friendly agriculture. The objectives of site specific farming are increasing yields, together with decreasing environmental impacts. Furthermore, growing interest in automated data acquisition and information processing is going to form another milestone towards improved farm management and an overall traceability in agricultural food production. The benefit and effectiveness of using precision farming techniques is highly dependent on the capabilities of the utilised technology. The investigations of the presented project concentrate on an overall quality-related evaluation of precision farming applications.

Major components such as the positioning system or the standardised CAN-based communication are subjects of these investigations. Delays and errors of single system components, and their relating variance, have main influence on the acceptance and adoption of precision farming systems. Beyond this, knowing all strengths and weaknesses of actual systems will be the base for developing enhanced precision farming equipment in future.

System analysis

Precision farming equipment consists of various components and tools, which jointly comprise the useable system. Each single component is imperfect and has therefore certain influences on the overall system capability. The first requirement was to analyse and split up the universal system into logical and workable parts. Three main sections with several independent components and interconnections were identified (Figure 1).

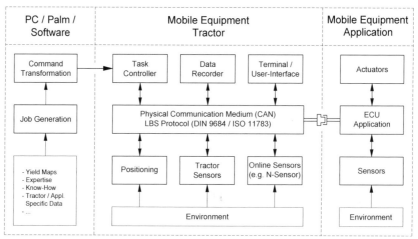

Figure 1. System analysis and segmentation.

A personal computer (PC) with precision farming software represents the first section (Figure 1, left). Farm specific data, preceding yield maps and other important information are stored and accessible an appropriate database. The generation of location based jobs comprehends several sources of errors like algorithm impreciseness, which obviously influences the quality of the real application. However, the vast amount of proprietary software tools makes a scientific evaluation not reasonable today. Therefore, the software section is not considered within this project.

The tractor, representing the second section, is provided with essential precision farming equipment (Figure 2, middle). Two deployed key technologies are the Global Positioning System (GPS) and the communication bus based on Controller Area Network (CAN). On this bus, modern tractors provide important sensor signals like speed as broadcast messages, which are conform to the Agricultural Bus System standard DIN 9684 (LBS) or ISO 11783 (ISOBUS). The satellite positioning system GPS, as well as the bus communication between the single components are expected to have the most significant influence on precision farming quality at this system level.

The third logical unit are all parts related to the application (Figure 2, right). These are usually the Electronic Control Unit (ECU) of the application implement (drill, planter, spreader, sprayer) as well as interconnected sensors and actuators. Response times of various actuators and control algorithms are considered to have primary influence on precision farming quality.

By means of system analysis, positioning and the communication among ECU's were identified to have major influence. Also, response times of involved actuators and their control algorithms are essential quality parameters. The following investigations focus on Differential GPS (DGPS) and the standardised communication protocol LBS respectively ISOBUS.

Positioning system quality

In the last decade, GPS has won the recognition of being the standard positioning system in agriculture. For reasonable accuracy, the additional service DGPS is utilised. The investigations were focused on the absolute positioning accuracy as well as the computation time (latency) of DGPS receivers. The method was to setup a highly accurate reference measurement system. For this purpose, a position reference, a time reference and a real-time data recording and synchronisation system are necessary and have been established and evaluated.

Geodetic measurement instrumentation differ considerably in measurement frequency, availability and accuracy. Therefore, the position reference consists of a Real Time Kinematic DGPS (RTKDGPS) and a Terrestrial Positioning System (TPS). Special fitted GPS receiver deliver a Pulse Per Second (PPS) event, which served as precise time reference. Both references fulfilled the postulated accuracy (position ± 3 cm, temporal resolution 1 ms).

The main work was the design of a PC based real-time (1 ms) data recording and synchronisation system, which also serves as interface for all different subsystems. The software was designed for Windows NT/2000 Operating System (OS), where a programmable temporal resolution of 1 ms is available.

The RTKDGPS receiver, the TPS system and the DGPS test receiver output their data via serial port (RS232). The PPS event generates a square pulse, which can also be detected via RS232 interface.

The final data recording and synchronisation system consists of a PC based platform, running Windows 2000 OS and using four serial ports (Figure 2).

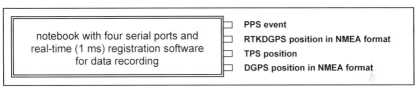

Figure 2. Schematic setup of measuring system for real-time data recording

One important result is the latency, that a certain receiver needs for position calculation and output. This delay entails a position displacement, dependent on the speed of motion. The latency of the tested receiver (Trimble AgGPS 132 with Omnistar DGPS correction service) as well as the reference system (Leica SR530 in RTK configuration) in different configurations and measurement modes are illustrated in figure 3.

The number of satellites used for position calculation has major influence on the latency. For increasing satellite number, the algorithm has to solve more and more unknowns, resulting in extended computational periods. The activation of filter algorithms for smooth position output always caused additional discrete offsets. Firmware updates generally resulted in increasing delays.

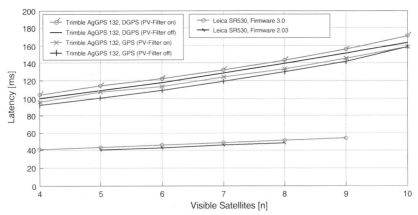

Figure 3. Receiver delay in dependence of satellite number.

The absolute position variance of the tested receiver (Trimble AgGPS 132 with Omnistar DGPS correction) was in the range of ± 3,5 m, whereas 60,8 % of 12.500 recorded points resided within an three-dimensional offset of 1,0 m. 79,3 % were within 1,5 m and 91,6 % were better than 2.0 m at maximum.

Extensive measurements with different receiver settings and operational modes have been made. This comprises GPS / DGPS operation, different correction services or different azimuth and elevation settings. Also, various receivers from low-cost (Garmin ETrex) to high-end have been investigated and compared.

Communication quality

The communication within an open system, as the LBS and ISOBUS protocols are, is very complex and allows the connection and communication of an unknown number of different components. Therefore, reasonable investigations require the definition of sensible boundary conditions. The design of a communication driver library, which fulfils all parts of the standard, builds the base for all investigations. This communication driver is a hardware independent software, which resides in each single ECU of the network and handles all communication among the ECU's as the standard claims.

Such a communication driver was developed within another project at the Technische Universitaet Muenchen, Crop Production Engineering under the terms Open Source Library LBS$_{lib}$ which supports the LBS protocol [Spangler et al., 2001]. Main effort is now spent on the further development of the communication driver to facilitate the conformity to ISOBUS, which is the preferred protocol for all examinations.

By means of the LBS$_{lib}$, a test environment is in preparation with a variable number of components like tractor ECU, user interface (terminal), GPS, data recording ECU and one or more implement controllers. A specific PC software will be the main tool for exploring the system in various conditions.

Considering examinations about communication quality, it is necessary to define the meaning of quality or quality indicating factors in this context. The following facts were identified of being important qualitative parameters.

- Compatibility:
  Full compatibility is the premise that a certain standard wins recognition and gets supported by industry. In case of an intelligent communication network, it must be ensured that each participant functions strictly according to well defined rules.
- Availability:
  The rigor environment, seasonal capacity overloads or weather dependency approve availability of agricultural equipment a high significance.
- Efficiency:
  The term efficiency covers factors like response times, minimal message size or rational data exchange intervals. Furthermore, a restricted bandwidth makes high demands on efficient communication.
- Security, Safety:
  In all sectors, where human live is endangered, safety considerations and precaution becomes important. An open bus structure facilitates the communication of several independent and unfamiliar components. Such a system allows several methods of control and therefore different levels of safety. Has the user a permanent emergency stop possibility? Are there any latent ambiguities, causing critical malfunctions?
- Intelligence, Flexibility:
  LBS and ISOBUS already imply a great flexibility and system intelligence as they are

designed to support highly complex system structures with a vast amount of participants, download functionality or commonly used components. The software design will have great influence on future demands as cruise control or implement controls tractor scenarios. This requires top implementation levels and a maximum flexibility of all interlinked components.

These and additional parameters such as malfunction, error management, response times, diagnosis or others should be investigated and quantified according to a predefined test scheme (Figure 4).

Layers	Quality features	Measurement ranges
Task / Application (e.g. sprayer control software)	security, safety	different levels
	malfunction	possible (y/n)
	error management	implemented (y/n), different levels
	response times	latency [ms]
	diagnosis	implemented (y/n), different levels
	...	
Communication driver (software) LBS$_{lib}$	compatibility	different levels
	efficiency	code size, run time [ms], ...
	flexibility	adaptation to different hardware, ...
	intelligence	dynamic storage management (y/n),...
	error management	implemented (y/n), different levels
	...	...
Hardware (CAN chips, electronics, connectors,...)	availability	breakdown (y/n), watchdog reset, ...
	temperature range	range sufficient (y/n)
	bus load	bus off risk (y/n)
	...	...

Figure 4. LBS / ISOBUS investigation test scheme.

Conclusion

The successful use of precision farming technology for automated data acquisition, site specific farming, fleet management and field robots makes it necessary to have a detailed knowledge on the accuracy of all utilised equipment. Two key technologies, GPS and LBS / ISOBUS communication have been identified to play a major role. Therefore, a system to investigate the performance of GPS receivers has been developed. The measurement system is able to quantify absolute position errors as well as calculation times and offsets. Several receivers have been investigated and in further proceeding, new receiver hardware and firmware will be tested in the same manner for generating comparable findings.

The communication on agricultural bus systems (LBS, ISOBUS) has to be explored in various conditions. The LBS protocol is already realised within the LBS$_{lib}$, whereas current work focuses on the development of ISOBUS conform code and a specific simulation and evaluation software. A test environment is in preparation and the first layout of a test scheme is already available.

Expected results are an universal evaluation of important precision farming equipment, allowing identification of limiting factors and formulation of general design rules for a new and higher accurate generation of agricultural control systems.

References

Auernhammer, H., Demmel, M., Ostermeier, R. and Weigel, R. 1996. Bus Configuration and Bus Load in a Tractor Fertilizer Spreader System (LBS by DIN 9684). In: Proceedings of the International Conference on Agricultural Engineering AgEng'96 Madrid, 207-208.

Demmel, M., Rothmund, M., Spangler, A. and Auernhammer, H. 2001. Algorithms for Data Analysis and first Results of Automatic Data Acquisition with GPS and LBS on Tractor Implement Combinations. In: Third European Conference on Precision Agriculture (vol. 1), 13-18. Montpellier, France.

Hoffmann-Wellenhof, B., Lichtenegger, H. and Collins, J. 2001. GPS Theory and Praxis. 5th ed. Vienna., Austria: Springer Wien NewYork.

Spangler A., Demmel, M. and Auernhammer, H. 2001. LBS$_{lib}$ als Open Source fuer Jedermann. In: Tagung Landtechnik 2001, VDI Berichte 1636, 101-106. VDI-Verlag. Duesseldorf, Germany.

Stempfhuber, W. and Maurer, W. 2001. Leistungsmerkmale von zielverfolgenden Tachymetern bei dynamischen Applikationen. In: DVW-Schriftenreihe 42/2001, Qualitaetsmanagement in der geodaetischen Messtechnik, 189-205. Wittwer Verlag. Stuttgart, Germany.

Stempfhuber, W. 2001. System Calibration in Precision Farming. In: Optical 3-D Measurement Techniques V, 366-376. Gruen and Kahmen, eds. Vienna, Austria: FIG Commision 5 and 6, IAG Special Comission 4, ISPRS Commission 5.

Acknowledgement

This project is part of the FAM research project and sponsored by the German Bundesminsterium fuer Bildung und Forschung (BMBF).

3D geographic information technology used for site-specific agricultural management

Sabine Grunwald
Soil and Water Science Department, Institute of Food and Agricultural Science, University of Florida, 2169 McCarty Hall, PO Box 110290, Gainesville FL 32611-0290, USA.
SGrunwald@mail.ifas.ufl.edu

Abstract

Optimized site-specific agricultural management depends on reliable and high-quality information describing the spatial distribution of soils, parent material, topography, and land cover. Our goal was to develop a methodology for quantitative, spatially-explicit 3D modeling which describe above and below ground soil and landscape properties. The study area was located in southern Wisconsin, USA. Geostatistical methods and 3D scientific visualization tools were used to reconstruct and visualize the study site. Validation showed a good fit between predicted and measured depths of soil material. The models are beneficial for farmers to optimize their land use management accounting for spatially-explicit land resource data.

Keywords: 3D, spatially-explicit models, soil-landscape models, site-specific management

Introduction

Optimized site-specific agricultural management depends on reliable and high-quality information describing the spatial distribution of soils, parent material, topography, and land cover.

Subsurface attributes have in common that they vary through three-dimensional (3D) space and through time. Commonly, two-dimensional (2D) maps are used to describe and visualize the spatial distribution of subsurface patterns (Pennock and Acton, 1989; Osher and Buol, 1998). In the U.S., soil information systems provide soil data in 2D spatial data formats (e.g. ArcView shapefiles) such as the State Soil Geographic Database (STATSGO 1:250,000) and Soil Survey Geographic Database (SSURGO 1:24,000). Other soil-landscape representations used a 2½ D design, draping soil and geologic data over a digital elevation model (DEM) to produce a 3D view (Su et al., 1996; Hogan and Laurent, 1999). Since this technique describes patterns on 2D landscape surfaces rather than the spatial distribution of subsurface attributes (e.g., soil texture, soil horizons), it fails to address 3D soil-landscape reality. Soil information systems such as SSURGO and STATSGO use the *entity geographic data model* based on the vector polygon as the geographical primitive. Each polygon is associated with an attribute data record. These crisp sets are too inflexible to take account of genuine uncertainty (McBratney and de Gruijter, 1992). The entity geographic data model is practical; however, it ignores spatial variation in both soil-forming processes and in the resulting soils. The spatial variation of soil properties within these polygons is unknown. In contrast, the *field view* displays the real world as a set of pixels or voxels (volume cells) (Burrough and McDonnell, 1998). The *field view* is used for modeling of natural phenomena that do not show obvious boundaries (e.g. soils) (Peuquet, 1988; Goodchild et al, 1992).

Discrepancies between spatial resolutions of below ground (soils) and above ground (topography, land cover, land use) properties are increasing at a rapid rate. For example, in Florida the average polygon size of SSURGO soil map units is 605,176 m^2 , which mismatches land cover and topographic data derived at spatial resolutions of 1 to 10-m (e.g.

land use derived from hyperspectral scenes such as IKONOS and topography derived using LIght Detection And Ranging (LIDAR) technology) to 30-m (e.g. land use derived from Landsat ETM7+ scenes and the U.S. Geological Survey National Elevation Dataset). Quantitative spatially-explicit soil datasets are lacking in the U.S. to address the need for precision farming, optimized land use management and soil and water quality assessment. Geostatistical techniques (Goovaerts, 1997; Burrough and McDonnell, 1998) facilitate production of spatially-explicit predictions which describe gradual changes of soil properties using the *field view* geographic data model. Yet most applications are still limited to 2D. Emerging geographic standards for 3D visualization such as the Virtual Reality Modeling Language (VRML) (McCarthy and Descartes, 1998) and X3D (eXtensible 3D) (W3D Consortium, 2003) are developing at a rapid rate but only few agronomists and soil scientists have adopted them.

The goal was to overcome these current limitations and to develop spatially-explicit 3D soil-landscape models which (i) describe the spatial distribution of subsurface attributes in the x, y, and z direction (3D), topography and land cover/use/management in 2D, (ii) integrate subsurface and surface attributes into one coherent model, and (iii) employ the *field view* geographic data model to capture gradual changes of subsurface and surface attributes over a study domain. The objectives were to reconstruct and visualize soil-landscapes in 3D using an object-oriented and multi-variate approach utilizing geostatistical methods and 3D rendering. The long-term goal was to improve the quality of agroecosystem representations and to optimize management to fit soil-landscape variability.

Materials and methods

Concept

Management data collected at the University of Wisconsin-Madison Agricultural Research Station West Madison (UW-AgWM) in southern Wisconsin, USA, indicated that on loess soils, the average corn yield was 9,406 kg/ha and on soils formed in shallow loess overlying thick sandy-loam glacial till corn yield averaged 5,640 kg/ha (internal statistics UW-AgWM, 2000). Soils were formed in loess and sandy-loam glacial till material which can be distinguished by soil characteristics such as bulk density, soil texture, soil water content, and organic matter content. It was hypothesized that penetration resistance (PR) can be used as a surrogate to characterize soil materials such as loess and glacial till showing contrasting soil characteristics. Mapping the 3D spatial distribution of penetration resistance infused with a digital terrain model could then be used to derive a stratigraphic 3D model showing the spatial distribution of loess and glacial till.

Dataset

The study site was a 2.73-ha field located on the UW-AgWM. Soils were classified as fine, mixed, mesic Typic and Mollic Hapludalfs and Typic Argiudolls which are highly productive. We used a Trimble 4600 LS differential global positioning system (dGPS) (Trimble, Sunnyvale, CA) to georeference sampling locations. The dGPS was mounted on an all-terrain vehicle to conduct a kinematic survey using a second receiver stationed close to the field to derive a DEM with 1-m grid size (vertical resolution error +/- 8 cm). A truck-mounted constant-rate profile cone penetrometer with 60° cone angel and 2-cm diameter surface area was used to measure cone index (CI) at 273 locations at depths down to 1.30 m on a 10 x 10 grid.

Twenty one locations were sampled; from each location and horizon a soil sample was collected, approximately simultaneously with penetration measurements. Each horizon was analyzed for soil texture, bulk density, and water content. For validation, soil cores at 77 sampling locations were collected using a targeted sampling design (dense sampling in topographic heterogeneous area). Soil cores (4.3-cm diameter) were analyzed at 5-cm depth increments for bulk density resulting in 1831 total samples and soil horizons. Elevations ranged from 321.3 to 329.6 m above mean sea level. Land use was a corn (*Zea mays*) - alfalfa (*Medicago sativa*) rotation.

Reconstruction and visualization

The flowchart below lists all major steps to reconstruct and visualize the quantitative spatially-explicit models (Figure 1). The logical data model of Roshannejad and Kainz (1995) was extended to represent data evolution as a sequence of spatio-temporal object quadruplets {(ID, s, t, c)} including an identification number (ID), spatial (s) and temporal (t) coordinates, and attribute values (c). Reconstruction comprised computation of quantitative spatially-explicit models using 2D and 3D ordinary kriging described by Goovaerts (1997). The SPlus (MathSoft Inc., Surrey, UK) and EVS-PRO (CTech Development Corporation, Huntington Beach, CA) software was used for spatial modeling. Model types comprised point, field view and stratigraphic models depending on the attribute being modeled. Innovative was the infusion of above-ground and below-ground attributes into one coherent 3D model. Measured

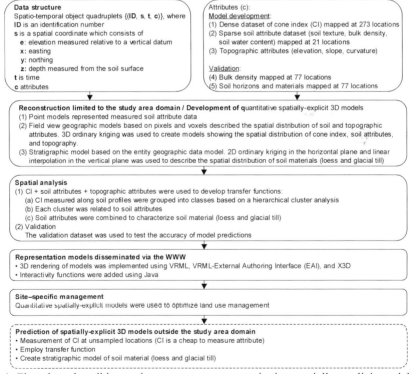

Figure 1. Flowchart describing major steps to create quantitative spatially-explicit models.

and predicted attribute values were visualized using virtual reality techniques. We used the Virtual Reality Modeling Language (VRML), eXtensible 3D (X3D), augmented with Java to visualize 3D models. To make results globally accessible and share spatially-explicit 3D models with farmers, land owners, soil scientists, and other users the models were posted on the Internet. Clients do not need any specialized software. They can download the models and render the 3D models using web browsers and a VRML plug-in available at no cost.

Results and discussion

The stratigraphic soil material model is shown in Figure 2. Our hypothesis was drawn in analogy to other studies (Hilfiker and Lowery, 1988; Unger, 1996; Tessier et al., 1997) in which PR was used to measure response to soil-factor combinations induced by different tillage systems. Glacial till and loess impart major control on crop growth and yield as well as the transport of agrichemicals. Stratigraphic detail helps to gain insight into complex phenomena such as the spatial distribution of soil attributes and soil materials. Validation showed a good fit between predicted and measured depths of soil material with a root mean square error of 3.81 for the top layer, 4.18 for the loess layer, and 4.98 for the glacial till layer. The models are beneficial for farmers to optimize their land use management accounting for spatially-explicit land resource data.

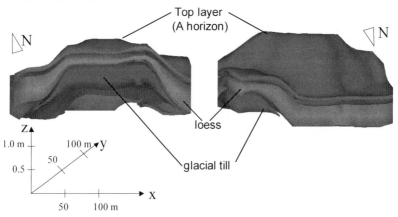

Figure 2. Stratigraphic model showing the spatial distribution of soil materials.

Conclusion

Site-specific analysis of soils, topography and land use/management requires to consider the 3rd dimension. The adoption of 3D spatial modeling and visualization techniques enables us to develop quantitative spatially-explicit soil-landscape models which are beneficial for precision farming.

Acknowledgement

I wish to thank my collaborators in various aspects of the research reported here: K. McSweeney and B. Lowery for financial support; P. Barak for collaboration on 3D modeling; G. Hart , A.I. Malik and D.J. Rooney for support to collect data. University of Wisconsin-Madison, Soil Science Department.

References

Burrough, P.A. and McDonnell, R.A. 1998. Principles of Geographical Information Systems – Spatial Information Systems and Geostatistics. Oxford University Press, New York.

Goodchild, M., Sun, G. and Yang, S. 1992. Development and test of an error model for categorical data. Int. J. of Geographical Information Systems 6 (2) 87-104.

Goovaerts, P. 1997. Geostatistics for natural resources evaluation. Applied Geostatistics Series, New York, Oxford University Press: 483.

Hilfiker, R.E. and Lowery, B. 1988. Effects of conservation tillage systems on corn root growth. Soil Tillage Research 12 269-283.

Hogan, M. and Laurent, K. 1999. Virtual earth science at USGS (U.S. Geological Survey). Available at: http://virtual.er.usgs.gov/

McBratney, A.B. and de Grujiter, J.J. 1992. A continuum approach to soil classification by modified fuzzy k-means with extragrades. Journal of Soil Science 43 159-175.

McCarthy, M. and Descartes, A. 1998. Reality Architecture – Building 3D worlds with Java and VRML. Prentice Hall, New York.

Osher, L.J. and Buol, S.W. 1998. Relationship of soil properties to parent material and landscape position in eastern Madre de Dios, Peru. Geoderma 83 143-166.

Pennock, D.J. and Acton, D.F. 1989. Hydrological and sedimentological influences on Boroll catenas Central Saskatchewan. Soil Science Society of America Journal 53 904-910.

Peuquet, D., 1988. Presentations of geographic space: towards a conceptual synthesis. Annals of the Association of American Geographers 78 (3) 375-394.

Roshannejad, A.A. and Kainz, W. 1995. Handling identities in spatio-temporal databases. In: Proceedings of ASCM/ASPRS Annual Convention and Exposition. Tech. Papers, 1995.

Su, A., Hu, S.-C. and Furuta, R. 1996. 3D topographic maps for Texas. Available at:http://www.csdl.tamu.edu/~su/topomaps/

Tessier S., Lachance, B., Chen, Y., Chi, L. and Bachand, D. 1997. Soil compaction reduction with modified one-way disker. Soil Tillage Research 42 63-77.

Unger P.W. 1996. Soil bulk density, penetration resistance, and hydraulic conductivity under controlled traffic conditions. Soil Tillage Research 37 67-75.

W3D Consortium 2003. Available at: http://www.web3d.org/.

An interactive tool to determine consitent within field patterns based on satellite imagery

F. Lahoche, J. Bergerou, L. Layrol and D. Lepoutre
GEOSYS, 20 Impasse Couzinet, 31500 Toulouse, France

Abstract

Management zones offer great promises in terms of improved bottom lines for farmers, and decreased pressure to the environment. This paper outlines how a combination of wavelet and segmentation algorithms can be used to define meaningful management zones solely based on satellite imagery. The input dataset was composed of thirty-eight archived satellite images (Hyperion, SPOT, IRS and Landsat TM / ETM) acquired across 10 growing seasons over Blue Earth, USA. A stack of the various wavelet transformed images was then used to obtain a vector-valued image representative of the within-field variations over the last decade. A segmentation scheme was then applied for an an efficient clustering of each pixel of interest into corresponding management zones. Results and validation are given on Blue Earth and also on Baziege site, in southern France.

Keywords: management zones, wavelet, satellite image segmentation, precision farming.

Introduction

The interest for management zones has grown in recent years as an easy way for the farmer to manage within-field patterns of variability. Management zones are commonly understood as areas within a field that show relatively little variability in growth conditions. The variation within a management unit is considered to be negligible for management purposes (Van Alphen and Stoorvogel, 1999), which allows reducing the theoretically infinite variation of growth conditions in the field to a limited set (Doerge, 1999). Adapting the management practices to the specifics of each zone could prove financially and environmentally beneficial. Increased input savings (e.g. variable rate seeding) and decreased nitrogen loss to ground water (variable rate fertilizer application) are among the most anticipated benefits from identifying these zones. The usefulness of remote sensing in precision agriculture has been widely examined in recent years (Robert, 1996). The majority of the previous research focused into the use of multiple satellite images for land cover classification and change detection (Coops and Walker, 1996; Lee and Marsh, 1995). The objective of our research project was to determine whether archived satellite images could also be used to determine management zones. Remote sensing is a cheap and valuable source of knowledge / observed in the definition of management zones mainly because radiometric measurements are closely related to crop behavior. The continuous spatial coverage of the fields at a fine spatial resolution (Doerge, 1999), and the speed of the zone determination using archived images make this technique superior to survey and yield based zones. A main objective of the present research was to assess whether wavelets and segmentation algorithms could be used in determining meaningful management zones. An additional objective was to assess which bands among our extensive database were bringing most of the information relevant to the segmentation process.

Materials and methods

Our research focused on the definition of management zones in corn and soybean fields around Blue Earth, Minnesota based on satellite imagery. Thirty-eigth images (Landsat, Spot,

IRS, Hyperion) were included in this study. All images were acquired across 10 growing seasons over Blue Earth, USA. Images were re-sampled to bring all of the resolution to 20 meters (down from 30 meters for Landsat).

Wavelet theory and use

Wavelets are mathematical functions that partition data into different frequency components. They represent a similar approach in signal processing as Short Term Fourier Transforms (STFT, a.k.a. Gabor), but with a variable window size (Matlab, 2002).
Wavelets decompose a 1D or 2D signal into a series of approximation (low frequency) and details (high frequency) functions, each sharing the same mathematical function (a.k.a. mother wavelet) but undergoing a specific dilation and translation.
A wavelet decomposition was used for each one of the 37 images based on an "a-trou" algorithm, as it offered the advantage to keep the same spatial resolution than the input image (Dutilleux, 1987). Isotropic filters have been used to avoid giving more importance to any direction during the decomposition. Wavelets transforms are very useful to consider images in a multi-resolution scheme, and therefore discern regions more or less large. On Figure 1, we display the result of four levels of wavelet decomposition on a uniform area (representing the field) with two separated randomly points distribution (representing within field zones). We are interested in segmenting the field into 3 zones, but not in segmenting the texture itself of the within field zones. Wavelet decompositions are then used in order to make a pre-segmentation tool. It becomes easier to segment the 4th level than the raw image.

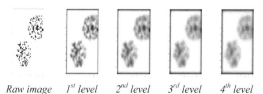

Raw image   *1ˢᵗ level*   *2ⁿᵈ level*   *3ʳᵈ level*   *4ᵗʰ level*

Figure 1. Four levels of wavelet decomposition on a synthetic image.

Each pixel was then represented by a whole vector (approximation of image from the wavelet transform), which translated in a vector-valued image of N values per pixel such as: N = (number of bands) * (decomposition at last level)
The next step consisted in the segmentation of the vector-valued image into regions or clusters. In order to do this, the vector-valued image was stored as a feature, and a region growing algorithm was used to segment the image. This entire process resulted in the classification of all the pixels within the area of study, which, in this application is a specific field.

Band selection

The automatic selection process described here had for objective to assess which bands among the 189 at our disposal had the most weight in determining the zones output. We compared the segmented zones using all of the bands at our disposal with the segmented zones using all of the bands but one for 36 different fields.
To perform an unbiased comparison, the mathematical derivative of both segmented images was computed, creating two binary gradient images of the zones. The difference between pixels on stable zones (derivative of zero) and pixels on transition zones (derivative other than

zero) outlined the limits of the zones within each image. Linear correlations were then run taking the binary image based on the entire database as our reference image. The greater the difference between these two final images (lower correlation coefficient), the more important the eliminated band was in determining the limit of the zones. There were as many comparisons as bands to compare. After all comparisons were finished, the image that differed the least from the reference image was definitely pulled out from the database.

A new round of comparisons was launched after the elimination of the less informative band, which resulted in the elimination of the second less informative band. One new band was pulled out after each loop of computation, until the segmented image using the remaining bands reached a correlation rate of less than 0.8 with the reference image, or there was no more bands left to eliminate.

Validation process

Also included within the database were yield data acquired between 1996 and 2001 using combine yield monitors. We obtained at least one year of yield data on over 80 fields in the Blue Earth, MN area, which represented around 3000 acres per year. Yield data was corrected for edge and time lags, and normalized to allow for comparisons between crops and years. Using an ISODATA classification, each pixel was assigned to a pre-determined number of classes. Depending on the available yield data, these yield classification maps were based on 2 to 5 years of yield data for 53 fields (no classification for fields with only one year of yield data). These maps were used as ground-truth data to verify the validity of the zones resulting from the satellite images segmentation. Most importantly, discussions with farmers brought significant input on the relevance and possible use of these zones. The segmentation process was performed in all of the 82 fields cultivated by the farmers involved in that study, and presented to them for feedback on the validity of our algorithms.

Also, on another area (Baziege, in Southern France) we compared the management zones results od our process with ground conductivity maps obtained with use of a Geonics EM38 sensor. Actually, conductivity measurements can be used to identify relatively homogeneous time invariant areas with regard to soil texture.

Results

Band selection

As explained in the Materials and Methods section, the algorithm removes one band after another from the original pool of 7 bands, until the segmented image achieves a correlation coefficient of less than 0.8 with the reference image or that there is no more bands to remove. The relative importance of each band in determining the overall management zones was outlined by whether or not they tended to be eliminated frequently. The range was quite wide, from the green band that was eliminated for almost every field, to the near infrared band that was eliminated for only 7 of the 36 fields when considering a single date image. Overall, the relative relevance of the seven bands studied was:

Near Infrared > Mid Infrareds >> Blue and Far Infrared > Red and Green

We also tested the segmentation process for some band combinations against the segmented image based on 7 bands (a.k.a. reference image). When only one band was used, the near and mid infrared bands achieved the best match with the reference image. Logically, the best result with several bands was achieved when combining all of the near and mid infrared bands. Also, the red and green bands did not bring significant information to the segmentation process. As a consequence, satellite images within our database that were acquired by the Landsat satellites were much more informative than images acquired by the Spot or IRS satellite which have no mid-infrared bands. The near and mid-infrared spectrums appear to carry the most valuable information for detecting management zones.

A similar use of Hyperion hyperspectral data (198 bands in visible, NIR and MIR bands) has been made and allowed a similar conclusion. The 198 bands have been combined into six main blocks: VIS (428-712 nm), NIR (722-943 nm), MIR1 (953-1134 nm), MIR2 (1145-1417 nm), MIR3 (1427-1911 nm) , MIR4 (1921-2396 nm). The segmentation process based on the entire 198 bands was then correlated to the segmentation based on some type of combination of these six blocks. Again, the central role of the near infrared spectrum is once again confirmed by this analysis, as the highest correlation for individual blocks was observed for the NIR block). The removal of the NIR block had a much stronger effect on the correlation level than the removal of any other band (e.g. the absence of the visible spectrum decreased the correlation in a small amount).

Validation process

Comparison with isodata multitemporal yield classification, and feedback from end users
When the segmented images were presented to the farmers, three different factors were advanced to explain the zones outlined by segmentation process (see examples on Figure 7). For at least one of the management zones in 72% of the studied fields, farmers considered the zones to be reflective of the water flow (or absence of flow) across the field. Thus, topography and drainage explained most of the zones observed. Although less common, differences in soil type and the resulting variations in yield potential was used to explain at least one of the zones in 42% of the studied fields. Finally, management practices (such as repeated manure application, tillage practices, crop rotation patterns…) were assessed as a leading cause for one of the zones in 16% of the studied fields.

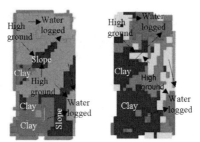

Figure 3. Results from the our process (on the left) and yield ISODATA classification for one field, with explanatory factors outlined by farmers. Arrows represent water flow.

The zones solely derived from satellite images were also compared with the zones issued from the multi-temporal yield-based (ISODATA) classification on 53 different fields. In more than 85% of the cases, farmers deemed the segmentation process as much or better representative of the actual within-field variability than the ISODATA yield classification. Figure 8 presents an example of how close the zones based on these two different methods can be. When yield-based zones were derived from only 2 or 3 years of data, farmers were always more confident in the zones based on image segmentation rather than yield classification. Following these results, farmers decided to customize the seeding rate to the characteristics of each zone determined through image segmentation on 16 of their fields.

Comparison with EC maps

The ground-truth data (electrical conductivity), as well as the images for two fields were acquired as part of a separate research project. Fields are located at Baziege, near Toulouse, in Southern France.

Electrical conductivity was measured with a Geonics® EM38 pulled by a quad machine. This non-invasive tool measures an induced electro-magnetic current between two electrodes. EM38 measurements have been taken during the summer of 2001, just after harvest to have bare soil conditions. The weather conditions have not been the same for all four fields, and the ground was more or less dry depending on the rains. That constraint could have perturbed acquisitions.The following image database was composed of the following characteristics: Spot, (Panchromatic, 10 meters resolution, acquired 05 Apr. 1996), IRS (Multispectral, 5 meters resolution, acquired 23 Sep. 2000), Landsat5 (Multispectral, 30 meters resolution, acquired 11 Feb. 2002), Landsat7 (Multispectral, 30 meters resolution, acquired 31 Aug. 1999), Aerial photography (IRC (3 bands with IR), 1/10.000 resolution, acquired 30 July 2001.

All images were brought to a 5 meters resolution (based on a bilinear method) to allow for the stacking of the bands from all satellites into one single matrix. Histogram equalization was applied on each band to better outline the within-field contrasts. We display on Figure 4, and over two fields, the obtained segmentation for these fields when the entire matrix of images is used as input to the algorithm. Limits of the zones are overlaid on top of EC maps.

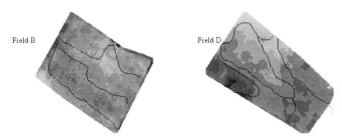

Figure 4. EC maps with overlaid management zones based on our segmentation algorithm for fields A and D. (whote: low value of EC, black: high).

Results are quite satisfactory. The determined zones appear to correspond to some degree to soil variability. The differences between the observed and predicted within-field variations could be in part due to the type of meteorological conditions encountered during the EM38 acquisitions.

An interactive tool

Our process has been written in C++ to allow fast processing and use over PC platforms. The C++ code has been then compiled into a dll that can be called by other applications, like web services.

A web interface prototype has been created for applying SAMZ algorithms to archive images in real time, based on JAVA technologies and Mapserver open source developed by

University of Minnesota. Several tools allow the navigation in the image and the selection of the area that will be processed for management zones delineation. Shapes layers for road and rivers are also displayed to make the navigation easier in the Blue Earth area.

Conclusions

Vector valued images issued from wavelet decompositions and segmented into one overall image reveal management zones. These zones are consistent with the farmer knowledge of the within-field variability and with ground-truth yield data. The information provided by the segmentation process was used for zone-specific management. The near and mid infrared bands seemed to contain most of the information relative to the within-field variability but still need to be combined.with other bands at different time period for a maximum representation of the total variability. Rules to select optimal combination of images to define management zones over any given area is the next step of this research program.

Acknowledgements

This project was funded by NASA, contact number NASA-NRA-98-OES-09.

References

Coops, N.C. and Walker, P.A. 1996. The use of the Gower metric statistic to compare temporal profiles from AVHRR data: a forestry and agriculture application. International Journal of Remote Sensing, 17 (17): p. 3531-3537.

Dutilleux, P. 1987. An implementation of the "algorithme à trou" to compute the wavelet transform. In J. M. Combes, A. Grossman, and Ph. Tchamitchian, editors, Wavelets Time-Frequency Methods and Phase Space, pages 2-20, Berlin Heidelberg, 1989. Springer-Verlag.

Doerge, T. 1999. Defining management zones for precision farming. Crop Insights, 8 (21). Available on the internet at:
www.pioneer.com/usa/technology/precision_farming_management_zones.htm

Matlab, 2002. Wavelet Toolbox built-in help. Matlab\Help\Contents\Wavelet Toolbox.

Robert, P.C. 1996. Remote sensing requirements for precision agriculture. Multispectral Imaging for Terrestrial Applications. In Proc. International Society for Optical Engineering, 8-9 August 1996, Denver, Colorado. Vol. 2818: pp. 54-58.

Paz, J.O., Batchelor, W.D., Colvin, T.S., Logsdon, S.D., Kaspar, T.C. and Karlen. D.L. 1998. Analysis of water stress effects causing spatial yield variability in soybeans. Transactions of the ASAE. 41, pp. 1527-1534.

Van Alphen, B. J. and J. J. Stoorvogel. 1999. "A Methodology to Define Management Units in Support of an Integrated, Model-Based Approach to Precision Agriculture." In Proceedings of the Fourth International Conference on Precision Agriculture, 19-22 July 1998, Minneapolis, MN, ASA-CSSA-SSSA, Madison, WI, 1267-1277.

Mapserver
http://mapserver.gis.umn.edu

Student design contests promote hands-on learning and innovation in precision agriculture

J. Müller, J. Smit, J.W. Hofstee and D. Goense
Wageningen University and Research Centre, Mansholtlaan 10-12, 6708 PA Wageningen, The Netherlands
Joachim.mueller@user.aenf.wag-ur.nl

Abstract

Student numbers in Agricultural Engineering are decreasing. The high-tech approach of Precision Agriculture with its strong involvement of Information Technology is certain to improve the public image of Agricultural Engineering and to attract the interest of high school students. Hence, Wageningen University and Research Centre (The Netherlands) is organizing a two days field robot event. The combination of a student design contest and a scientific fair generates high academic, agricultural and publicity value.

Keywords: field robot, design contest, education, students

Introduction

As a global trend in Europe and the U.S.A., student numbers in Agricultural Engineering are decreasing in spite of a good job market for graduates. Technically oriented high school students prefer studies that are more related to high-tech disciplines and Information Technology. Precision Agriculture includes both aspects and offers a good opportunity to attract more students. Especially the combination of "serious" and "playful" aspects of robotics are appealing to the upcoming student generation. Therefore, Wageningen University and Research Centre (WageningenUR) is organizing a field robot event, including a student design contest as well as a scientific fair.
The objectives of the WageningenUR Field Robot Event are:
- Improving the public image of Agricultural Engineering
- Attracting interest of high school students for Agricultural Engineering
- Creating a platform for exchange of knowledge on field robots
- Harnessing students creativity to promote the development of field robots
- Promoting off-curriculum skills like communication, teamwork, time management and fundraising

Materials and methods

The WageningenUR Field Robot Event is composed of various academic, scientific and publicity elements (Figure 1). The pedagogic elements are derived from the design competition of the Massachusetts Institute of Technology (MIT). As design object, field robots have been chosen, i.e. the scientific element is originated in precision agriculture. Publicity elements such as open-air atmosphere and agricultural context are in common with events such as tractor pulling. These elements will be explained in more detail.

MIT 6.270: academic archetype of the WageningenUR Field Robot Event

At the Massachusetts Institute of Technology, design projects based on Papert and Harel's (1991) educational theory of constructionism have a two decade tradition. According to the theory of learning-by-creating, people learn most effectively when they create an external artefact in the world and therefore have to develop new ideas without being obstructed by existing solutions.

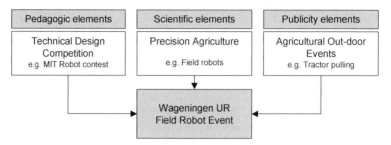

Figure 1. Academic, scientific and publicity elements of the WageningenUR Field Robot Event.

One of such design projects with course number 6.270 started in 1987 as a programming competition in which students wrote computer games simulating robots that tried to find and destroy other robots. Two years later, the course was developed to a hands-on class where students designed and built a real robot that played in a competition at the end of the course. The students worked in teams of two or three. Each team was given the same kit containing various sensors, electronic components, batteries, motors, and LEGO. The kits were handed out three weeks before the competition. The objective for the students was to design an autonomous machine that would be able to navigate its way around the playground, recognize opponents, and manipulate game objects without human interference. Teachers of 6.270 found, that students *"...can learn everything they need to know by working with each other, being introduced to some material in class and, mostly, by hacking on their robots"* (Anon, 2003).

The game idea and playground are developed afresh every year. Figure 2 shows as example the playground of the contest in 2001 together with a robot in action.

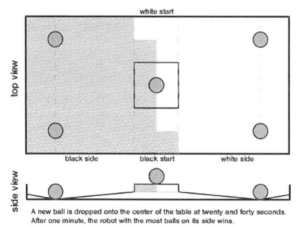

Figure 2. Playground of the 6.270 MIT's Autonomous Robot Design Competition 2001 "Master of the Universe" (Anon, 2003).

Meanwhile, the MIT competition became very popular and similar competitions have been organized worldwide at many technical universities. Winning teams are sent to a final international contest.

Field robots: scientific object of the WageningenUR Field Robot Event

Technologies in Precision Agriculture concerning GPS-navigation and sensor systems have now reached a stage which seems to make autonomous vehicles for open field operations a realistic option. Recently, the number of research groups presenting their field robots at congresses or in scientific journals has been increasing. The field robots can be classified into auto steering tractors, super-canopy robots and sub-canopy robots. Examples are presented in Table 1.

Table 1. Examples for different types of field robots.

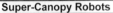

Auto Steering Tractors	
	Auto Steering Tractors are based on conventional tractors but equipped with navigation systems and other sensors. Example: Cemagref, LASMEA, France (Thuilot et al., 2002)
Super-Canopy Robots	
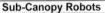	Super-canopy robots are small platforms, which are straddling one or more rows and surmounting the crop canopy. Example: Technical University of Denmark (Madsen and Jakobsen, 2002)
Sub-Canopy Robots	
	Sub-canopy robots are small enough to move in-between the plant rows. Example: Wageningen University, The Netherlands (Claessens et al., 2002)

Auto steering tractors are based on conventional tractors that are equipped with high quality GPS-navigation systems and various sensors to keep track on a programmed path without the intervention of a driver. Super-canopy robots are autonomous vehicles without driver cabin. These platforms are smaller than conventional tractors, navigating on a programmed path or along the plant rows with the help of sensors. Super-canopy robots straddle one or more plant rows, riding over the crop canopy. Sub-canopy robots are small enough to move in-between the plant rows, i.e. below the crop canopy. High quality GPS receivers are presently still too heavy for sub-canopy robots; navigation is based on machine vision or other sensor systems.

Results

The Wageningen UR Field Robot Event combines academic, scientific and publicity elements, inviting university teams European-wide to contribute to a two days inter-campus event on autonomous vehicles in agriculture, see Figure 3.

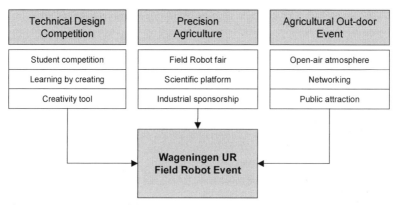

Technical Design Competition	Precision Agriculture	Agricultural Out-door Event
Student competition	Field Robot fair	Open-air atmosphere
Learning by creating	Scientific platform	Networking
Creativity tool	Industrial sponsorship	Public attraction

Wageningen UR Field Robot Event

Figure 3. Composition of the WageningenUR Field Robot Event.

A student design contest comparable to the MIT's Autonomous Robot Design Competition is integral part of the event. Instead of developing new robot playgrounds every year, as practiced in the MIT contest, a challenging playground is provided by the natural degree of non-uniformity of a maize field. Navigating in the plant rows, turning at the headland and treating obstacles are program disciplines. Scouting capabilities are performed in freestyle and rated by an expert jury. In contrast to the MIT contest, there is plenty of time to design a robot during spare time and the choice of components is not limited to a standard kit. This allows unhampered creativity of the participants. In return, the jury assesses the hardware costs of the robot, based on the bill of material, admitting economy to an additional selection criterion. By accepting freestyle robots that are built at the home universities, the contribution of non-student resource persons cannot be ruled out. This is not a disadvantage but an increase of value, because identifying and motivating resource persons as well as fund-raising are skills which students are also learning from the project. Furthermore, the system is more promoting the spirit of university competition than individual competition, turning loose additional resources and creative energy. Parallel to the student teams, research teams are also invited to demonstrate the performance of their autonomous agricultural vehicles in the contest, participating in a separate, professional class.

All student and researcher robots are presented together with posters and additional information material in a Field Robot fair that fills half a day of the program. The fair is forming the scientific platform of the event, showing the state of the art in agricultural field robotics and promoting the exchange of knowledge amongst researchers and students. Taking into consideration the creativity, curiosity and IT-skills of the e-generation, the flow of knowledge between scientists and students is not limited to a one-way direction, inspiring the idea of the university as *universitas magistrorum et scholarium*.

The Field Robot fair is not only a scientific platform but also a market place for innovative companies in agrotechnology, to demonstrate advances in mechatronics and to recruit high-potential students for their R&D-departments. As the invitations for the event are sent to all relevant university departments in Europe, company sponsoring is expected to pay off well.

Finally, using a maize field as playground offers not only a real-world scenario in terms of weather, plant and soil condition but also creates an agricultural open-air atmosphere, attracting additional visitors from the surrounding. Media reports about the event are reflecting an innovative high-tech image of education in Agricultural Engineering and attract potential first-year students.

Discussion

In Figure 4 the WageningenUR Field Robot Event is ranked according to its academic, agricultural and publicity value in comparison to tractor pulling, scientific congresses on field robots and the MIT's Autonomous Robot Design Competition.

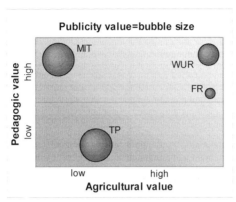

Figure 4. Portfolio of agricultural, academic and publicity value of events such as tractor pulling (TP), scientific congresses on field robots (FR), MIT's Autonomous Robot Design Competition (MIT) and WageningenUR Field Robot Event (WUR).

The MIT's Autonomous Robot Design Competition as prime example of constructionism is of high academic value. Also the publicity value is high due to the standard rules and the high worldwide popularity. However, the MIT's Autonomous Robot Design Competition is an indoor event, focused on mechanical engineering and therefore the agricultural value is low. Tractor pulling events have developed to sport-spectacles with high publicity and professional teams. The reference to its agricultural origin is decreasing. Apart from special tractor pullings like the ASAE Quarter Scale Tractor Competition for students, the academic value of typical tractor pulling events is low. Field robot sections in scientific congresses are innovation motors for Precision Agriculture and hence of high agricultural value. There is a medium academic value for students who are visiting such congresses. The pedagogic value increases if students are enabled to contribute to the program. Field robot sections are focusing on circles of experts; therefore the publicity value is low. Combining elements of the events mentioned, the WageningenUR Field Robot Event is ranking high in the three chosen scales: letting the students create own field robots is of high academic value that is increased by meeting leading researchers during the event. As the WageningenUR Field Robot Event is focusing on field robots and includes also a Field Robot fair the agricultural value is also scoring high. Being the first event of this kind, the publicity value of the WageningenUR Field Robot Event is still medium, but could increase if expanded into an annual or biannual event.

Conclusions

Due to the quickly growing knowledge in navigation and machine vision, various groups of scientists are working on the development of autonomous vehicles for precision agriculture. Most of these activities are based on national research projects that are not linked on an international level. A competition of such field robots is a welcome occasion for the researchers to get together in a sportsmanlike atmosphere for personal exchange of knowledge and a unique opportunity for the audience to see a variety of autonomous vehicles at work. Opening the competition for student teams gives a fresh impulse of creativity in Agricultural Engineering. Furthermore, hands-on learning by designing own field robots is a strong motivation for students, also addressing off-curriculum skills like communication, leadership, teamwork and fundraising. The publicity effect of such contests might increase the interest of high school students to get involved into the subject. Like the MIT's Autonomous Robot Design Competition, the WageningenUR Field Robot Event should become an annual or biennial event and the design process should be embedded in the curriculum according to the educational theory of constructionism. Due to the high costs of the required components, periodical sponsorship will be required in the long run. With respect to the growing mechatronic market and its need of skilled and innovative employees, companies would benefit from sponsoring such student design contests.

References

Anon. 2003. 6.270 MIT's Autonomous Robot Design Competition. Http://web.mit.edu/6.270, viewed 7/1/2003.

Claessens, L., Huizinga, P., Huls, M. Miltenburg, A. Thelen, J. and Vanthoor, B. 2002. Robot ants in micro-mechnanisation of plant production. Proceedings of AgEng2002, Budapest, Paper-No 02-VE-001, 8 pp.

Madsen, T. E., Jakobsen, H. L. 2001. Mobile Robot for Weeding. Master Thesis Project at the Department of Control and Engineering design of the Technical University of Denmark, pp. 160.

Papert, S. and Harel, I. 1991. Constructionism. Ablex Publishing Cooperation, 519 pp.

Thuilot, B., Cariou, C., Martinet, P. and Berducat, M. 2002. Automatic Guidance of a farm Tractor relying on a single CP-DGPS. Autonomous robots 13 53-71.

Spatial dependence of available phosphorous in grazed silvopastoral system using semivariogram models

L.C. Nwaigbo[1] and G. Hudson[2]

[1]Faculty of Agriculture and Veterinary Medicine, Abia State University, P.M.B 2000, Uturu, Abia State, Nigeria.
[2]Macaulay Land Use Research Institute, Craigiebuckler, Aberdeen AB15 8QH, UK.

Abstract

Semivariogam models were used to study the spatial dependence of soil (0-20 cm depth) available phosphorus (P) at tree scale in a silvopastoral system site at Glensaugh, Northeast Scotland. Treatments were plots of about 0.8 ha each planted with Hybrid larch (*Larix eurolepis* Henry) at 5m x 5m (HL400), 7m x 7m (HL200) and 10m x 10m (HL100), spacings (stems ha-[1]) on a predominantly rye grass (*Lolium perenne*) pasture grazed by sheep yearly from April to October. The randomised complete block design also included grazed pasture plots without trees (Agricultural control, AC). All treatments were replicated three times. Soil samples collected at 1m intervals around two randomly selected trees in each plot and from the two hypothetical tree positions in the AC plots were used for spatial dependence analyses using semivariogram models. Three geostatistical models were tested for the spatial distribution of soil available P at tree sacle. Spatial dependence of soil available P was observed only in the AC and HL100 treatments but while the AC showed isotropic (similar) spatial dependence the HL100 had anisotropic (dissimilar) spatial dependence in the North-south and East-west directions.The power model based on least squares technique weighted by number of sample pairs divided by sample distance had the best fit of 82.4% and 67.5% for the AC and HL100 treatments respectively. The estimated range of spatial dependence was between 5 and 6m in both treatments. There was no spatial dependence of avialable P observed in the HL200 and HL400 treatments. This indicates a random distribution pattern of available P in the HL200 and HL400 plots. It may suggest the effect of increasing tree density and the low animal: tree ratios of the treatment plots and will improve our understanding of the spatially variable fertiliser applications that may be required.

Keywords: phosphorous, spatial dependence, trees, grazing, semivariogram

Introduction

The phosphorous status of many European agricultural soils has risen steadily during the last decades. This is due to generous fertilising needed to improve poor P plant uptake efficiencies. In a derived agro ecosystem such as the silvopastoral system, some soils especially those common to acid upland areas, still have relatively low amounts of available P. Human activities such as agriculture influence P concentrations and plant - animal interactions can influence the soil nutrient distribution pattern in silvopastoral systems (Nwaigbo, 1996). However, it has been found that soil variables can exhibit some degree of spatial dependence - values at sites that are close together are more similar than those further apart (Webster & Oliver, 1990). The semivariogram is an important tool of geostatistics (Webster & Oliver, 1990) that can be used to model, predict, estimate and eventually map spatial patterns of soil physical and chemical properties. Geostatistical methods are based on a fundamental assumption that points situated close to one another in space share more similarities than those farther apart. This means that the squared difference in values between two points is a quantity that depends only on the distance between them and not on their absolute positions (Bringmark & Bringmark, 1998). The objective of this study is to assess the spatial dependence of soil available P around the trees in grazed silvopastoral system at Glensaugh northeast Scotland using semivariogram models.

Materials and methods

The soil samples for available P determination were collected (0–20cm depth) in N-S and E-W directions at 1 m intervals from around two randomly selected trees in each grazed silvopastoral treatment plot planted in April 1988 with Hybrid larch (*Larix eurolepis* Henry) at 5m x 5m (HL400), 7m x 7m (HL200), and 10m x 10m (HL100) spacings (stems ha-[1]) on an upland ryegrass (*Lolium perenne* L) pasture site in NE Scotland. Included in the randomized complete block design as control treatment were grazed pasture plots without trees (Agricultural control, AC) from which soil samples were also collected from two hypothetical randomly selected tree positions. The treatments were replicated over three blocks of land. The plots were grazed by sheep yearly from April to October and received an annual application of 160 kg N ha-[1]. Soil samples around each tree were collected up to mid point of the separation distance between the trees. The soil data were used to determine semivariograms of available P for each treatment. The considered resolution was for 8 lag distance which was more than half the maximum distance in any direction. The experimental semivariogram was obtained by the relationship (Lascano & Hartfied, 1992).

$$\gamma^*(h) = 1/2N(h) \sum_{i=1}^{N(h)} [Z(x_i + h) - Z(x_i)]^2$$

where γ

$\gamma^*(h)$ semivariogram estimator for lag distance class h;

$Z(x_i)$ measured sample value at point x_i;

$Z(x_i + h)$ measured sample value at point $x_i + h$;

$N(h)$ total number of pairs of observations $[Z(x_i), Z(x_i + h)]$ separated by h.

The fits of three semivariogram models were tested for the soil available P data:

(i)   Power model: $\gamma(h) = C_0 + bh^a$, where $0 < a < 2$ and $b > 0$. Also where b is the slope and C_0 is the intercept (nugget effect);
        A power variogram indicates a linear spatial gradient.

(ii) Spherical model $\gamma(h) = C_0 + C_s[(3h / 2a) - (h^3/2a^3)]$ where $0 < h < a$ ..... (a)
                $\gamma(h) = C_0 + C_s = C$ where $h > a$ .... (b)
where C_0 is the nugget component, C_s is the spatially dependent component, a is the range of the semivariogram beyond which a soil nutrient is no longer spatially correlated; and

(iii) Exponential model $\gamma(h) = C_0 + C_e [1 - \exp(- h/a)]$ where C_e is the exponential component.
The above models were fitted to the experimental semivariogram data in three ways:
(a) no application of weight;
(b) weighted by the number of sample pairs;
(c) weighted by the number of sample pairs divided by sampling distance (lag distance).
The fitting of various models and estimation of the parameters of the models were carried out using least squares procedures (Cressie, 1985).

Results and discussion

The semivariogram indicated spatial dependence of availabe P in the HL100 and AC treatments (Figs 1 & 2) respectively. The estimated range of spatial dependence of soil available P was between 5 and 6m in both HL100 and AC treatment plots. There was no spatial dependence of available P observed in HL200 and HL400 treatments and this suggests a random distribution pattern of available P around the trees in these treament plots. The power model for all directions weighted by the number of pairs divided by the sampling distance gave the best fit for the soil available P data. Grazing of pasture by sheep was common to all the treatment plots but the best fit was in the AC plots which had no trees ($r^2 =$ 82.4) while the HL100 which had the least tree density was next best with $r^2 = 67.5$ (Table 1). The Power model semivariogram for the HL100 showed anisotropic (dissimilar) spatial dependence in N-S and E-W directions (Figure 1). That is to say that the semivariogram of available P in the N-S direction is different from that in the E-W direction. The AC exhibited isotropic (similar) spatial structure in N-S and E-W directions (Figure 2). In these AC plots there were no trees to serve as rallying points for the sheep as was the situation in the HL100, HL200 and HL400 treatment plots. The difference in the spatial pattern of available P in HL100 and AC plots may be due the impacts of sheep around the trees which serve as rallying points for the sheep which seek shelter and shade under trees in silvopastoral systems (Sibbaid et al., 1995). The absence of a defined spatial pattern of available P in the HL200 and HL400 treatment plots as was the situation in the HL100 plots may suggest the effect of increasing tree density and the low animal: tree ratios of the treatment plots. Animal - tree interactions in silvopastoral systems, therefore can cause significant redistribution of soil nutrients around the trees. In the absence of trees in grazed pastures the spatial pattern of available P was similar in all directions. This pattern varied in directions with the presence of trees at HL100 and the pattern became a random one at HL200 and HL400 stems ha-[1]. At higher tree density ha-[1] it was likely that the animal impacts were spread over many trees coupled with the influence of the trees on soil available P pool. The HL100, HL200, HL400 and AC plots were about the same size and had the same number of sheep of about the same weight and were therefore expected to have been influenced equally in every way by the sheep. This was not the case and the reason may be due to the presence of trees and the differences too in the number of stems ha-[1] of the plots (Nwaigbo, 1996). The impact of these on the spatial variability of available P around the trees in grazed pastures has an effect on the subsequent need for application of fertiliser inputs in grazed systems with trees.

Table 1. Power model estimated parameters of P for HL100, HL200, HL400 and AC (Control) treatments.

Treatment	Weight	Fit	Nugget	Slope
HL100	Number of pairs/ lag distance	67.5	0.467	0.146
HL200	Number of pairs/ lag distance	-	1.041	0.004
HL400	Number of pairs/ lag distance	23.9	0.695	0.122
AC	Number of pairs/ lag distance	82.4	0.425	0.166

Figure 1. Fitted Power model weighted by number of pairs/lag distance for anisotropic P in the N-S and W-E directions in HL100.

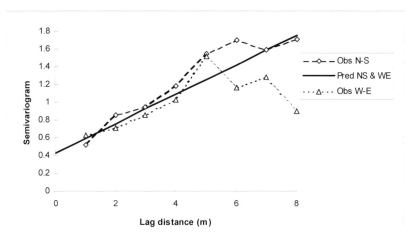

Figure 2. Fitted Power model (weighted by number of pairs/ lag distance) for isotropic P in the N-S and W-E directions in AC.

Conclusion

It is an acceptable fact that animals in silvopstoral systems seek shade and shelter around trees. This interaction can cause soil nutrient redistribution in such systems. In the grazed pastures without trees spatial dependence pattern of available P was definitive and similar in all directions. The presence of trees, animal and the interactions between such components of the silvopastoral system redefined this pattern and at a certain level of tree density the spatial dependence pattern became random. This variation can only be attributed to the presence of trees.

Acknowledgements

LCN is grateful to the Association of Commonwealth Universities, Commonwealth Scholarship Commision, UK, for funding this study. Also the kind permission of the Director Macaulay Land Use Research Institute, Aberdeen, to use the agroforestry research site at Glensaugh, Scotland, is appreciated.

References

Bringmark, E. & Bringmark, L. 1998. Improved soil monitoring by use of spatial patterns. Ambio 27 (1) 45-52.

Cressie, N. 1985. Fitting variogram models by weighted least squares. Mathematical Geology 17 563-586.

Nwaigbo, L.C. 1996. Spatial variation of tree growth and site factors in a silvopastoral system in northeast Scotland. PhD Thesis, University of Aberdeen, Aberdeen, UK.

Shand, C., Edwards, A.C. and Macklon, A. 1994. Nutrient availability: Phosphorus. Macaulay Land Use Research Institute (MLURI), Aberdeen.

Sibbald, A.R., Dick, J and Iason G.R. 1995. The effects of the presence of widely spaced trees on the behaviour of sheep. Agroforestry Forum 6(2) 22-25.

Webster, R and Oliver, M.A. 1990. Statistical Methods in Soil and Land Use Resources Survey. Oxford University Press, Oxford. 316pp.

Application of simple technologies for a precision farming model in mexican agriculture

H. Ortiz-Laurel and D. Rössel
Campus San Luis Potosi, Postgraduate College, Iturbide 73, Salinas de Hgo., S.L.P., C.P. 78600, Mexico
edietmar@colpos.mx

Abstract

The aim of precision agriculture is to provide for each different unit or plot an optimal treatment in space and time by using advanced and integrated technological system components. In Mexico, farmers showing great interest in precision farming are to some extent unwilling to accept it because of the cost and complexity. This is also due to: a) The reduced size and irregular shape of plots, b) High investment for the monitoring equipment and associated devices, c) Lack of training for using information technology and advanced electronics, d) Limited availability of specific software and e) Difficulty in explaining how to interpret production data. However, some type of simplified precision agriculture will have to be aimed at for Mexico, where the differentiated management of the entire set of production factors is exercised through precise organisation and direct intervention. This reports ends with a very cautious assessment of the prospects for the development of precision farming in Mexico.

Keywords: information technology, advanced electronics, control process, small farms.

Introduction

Mexico covers an area of about 2 million sq. km and has a population of about 100 million people. The economically active population is around 36.6 million and around 18% are engaged in agriculture. Of 31.1 Mha of agricultural land, 5.6 Mha (18.1%) are irrigated and 25.5 Mha (81.9%) are rainfed (Table 1). According to farm size, most farming units are small and as agricultural mechanisation is concerned it can be said that it still ranges from hand tools, tools for draught animals to mechanised equipment.

Table 1. General situation of Mexican agriculture.

Agricultural land, Mha	31
Cultivated land, Mha	23
Irrigated land, %	18
Agricultural farms, million	2.8
Average size of farmland plots, ha	12.1
Number of tractors	170 000
Number of combines	20 000
Tractorization index	0.3501

Mexican agricultural production has increased, and concentrates on cereals mainly for domestic consumption. This is achieved in very concentrated and defined zones, and by using big machines, tillage implements and farm equipment, harvesting operations fully mechanised and efficient use of good quality seeds, fertilisers, pesticides and irrigation. On average the mechanisation index is 0.443 kW/ha of energy applied on farming activities (Lara Lopez, 2000).

In Mexico there is a wide contrast of farming activities. Much of the land (5 574 770 ha) is cultivated by small farmers (2 620 399 holdings with < 5 ha) but there are also areas of large farms which are already highly mechanised. To meet the national needs for increased crop production, it is necessary for the small farms to adopt improved technology including mechanisation of farming operations, and for the larger farms to keep abreast of current developments in agricultural engineering technology such as precision farming techniques.

Larger-scale Mexican farmers are aware that they can benefit from increased speed, accuracy and efficiency through the use of precision agricultural tools. Farmers adopting precision farming understand that they will be doing their own research as they adopt and demand any upgraded technologies for improved management of crop-production systems. At this stage, great emphasis is required to design as simple as possible research approaches to help farmers to learn the complex interactions between the multiple factors affecting crop growth and decision making, and they should know that managing this technology has a multidisciplinary character. The degree of adopting and using the various levels of precision farming technology may be highly variable. In fact, the need and appropriateness of this technology will be dependent on the field conditions and producers' management abilities.

Farmers are reluctant to adopt any new farming technologies if costs are increased or profits reduced. Although herbicides have assisted the economic viability of farmers and helped reduce the risk of soil erosion, there are many situations where weed control with sprays is more expensive than cultivation, and their use is seen as a potential ecological hazard. By spotting and spraying only weeds can reduce overall use of pesticides and increase adoption of sustainable tillage practices by farmers.

In this work, Mexican agricultural technology had a brief assessment and focus was trained on the advantages provided by new machines equipped with electronic and control systems able to reduce the consumption of pesticides and herbicides as well as of mineral fertilizers. In this connection, the development of technological systems for the concrete application of precision agriculture is especially interesting, specifically as regards the use of fertilisers and pesticides as well as the dissemination of innovative practices in soil cultivation with the right machinery.

Materials and methods

The general objective in agricultural production is to have a highly efficient agriculture, market oriented and environmentally sustainable. This mission can be achieved by means of different effects and phenomena, such as; increase of accuracy, synergy application, more complex modeling, etc. Precision agriculture is based on accuracy to carry out each process, therefore, it requires control systems to support the technique. At present, agricultural production is the science of technology with key elements from the technique, the biological process and process management by using equipment handled by man. However, it is highly expensive to achieve the greatest accuracy.

The average size of farms in Mexico is 12.1 ha. In these small farms, agricultural activities must be fully organized in order to be efficient and the work the machines are used for organized appropriately. There are suitable ways and conditions for managing crop production and its inputs, such as concentrating the production as a mean of organizing the production; climatic conditions; a suitable market; etc. The interaction between these elements is important in order to get a technique with quality and quantity, for the nature biological process and the human being.

A successful area for precision agriculture is agrochemical application. In Mexico, for instance, in the last 30 years there has been an increase of 1000% in the amount of pesticides sprayed on crops (Ortiz-Laurel & Rössel, 2002). It is possible to increase spraying accuracy through a high technical control and as well as, through applying technical principles in order to have adaptable equipment, the adequate mix concentration, the right time for spraying, etc. The latter ones are less expensive compared to an overall technical control and also it ensures profits and reduces negative environmental impacts. Using a local sensor-based system, weeds can be identified so overall treatment rates can be reduced to the easy-to-kill targets. It is also expected an imminent production of distributing machines with vision systems to ensure a reduction of up to 80%, of the dispersion of pest control products.

In practice, dosage of fertilisers are distributed on the land in relation to how much the land itself is able to produce in present conditions, without attempting to modify it through corrective measures or improvements. It is now possible to apply fertiliser on the basis of plant colour. A chlorophyll meter device can be used to assess the crop's nitrogen stress level, meaning a real-time counter-measure taken accounts of the plants´ needs becomes possible. A differentiated distribution of nitrogen includes detection, through specific sensors of the chlorophyll content of the crop to be fertilized and in relation to this content, intervenes in real time to regulate the spreader (Balsari, 1999).

The same situation is found in livestock management. In this case, the high number of animals in the farm helped to control effectively the process for supplying forage, climate control (temperature, relative humidity, ventilation, etc.), cleaning, organization, infrastructure, collecting the produce, reproduction and waste management. It can also be achieved by having in mind the impact on the environment and awareness of the new findings, such as: modeling, application of open models, artificial intelligence, fuzzy logic, etc. In this example, it is possible to achieve the accuracy necessary on all aspects of livestock production.

Finally, there is a great concern for an adequate seedbed preparation, where the main purpose is to achieve a good germination and reduce the rapid growth of weeds. Precise depth control in seedbed preparation is very important, as it conserves moisture, incorporates chemicals and creates a high quality soil structure for seeding. It is possible to facilitate the control of seedbed preparation by changing the technology and by combining some processes as well as to reduce costs, such as practising minimum tillage and organic farming. Through applying a mulch layer the process can be easily controlled by getting a reduction on the variability of field conditions, practising minimum tillage, direct drilling and achieving a reduction of weeds due to the mulch layer.

Discussion

In spite of initial interest and enthusiasm showed by farmers, it seems that such a limited vision of precision agriculture would be very difficult to apply to the Mexican agriculture picture, mainly because of the extreme fragmentation of their lands. On the other hand, the irregular shape of Mexican farmlands makes difficult to get uniform distribution of seed, herbicides and fertilizer near the headland. The waste of these products has run up from 10%-20% of the total distributed.

Setting up a precision farming programme for sceptical farmers should deal with the technical aspects, but also with economic analysis so that farmers get answers to their monetary as well as their practical concerns.

In micro economic terms, precision agriculture is advantageous when the value of the lower variable costs is greater than the increase in the fixed costs of the investment required plus the losses in revenues. Although, reduction in pesticides and fertiliser used is a key element in the cost reductions stemming from precision farming, it is not in itself sufficient justification for its adoption. That at least is the case in purely business terms. However, on environmental terms there is a different panorama taking into account the reduction in the use of chemical fertilisers and pesticides.

Certainly, of major importance, as explained in the above examples is the considerable wealth of multidisciplinary "know-how" required for the application of these new techniques to our own farming situation, "know-how" that we do not yet fully possess. Thus we probably need to look for specific solutions of our own, many of which still need to be trailed. What appear to be of particular interest are applications that reduce the development of variables without reducing output and that allow for the direct distribution of variables.

Conclusions

It is possible to develop a new agriculture for Mexico over the medium term as shown by the examples in this work, incorporating electronics and information technology for controlling the machine functions in real time through open models and fuzzy logic. Its goals in large part, identical to those of the American precision agriculture model. Research is being undertaken to examine areas of implement control, which might lead to higher levels of accuracy being achieved. Soil tillage for seedbed preparation, weed management and crop fertilisation appear to offer the opportunity for the development and spread of this technology.

References

Balsari, P. 1999. Present technologies available to precision agriculture. Machinery World. VIII (9) 8-17.

Lara López, A. 2000. Situación de la mecanización agrícola en México (Status of agricultural mechanization in Mexico. Proceedings of the Congreso Latinoamericano de Ingeniería Agrícola (Latin America Conference on Agricultural Engineering) - CLIA 2000 and X Congreso Nacional de Ingeniería Agrícola (National Conference on Agricultural Engineering) - AMIA 2000. CM-06. pp. 53-62.

Ortiz-Laurel, H. and Rössel, D. 2002. Present situation and future prospects of agricultural mechanisation in Mexico. Proceedings of AgEng 2002. 02-RD-003. Budapest, Hungary. 11 p.

Realization of an in-field controller for an Agricultural BUS-system based on open source program library LBS-Lib

R. Ostermeier[1], H. Auernhammer[2] and M. Demmel[2]

[1]Technische Universitaet Muenchen, Department of Bio Resources and Land Use Technology, Crop Production Engineering, Am Staudengarten 2, Freising-Weihenstephan, Germany
[2]Technische Universitaet Muenchen, Department of Bio Resources and Land Use Technology, Crop Production Engineering, Am Staudengarten 2, Freising-Weihenstephan, Germany

Abstract

Site specific plant production is one main focus in Precision Farming. Current Agricultural BUS-systems (ISO 11783, DIN 9684) use primarily the "mapping approach" for process control in mobile application systems. The "sensor approach", gaining in importance, is implemented almost only proprietary. For the comprehensive approach "Real-time approach with map overlay" a definition is missing until now. The new definition of an „In-field Controller", a better integration of on-line sensors and a modification of process data handling, enable modular and scalable process control systems in compliance with Agricultural BUS-systems. The In-field Controller can supervise, control and document "In-field activities" based on on-line sensor measurements, "overlay maps" and the inclusion of an expert system. The further decisive workstep will be, especially for spatial variable fertilization, the formulation of knowledge and rules for the "mobile expert system", which must be real-time capable. The Open Source program library LBS-Lib offers a solid basis for the realization and test of an In-field Controller.

Keywords: ISO 11783, DIN 9684, Site specific fertilizer application, In-field Controller, Sensor based fertilizer application, Open Source program

Introduction

Site specific plant production, e.g. spatial variable fertilization, is one main focus in Precision Farming. Three different system approaches determine the process control in mobile application systems. These are "mapping systems" ("mapping approach"), real-time sensor actuator systems ("sensor approach") or the combination of both ("Real-time approach with map overlay"). Positioning equipment (e.g. GPS) is used for the geographic reference at the mapping approach. The input information is gathered automatically or manually. Based on these, the desired application maps are generated for field work like site specific fertilization, plant protection, seeding or irrigation. Real-time systems require neither positioning nor soft- and hardware for the creation and processing of (application) maps. The decisive parameter is acquired with a sensor and compared with an expectancy. Thus, the application action is derived and immediately executed. Both approaches show system conditional different weak points. Data which were collected in the past ("mapping approach") describe static conditions and therefore can't show the complete weather conditional variability within the growth period. The exclusive analysis of the current sensor value neglects the knowledge about the fundamental growth and yield potential as well as an environmental protection specific restriction on the current partial area (topological unit). "Real-time systems with map overlay" combine the "sensor approach" with the "map approach" and can overcome the disadvantages of both. The technical realization of this most sophisticated approach requires an effective forward-looking agricultural engineering implementation. With the availability of Agricultural BUS-systems (ISO 11783, DIN 9684) in mobile agricultural equipment, efficient distributed electronic systems can be realised and the previous isolated solutions based on "mobile agricultural computers" can be displaced. Agricultural BUS-systems permit multiple use of tractor sensors, of GPS with the current position and time information, of data output and input devices for the management as well as the user terminal is only needed once. By the integration of additional sensors further "on-line available" information can be integrated into the system. But the integration is neither defined nor standardized till now. Differentiated suggestions on special services in BUS-systems with algorithms of superposition for on-line sensor technology and "overlay maps" are just as little known. Therefore, such systems are realized proprietarily in Agricultural BUS-Systems till now, if at all.

It is the aim of this work to show a way, on the one hand, how "Real-time systems with map overlay" can be realized ISO 11783 / DIN 9684 compliant and not in a manufacturer specific way. On the other hand, this solution shall make it possible that process control can individually be adapted and scaled by the user to the desired degree of complexity (from a manual control up to the support by expert/decisicon support systems on the mobile sytem) in application systems.

Materials and methods

Precision farming maps

In principle, maps are used as input information in decision support systems, as data storage and as output information of processes of decision. The bandwidth of mapping is extensive and runs from yield mapping, weed mapping, soil mapping (e.g. soil sampling, EM38, AWC), surveying (e.g. borders, obstacles, digital terrain models), remote sensing maps to application maps (fertilization, plant protection, seeding).

On-line sensor technology:

The functionality of on-line sensor technology can go beyond the pure measurement determination and include the generation of an application setpoint within. The setpoint derivation is based on a difference comparison of current sensor measurements with a predefined expectancy. Primarily following attributes are an input parameter. On the one hand these are the soil parameters, like structure, moisture, available water capacity and (available) nutrients, on the other hand the plant parameters, like stress, nutrient supply and health state. The technical basis for the sensor technology is primarily based on spectroscopy, electromagnetic induction, electrical conductivity, ground penetration radar, optoelectronics, ultrasound, infrared radiation, image processing and force-impulse measurement. The development of on-line sensor technology has progressed furthest for nitrogen fertilization and weed control. Current plans on fertilizer systems based on laser induced chlorophyll fluorescence or the Pendulum meter are very promising.

Agricultural BUS-systems (ISO 11783 / DIN 9684):

The standards ISO 11783 and DIN 9684 specify a serial communication system for the safe, reliable and compatible information exchange between the electronic units of tractor-implement combinations and the data interchange with the stationary Farm Management Information System (FMIS). The current 11/4 parts of ISO 11783 / DIN 9684 define the complete communication functionality according to the OSI layer model, in detail, the Physical Layer, the Data Link Layer and the Network Layer, and cover further-reaching requirements of tractor-implement combinations with definitions in the area of the

Application Layer. These are functions as important as the "Virtual Terminal", the "Implement messages application layer", the "Power train messages application layer", the "Tractor ECU", the "Task Controller and FMIS data interchange" and the "Mobile Data Element Dictionary". Nevertheless, all desirable features aren't covered yet. With the current add-on of two "New Work Items", namely "Diagnosis" and "File Server" (NWI ISO 11783-12&13), a first step was made. However "on-line sensor technolgy" or "map overlay" don't get sufficient attention or are left to the area of proprietary subsystems.

Results and discussion

The realization of the approach "Real-time systems with map overlay " for ISO 11783 / DIN 9684 means that parts of the FMIS-functionality, namely the derivation of setpoints, must be provided in the Mobile Implement Control System (MICS) and on-line sensor technology must be integrated standard compliant. The current definition of the Task Controller and of the data interchange with the FMIS doesn't include this. The Task Controller works on management tasks, i.e. the handling of predefined tasks. The solution of this problem is either the upgrading of the fundamental definition of the Task Controller and the data interchange with the FMIS or the choice of a more universal solution. For this purpose an additional service is defined, the "In-field Controller" (IFC), which can supervise and control the "In-field activities". The In-field Controller then would receive the starting signal and further organizational task data from the Task Controller after task selection, receive or request the current position, derivate the local setpoint based on on-line sensor values, "overlay maps" and by means of an expert system and send this setpoint to the implement controller. Besides the setpoint generation, the documentation of the local information is a second important task, as far as this isn't done by the Task Controller already or can't be commissioned to the Task Controller. The functionality of the In-field Controller at a glance:

- *General specifications:* ISO 11783 / DIN 9684 compliant
- *Take-over of input information:* Set point curves, differentiated "overlay maps" which are time and position referenced and expert knowledge
- *Data acquisition* of several on-line sensors real-time compliant
- *Derivation of the local setpoints:* Inclusion of expert knowledge and an expert system; Consideration of exclusion areas with a reduced or not permitted application rate, of section widths at the field and border area, of driving on the same track repeatedly; Smoothing for zonale transitions based on the map overlay
- *Documentation* of local measurement values and setpoints
- *User interface:* work and configuration menu, possibility of manual intervention

If an already existing process control system for application shall be extended in an open scalable way, the idea that one controller (Task Controller or on-line sensor technology) is the only source for the setpoint generation and of a virtual point to point connection with the implement controller is no longer suitable. The problem that several competing setpoints can exist, namely of Task Controller, on-line sensor technology, In-field Controller and the operator must be solved. Methods from the topic "(networks of) autonomous agents" would make the handling of competing setpoints in Agricultural BUS-systems (ISO 11783 / DIN 9684) possible. It is a decisive basic concept that every single implement controller has "intelligence" to derive a solution from competing setpoints with regard to their value and time of arrival. Besides the necessity of a priority algorithm, it is important to increase the degrees of freedom in the system for this derivation, i.e. sending setpoint intervals instead of exact values, whenever possible.

Three key aspects emerge at the realization of the suggested system architecture (Fig. 1) in ISO 11783 / DIN 9684:

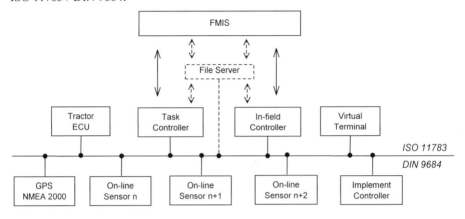

Figure 1. System architecture „In-field Controller" - ISO 11783 / DIN 9684.

A, Integration and definition of an "In-field Controller"

The realization of the In-field Controller with its complete functionality requires a clear network identity as a basis of a system compliant participation in the Agricultural BUS-system and the ability for process data interchange as well as a standardized data interchange with the FMIS. In ISO 11783 network identity can be ensured by a „Device Class"- and „Function"- definition for an In-field Controller according to the „naming structure (NAME)". The classification of the „Adress Configuration Type" with „self-configurable Adress" permits a straightforward address allocation. The definition of the "ECU type" with "Standard" and the "Industry Group" assignment with "Agricultural and Forestry equipment" is obvious. The In-field Controller must be able to exchange the following data types with the FMIS: Set point curves and functions, "overlay maps", (update of) expert knowledge and the documentation data. After completion of the standardization work to the "File Server", the data transmission could be performed via this service. Further specifications must be made with respect to the data structure and interchange format. As far as possible, definitions of data interchange between Task Controller and FMIS should be inherited and adapted. Since with the In-field Controller a part of the FMIS is relocated on the MICS, it seems appropriate to take into account possible existing standardizations for data interchange between different FMISs. This aspect wasn't processed here yet and therefore cannot find sufficient consideration yet. "Overlay maps" demand basic GIS functionality. The data must be in the position to be exchanged in the grid format or vector format provided with spatial and temporal references. The support of a 2D-system is necessary, a 3D-system is desirable. ISO 11783-10 / DIN 9684-5 have integrated the grid ("Grid") and the vector format ("Polygon"). Starting from a newly defined basic data object "Map" the necessary extension can be made for "overlay maps" according to the schemes of "TreatmentZone", "ProcessDataSetpoint" and "ProductSetpoint". However it is important that restriction to pure setpoints is canceled. The temporal reference is given for tasks, comments and data logging values in ISO 11783 / DIN 9684 only, so it must be defined newly for "maps". For documentation, the definiton in ISO 11783-10 / DIN9684-5 for data logging ("DataLogValue") can be used almost without modification. With regard to the data structure for the data interchange of expert knowledge, the question arises, how far the knowledge can be treated as a stand-alone unit and can be transfered or whether the upgrade of the "mobile expert system" must be executed by the diagnosis and maintenance functionality (perhaps proprietary).

B, Integration of on-line sensor technology

Analogously to the In-field Controller, the integration of on-line sensor technology requires a clear network identity as well as the identification of on-line sensors process data as Mobile Data Dictionary Elements. Network identity can be ensured according to the „naming structure (NAME)". On-line sensors can be source and sink for all four process data types. The identity of the process data is given by the entry in the Data Dictionary. Unfortunately only 16 implement types with one accompanying matrix table each with 16x16 data objects are defined. A definition of identical on-line sensors process data in reference to different implements types (e.g. tractor, fertilization, seeding etc.) leads inevitably to data redundancy and thus, possibly to difficulties in the data consistency. As indicated already in the current ISO standard draft (and in DIN 9684), on-line sensors should get "implement types" of their own. At the moment, vacant space in the INST/WERT matrix table of the implement type "basic attributes" is reserved for weather data, soil and plant attributes. Because the vacant space of this matrix table is already very bounded, the use of the „Reserved Bit" in data byte 1 of the process data telegram definiton should be taken into consideration however for a second set of implement types with a potential of further 16x16x16 Data Dictionary elements. For DIN 9684 this is equivalent to the definition of a new "LIS" list.

C, Modification of the mode of operation with process data

The previous definition of process data must be adapted to the handling of competing setpoints with regard to their value and time of arrival at the receiver. A "setpoint interval" has to be attached to the existing definition of exact-, minimum- and maximum- setpoints. An extended protocol for request, allocation, information, confirmation and rejection of setpoints must also be established. This is reached by a protocol specification in the data bytes-part of the process data telegram with function classes and parameter values (cf. "Virtual terminal", COMMAND and PARAMETER) and an alternative number representation. The suggestion for modification of process data definition (ISO WG1 N275) would relatively simply allow a new specification with its 1 byte long "Command"-word since it doesn't "offend" the 4 bytes for number representation. To be able to use and to realize these new possibilities of the process data structure, implement controllers must be equipped with more "stand-alone intelligence". The user must be able to select priority ranking by a configuration menu of the implement controller.

Realization and test of an In-field Controller for ISO 11783 / DIN 9684:

The theoretical considerations will be realized practically and checked for feasibility in a test stand, especially for a process control system based on on-line sensor technology and "map overlay" for intensive nitrogen fertilization. The Open Source program library LBS-Lib (Spangler et al., 2001) is a solid base for the implementation of the complete ISO 11783 / DIN 9684 - framework for the In-field Controller.

Conclusion (and outlook)

With the definition of an In-field Controller, the implementation of "Real-time systems with map overlay" in the context of Agricultural BUS-systems (ISO 11783 / DIN9684) can be made in an open, standardized and scalable way without having to decide for a (limiting) manufacturer specific solution. However the current standards must be extended by the definition of an In-field Controller, a better integration of on-line sensor technology and a modification of the process data specification for handling of competitive setpoints. The Open Source program library LBS-Lib offers a solid basis for the implementation of an In-field Controller. Questions on the topic "mobile expert system" will be decisive further worksteps. From the application oriented (agricultural) view, the definition of the expert knowledge, rules and overlay algorithms (especially for the nitrogen fertilization) is the main focus. For this the interdisciplinary cooperation of different departments in the IKB Duernast project is of great importance. The information technological challenge is the derivation of the real-time requirements for the process control and finding an expert system architecture which fullfils the real-time requirements and is suitable for the formulated knowlegde and rules.

References

Auernhammer, H., Schueller, J.K. 1999. Precision Agriculture. In: CIGR-Handbook of Agricultural Engineering. Vol. III: Plant Production Engineering. ASAE, St. Joseph, pp 598-616.

Auernhammer, H.: Precision farming - the enviromental challenge. 2001. In: Computers and Electronics in Agriculture, Millenium Special Issue: Past Developments and Future Directions. Elseviere Science 30 (2001), Amsterdam, pp. 31-42.

DIN - Deutsches Institut für Normung: Schnittstellen zur Signalübertragung, DIN 9684/2-5. Beuth Verlag, Berlin.

ISO - International Organization for Standardization: Tractors, machinery for agriculture and forestry - Serial control and communication data network. Part 1 to 11. Geneve: ISO/TC23/SC19/WG1, ISO 11783 (under development and partly published)

Spangler, A., Auernhammer, H., Demmel, M. 2001. Stimulating use of open communication standards in agriculture (DIN 9684 and ISO 11783) with capable Open Source Program Libary as possible reference implementation. In: Proceedings of the "Third European Conference on Precision Agriculture" edited by G. Grenier, S. Blackmore, agro Montpellier, Vol 2, pp. 719-724.

Acknowledgement

This work is part of the research project "Information Systems Precision Farming Duernast" IKB Duernast (http://ikb.weihenstephan.de) which is promoted financially by the "German Research Council DFG".

Decision support tool for extension and sales purposes to evaluate the economic effects of introducing Precision Farming technology to a farm

Susanne Otter-Nacke
Agrocom GmbH & Co. Agrarsystem KG, 33719 Bielefeld, Germany
otter-nacke@agrocom.com

Abstract

Adoption of precision agriculture technology has been slower than expected. In part this delay is due to the site-specific nature of benefits. In individual cases, a farmer intending to invest in this technology needs evidence that it will pay. Evaluation on an individual farm basis is required to enable this decision. Since a range of technical solutions is available, the farmer needs assistance in determining which system would be suitable given price relationships, crops produced, management ability, farming practice and heterogeneity of the site. Defining realistic scenarios for the individual farm enables calculation of potential benefits from application of PA. A set of Excel-based worksheets was developed allowing these individual calculations and thus a customized cost / benefit analysis based on varying scenarios. The tool, which is designed to be operated by sales personnel or advisors, supports the decision making process and helps the farmer to identify his best approach to PA.

Keywords: economics, decision support tool, cost/ benefit analysis

Introduction

Farmers are continuously forced to find new strategies to reduce costs and maximize profits. Therefore, before investing in new technology it is vital to get as much information as possible about the economic effects of such a decision. When trying to evaluate whether the step into the information age technology will pay for a future user it is most important to employ relevant figures.

Adoption of PA technology is slow because the benefits are not generally achieved and are not obvious in particular cases. According to Kloepfer et al. (2001) 'a broad acceptance of PA is inhibited above all due to the huge investments to make, the not obvious benefits and technical problems'.

A typical situation for discussion about the economics of precision agriculture occurs in connection with planned investments in precision agriculture technology. Sales personnel could make use of a guided worksheet which helps to introduce the farmer to the different approaches to and potential benefits of Precision Agriculture. As Swinton and Lowenberg-DeBoer (2001) pointed out, farmers are still reluctant to adopt PA technology. One of the reasons that they mention is that 'use of the technology requires acquisition of a whole unit that is not easily subdivided.'

Materials and methods

The farmer has to decide about investing in a relatively costly system which also requires new farming strategies and corresponding skills. This causes uncertainty which adds to the questions about economic benefits.

Answers are sought to the following questions:
- What are the potential benefits of PA on my particular farm?
- How much do I have to invest in this technology?

- How to quantify the actual potential profit or benefit needed to make up for the additional fixed costs caused by the investment into this technology?

To meet these requirements, an Excel worksheet tool was developed to be used by sales personnel or possibly advisors or any other skilled person to answer these questions for the individual case. The tool guides the user through a set of worksheets which require the input of real farm data. The sheets are related to:
- farm description
- specifying the technical details for a system suitable for the farm and corresponding to the farmer's needs/ skills/ expectations
- calculation of fixed costs (per year, per ha)
- evaluation of potential benefits
- cost / benefit analysis
- results (printable for the user to take along)

The worksheets may be operated using average values, individual farm values that the farmer has at hand or data extracted from a field record program. After specifying the technical details of the system desired, the fixed costs are calculated for that scenario based on the user's own details for depreciation, interest and insurance. The annual fixed costs are then related to the area of arable land and area to be managed according to PA principles.

Different levels of PA may be recognized and taken into consideration. Based on a scheme being developed by Kottenrodt (2002) as part of the preagro project, it is possible to determine different levels of intensity of PA technology. **Level I** corresponding to low technical intensity is recommended if the heterogeneity of the site is determined to be low.

Following Kottenrodt (2002) the elements listed below can be considered to be the basic equipment needed to get started:
- soil property maps
- yield mapping system (e.g. GPS based)
- PDA plus software used for field scouting
- basic software: field record system

These few requisites enable the farmer to make use of several data layers. Information may be condensed and combined with the farmer's long term experience on his fields. Determination of different management zones may result and amounts (of fertilizer or chemicals) to be applied may be varied manually or by adjusting the driving speed.

Recommendations for **level II** at medium heterogeneity (see Kottenrodt (2002)) include:
- DGPS receiver
- yield mapping system capable of producing yield maps as a basis for generating application maps (e.g. AGRO-MAP Standard)
- job computer and controller for implements (sowing machine, sprayer etc.)
- (in the longer term) on-line measurements (N status and/or degree of weed coverage)
- additional data layers such as aerial imagery, elevation maps
- additional GIS-based software

The highest intensity of PA technology is at **level III** which is advisable either for very heterogeneous sites or as the ultimate stage of PA. At this level, all farming operations will be site specific. Thus the following investment is required (if not already present):
- DGPS receiver
- yield mapping system
- job computer and controller for all implements

- LBS control system for tractors
- on-line measurements (N status and/or degree of weed coverage)
- additional data layers such as soil analysis maps, official soil property maps
- GIS-based software.

To add even more utility to the calculations, system specifications within this tool are based on the particular range of products of Agrocom including price information. However, the calculations performed are generally applicable and transferable. Having decided on a certain system, the user may start calculation of fixed costs. As an alternative, the same calculations may be performed for investments at a different level and results may be compared to each other.

Results and discussion

In order to demonstrate the working mode, representative examples are given. Figure 1 refers to system specifications for a level III approach to PA according to the suggested list (see above) for a given farm situation. The corresponding fixed costs calculations are displayed in Figure. 2.

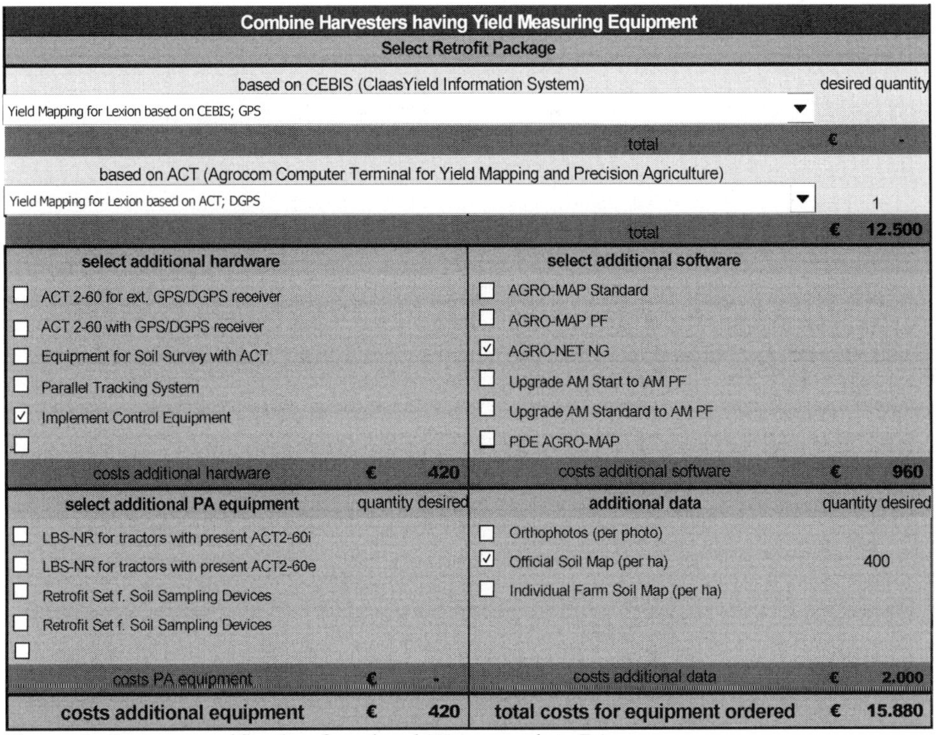

Combine Harvesters having Yield Measuring Equipment			
Select Retrofit Package			
based on CEBIS (ClaasYield Information System)			desired quantity
Yield Mapping for Lexion based on CEBIS; GPS ▼			
		total	€ -
based on ACT (Agrocom Computer Terminal for Yield Mapping and Precision Agriculture)			
Yield Mapping for Lexion based on ACT; DGPS ▼			1
		total	€ 12.500

select additional hardware		select additional software	
☐ ACT 2-60 for ext. GPS/DGPS receiver		☐ AGRO-MAP Standard	
☐ ACT 2-60 with GPS/DGPS receiver		☐ AGRO-MAP PF	
☐ Equipment for Soil Survey with ACT		☑ AGRO-NET NG	
☐ Parallel Tracking System		☐ Upgrade AM Start to AM PF	
☑ Implement Control Equipment		☐ Upgrade AM Standard to AM PF	
☐		☐ PDE AGRO-MAP	
costs additional hardware	€ 420	costs additional software	€ 960
select additional PA equipment	quantity desired	additional data	quantity desired
☐ LBS-NR for tractors with present ACT2-60i		☐ Orthophotos (per photo)	
☐ LBS-NR for tractors with present ACT2-60e		☑ Official Soil Map (per ha)	400
☐ Retrofit Set f. Soil Sampling Devices		☐ Individual Farm Soil Map (per ha)	
☐ Retrofit Set f. Soil Sampling Devices			
☐			
costs PA equipment	€ -	costs additional data	€ 2.000
costs additional equipment	**€ 420**	**total costs for equipment ordered**	**€ 15.880**

Figure. 1. System specification for a level III approach to PA.

Fixed Costs Calculation

Retrofit based on		...CEBIS	...ACT	Prec. Farming
		Level I	Level III	Equipment
Basic Price	€	-	€ 15.880	
Chip card reader present ☐	€	-	€ -	
List Price	€	-	€ 15.880	€ -
Net Price	€	-	€ 15.880	€ -
Implement Control	€ ▲▼	-	€ ▲▼ 700	
Value at End of Year 8	€	-	€ 4.518	€ -
Average Annual Depreciation	€	-	€ 1.508	€ -
Average Annual Interest Costs	€	-	€ 715	€ -
Insurance Costs	€	-	€ 127	€ -
Annual Fee DGPS-signal	€	-	€ 600	
Total Fixed Costs/year	€	-	€ 2.949	€ -
Fixed Costs/ha arable/a	€	-	€ 5,67	€ -
Fixed Costs/ha SSF/a	€	-	€ 9,22	€ -

Type of Depreciation:	degressiv ▼
Number of Years:	8 ▲▼
Depreciation Rate:	15,0% ▲▼
Interest incl. Fin. Charges:	4,5% ▲▼
Insurance, % of Costs	0,8% ▲▼
Arable Land [ha]	520
Arable Area respondg. to SSF	320

Figure 2. Fixed costs calculation.

Cost/Benefit Calculation

Annual Marginal Profit due to SSF

Marginal Profit resulting from Factor Input Savings

	Seeds	N fertilizer	PK fertilizer	Herbicides	Fungicides	Machine cost	Labour	other costs	total
% Savings	0%	5%	0%	0%	0%	0%	-2%	0%	
Aver. Input [€/ha]	0,0	102,0	0,0	0,0	0,0	0,0	35,0	0,0	
Marg. Profit per ha	€ -	€ 5,10	€ -	€ -	€ -	€ -	€ (0,70)	€ -	€ 4,40
Area SSF [ha]	0,0	320,0	0,0	0,0	0,0	0,0	320,0	0,0	
Total Marg. Profit	€ -	€ 1.632,0	€ -	€ -	€ -	€ -	€ (224,0)	€ -	€ 1.408,0

Marginal Profit from Yield Increase	Total Profit Factor Input Savings plus Yield Increase	Total Fixed Costs p.a. Alternative I	Total Fixed Costs p.a. Alternative II	Total Fixed Costs p.a. SSF Equipment
% Savings	0,6%		ng for Lexion based on CEBIS; GPS	ng for Lexion based on ACT; DGPS
Aver. Revenue [€/dt]	800			
Marg. Profit per ha	€ 4,80 per ha	€ 9,20 per ha	€ - per ha	€ 9,22 per ha	€ -
Area Responding	320,0			
Total Marg. Profit	€ 1.536 p.a.	€ 2.944 p.a.	€ - p.a.	€ 2.949 p.a.	€ -

Figure 3. Example of cost / benefit analysis.

Customized fixed costs calculation is based on the investment required for this solution and compared to potential benefits as envisioned by the farmer (see Figure 3). In this case, the scenario for a 520 ha farm is based on the assumption that the most promising approach would be to apply N in cereals (320 ha) using N application maps. Five percent savings in fertilizer costs seemed to be possible through site-specific application of nitrogen (see figures reported by Wagner (2000)). This management measure would cause higher expenses for labour (estimated at 2%), but also a 0.6% increase in net return due to higher yields and better quality. Meeting these goals would mean balanced economics of introducing PA on this farm.

In Figure 4, investments needed for levels I and II are displayed as alternatives. This scenario is for a 250 ha farm typical for Western Germany growing 140 ha of cereals where PA could be initiated by site-specific application of herbicides. Annual fixed costs are considerably higher for the level II approach, especially when applied to a small area. This effect is increased through the annual fee for the DGPS signal which may not be relevant in the future. This difference in fixed costs should be discussed with regard to the additional benefits the farmer may realize if he decides to step from level I to II. In that case application of PA technology to a higher percentage of the arable land would be advisable in order to reduce the annual costs per ha.

Fixed Costs Calculation			
Retrofit based on	...CEBIS	...ACT	Prec. Farming
	Level I	Level II	Equipment
Basic Price	€ 6.980	€ 17.150	
Chip card reader present ☐	€ -	€ -	
List Price	€ 6.980	€ 17.150	€ -
Net Price	€ 6.980	€ 17.150	€ -
Implement Control	€ -	€ 700	
Value at End of Year 8	€ 1.902	€ 4.864	€ -
Average Annual Depreciation	€ 635	€ 1.623	€ -
Average Annual Interest Costs	€ 314	€ 772	€ -
Insurance Costs	€ 56	€ 137	€ -
Annual Fee DGPS-signal	€ -	€ 600	
Total Fixed Costs/year	€ 1.005	€ 3.132	€ -
Fixed Costs/ha arable/a	€ 4,02	€ 12,53	€ -
Fixed Costs/ha SSF/a	€ 7,18	€ 22,37	€ -

Depreciation settings:

- Type of Depreciation: degressiv
- Number of Years: 8
- Depreciation Rate: 15,0%
- Interest incl. Fin. Charges: 4,5%
- Insurance, % of Costs: 0,8%
- Arable Land [ha]: 250
- Arable Area respondg. to SSF: 140

Figure 4. Comparison of fixed costs for level I and II investments.

Analysis for this example reveals that the necessary increase of yields (%) for break-even would be 1.1 % at level I and 2.8 % at level II. Or, considering only factor inputs (e.g. herbicides), necessary reduction for break-even are estimated to be 10.5 % and 28 % resp. These figures are within the range of results reported by Wagner (1999) and Dabbert and Kilian (2002). Applying this technique to a larger area means notably smaller necessary reductions for break-even showing the effect of large-scale farming also for SSF strategies.

Conclusion

The worksheet application described above is a handy tool for sales personnel, advisors or other skilled persons helping to structure the decision making process and providing support by using real farm figures to evaluate the economic implications of the decision. The tool offers not only to evaluate the situation for different levels of PA technology but also allows comparison of the economics of these different approaches. The user may then play with different scenarios of profit improvements from PA in order to get additional information about the risks implied with the decision. This should be valuable information for farmers to enable him to make decisions about the level of PA that is practicable and encourage the adoption of PA technology.

References

Dabbert, S., Kilian, B. 2002, Ökonomie (TP IV-1) (Economics), Precision Agriculture, KTBL-Sonderveröffentlichung Nr. 38, S.423-446.

Klöpfer, F., Kottenrodt, D., Weltzien, C., 2001, Implementation and Acceptance of PA by farmers and agricultural contractors within the research project 'PreAgro'. In: Proceedings of the 3rd European Conference on Precision Agriculture ECPA 2001, Eds.: G. Grenier, S. Blackmore, Montpellier, France, p. 527-532.

Kottenrodt, D., 2002, Kompendium Teilflächenmanagement (Compendium site specific management), Pre Agro Projekt, unpublished information.

Swinton, S.M., Lowenberg-Deboer, J., 2001, Global adoption of Precision Agriculture Technologies: who, when and why? In: Proceedings of the 3rd European Conference on Precision Agriculture ECPA 2001, Eds.: G. Grenier, S. Blackmore, Montpellier, France, p. 557-562.

Wagner, P. , 1999, Produktionsfunktionen und Precision Farming (Production functions and Precision Farming), (www.weihenstephan.de/ui/veroeff/pfupf.htm).

Wagner, P., 2000, Problems and Potential Economic Impact of Precision Farming, In: Proceedings of the 7th Int. Congress for Computer Technology in Agriculture, ICCTA, Florence, Italy, p.241-249.

Use of guidance systems in precision agriculture

K. Persson and H. Skovsgaard
Danish Institute of Agricultural Sciences (DIAS), Research Centre Bygholm, P.O. Box 536, DK-8700 Horsens
Krister.Persson@agrsci.dk

Abstract

Guidance systems are rapidly becoming important tools for the farmers in North America were foam markers until recently have been used in order to make it possible to find the correct positioning in the field. The guidance systems are used for manually steering but also becoming important in relation to auto-steering.
I Europe foam markers have not been in used for a long period but instead substituted by fixed tramlines laid out during the sowing. This system is efficient but on the other hand time consuming and limits the farmer in only using machines having fixed working widths. Guidance systems are easier to use and allow the farmer to change working width of the used machines during the season.
The accuracy of the guidance systems is important as it have to be as accurate as the tramline system in order to satisfy the European farmers. One guidance system has been tested at Research Centre Bygholm and the first results using straight line guidance shows that the average error (pass to pass) was found to be 0,12-0,27m when testing tramline from 3 to 24m in steps of 3m. The standard deviation was found to be 0,25-0,45m when using Omnistar differential correction signals.

Keywords : Guidance, straight line, RTK/GPS

Introduction

Several methods for optimising the field work have been developed over time. In Europe the tramline system is the most recent method for securing the correct distance in between each single run in the field that will optimise the use of the machines. Too small or too big distance in between each run in the field causes a decrease in utilisation of expensive machines or other input. Fixed tramlines may be very accurate if the establishment of the tramline is done carefully. The fixed tramline makes it possible to drive more or less 24 hours a day but as the crops grow it may be more and more difficult to find the tramline. Further more the fixed tramlines limits the farmer to use machines having a different working width from the width laid out in the field..
In North America the same problem until recently have been solved by using foam markers. The advantage of the foam marker is that it may be adapted to the actual working width of the machine. The disadvantages are that the foam may disappear after some time and even roll away from the spot were it has been placed. When using machines having a wide working width it may even be difficult to adjust the distance from pass to pass on basis of the foam.
The Global Positioning System (GPS) has opened a new technology which makes it possible to leave the fixed tramlines as well as the foam markers behind. Based on GPS the guidance systems has been developed and especially in North America been adapted. A guidance system is a system that allows tracking of invisible tramlines in a pre-set distance from the actual position. In the literature it is stated that the guidance system among others allows higher operational speeds, minimizing the over- and under-lap in adjoining swaths and increased compatibility between implements of different working width etc. Buick (2002) states as well

that the use of the guidance system improve the in-field productivity, improve crop yields, extend the hours in the field at critical times (e.g. planting) and at last reduced operator fatigue. The guidance systems are on their way to Europe and already in operation some countries like the UK. In Denmark the systems are introduced only a year ago and from this reason DIAS has started an evaluation of a guidance system. The aim of the evaluation is to clarify if the guidance system is competitive to the traditional tramline system that has been used for years.

Methods

The work has been carried out at Research Centre Bygholm as a part of a research project concerning techniques related to precision agriculture – fertiliser distribution. The project concerns precision in fertiliser application and accuracy of GPS/DGPS receivers.

The work, which has been carried out in relation to guidance systems, covers development and test of methods for determination of the precision of guidance systems. The proposed methods cover tests of guidance systems used in fields where the working directions are either north-south, east-west or diagonal. The reason for testing systems in different directions is that earlier tests have proved different accuracies of the GPS receivers in those directions. The tests cover operations at different working widths, i.e. 3-36 m in steps of minimum 3 m, in order to show differences in accuracy depending on the desired working width. The tests covers investigations of straight line guidance .The test-methods are based on registration of actual positions in the field by use of RTK/DGPS (Topcon with a reference station within 2.5 km) with accuracies of ± 0.02 m. This method has been used by Han et al. (2002) as well.

By processing the data collected from the RTK/DGPS information's as the *pass accuracy* (the deviation from the actual track), the *pass-to-pass accuracy* (distance between each track) and the *standard deviation (*the error of the mean value) was calculated..

In the tests that have been carried out until now an OutbackS from RHS in USA has been used.. OutbackS includes an antenna and a combined controller and display (figure 1).

The controller has a number of buttons for choosing the setting of the unit as well as the type of guidance to be used. The Outback give the possibility to make straight line or contour guidance. By straight line guidance the first tramline in the field is fixed by pressing the activating knob at the A-point (starting point) and at B-point (end point) in the tramline. Depending on which working width that is chosen the indicating lamps will start giving instructions about position and directions when the tractor leaves the B point. By contour driving a first tramline e.g. along the field edge will be registrated by the controller and immediately after finishing the first run the system starts to guide the driver around in the field.

Indicator showing the direction

Indicator showing the position

Figure 1. The OutbackS controller and display placed in-side the tractor. The buttons in lower part of the unit are used for setting the equipment and choosing the way of running the guidance in straight line or in contour mode.

The actual system (OutbackS) used the Omnistar differential correction signal as standard but when running in Europe the Egnos signal may be chosen as well. Both types have been tested. Also the sensitivity (=what are the allowed deviation from calculated correct position before signal for correction are given) may be set on the Outback. In the tests generally the highest sensitivity has been used. During the tests in the field a forward speed of approximate 10 km/h were used. While testing, the system were challenged by running the tractor from side to side in the tramline until a signal were given to the driver to adjust the direction. By driving from side to side the deviation that the systems allow when running in the field were found.

Figure 2 show the how the OutbackS and the Topcon antennas were mounted on the roof of the tractor. The reference signals were transmitted from the base station to the tractor by radio signals. The radio antenna was mounted at the back of the tractor as well. The RTK positions were received with a 5 Hz frequency and stored on a lap top.

Figure 2. Antannas for the Outback and RTK units mounted on top of the tractor cab. Distance in between the antennas is 0.6m. Antenna receiving correction signal mounted in the right back corner of the tractor cab.

Results

From the data collected from the RTK/DGPS receiver results as shown in figure 3 and 4 and table 1 and 2 were calculated.

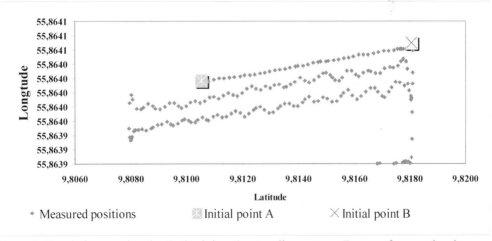

Figure 3. Result from testing the Outback in a 3m tramline system. Egnos reference signals. Sensitivity : Low.

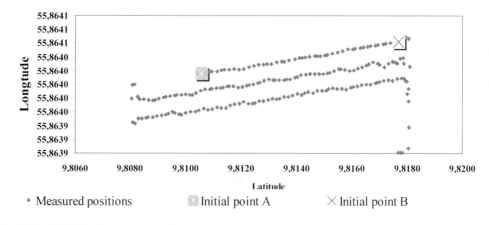

Latitude

• Measured positions   ⊠ Initial point A   ✕ Initial point B

Figure 4. Result from testing the Outback in a 3m tramline system. Egnos reference signals. Sensitivity : High.

The recorded positions shown in the figures indicate the influence of the set sensitivity. I figure 3 the low sensitivity were used and it can be seen that rather big deviations from the average tramline were allowed. In figure 4 the high sensitivity were used and the allowed deviations from correct tramline position is much lower.

In table 1 the performance of the Outback using Omnistar and Egnos differential correction signals are compared at the 3 sensitivity levels. The pass accuracy in this case is good but covers an average value as the tests have been carried out by driving from side to side. The standard deviation on pass to pass accuracy express how far away from the true middle of the tramline the guidance system will allow the drive to go. The standard deviation on the pass to pass accuracy from this tests show a decreasing tendency by increasing the sensitivity level which was expected. The results show that the quality of the Omnistar and the Egnos signals on the day when the tests were carried out were of the same level.

Table 1. The pass accuracy and the standard deviation of the pass to pass accuracy when testing different combinations of correction signal and sensitivity of the guidance system. OutbakS guidance system. Straight line guidance. 3m working width.

Set sensitivity	Omnistar correction signal		Egnos correction signal	
	Pass accuracy, m	Standard deviation on the pass to pass accuracy, m	Pass accuracy, M	Standard deviation on the pass to pass accuracy, m
Low	0,05	0,38	0,04	0,40
Medium	0,09	0,26	0,06	0,24
High	0,17	0,29	0,04	0,21

Table 2. The pass to pass accuracy and the standard deviation of the pass to pass accuracy when testing different working widths. OutbakS guidance system. Straight line guidance. Omnistar correction signal. Sensitivity : High

	Working width, m				
	3	6	9	12	24
Pass accuracy, m	0,23	0,16	0,12	0,16	0,26
Standard deviation on pass to pass accuracy, m	0,26	0,38	0,32	0,41	0,44

The results shown in table 2 indicates that the pass accuracy by different working widths are more or less equal but the standard deviation on pass to pass accuracy seems to increase by the working width.

By comparing the results from testing 3m tramlines in table 1 and 2 it can be seen that the pass accuracy is not the same. The reason for this is not clear at the moment but the quality of the DGPS signals may influence the results. The results from the trials that have been examined are not from the same day and the experience is that the results may vary from day to day.

Conclusions

The initial tests that have been carried out show that the possible pass accuracy when using guidance systems may be within a range of 0,05– 0,26m depending on the working width and the sensitivity of the system. Also the GPS constellation may influence the accuracy as the recorded positions vary over time.

The test show a need for a standardized methods including guidelines for testing over time for these equipments.

The guidance system may in the future substitute the foam markers or the fixed tramlines but the calculated standard deviations indicate a tolerance higher than expected for fixed tramlines may be expected.

References

Buick, R. .(2002). GPS guidance – Making an information decision. Proceeding on the 6th. International conference on Precision Agriculture, Minneapolis, MN, July 2002. pp25

Han, S., Noh, H.K., Zhang, Q. and Shin,B-S.. (2002) Dynamic performance Evaluation of DGPS Receivers for Parallel Tracking. 2002 ASAE Annual meeting. Chicago. pp 17

OutbackS owners manual (2002) .Outback guidance, dev. of RHS . pp28

GIS-supported simulation of the economic risk of different site-specific nitrogen fertilisation strategies

Antje Reh
AGRO-SAT Consulting GmbH, Schulstrasse 3, 06388 Baasdorf

Abstract

At present site-specific fertilisation strategies are discussed as a possible way to increase profitability in agriculture. Beside human influences spatial differences in soil fertility lead to crop yield variation within a single field. For crop yield level causal relations between soil, climate and crop management are characteristic. So far in Germany only very few empirical data about input-output relations between crop inputs and yields is available. In order to compensate this data lack empirical data from the USA was used as a basis for stochastic production functions. In order achieve economic and also ecological advantages, crop input intensity must be adapted to the site specific yield potential areas within a field. For the determination of the economically optimal input rate of nitrogen under uncertainty and evaluation of the environmental risk under changing climate conditions, GIS tools like supervised classification and simulation techniques like Monte Carlo methods were evaluated.

Keywords: GIS, Monte - Carlo - Simulation, Nitrogen Fertilisation, Economic and environmental risk

Introduction

In the more effective use of resources is gaining more and more importance. Imposed specific environmental regulations and tightened agricultural price policies increase the economic and also ecological pressure on the agriculture. One way of farmer's reaction could be the more efficient use of crop inputs (optimal special intensity) during the production cycle.

However spatial differences of the soil fertility within a single field require the evaluation of different crop input strategies. Among others site-specific nitrogen fertilisation are discussed strategies as one possibility to increase the profitability in crop production. Farmers presently fear that these methods involve the risk to supply insufficient quantity to certain areas within the field and to loose possible profits. Beside nitrogen input crop yield is significantly depending on the interaction between climate, soil fertility and cropping system. As farmers cannot predict climate and product price at harvest they have to make decisions under uncertainty. To investigate different nitrogen strategies according to yield potential within a field Monte Carlo methods are used to assess simultaneously economic and environmental risk.

Material and methods

The use of a geographic information systems (GIS) enables the spatial demarcation of inhomogeneous soil fertility within a field. These areas can be derived by processing different information sources. Remote Sensing data, yield and soil maps serve as data for appropriate processing.

For the estimation of the relative site specific yield potential within a field among other procedures complex classification algorithms was used. Supervised classification procedures like Maximum Likelihood Classification are powerful tools used to extract quantitative information from different data by transforming multispectral images into thematic information classes.

To perform it one first determines the classes they wish to obtain from the different images. For each class, a sample of pixels that correspond to it, is selected to allow a reasonable estimate for the range of pixels in each class. These ranges, called training sites, are saved in a vector file which is then used to create a signature or spectral response pattern for each class. In supervised classification the identity and location of training sites (feature classes) has to be well known. These signatures are then used to classify the full image.

For a case study empirical data like yield maps, multitemporal aerial images and soil survey maps were incorporated into a supervised classification. The result is a thematic input management map presented in picture 1.

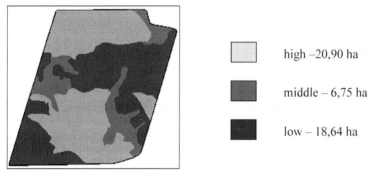

high –20,90 ha

middle – 6,75 ha

low – 18,64 ha

Picture 1. Crop input intensity map.

The input intensity map describes the subdivision of the field into different yield potential areas. The red areas represent high yielding areas, the green areas average and the blue areas low yield zones.

A further step of this work is the development of an input-output model, which takes soil fertility, the influence of climatic conditions and economic parameters like product price and variable costs of crop production into account. A site specific crop input-output-model requires site specific production functions, which reflect the different yield potential in each area adopted to favourable or unfavourable weather periods.

There are principally two different ways to get realistic equations for this part of the simulation model. Beside the estimation of production functions by appropriate crop growth simulation models (CERES and others), empirical data could be statistically analysed for the estimation of production functions. In Germany up to now there is no sufficient data available to calculate site specific production functions for nitrogen input.

For the purposes of this work American trial results in summer wheat were used. In this experiment yield of summer wheat in different zones with low, average and high yield potential under different nitrogen input has been studied over several years (1). By regression analysis three different production-functions were derived.

Table 1 shows the equations which build the basis for the calculation of cereal yield under different nitrogen inputs. In order to adopt the American production functions approximately to German input output relationships in winter wheat, the equations which originally were calculated on bushel/acre where transformed directly into German metric system by converting regression parameters one to one. Regression coefficients were checked by comparing with theoretically calculated and practically measured input-output relations in an existing field in Saxony-Anhalt, Germany.

Table 1. Site specific production functions for winter wheat.

Zone	Yield Potential in USA	Production Function in Germany (formula)
1 (blue) Low (< 36 bu/acre)		yield dt/ha = $21.0 + 0.18\,N - 0.00051\,N^2$
2 (green) Middle ($36 - 56$ bu/acre)		yield dt/ha = $24.4 + 0.36\,N - 0.00113\,N^2$
3 (red) High (> 56 bu/a)		yield dt/ha = $30.2 + 0.31\,N - 0.0003\ \,N^2$

Source (1)

Table 2 shows the basic design of the simulation model. For the three yield potential areas different nitrogen input amounts are given as a strategy (left column). The described production functions calculate the correspondent yield. The independent variables like climate, variable costs and product price are stochastic and are calculated with support of the software package "@RISK". @RISK uses simulation, sometimes called Monte Carlo simulation, to do a risk analysis. Simulation in this sense refers to a method whereby the distribution of possible outcomes is generated by letting the computer recalculate the model over and over again, each time using different randomly selected sets of values for the probability distributions for the stochastic variables (2).

	N-Input kg/ha	yield dt/ha	factor weather	factor draught stress	price wheat EUR/dt	price per N kg	sales EUR/ha	N cost EUR/kg	other costs EUR/ha	subsidy (1) EUR/ha	C M EUR/ha	N balance kg/ha
distribution			Norm(1/0,0667)		Tri(7,8,10)				Tri(380,400,430)			
yield pot.												
average	130	52.1	1.00	1.00	8.3	0.5	434.19	65.00	403	353	**319**	-5.47
low	105	34.3		1.00			285.64	52.50			**183**	15.88
high	185	77.3		1.00			644.02	92.50			**501**	-15.93
average	**145**										average	**-1.60**

Calculation of contribution margin (C M) for field "Finkenherd"

yield potential	GIS area Ha in zone	C M in zone EUR
average	6.75	2152
low	18.64	3408
high	20.9	10475
total	46.29	16034.60

(1) Agenda 2000, Bundesministerium für Verbraucherschutz, Ernährung und Landwirtschaft
http://www.verbraucherministerium.de/landwirtschaft/agenda-2000-pflanzen.pdf, Ausgabe 2002

Picture 2. Simulation model to the calculation of the risk of different N-strategies.

The site specific different negative influence of weather conditions to yield is modulated by a second factor (yield multiplier) which follows a normal-distribution with a mean of 1.0 and different standard deviations according to yield potential zone. Bad weather conditions are expressed by values lower than 1.0, favourable conditions lead to a factor grater than 1.0. Within areas of low yield the factor can reach a minimum value of 0.65 due to advanced dry-stress effects, in high yielding areas this factor goes only down to 0.8.
The price winter-wheat is simulated between 70 and 100 Euro per ton following a triangle probability distribution with the most frequent value at 80 EUR per ton. Variable costs are simulated between 380 and 430 EUR per hectare following a triangle distribution with 400

EUR as the most frequent value. The price for nitrogen remains unchanged at 0.5 EUR/ kg as well as the payment of the direct subsidies with 353 EUR per hectare. For the economic analysis the model calculates within each simulation iteration the contribution margin of the total field.

For the environmental analysis the balance of supplied fertiliser according to N-strategy and theoretically consumed fertiliser according to achieved yield is calculated. It is assumed that the production of 100 kg of wheat consumes 2.6 kg of nitrogen including the amount for biomass (2).

By starting the simulation process the above described spreadsheet is calculated over and over again – where each recalculation is an "iteration" delivering the correspondent contribution margin and nitrogen balance - with the new set of sampled values for climate factor, variable costs and product price. As more as iterations are run, output distributions become more "stable", therefor it is important to run enough iterations so that the statistics generated on the outputs are reliable (3). Test have shown that 1000 iterations are sufficient to achieve stable results.

Results and discussion

Table 2 describes four different nitrogen strategies to be analysed. Number 1 represents the economic optimal site specific N fertilisation under average climatic conditions based on the production-functions for each of the 3 yield potential areas. Strategy 2 reflects a site specific nitrogen distribution which was used in practise.

Table 2. Description and economic results of different nitrogen Strategies.

Management zone	Strategy 1 N amount in kg/ha	Strategy 2 N amount in Kg/ha	Strategy 3 N amount in kg/ha	Strategy 4 N amount in kg/ha
Green (middle)	130	150	130	185
Blue (low)	105	120	130	185
Red (Good)	185	180	130	185
Average N amount in kg/ha	145	151	130	185
„80%-value" in EUR	>10.569,	>10.384	>9.366	>9.892
„33%-value" in Euro	>16.442	>16.147	>14.535	>15.587

Strategy 3 is a strategy considering ecological aspects with the goal to minimise the risk of nitrogen surpluses by fertilisation. With the uniform application of 185 kg/ha in strategy 4 farmers intend to maximise possible profits in the highest yielding area.

The 80%-value in table 2 refers to the contribution margin of the whole field which can be expected at least with a probability of 80% under the above described assumptions. The higher this value is, the better and the less risky is the chosen nitrogen strategy from the economic point of view.

It is evident, that strategy 1 and 2 are economically very similar. The "ecological strategy" (no. 3) offers the lowest expectable profit. In practise many farmer believe that strategy 4 could give them at least an additional chance to gain higher profits than in site specific strategies. The results of the simulations showed very clearly that under our assumptions the chance to achieve higher profits is even better with site specific strategies compared to

132

number 4. With a probability of 33% strategies 1 and 2 offer higher profits (>16.400 or 16.100 EUR) than strategy 4 (>15.600 EUR). The detailed distribution of the simulation results are shown in figure 1.

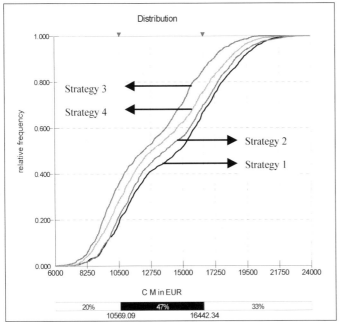

Figure 1. Distribution (relative frequency) of contribution margin according to N-strategy.

A further goal of this work was the risk assessment of potential N-excesses. As basis for N-consumption served a general factor described in the literature (2) assuming that the production of 1decimal ton of wheat grain requires 2.6 kg N. As one could expect the highest risk of N-surpluses occurred in the low yielding areas of the field. in the respective yield potential area. It is evident that the site specific N-strategy no. 1 involves the lowest risk of greater N-surpluses. In 50% (500 iterations) the positive N-balance of this strategy was lower than 20 kg N per hectare. Strategy 3 provided in nearly 50% of the iterations N-excesses between 40 and 60 kg per ha. The high input strategy no. 4 is clearly the worst solution. In more than 50% of the cases N-surpluses of more than 110 kg in the low yielding area were simulated.

Figure 2 shows the result of the simulation by the absolute frequency of calculated N-excesses.

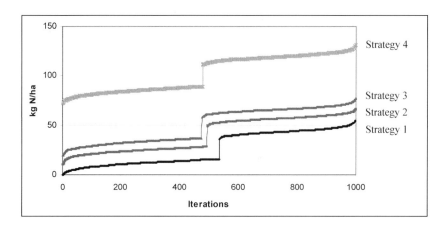

Conclusions

The combination of GIS-techniques and Monte Carlo simulations proved to be a useful tool to analyse the risk of different N-strategies taking into consideration uncertainties caused by climatic factors and economic parameters. For the described case for the economic point of view it could be shown, that site specific N-strategies were less risky and do not reduce the chances for high profits. The analysis of the environmental risk of different N-strategies demonstrated that the economically best strategy proved also to be the one with the lowest risk of considerable N-surpluses. However the model is not capable to evaluate if the calculated N-surpluses are really a danger for the environment or mankind by leaching down into the ground water.

Even though the used tools proved to be very powerful for such kind of analysis the lack of appropriate empirical data became evident. Particularly for Germany more extensive investigations are required to quantify more precisely site specific (soil related) input-output relation between usual crop inputs and crop yields.

References

1) Grant Jackson, Predicting Spring Wheat Yield and Protein Response to Nitrogen, 1998
   http://landresources.montana.edu/FertilizerFacts/17_Predicting_Spring_Wheat_Yield.htm
2) http://www.uni-hohenheim.de/~kulaholo/public_data/nkey0305www.pdf
3) Risk Analysis and Simulation – Add-In for Microsoft Excel, Guide to Using @Risk, Palisade 2001

Acknowledgement

Hereby the author would like to acknowledge Prof. Dr. B. Dohmen at Anhalt University of Applied Sciences, Germany for his invaluable discussions and comments.

Site-specific crop response to temporal trend of soil variability determined by the real-time soil spectrophotometer

S. Shibusawa[1], C. Hache[1], A. Sasao[1], I. Made Anom S.W.[2] and S. Hirako[3]

[1]*Faculty of Agriculture, Tokyo University of Agriculture and Technology, Fuchu, Tokyo 183-8509, Japan*
[2]*Department of Agricultural Engineering, University of Udayana, Denpasar, Bali, Indonesia*
[3]*Omron Institute of Life Science Co. Ltd., Kyoto 615-0084, Japan*

Abstract

Using the real-time soil spectrophotometer, temporal changes in spatial variability of soil parameters on a small field (155 m long and 30 m wide) were evaluated during the two crop seasons in 2000 to 2001, and then the corn plant response was traced in the following season of 2001. The field was divided into 8 plots by soil treatment: manure, fertilizer and manure-fertilizer mixed. The spatial trend and temporal stability maps of moisture content, soil organic matter (SOM) content, NO_3-N, pH and electrical conductivity (EC) were obtained. The maps showed that high level of temporal variability came out for moisture content and NO_3-N, and that areas with stable high levels of NO_3-N were observed in fertilizer plots while areas with stable low level or unstable of NO_3-N in manure plots. During the crop growing period of 15 weeks, SPAD values of crop were fluctuated in one fertilizer plot giving stable high level of NO_3-N and in a manure plot with stable low level of NO_3-N.

Keywords: real-time soil sensor, soil mapping, temporal variability, crop response

Introduction

Spatio-temporal variability of the field is indispensable knowledge for site specific management. Simard *et al* (2000) observed the spatio-temporal variability of anion exchange membrane P and Hoskinson *et al* (1999) reported temporal changes in spatial variability of soil nutrient, for example, and they have emphasized the importance of the spatio-temporal trends of soil parameters. On the other hand, Shibusawa *et al* (2001) have been developed a real-time soil spectrophotometer for in-situ measuring several soil parameters, which can be useful to determine the spatio-temporal variability of the field.
The objectives of the work were to observe the temporal changes in spatial variability of soil parameters of a small field using the real-time soil spectrophotometer, and to trace the crop response to the soil variability.

Materials and methods

The real-time soil spectrophotometer mounted on a tractor with an RTK-GPS, collected soil reflectance in the range of 400 to 950 nm wavelengths with a resolution of 3.6 nm and the range of 950 to 1700 nm with a resolution of 6 nm, respectively (Shibusawa *et al* 2001). Renewed devices were an EC electrode and a load cell to detect soil compactness, as shown in Figures 1 and 2.

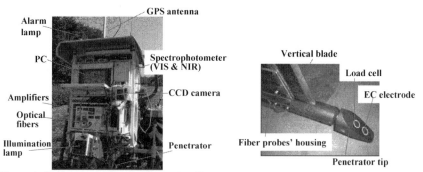

Figure 1. Soil spectrophotometer and soil penetrator.

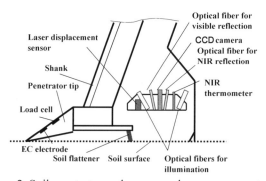

Figure 2. Soil penetrator and sensor probes arrangement.

The experimental field was located on the experimental farm of Tokyo University of Agriculture and Technology (TUAT), which had been divided into eight plots with three different types of soil management (manure, chemical fertilizer and chemical fertilizer-manure mixed) for more than ten years (Fig. 3). The soil type was Andisol and its texture was 20% clay, 28% silt, and 52% sand. Two-year three-crop rotation: wheat-corn-wheat-bean had also been performed. Soil sensing experiment was conducted for three times from June 2000 to May 2001. The sensor speed was 30 cm/s and the measuring depth was about 15 cm. Three segments of the field at a scanning interval of 1 m in longitudinal direction and 5 m spacing in lateral direction, resulted in a total of 360 scanning points. The crop observation was performed every week during the growing period of 15 weeks, at 5 measuring points with 6 plants in each plot. The crop was dent corn (Gold Dent KD772D, Kaneko Seeds Co.) and crop growth parameters were SPAD, height and number of leaves.

To calibrate the soil reflectance, about 75 samples were taken every 5m at the same depth and location as the corresponding sensor's scanning point. These samples were analyzed in laboratory for moisture content (MC), nitrate nitrogen (NO_3-N), soil organic matter content (SOM), EC and pH. Fifty samples were used for calibration and 25 for validation of the prediction model. For calibration procedure, the moving average smoothing method, Kubelka-Munk transformation method, multiplicative scatter correction, and stepwise multilinear regression analysis were performed. Using the kriging method, interpolated maps were created and spatial variability was presented in both soil and crop parameters.

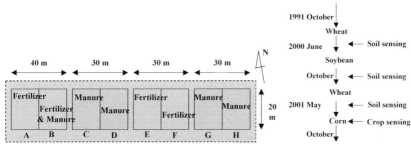

Figure 3. Soil treatment in TUAT field (left) and crop rotation (right).

Results and discussion

Temporal changes in spatial variation of NO_3-N are presented in Fig. 4 (I Made Anom *et al* 2002). Dividing the treatments with their respective averages normalized the data. The pattern of variability was changeable, for example, areas with higher NO_3-N were observed along the southern part of the field in June 2000 (before soybean planting), on the west part in October 2000 (after harvest) and on the west part in May 2001 (after wheat harvest). It might be seen that soybean cultivation changed the spatial variability of NO_3-N and wheat cultivation did not.

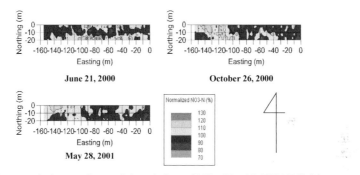

Figure 4. Temporal changes in spatial variation of NO_3-N with TUAT field.

Next, the average and standard deviation of the normalized data series were calculated in time direction, and the coefficient of variation (CV) was obtained. In this case only three sets of time series were provided. The averages of normalized data gave spatial trend maps and the CV data implied temporal stability maps. Figure 5 shows the spatial trend and temporal stability maps of soil parameters (I Made Anom *et al* 2002). Moisture content and NO_3-N fluctuated temporally in high levels on several areas while SOM, EC and pH showed temporally stable.

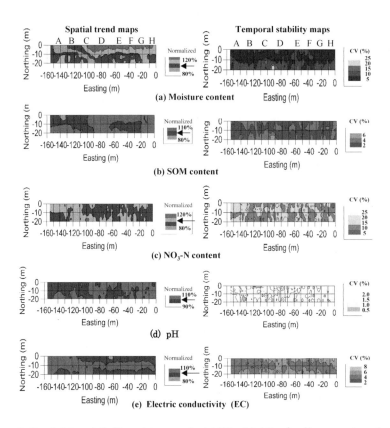

Figure 5. Spatial trend (left) and temporal stability (right) of soil parameters during three seasons in TUAT field. Data were normalized to the respective averages over the whole field during three seasons.

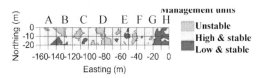

Figure 6. Management unit map for NO3-N of TUAT field.

With the spatial trend and temporal stability maps, management unit maps can be calculated as three levels of management unit, where "stable" implies CV values below 10 % and "High" implies the temporal averages above 100%. Figure 6 shows the case of NO$_3$-N (I Made Anom *et al* 2002). It can be used for understanding the feature of field, for example, stable high area was found in plots A and F, which were chemical fertilizer plots, stable low areas were observed in plot H, manure plot, and unstable area came out in plot D, manure plot. In order to make a correct recommendation a lot more time series data were required.

Figure 7 shows temporal changes in spatial variability of SPAD, number of leaves and plant height of the corn plant during the growing period (Hache *et al* 2002). The second week after planting showed a similar spatial pattern of three parameters, though the patterns were different in the 7th week. The correlation among the three parameters was changeable in time. Areas with relatively high CV were patched over the field for three parameters, and an interesting fact was that plot A with stable high of NO_3-N and plot H with stable low tended to give higher CV of SPAD and number of leaves. More data on soil and crop as well as on farm work conditions were required for detailed consideration.

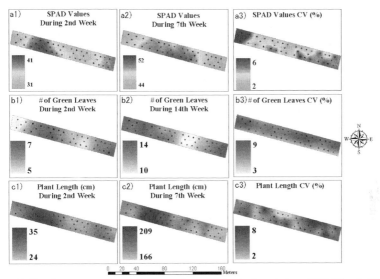

Figure 7. Spatio-temporal variability and coefficient of variation (CV %) of SPAD, number of leaves per plant and plant length (cm).

Conclusion

The real-time soil spectrophotometer developed was useful for evaluating the temporal changes in spatial variability of soil parameters in a small field. During the two crop seasons in 2000 to 2001, spatial trend and temporal stability maps of moisture content, soil organic matter (SOM) content, NO_3-N, pH and electrical conductivity (EC) were obtained. Results showed that high level of temporal variability came out for moisture content and NO_3-N. A management unit map was introduced for NO_3-N, and stable high area was found in fertilizer plots while stable low area or unstable area in manure plots. During the crop growing period of 15 weeks, SPAD values of corn plant were fluctuated in one fertilizer plot giving stable high level of NO_3-N and in a manure plot with stable low level of NO_3-N.

References

Hache, C., S. Shibusawa, Y. Kato, A. Sasao, A. Otomo, K. Matsuzaki. 2002. Site Specific Corn Response to Soil Spatial Variability in Different Fertilizer Treatments. *Proceedings (on CD-ROM) of the 6th International Conference on Precision Agriculture*. July 14–17, Bloomington, Minnesota, USA.

Hoskinson, R. L., J. R. Hess, R. S. Alessi. 1999. Temporal Changes in Spatial Variability of Soil Nutrients. *Proceedings of the 2nd European Conference on Precision Agriculture*, 11-15 July, Denmark.

I Made Anom S. W., S. Shibusawa, A. Sasao. 2002. Temporal Changes in Spatial Variability of Soil Parameters within a Small Paddy Field and Forage Field. *Proceedings (on CD-ROM) of the 6th International Conference on Precision Agriculture.* July 14–17, Bloomington, Minnesota, USA.

Simard, R. R., A. N. Cambouris and M. C. Nolin. 2000. Spatio-temporal Variation of Anion Exchange Membrane P in a Corn Field. *Proceedings (on CD-ROM) of the fifth International Conference on Precision Agriculture.* July 16–19, Bloomington, Minnesota, USA.

Shibusawa, S., I Made Anom S. W., S. Sato, A. Sasao, S. Hirako. 2001. Soil Mapping Using the Real-time Soil Spectrophotometer. *Proceedings (on CD-ROM) of the 3rd European Conference on Precision Agriculture*, 18-20 June 2001, Montpellier, France.

Development of production level management zones for nitrogen fertilization

D.G. Westfall*[1], W.J. Gangloff[1], R. Khosla[1], R.M. Reich[1], A. Hornung[1], D.F. Heermann[2], H.J. Farahani[2] and K. Fleming[1]
[1]*Colorado State University, Fort Collins, CO 80523, USA*
[2]*USDA-ARS Water Management Unit, Fort Collins, CO, 80523, USA*

Abstract

Grid soil sampling has been used to develop nutrient maps to guide precision fertilizer applications since the inception of precision agriculture. However, the cost and labor associated with collection and analysis of soil samples to accurately describe spatial properties of fields can be prohibitive. Consequently, the use of production level "management zones" (MZ) to allow implementation of precision agriculture appears to be a good alternative. The objective of our research was to evaluate four methods of developing production level MZ (low, medium, and high) for precision N fertilizer. Two-way analysis was used to compare accuracy of MZ delineation techniques compared to actual crop yield. The average accuracy ranged from 24 to 55% and all were statistically significant based upon the Chi^2 goodness of fit, indicating these techniques are identifying yield patterns in the fields.

Keywords: precision agriculture, management zones, nitrogen management

Introduction

Management zones are defined as a sub-region of a field that expresses a homogenous combination of yield limiting factors for which a single crop input is appropriate to attain maximum efficiency of farm inputs. Several studies have indicated that site-specific MZ could be used as an alternative to grid soil sampling to develop nutrient maps for variable rate fertilizer applications (Fleming et al., 1999; Khosla et al., 2002). Different MZ delineation techniques have been proposed. Some methods are the use of soil survey maps (Franzen et al., 2000), utilization of topography or landscape positions (Kravenko and Bullock, 2000), the apparent soil electrical conductivity (ECa) (Sudduth et al., 1998) and soil color and the framers production experience (Fleming et al., 1999). The objective of this study was to compare four techniques for delineating MZ.

Materials and methods

Irrigated maize (*Zea mays* L.) fields in Colorado, USA that were managed by the farmer were utilized. Management zones were delineated to identify areas of high, medium, and low-yield potential using four techniques. Two N rates were evaluated, 0% and 100% recommended N. The techniques evaluated, from the least to most technical are: Technique 1; a commercially available technique that utilizes panchromatic bare-soil imagery and the farmers past management experience (Fleming et al., 1999) (SCMZ). Technique 2; a technique that utilizes Veris®E.C. cart to measure apparent soil conductivity (ECaMZ). Technique 3; a technique that uses multi-spectral bare-soil imagery; soil organic matter; cation exchange capacity; texture (sand, silt and clay content); and previous years' yield monitor map as the data layers (YBMZ). Technique 4; a technique that utilizes bare-soil imagery to develop a stratified, cluster soil-sampling design followed by soil analysis (RCMZ).

Results and discussion

The average comparisons of grain yield zone classification (low, medium, and high) zs MZ delineation across sites were evaluated using a two-way comparison. The average agreement was 38, 45, 32, and 32% at 0% recommended N and 41, 34, 31, and 37% at 100% recommended N for the RCMZ, ECaMZ, SCMZ, and YBMZ techniques, respectively. The Chi^2 overall goodness-of-fit tests were all significant. This indicates the delineation techniques are "picking-up-patterns" in the field. A more meaningful comparison is the relationship between a production level MZ (low, medium, or high) and the observed yield zones classifications (low, medium or high). The average agreement between the observed yield zone classification and the MZs for the 100% N rate were 38, 36, 31, and 36% RCMZ, ECaMZ, SCMZ, and YBMZ techniques, respectively. The agreement between individual yield zones within individual fields and production level MZs ranged from as high as 82% to as low as 0%.

Conclusions

The delineation of production level MZs holds promise as a method for economic implementation of variable rate fertilizer management. However, other factors influence yield, which we were not able to evaluate. Consequently, this impacted our results. Fertilizer management using the MZ concept is more economical and less labor intensive than grid sampling based management. There was an acceptable level of agreement between MZ delineation and yields when one considers the data "smoothing" that occurs in MZ delineation and yield values to make it possible to use large commercial application equipment for fertilizer additions. We conclude that the MZ concept is a viable method of implementation of precision fertilizer management. However, the final model has not been quantitatively identified.

References

Fleming, K.L., D.G. Westfall, D.W. Wiens, L.E. Rothe, J.E. Cipra, D.F. Heermann. 1999. Evaluating farmer developed management zone maps for precision farming. In P.C. Robert, R.H. Rust and W.E. Larson (eds.) In Proceedings 4th International Conference on Precision Agriculture.

Franzen, D.W., A.D. Halvorson and V.L. Hofman. 2000. Management zones for soil N and P levels in the Northern Great Plains. In P.C. Roberts et al. (ed.) Precision agriculture. Proc. Int. Conf., 5th, Bloomington, MN. 16-19 July 2000. ASA, CSSA, and SSA, Madison, WI.

Khosla, R., K. Fleming, J.A. Delgado, T. Shaver and D.G. Westfall. 2002. Use of Site Specific Management Zones to Improve Nitrogen Management for Precision Agriculture. J. Soil and Water Conservation. Vol. 57 (6) 513-518.

Kravenko, A.N. and D.G. Bullock. 2000. Correlation of corn and soybean grain yield with topography and soil properties. Agron. J. 92:75-83.

Sudduth, K.A., N.R. Kitchen and S.T. Drummond. 1998. Soil conductivity sensing on claypan soil: comparison of electromagnetic induction and direct methods. p. 979-990. In P.C. Roberts et al. (ed.) Precision agriculture. Proc. Int. Conf., 4th, St. Paul, MN. 19-22 July 1998. ASA, CSSA, and SSA, Madison, WI.

Precision agriculture and mechanical soil protection

Peter Weisskopf[1], Claus Sommer[2] and Eugen Kramer[3]

[1]*Federal Research Station for Agroecology and Agriulture, FAL Zürich-Reckenholz, CH-8046 Zürich, Switzerland*
[2]*Institute for Production Engineering and Building Research, Federal Agricultural Research Centre (FAL), D-38116 Braunschweig, Germany*
[3]*Agricultural Engineering and Safety, Agricultural Centre Strickhof, CH-8315 Eschikon, Switzerland*
peter.weisskopf@fal.admin.ch

Abstract

Mechanical subsoil protection is an important issue for sustainable soil cultivation. Actual trends in agricultural vehicle development are leading to an increased risk of damages to soil structure. Legislative approaches to mechanical soil protection tend to simplify the interactions between vehicles and soils and could lead to considerable restrictions for agricultural mechanisation.

This paper presents ideas for a precision agriculture approach to prevent mechanical soil compaction. This approach tries to consider the complex interactions between vehicle and soil, the changing soil conditions and the practical needs of everyday field operations. Possible ways to realize this concept, the availability and requirements of technology and knowledge are discussed.

Keywords: field traffic, vehicle properties, soil compaction, mechanical soil protection, precision agriculture.

Agricultural mechanisation and mechanical soil protection

The development of agricultural field mechanisation during the last decades resulted in efficient and powerful but also heavy vehicles (Weißbach, 2001). As a consequence of the increasing weight of tractors, self-propelled harvesters and transport vehicles, the risk of damages to soil structure, in particular to persistent subsoil deformation under wet soil conditions ("subsoil compaction") has intensified (Arvidsson et al., 2000).

Preventive means to control the risk of damages to soil structure are in the interest of the farmers in order to maintain soil quality as well as to minimize yield losses and expenses for soil tillage. Likewise the maintenance of ecologic soil functions lies in the public interest to preserve natural resources and to guarantee food safety and safe living conditions. These intentions are leading more and more to legislative regulations concerning soil protection (e.g. the German "Bundesbodenschutzgesetz" (Anonymous, 1998; Sommer et al., 2001)).

To improve soil protection against mechanical stresses, threshold values for vehicle parameters (e.g. the axle load) have been put forward (Anonymous, 2002). However these propositions raised discussions about unforeseeable technical and financial consequences for agricultural mechanisation.

In order to reduce the risk of soil compaction by wheeling, it is helpful to analyze the causes and processes leading to damages of soil structure.

Fundamentals of mechanical soil protection

Potential solutions have to base on prevention and should concentrate on subsoil protection because regeneration of a damaged subsoil structure proceeds very slowly, leading to long lasting yield losses and higher costs for agricultural soil cultivation (Alakukku, 1996). Nevertheless, damages to soil structure can't always be avoided because of market pressures to harvest at a given time and with prescribed product quality standards.

The extent of damages to soil structure depends on the intensity and duration of the effective stresses exerted by vehicles and on the mechanical stability of the soil structure, which again is defined by the composition, structure and moisture of a soil at a given time.

Preventive mechanical soil protection requires an adjustment of the soil stresses caused by vehicles to the structural stability of a soil, i.e. its mechanical strength. For that purpose, a risk assessment is necessary considering intensity and duration of damages to soil structure in relation to economic and ecologic criteria.

Assessing the risk of damages to soil structure requires the following information (Figure 1):

1. extent of soil structure deformation caused by the mechanical stresses transferred by the wheels (or tracks) of a vehicle to the soil; this can be done by directly measuring or by predicting (modelling) soil deformation.
2. agricultural and ecological significance of soil structure deformations;
3. regeneration potential of a deformed soil structure at given site conditions.

Based on the intensity, relevance and duration of soil structure deformations, a target value for maximum tolerable soil deformation can be derived. Predicting deformation of a soil structure by modelling the interactions between vehicle and soil requires the understanding of

a) the stresses transferred by wheels or tracks of a vehicle to the soil (at least for a static situation);
b) the stress distribution in a soil (depending e.g. on soil conditions);
c) the deformation characteristics of a layered soil (for subsoil simplified to a stress-strain relationship).

Finally risk assessment is based on the comparison between deformation of soil structure and the corresponding target value for maximum tolerable soil deformation. Because normally absolute values for soil structure deformation and maximum tolerable soil deformation are not available, often a conservative approach is used, which assumes that no change of soil structure is acceptable in subsoil.

Threshold values for vehicle properties?

Threshold values for vehicle parameters are always related to individual wheeling situations, i.e. defined constellations of mechanical stress and structural soil stability. If soil stability changes, the threshold value for a vehicle parameter should change as well in order to maintain a given probability level for damages to soil structure.

Generally valid threshold values for vehicle parameters are easy to execute administratively, but with the drawback that the judgement based on these threshold values is either too generous or too restrictive, depending on whether the values have been adjusted to the upper or lower boundary of soil stability.

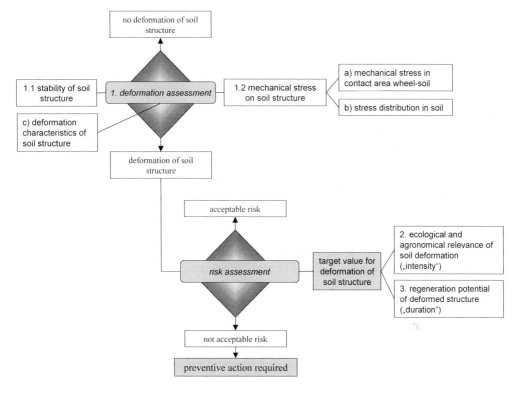

Figure 1. Most important components of site and vehicle properties for a risk assessment of damages to soil structure.

Scientifically, preventive soil protection would rather demand vehicle parameters be adapted to the given wheeling situation, i.e. to the differing structural stabilities of different soils (and of the same soil in different conditions). Up to now the main drawbacks of such a concept were the missing soil and vehicle information and the missing possibilities to adapt vehicle properties during operation.

Focussing on agricultural engineering, three approaches to adapt vehicle properties to site restrictions are available which can contribute to a sustainable agricultural soil cultivation:
1. Soil respecting design and engineering of agricultural vehicles
2. Site-specific choice of agricultural vehicles
3. Situation-specific adaptation of agricultural vehicles

Precision agriculture: adaptation of vehicle properties to site restrictions

The fundamental idea for this precision agriculture approach is shown schematically in Figure 2. Several interacting vehicle and tyre properties are determining the extent of stress in the contact areas between tyres and soil. Stress distribution in soil depends not only on vehicle properties, but also on soil characteristics, and is finally decisive for the extent of stresses on soil structure. Together with the actual stability of soil structure, these stresses are determining soil deformation and the subsequent change in soil functionality.

Figure 3 shows the required vehicle and soil parameters to judge the actual risk of soil structure deformations. There is a distinction between

- basic data, registering general properties of vehicle and tyre design, which can be stored in a database, and
- variable data, including changing vehicle and soil properties, which have to be measured or calculated permanently during operation.

Moreover, the figure shows possible control schemes to judge the risk of soil damage and possibilities for the subsequent adaptation of the running vehicle to the actual soil stability.

These parameters allow the continual coverage of the impact of wheeling on soil structure and the consecutive comparison with a target value for acceptable soil deformation, thereby assessing the risk for soil deformation. Depending on the technical possibilities, soil structure deformation is either directly measured or indirectly modeled. A directly measured parameter for soil structure deformation could be rut depth, a modelled parameter could be calculated subsoil stress.

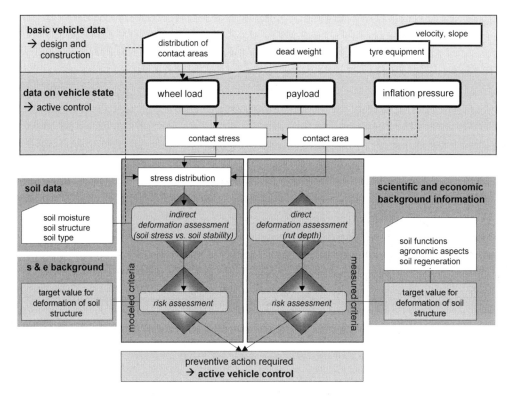

Figure 2. Required information to judge the risk of structural soil deformations. Apart from basic data about general properties of vehicle and tyre design, variable data related to changing soil and vehicle properties must be available and have to be measured during field operation.

If the vehicle is not equipped with active control devices, the conclusion of this risk assessment can only be whether maximum tolerable soil deformation or soil stress is reached. This would indicate to the driver when his vehicle has reached its load limit under the actual soil conditions.

If control devices are available to influence load distribution and inflation pressure, vehicle properties can be adapted to the actual soil conditions. For vehicles used in field and road traffic, such an adaptive control system allows the optimization of contact area properties to the totally different requirements of wheeling on a soft soil and on a hard road surface respectively.

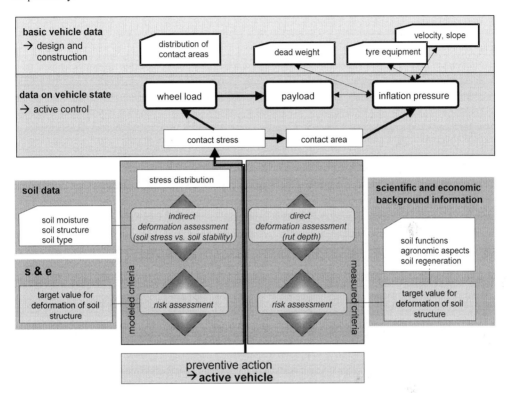

Figure 3. Possible ways to control vehicle properties relevant for mechanical impacts on soil during field operation.

The procedure to adapt a vehicle properties continually to the stability of the soil could be the following:
- continual load balancing of wheels in order to distribute loads among wheels, according to available contact areas and load capacities of the tyres;
- additional adaptation of inflation pressure of the tyres according to the actual stability of soil structure as well as to the load capacities and velocity restrictions of the tyres;
- information of the driver when the technical limits of vehicle and tyre equipment determining its maximum field-transport capacity are reached; consequently the driver can stop field operations (i.e. to empty the tank) in order to limit the risk of damages to soil structure.
Together with a GPS application, it is possible to access basic site data more specifically and to document the result of this prevention strategy geografically. For a contractor this could be a way to give a proof of his working quality to the customer.

First experiments to assess soil deformation during vehicle operation by measuring rut depth were carried out at the FAL Braunschweig-Völkenrode (Brunotte et al., 2000). The technical basis for this precision agriculture approach exists - sensors and devices to analyse data and control vehicle parameters during operation are available. Urgently needed are data describing properties of agicultural vehicles and tyres relevant to mechanical impacts on soil. Furthermore, data and models are required describing the relation between soil stress and soil deformations as well as between soil structure and soil functions, including physiological requirements of plants. Based on this know-how target values for risk assessment can be derived.

The described precision agriculture approach could be a valuable contribution to preventive mechanical soil protection, adapted to the possibilities and needs of farmers and usable in everyday vehicle operations (Alakukku et al., 2003; Chamen et al., 2003). It requires the cooperation between soil scientists, agronomists, agricultural and mechanical engineers.

Conclusions

Preventive means to protect soil structure, especially in the subsoil, are important for sustainable agriculture. As the soil deformation process is determined by many factors and multiple interactions, it is difficult to develop simple solutions for mechanical soil protection. Legal approaches to solve the problem in detail tend to simplify and generalize the situation, so either soil protection goals are missed or agricultural soil management is restricted more than necessary. The basic requirement of preventive mechanical soil protection is to adapt vehicles during field operations to the limited stability of soils. Agricultural engineering could contribute to preventive actions with new approaches related to vehicle design and vehicle adaptation. As part of the precision agriculture strategy, active vehicle control based on the registration and processing of actual information on soil and vehicle properties could lead to a better adaptation of vehicles to site restrictions, optimizing demands of preventive soil protection and agricultural soil cultivation at the same time.

References

Alakukku, L., 1996. Persistence of soil compaction due to high axle load traffic. I. Short-term effects on the properties of clay and organic soils. Soil Tillage Research 37 211-222.

Alakukku, L., Weisskopf, P., Chamen, W.C.T., Tijink, F.G.J., van den Linden, J.P., Pires, S., Sommer, C., Spoor, G., 2003. Prevention strategies for field traffic-induced subsoil compaction. Part 1. Machine/soil interactions. Soil Tillage Research (in press).

Anonymous, 1998. Gesetz zum Schutz des Bodens (Bundesbodenschutzgesetz) (Law on the protection of soil (Federal German Soil Protection Act)). Bundesgesetzblatt Nr. 16 vom 17. März 1998, 502-510.

Anonymous, 2002. Gefügestabilität ackerbaulich genutzter Mineralböden (Structural stability of arable mineral soils). ATV-DVWK-Regelwerk Teil III, Merkblatt M 901, 16.

Arvidsson, J., Trautner, A., van den Akker, J.J.H., 2000. Subsoil compaction – risk assessment and economic consequences. Advances in GeoEcology 32 3-12.

Brunotte, J., Weissbach, M., Rogasik, H., Isensee, E., Sommer, C., 2000. Zur guten fachlichen Praxis beim Einsatz moderner Zuckerrüben-Erntetechnik (Good agricultural practice using modern sugar-beet harvesting equipment). Zuckerrübe 49 (1) 34-40.

Chamen, W.C.T., Alakukku, L., Pires, S., Sommer, C., Spoor, G., Tijink, F., Weisskopf, P., 2003. Prevention strategies for field traffic-induced subsoil compaction. Part 2. Equipment and field practices. Soil Tillage Research (in press).

Sommer, C., Brunotte, J. und Kloepfer, F., 2001. Gute fachliche Praxis zur Vermeidung von Schädigungen des Bodens (Good agricultural practice to avoid damages to soil). In: KTBL-Schrift 400, 69-82.

Weißbach, M., 2001. Bodenschonende Reifen für Grossmaschinen und Schlepper (Soil respecting tyres for heavy machinery and tractors). Rationalisierungs-Kuratorium für Landwirtschaft (RKL), Rendsburg/Deutschland, 339-395.

Abstracts from oral presentations for ECPA

On-the-go mapping of soil properties using ion-selective electrodes

V. Adamchuk, E. Lund, A. Dobermann and M. Morgan

The main objectives of precision agriculture are to increase profitability of crop production and reduce the negative environmental impact by adjusting application rates of agricultural inputs according to local needs. Variable rate management of soil pH and macronutrients has been one of the most promising strategies involving precision agriculture technology. It has become a common practice to integrate different layers of spatial field data to develop application algorithms, known as prescription maps. Results of systematic soil sampling followed by standardized laboratory analyses usually serve as the best estimate of soil nutrients and pH distribution throughout the field. However, implementation of variable rate application of fertilizers and lime frequently do not result in a positive economical impact. The two main limitations of current methods are: 1) inadequate recommended application rates which were developed for macro scale agricultural practices, and 2) high cost of soil sampling and analysis resulting in maps with inadequate spatial resolution. Alternatively, high-density soil data layers (bare soil aerial photography, electrical conductivity maps, digital elevation models, etc.) have been utilized to modify soil-sampling patterns and develop management zones requiring similar soil treatments. However, a methodology for reliably obtaining higher density direct measurements of soil properties could provide capabilities for resolving both limitations of variable rate technology. An on-the-go system for mapping soil pH and macronutrient levels could be used to either prescribe variable rate application maps or conduct research studying local yield response to certain soil treatments. Although, no promising techniques to rapidly measure soil phosphorous content have been discovered at this time, soil pH, available potassium and residual nitrate contents can be mapped using an automated soil sampling system. In this work, combination, flat-surface, ion-selective electrodes were used to determine ion activities of H+, K+, and NO3- in naturally moist soil samples on-the-go. A prototype automated soil sampling system has been developed to conduct simultaneous measurements of soil pH, available potassium and residual nitrate contents at a constant depth while moving across the field. The location of each sample is geo-referenced. The experimental work has shown that a relatively stable output of ion-selective electrodes in direct contact with tested soil samples can be achieved in 10-15 seconds. Although, individual measurements have lower repeatability compared to standardized laboratory tests, a significant increase of sampling density results in higher overall quality of the produced soil maps. While traveling at 5 km hr-1 with 30 m between passes, a single automated soil sampling system provides approximately 20 samples per hectare. Laboratory experiments, performed on 15 Nebraskan soils, showed that standard errors between 0.2-0.4 logarithms of measured ion concentration (pH, pK or pNO3) should be expected when measuring unsaturated soil samples. Results were compared to standard soil analysis performed by commercial soil testing laboratories. The major drawback of the proposed methodology is that the values measured on-the-go represent snapshots of nutrient availability with little information regarding soil buffering and nutrient release over time. Thus, current lime and fertilizer recommendations cannot be directly applied to the generated maps. However, secondary spatial information (texture, organic matter content, cation exchange capacity, etc.) of immobile soil buffering characteristics can be integrated with the maps obtained using the automated soil sampling system to estimate the amount of fertilizer or lime to be applied. Commercialization of the proposed methodology is currently being pursued.

Strategy for implementation of precision irrigation in irrigation water management in Jordan

Esmat Al-Karadsheh, Heinz Sourell and Abdel-Nabi Fardous

At the start of the new millennium, nations are still facing the crucial challenge of ensuring food supplies and sustainable utilization and development of agricultural resources. An ever-increasing population, resource shortages and degradation of the ecological environment have added even greater pressure on countries. The use of precision farming for irrigation water management/scheduling, known as precision irrigation, in order to apply water in the right place with the right amount at the right time is just beginning to be explored and still in the development stages and requires more experimental works to determine its feasibility and applicability. This study has focused on an approach for establishing a strategy for implementing precision irrigation in irrigation water management towards improving the overall efficiency that is beneficial to the farm, both economically and environmentally. Fields that show spatial variability in available water content (AWC) would benefit from a precision irrigation system which has the ability to vary the amount of water applied. Also, it may be desired to vary depth of irrigation under semi-mobile and reel systems to account for uniformity, and/or to avoid application to unproductive zones such as rock outcrops or waterways. Results indicated, mapping soil ECa using EM38 can be a good surrogate measurement for spatially variable factors, that are not easy to sense and map such as soil type and moisture content. Variable rate application (VRA) refers to the application of agricultural inputs in specific and changing rates throughout the field. As proposed in the suggested strategy, the VRA of water could be fulfilled by using some available technologies, such as speed control systems and regulating discharge rate. Results indicated, speed control systems succeed in adjusting travelling speeds of reel machine at acceptable accuracy which, in turn, change the amount of applied water exponentially. Also, presented results justified the modifications established to the commercial centre pivot using solenoid valve and controlled with programmable logic control (PLC) to produce variable-rate water application system to irregular-shaped areas. The PLC obtained the positional information and open/close the addressed solenoid valve(s) to obtain the target depth and succeed in varying the amount of water throughout the field, using the concept of pulse irrigation. The PLC was not adequate in its current configuration for the experiments which were to be performed, including fertilization and chemication. The agricultural sector in Jordan uses about 70% of the available water resources, and the rest is divided between the other sectors (industrial and urban). Due to the increases in industrial and urban activities, large amounts of irrigation water are being shifted towards the non-agricultural activities. Accordingly, new management systems for water must be introduced in order to save on the amount of water being used to meet the requirements for increasing the agricultural production. For wide implementation and acceptance of precision irrigation technologies, the starting point that should be considered includes preparing a nation-wide agricultural development program. Evaluation of the expenses of this system is beyond the scope of this work, and is another important direction toward which this research can be profitably extended, but a brief description for the extra technical costs was conducted. Because of this, a full economic analysis cannot be conducted until full results from research programmes are available and also, because this management system is still in the beginning stages, many industrial accessories are expensive.

Nitrate leaching, yield and soil moisture prediction by CERES-Maize of DSSAT model under tropical conditions

M.E. Asadi and R.S. Clemente

Abstract Transport of nitrate and water in the soil under field conditions represents both an economic loss to farmers and an environmental pollutant for groundwater. Computer simulation models are useful management tools for estimating NO3-N leaching, but they need to be calibrated and validated before use. The CERES-Maize of DSSAT v3.5 model was used to simulate nitrate leaching, nitrogen uptake, corn yield and soil moisture content in an acid sulphate soil in the central region of Thailand. The validation data was obtained from a two-year study with conventional tilled corn (Zea mays L.) during 1999 and 2000 at the Asian Institute of Technology, Bangkok, Thailand. Nitrogen source was urea and there were four N fertigation treatments which include 0 kg N ha-1 (control), 100 kg N ha-1, 150 kg N ha-1, and 200 kg N ha-1, each having three replications arranged in a randomized complete block design (RCBD). The soil was irrigated to field capacity at 50% available soil moisture depletion regime throughout the season. Nitrate leaching losses were measured with soil water sampler and soil moisture content was measured by the gravimetric method. Inputs to the model included site information, daily weather data, soil properties, soil initial conditions, irrigation and fertilizer management and crop performance data. The weather data for 1999 and 2000 were collected from a local weather station. Simulation showed that, the discrepancies between observed and simulated values are generally small in both years. The model overpredicted corn grain yield slightly for some treatments, generally overpredicted total N uptake and underpredicted total N leached and soil moisture content. The lower observed corn grain yields at high rates of nitrogen application was due to experimental factors, which the model is not sensitive. Low values of nitrate leaching were predicted by the model because of low drainage. The relation between results obtained from experiment (Yo) and simulation (Ys) was expressed by the equation Ys= 1.058Yo with R2=0.97 for grain yield, Ys=0.7396Yo with R2=0.86 for nitrate leaching, and Ys=1.1103Yo with R2=0.99 for total N uptake. Validation results showed the reliability and good performance of the model. The study also showed that the CERES-Maize of DSSAT model may be applied with confidence to study effects of N and irrigation management on maize yield, nitrate leaching and N uptake under irrigated tropical conditions.

The detection of relative height of weeds as a main determinant of an early competition index estimate in the field

L. Assémat and M. Chapron

Height has been constantly suggested as one of the key parameter for driving light competition and most competition models have to posses a module that simulate expected height growth in relation to time and/or development stage of species. Whether height differences between species were the result of emergence differences, whether only species growth characteristics and canopy architecture explain these differences, almost no remote sensing at any scale in the field of this variable has been attempted. However, small initial differences at these early stages - where control decisions have to be undertaken - are known to explain a bigger part of variability in the outcome of competition. In the context of Precision Agriculture, where localized control of weeds is the aim, a better evaluation of local competitive pressure should be devised.

Our demonstration is in 2 parts. Image analysis of color photography of weed cover at seedling stage and simulation of light competition between species.

First, we use a methodology recently developed (Chapron et al., 2nd European Conference on Precision Agriculture, 1999) in order to analyse stereo images of experimental plots of weed seedlings. In this method, an initial classification of pixels of stereo images in two classes (vegetation, soil) is performed from the red, green, blue and near-infrared images using the Principal Component Analysis. Then, an adaptive cross-correlation is carried out on the contour pixels of leaves in the left and right images. This provides the disparity and correlation coefficient images. The correlation coefficient represents the reliability of the stereo-matching and the disparity gives the height through the calibration parameters which model the stereovision system. The intrinsic and extrinsic calibration parameters model the projections of the 3D world on 2D image planes. The heights inside the contours of leaves are obtained by a diffusion technique based upon a partial differential equation. The results of this stereovision method are compared with the true values of measured heights of leaves.

Second, a 3D simulation model for radiative transfer balance in a multi-species row structure is constructed (from Sinoquet & Bonhomme, Agric. For.Meteorol., 1992) in order to estimate what is the importance of the height information of plants on the light efficiency (part of incident light intercepted) of each species. Competitive pressure at this local scale is then estimated by the ratio of actual light efficiency over light efficiency of the same species when all (or part of) other species have been removed.

We present the results for experimental plots measuring 48 x 84 cm and identify possible sources of errors for all steps from picture shot in the field to the computation of competition indices.

Finally, we try to estimate what would be the importance of plant height knowledge in comparison with the identification and estimation of plant cover of the species only, for a correct estimation of potential weed competition at this local scale in the field. The discussion concerns how much effort we should continue to put in a 3D description of weed-crop cover, relative to a 2D description much simpler, yet difficult. One conclusion will be that height can also be a good criteria for species identification where all other attributes (color, texture, shape, ..) have failed. The spatial nature of weed competition concerns the horizontal patchy distribution of weeds in the field, and the vertical distribution of species hierarchy : both control the strength of competition with the crop and ultimately the economic and environmental benefits of any future precision weed management.

The use of radiative transfer models for remote sensing data assimilation in crop growth models

H. Bach, W. Mauser and K. Schneider

The application of remote sensing in precision agriculture is still in an early stage. Often images acquired form airborne or spaceborne platforms are only used for simple interpretation of the heterogeneity within a field. This interpretation is very much depending on the knowledge of the analyser and thus the results obtained are subjective. In other attempts empirical regressions are developed that describe the relationship between a spectral parameter (e.g. NDVI) and an agricultural parameter (e.g. leaf area, yield, soil water status). The nature of these regressions is that they are empirical and hardly transferable from one region to another and from one scene to another. Changes of sensor characteristics and observation geometry can be reasons for this. To overcome this problem a new approach is developed that allows the evaluation of remote sensing data in a quantitative way. The baseline idea is to use radiative transfer modelling to understand the physical background of the signal measured by optical sensors and to assign plant physiological parameters like leaf area, wet and dry biomass, chlorophyll and water content to measured reflectance values. The radiometric calibration of the remote sensing data is a prerequisite for this. The retrieved plant physiological information is then assimilated in a crop growth or plant production model in order to be able to model plant development with time and parameters that can not directly be seen by a remote sensing instruments, like grain yield. Thus first, the radiative models convert the reflectance spectra into spatial fields of land surface parameter values. In a next step they can be used as spatially distributed inputs to the process models. The radiative transfer model developed for this approach is called GeoSAIL. It is a two-layer version of the canopy reflectance model SAIL and incorporates a submodel for the soil reflectance. In GeoSAIL brown leaves are separately considered in order to mimic the vertical leaf colour gradient within a canopy. The spectral information on the leaves are calculated using the PROSPECT model. By combining the models PROSPECT and GeoSAIL, the major aspects related to optical observation of vegetation and realistic modelling of directional canopy reflectance spectra are covered. In a third step the coupling of the surface reflectance model to an atmospheric radiative transfer model needs to be addressed. Using the model MODTRAN4 realistic computing of absorption and scattering in the atmosphere at high spectral resolution over the solar-reflective spectral range is obtained. As a result the signal observed by a sensor is simulated. These simulations can be performed for any optical sensor or acquisition scenario by adapting the sensor characteristics in the model environment. To fully exploit the information content an inversion of the model is required. The values of the parameters used in order to describe the radiative transfer within a canopy (e.g. LAI, chlorophyll content) are adjusted until model outputs and measurements coincide. Best results can be achieved when the radiative transfer model is coupled with a crop growth model, which can suggest a plausible range of parameter values from plant growth modelling based on the weather and soil conditions. An example is shown, where the radiative transfer model GeoSAIL is coupled with the plant growth model PROMET-V to successfully determine the heterogeneous distribution of biomass, carbon flux and nitrogen leaching. Multispectral data (e.g IKONOS, SPOT, TM) were used in this example. Additionally results using data of the hyperspectral airborne sensor AVIS will be shown. Hyperspectral data offer the possibility to analyse specific absorption features of substances like water or chlorophyll to determine their abundance in the fields. With 2 to 4 m spatial resolution the AVIS sensor is an ideal source of information for precision agriculture.

A metadata profile for Precision Farming based on ISO-19115-standard

M. Backes, D. Dörschlag and L. Plümer

A new metadata profile for Precision Farming based on ISO-19115-standard and a tool to administrate the metafiles will be presented. Within Precision Farming the survey of space-oriented data (geodata) is an indispensable condition. Because of the very high spatial and temporal resolution of geodata in Precision Farming the systematic data management is important, when offering efficient ways to operate with geodata. One important pre-condition for systematic data management is the creation of "data about data" thus metadata. This metadata is only reasonable, when there are standards being stable over a long period of time and therefore providing access to a huge number of users. There exist proper standards for geospatial metadata. Those standards have been established from the ISO (International Organisation for Standardization) the ISO-19115 and from the FGDC (Federal Geographic Data Committee) the Content Standard for Geospatial Metadat, which are both popular in the geoinformation community but they provide much more possibilities than needed and/or reasonable for Precision Farming. A necessity for limitation of amount and complexity follows from the question what an originator of geodata is willing to invest in the creation of metadata. Especially the amount of basic essentials that are required is dependent on the group of the originators and what you can expect of them. As the concept of the ISO-19115-standard already includes the possibility of creating profiles for differend user groups, in this place there will be defined such a profile for precision farming. An corresponding realisation based on FGDC standard will not be realised, because the common systems in geoinformation have already an implemented an interface based on ISO-19115-standard, or will do so soon. Precision Farming is dominated by three major groups of interest. There are farmers collecting data, sometimes using it for personal analysis. Agricultural service providers also collect geodata using it for management or documentation purposes. The 3. group of interest is the scientific community, which has high demands on precision and volume of data. Consequently this last group is able to collect, analyse and document data without having any restrictions in time and money. Farmers have the least time to invest in collecting metadata thus the concept of the agricultural profile for metadata should be appropriate for this user group. Nevertheless with respect to the other user groups having different aims the sustainability of geodata should be assured by this metadata profile as well. As most important descriptive elements of the ISO-19115-standard we identified the Data Identification, Reference System Information and the Citation and Responsible Party Information elements. They will be used as mandatory elements in the new profile, which is presented here. For cost/benefit relation reasons all other elements will not be provided. Another advantage of the ISO-19115 is its being based on the XML (extended mark-up language) standard of the W3C (world wide web consortium). The XML-standard provides a schema definition which enables the validation of metadatafiles and the possibility to check wehter they are conform with the profile. A definition is integrated in the ISO-19115-documentation. Due to its origin in ASCII it con be apllied to different operating systems. For editing geospatial metadata in XML-format there are several tools available partially implemented in commercial GI-software, which are too complex for use in Precision Farming applications. Therefore a new user group oriented tool was realised, which will be demonstrated. By introducing geospatial metadata profiles in Precision Farming it should be possible to merge the collected data from different user groups and different periods of time to improve and evaluate modern management in Precision Farming. For all participating user groups the amount of usable data will be enormously extended and the costs of data collection will be reduced by increasing the reuse of collected data.

Temporal stability of sward dry matter and nitrogen yield patterns in a temperate grassland field and the potential for site-specific management

J.S. Bailey, J.P. Frost and H.W. Xu

Introduction Little is known about the nature, extent or temporal stability of within-field variability in forage grass production, and whether or not it would be feasible to manage such variability in a site-specific way. In the absence of automated yield-mapping equipment, Bailey et al. (2000) have developed a manual procedure for mapping variability in forage grass DM yields within individual fields. Using this manual procedure, spatial distributions in sward DM production were mapped in 1999 and 2000 on a forage grass field. A third season of yield mapping is required, though, before any robust assessment of temporal stability in yield patterns may be made. From the results to date, it would appear that variability in DM yield is greatest at cut 1, and due to differences in net soil N mineralization across the site. The objectives of the present study were to quantify the temporal stability of DM yield patterns in the field at 1st cut, produce classified management maps showing areas of consistently high or low yields, and determine if site- specific applications of N could be used to achieve uniformity in DM production across the site. Materials and methods The study was conducted on a 7.9 ha silage field in County Down between 1999 and 2001. Fertiliser N was applied at a rate of 100kg ha-1 in March, and first cut silage crops were taken in late May. 101 grid-sampling points were identified and located using a GPS. Before harvest, grass samples were collected at each sampling point, chemically analysed, and the results used to predict DM and N yields (Bailey et al., 2000). Values for DM and N yields at unsampled locations were estimated using ordinary punctual kriging. Spatial trends in DM and N yields were calculated as the mean values at each sampling point over the 3-year period, and temporal stability's of the yield distributions were assessed by calculating coefficients of variation over time at each point. The results of these analyses were then combined to produce classified management maps showing areas of the field that were high-yielding and stable, low-yielding and stable, or unstable (Blackmore, 2000). To investigate if site-specific applications of N could be used to achieve uniformity in crop yields across the field, twin plot experiments, each with 40 plots, were established on the high yielding and low-yielding parts of the field, in 2001. Fertiliser treatments were 0, 50, 100, 150 and 200kg N ha-1 (with 8 replicates). DM and N yield measurements were made at 1st cut. Results and discussion The temporal stability analyses showed that the patterns of DM and N yield in the field at 1st cut were relatively stable, with coefficients of variation at most points less than 20%. Accordingly, almost 95% of the field was classified as stable for both DM and N yields, with 48% categorised as stable and high yielding and 47% as stable and low yielding. Furthermore, the high-yielding and low- yielding areas formed two distinct blocks, each of which might be managed as a separate sub-unit. The results of the twin plot experiments showed that maximum DM yield on the high-yielding area could be matched on the low-yielding area by applying 50kg ha-1 more fertiliser-N. It was concluded therefore, that uniformity in DM production could indeed be achieved by adopting a site-specific approach to N fertilisation. References Bailey JS Higgins A; Jordan 2000. Empirical models for predicting the dry matter yield of grass silage swards using plant tissue analyses. Precision Agriculture 2, 131- 145. Blackmore S 2000. The interpretation of trend from multiple yield maps. Computers and Electronics in Agriculture 26, 37- 51.

Using site specific weed management for control of winter wild oats in Spain: an economic evaluation

J. Barroso, C. Fernandez-Quintanilla and B. Maxwell

One of the most important implications that arises of the aggregated nature of weed infestations is the possibility of reducing substantially the quantity of herbicide used by applying the herbicide only to patches where weed densities are above a given threshold This is, basically, the concept of site specific weed management (SSWM).The economic benefits of using SSWM for controlling winter wild oats (Avena sterilis ssp. ludoviciana) are likely to be related to the proportion of the field that is weed-free (or below a given population threshold), the size of weed patches and the spatial resolution of sampling and spraying technologies. In this paper we simulate various types `on/off` patch spraying techniques considering different combinations of the above mentioned factors. In the context of Spanish winter barley crops, patch spraying was found to be economically superior that spraying the whole field under most of the spatial distribution conditions tested. The profitability of SSWM systems increased as the proportion of the field being infested by this weed decreased. On the other hand, and independently of the proportion of the field infested, profitability increased when the patch distribution was more concentrated (patch size was larger). Although the site specific strategy was superior to the standard strategy in most cases, positive net returns of controlling winter wild oats with this approach were only obtained when the weed-free area was smaller than 70%. Regarding patch mapping and spraying resolution, the critical parameter which determined the viability of the various options was their technological costs. The spatial pattern of weed infestation also determined the relative profits of the various systems. Simulation results indicate that a few large patches can be mapped with a 12 m x 12 m resolution and sprayed with standard equipment with a reasonable technology cost (9 e ha-1). The presence of smaller patches may require using 6 m x 6 m resolution patterns and spraying independently with the two boom sides, increasing the technology cost (13.5 e ha-1) and decreasing the profitability of the system. Changing the mapping resolution to 3 m x 3 m and spraying independently 1m segments was found to be not economical due to the high technology cost (27 e ha-1). The maximum technology costs that can be justified economically (`brake point`) for different patch sizes and resolution levels was estimated. An analysis of the errors and costs associated with using SSWM systems under various types of conditions show that mapping and spraying errors increase gradually as the proportion of the field weed infested decrease. This fact results in subsequent increases in the yield losses produced by residual weed populations (weedy areas not sprayed) and in herbicide waste (weed-free areas sprayed). Spraying error also increased as patch size and spraying resolution decreased, reaching a maximum error of 40% when spraying 400 m2 patches with a 12 m x 12 m resolution. We can conclude that site specific management of winter wild oats offers a practical and economical alternative for the control of this weed in low-yield Spanish cereals.

Assessing and simulating soil water and nitrogen effects on spatial variability of yield and grain quality of durum wheat

B. Basso and P. De Vita

Spatial variation in soil water and nitrogen are often the causes of crop yield spatial variability due to their influence on the uniformity of plant stand at emergence and for in-season stresses. Natural and acquired variability in production capacity or potential within a field causes uniform agronomic management practices for the field to be correct in some parts and inappropriate in others. To achieve the ultimate goal of sustainable cropping systems, variability must be considered both in space and time because the factors influencing crop yield have different spatial and temporal behaviour. Recent advances on the resolution and availability of remote sensing imagery, coupled with a decrease in its associated costs, have allowed the collection of timely information on soil and crop variability by examining spatial and temporal patterns of vegetation indices. The objectives of this study were to assess spatial and temporal variability of soil water content and nitrogen availability across the field; to assess causes of yield variability at field scale; to understand water/nitrogen interaction on grain quality across the field and to spatially validate CERES-Wheat for simulating interactions of stresses on crop growth and development occurring throughout the season. A field-scale study on durum wheat was carried out at the Experimental Institute for Cereal Crop located in Foggia-Italy (41° 28" N, 15° 32" E and 75 m asl). A regular grid sampling scheme (25 m) was imposed on the field to select points of measurements. Soil physical, chemical and biological properties were measured on each grid point a depth increment of 15 cm. Measurements on plant density and difference in emergence were recorded for the field. Data on soil water content, soil nitrogen were taken throughout the season every two weeks. Plant canopy reflectance was also determined every two weeks using a hand-held multispectral radiometer. The vegetation indices used to assess the crop performance were (RI, GVI, NDVI, WDVI, and TSAVI). Results of field experiment showed high spatial variability of soil water and nitrogen dynamics. Due to a severe droughts occurred at different phonological stages soil water had a major effect on spatial variability of yield. Soil water stress occurred at grain settings in a part of the field and at grain filling in another part of the field. Durum wheat yield resulted quite low due to the severe water stress, ranging from 0.4 tons ha-1 to a maximum of 3.5 tons ha-1. Grain quality was also variable throughout the field ranging from 12% in areas of higher yield to 23% in areas of lower yield. The weight of 1000 seed, text weight and harvest index were also determined for each grid point measured. The CERES-Wheat crop simulation model was applied on each selected point and on uniform zones based on similar crop performance within the field identified by vegetation index analysis. A reclassified NDVI map was used to identify spatial patterns, to select uniform zones across the field to execute the model on. Simulation results were compared to measured data for soil water, soil nitrogen, yield and grain quality using RMSE. The model was able to closely predict the measured variables when correct inputs were used.

Evaluation and simulation of variable depth tillage effects on yield and soil physical properties

B. Basso, L. Sartori, M. Bertocco and G. Oliviero

In Mediterranean areas the adoption of an appropriate tillage system is important for conserving soil water, increasing soil fertility, improving environment quality and reducing production costs. Conservation of natural resources is the base for a long-term sustainability of agricultural and natural ecosystems. Due to spatial variability of topography, soil type and soil depth, a tillage operation at a certain depth may be correct in some areas of the field and inappropriate in others. Soil compaction is a complex phenomenon that is in many cases responsible for yield losses and environmental impact due to the limited exploitable root volume and to the decrease in soil fertility. Soil compaction is not only a result of physical, chemical and biological soil properties but also of tillage intensity and it occurs with a high degree of spatial variability within the field. The objective of this research was to assess spatial and temporal variability of soil compaction throughout a 20 ha field and to prescribe a map of tillage (chisel plow) at variable depth based on soil compaction data collected with a penetrometer during the season. A secondary objective was also to simulate the long term effect of tillage depth on the yield of a maize crop (Zea Mays L.) using the SALUS model. The study was carried out in 1999-2000 on a middle-mixed silty farmers field (Azienda Agricola San Basilio, 44°57' N and 12°10' E, 0 meters a.s.l.) in the Po Valley, near Rovigo (Italy). During the trials, functional (work rate), energetic (fuel consumption) and agronomic (plant density) parameters were compared between tillage at constant depth (35 cm) and tillage at variable depth. The average rate of work increased at variable depth, while average unit fuel consumption was lower during tillage at variable depth compared to constant depth. Tillage at variable depth reduced the unit cost compared to tillage at constant depth by 12 €/ha (- 31 %); the economic and energy results are correlated to the spatial variability and entity of soil compaction within the field. A long term simulation was performed to evaluate the effect of weather variability on changes of bulk density, maximum ponding capacity and saturated hydraulic conductivity. The SALUS (System Approach to Land Use Sustainability) model requires data on soil, weather, genetics and management to model continuous crop, soil, water and nutrient conditions under different management strategies for multiple years. Simulated and measured yield maps were compared using RMSE. Probability distribution function of yield was evaluated for variable depth tillage across the field and compared with uniform tillage depth. The economic benefit threshold area using variable depth tillage resulted in an annual tilled area of 165 hectares. The threshold was determined by analysing the unitary cost of soil tillage. The adoption of a variable soil tillage depth system could be profitable in order to improve economic and energy aspects, however the profitability of a variable depth tillage system is function of the spatial variability of soil compaction. Plant density was not influenced by tillage operations adopted within the field. The results observed between tillage depths, simulation and economic analysis were satisfactory and encouraging.

Estimating potential economic return for site-specific soybean variety management

W.D. Batchelor, J.O. Paz and J.W. Jones

Crop growth models have proven to be a useful tool for identifying causes of spatial yield variability in soybean fields in the Midwestern United States. Through model analysis, we have found that water stress is one of the leading contributors to spatial yield variability for soybeans. Producers have little control over spatial water stress in non-irrigated fields. Several alternatives are available to manage spatial water stress. These include planting different varieties within the field, altering seeding rates to conserve water in dry areas and using variable tillage practices within a field. Producers in the midwest are interested in matching varieties to specific locations in a field to capitalize on differences in variety response to water stress. Some producers have implemented strip variety trials in their fields to determine where different varieties outperform other varieties. One problem that exists is that the best variety for a particular location in a field depends upon the weather than occurs during the season. Thus, there is a risk that the variety prescription applied at planting may not be the best prescription for the weather that occurs during the season. Thus, risk analysis techniques must be used to develop variety prescriptions. Process-oriented crop models can be useful for determining optimum variety prescriptions that maximize return under uncertain weather information. This paper describes the use of a crop growth model in the development and analysis of soybean variety prescriptions for two 20-Ha fields in central Iowa. The Home and McGarvey fields were sub-divided into grids 0.2-Ha in size, and the CROPGRO-Soybean model was calibrated to fit three seasons of historical yield data in each grid. Yields for 70 soybean varieties were simulated for each grid using 34-years of historical weather data. This resulted in 34 estimates of yield for each variety and grid. Several variety prescription scenarios were developed and evaluated. Prescription A was developed to estimate the maximum net return possible for a variable variety prescription. It assumes that we always plant the variety that gives the maximum yield for a given year of weather, and that we have a priori knowledge of weather. Prescription B was developed to estimate the realistic expected net return that a producer could expect from a variable variety prescription. In this prescription, we selected the variety that maximized the net return for a grid over 34 seasons of historical weather conditions. Thus, this prescription does not rely on a priori knowledge of future weather. Prescription C was developed to estimate the net return for a single variety (maturity group II) planted in each grid. Not surprisingly, Prescription A produced higher yields and net returns compared to Precriptions B or C. Prescription B is a viable option for farmers to use without the burden of having a priori knowledge of weather information associated with implementing Prescription A. The difference between Prescriptions B and C is the increase in net return if the producer moves from a single variety to multiple varieties in the field. Our study also supports the idea of growing a single top-yielding variety across the whole field. This management strategy produced higher net returns than planting a single uniform variety (Prescription C), and roughly the same level of profitability as that of Prescription B.

Simulating the spatial variation of soil mineral N within fields using the model SUNDIAL (Simulation of Nitrogen Dynamics In Arable Land)

S.J. Baxter, M.A. Oliver, J. Gaunt and M.J. Glendining

Where water and other nutrients are not limiting, nitrogen (N) has the largest effect, on the growth and yield of cereal crops. There has been interest in variable-rate application of N fertilizer because of environmental and financial benefits that can accrue. To benefit from this technology the spatial variation of N supply within fields must be known in order to adjust fertilizer rates accordingly. Appropriate cost effective methods to predict N fertilizer requirement accurately and rapidly are still unavailable. This is because N transformations are complex within the crop and soil system. This makes regular measurements of N supply within fields difficult to achieve practically. An alternative approach is to simulate N supply using a model that describes these transformations and pools of N. If N transformations and pools could be modelled reliably using ancillary information, such as weather and crop data, and those soil properties that do not change markedly from year to year, such as texture, then modelling might provide a practical solution for site-specific management of N fertilizer applications. SUNDIAL is a simple process-based model. It comprises sub-models for the major processes governing the N cycle in arable soil, i.e. mineralization of soil organic matter, immobilization of inorganic N, denitrification, ammonia volatilization, nitrate leaching and crop uptake. The processes involved are described by a set of parameterized zero and first-order equations. The user usually enters readily available information for a field to simulate single values of N turnover. However, it has been used here to simulate more than 100 values of N within fields to compare the simulated results with actual measurements. The soil was sampled during 2000 or 2001 in the spring at over 100 sites in each of four fields: BM, WAR, CM and FF in Bedfordshire, UK. Mineral N measurements (one indicator of N supply) were made together with loss on ignition and particle size fractions. For two fields, CM and FF, mineral N was measured again in the summer: CMjuly and FFmay. To test the performance of SUNDIAL we compared the predicted and actual soil mineral N within the fields at different places. The parameters entered into the model at each location included: available water capacity, total active carbon, the fraction of incoming substrate converted to soil microbial biomass carbon and to humus carbon, and the minimum level of mineral N at each sample site. Other than mineral N these parameters have not been measured within the field, but were derived from other measured properties. For example, total active carbon was estimated from loss on ignition and available water capacity was estimated from soil texture. The results show that the simulated values of mineral N correspond to the measured values for BM, CM, CMjuly and FF, but those for WAR and FFmay did not. The model is based on assumptions that take into account the amount of organic matter, clay content and available water in the soil. The correlations between loss on ignition and mineral N were greater than 0.6 for BM, CM, CMjuly and FF, but zero and 0.3 for FFmay and WAR, respectively. It appears that the stronger the relationship of mineral N with organic matter, clay content and available water the more accurate the predictions from simulation will be. Total active carbon had a greater influence on the accuracy of the simulations than the other within-field variables. These results suggest that it is possible to use SUNDIAL together with ancillary information to provide information for site specific management and this will be described and illustrated with examples.

Development of a small autonomous research tractor

Simon Blackmore, Henning Nielsen and Hans Werner Griepentrog

A small tractor was modified to accept computer control as part of an ongoing research program to develop an autonomous research tractor that can behave in a sensible manner, for long periods of time, unattended, in a semi natural environment. A platform with these characteristics can allow a new range of mechanisation options that can reduce the scale of Precision Farming from sub field to potentially plant level, or Phytotechnology. The initial required behaviours of the tractor have been identified in (Blackmore et al. 2001) and a suggested system architecture has been proposed to enable these behaviours to emerge. (Blackmore et al. 2002) Part of the architecture proposed that the tractor be 'encapsulated' within a Hardware Abstraction Layer (HAL). This paper describes the internal elements to the HAL and how it will interface to the rest of the architecture. The tractor used was a 26hp grounds tractor called the Hakotrac 3000 from Hako in Bad Oldesloe, northern Germany. (www.hako.com) The main tractor control systems were modified to allow manual, electronic and computer control in four operational modes. (Off), Manual, Start, Remote and Automatic. Manual mode is were the tractor has all the actuators removed or disabled to allow a driver to take normal control of the vehicle. Start mode enabled the starter and fuel provided that the vehicle was safe to start. Safe conditions required that the transmission was in neutral and the safety system was enabled. The tractor will not start unless the driver also depresses the 'clutch' pedal. Remote mode enables the steering, engine speed and Continuously Variable Transmission (CVT) to be controlled from a hand held remote control box connected to the tractor via an umbilical chord. This remote control box is not used for driving the tractor but to take immediate control of each of the systems independently. Therefore it does not incorporate a computer system but utilises analogue voltages directly from potentiometers. Automatic mode is where a computer system can take complete control over all the systems. This will be achieved through the use of the GT2000 in-cab terminal PC from GEOTEC (www.geo-tec.de), or another computer that holds the route plan. The ESX job computer from Sensor-Technik Wiedermann handled the control loops on the tractor and the CAN interface from the GT2000. The operational modes were requested by the operator turning a multi position switch that inputted the request to a dedicated microprocessor dealing with the 'glue logic'. This configured the tractor into the new mode if conditions allowed. To move from Start to Remote required no special conditions but Remote mode disabled the starter. To move from Remote mode to Automatic mode, the CVT must have been in neutral and the controlling computer must show its active status. The Hako hydrostatic steering system was modified by Sauer Danfoss (Nordborg, Denmark) to accept their electro hydraulic proportional valve unit with a modified servo actuator to deal with the small oil flows for this tractor. An electrically operated switchover valve to enable manual or electrical control was also fitted. The engine speed was controlled remotely by adding a Linak LA12 (Silkeborg, Denmark) linear move actuator with positional feedback to the diesel injection pump via a Bowden cable. Setting the swash plate on the CVT with a second Linak sytem controlled the transmission ratio. Safety has been a priority throughout the design, building and operation of the tractor. To this end not only did the tractor have multiple stop switches, but also had many failsafe electronic devices and clearly identified operational procedures. Tests have shown that the tractor stopped within 2m after pressing the emergency stop switch at full speed.

Digital soil maps: A requirement for precision agriculture

J. Boess, H.-J. Heineke and J. Kues

Within the scope of the "pre-agro" project, NLfB has conducted a detailed mapping of the soil of a farm near Helmstedt. The resulting soil map can be used as a basis for precision agriculture utilizing the satellite GPS system. The locations of boundaries on this map must be determined with high resolution, for which the GPS system must also be used. In addition to the conventional sources of information (geological maps, relief maps and soil assessment maps), new sources of information, such as aerial photographs, electrical conductivity maps, ground penetrating radar (GPR) records and maps of crop yield, were tested to determine their efficiency and usefulness for obtaining the required high resolution. Whereas the geological and soil assessment maps are based on mapping with a soil auger, the crop yield maps and geophysical records are derived from areal measurements (GPR) or point measurements with a spacing of less than 10 m (geoelectric, g-spectrometry, soil assessment maps).

In the case of a sandy subsoil, georadar provides a good picture of the structure of the top layers. If the traverses are closely spaced, a quasi three-dimensional picture is obtained. Boundaries can be determined with high precision. Cohesive soils, however, attenuate the reflections. In glacial soils, georadar can be used to determine boundaries between till, glaciofluvial sand, and periglacial sand and gravel.

The apparent electrical conductivity parameter sums the electrical conductivity over the entire depth of penetration, which is determined by the water and clay contents.

The natural gamma-radiation of the soil is determined by g-spectroscopy. Tests show a very good correlation between soil textures in the top 30–40 cm and and the gamma-activity. Intersection of maps of electrical conductivity and gamma-activity permits changes in soil material. Because soil texture is an important parameter in many evaluation methods, these two methods are important mapping tools.

Patterns and structures identified by evaluation of the data were classified, tested by taking soil samples with a soil auger, and interpreted. Statistical methods were used to evaluate the contribution of the individual sources of data. The resulting digital soil map shows the heterogeneity of the soil with the high-resolution needed for optimum tillage, planting density, and use of fertilizers and pesticides. The main parameters on the soil map are soil texture, relief, humus content, effective capacity, and soil group. The data were evaluated using the NIBIS methodsbase of the NLfB. New methods were developed and tested in cooperation with other institutions.

Of the new geophysical methods, geoelectrics is available on the commercial market at favorable prices. Gamma-spectroscopy is not yet established in the soil sciences. A relatively small apparatus has been used for studies up to now, making the method rather time-consuming for soil mapping. A gamma-spectroscopy survey with a large instrument would take less time, possibly in combination with conductivity measurements. In this case, it can be expected the this method can be made available at acceptable prices.

A statistical comparison of the digital soil maps and the soil assessment maps shows that the digital soil map prepared with the above-described parameters has a distinctly higher resolution. The geophysical methods make a large contribution to the precision of boundaries on the digital soil map.

Prediction of protein content in spring barley and winter wheat using handheld sensor from Hydro and Hydro N-Sensor

Thomas Börjesson, Mats Söderström and Svenska Lantmännen

Within-field differences in protein content of cereal crops can be extensive and the pattern may vary between years (Stafford, 1999). It could be advantageous for farmers if the differences could be retained and not get mixed up at harvest, possibly missing the protein target for the whole lot. To retain the differences, fields could be divided into management zones that are harvested separately based on information collected during the growing season. Hansen et al. (2002) used canopy reflectance at Zadok stages 28 to 50 to predict yield and protein content in spring barley and winter wheat. The prediction of protein content in wheat was quite successful but not in barley. The regression coefficient for the relation between predicted and measured protein content was quite high in wheat ($r2 = 0.57$) but lower in barley (0.21). One problem that is discussed is that environmental factors late in the growing season could influence the protein content and that a later measurement might be beneficial. During 2001 we have made measurements in field trials in 6 different locations in Sweden with a handheld sensor from Hydro at different Zadok stages of spring malting barley and winter wheat; 31 (barley only), 45, 69 and 87. Two varieties of each cereal were used and three (barley) or two (wheat) fertilization levels. The handheld equipment used is similar to the tractor-borne N-Sensor used for site specific spreading of nitrogen fertilizers and is designed for research purposes. Precipitation data were collected from the field trial locations. Models were constructed based on partial least squares algorithms and the ability to make predictions for protein content at harvest using sensor data collected at different stages were compared. The models were validated both using crossvalidations and by using them to predict protein content in 6 different fields, 3 spring barley and 3 winter wheat fields in 2002. In this case, tractor-borne N-Sensors were used and only one measurement was performed (in Zadok stage 69). Management zones were constructed with different predicted levels of protien content within the zones. Crossvalidations in 2001 showed that both for wheat and barley a measurement after heading, Zadok stage 69, gave the best predictions, $r2 = 0.49$ for barley and 0.59 for wheat. Predictions were improved for barley by adding precipitation data for the three summer months ($r2 = 0.69$). For wheat, no such improvement was obtained, but variety specific models gave better results. With a model based on Tarso alone, an $r2$ of 0.76 was reached. Validations in 2002 were successful for barley, i.e. the correlation between predicted protein content in the zones where samples were taken and measured protein content at harvest was high. In wheat, the prediction was only successful in one of the fields. In the rest, the results were reversed, i.e. in zones with low predicted protein contents, high protein contents were encountered and vice versa. The reason might be that 2002 was drier than 2001 and that maturing-related colour changes that starts earlier in weak, high protein parts of the fields affected the reflectance. In conclusion, the study has shown that canopy reflectance can be a valuable tool for prediction of within-field differences in protein content. However, the method needs to be combined with data describing environmental conditions specific for the season. Variety specific models seem to be beneficial also. References Hansen, P.M., Jörgensen, J.R. and Thomsen, A. 2002. Predicting grain protein content in winter wheat and spring barley using canopy reflectance measurements and partial least squares regression. J. Agric. Sci. 138:1-12. Stafford, J.V. 1999. An investigation into the within-field spatial variability of grain quality. Proceeding of the 2nd European Conference on Precision Agriculture. 353-361.

Integration of expert knowledge in precision agriculture management systems - practicable use of scientific results

Stefan Böttinger

As in other industrial sectors also in agriculture a management software will be the backbone of upcoming business decisions. Due to the central role of software in a management system especially for the use in agriculture great demands concerning the functionality and the user interface are made. The corresponding tasks for the software are derived from the necessity to collect data of different qualities (precision and geographic resolution), different formats (from onboard computers, laboratories or Internet) and to calculate with these data. Especially with geographic data the processing and the presentation of the results has to be in corresponding graphics and maps. Only summaries can be outputted as text in a table. To support further the introduction of precision agriculture in practice it is necessary to make scientific results available for every farmer. Therefore expert modules are developed by scientist and integrated via a common interface in a standard precision agriculture software (main software). For the development of this interface two main topics have to be taken into account: - To ensure the responsibility and the service for the expert knowledge independent from the used main software - To enable the use of expert knowledge by farmers by the means of an easy to use and adapted user interface By the use of OLE for the development of the interface between main software and expert module and the realization of the module in a DLL several advantages are realized: - The definition of the responsibility for the content and the functionality of the expert module and the standard precision agriculture software is clear. By this the further development of module and main software is separated, mainly independent and could be realized by different persons or companies. - The user is free to choose the main software of his favourite supplier. In future the user will also have the possibility to choose between several expert modules with similar functions from different institutions or companies. Regarding the user interface the broad spectrum of the skills of the users has to be taken into account. This spectrum ranges between beginners in Precision Agriculture and with PCs, who are using such a software only sometimes, and professional users who are working daily with such a software and having high demands to the automation of the data processing. Based on the experiences of software manufacturers, developers and on the discussions with advanced users several design rules for the user interface can be derived: - Users are demanding understandability of the background of expert models, they need white box models. - Provisional results should be visualized and stored for e.g. later comparisons with results of calculations using other parameters or to support the understandability. - Data input from the user has to be minimized, e.g. by automatically using values from stored lists and tables and / or by storing the user's input to use it as preset for further usage of the model. Under consideration of these results regarding the software interface and the user interface first expert modules are developed and integrated to main software packages.

Assessment of spatial correlation between wheat yields and some physical and chemical soil properties

Hocine Bourennane, Bernard Nicoullaud, Alain Couturier and Dominique King

Topics: Methods and tools to analyze spatial variability Or Statistics, data analysis and research design Abstract The objectives of this study were to (i) quantify spatial correlation between wheat yield and some physical and chemical soil properties at different spatial scales and (ii) examine the temporal persistence in spatial pattern and correlation structure between wheat yield measurements conducted in two growing seasons and soil properties. The general aim of this experiment is to examine the potential value of yield maps, which are easy to obtain by many farmers, to infer variation in soil properties and thus provide a basis for managing lands in precise way. Soil samples from the 0- to 30-cm depth were collected at 104 auger holes in the southwestern Beauce Plain (France). For each sample 15 soil properties were measured as their simplicity analyses and presumed effects on wheat yield. Moreover based on the geographic coordinates of the soil observations in the field, each sample has assigned a value of available water content (AWC) and wheat yield values in 1998 and 2000. Wheat yields were measured using Proserie-system from RDS technology. AWC map is result of previous study over the study area in the frame of a precision irrigation project. Factorial Kriging Analysis (FKA) was used to describe the coregionalization of soil properties and wheat yield. The major steps of this geostatistical technique are described below. 1. The analysis of the coregionalization of a set of variables leading to the definition of a linear model of coregionalization (LMC) The p (p+1) / 2 experimental direct- and cross-variograms of the p variables requires a prior modeling using a linear combination of the same set of Ns variograms standardized to unit sill. 2. The analysis of the structural correlation coefficient Each coregionalization matrix describes the relationships between the chosen variables at the particular spatial scale, defined by the basic variogram functions. However, the structural correlation coefficient is more revealing as it is a unit-free measure of correlation between any two variables at the different spatial scales defined when modeling the coregionalization. However, depends entirely on the modeling of the coregionalization between each variable pair. 3. The principal component analysis on the coregionalization matrices and the cokriging of specific factors at different spatial scales A principal component analysis (PCA) was applied to coregionalization matrices, which are the variance-covariance matrices describing the correlation structure of a set of variables at different spatial scales. Unlike PCA performed from the classical variance-covariance matrix, the PCA on the coregionalization matrices yielded sets of spatial components (regionalized factors) for each spatial scale. Finally, the scores of the regionalized factors can be estimated at different spatial scales and over the whole study area using cokriging systems. The cokriged maps summarize thus the behavior and the relationships among variables at different spatial scales. Geostatistical mapping of wheat yield (cokriged map) was thus performed accounting for scales of spatial dependence between variables. The performances of cokriged yield map were compared to wheat yields measured using Proserie-system. The results show that the structure of the variation in yield appears to be relatively stable for two years. In addition the scale of variation of crop yield can be related to that of the soil properties. Stable areas suggest that some form of differential management might be feasible. Areas of low temporal variability would prove the easiest to determine management regimes, with areas expressing high temporal variability more likely to be identified as high risk zones and treated accordingly. Keywords: Soil properties, Wheat yield, Factorial Kriging Analysis, Yield temporal variation

Non-contacting chlorophyll fluorescence sensing for site-specific nitrogen fertilization

C. Bredemeier and U. Schmidhalter

Map- and sensor-based approaches are basic methods of implementing site-specific management (SSM) for the variable-rate application of crop inputs. The majority of available technologies in SSM utilizes the mapping approach based on sampling, map generation and variable-rate application. On the other hand, sensor-based methods of plant analyses could detect nutrient needs on-the-go and simultaneously apply fertilizer rates based on those needs, avoiding costs for soil sampling and analysis and data management. One technique to observe the N-status of plants by non- contacting methods is the detection of chlorophyll fluorescence. Laser-induced fluorescence is the optical emission from chlorophyll molecules that have been excited to a higher energy level by absorption of electromagnetic radiation. The role played by N in chlorophyll synthesis suggests that N-deficiency can be detected based on changes in the plants fluorescence spectra. The main advantage of active laser-based fluorescence sensors in relation to passive sensors is the possibility of measurement almost independent of light conditions and sun position, perhaps even during the night. In addition, the fluorescence signal has a very low background, since the signal comes mainly from the green plant parts. Experiments with corn and wheat have been carried out under controlled environmental (three years) and field conditions (one year). The treatments consist of different N- fertilization levels applied to the plants and different sowing rates. The excitation wavelength of the laser is 640 nm and the chlorophyll fluorescence is detected at 690 nm and 740 nm. We use different sensors for growth chamber (portable sensor) and field evaluations (tractor-mounted sensor). After the measurements are done, biomass production, N content and chlorophyll content (using the chlorophyll meter SPAD-502â) is determined. The effect of light intensity and temperature on the fluorescence ratio F680/F740 was also studied. The fluorescence intensity was measured in different light intensities (from 10 to 840 umol m-2 s-1) and air temperatures (from 5oC to 25oC). The fluorescence ratio F690/F740 in corn under field conditions was good correlated with SPAD-values measured on the latest fully developed leaf ($r2=0.78$). The ratio F690/F740 varied from 1.157 (in plants without N- fertilization) to 0.904 (in plants receiving 145 kg N/ha), while the SPAD-values varied from 34.7 to 51.5. The chlorophyll fluorescence intensity in wheat at 740 nm increased ($r2$ between 0.69 and 0.86) with increasing dry biomass production under controlled environmental conditions. This result indicates that chlorophyll fluorescence measurements can be influenced by both chlorophyll content and biomass quantity. The fluorescence ratio F680/F740 decreased from around 2.0 to 1.8 with an increase in the dry biomass from 5 to 9 g/pot ($r2=0,78$). The ratio F690/F740 was also correlated with chlorophyll meter (SPAD-502) measurements ($r2$ between 0.33 and 0.44). The ratio F680/F740 was little sensitive to light intensity between 10 and 840 umol m-2 s-1, suggesting that this system can operate with the same accuracy even at low light intensities, like at dawn and dusk. This feature of the ratio F680/F740 represents a great advantage compared with reflectance measurements, since reflectance varies greatly depending on the light intensity. The fluorescence ratio F680/F740 was slightly affected by air temperature. The ratio F680/F740 was linearly correlated with the air temperature ($r2=0.97$), decreasing approximately 16% with a decrease in the air temperature from 25oC to 5oC. N-fertilization levels in wheat under field conditions could be differentiated by means of fluorescence ratio F690/F740 measurements. The ratio decreased from approximately 0.79 in plants receiving 90 kg N/ha (SPAD- reading around 46.0) to 0.73 in plants receiving 210 kg N/ha (SPAD-reading around 50.0). This system is proving to be reliable to detect the chlorophyll content and the N-status of wheat plants under different weather conditions and almost independent of light intensity.

Estimating humus content of bare soils using digital aerial photographs and topographic data

Niels H. Broge, Mogens H. Greve and Rene Larsen

Knowledge of plant growth potential is important for farm managers seeking to optimise agricultural production at minimal economical and environmental costs. Hence, soil textural parameters, especially clay content and humus content, play an important role in the complex water and nutrient exchange processes occurring in the root zone. Both clay and humus is highly correlated with ground conductivity, which may be measured with mobile equipment in the field. However, in order to separate the contribution from clay and humus, an independent measure of one of the other is needed. Soil organic matter content is known to have a strong influence on soil reflectance (Baumgardner et al., 1985). Several studies have shown that spectral reflectance can be used to estimate surface organic carbon levels of soils. Coleman and Montgomery (1987) used a field-portable multiband radiometer, with a band configuration resembling the Landsat TM satellite sensor, to differentiate between soil types and soil properties including organic matter. Wilcox et al. (1994) showed that a linear relationship can be established between transformed Landsat TM band ratios and soil organic carbon. Hummel et al. (1996) found that SOM appears to be more correlated with the general shape and level of the soil reflectance curve than with reflectance at specific wavelengths. The results of Ingleby and Crove (1999), who measured spectral reflectance for soil samples of five different soils representing a range of organic carbon levels seems to support this argument. Topography is known to have an impact on the composition of the topsoil composition for cultivated fields. The A-horizon is typically thinner at hilltops due to tillage practice, resulting in accumulation and thickening of this horizon in field depressions. Further, humus concentrations tend to be higher in local depressions in the field, probably due to weathering effects. On this background, we decided to also include a topographic measure known as the wetness index (Burrough and McDonnell, 2000) as a predictor in the analysis. The potential use of spectral reflectance data and surface topological measures was investigated with the specific aim of estimating soil organic matter (SOM). Intensive measurements of soil texture and topography in two spatially and geologically independent Danish fields combined with digital aerial photographs form the basis of the analysis. Analyses of semi-variograms indicate similarity between spatial variations of spectral, topographic and textural measures. Multivariate statistics were applied to investigate the potential of spectral reflectance data and topological data for prediction of SOM. The results show that spectral reflectance data obtained at visible and near infrared wavelengths is correlated with humus content in one of the fields, and that the spectral data are highly co-linear between bands. The spectral reflectance data proved to be better than topological data for estimation of SOM in the two fields. However, none of the spectral indices or topological measures were able to properly predict SOM fractions larger than 5%. Combining spectral and topological data did not improve SOM estimates considerably.

Methods for site-specific nitrogen fertilization - results from German precision agriculture project Preagro

S. Brozio, R.I.B. Gebbers, D. Barkusky, G. Verch and K.-O. Wenkel

Site-specific nitrogen fertilization is one of the most important options in Precision Agriculture. This is due to the availability of the technology (fertilizer spreaders) and the importance of nitrogen for crop growth. Buying and using a VRT (variable rate technology) spreader is usually the first step in active precision agriculture management after adopting global positioning system (GPS) survey and yield mapping. An important prerequisites for variable rate nitrogen technology are user friendly methods for the calculation of optimum site-specific nitrogen rates. Such algorithms were the objectives of our working group within the project "preagro". The aim was the development of methods and algorithms for site-specific nitrogen management that improves economic and environmental benefits. To achieve this, results of the latest research, recent developments of industry as well as the official German guidelines have to be included.

The offline preagro-N model is designed as a balance-algorithm with "white box" character. The general recommendations of nitrogen are determined by yield goals, variety of winter wheat and proposed protein content. Spatial and temporal variations of the recommendations depend on the current situation (plant density and nitrogen content of the plants), soil types, relief, organic fertilization and weather conditions. Ecological and economical restrictions can be included as well. A version of the algorithm as an ArcView-based nitrogen recommendation system for winter wheat is available. The algorithms for crops winter rape, winter barley, triticale, and sugar beet exist as an Access data base, but are not yet implemented in ArcView. During the project, over three years a large number of field trials have been conducted. The type of trials vary from large field on-farm experiments, to strip trials and to small scale plot experiments. There have been nearly 50 large field experiments on the preagro research farms (16 per year), five strip trials, and one plot experiment at the ZALF experimental stations. Within the large field experiments constant rate and variable rate (using preagro-N algorithm) were compared. The results of these experiments are presented and discussed in detail. The strip trials were accomplished by the ZALF experimental station in Dedelow. These strip trials include tests of nitrogen recommendation of preagro-N algorithm, real time sensor based application rates (Hydro Agri N-Sensor), recommendation of mapping approach, sensor recommendation with map overlay, constant rate application and reduced nitrogen fertilization. The employment of thematic maps in combination with sensor were tested. The mapping approach and map overlay were based on ancillary data. Thematic maps were derived from yield maps of last years or potential yields, and relief units. The technology of sensor with map overlay is practicable and easy but not often used in practiced. Restrictions can simply be combined with real time sensor application. Special on-farm experiments were established which investigate the performance with the Hydro-N sensor. The sensor could not achieve higher yields, as far as our experiments are concerned. A higher nitrogen efficiency of the sensor could also not be proven as well. But this may be due to technical problems, which appeared sometimes during the storage of the application data. In this cases, the N-balance could not be calculated, leaving only a very few experiments where an analysis of nitrogen efficiency was possible. Using the plot trials at the ZALF research station Müncheberg, we are trying to compare the local application of the preagro-N algorithm with other methods of nitrogen fertilization strategies and techniques more accurately. The preagro-N algorithm was successfully conducted on seven farms, which are located in different regions of Germany. Results of our fertilization experiments show that site specific recommendation tend to have a higher nitrogen efficiency or a lower recommendation rate compared to uniform application.

Economic evaluation of the application of hyperspectral imagery as an aid to whole farm crop management through precision agriculture techniques in the UK

R.J. Bryson, R. Reeves and H. Poilve

Sustainability of the arable sector requires farm managers to be profitable whilst at the same time addressing the concerns and demands of the consumer as well as being mindful of the environment in which they operate. Crop production and whole farm management requires tight controls on both fixed and variable costs. The application of precision agriculture techniques may offer reductions in input costs through reductions in nitrogen, PGRs and herbicides inputs but it is essential that this is not at the expense of yield and fixed costs through increased machinery costs, time management and consultancy. A series of studies in 2001/02 evaluated the application of airborne hyperspectral imaging techniques developed by Astrium and ITCF in France using crop growth models to produce maps of leaf area index (LAI), tiller density, nitrogen requirement and yield prediction. A total of 850 ha over 7 farms and consisting of 51 fields were imaged at three key growth stages, GS32-33, GS37-39 and GS59-61. Management decisions for the crop area were based on both the imagery and the local farmer best practise and then compared. Variable inputs were applied using existing farm machinery either as zones, tramlines or split fields as appropriate. In order to determine gross margins, all experimental fields were yield mapped at harvest and grain quality parameters recorded. At Vine farm in Cambridgeshire the whole farm area of winter wheat, 20 fields in total (2 second wheat crops) and giving an area of approx. 300ha, were imaged and managed accordingly throughout the season. On all farm sites the imagery was ground truthed using established crop physiology techniques at each growth stage, good correlations were found between imaged and actual tiller density ($R2 = 0.92$) and LAI ($R2 = 0.87$). However, the relationship was more variable with crop nitrogen requirement and yield potential which was believed to be due to differences between the behaviour of some varieties under UK and French conditions. Within-field comparisons of input management using precision agriculture techniques, imagery and targeted N and PGR applications, with farmer best practise generally increased gross margin in the majority of cases with increases of up to £40/ha. Situations where no increases in gross margin were achieved generally reflected cases where few changes based on the imagery, away from best practise, had been made. In the whole farm study (300ha) the hyperspectral imagery not only helped input management decisions but helped to target field walking over the 1,500ha farm enterprise, confirmed general management practises based on indigenous farm knowledge (such as the use of take-all treatments) and helped to prioritise farm operations. Farm management time in most cases was increased due to the need to become equated with imagery interpretation, potentially increasing fixed costs, however farm managers felt that this would eventually reduce the need for farm consultant time and was not felt to be a negative effect. Added benefits of the precision agriculture techniques employed in this study were also found to include increased traceability to justify input decisions, identification of more or less productive areas and targeted crop walking for newly acquired land. This study is being extended for the 2002/03 season to cover an area of approximately 20,000 in the UK, it is hoped that a more comprehensive cost:benefit analysis, on a whole farm basis will be possible as well as an evaluation of the potential environmental benefits of applying these techniques.

Regression tree analysis to identify factors affecting yield variability

G. Buchleiter

Progressive crop producers interested in precision farming (PF) are using new tools such as global positioning systems (GPS) with yield monitors and soil electrical conductivity (EC) equipment, to collect large amounts of data about their fields. Some of their prevalent concerns are: (1) how to determine the production factors that are limiting yield, (2) which data are most useful and affordable to aid in modifying management. One hindrance to effective use of the large amounts of collected data is a clear understanding of how various production factors affect yield in various parts of the field. The objective of this research was to identify and quantify the various factors affecting yield variability to identify and implement the PF practices that could have the greatest economic impact. Data collected by a multidisciplinary research team on two sprinkler irrigated maize fields in northeastern Colorado USA for 4 consecutive years, have been used to develop an autoregressive model to identify various soil, water, fertility, and pest factors that affect yields (Heermann, et al., 2000). However, this multiple regression approach does not consider the possibility that factors most affecting yield can vary by location within a field. Another concern of these types of analyses is that various data parameters contain overlapping information e.g. % sand is reflected in part by EC, minimum elevation is related to irrigation amounts as well as potentially EC. A regression tree analysis was used on this 4 year dataset to map areas (76 m x 76 m grid) within fields where yield correlated best with different factors. This technique sorts the entire dataset into successively smaller, non-overlapping, more homogeneous yield classes having the smallest within-class deviance and the largest between-class deviance. Results indicate that although factors and the locations within a field can change from year to year, soil electrical conductivity (EC) and to a lesser extent applied water explains statistically much of the yield variability. Since regression trees can be constructed to explain successively smaller amounts of the variance, it is important to 'prune' the tree and not overfit the data. Generally the top 4 factors explained 75% of the yield variance. Regression trees can accommodate disparate data ranging from qualitative to quantitative, and can structure non-linear and interacting relationships. Mapping the locations of the data in various subclasses suggested that different parameters influence yield in different parts of the field. Nutrient levels were generally adequate across the field so there was little correlation with yield. Since salinity is very low and not problem in these sandy fields, the EC data were indicative primarily of the soil texture and the percentage of clay in the soil profile. Thus for these two fields, it appears that the water holding capacity of the soil and the variability of applied water had significant impacts on yield particularly in the sandier areas by statistically explaining the majority of the yield variability. Other factors statistically accounted for smaller amounts of yield variability. In one of the fields, Western bean cutworm levels were correlated with lower yield especially in the areas with low EC. In the other field, lesion nematodes and yield correlated best in the higher EC areas.

Retrospective analysis of water supplies to grain crops using site-specific protein maps

D. Butler, R. Kelly, W. Strong, T. Jensen and D. Woodruff

In northern Australia, yields of dryland grain crops are most likely to be limited by supplies of either nitrogen (N) or water. Much of the cropping soils in this region has a high proportion of clay (60-80%) which enables these soil profiles to feasibly support a winter crop on stored summer rainfall. Risk-averse farmers prefer to sow winter crops on a full moisture profile, but scope exists for others to sow on reduced moisture on the chance that in-crop rain may be sufficient. Coincidental protein/yield outcomes for dryland grain crops provide an indicator of the balance between N supplies and antecedant supplies of soil moisture. For example, wheat grain with a protein content of >12.5% and a relative yield of <25% is likely to have been yield-limited by supplies of moisture rather than of N. An interactive simulation model, known as Wheatman, has been developed over the past decade in the region to assist with N budgeting, timing of sowing, scenario analyses, and to explore opportunities for enhanced crop management. The model has proven to be robust in predicting grain yield/protein outcomes on the basis of antecedant soil moisture and initial N supplies. We wished to apply Wheatman to retrospectively assess what antecedant supplies of moisture ought to have been available for crop uptake. A barley crop was sown in May 1999 across two previously separated fields which had different cropping histories: 1/.long-fallow from the previous summer 2/.cropped with mungbeans. In November 1999, the crop was harvested with a John Deere header (AgLeader monitor) along permanent traffic lines. Samples of grain were simultaneously captured at harvest using a custom-built sampling device (ever 20-30 m). Consequently, 569 samples were taken within a 40 ha paddock. These samples were analysed for moisture and protein content using a NIR spectrophotometer. Each data layer was interpolated to a common 10 m grid for coincidental interrogation. We used algorithms within Wheatman to model the antecedant moisture for any given yield/protein output. Yield/protein outputs that were optimal for barley were used to develop a water use efficiency derivative for the field. Fallow and in-crop rainfall were used as inputs to develop water use efficiencies in the remainder of the field. In this way, yield/protein outputs could then be used to derive antecedant moisture supply maps on a 10 m grid resolution. Grain yield varied between nil and 7.7 t/ha. Mean yield was 2.17 t/ha, and the coefficient of variation (CV) was 28.2%. Yield values were distributed in two major zones within the field which appeared to follow the pattern of previous crop rotation. As expected, lower yields (relative yield <50%) were largely found in the region where double-cropping had taken place. Grain protein varied between 7.44 and 15.2%. Median protein was 12.98% with the distribution tending to be skewed to the upper limit (i.e. 25% quantile 10.96%, 75% quantile 13.75%). The coefficient of variation was 15.7%. Based on the interpolated data sets, grain yield was inversely related to grain protein (r=-0.89; n=5298). Antecedant soil moisture varied from 35 to 210 mm with the lower values tending to be found where double-cropping had taken place. In conclusion, the technique developed will enable farmers and managers to: · estimate antecedant supplies of soil moisture · predict the location and impact of factors, such as salinity or nematodes, that may reduce water use efficiency below that anticipated for any grain crop in a given season · determine if variable supplies of moisture are related to stable factors (proximity to run-off sites or field depressions), or unstable/unmanageable factors (patchy rain events etc.), · determine what supplies of moisture remain in the profile for successive crop access. It is likely, although not tested in the above study, that these techniques could also be utilised in irrigated crops thus highlighting where irrigation outlays were inadequate or faulty.

Mapping soil compaction measuring cone penetrometer resistance

M. Carrara, A. Comparetti, P. Febo, G. Morello and S. Orlando

The traffic of agricultural machines can cause soil compaction, reducing soil porosity and creating obstacles to air and water movement and to root penetration. Soil compaction can be measured in several ways, most of which require the drilling of core soil samples and time consuming laboratory tests, or lengthy field preparation, where trenches have to be dug for burying probes and load cells. Probably the quickest way for monitoring soil compaction is measuring cone penetrometer resistance. The literature is full of examples where cone penetrometer resistance measurements have been used for monitoring soil compaction. On the basis of these experiences, it can be claimed that cone penetrometer resistance measurements can only monitor the changes, which have occurred after compaction in those particular soil conditions (texture, water/moisture content, etc.) of the field used for the tests. They cannot give any absolute measurement that can be compared with the state of compaction of other fields and/or soil conditions. However cone penetrometer resistance measurements can be used to quantify and evaluate the damage caused by the traffic of agricultural machines and/or the action of tillage implements. Such an evaluation can be useful for choosing the most suitable size/mass of tractors and machines, type and inflation pressure of tyres, and shape and working depth of implement tines. Soil compaction in a field is significantly spatially variable, depending on the amount and distribution of the traffic of agricultural machines, and of the action of tilling tools. Therefore, it would be useful to measure soil compaction on a localised basis, in order to produce a map of the compaction occurred in the field. The aim of this paper is to present an instrument developed at Dipartimento I.T.A.F. (Department of Engineering and Technologies in Agriculture and Forestry) of Palermo University and the method used for mapping soil compaction. An electrically driven penetrometer is fixed to a frame, which can be mounted on either a tractor or a trailer, or on a self-propelled machine. A DGPS mobile receiver and the relative satellite antenna are also fixed to the frame. The penetrometer and the DGPS mobile receiver are connected to a portable computer, which logs the values of cone penetrometer resistance and the related DGPS positions. GIS software allows the production of maps of the cone penetrometer resistance measurements. The idea is to map the measurements before and after the traffic of machines on the field and to compare the two maps in order to quantify and evaluate the intensity of soil compaction, creating a sort of soil compaction field map. The traffic itself will also be mapped using the DGPS mobile receiver and the GIS software. Then the two maps will be superimposed, in order to find ways of avoiding, or at least reducing, the amount of compaction. This could be done either by choosing the most suitable tractors, machines, tyres/tracks and tyre pressures, or by studying traffic systems that will confine the compaction in places where it can be less damaging or more easily controlled. The entire system is at present being tested for its capability and accuracy in different field and soil conditions

Field-based multiangular remote sensing for plant stress monitoring

R. Casa and H.G. Jones

Successful remote sensing of plant stress should open up a number of applications in precision agriculture, which would allow the correction of the stresses detected to optimise production and quality. Cost-effective monitoring by remote sensing would be useful in the development of technologies for the optimisation of the localised control of irrigation and fertilisation. In general any stress, almost by definition, leads to sub-optimal growth. Such sub-optimal growth can be detected remotely through the resulting reduction in leaf area index (LAI); this reduction in LAI can it turn be determined through the use of one of the many spectral vegetation indices (VI) that have been developed over recent years. However these vegetation indices, eg. NDVI, are used to derive statistical relationships with vegetation properties that do not have a general validity. In addition they are influenced by leaf and soil spectral properties, canopy geometry, illumination and view zenith angles. Moreover a saturation in the VI-LAI relationship can occur at LAI values higher than 3. Improvements in the methodologies are therefore needed before routine remote sensing estimation of LAI can be carried out. Multiangular remote sensing, as compared to the typical use of measurements from only one view angle, offers the possibility of exploiting the directional information, which has been shown to be especially influenced by the geometric properties of target. The availability of multiangular radiation models that incorporate simulation of leaf spectral properties into canopy reflectance models, have stimulated in the last decade a number of studies in which crop canopy properties are inferred from remote sensing measurements by model inversion. However in in many circumstances the theoretical capability of a model inversion procedure to retrieve the correct parameter values has not been matched by the practical applicability using measured data, in which large experimental errors are usually included. In addition it has been shown that the same multiangular reflectances or spectra can be modelled by different combinations of input parameter values. The main goal of our research was the development of unexpensive methodologies for the retrieval of LAI from field based multiangular remote sensing data and inversion of physical models of the vegetation. Our approach was based on the exploitation of the angular variation of image fraction components, i.e. the fractions of images occupied by sunlit and shaded leaves and soil. For this purpose a colour-infra-red digital camera (Dycam ADC) was used to acquire an extensive data set of multiangular red/NIR images on potato canopies subject to contrasting irrigation and nitrogen fertilisation treatments. Supervised classification of the images, into shaded and sunlit foliage and soil classes, was then carried out. Multiangular image fraction data were used for the inversion of a simple ray tracing canopy model developed for the purpose. The latter is not a canopy reflectance model, but just an application of ray tracing algorithms to a geometrical optical problem. In this way the model parameters were kept to a minimum, being only related to the canopy geometric properties. The inversion of the model was performed using a look-up-table approach. This allowed to estimate successfully the main canopy structure parameter, LAI, from the measurements carried out on potato canopies.

Stone density of soils by image processing

Michel Chapron and Philippe Huet

This work is concerned with the stoniness of soils. This consists in computing the density of stones on the soil and extracting some features of stones. The properties of soils are the following : texture (stony, argillaceous, sandy, ...), the structure (fine, coarse, medium, ...) and the consistency (friable, sticky, damp, ...). Herein, we try to characterize the stony soils by image processing, which permits to predict the ability of the soil to stock water and determine the well adapted crops on the observed soil with other agronomical parameters. The image treatments are carried out on grey-level images because the colour provides almost nothing. Usual techniques applied for classifying different textures on images can not be easily utilized for separating stones from soil because the stones are too different to separate them from each other. The proposed method is the following. The first step of the method is to filter the image in order to better distinguish the stones from the soil. For reaching this aim, different processings have been tried : median filter, filterings based upon the coefficient thresholding of the different types of wavelets and packet wavelets, filter founded on the total variation principle. The best results are obtained from the last filter which is applied several times. This process well preserves the stone contours. During the last few years, there have been more and more applications of partial differential equation based techniques (PDE) in image processing (see the works of Morel and Alvarez). These new methods are motivated by a more systematic approach to restore images with sharp edges, this can also be the previous step of the segmentation stage. The soil and stones are denoised according to a non-linear anisotropic diffusion PDE, designed to smooth less near the edges. The main drawback of using quadratic regularization functional is the inability to recover sharp discontinuities (edges). To improve the Tikhonov's regularization with quadratic regularization functionals, Rudin, Osher and Fatemi proposed to use the total variation as a regularization functional. From the filtered image, it is easy to calculate the segmentation of the image which permits to compute geometrical parameters such as area, perimeter, compactness factor and invariant moments (translation, rotation and scaling) of the stones, these parameters characterize them. Then, the stone density is computed. A second work consists in characterizing the stone contours by using a multi-scale technique based on the wavelet based multi-resolution representation of the image. The estimation of the coefficients of regularity of Lipschitz a on significant pixels of stone contours is performed by following the modulus maxima obtained from Canny's edge detector through different resolutions. This can be represented in a scale space plane and the estimation of a use the regression technique in this plane (Mallat). When following the modulus maxima through the different resolutions the orientation of the contour in the neighbourhood of these singular points is taken into account. Short sequences of modulus maxima do not permit to compute the coefficient a with high reliability. This technique can be implemented in a real time way. This method gives good results on different types of soil.

Is precision agriculture irrelevant to developing countries?

S.E. Cook, R.J. Corner, R. O'Brien and T. Oberthur

A commonly-stated reason for low adoption rates of precision agriculture in developed countries is that its benefits are insufficient to justify the additional costs. Ostensibly, this seems to preclude any possibility of PA for developing agriculture, which suffers much lower profitability than virtually any sector of agriculture in developed economies, and has virtually no capability to pay for the technology used. We question this assertion, and postulate that the basic purpose of PA - namely to reduce the uncertainty introduced by spatial variation- far from being a luxury, could become a necessity for the survival of farmers in many parts of the developing world, even if it appears there in a completely different form. Using examples from Latin America and elsewhere in the developing world, we examine this question in relation to four key topics: 1 The value of information: What is the relative cost of ignorance? How does ignorance make poor farmers poorer? 2 Cost of information: How realistic is it to consider bridging the digital divide in developing countries? What alternatives exist to provide the information required? 3 Scale of application: At what geographic scales is information likely to be used? 4 Organization: How could the flow of information be organized at these scales to support development processes? Precision Agriculture is most often used to manage farm inputs more efficiently by managing spatial variability. In developing agriculture not only is information on these inputs often not available, neither is the technology commonly used for analysing spatial variability. However smallholder farmers, precisely because of lack of technology, manage spatial variability on their farms by tending their fields on foot or using draft animals. Therefore it is both possible and relatively easy for them to apply the PA concept of spatially variable management. In order to survive, farmers in the developing world need to respond to changes in the global economic climate and also manage the temporal and spatial uncertainty associated with short and long term cycles of climate change. Due to the small scale near subsistence level of some of these operations, their agricultural practices need to be extremely resilient. Therefore decision support needs to be highly flexible and locally targeted. We conclude that although PA - as seen in Europe and North America - is likely to be largely irrelevant in developing countries, the need for spatial information here is actually greater, principally because of stronger imperative for change and lack of conventional decision support. The acquisition and delivery of spatial information in developing countries is a major impediment to progress, and here, we suggest alternatives to expensive data-gathering technology which build on available information at regional and local scale. In addition, readily-available spatial data, although often at a scale seen as too coarse for PA, can often be updated using local knowledge and made usable without the need for expensive GPS data collection. Such methods would eliminate the opportunities for technology suppliers, who have been key to much of the progress of PA in the developed world, so we propose an organizational structure which could implement change without their participation.

Error propagation in agricultural models

R. Corner, D. Purnomo and M. Adams

Spatially distributed agricultural models are being used for the routine delivery of agronomic advice to growers. These models are generally data intensive and may involve data collected at a number of different spatial resolutions and accuracies. Some data sources, such as grain yield, are collected from closely spaced samples whilst others are interpolated from a small number of more widely spaced samples. This means that we have error in measurement compounded by error in interpolation. When a number of data sets are combined using map algebra in a spatially distributed model, these errors are propagated through the algebraic functions. Depending on the form of those functions this may cause serious uncertainty in the resulting recommendations. A method is therefore required to determine the magnitude of this propagated uncertainty. This case study looks at a model to predict the Nitrogen fertiliser required to obtain maximum economically achievable wheat yield. There were two stages to the work. The first looked at errors in the interpolation of soil property surfaces and the second addressed the question of error propagation through the model itself. The study area comprises two paddocks on a wheat and sheep farm near Geraldton, Western Australia. More than 150 soil samples were collected on a 100 meter and 25 metre grid format. These samples were then analysed in a commercial laboratory for a standard range of properties. An experiment was carried out in order to investigate the effect of different sampling densities on the reconstruction by interpolation of a soil surface. No very-high resolution soil sample data was available for an extensive area but an airborne gamma-radiometric dataset for the entire farm was available. The thorium channel of this dataset was chosen as being an integrator of a number of soil properties and therefore as representing the spatial variability of those properties. The thorium surface was then sampled at a range of intervals from 25 to 200m on a regular grid and a variety of interpolation methods (Kriging, Inverse Distance Weighting and Spines) were used to recreate the original surface. The most suitable interpolation method and data density chosen based on comparison on the root mean square of the differences between the predicted value and its actual value. The best compromise between fidelity of surface reproduction and sampling cost was provided by a sampling interval of 100m and a spline interpolator The actual samples were then used to create surfaces of the soil properties required for the Nitrogen requirement model. Cross validation techniques were used to estimate the interpolation error at each data point. These error estimates were themselves interpolated to form an error surface for each soil property. The initial N calculation was modified to maximise the profit and output a nitrogen recommendation for maximum profitability. Error propagation routines were written using both First and Second order Taylor Series in order to track the propagation of error through the nitrogen requirement model. The interpolation error surfaces were used as well as error surfaces that assume percentage errors caused by measurement error. Results show that the errors in estimations of nitrogen application required for maximum profitability had mean values between 33% and 43% depending on the methods used to estimate the individual data layer errors. The 33% figure is the result of propagating conservative estimates of individual data layer error. Further work in continuing on a Potassium model whose algebraic form suggests that error propagation will be more severe.

Up scaling of soil data using ancillary data and probabilistic inference

R. Corner, N. Schoknecht, W. Verboom and D Purnomo

The effective implementation of precision agriculture requires a supply of soil data at a high spatial resolution. Ideally this should describe soil properties rather than soil types. Australia, with large paddocks exhibiting considerable variability is well positioned to take advantage of the benefits of precision agriculture. However it suffers from a paucity of soils information. In Western Australia for example, soils are mapped at a soil landscape level rather than at a soil type or soil property level. It is of course possible to collect base soil data by carrying out detailed surveys and soil properties may be derived from sampling. However this is often economically marginal, particularly in areas with large paddocks and relatively low returns per hectare. This paper looks at the opportunities for using soil landscape and soil type maps as a base for an up-scaling process using Bayesian inference. This is particularly applicable in a case where the coarser scale spatial data is linked to non-spatial data that describes the statistical distribution of soil types or properties at a finer degree of differentiation. The study site was an area of 8000 ha in the West Australian wheatbelt for which airborne gamma radiometric data had been flown. Individual paddock in the area are typically 80-100ha in size and may contain four or five soil types. The radiometric data was collected as part of a pilot study to determine the value of collecting such data over the whole of the agricultural area of the State. A soil landscape survey of the area carried out a nominal scale of 1:250,000 was available in digital form. Attached to this data set is a table describing the proportions of soil types occurring in each of the soil landscape units. In addition a high resolution digital elevation model (DEM) of the area was available. A number of terrain attributes were synthesised from this DEM. An adaptation of the Expector soil attribute mapping method was used. This is a Bayesian method originally designed to update prior probabilities of soil property occurrence using diverse related data. The initial version of Expector set those prior probabilities entirely from sample data and assumed that they were ubiquitous throughout the area. The modified version uses the proportions in the soil landscape map as priors. That is, it starts with spatially distributed priors It is suitable for the updating and up-scaling of any map in which there is a hierarchical regime of classification. For example it can be applied to up-scaling soil-landscape maps to soil type maps, or soil type maps to soil property maps. In each case it requires the ability to develop relationships between spatially extensive environmental variables, such as terrain attributes, and members of the lowest target stratum of the hierarchy. This is achieved by using sample data and expert knowledge. The system is able to account for uncertainty in the input data by weighting each class in an input map with a confusion matrix for that map. It is also possible to weight entire data layers in accordance with an expert opinion of their predictive strength. An interesting point raised by this research is that of incompatibility in data sets. This occurs, for example, if the base map to be updated does not allow the possibility of a particular soil type existing in an area, yet expert knowledge and ancillary data does. Methods are being developed to resolve these ambiguities.

Rows detection in high resolution remote sensing images of vine fields

J.P. Da Costa, C. Germain, O. Lavialle and G. Grenier

Variability of soils is the basis for French Appellation d'Origine Controlée production areas delineation, so wine growers are very receptive to Precision Agriculture concept, and most of them are thinking that PA technologies could be useful for a better management of vineyard and for improving wine quality. Unfortunately, adoption of Precision Viticulture technologies is more limited by the lack of appropriate tools than by wait-and-see attitude of wine growers : yields maps are not useful for this purpose because yields are voluntarily limited – on a legal basis defined region by region – by way of pruning, and designing of soils maps by way of electro or magneto-resistivity methods must be done – at this time – only before vine planting. In this context, remote sensing imagery could be a great help for the better understanding of in-field variability and to access to a Precision Viticulture management. But, due to the small size of vine fields (in France mean size is around 0.3 Ha), high resolution images must be used in order to obtain an accurate zones delineation. The particular spatial arrangement of the vine induce the need for new methods of images processing : rows of 0.4m wide are regularly spaced (depending of the region and management choices, this spacing out could be from 1m to 3m). In high resolution images, rows of vine must be separated from soil or grass located between rows. The goal of this work is to achieve a reliable tool for detecting rows in this type of images and to draw up a map of vine sturdiness row by row. This map could by used in a second step to split the field in different zones according to a sturdiness index. For rows detection we used a FFT method : despite the images are taken on natural scenes we obtain two definite peaks. With these peaks it is easy to calculate coarse orientation and spacing of rows. These data are used for positioning on the image a network of parallel lines regularly spaced, using pivot points put on a line perpendicular to the direction of rows. These lines are deformable shape models, and the following step of this process is the calculation of the better orientation and spacing of lines according to a cost function which must be minimized. This cost function is calculated on the entire line, based on the sum of grey levels for all pixels on the line. The line is little by little moved right and left from is original position, and also rotated from is original direction. Results sow that it is possible to obtain a very good detection of each row, even if some of them are interrupted (missing plants), or when the spacing of rows is irregular (geometric distortion in the image, irregular planting spacing along the row,..). At the end of calculation, the line appear to be correctly positioned on the squeletton of each row (even if there is some holes in the row). Finally a sturdiness index is calculated using a rectangular window centred on each line, the size of this window is row spacing in wide and line length. Work is going on for designing a reliable method for high resolution remote sensing images segmentation and zones delineation.

Comparison of EMI and DC methods for soil mapping in precision agriculture

M. Dabas and A. Tabbagh

Electrical Methods, named Direct Current or Galvanic (DC) and ElectroMagnetic Induction Methods (EMI) are powerfull tools for non invasive measurements of soils. These methods have been used for years and are known to be important tools for cost-effective site assessment. These two classes of methods measure the same physical parameter: Electrical Resitivity (ER) for DC methods or its inverse, Electrical Conductivity (EC) for EMI methods. ER and EC play a major role in the delimitation of zones in Precision Agriculture (PAg). It is a direct consequence of existing relationship between agronomical important parameters (like clay or available water content, porosity and texture) and ER/EC. It is also a consequence of both the existence of easy-to-use instruments in the market and a wide dynamic range of ER and EC in soils. Moreover strong correlation are sometimes observed between yield and ER/EC for specific years. Definition of management zones from the knowledge of yield maps, even with several years, still remain a challenge. It is now often stated that ER or EC maps are the first step in PAg. They serve as a basis for a precise location of soil samples and for a delimitation of homogeneous zones needed in the definition of the management zones. They also bring a spatial accuracy which is compatible with the need of precision agriculture machines. The aim of this paper is two fold: from a theoretical point of view (implementation of Maxwell's equations), we have performed 1D (one dimentionnal) and 3D numerical simulations to test the responses of available instruments on the market (EM38 (Geonics), Veris3000 (Veris) and ARP01 (Geocarta)). On the other hand, these instruments have practical limitations in terms of their design, calibration, etc. They were compared in the field by surveying the same plot with the same conditions. For both methods, we have tested: 1 : The influence of the height of the coils above the soil surface in EM surveying and tested it with the published curves from Geonics 2 : The influence of the LIN (Low Induction Number) hypothesis for absolute determination of soil conductivity. 3 : The vertical response of instruments (depth sensitivity by simulation of the effect of a very thin layer) 4 : The investigation depth of the instruments (effect of a two layer soil) This was tested both for conductive and resistive targets. In a second time, a test plot of 30 acres was surveyed with the different geophysical techniques and comparisons were drawn from the derived maps. Tha advantages of using EMI methods were found to be the ease of making a measurement and cost-effective instruments available on the market. The drawbacks are: important thermal drift and time drift of the electronics, absolute calibration nearly impossible in normal conditions, a low signal for high resistivities, an investigation depth which is not controlled, a time lag of the response for available instruments. The drawbacks of using ER methods were found to be: labour intensive measurements and problems when high contact resistances are encountered (dry soils). The advantages are: absolute measurements with no drift, robust instruments , several depth of investigation at the same time by use of multiple electrodes, a very quick response time which makes data collection at high speed possible with no spatial lag. From this comparison wc found to be in favour of electrical methods.

New innovation profitability: precision agriculture adoption by sugar beet producers in the Red River Valley of the U.S.

S. Daberkow, W. McBride and M. Ali

Precision agriculture technologies, such as grid soil tests, yield monitors, remote sensors, and variable rate applicators, are tools to manage sub-field spatial and temporal variability to enhance economic and/or environmental benefits. However, the adoption level for many of these tools is modest for most commodities. There is some evidence that adoption of these tools is more profitable for higher value crops, such as sugar beets. This analysis explores the financial performance and nitrogen fertilizer use of sugar beet producers in the Red River Valley of the U.S. who have adopted precision agriculture technologies relative to those who have not. Nitrogen fertilizer management is a key factor in the profitability of sugar beet production. Nitrogen is not only the most important yield-limiting nutrient but also its management is critical for producing high quality sugar beets. Production of high quality sugar beets is important in a quality-based payment system because growers are being paid on the basis of extractable sucrose content of their sugar beets. Nitrogen sources for sugar beet production include soil organic matter, which often varies spatially across fields, and commercial nitrogen fertilizer, which can be applied using variable rate applicators. Adequate nitrogen fertilizer use normally increases yield of both roots and sugar; however, excessive use of nitrogen fertilizer decreases the sucrose content in the root. The use of variable rate applicators is intended to make more efficient use of nitrogen by applying fertilizer to meet plant needs that vary spatially across the field. This study will examine the impact of using variable rate technology (VRT) on nitrogen fertilizer application rates and on farm financial performance. The Red River Valley of North Dakota and Minnesota was chosen as the study area for this analysis because of its importance to U.S. sugar beet production, and because of the significant adoption of precision farming technologies in this area. Over a third of the sugar beet farms in the Red River Valley had soil properties mapped and geo-referenced while over half had remote sensed images made of sugar beet fields. About 17 percent of the farms used this information for variable rate fertilizer applications. Data for the analysis comes from a detailed survey of U.S. sugar beet operations conducted in 2000 as part of USDA's annual Agricultural Resource Management Study (ARMS). Each farm sampled in the ARMS represents a known number of farms with similar attributes so that proper weighting provides a basis for calculating estimates for the target population. The target population in the sugar beet survey was operations planting one or more acres of sugar beets in 2000. The survey collected information about sugar beet input use, production practices and yields; costs of production; and farm financial status. The survey also collected specific data on the extent to which various precision farming technologies were used in sugar beet production. The model used to assess the impacts of VRT adoption on sugar beet farms is estimated in two stages. The first stage is the estimation of an adoption equation for VRT using a binary probit model. Predictions from the adoption model are specified as an explanatory variable in OLS regressions relating farm and operator characteristics to fertilizer use (i.e. nitrogen application rates) and to measures of farm financial performance (i.e. net farm income and sugar beet returns above variable costs). This specification is used as a means of correcting the self-selection bias inherent in estimating the VRT adoption impacts based upon farmers themselves deciding whether or not to use the technology.

Sensor-controlled variable rate real-time application of herbicides and fungicides

Karl-Heinz Dammer, Gerhard Wartenberg, Hartmut Böttger and Detlef Ehlert

Weeds and plant diseases within cultivated fields are not equal distributed; they occur often in patches. A variable rate application of herbicides and fungicides according to the weed and disease occurrence helps to optimise the input in means of production and reduces the input of biocides into the environment. Because a weed and disease assessment in the field by walking is time and labour consuming, at the Institute of Agricultural Engineering (ATB) sensors were developed. They are used in combination with a field sprayer for the real-time application of herbicides and fungicides.

Because of general difficulties in discrimination between weeds and the cultivated crop within the same standing area, an optoelectronic sensor detects the weeds within the tram lines of narrow row crops, e.g. cereals and legumes. Based on the found relationship between the weed counts in total and the sum of the calculated yield loss due to certain weed species there is no discrimination between weed species. The application rate is than adjusted according to the expected yield loss.

Since there are no sensor-based technologies on the market for the direct detection of plant diseases, an alternative method has been developed to optimise fungicide application. The strategy is to apply the same concentration of active fungicidal substance per unit of plant surface area (leaf area index). Small-scale information about the present plant surface of the cultivated crop can be obtained indirectly by the pendulum-meter, a mechanical sensor for scanning biomass.

Field experiments for testing the procedure of the sensor-controlled variable rate real-time application of herbicides and fungicides were conducted in commercially cultivated fields of cereals and pea.

For evaluation of the efficiency of the real-time applications weed and disease occurrence as well as yield were recorded in uniform compared to adjacent variable rate treatment plots. Weed counting and disease assessment were done approximately two months after application. The experimental fields were harvested by a combine harvester with a yield logger. From the local yield measurements relative yield levels between the variable rate treatments and the uniform treatments were calculated for 9 meter distances. Because of the resulting high sample size, for yield comparison frequency plots of the relative yields were used. Three year results are presented.

In areas with low application rates, as well as in areas with the maximum application rates, there were no differences between weed and disease occurrence. Local yield reductions, as well as yield increases were obtained, but at mean there were no yield reduction due to the variable rate real-time application.

Automated methods for mapping patterns of soil physical properties as a basis for variable management of crops within fields

P. Dampney

Many previous projects have demonstrated that the nature and variability of soil physical properties within fields are of key importance when assessing the potential for variable rate management of some crop inputs and for delineating within-field management zones. This paper reports key results and conclusions from recently completed projects which have investigated two alternative approaches for mapping soil physical properties using automated, cost- effective methods. Firstly, the use of yield map sequences to predict soil patterns; secondly the use of Electro- Magnetic Induction (EMI) non-intrusive sensing measuring the apparent soil electrical conductivity (Ec). Using cluster and spatial analysis statistical techniques, yield map sequences from over 30 fields in England on contrasting soil types have been analysed to identify the occurrence, scale and magnitude of variation of persistent sub-regions within each field. For each field, an independent pedological survey of the field has been carried out. From this, a series of spatial and non-spatial indices of spatial variability have been identified and related to field and associated laboratory soil analyses. In addition, a 5 class agronomic assessment (the "A" rating) of the potential economic benefit from variable rate application of crop inputs has been made for each field based on the pedological soils information. Results to date (see note below) show that yield map data have a good relationship with the "A" rating and can thus provide useful information about the underlying soil variability of a field. that is of practical significance for crop management purposes. This information can be used to allow provisional mapping of soil patterns. Between 2000 and 2002, four fields of contrasting soil types, including sandy, clayey and shallow soils over rock, have been investigated using an ATV (All Terrain Vehicle) drawn EM 38 model to determine i) the soil factors influencing Ec, ii) the stability of Ec maps under contrasting measurement conditions, and iii) the optimum method for practical use. In each field, horizontal and vertical dipole measurements were carried out under dry, wetting, wet and drying soil moisture regimes and also at 6 metre between-track intervals. Associated soil moisture and soil physical measurements were carried out in multiple transects within each field. Linear modelling, ordination analysis and robust geo-statistical analysis of the data have provided the following conclusions. Soil Ec is effective at distinguishing between soil types, Soil moisture, organic carbon and clay content are the main factors influencing soil Ec Although the absolute values of soil Ec change significantly according to the season and the different levels of soil moisture, the spatial patterns of Ec are generally stable; Soil Ec data based on 20 metre transects across a field can give reliable maps of soil Ec The advantages and disadvantages of the different approaches for mapping patterns of soil physical properties within fields will be discussed. Note to Program Committee We are in the final stages of data analysis for these projects which will be completed by November 2002. If this abstract is accepted, the key statistical and interpreted data will be presented in the full paper which we do not yet have available.

Weed leaves recognition in natural complex scenes by model-guided edges pairing

B. De Mezzo, G. Rabatel and C. Fiorio

New weeding strategies for pesticide reduction rely on the spatial distribution and characterisation of weed populations. For this purpose, weed identification can be done by machine vision applied in the field. Due to the scene complexity, *a priori* knowledge on the searched shape is valuable to enhance the image segmentation process. We propose here an approach based on a primary analysis of object boundary pieces in the image. This analysis relies on shape modelling, and leads to the generation of hypotheses about actual leaves in the scene. Initial results are presented, and further developments are proposed.

Planning the infrastructure for optimisation of sugar beet production in transborder farming

M. Demmel, J. Feldmann, M. Rothmund, H. Auernhammer and T. Rademacher

Sugar beet production delivers the highest profit of crops in Western Europe's agriculture. Nevertheless there are further potentials to increase the profit from sugar beet growing. Because changes in the market system of sugar and sugar beet are discussed the possibilities of optimising sugar beet production again will be discussed more intensively. Beside biological and cultural improvements of the sugar beet production also the very important area of organisation of the sugar beet harvest and transport seems to have potential for improvement. Growing sugar beet delivers the biggest mass of harvesting material (60 – 100 t/ha) of all crops. The processing of sugar beet is concentrated on a small number of plants. Therefore the beets have to be transported over long distances to the plants. These factories are working 24 hours a day, 7days a week through the short campaign of 3 – 4 month. During this period all the time enough beets have to reach the factory that it won't be forced to stop processing. To hold the factory running under all circumstances they normally have big piles of beets laying in front of the plant. Optimisation from the position of the sugar factory means reducing this stock and realising a jus in time delivery of the sugar beets. But on the other hand the community of sugar beet producers have real problems to organise just in time delivery. The structures of sugar beet producing farms are often small size , especially in Western Germany. In many regions the average sugar beet field size is between 1 and 2 hectares. To improve this structure the land consolidation is the normal (official) but long lasting way. Virtual land consolidation by transborder farming is a new, fast to realise and flexible way to improve the structure of farms (Auernhammer et al. 2000, Rothmund et al 2002). The realisation of transborder farming needs the tools of precision farming and integrates the strategy of site specific farming into structure improving measures. Small fields are farmed together over the borders (transborder) with the best available technology and knowledge. On the larger fields than spatial variable plant production can be established. For sugar beet production within transborder farming further improvements regarding harvesting and transport are possible. Beside larger fields created by ransborder farming, the virtual land consolidation gives the chance to locate sugar beet fields beside roads where loading and transport (by trucks) is possible without disturbing the public traffic or without the risk that trucks will slip of bad roads. This specific location of the sugar beet fields and its planning is difficult because it has to consider whole rotations. Therefore planning tools in a GIS environment are needed. It also can be considered that the beets will not be stored in piles on the headland of each sugar beet field but on a number of centralised beet collection places where highest loading capacities can reached because the cleaner-loader does not have to change fields and from which the transport is very easy. The proposed presentation will describe the planning procedure, the planning alternatives and possible results of the measures of optimisation of the sugar beet production in transborder farming. The planning is made for the Lower Franconia county Zeilitzheim where the practical realisation of transborder farming is investigated within the integrated research project PREAGRO (not regarding optimisation in sugar beet production).

An artificial intelligent approach to identify complex associations among soil properties and weed density

B. Díaz, A. Ribeiro, D. Ruiz, J. Barroso and C. Fernandez-Quintanilla

Although the selection and application of weed control practices is generally based on overage field conditions, weed abundance varies spatially across a field in non-random spatial patterns defined as patches. On the other hand, it is well known that site properties vary spatially across fields. The associations of site properties and weed abundance within a field are not well understood, in spite of the considerable research work done in this direction. The causes that generate a higher weed density in some areas of the field are multiple and probably interrelated. Biological factors, weed management practices, field traffic patterns, and temporal weather variations could influence the dynamics of weed patches. The approaches that have been used in the past to analyze the effects of these factors assume accurate measurements. However, in real-world scenarios there is always a degree of uncertainty. Consequently, obtaining a precise analytical model for weed dynamics is a very complex, probably unreachable, task. A solution could be to find a qualitative and approximate model; such as, a set of rules that explain input data in the same way that an expert farmer: IF (phosphorus_concentration IS high) AND (clay_content IS low) AND ... THEN (weed:density IS high) (1) The objective of this study is to assess how an artificial intelligence approach -a rule induction method- could be used to identify complex association among site properties and the abundance of weeds within agricultural fields. The proposed approach has been tested finding out a rule set that explains the abundance of winter wild oat (Avena sterilis L.) in regard to eight soil parameters. Site properties and weed abundance data have been collected in four barley fields located in Madrid surroundings (Spain). To reach general conclusions, the fields have been selected to cover a wide diversity of situations. Eight soil properties have been considered: pH, organic matter (OM), Nitrogen (N), Phosphorus (P), Potassium (K) and sand, silt and clay percentages. The considered soil properties and the wild oat abundance data have been collected at each grid intersection point of quadrangular sampling grids. Wild oat counting has been performed around of the center of each grid intersection point in a 0.1 m2 sampling area. The proposed approach consists in three stages. First, the inter-fields data are homogenized, making them comparable, by a linear scaling technique, with all the input data taking their values in the range 0-1. In the second stage, the method used transforms the input data into labels with an associated semantic. The range of values of the site properties is split into three equal intervals that correspond to the Low, Medium and High labels respectively. Also the range of values for the weed abundance is split into two intervals, Low and High, but in this case with the threshold defined in 0.2 value, since the short weed presence in the farmer-managed fields forces to define a low threshold. The two labels of the weed abundance divide the input data into two sets, namely training sets. The third step is a learning process, where a Genetic Algorithm (GA) searches a set of rules as (1), which explain the two training examples. The validity of the rule set produced by the GA is measured as the percentage of input data that are correctly classified. The firsts results are very promising. The proposed method explains, with a small set of rules, about a 70% of the input data.

Spatial and temporal dynamics of weed populations in crop rotations under the influence of site-specific weed control

D. Dicke, P. Krohmann and R. Gerhards

Field experiments were conducted on five arable fields to study the spatial and temporal dynamics of weed populations in a crop rotation under the influence of site-specific weed control at Dikopshof Research Station in Germany from 1997 until 2002. One of the five fields was planted to maize in all six years; on the other four fields maize, sugar beet, winter wheat and winter barley were rotated. For the manual weed mapping, a regular 15 * 7,5 m grid was established in all fields. Weed seedlings were counted in a 0,4 m² quadrate frame placed at all grid intersection points. Linear triangulation interpolation was used to estimate weed seedling density at unsampled positions to create a continuous map of weed density. Weed control methods were applied site-specifically in all four crops using a GPS-controlled patch sprayer. A decision algorithm based on weed thresholds was used for patch spraying. Weed distribution maps were reclassified based on the weed threshold model to obtain weed treatment maps. Classes were defined as no, low (60%), medium (80%) and high rate (100%) of post-emergence herbicide. In sugar beet the dominating weed species were Chenopodium album, Fumaria officinalis and Viola arvensis. In maize, major weed species were Chenopodium album, Polygonum arviculare and Viola arvensis and in winter wheat and winter barley Veronica hederifolia, Galium aparine and Alopecurus myosuroides.The major weeds in continuous maize were Galinsoga parviflora, Solanum nigrum and Echinochloa crus-galli. For most species, spatial distribution and density of weeds in the rotation fields were very stable over the time of this study. Areas of the field that were not sprayed according to the weed threshold model and areas that received a reduced rate of post-emergent herbicides did not increase in weed seedling populations within the six years of study. Patches with high density weed infestations persisted despite of intensive weed control in all six years. In contrast to the rotation fields, weed seedling density and size of weed patches have increased in continuous maize during the period of study. Weed seedling density was counted before and after post-emergent herbicide application and before harvest at the same sampling points in the field and efficacy of weed control was calculated. Seed mortality by predation and fatal germination was assessed and the number of seeds was counted for those weeds that survived weed control and crop competition. These data were collected in all ripe crops of the rotation and in continuous maize and entered in a weed population model published by ZWERGER and HURLE (1990). The objective of the study was to find out if the model can explain the persistence of high density weed patches. In addition to that, the model is used to predict the dynamics of weed populations for the followings years based on the results of the six-year study in the rotation and in continuous maize. If the predictions of the model correspond to the actual weed seedling distribution in the fields, recommendation on site-specific weed management can be made based on previous years data. This would significantly reduce the time and costs for assessing weed distributions in the field. The knowledge of weed seedling populations dynamics could also be transferred to fields where Precision Farming has not been practised yet.

190

Optimal path nutrient application using variable rate technology

C. Dillon, S. Shearer and J. Fulton

Variable rate technology has been accompanied by unique opportunities and difficulties. One of these difficulties is the error of actual versus desired amount of nutrient applied as an operator traverses a field and the desired amount changes. Engineering research has determined that application errors exist and, furthermore, is a function of both direction and amount of change in the prescribed amount on nutrient applied. There is a need to reflect this degree of accuracy in application decisions. Specifically, this error can influence the decision of the optimal path an operator uses during application. An economic decision-making model is herein proposed to account for rate change error during variable-rate application of granular fertilizers. Optimal path mathematical programming models are available in the form of the classical traveling salesperson model in which the best route is determined while still visiting all desired locations. This formulation lends itself to addressing a minimization of application error in a variable-rate fertilizer application setting. Although widely researched in business, very few applications of this technique have been made in agriculture. A contribution of this research is the innovative use of this technique for this problem. The purpose of the proposed research is to provide insights regarding improved profitability by actively considering the path of nutrient application. Objectives are to 1) Develop an economic optimization model that determines the optimal application path and 2) Provide an empirical application of this model. An integer programming model which incorporates hexagonal decision cell binary variables will be modeled to reflect six possible exit and entry points for a given location. Data (from both published and unpublished sources) regarding the level of accuracy in nutrient application level dependent upon previous application rate will be used in ascertaining an optimal entry and exit point for each location. In combination, this will determine the optimal path of fertilizer application. Two possible objective functions are envisioned: one which minimizes application error and another which minimizes cost of error including such factors as application cost, time of application and therefore turning time, yield differentials, etc. Binary decision variables for application will be incorporated and include information on location and travel direction. The possibility of traveling on a given cell more than once will be considered while still reflecting the appropriate economic evaluation. Constraints will reflect logical feasibility factors such as the fact that a given cell, once entered, must be exited. The potential problems of subtours described in traveling salesperson literature wherein the optimal solution is for application of portions of the field without feasible connection will be eliminated using traditional established constraints should the need arise. These constraints will be excluded unless needed for simpler, more easily solved formulation. This study has potential implications for enhancing the value of variable rate technology for dry preplant fertilizer application and beyond. The economic optimization model will provide a decision-making framework suitable to optimal path determination for other machinery operations as well if supported by the appropriate underlying physical data regarding accuracy as influenced by varying rate of application. As the development of automatic guidance creates a need for computerized instruction of the path of operation, this research displays enhanced potential benefits.

An economic optimization model for management zone configuration

C. Dillon, T. Mueller and S. Shearer

One of the most fundamental issues associated with variable rate technology of precision agriculture is how to optimally configure management zones. A greater degree of accuracy with respect to spatial information regarding the optimal level of input (e.g., fertilizer) is desirable on as fine a scale as feasible. However, the fixed costs per zone (e.g., soil sampling) associated with this greater accuracy will not, at some point, justify the additional costs of this fine scale of management. The decision of how to appropriately delineate economically optimal management zones represents a great opportunity to assist producers in achieving profit maximization. Nonetheless, this decision likewise presents a daunting and complex problem that challenges researchers, extension specialists, industry leaders and producers alike. Consequently, producers are left facing the difficult decision of how to delineate management zones without consistent guidance. The most appropriate method of responding to this research question is through a multidisciplinary framework representative of key elements of the decision-making environment (including economic, agronomic and engineering aspects). While the specific focus of the empirical application of this project is upon variable rate seeding for a Kentucky corn producer, the techniques developed here will be suitable for a broader audience. This research is especially relevant to all producers who use variable rate technology in that it provides a missing key element for properly assessing alternatives regarding variable rate input application. Consequently, the long term goal of this proposed research project is to assist crop producers enhance their profitability by providing procedures and information that will assist them in determining how to establish management zones. Accomplishing this goal can be begun by meeting three objectives: 1) Develop an optimization procedure which will accurately and definitively ascertain the economically optimal delineation of production management zones based on complete data, 2) Provide an empirical application of this model, and 3) Perform sensitivity analysis to ascertain how optimal management zones change/respond to fluctuations in the economic decision-making environment. A mathematical programming (specifically mixed integer) model embodying the economic decision framework of a representative Kentucky crop producer will be formulated. The objective function will be to maximize net farm returns above selected relevant costs. Decision variables will include variable seeding rates and seed type by management zone as well as selection of the number, position and size of management zones using soil test information from each management zone. Constraints modeled will include land available, input purchases, commodity sales and management zone calculations. Published yield response data to seeding rates for soil with varying topsoil depth will serve as the agronomic setting. The model will therefore determine the management zone that each individual cell should optimally be allocated within. Given the cost of a soil test for each zone, additional costs for increasing the number of zones is weighed against the yield result differentials for added accuracy. Sensitivity of management zones with respect to changes in the economic decision-making framework will be examined. This study aims to establish a fundamental framework which will permit analysis of a very germane and basic question currently plaguing the successful implementation of precision agricultural variable rate technology: How does a producer identify the optimal management zone? The innovative model formulation proposed within this research project permits the appropriate economic analysis to be conducted for comparison to the less data intensive and farmer friendly management zone delineation procedures presently being conceived (such as delineation by soil properties or electrical conductivity).

Processing of yield maps for delineating yield zones

A. Dobermann, J.L. Ping, G.C. Simbahan and V. Adamchuk

On-the-go yield mapping using combine-mounted yield monitors has become one of the most widely used precision farming technologies. However, the analysis and interpretation of yield point data has lagged behind the developments in yield monitoring systems and their adoption by farmers. Yield maps contain numerous systematic and random sources of measured yield variation, including (i) natural yield variability related to climate and soil-landscape features, (ii) seasonal, management-induced yield variability, and (iii) measurement errors. Therefore, a single-year yield map is useful for posterior interpretation of possible causes of yield variation, but of limited value for more strategic site-specific management decisions over medium to long-term periods. Procedures must be developed to correct or eliminate erroneous yield monitor data, create and understand annual yield maps, and integrate multi-year sequences of yield maps into classes of different yield performance or yield goals. The objectives of our research were to (i) develop a general algorithm for post-processing of yield monitor data and (ii) compare different procedures for classifying multiple-year yield maps into categories or zones of different average yield and its variability among years. Yield monitor data measured for five to six years in two production fields in Nebraska were used as examples. The proposed yield data cleaning algorithm screens and deletes six basic types of erroneous values: (1) header status up, (2) start and end pass delays for both headlands and stop-and-go segments within the field, (3) short segments, (4) frequency distribution outliers of distance traveled, grain flow, and grain moisture, (5) co-located yield records caused by GPS drift, and (6) small patches of local neighborhood outliers and narrow low yield strips. The latter are removed based on a local neighborhood test performed for each yield data point following the movement of the combine through the field. The cleaning algorithm typically removed about 20% of the original yield monitor records, improved the frequency distribution, and removed outliers in the resulting yield maps. Normalized and interpolated yield data for five (Site 1) and six (Site 2) years were used to evaluate the performance of empirical, hierarchical (Ward's method), and non-hierarchical (k-means, fuzzy k-means, ISODATA) clustering techniques. All methods were evaluated based on the relative yield variability explained and landscape pattern metrics describing the spatial fragmentation of the yield classes maps. Empirical yield classification methods resulted in less fragmented maps, but frequently explained less than 40% of the total spatial and temporal yield variability. These results were sensitive to subjective decisions that define the yield classes. In contrast, clustering techniques explained up to about 70% of the yield variability, but resulted in more disperse maps of yield classes. All cluster analysis procedures produced relatively consistent results if used as univariate classification of mean yields or bivariate classification of mean and standard deviation of yield across years. K-means clustering showed little sensitivity to the choice of data source, but the fuzzy k-means method was most sensitive to both choice of input data and the number of classes selected. The clustering procedures focused on maximizing the variance between while minimizing the variance within classes, without constraints to form spatially contiguous management zones attributed to the same yield class. We discuss the use of spatially constrained multivariate classification, numerical spatial aggregation, and application of post-classification smoothing filters to obtain spatially congruent yield classes while at the same time accounting for a high level of yield variability. In addition to statistics describing the yield variability explained, landscape pattern metrics that quantify map composition in terms of patch size and density, dispersion, and interspersion are useful criteria for evaluating spatial classification procedures in site-specific crop management. The techniques discussed will allow more objective mapping of yield goal zones, which is an important data layer required by many algorithms for prescribing variable rates of agricultural inputs.

Detecting management zones in respect to yield empirically

H. Domsch, T. Kaiser, K. Witzke and O. Zauer

A practicable approach to start precision agriculture in a farm now is the mapping approach according to Werner et al. (Werner et al. 2002). The measures of precision agriculture are planned on the basis of the average site conditions which are summa-rised in the term yield potential. The yield potential is primarily a function of the soil including the depth of the water table and the relief. Farmers are able to collect cereal yield data for several years but they do not know how yield potential maps can be developed from them. Thus they need temporarily an empirical way to delineate management zones from yield data. The yield potential can be estimated by using the yield level in a single or several years if two conditions were fulfilled: average weather in this year and a crop man-agement according to the newest standard of knowledge without mishaps in techni-cal measures. We can assume that boundaries of yield zones are boundaries of yield potential zones if the boundaries of yield zones can be explained by soil parameter or relief parameter boundaries. Without scientific support but with support of service providers farmers are able to acquire additional to annual yield data, soil electrical conductivity (EC) data, digital terrain models (DTM), time-series aerial images, and weather data. The farmer will not have the tools to remove all the errors existing in the data. However by using several different data levels simultaneously their influence will be reduced. The aim of the paper is these data levels to compare for fields of a farm that partici-pates in a project promoted by the Federal Ministry of Consumer Protection, Food and Agriculture. The comparison is to take place only visually using filled contours and contourlines to display two data maps on the screen in the same time. Coinci-dent contourlines in different data levels or different years attest that this contourlines are natural boundaries that can be used as yield potential boundaries. Results Data are available from the years 2001 and 2002. The precipitation was somewhat above average in the first of the year 2001 and was somewhat lower in the May. The precipitation in the year 2002 corresponded to the average. The DTM was bought by the mapping agency Saxony-Anhalt as raster data with 10 m grid size and an accuracy of ±0.5 m. The soil electrical conductivity was measured using the EM38 instrument and referred to a soil temperature of 25 °C (EC25). Two aerial photographs per year were taken using a Cessna and a digital camera and the visible atmospherically resistant index were computed from them. Yield data were collected with John Deere combine harvesters by the farmer. The data of the year 2002 are not very reliable due to the unfavourable climatic conditions during the harvest. The raw data were only removed from obviously erroneous data caused by the starting phase and incorrect crop cut width before mapping. In a relatively even field yield pattern with a wide range between 1000 and 12000 kg/ha corresponded well with the EC25 pattern indicating that the available water limited the yield. In a hilly field yield pattern were caused by the EC 25 pattern too however this de-pendence was overlaid by the relief. Conclusion Delineating of sub-units in a field by comparison of canopy pattern with EC pattern and the digital terrain model is an empirical approach that the farmer enables to start PA. In the investigated fields, the yield pattern could be explained by one of the other data. The empirical approach can bridgeover the time till a software aided approach will be available. References Werner A. et al. 2002: Yield potentials of sub-units within fields as a key input for crop management in precision agriculture. In: Precision Agriculture, Tagungsband Preci-sion Agriculture Tage 13.-15. März in Bonn, S. 197- 200.

GIS-Based farm machinery logistics optimization: A modelling approach for precision agriculture

J.P. Douzals, T. Castel, P. Zuchetto, G. Trouche and J.-L. Maigrot

Agriculture has to deal with an increase in both stand and farm surfaces. The two main consequences lie on the development of larger farm implements and a substantial time loss and a lack of productivity during transportation. Those are coupled with changes on agricultural practices toward the implementation of various time-saving cultivation methods such as simplified cultural techniques. This new context involves technical, economical and environmental consequences on farm management. From there, the modelling of displacements (i.e. transportation) within the farm area may be seen as a good indicator (integrative variable) of the system functioning. Hence, the evaluation of the 'real' transportation cost is of primary importance for decision making. In order to solve such complex transportation problem, new approaches related to precision agriculture - as Differential Global Positioning System and Geographic Information Systems for Transportation (GIS-T) - offer powerful tools. However if many applications address public, commercial or emergency travel and transportation management, few are specific to the agricultural domain. Indeed, from our knowledge the integration of geographic parameters as road network, field plot, field relief are not generally integrated into logistic programs dedicated to agriculture. This paper presents the capability of GIS-T to address agricultural transportation issues for which network representation of the farm system is needed. It focuses on a preliminary analysis of the two main i) Geo-relational and ii) Object-oriented data modelling techniques in the context of precision agriculture. A first comparison of the data models is conducted in an optimization context in order to minimize time losses and to improve implements and manpower efficiency as well as productivity. The experimental data used in this study concern a farm association (5 initial farms) with a total arable surface of about 1500 hectares in the northern east part of France for which only two machinery storages are used. The maximum distance between storage and fields is about 25 km. Soil tillage is one the agricultural practices that involves major agronomical and economical outcomes. By the shift from ploughing onto simplified cultivating techniques among the whole wheat crop fields, logistic problems appeared. It was then decided to complete the initial study with an extension to other agricultural practices as seeding, fertilizing and sprayding. It is assumed that specific agricultural practices as harvest will be integrated into the model. The results illustrated the potential of both data modelling and database design for the analysis of the cost of the farm machinery logistics. Such information leaded us to propose a conceptual approach for the development of a simplified dynamic model within GIS environment for the optimization of the farm machinery transportation of our farm system. Finally it discusses how such spatial data modelling as well as spatio-temporal data models may be relevant for precision agriculture. For example, the integration of in-field agronomic variability would have a significant effect on the logistic management of fertilizers or pesticides. Initially based on the 'real' description of farming practices, the final aim is to allow simulation before implements acquisition. Possible extension of the modelling is planned in the domain of cooperative farm machinery management structures and private entreprises for agricultural management.

Strategies for site-specific nitrogen fertilization with respect to long-term environmental demands

Thomas Ebertseder, Reinhold Gutser, Ulrich Hege, Robert Brandhuber and Urs Schmidhalter

Strategies for site-specific nitrogen fertilization usually focus on reducing spatial heterogeneity of crops. Because in most cases plant growth is limited by water availability, fertilizing systems which are more adapted to site specific yield potential or soil characteristics should fit better with environmental demands. So far, no results are available on long-term effects of any strategy on site-specific yield, soil fertility or nitrogen losses. Therefore a number of long-term field trials were implemented. In this paper the fertilizing strategies applied will be discussed by presenting results from the first year of these experiments. The tested strategies are: (I) uniform fertilizer application (II) site-specific fertilizer application according to mapped yield zones (mapping approach). The amount of nitrogen applied depends on expected yield potential (low input on sites with low yield expectations and vice versa). For the delimitation of yield zones several years yield maps, soil maps, remote sensing or tractor based crop sensors are used. (III) exclusively on one site: sensor based fertilizer application (Hydro N-Sensor, online approach) The trials (static strip plots) are located in three regions, differing in climatic conditions and soil properties. The crops change from year to year due to farm specific crop rotation. The results presented in the following all came from winter cereals. The nitrogen status of crops was mapped several times in the growing period by Hydro N sensor. On certain sites of each field yield and nitrogen uptake were determined by hand cuts and yield maps were recorded additionally. As supposed, strategy (II) amplified spatial heterogeneity of crops to a certain degree, when compared to uniformly applied nitrogen (strategy I). This was detected by sensor measurements as well as remote sensing and corresponded to yield either (strategy II: decreased yield on zones with low yield expectations up to about 1 t ha-1, increased yield up to 1 t ha-1 on deep soils). The effect on yield was attributed to several yield components, particularly crop density. Despite of decreased yield, on most sites with low yield potential nitrogen efficiency was increased by strategy II, varying with the amount of N applied, plant species, and soil characteristics. When looking on the whole fields no or only little differences in yield (depending on site, plant species, and extent of different yield zones) could be detected between strategy I and II. Online nitrogen application by Hydro N sensor (strategy III) tended to equalize crop appearance (decreasing heterogeneity as detected by the sensor in the period of ripening). Mainly this was not due to spatial variation of total nitrogen applied (same amount as with strategy I on most parts of the field) but rather due to temporal variation (different doses in individual growing stages). In most parts of the field yield obtained by sensor based nitrogen application differed hardly from yield produced with strategy II, although there were some changes in yield structure (crop density, grain mass). N efficiency was lower than with strategy II, especially on sites with low yield potential. For the whole field yield was almost the same as for strategy I and II. The results show that temporary there is a potential for more efficient fertilizer use by taking into account maps of site-specific yield potential. The effect of the N sensor seems to be mostly a homogenization of crops. What is happening in the long-term on different sites of the fields (yield potential as well as potential for N losses) will be evaluated in the following years.

AVIS - a new sensor for environmental monitoring and precision farming

Rainer Efinger

Introduction: For the purpose of environmental monitoring and precision farming with remote sensing systems, the choices of system selection are limited. The spectral resolution of classical systems like Landsat TM is not accurate enough for the derivation of parameters like chlorophyll content or the supporting of fertilization or pest control. On the other hand there are no spaceborne Imaging Spectrometer systems available, and airborne Imaging Spectrometers such as DAIS, AVIRIS and HYMAP are expensive and difficult to obtain for more than one or two missions per year. Therefore a cost-effective system is needed which is affordable for official institutions and communities to monitor the environment over a longer time period. The airborne imaging spectrometer AVIS (Airborne Visible near Infrared Imaging Spectrometer) was developed at the Institute of Geography, University of Munich, as an attempt to close this gap. Sensor description: AVIS is a line imaging spectrometer consisting of two main components. On the one hand the camera unit with camera, spectrograph, lens and filter and on the other hand the storage unit with PC, frame grabber and GPS. The spectrograph is used with 64 resp. 128 bands and covers a spectral range from 400 to 870nm. The spectral resolution is 8nm. It is mounted between an objective and an IR CCD black and white camera. Due to the spectrograph slit each recorded image is a line of 1280 resp. 640 pixels with a spatial resolution from 2 to 8m depending on the aircraft altitude. The resulting view angle is 57 degrees. The AVIS – sensor can be installed in a chassis which fits onto a standard aircraft camera mount. It is connected to a computer via a 14 bit frame grabber card for image data capture and processing which provides near-real-time data. The connected monitor enables the captured images to be supervised during the flight. Auxiliary DGPS data including data, time, geographical position, and camera data is recorded. The exact position of the aircraft or better the behavior of the aircraft in motion is registered by an IMU-System (Inertial Measurement Unit), which records the roll, pitch and yaw angles, 40 times a second. Application: Hyperspectral remote sensing enables the spatial monitoring of vegetation and the derivation of individual plant components such as chlorophyll and nitrogen content additionally to the standard vegetation indices like NDVI, SAVI or others. These parameters can be taken to optimize agricultural management by spatially differentiated fertilization and vegetation yield modeling. The spectral behavior of chlorophyll a and b with their specific absorption features as well as their respective effects on vegetation reflectance spectra can be derived only by hyperspectral data. The chlorophyll content can be obtained by the CAI (Chlorophyll Absorption Integral), which is calculated by a specific spectral area within the red edge region. The CAI correlates with the LAI (Leaf Area Index), therefore AVIS can be used for LAI monitoring. The nitrogen within the plants is not detectable in the visible and near infrared wavelength range because there are no specific absorptions or reflection patterns. But there is a very tight relationship to the chlorophyll, because every chlorophyll molecule contains four nitrogen atoms. So the content of nitrogen can be derived by the amount of chlorophyll. The possibility to map the spatial distribution of nitrogen can be used for the monitoring of maturity stages for the determination of the most favorable harvest date.

Advanced sensor pendulum-meter

D. Ehlert, S. Kraatz and H.-J. Horn

Information about the distribution of plant mass growing in a field is a prerequisite to precision agriculture. For time-efficient, non-destructive and labour-saving measurements, sensors are needed which can determine and predict - preferably on-line - the plant mass and yields of crops. Determination of spatially variable plant mass is important for optimising inputs of agro-chemicals, and improving management and environmental protection. Surveying plant mass distribution is possible by manual methods, aerial photography, and vehicle based methods. A vehicle-based mechanical sensor (pendulum-meter) was developed to measure the plant mass of crops. In the first stage of development, a dynamometer with pendulum-meter, sensory and electronic equipment was developed for testing the sensor under defined conditions. Parameter trials were done for the assessment and optimising the pendulum meter. The correlations of functional relationship between plant mass and pendulum angle measurements in winter wheat, winter rye, grass and rice were $r^2 = 0.89$. Based on this positive result, in the recent years the sensor was developed stepwise to meet the demands for practical use under different und hard conditions. Technological solutions for the compensation of the depth of tram lines and slopes and for the automatic change from sensing into transport position and vice versa involving market available components were found. The sensor has a total weight of about 40 kg and is mounted in front of the basic vehicle (tractor, tool carrier). The sensed stripe has a width of 1 m and is arranged between the both tram lines. Depending on the crop and soil conditions, the sensor can be varied in height of pivot point, the pendulum length and mass of the cylindrical sensing body. The low level electrical energy (12V, maximum 50 W) for all sensor functions is supplied by the generator of the the basic vehicle. To minimise the expense for sensor calibration different methods and the accuracy were investigated. Based on results of statistic calculations additional co-relation between plant mass and electrical soil conductivity and yields for some fields in different regions will be presented. The research works for developing and optimising the pendulum-meter were finished in the growing season 2002 mainly. In the present stage of development the sensor can be used for the measurement of plant mass, depth of tram lines and slopes. Measuring values for these parameters for some crops and sites will be presented and discussed. The sensor pendulum-meter give the farmer valuable information to assess the variability of soil and crops for first steps in the realisation of precision agriculture. To demonstrate the practical potential of the sensor, variable rate applications of nitrogen fertiliser, growth regulators and fungicides in off line and in real time at different farms were performed. For this purpose the sensor was mounted on the three point lift in front of tractors in combination with a spreader and field sprayer. Examples and experiences for the pendulum-meter supported variable rate technology will be presented. Based on the obtained results, it can be concluded that the sensor pendulum-meter is ready for the stage of commercial manufacturing and comprehensive use in precision agriculture.

Quality of agricultural positioning and communication systems

M. Ehrl, W. Stempfhuber, M. Demmel and H. Auernhammer

Precision farming becomes more and more an accepted way of crop production and helps to achieve a sustainable environmental friendly agriculture. The objectives of site specific farming are increasing yield, together with decreasing environmental impacts. The effectiveness is close-knit with the accuracy and response speed of the utilised technology. Precision farming equipment consists of various components and tools, which are jointly generating the useable system. Each single component is rather not perfect and has therefore certain influences on the overall system capability. A thorough investigation allows to determine the quality and accuracy of major precision farming components. The positioning and navigation system, as well as the communication between the single components were identified to have the most important influence. Here, the "Differential Global Positioning System" (DGPS) and the standardised "Controller Area Network" (CAN) communication protocol, well known as LBS (DIN 9684) or the upcoming ISOBUS standard (ISO 11786), are used. The objective is to identify and quantify all sources of quality-influencing factors in these systems. The investigations regarding DGPS have been focused on the absolute positioning accuracy as well as the computation time. A highly accurate hybrid reference system was established, consisting of a "Real Time Kinematic DGPS" (RTKDGPS) and a "Terrestrial Positioning System" (TPS). The "Pulse Per Second" (PPS) event of special fitted GPS receivers served as precise time reference. The main work was the design of a PC based "real-time" synchronisation and data recording system, which also served as an interface for all different subsystems. With this set-up, various different receivers from low-cost (handheld) to high end range have been investigated and compared. The communication within an open system, as the LBS and ISOBUS protocols are, is very complex and allows the connection and communication of an unknown number of different components. Therefore, sensitive investigations require the definition of reasonable boundary conditions. The development of a communication driver library, which fulfils all parts of the standard, builds the base for these investigations. This communication driver is a hardware independent piece of software, which resides in each single hardware component of the network and handles all communication between the components as the standard claims. A testbed with a variable number of components like tractor electronic control unit (ECU), user interface (terminal), GPS, data recording ECU and one or more implement controllers is the target environment for all investigations. A specific PC software, listening to the traffic, will be the main tool for exploring the system in various conditions. All quality relating factors such as compatibility, intelligence, flexibility, malfunction, error management, response times, safety or others should then be investigated and quantified due to a well defined test scheme. Also, the simulation of real bus load scenarios or extreme bus situations are considered to give a close insight how the overall system reacts. Expected results are an universal evaluation of important precision farming equipment, allowing to point out limiting factors and formulate general design rules for future developments. The results are significant for the design of a new and highly accurate generation of agricultural control systems. Increasing accuracy in combination with standardised bus communication will be the next milestone to increase the acceptance of precision farming.

A precision agriculture study for Ohio strawberry production

M.R. Ehsani, D. Saraswat and M. Sullivan

Information technology is playing an increasingly important role in today's agricultural production systems, regardless of operation size, type of commodities and management approach. The rapidly changing market requires new farming strategies. Precision agriculture can potentially increase crop profitability by reducing over-application of crop inputs, such as water, fertilizers, and pesticides. In theory, the application of precision agriculture to higher value crops would show more pronounced profits than for lower value crops because fruits and vegetables consume more inputs such as pesticides, and fertilizers and may benefit more from variable rate application of inputs. The long term goal of this research was to develop low-cost tools and methods to enhance the profitability of high-value crop production. The specific objectives, with strawberries selected as the high-value crop, were: 1) To find a low cost yield-monitoring system for strawberries in small farms. 2) To quantify the magnitude of yield variability in a strawberry field. 3) To determine the relationship between soil and plant characteristics and yield by conducting field experiments. 4) To find the main causes of yield variability for a strawberry field. 5) To investigate the relationship between soil electrical conductivity and yield. 6) To determine the relationship between NDVI and yield. For yield monitoring of strawberries, one indirect method and two direct methods were investigated. For the indirect method, a computer vision method was studied. The results of indirect method indicated that although the contrast between strawberries' red color and its surrounding can be used to predict yield, a high resolution camera and more sophisticated algorithms are needed to provide the desired accuracy. In direct yield monitoring, two different techniques of weighing the strawberries were used. In the first technique, a batch type yield monitor for semi-mechanical harvester was designed and tested. The batch type yield monitor developed in this study was a relatively low-cost yield monitoring system and showed very good results under laboratory test conditions. The second technique was based on direct weighing using a commercial scale with RS232 output. A computer program was written to collect weight and GPS data and store them on the computer. For practical purposes, we decided to use this method for our field study. The field study was conducted at a one-acre strawberry farm in London, Ohio. Composite soil and plant samples were collected at the beginning of the season on a 6.1 m. X 6.1 m. (20 ft. X 20 ft.) grid. The soil samples were analyzed for nitrate, available phosphorus, exchangeable- K, calcium, magnesium, Cation Exchange Capacity (CEC), organic matter, and pH. Moreover, soil compaction, moisture and electrical conductivity, were measured and mapped. Plant samples were analyzed for total N, S, P, K, Mg, Ca, Na, B, Mn, Zn, Fe, Cu, and Al. Two multi-spectral aerial images (early in season and at the end of season) were taken from the field and the NDVI index was calculated. At the end of the season, yield was measured for each grid point and a yield variability map was created. The yield map showed significant yield variability throughout the field. The input -output relationship between yield, soil and plant data was studied. Data were analyzed using a multiple linear regression technique and also a new wavelet approach, which is a time-scale analysis, to find the input-output relationship. The results of the data analysis are presented and discussed in the paper.

Developing recommendations for site-specific nitrogen management of irrigated maize

Richard Ferguson, Achim Dobermann, Charles Shapiro, Daniel Walters, Jürg Blumenthal, Charles Wortmann, David Tarkalson, Gary Hergert and James Schepers

Groundwater nitrate contamination associated with intensive irrigated maize production in Nebraska in the US has been an issue of concern since the 1960's. Research efforts have produced a variety of fertilizer and irrigation management practices that reduce the potential for nitrate leaching. Intensive educational efforts have resulted in the adoption of many of these practices by Nebraska producers. However, N use efficiency by irrigated maize for many producers remains relatively low – in the range of 35-40%. Research was initiated in Nebraska in 1993 to investigate variable rate nitrogen (N) fertilization as a means of increasing N use efficiency of irrigated maize, and to further reduce the potential for nitrate leaching to shallow aquifers. Two separate studies, encompassing both furrow and sprinkler irrigation, were conducted over 13 site-years from 1993 through 1997. At all site years, variable rate N was compared to uniform N in field-length, replicated strips. Variable N rates were determined from grid-sampled soil organic matter, residual nitrate-N, and uniform expected yield based on the University of Nebraska N recommendation algorithm for maize. Uniform N rates were based on mean values of these parameters for each field. Spatial grain yield was determined by yield monitor equipped combine harvesters. Treatment effects on soil residual nitrate-N were determined by soil sampling each year in the original grid pattern to a depth of 0.9 m. Over the 13 site-years, the average fertilizer N rate for the uniform treatment was 142 kg ha-1; the average rate for the variable treatment was 141 kg ha-1. For most site-years, there were no significant differences in grain yield between uniform and variable N treatments. Mean grain yields ranged from 4.5 to 13.9 Mg ha-1 among site-years. Variable rate N significantly reduced grain yield compared to uniform application at two site-years; uniform application significantly reduced yield compared to variable application at one site-year. There were no significant differences in soil residual nitrate-N to a depth of 0.9 m between uniform and variable N application, with mean NO3-N concentrations ranging from 2.7 to 14 mg kg-1. We concluded that the current N recommendation algorithm for maize, while in most cases sufficiently accurate for field to field recommendations, is not sufficiently sensitive to spatial differences in soil N supplying potential and crop N demand to use in generating variable rate N prescriptions. Two additional studies, using similar approaches conducted from 1995 through 1998 over an additional 13 site-years for irrigated maize, found similar results. In 2002, a three-year study was initiated to refine nutrient recommendations (N, P and K) for irrigated maize in Nebraska, accounting for climate and soil differences across the state. The range in Nebraska of cropping heat units (expressed as Growing Degree Days [base 50o F]) is 2000 to over 3700. Mean annual water balance ranges from 150 to -350 mm. Root zone available water holding capacity for soils range from 25 mm to 300 mm. One goal of the project is the development of region-specific N recommendation algorithms for irrigated maize. Refined N recommendation algorithms may then be sufficiently sensitive to soil N supply/crop N demand that they can be used to develop variable N rate fertilizer prescriptions. Twelve site-years of this study were conducted in 2002. While results from the first year of this study are unavailable at the time this abstract was prepared, they will be summarized in the proceedings paper and presented at the conference.

Information sources in precision agriculture in Denmark and in the USA

S. Fountas, D.R. Ess, S.E. Hawkins, J. Lowenberg-DeBoer and S. Blackmore

Precision farming is information-intensive. Farm managers acquire a vast amount of data for the spatial and temporal characteristics of their fields and face many challenges on how to interpret these data as the basis for decision making on crop management. The transparent transformation from data to useful information that can effectively be used in the decision-making process is still not available. Therefore, it is important to gain more knowledge on the experiences of farm managers, who currently practise precision farming. Such knowledge would assist in identifying the problems and obstacles of further expansion in precision farming and, consequently, form the basis for guidelines on modification of existing technologies and methods or introduction of new ones. To address the problems with information management in precision farming, two mail surveys were carried out, one in Denmark in January 2002 and one in Indiana, USA in July 2002. The Danish survey was administered through the Danish Advisory Center, the Danish Institute of Agricultural Sciences and the Royal Veterinary and Agricultural University. The American survey was administered through the Purdue Site Specific Management Center (SSMC) and Purdue Extension Service. Both of the surveys were targeted to farmers who had been used precision farming practices. The surveys included self-administered questionnaires that included close-ended and open-ended questions. A total of 580 received the survey in both countries. The respondent names were identified by commercial companies and extension educators. In both cases, an incentive related to precision farming was offered to ensure a high response rate. The surveys focused on information management issues. Firstly, issues regarding the status of the various precision farming applications used as well as equipment and software. Secondly, the use of Internet and e-mail as information sources in agriculture and precision farming. Thirdly, the information source preference for investing in precision farming and practicing it, as well as the satisfaction level from the different provided services in the area of precision farming, in both countries. Moreover, issues regarding the impediments in data handling and data interpretation as well as the ways of data storage and data ownership were addressed. Furthermore, usefulness of the yield and soil maps for making farm management decisions and potential changes in management practices due to precision farming were communicated. Finally, the desired further information and services regarding precision farming and the next step in this domain were addressed. A detailed overview on information management aspects both in field practices and in the farm office was obtained. Comparisons were also made on different adoption trends, information preference sources and precision farming applications, equipment and software in both countries. These comparisons were accompanied with cultural and operational differences in agriculture in Denmark and in the USA. Furthermore, the main obstacles in data handling and interpretation are identified and selective recommendations for more efficient use of precision farming data and interpretation are provided. As a result, there are suggestions for the role of crop consultants, co-ops, university specialists, extension educators and software and hardware vendors for their role in precision farming to enhance the adoption and ease the difficulties in the current practice of precision farming.

Exploring the spatial relations between soil properties and ElectroMagnetic Induction (EMI) surveys using statistical and geostatistical methods and the implications for management

Z.L. Frogbrook and M.A. Oliver

Site specific management in agriculture depends on the accurate determination of the within field variation of soil properties. This inevitably involves considerable sampling effort, which can rarely be afforded for assessing input needs. There is a need, therefore, to sample the soil in such a way as to maximize the amount of information gained for management and to use ancillary data where possible. One such kind of ancillary data is from ElectroMagnetic Induction (EMI) sensors, which measure the electrical conductivity of the soil. They provide large amounts of data rapidly and fairly cheaply. Such surveys could provide a guide to targeted sampling of the soil for management. This paper describes a survey of a 25 ha field in Berkshire, England. The soil was sampled in March 2002 at over 100 locations on a 30-m grid. Soil samples were taken from two depths, 0-15cm and 30-60cm. The soil was analysed for pH, organic matter and the percentages of clay, silt and sand. Soil bulk density was measured for the two depths and depth to parent material was also recorded. Soil moisture readings were taken in October 2001 and May 2002. Measurements of electrical conductivity were made in March 2002 along transects approximately 15m apart, resulting in over 3000 values. These measurements were made with the EMI sensor in the vertical position. The relations between the measured properties were analysed initially using a variety of statistical methods, including the product moment correlation coefficient, principal component analysis and partial regression analysis. The correlation coefficient was computed for the field as one unit and also for moving windows. The latter allows the change in the relations across the field to be examined. The correlation results show a strong relation between EMI and the soil particle size distribution. Sand shows a negative relation while silt and clay a positive one. Soil bulk density is negatively related to EMI at both 0-15cm and 30-60cm. Soil moisture shows a positive relation to EMI, with a stronger relation for moisture in May than in October. Geostatistical methods were used to analyse the data further. Variograms were computed for the soil properties at both depths and for the EMI data. The variograms for the majority of soil properties show some structure, the exceptions are soil moisture, October 2001, and soil bulk density, 0-15cm. The variogram for EMI is nested and the long-range component is similar to the range of spatial dependence for most soil properties. Ordinary kriging was used to predict values of soil properties and EMI values at unsampled sites for mapping. The map for EMI shows several distinctive features that are also evident in the maps of soil bulk density, and sand, silt and clay contents. Cross variograms were computed to examine the coregionalization between those soil properties that were correlated with EMI. This analysis will provide additional information to determine which soil properties and soil depths relate most closely to electrical conductivity. The EMI data can be used to target soil sampling only if there is a strong link with the soil properties of interest. The strength of the spatial relations between EMI and the soil properties will be discussed and the implications this has for precision agriculture and future soil sampling strategies.

Spatial dependency of soil samples and precision farming applications

W.J. Gangloff, R.M. Reich, D.G. Westfall and R. Khosla

Variable rate application maps are often based on a grid sampling approach. Problems with this approach are primarily due to large separation distances among samples. The result is a lack of appropriate sample numbers to develop sound spatial predictive models for variables of interest. In most cases, scales-of-pattern associated with soil variables occur at separation distances much smaller than separation distances observed with a grid sampling approach. This can result in data sets that are spatially independent and can preclude proper application of interpolation techniques. Hence, this approach may generate models and prescription maps that are questionably accurate. The objective of this study was to illustrate and quantify problems associated with grid sampling. Two fields were sampled on a 76 x 76 m2 coarse grid (CG), and a portion of each field was sampled on a 15 x 15 m2 fine grid (FG). Samples were analyzed for P, K, organic matter, pH, NO3-N, NH4-N, and Zn using standard soil testing procedures. Sample locations were randomly located within each cell and recorded with a global positioning system. The CG data was tested for spatial autocorrelation using Morans I statistic. Morans I statistic is analogous to a weighted correlation coefficient between possible pairs of n observations. Morans I is a dimensionless statistic that ranges from -1 to 1. The point autocorrelation coefficient (Ii) was also determined for each soil variable at each sample location. This statistic can be used to identify outliers, extreme data or local anomalies. The point autocorrelation coefficient is calculated by decomposing Morans I statistic to obtain the relative contribution of each sample point to the overall Morans I statistic. Data collected from the CG was used to generate empirical variograms. Gaussian, exponential, and spherical variogram models were fit to empirical variograms. Effectiveness of models was tested with cross-validation and goodness-of-prediction statistic (G). The G-statistic is a measure of how effective a prediction might be relative to that which could have been derived by the sample mean. The FG data was analyzed with correlograms and cumulative correlograms. Correlograms and cumulative correlograms were constructed with Morans I statistic. Soil samples collected from the CG had significant spatial autocorrelation for all variables, except P, based on Morans I statistic. This suggests these data are good candidates for modeling techniques such as kriging. However, based on G-statistics, the kriging models did not perform well. At least two factors are contributing to poor performance. First, based on evaluation of point autocorrelation coefficients it appears some sample locations may exhibit anomalous characteristics that influence the overall value of Morans I statistic. These anomalous samples may lead to the conclusion of significant spatial autocorrelation when in fact the data set is spatially independent at local scales. Morans I statistic is a composite statistic that evaluates the spatial dependence of a variable at several scales. In some cases, techniques such as kriging may not be applicable for a data set despite a significant Morans I. Second, based on evaluation of cumulative correlograms, separation distance among samples was too large. Cumulative correlograms indicated the maximum scale of spatial dependency was 20 to 30 m. However, CG data sets had median separation distances in the range of 70 to 75 m. Finally, when standard and cumulative correlograms are compared only cumulative correlograms are effective for elucidating scales-of-pattern in soils. To optimize precision farming applications, sampling designs should focus on properly capturing spatial variability and dependency before developing models. Alternative sampling designs that capture scales-of-pattern may include double sampling and/or regression approaches that utilize auxiliary data such as soil color or topography.

Use of satellite SAR and optical data in order to characterise within-field variability

F. Gasc, A. Gachet and D. Boisgontier

The determination of management zones within a field is the first step towards precision agriculture, the two others being the interpretation of these zones, and the agronomic decisions derived from the analysis of these zones. This paper outlines how satellite SAR (Synthetic Aperture Radar) and optical data can be combined in order to reveal long-term homogeneous zones within a field. Such zones are, in precision agriculture, the basis of zone management. Among all the agricultural parameters, soil is one of the most stable over time, and soil properties have an important place in terms of agricultural decisions. Hence, it is an important factor to take into account in the management of zones for precision agriculture. Many studies, in the past decades have shown that satellite images could be used to finely delineate surface soil variations or characterise particular soil properties. However, the whole range of sensors onboard commercial satellites now offers the opportunity to use and compare data from different sensors, and thus from different wavelengths and characteristics. Moreover, the great improvement in spatial resolution of the sensors onboard satellites, enables the use of satellite data in order to determine very fine spatial variability. Remotely sensed optical, radar or hyperspectral data are now commonly used in soil characterisation. On one hand, in the scope of soil properties, data issued from Visible/Near Infra Red bands, have been widely used to determine soil variation, for example through soil brightness. On the other hand, radar, which can penetrate soil under certain circumstances, has demonstrated to be very informative on soil moisture and soil texture. Thus, radar and optical data complement each other for information on soil. The first objective of this project is to study how precision agriculture could benefit of a satellite SAR image. The first step is, therefore, to extract from the radar image the information in terms of soil characterisation. The second step is to interpret the zones at the scale of a French field and find out the relevance of each zone in the agronomic point of view. The study sites were chosen to in typical cereal basins in France. Each field site was as representative as possible of its basin in terms of pedoclimatic environment. The size of the studied fields is about 10 ha. In order to interpret correctly the response of SAR backscatter to soil conditions, soil properties of the studied areas were sampled at a density of ten points per hectare. Sampling analysis included chemical analysis, physical data, depth, stoniness, texture, organic matter, colour, and electric resistivity. Other data were gathered over the fields, such as meteorological data, in-field notations and any other farmer knowledge of the field. The second objective of this study is to characterise within-field variation in terms of soil properties using combined satellite SAR and optical data. The combination of radar and optical data is a powerful method to distinguish typical homogeneous zones within a field. Images were acquired over the study sites in both radar and optical sensors. In order to interpret the homogeneous zones issued of the combination of SAR and optical data, the same ground data collected over the fields were used.

Approach to accuracy assessment of site-specific fertilization

R. Gebbers, K. Persson and C. Weltzien

While the accuracy of some of the components and methods of site-specific fertilization has been thoroughly analysed (e.g. GPS-receiver), accuracy assessment of the whole system is rare until to date. Analysing the error propagation within the system of site-specific fertilization is the objective of this work. This can help to understand the effects that can be achieved by site-specific fertilization. It will show which component has the greatest influence on the overall accuracy, and points out parts of the system were improvement of accuracy is most efficient. The accuracy assessment presented here is focussed on basal dressing. Examples are drawn from the application of phosphorus. Looking at the system of site-specific fertilization we can identify components and associated procedures:
*Input data (soil sampling, soil testing, yield mapping, estimating yield goals), *GPS-receiver (positioning, navigation), *GIS (spatial operations like interpolation, coordinate conversion), *fertilization algorithms (calculation of the recommendation), *board computers and controllers (digital-to-analogue conversion), *fertilizer spreaders (application of the fertilizer).
The related statistical processes can be described as random functions. They can be linked by the law of error propagation using analytical approaches (e.g. first order Taylor method) or stochastic simulation (Monte Carlo method). Having to regard the spatial dimension of the statistical processes makes this analysis complicated. Spatial accuracy of the input data is analysed and described by geostatic methods like variography, calculation of the kriging variance, and conditional simulation. Accuracy of fertilizer spreaders and GPS-receivers has been investigated by the DLG (German Agricultural Society) and DIAS (Danish Institute of Agricultural Science). A simulation procedure was derived that describes site-specific fertilizer application under field conditions. This simulation covers the technical aspects of fertiliser distribution. It is assumed that the recommendation map used is representing exactly the actual requirements of fertiliser. Concerning GIS, especially interpolation and resampling are spatial operations that influence accuracy. Computers can originate numerical errors. Fertilizing algorithms can contribute to inaccuracy by rounding and classification. A first approach to whole system accuracy assessment of site-specific fertilisation is presented, showing results from calculations based on real data from preagro fields and spread patterns from the Test Centre in Bygholm. Assuming a variogram for phosphorous with a range of 66 m, a nugget of 0.06, a sill of 2.68 and spherical model, soil nutrient maps have coefficients of variation (c.v.) higher than 20 %, even if they are sampled on a 1 ha grid. Accuracy of soil nutrient maps depends on the sampling density, the sampling scheme, and the interpolation algorithm. Other, even more important input variables are the estimated yield goals. Their contribution to the recommendation of basic fertilizers can be very high when fertilizers are given in advance for a whole crop rotation. But little is known about the accuracy of yield goals, statements vary from 10 to 50 % c.v. Calculating the error propagation of the fertilizer recommendation for a typical crop rotation (four years; oilseed rape, wheat, wheat, barley), we can show, that the variance of the yield goals can have a high influence on the overall accuracy of the application. This influence is only slightly modified by the correlation between the yields of different crops. Fertilizer spreaders contribute to the errors of fertilizer application as well. The deviation from the recommended rates can easily be more than 10 % c.v. Developing an approach to accuracy assessment of site-specific fertilization can provide means to incorporate uncertainty in the decision making process of nutrient management. Future algorithms for site-specific fertilizer recommendation should include routines for accuracy assessment. This will help to decide where improvement of accuracy is most efficient or whether site-specific application is beneficial at all.

Precision farming in weed control - system components and economic benefits

R. Gerhards and M. Sökefeld

Weed seedling distributions have been found spatially and temporally heterogeneous within agricultural fields. They often occur in aggregated patches of varying size or in stripes along the direction of cultivation or along the field borders. The spatial distribution of weeds has so far been ignored in weed management practices. Nevertheless, there is a general agreement that herbicide use can be reduced significantly when the variation in weed population is accounted for weed control strategies.

Precision Farming in weed control requires the use of modern information technologies. These include a digital image analysis system for online weed species identification and GIS-Software to generate georeferenced weed distribution maps. Secondly, a decision support system is needed to estimate the local yield loss caused by weed competition and to select the optimal type and dosage of herbicide. Third, a patch sprayer is essential that allows the separate control of boom sections, variable rate application and variable herbicide mixture. Prototypes of all three components for Precision Weed Control exist and will be presented in this paper.

A cost – benefit analysis for Precision Weed Control was carried out based on experiments in 18 cereal- , 21 sugar beet- and 7 maize-fields in West- and East Germany. The average costs for Precision Weed Control in cereals were reduced by 36 Euro per ha, in sugar beet by 79 Euro per ha and in maize by 8 Euro per ha compared to a broadcast application of a herbicide mixture.

Site-specific analysis of corn nitrogen status using reflectance measurements

Simone Graeff and Wilhelm Claupein

Site-specific N management practices offer the potential to increase yields, improve N use efficiency by crops, and minimize NO3-N leaching. Yet, because tillage practices, climatic variability, and the dynamics of crop growth all influence various components of the N cycle, predicting plant nitrogen requirements with great accuracy is difficult. Up to now soil must be sampled each year to determine the variation in soil N. The expense and the time to sample is high and cost prohibitive. In spite of the dynamics of N in agricultural soils, effective diagnostic tools and procedures have been developed that can help farmers to make site-specific N management decisions. Especially, in the last few years remote sensing techniques have become readily available and may be helpful to determine management zones differing in N availability. Remote sensing of agricultural fields has been used for a variety of applications ranging from assessment of water or nutrient status to detection of weeds and insects. Current applications of remote sensing involve comparing the reflectance of a surface to the reflectance of a known target surface and thus identifying crop stresses. Spatial variability can be evaluated by comparing reflectance with that from a reference of choice. The objective of our study was to test the use of reflectance measurements to identify the spatial and temporal variation of corn N status in a field during a growing season, in order to adjust site-specific N application rates. A field trial was conducted at the experimental station Ihinger Hof (48° 44' N 8° 56' E, 687 mm, 7.9 °C), of the University of Hohenheim, Stuttgart, Germany. An inhomogeneous trial site was chosen, representing different soil types and different previous crops. Corn [Zea mays L. cv. Tassilo] was planted on 05/02/01 at a rate of 90000 kernels ha-1. In order to determine the natural induced variability of soil nitrogen, no fertilizers were applied. Reflectance measurements were conducted once a week during the growing season (May – July) to document spatial and in-time variability of corn N status. Reflectance measurements were carried out every 2 m along a transect of 50 m at the trial site. Reflectance measurements were performed with a digital, light-sensitive (ISO 200 – 2400), high resolution (5140*5140 pixel) Leica S1 Pro camera (Leica, Germany) in conjunction with a light source (HMI 21 W/D ~10 W m-2, Sachtler, Germany) of total daylight spectrum. Total daylight spectrum was split into various wavelength ranges using long-pass filters (Maier Photonics, Manchester, USA), active at wavelengths longer than 380 nm, 390 nm, 430 nm, 470 nm, 490 nm, 510 nm, 516 nm, 540 nm, and 600 nm, respectively. Scans were carried out with the software SILVERFAST V. 4.1.4 (LaserSoft GmbH, Germany) and analyzed with the Software ADOBEÒ Photoshop 5.0 in the L*a*b*-color system (CIE, 1976). The results showed that leaf reflectance of corn plants varied along the chosen transect of 50 m. Reflectance changes could be closely correlated with chemically determined corn nitrogen status. As it has been shown by Graeff et al. (2001) reflectance patterns are different for various crop stresses. Thus, using different wavelength bands other stress factors, which might have an influence on reflectance, could be excluded. Nitrogen deficient zones in a field could be clearly detected. Reflectance measurements might be a useful tool for the estimation and optimization of fertilizer requirements of corn plants during a growing season. The results support the need for on-the go measurements of soil properties and plant responses. Plant surveys could be used in conjunction with soil surveys to identify management zones and to optimize fertilizer applications. GRAEFF, S., STEFFENS, D., SCHUBERT, S. (2001):Use of reflectance measurements for the early detection of N, P, Mg, and Fe Deficiencies in Zea mays L. J. Plant Nutr. Soil Sci. 164 (4): 445-450.

Understanding the spatial variability in the field dissipation of acetochlor and isoxaflutole

C.D. Graff, W.C. Koskinen, J. Anderson, T.R. Halbach and R.H. Dowdy

Current research in precision agriculture has focused on solving problems dealing with soil fertility and crop yield. Specifically, geostatistical methods are being used to link the spatial distribution in soil physical and chemical properties to long recognizable spatial patterns in crop yield. Because soil properties such as organic carbon content (OC), percent clay, and pH can affect the field dissipation of certain herbicides, and they can significantly vary spatially across the landscape, it is possible that herbicide dissipation rates, and subsequent weed control, have spatial structure as well. Experiments at the Rosemount Experiment Station in Dakota County, MN were conducted to test whether or not principles guiding research into the precision management of fertilizers can be used to study soil-herbicide interactions. The spatial variability of surface soil properties was characterized for a watershed cultivated in corn and displayed on a digital elevation model (DEM). Soil pH ranged from 5.5 to 7.5 and total organic carbon varied between 1.18 and 4.15. Textural fractions of % silt and % clay ranged from; 39 to 64 % and 12 to 27 %, respectively. Utilizing Arc/Info, terrain attributes such as slope, aspect, plan and profile curvature and flow accumulation were derived from the finely sampled DEM in order to determine if such attributes enhanced prediction of soil characteristics and/or herbicide residues. Acetochlor was applied at the uniform recommended rate of 2.2 kg ha-1 for both the 2000 and 2001 growing seasons. Isoxaflutole was applied at a rate of 80 g ha-1 to the center 4 ha of the field in both years as well. At 2 days, 2, 6 and 11 weeks, soil samples were taken to a depth of 70 cm at numerous georeferenced locations having a range in soil properties representative of the site. Soils were analyzed at 0-10, 10-20, 20-40, 40-60 and 60+ cm for herbicide residues. Surface field dissipation rates (k) for both herbicides during each field season were calculated at each georeferenced location where there were enough points to do so. The spatial distribution of k and herbicide concentrations at each sampling time were modeled in Vesper and displayed with Surfer. Various regression and spatial techniques were applied to correlate observed spatial structure in the raw data to the distribution of soil properties and terrain attributes across the field. Where significant correlation was found (30%), regression kriging was performed. From this procedure, an interpolated map of k or herbicide concentration across the field could be obtained utilizing only existing soil property or terrain attribute data. Due to the limited number of data points which eliminated the existence of a validation set, a boot strap method was employed to validate the interpolated maps. These analyses were performed on both herbicides and in both years to understand how the relationships might vary between different years and different climactic conditions, and to determine how different herbicides chemistries react under the same environmental inputs. Understanding how herbicide dissipation rates vary with soil characteristics could lead to reducing off-site transport via precision management, where soil properties and terrain attributes can be used to describe the distribution of the field dissipation rates.

Investigations on the use of airborne remote sensing for variable rate treatments of fungicides, growth regulators and N-fertilisation

Görres Grenzdörffer

Remote sensing data for variable rate treatments may be separated into two categories: spatially static base information of soil properties or yield potential (= site (potential) map) and spatially dynamic information of the canopy development, the soil water dynamics or the quality of recent crop management decisions (= status map). Remote sensing data for dynamic applications, e.g. fungicides growth regulators, N-fertilisation, contributes to provide current crop status information, e.g. biomass, nitrogen content, which are of great importance for several treatments. Airborne or satellite remote sensing is able to deliver current and large scale crop status information. Due to the importance of the turn around time (the time between image acquisition and delivery) and flexible image acquisition digital airborne sensors are best suited. To fulfil the special requirements of precision agriculture, a digital remote sensing system (PFIFF) has been developed at the institute for geodesy and geoinformatics. The system is temporarily installed for a photo flight in a Cessna 172 with a small ground hole. The core of the image acquisition system PFIFF is a high-resolution digital colour camera. At an altitude of 2.500 m - 3.000 m a ground resolution of 0.75 - 0.9 m is reached. With the digital workflow and image processing procedures it is possible to preprocess and geocode the images within a few days. The transformation of remote sensing information to site-specific treatments or for combined utilisation in a GIS requires an exact and quick geocoding of the imagery. Due to this reason a GPS-AHRS system was developed and different photogrammetric procedures which require none or very few ground control points were investigated and used for automatic aerotriangulation. With respect to crop development, information delivered by remote sensing imagery maybe outdated quickly. For applications such as the 2nd and 3rd N-dressings, relevant derivatives must therefore be extensively automated and generated objectively. Due these reasons digital image analysis procedures such as the Visible Atmospherically Resistant Index (VARI) was tested successfully for winter wheat and winter barley during two vegetation periods. This index relies only on the spectral bands in the visible spectrum and is highly correlated to the vegetation fraction (= crop density, biomass). Based upon the remote sensing data the application map is generated in three steps: 1. Generation of a conceptual map: the geocoded imagery is either classified or separated into a few classes based on a vegetation index. The minimum mapping unit is 0.3 - 1 ha. 2. Field survey: based upon the conceptual map typical areas and special features are investigated in the field. 3. Application map: The application map is generated upon the previous information. Thereby the relative amount of growth regulators etc. is determined. The precise amount is determined in the common way by models and based upon the current moment of spraying. Using conventional sprayers, which are extended for variable rate treatments with modern electronics and GPS, several questions have to be answered in order to generate the appropriate application maps for the machinery, e.g.: 1. How accurate does a sprayer work under conventional and variable rate conditions ? 2. How long is the time lag between the controller and the answer at the sprayer ? 3. What is the size of the minimum treatment unit ? 4. How many steps are reasonable for an application map (3,5,7) ? To answer these questions practical tests with two sprayers of the KSG Kassow GmbH where undertaken: a Rau sprayer and a Danfoil sprayer with air support. Due to the different technological approaches different reactions of the two sprayers will be presented. The size of the minimum treatment unit has a large influx of the profitability of site specific applications. In order to access the economic return of a variable treatment of fungicides different minimum treatment units were generated and compared.

RTK-GPS for single plant care in cropping systems

H.-W. Griepentrog and M. Nørremark

Agriculture has benefited in the past from the success of technological developments that have brought greater productivity and economic efficiency. Historically, the emphasis of these developments has been on the mechanisation of field operations to increase work rates achievable by individual operators. Today, however, the trend of increased efficiency through the use of larger and more powerful machines becomes more crucial due to environmental hazards as soil compaction and high chemical and fuel inputs. Large scale machinery also seems to have drawbacks to match the general requirements for Precision Farming. The trend of increased machinery size and weight may be substituted by newer information based technologies that may ultimately enable reliable autonomous field operations. This scale-reduction process, embodied in Precision Farming, may lead to the possibility of individual plant care cropping systems. These cropping systems require accurate information about the position of the crop plants. The objective of the project is to provide high accuracy seed position mapping of a field of sugar beet. This highly accurate information enables several automatic controlled field operations as - guidance of vehicles (e.g. parallel to row crops), - guidance of implements or tools (e.g. inter- and intra-row weeding), - application of fluids or granules to individual crop plants (e.g. insecticides, fungicides, fertilsers etc.) and - measuring growth status of individual plants (e.g. multi-spectra, shape etc.). A six row precision seeder for sugar beet was retrofitted with a real-time kinematic GPS positioning (RTK-GPS) and a data acquisition system. Six optical sensors (one per seeder unit) were mounted directly above the coulters and detected the seeds as they dropped into the furrows. In order to correct the tilt of the seeder and the attached GPS-antenna an inclinometer was added to log the tilt information. The data logging system stored the GPS time and the UTM coordinates 20 times every second. The data logger also monitored the optical sensors and the time-tagged seeds were also stored in the memory. The overall objective was to map the placed seeds in the soil but also to get the information about where the individual plants are. Therefore the total mean deviation between estimated seed position and true plant position were determined. Error sources for plant position determination are - accuracy of the positioning system (RTK GPS) - movement (play) of sowing devices relative to the positioning reference point - displacement of seeds in the furrow after passing the optical sensor - deviations between seed position and plant position (effect of seed bed structure, seeding depth etc.). Field tests were conducted not only to check the performance of the seeder. A field experiment was set up to investigate the effect of the seed bed quality and the soil type on mean deviations between seed position and true plant positions. In a different experiment the seed spacing and vehicle speed were altered to check its influence. The results showed that the overall accuracy of the estimated plant positions is acceptable for the guidance of vehicles and implements. For subsequent individual plant cultivations the deviations were not in all cases small enough to ensure an accurate targeting of treatments. Perspective: The developed seed mapping technology will be improved and utilized within a new funded danish research project called 'Robotic Weeding'. The mapped seed positions will give a prior information about a field for subsequent more accurate scouting of not only the true crop plants but furthermore also the weed plants. A planned cultivation consisting of mechanical operations (intra-row weeding) or chemical treatments (micro spraying) are part of the project contents.

Stability of spatial structure in successive grid samplings of unchanging soil properties in fields under no-tillage management

J.H. Grove and E.M. Pena-Yewtukhiw

It is well understood that the long-term value of a mapped soil property becomes more questionable as the temporal instability of the property increases. However, growers have also questioned whether maps of relatively unchanging soil properties (texture, organic matter) will be relatively similar across sampling events. An evaluation of this question was the objective of this study, conducted for three years (1999, 2000 and 2001) on three fields under no-tillage soil management. All fields were planted to a corn, wheat-doublecrop soybean, and corn rotation for the years under study. No-tillage management has been used in these fields since the 1991 season production season. Fields were approximately 20 ha in area. The soils in these fields were largely Paleudalfs, Hapludults and Fragiudalfs. One field had a history of uniform chemical fertilizer application, while the other two had a history of liquid swine manure and chemical fertilizer N applications. Soil samples (8 cores to a 10cm depth) were taken on an approximate 54m by 60m grid in each field, after each crop, resulting in four grid sampling events for each field over the course of the study. Grid positions were different with each sampling event and locations were recorded using GPS. A digital elevation map was developed for each field using data taken on a 7m by 7m grid interval with a real-time kinematic GPS system. Slopes averaged about 2% in all these fields, but individual fields evidenced very different ranges in this parameter. Relatively unchanging soil properties like water pH, soil organic carbon (dry combustion method) and soil texture (pipette method) were determined on each soil sample. Soil pH values were nearly 6.5 in all fields. Average soil organic matter levels were very different among fields, averaging as little as 2.5%, or as much as 3.5%. Soil texture behaved in a manner similar to slope. All fields had a similar average clay content, about 20%, but individual fields exhibited very different ranges in this textural parameter. Localized variograms were used to study the spatial structure of these soil properties. Ordinary point kriging interpolation was used to obtain values at unsampled locations. Grid soil sample information was examined for normality, and the resulting semi-variograms examined for anisotropy and stationarity. For each individual sampling event, and for each sampled property, descriptive statistics were performed. All sampling events were grouped together and descriptive statistics were again performed. The population parameters for the different groups of samples were compared (individual events versus grouped events). The same grouping procedure was used for spatial analysis. For each individual sampling event, the spatial structure of the soil properties was analyzed, and then a new spatial analysis was performed with the events grouped together. The results from each individual event and the pool of events were compared. In general, the variogram model fitted to the soil properties was spherical with a nugget effect. The nugget/sill ratio ranged from 0 to 15%, depending on the soil property and the field considered. One field evidenced poorer spatial structure for all soil properties, contributing to considerable map instability with individual sampling events.

3D geographic information technology and site-specific agricultural management

S. Grunwald

Optimized site-specific agricultural management depends on reliable and detailed information to describe the spatial distribution of soils, parent material, topography, and land cover. Land resources are useful indicators of terrestrial change because they absorb the impacts of land use and thereby accumulate and consolidate information. Yet to address the current status of land resources we need to acquire detailed and spatial explicit information about soil-landscapes. Subsurface attributes have in common that they vary through three-dimensional (3-D) space and through time. Our goal was to overcome current limitations of geographic information systems to manage, analyze, and visualize geographic land resource data in two dimensions (2-D). Our objectives were to reconstruct and visualize soil-landscapes in 3D using an object-oriented and multi-variate approach utilizing geostatistical methods and 3D rendering. Commonly, two-dimensional (2D) maps are used to describe the spatial distribution of soils, topography, land use, geology, hydrology and others. In the U.S., soil information systems provide soil data in 2D GIS formats (e.g. ArcView shapefiles) such as the State Soil Geographic Database (STATSGO) and Soil Survey Geographic Database (SSURGO). Other soil-landscape representations use a 2½D design, where soil and land use data were draped over a digital elevation model (DEM) to produce a three-dimensional (3D) view. Since this technique describes patterns on 2D landscape surfaces rather than the spatial distribution of subsurface attributes (e.g., soil texture, soil horizons) it fails to address 3D soil-landscape reality. Conventionally, soil information systems such as SSURGO and STATSGO contain polygons representing soil mapping units and associated representative soil attributes. Most digital soil data use this entity data model based on the vector polygon as the geographical primitive. This data model is practical; however, it ignores spatial variation in both soil-forming processes and in the resulting soils. This suggests to (i) develop 3D models, which describe the spatial distribution of subsurface attributes in the x, y, and z direction (3 dimensions) and topographic and land cover attributes in 2 dimensions, (ii) describe soil and landscape attributes continuously to capture gradual changes, and (iii) employ field view rather than entity (object) view to describe the spatial variability of soil and landscape attributes. We used geo-referenced subsurface, topographic, and land cover attributes from several sites in southern Wisconsin, northwestern Ohio, and northern Florida to demonstrate the capabilities of our approach. These sites are used for agricultural production of corn and soybeans (Wisconsin and Ohio) and potatoes, corn, pasture and other (Florida). Reconstruction was based on 2D and 3D ordinary kriging utilizing a suite of software such as the Environmental Visualization System (EVS) software, ArcGIS, and SPlus. The Virtual Reality Modeling Language (VRML) was used for 3D rendering of the models. VRML models are interactive and were posted on the internet (http://grunwald.ifas.ufl.edu). Three different types of soil-landscape models were distinguished: (i) Models representing subsurface attributes as points (ii) Models representing subsurface objects as polyhedrons or "volume models" and (iii) Block models consisting of voxels (volume cells). A server hosts the virtual soil-landscape models, which are accessible by multi-clients via an interface coded in HTML. These models are interactive, platform independent and enable users to analyze, explore and gain insight into the spatial distribution of surface and subsurface attributes in 3D geographic space. 3D soil-landscape models integrate a variety of land resource data including soils, topography, land cover, and potentially more (e.g. geology). Understanding the spatial variability and distribution of soil-landscape properties is a prerequisite for improved sustainable land resource management, land use planning, site-specific management, and more.

Site specific calibration of a crop model by assimilation of remote sensing data: a tool for diagnosis and recommendation in precision agriculture

M. Guérif, M. Launay and J.M. Machet

One of the main challenges of precision agriculture is to propose methods for variable rate application of the major cultivation practices according to the spatial heterogeneity of the agricultural fields. A lot of effort has been made in the last decade in order to measure different aspects of the within-field crop heterogeneity, but few methods have been proposed to handle the making of decision from this information. Crop models appear as interesting tools for simulating in real time crop behavior and thus help in diagnosis and decision-making. The model allows simulating the effects on the crop of the environmental conditions (soil and climate) of the crop - when accurately characterized, as well as those of the management options. Moreover, the model may be a way of integrating multiple-source information obtained during the growing cycle on crop status, by making a self-correction of its input variables or parameters, and thus providing a more accurate representation of the crop functioning. We propose here a method that allow the model to assimilate remote sensing data acquired during the growth cycle and thus to re-estimate some lacking information on input variables or some site-specific values for some model parameters. By coupling it to a radiative transfer model into the canopy (SAIL), through a canopy variable as the LAI, the crop model (SUCROS) is able to simulate reflectance in different wavelengths that can be compared to measured one at some dates by remote sensing acquisitions (SPOT or airborne measurements). The minimization of the differences between simulated and measured reflectance allow to re-estimate, with the spatial resolution of the images, input variables or parameters that permit to give a better simulation of crop functioning. This kind of site-specific re-calibration may allow therefore to realize site-specific simulations of crop behavior, according to different scenarios, and to evaluate some criteria based on these simulations in order to make decision. The proposed method is developed on 2 sugar beet fields grown on two years on a site devoted at INRA to precision agriculture programs. Very precise soil maps were established, with adequate pedotransfer functions that allow associating to each unit of soil its properties in terms of hydraulic properties for the crop model; initial values of soil water content were measured. SPOT images and airborne measurements with a CASI hyper spectral camera were performed at 4 times during the season (from April to July). Both measurements were transformed in surface reflectance data by performing atmospheric corrections. All these data were handled through a GIS and allowed to describe each field according to variable size grids (20m * 20m, 10m* 10m, 4m * 4m). A first step consisted in simulating for each grid cell the crop growth with the SUCROS model run alone. Then, the assimilation of the 4 sets of reflectance was performed into the SUCROS+SAIL model, according to different strategies: re-estimation of crop establishment parameters alone, of crop parameters plus root settlement parameters. The method provided maps of estimated parameters (status of LAI at 500°C.days after sowing, maximum rooting depth) and of simulated yields. These simulations were compared to yields measured with a yield monitoring system. Despite the large uncertainties on yield measurements, it was obvious that the assimilation method improved the simulation of yield made by the crop model alone. The limitations of the method in the case study presented are therefore discussed (number and dates of available images, ability of the model to represent crop growing conditions,). The potential of such a method for retrieving soil permanent properties and for decision-making are also debated.

Acceptance of precision agriculture in Germany - Results of a survey in 2001

E. Gumpertsberger and C. Jürgens

Since a few years Precision Agriculture offers a chance to consider site specific differences within a single field. To get necessary informations and to realize the consequences for the field work the farmer has to modify his machinery and adopt new computer technology. This is surely an obstacle that has to be overcome by the farmer. Considering this background one was interested to monitor how precision farming techniques enter the German market in respect of time and geographic location. Therefore surveys during the AGRITECHNICA fair were conducted in 1999 and 2001. German farmers were interviewed during the fair to find out their experience with the new technology and their attitude towards it. Those farmers who are not using precision farming techniques at the moment were asked for their reasons and if they would like to try it out during the next few years. As a result we can group all interviewd farmers into different types according to their knowlege and their farming background. Each group will be characterized by statistical analyis and important regional results will be presented on thematic maps that stem from GIS analysis. Some results are briefly summarized here: - More than 50 percent are familiar with the term „precision farming". Mostly they associate the following terms or techniques with it: GPS, Yield mapping and field area mapping. - Approximately seven percent of the interviewed farmers use precision farming techniques. - The reason for the farmers to use precision farming techniques are mainly economical. They expect a financial benefits. This expectation was fulfilled in most cases. Financial losses seem to happen rather seldom. - Very interesting was to analyze the techniques that are used in practice. The farmers use mainly techniques that provide information about the fields (e.g. yield mapping or soil sampling). Only a few farmers already use techniques that help to react according to the prior gained information (e.g. site specific seeding, or spraying). - The users seem to be happy with the techniques they apply. More than 80 percent would recommend precision farming to others. More than two thirds are willing to use additional techniques and approximately 60 percent are intrested to apply the techniques on a larger area. - Regarding the size of the farms one found out, that not only large farms (over 1000 ha) use precision farming techniques. Also small and medium farms use precision farming already. - The informed farmers who do not yet use precision farming have various reasons for their hesitation. Most of them say that the technique is still too expensive, many want to wait until they realize that the new technique has proved to be unproblematic. - From the regional distribution one can discover differences in the differnet groups. Most farmers who use precision farming are in the north and especially in the eastern part of Germany. This is probably because of the farm sizes there. However one can recognise a trend from the east to the west and also from large farms (probably of innovators) to medium and small farms (of adopters). As a final statement one could conclude that there seems to be a potential in Germany that presision farming will be applied by more and more farmers in the near future.

Hyperspectral image feature extraction and classification for soil nutrient mapping

Yao Haibo, Tian Lei and Kaleita Amy

Soil nutrient management is important for crop production. Soil nutrient can be managed via fertilizer application based on variable rate technology (VRT), which is a major type of precision farming practice. Successful implementation of VRT relies largely on acquiring the appropriate field nutrient information - or the soil nutrient map, upon which the application prescription map can be generated. The traditional way for nutrient mapping is using soil grid sampling with either 1.1 or 2.5 acre sampling interval. Soil nutrient map can be produced through spatial interpolation using the grid sampling data. However, results from the interpolation approach are affected by the interpolation methods (inverse distance weighting or geostatistical methods such as kriging) and may not accurately represent the real field condition in between the sampling locations. Remote sensing has shown its potential for many precision farming applications because it can provide detailed pixel by pixel spectral information. This paper will focus on using aerial hyperspectral imagery for soil nutrient mapping. A hyperspectral image has many narrow bands, normally several nanometers, in the same wavelength range as a conventional Multispectral image. With its abundant spectral information, hyperspectral imagery provides the opportunity to extract more detailed field information, as well as increases computational complexity. The nutrient properties under investigation include soil PH, potassium, phosphorus, and organic matter, which are relatively stable in the field over years. Two fields from the US Midwest region will be used in the study. As a comparison study with the interpolation method, this study will use both the soil sampling data and the hyperspectral image for nutrient mapping as compare with only use the sampling data for interpolation. In processing the hyperspectral imagery, a dimension reduction and feature extraction process will be used prior to image classification. This process is a supervised approach, which uses the soil grid sampling data as training data. The dimension reduction and feature extraction process first selects a subset of the original image bands and then implements a selective principal component analysis (SPCA). In searching for the best set of image band combination, an evolutionary algorithm will be used for avoiding a costly exhaustive search. The SPCA step will identify the best feature(s) in the principal component (PC) image space by calculating the class reparability, using statistical distance among different classes. The supervised image classification used both the maximum likelihood and the minimum distance classification methods depend on the number of feature(s) extracted. Past study has shown that there existed sensitive regions in the electromagnetic spectrum for different soil nutrient properties. The dimension reduction and feature extraction process used in this study will provide a way to identify the sensitive regions, which will also be benefit for future hyperspectral imaging sensor design for soil nutrient sensing. Both the hyperspectral image processing results and the interpolation results will be validated by additional dense soil sampling data (with 25 meter grid). Preliminary results showed that image classification could be implemented on a feature space using one or two features rather than on the original image space using all bands. It is also showed that by using the hyperspectral image, accuracy for phosphorous (P1 test) and organic matter mapping is better than using the interpolation approach.

Construction of management zones via a hidden Markov model approach

U. Halekoh and H. Nehmdahl

In precision farming, one attempts to adjust management activities or the rate of application of fertilizer to the local needs of the field. The needs are defined with respect to the optimization of yield or net return. The gradual change of the application may either be impractical or the costs for detailed mapping of the needs exceeds the benefit from gradual changes. Therefore, the field is treated equally on larger areas, called management zones, which should be rather homogeneous with respect to the needs. Additionally, their shape and size should allow the adjustment to be technically feasible. We consider the construction of management zones as a task of clustering the spatial locations based on variables like soil conductivity (EM38) or topological characteristics. These variables serve as easily and inexpensively measurable surrogote variables for the needs. Therfore, the ultimate utitliy of a clustering of a field depends on the relation between the surrogate variables and the needs. The method of KMEANS-clustering is a classical partitive clustering method. It has the disatvantage that it does not incoporate the spatial dimension of the data and therefore yields often not spatially sufficiently connected areas. We propose to use a Hidden Markov model approach for the clustering. The multivariate measurements at the locations belonging to the same zone are considered to be multivariate normally distributed with the same mean vector and the same diagonal covariance matrix. The spatial aspect is incorporated by modelling the cluster memberships via a Pott-Strauss Markov random field. Given the memberships of the neighboring locations, the model favours the membership of the majority of the neigbors as the membership of the central location. The strength of this relation is controlled by a single parameter. The parameters are estimated via an EM-type algorithm (Ambroise et al., 1997) where the memberships are formally considerd as missing data. For the expectation step we consider two algorithmic solutions (the iterative conditional maximization (ICM) and the marginal posterior means (MPM) principle. Both yield estimates of the class memberships and hence a clustering. Both algorithms turn out to be sensitive to the initial solution chosen, where the ICM approach seems to remain closer to the intitial solution than the MPM algorithm. A minimal size requirement for the zones is not automatically achieved with the simple Pott-Strauss model. Instead of incorporating such a restriction into the model we propose a simple aggregation of too small zones to neighboring ones. For some soil property variables we analyse how their spatial variability is related to a clustering based on average yield experience, soil conductivity and the wetness index, a topological oriented index for the water accumulation property to a location. For an example field it turns out that a clustering based on soil conductivity alone is not much different from that including the other variables with respect to the explained variability of clay content. For organic content the inclusion of wetness index improves the clustering considerably. Ambroise, C., Dang, M. and Govaert, G. (1997): Clustering of spatial data by the EM-Algorithm. In: A. Soares, J. Gòmez-Hernandez and F. Froidevaux(Eds.): geoENV I- Geostatistics for Environmental Applications. Quantitative Geology and Geostatistics, vol. 9, 493--504, Kluwer Academic Publisher.

Using aerial photography to detect weed patches for site-specific weed control - perspectives and limitations

Andreas Häusler and Henning Nordmeyer

Site specific weed control requires information on the spatial distribution of weeds. Weed mapping by field walking is very time consuming and costly. The aim of this study was to investigate the suitability of mapping techniques based on low altitude aerial photography for monitoring weeds in agricultural crops. The results were used to evaluate the possibility of integrating airborne remote sensing data in site-specific weed control. Several fields in conventional agricultural use located close to Braunschweig, Northern Germany, were chosen. Low altitude aerial photography was done in early growth stages of weeds and crops and some weeks before harvest. Aerial photos (colour and colour infrared slide film) were taken from 300 to 900 m above ground level using conventional cameras (6 by 6 cm and 24 by 36 mm format) mounted vertically in the floor of a fixed wing aircraft. After scanning, geocoding and rectification of aerial photos the sizes of ground pixels in digitised photos were between 0.15 and 0.85 m. In most cases only portions of the field were covered by single photos caused by low elevations in order to guarantee a high spatial resolution. Single photos of one field were combined in one mosaic of aerial photos using artificial and natural ground reference points. These mosaics were the basis for extraction of information on weed distribution and the comparison with ground truth data. The reference data on spatial distribution of weed populations were recorded by means of visual mapping shortly before or after the date of flight using the NAVSTAR DGPS (Navigation System with Time and Ranging - Differential Global Positioning System). In addition to grid-based monitoring of weeds with spacings between 5 to 6 m, the shape of selected weed patches with high plant or shoot densities was determined by DGPS-based mapping. Finally the results of ground-based mapping and a geographical information system (GIS) were used to verify the weed distribution data derived from mosaics of aerial photos. The aggregated occurrence of investigated weed populations confirmed a strong potential for spatially selective weed control. Furthermore with low altitude aerial photography it was possible to detect patches of important weed species before weed control in spring (May) until maximal 4 weeks after the sowing date of sugar beets. But only changes in total weed coverage were detectable. So, for example, only patches of Cirsium arvense with high shoot densities (50 shoots/m^2, BBCH-Code: 11-32) could be surely distinguished in sugar beet. Moreover high densities of an annual (Alopecurus myosuroides) and a perennial grass weed (Agropyron repens) could be extracted in aerial photos in early crop growth stages. Weed detection by aerial photography was also possible in the late growth stages of crops and weeds. Weeds (e. g. Apera spicaventi, Galium aparine, Cirsium arvense) taller and of different colour than the crop were differentiated in aerial photos some weeks before harvest. But again high weed abundances or biomass, respectively were necessary for successful weed detection by airborne remote sensing. It can be concluded that low altitude aerial photography can be used for assessing highly aggregated infestations of certain weed species. But the results also indicate that, if information on the weed distribution for the present weed control decision should be derived from aerial photos, low altitude aerial photography is only of restricted use due to the small time window at the beginning of the vegetation period, the dependence on weather conditions and the limited detection of weed densities close to threshold values.

The value of additional data to locate potential management zones in commercial fields

Dale F.Heermann, Kenan Diker, Gerald Buchleiter and Mary Brodahl

A multi-disciplinary Precision Agriculture research project has been conducted for the last six years (Heermann, D.F. et al., 2002. Interdisciplinary Irrigated Precision Farming Research. Precision Agriculture, 3. pp. 47-61). When presenting the results of our past five years to producers and crop consultants, they asked how much data is needed to identify zones that are candidates for variable rate application of inputs. Development of management zones is a dynamic process, with zones potentially changing with acquisition of more data and management changes. Even though extensive soil sampling maybe appropriate, it is very expensive and not acceptable to many producers and consultants desiring to use PA technology. The fewer years and easily sampled data would provide the producer and consultant a guide to adapting Precision Agriculture technology. The techniques need to be simple and easily understood by the producer and consultant. Results obtained with more complex statistical and analytical tools are not readily accepted by the user. Therefore, our objective was to develop a simple process of defining management zones using easily acquired data. Identifying potential management zones with limited years of data collection will also be evaluated. Research data were collected in cooperation with a commercial farmer on a 72 ha center pivot irrigated field in northeastern Colorado, USA to develop potential management zones. A multi-discipline research team collected data of the production inputs, physical conditions, soil, crop, pests and climatic conditions. Technical expertise included soil fertility, plant pathology, weed science, entomology, remote sensing, water use (ET), irrigation system evaluation and modeling, variable rate application of water and chemicals, hydrological and water quality modeling, statistics, and agricultural economics. University, government agencies, and industrial partners pooled their resources to fund the research. Soils, weed, and entomological data were collected on a 76m x 76m grid. Soil electrical conductivity, topography and yield data were collect continuously over the entire field. Limited aerial photographs were also taken. Results were presented at the first three European Precision Agricultural Conferences. Use of management zones (MZ) is a popular concept from the producer's viewpoint. Two concepts of spatial management are considered. One is to reduce inputs and lower the production costs on low yielding areas. The other is to increase the inputs on the low yielding areas in an attempt to have uniformly high yields over the entire field. We began with bare soil maps and farmer knowledge for the first estimate of management zones (Fleming, K.L., et al., 2000, Evaluating Farmer Defined Management Zone Maps for Variable Rate Fertilizer Application. Precision Agriculture, 2, 201-215). The potential MZs from yield data were based on whether each cell was above or below the within year average yield for the 72 ha field. The areas with consistently high or low yields were determined by counting the frequency that each cell (6.1m x 6.1m) exceeds or is less than the average yield. The data for the maximum years available were assumed to identify potential MZs. Results showed that the two MZ methods had total agreement on 49% of the area contained in each of the high, medium and low yielding areas. Only 7% of the area was in total disagreement where high and low yield areas intersected. The remaining 44% were areas of medium-low and medium-high intersections. Comparisons will also be conducted with alternative methods in which the yield average will be replaced with yield average +- ($\frac{1}{2}$ and $\frac{1}{4}$ standard deviation) for establishing high and low yielding areas and other ancillary data. The ancillary data such as soil electrical conductivity, topography and water application data will be examined for refinement of the yield base.

Development of a method for determining actual spread mineral fertilizer using digital image processing

Oliver Hensel

Application of mineral fertilizer is one of the most important measure in agricultural plant production whereby accuracy in distribution will influence success both in cultivation and economics. Since meanwhile machinery manufactures offer high precision flow indicators for their fertilizer broadcasters, farmers have exact information of the quantities leaving the machine but still do not have any information about its proper distribution. Therefore time costly evaluations have to be carried out by lining up trays to collect the fertilizer for weighing and calculating the accuracy of the machine's setup. Since such measurement can be done only in a consecutive way, no online data are available neither for a closed-loop control of the machine nor for a data storage to fit in actual applied fertilizer dose into application maps. This paper describes a new method for an automatic detecting and counting system for spread mineral fertilizer based on digital image analysis. Pictures of the soil surface set up with fertilizer where analysed and fertilizer particles were separated from soil components by specific differences in colour, shape an size. In the present paper first it is described how the rate of detection is influenced by environmental factors such as light, condition of soil and disturbing vegetation whereby samples are shown from both laboratory and on-farm situations. In addition, the effect of the specific fertilizers nature is described and figures are given which point out that the detection rate of the whole system is varying depending on the specific situation in the field: on clod free soil without biomass a detection rate of close to 100 % of the fertilizer particles was found, while mulch and vegetation decreased the accuracy. Fertilizer which was hidden by clods or vegetation could hardly be detected by the image processing system. In general, the conclusion can be drawn that the detection rate is depending less on the particles shape and size then on contrast between soil surface and the fertilizer particles. In the second part of the paper procedures are described which yielded high-contrast pictures, such as ultraviolet light based photography and thermal imaging. It is shown that by an advanced shooting technique the detection rate can be increased up to 100 % even under adverse conditions as they are found at on-farm situations. In the third part, it is shown that the collected data can not only be used to give the number of fertilizer particles but also the weight per hectare by planimetrical calculation using the digital image system and taking their size into account. Since in addition to the number the x-y coordinates of each fertilizer particle is known a spread pattern diagram of the machine can automatically be generated. At last, as practical application of the research work into farming practice samples are given such as a tractor equipped with a self regulating fertilizer spreading system where several camera modules observe the whole working width of a centrifugal broadcaster. Furthermore a specialized motorcycle equipped with GPS and the image processing system is presented which enables to fit in data of the actual distributed fertilizer into application maps.

Pest management module for tree-specific orchard management gis as part of the PRECISPRAY project

A. Hetzroni, M. Meron and V. Fatiev

Precision Horticulture encompasses variable-rate application of agrochemicals, nutrients, and water, according to specific needs of each individual plant. These plant-needs include water and nutrient status, pest and disease infestation, fruit growth, plant stress, and economic and horticultural considerations. Spatial variability of trees in the orchard is due to stationary and dynamic parameters. Static variables would be: yield potential, tree vigor, soil characteristics, foliage volume, and tree size. Dynamic variables would be: pest and disease infestation or pest-control measures. Customary applications of inputs are either aimed to provide for utmost needs which results with wasting by over-dose; or aiming to provide for the average needs - thus reducing waste, yet, resulting with under treating parts of the orchard. Variable rate treatment of the orchard by matching applied input to tree-specific requirements could increase economical efficiency. For this, reliable information regarding the individual plant is needed together with data-handling, decision system, and a suitable applicator. Integrated Fruit Production and ISO 1400 application for fruit quality requires data handling and information archiving. Horticulture management systems could comprise the requirements to maintain product records from tree to market as part of quality assurance, together with management of the necessary information for precision horticulture. Precision spraying by matching spray volume to tree size and foliage density is one of the high priority goals of R & D in the EC because of its potential to reduce pesticide load on the environment. The 5th Framework "PRECISPRAY" project was initiated to develop a horticulture precision spraying system. The system is made of a segmented boom cross-flow sprayer, where each segment can follow the outer contour of the tree and adjust airflow and spray volume according to foliage density and volume. The vehicle navigation system maintains tree shape and volume maps (TPV) obtained from aerial survey and photogrammetric processing. The PRECISPRAY spraying system and the Orchard Management GIS (OMGIS) are tied together at the applicative level. The pest management module of the OMGIS is the depository of the infrastructure and physical properties of the orchard, including the TPV maps. It is also the user interface to the pest knowledge base, infestation severity evaluation, and support for pest control decisions. It is used to communicate with field data such as pheromone traps and scouting findings, issue spraying commands and to store feedback data from the applicator. In order to serve as user interface in the farm level, it is designed to manage event location and time information, and to provide the user with simple and intuitive interface to farm data. The software prototype is being developed using standard software packages, thus enabling rapid application realization, certain level of documentation, and ease on further maintenance. The module is written in ArcView 3.x native AVENUE language and management data are stored in compatible tables. The content of the handbook of pest control is stored in MSAccess format. Separate modules of the package, i.e., TPV map generation module and applicator control unit, communicate with the central OMGIS via standard protocols, thus allowing independent development of the separate units.

Retrieving valuable crop parameter information from synthetic aperture radar

G. Holmes

Infoterra is leading a pan-European consortium in an EC Framework V research project "ISOCrop" (Integrating SAR and Optical products for Crop Management) over the 2002 and 2003 seasons. Through ISOCrop our ability to retrieve crop parameters using new generation space-borne Synthetic Aperture Radar (SAR) sensors will be established, along with the ability to integrate this information with existing services to growers and advisors based on optical sensors. Earth observation has for many years been acknowledged to have great potential for monitoring crop status and spatial variability for the benefit of growers, but it is still not in widespread use. The latter fact is attributable to a number of obstacles, not least of which is temporal reliability and the difficulty created by cloud cover for optical remote sensing. Radar has the distinct advantage of being able to penetrate cloud cover, but space-borne SAR instruments in the past have been limited by spatial resolution and information content. Now the next generation of SAR satellites (such as the European "TerraSAR" missions) promise to provide this cloud-penetrating data at much-improved spatial resolution (<10m) and with vastly improved information content (through dual-frequency capability and multiple polarisations). Previous research suggests it should be possible to retrieve useful information on crop and soil condition from this kind of data, particularly structural and dielectric properties. The ISOCrop project has been designed to prepare the way for exploiting TerraSAR for the benefit of growers, developing the necessary techniques for isolating useful crop and soil parameters in advance of its launch. Wheat has been selected as the initial focus of research due to its importance in terms of area and production in Europe. Trial sites have been selected in the UK and Spain to represent the diverse growing conditions and information requirements within Europe. Airborne image acquisition will take place over two growing seasons (2001/2 and 2002/3). SAR data are acquired by the E-SAR system operated by the German Space Agency (DLR). Near-simultaneous optical data are acquired by the AISA system operated by Infoterra Ltd. Both data sources will be used to simulate satellite-quality data. The flights are synchronised with the collection of intensive ground truth measurements in the trial fields. The fields are differentially managed to create the necessary variation in crop parameters for the development SAR retrieval techniques. Key research objectives are to retrieve parameters such as biomass, canopy moisture and surface soil moisture from the SAR data, and to analyse how these relate to parameters that can be routinely estimated from optical data sources, such as Leaf Area Index. These biophysical parameters can be coupled with field background information and agronomic models to generate anomaly maps and application recommendation maps. The integration of both SAR and optical information products may form the ideal basis for sustainable and reliable services to growers and their advisors. Imagery and results from the first year of research are presented. Ultimately the vision is that information derived from Earth Observation will become a valuable and widely used component in precision agriculture, providing a rich source of the spatial information that is needed to drive decision support systems and variable rate application strategies. By contributing a technology that will aid the optimisation of inputs, ISOCrop supports the EC objectives of the improvement of economic margins for European growers, and the reduction of environmental impact. ISOCrop is a shared-cost project (contract EVG1-CT-2001-00053) co-funded by the Research DG of the European Commission within the RTD activities of a generic nature of the Environment and Sustainable Development sub-programme (5th Framework Programme).

Evaluation of site-specific management zones: Grain yield, biomass, and nitrogen use efficiency

A. Hornung, R. Khosla*, R. Reich and D.G. Westfall

Precision agriculture fertility management practices, such as production level management zones, have been shown to account for soil's spatial variability. We are developing and evaluating a new technique, "site-specific management zones" (SSMZ) of delineating production level management zones. The objectives of this study were: (i) to compare the new SSMZ technique with a commercially available soil color management zone (SCMZ) technique, and (ii) to determine the optimum N-management strategy within the technique that most accurately describe the crop production parameters, grain yield, biomass, and nitrogen use efficiency. The study was conducted on two irrigated continuous maize (Zea mays L.) fields in northeastern Colorado, USA over two years. The GIS data layers used to delineate management zones based on new SSMZ technique were: multi-spectral bare-soil imagery; soil organic matter; soil cation exchange capacity; soil texture; and previous years' yield monitor map. Stepwise procedure was used to select a set of bands (Red, Green, Blue, and Near Infra-red) from multi-spectral imagery to describe the large-scale variability in field data using multiple regression equations. The errors associated with the regression equations were modeled using binary regression trees. These (regression equations and binary trees) were added together to create the surface maps of the field data. Finally, productivity level management zones (high, medium and low) were created using k-means clustering algorithm using all field GIS data layers. The commercially available SCMZ technique utilizes panchromatic bare-soil imagery, field topography, and farmers past management experiences as the three GIS data layers to delineate management zones. Five nitrogen management strategies were utilized to evaluate both techniques in this study. Those N strategies were: (i) uniform N application at two different rates (Recommended rate and 0.5 x recommended rate), (ii) grid soil sampling based N application; (iii) constant yield goal based N application; (iv) variable yield goal based N application; and (v) control, 0 N application. Recommended N rates for uniform N application were determined based upon N recommendation algorithm, which includes residual soil NO_3-, soil organic matter, and expected yield. Maize grain yield was harvested using a GPS equipped yield monitoring system. Spatial statistical methods were used to arbitrate treatment differences on grain yield, biomass, and nitrogen use efficiency. The data were analyzed using a nested design with a spatial autoregressive model. The results were compared to a classical nested design. When the spatial dependence among the sample points was ignored, no significant differences were observed between the two management zone techniques. In contrast, the spatial model, which accounts for spatial dependency, showed significant differences between techniques for delineating management zones. Detailed results from 4 sites years of data will be discussed.

Weed detection based on spectral imaging systems with CMOS cameras

S. In der Stroth, B. Ramler, A. Linz and A. Ruckelshausen

The further reduction of herbicide applications for weed control is mainly based on the availability of sensor systems with high selectivity and high-speed. So far developments with photo diodes [1], CCD-cameras [2] or multi-sensor systems [3] have been improved, however, prototypes or products in the market are still not available. In order to improve and speed up the process of transferring research results into the market, reliable online measurements of the percentage of the total weed population - excluding crop detection - have come into focus. In a first step of sensor-based weed control, the local application of herbicides has to demonstrate its ecological and economical advantages in commercially available systems. Recently, the authors have applied a commercially available optical system - ImSpector - for weed detection, which scans one line thereby splitting each detection point into a spectral pattern [4]. Thus a spectral analysis can be performed for each point. New digital CMOS imaging devices have now been applied for the detection of the two-dimensional information - position versus wavelength. Digital CMOS sensors allow the addressing and fast image processing of single pixel informations, thereby resulting in higher flexibility and speed as compared to standard CCD cameras. Moreover the choice of linear or logarithmic response as well as automatic gain control are important figures with respect to variations of the light intensitiy. Combined with the high sensitivity of the camera the whole spectral imaging system thus fulfils the specification for fast weed detection and allows the online detection of the percentage of the total plant population, being measured between the rows or in a tram line. Moreover, continuous spectral data are available which might be used in future sensor systems for crop-weed distinction as proposed by the authors [3]. Two different CMOS camera systems have been evaluated in combination with the optics (ImSpector): a) A commercially available PC-based camera system using a linear CMOS imager (Pixellink); b) A microcontroller-based development using the FUGA-1000 CMOS-sensor with logarithmic response. The measurement setup and a standard illumination (halogen lamp) are mounted to a vehicle for field tests. Plants even smaller than 1 mm have been detected. In first field tests both sensor versions can be applied up to 12 km/h. The sensor systems will be coupled to actors for weed control (sprayer or mechanical weed control). The project is supported by the Bundesministerium für Bildung und Forschung (BMBF), Germany, in cooperation with the ATB Potsdam-Bornim, SYMACON Bildverarbeitung GmbH (Magdeburg) and Müller Elektronik (Salzkotten). [1] Wartenberg, G.; Dammer, K.H. Teilflächenspezifische Unkrautregulierung im Echtzeitbetrieb, Tagung, Landtechnik 2000, VDI-Verlag [2] Sökefeld, M.; Gerhards, R.; Therburg, R.-D.; Nabout, A.; Jacobi, R.; Lock, R.; Kühlbauch,W. : Multispektrale-Bildanalyse zur Erfassung von Unkraut und Blattkrankheiten; Z.PflKrankh.PflSchutz, Sonderheft XVIII, 437-442 (2002), ISSN 0938-9938 [3] Ruckelshausen, A.; Dzinaj, T.; Gelze, F.; Kleine Hörstkamp, S.; Linz, A.; Marquering, J.: Microcontroller-based multi-sensor system for online crop/weed detection ; Proceedings of the International Brighton Conference Weeds, 1999, pages 601-606 [4] Borregaard,T (1997); Application of imaging spectroscopy and multivariate methods in crop-weed discriminations; PhD thesis, Department of Agricultural Sciences, The Royal Veterinary and Agricultural

Detecting fungal pathogens in wheat by use of image analysis

J. Jacobi, R. Gerhards and W. Kühbauch

For the search of new active agents thousands of new chemical compounds have to be tested by screening methods. Trained personal is involved in the process to estimate the fungi infestation on a set of plants. One problem in this visual assess is that the human eye recognizes and uses only the visible light of the spectral band (400 – 700 nm) and the assess cannot reliably be reproduced. The objective of this project was to investigate the options of digital image processing of multispectral images to evaluate fungal infestations. In the study wheat was grown in a greenhouse and infected with spores of Puccinia recondita, Blumeria graminis and Septoria tritici. At different stages of the fungal infection, images by multispectral CCD-camera were taken under controlled light conditions. The digital camera used performed simultaneous images in the blue, green, red and near-infrared spectrum at 1024*1024 pixels resolution and 8 bit color depth. With the multi-channel images various channel combinations including NDVIs have been tried to amplify the spectral differences between healthy and diseased plant tissue. A number of band combinations have been tested: In addition to the well known Red-NDVI, the Green-NDVI (replacing the reflectance values of the red by the reflexion in green wave band) and a Blue-NDVI (replacing the red by the blue spectral bands). Further, differences of reflected energy, NIR minus Red, NIR minus Green, NIR minus Blue, have been investigated to enhance the diseased areas of plant tissue. The image processing has been carried out with the following major steps:. First, transformation of RGB-images to HSI – Hue, Saturation, Intensity. Hereupon the pathogen lesions on the leaves were interactively marked in the image to show the corresponding HSI-ranges. This HIS-range is then copied into a mask. There after this mask has been applied to all images taken from healthy and diseased plant tissue with all spots automatically marked in the images according to the HIS-range of the mask. Because of the different spectral attributes for each pathogen a separate mask were necessary to apply. The marked area can be calculated in percent of the total leave area. At the same time the infestation may be evaluated according to size distribution of the lesions to indicate the pathogens stage of development. Disturbances of image processing by light reflections, e.g. at the stems and leaf edges, so far have not been eliminted completely. Therefore, in future investigations additional parameters should be considered to improve differentiation between damaged and intact leave area. Additional improvement of the results may be achieved by the variance of diseased area over time or form parameters of diseased tissue areas that may characterize the type of pathogen. On the other hand, with spectral reflectances of the crop canopy of winter wheat, diseased and intact crop stands hardly could be distinguished. Only in one wheat variety, extensively infested by Puccinia recondita, the investation has been recognized from the crop stand, and at a very late stage of infestation. It seems that multispectral image analysis has the potential at least to add to the visual assess of pathogen on plant leaves.

Capturing low-cost aerial images to use in site-specific agriculture

T. Jensen, A. Apan, F. Young and L. Zeller

One of the best methods of assessing the variability of a crop is to look at it from above. Cameras mounted on a remotely controlled aircraft are able to provide this information that, when combined with other site-specific data, enables farmers to make management decisions on a zone by zone basis. The desire to capture images with a remote control aircraft has arisen from the lack of useability of traditional methods. Satellite images have a set repeat cycle (every 16 days for Landsat TM), generally have a coarse resolution, may be influenced by cloud cover and cover vast areas. A much finer resolution is obtainable with aerial imagery which has the ability to be programmed and can better handle cloud cover. However, this imagery is costly when a dedicated mobilisation of the aircraft is required and the areas of interest are comparatively small (hundreds of hectares). Images have been captured from the remote control aircraft with two differing camera systems. Firstly, a digital still camera capturing traditional colour (blue-green-red) or infrared images to the PCMCIA card has been successfully trialed. The second camera system uses miniature analogue black and white video cameras to capture the red and infrared spectrum to a ground based digital video camera via a single radio down-link. Ground pixel resolution of 100mm is easily attainable with these cameras. The video footage is viewed and the required images are transferred to computer with a frame grabbing card. The images are then manipulated with image processing software. These images can then be overlaid with other site specific data to enhance the interpretation of yield maps. FUTURE WORK It is planed to use the remotely controlled aircraft with associated cameras to monitor cereal crops on a weekly basis to refine it's use as a agronomic tool. Areas that will be investigated include; · An evaluation of the minimum and achievable requirements to make the data useful. Parameters requiring investigation include positional accuracy, atmospheric/radiometric correction, geometric rectification, mosaicing and other parameters. · Determining the spectral requirements (what wavelengths of light to consider) for the adequate discrimination of crop cover features and their conditions. · Determining the optimal time (temporal resolution) to take images? Images will be taken weekly throughout the growing season to determine this. · Determining what spatial resolution (how small an object can be seen on the ground) is needed? Is it useful to see individual plants, or is one metre (1m) resolution (currently available from light aircraft mounted equipment) adequate? · Comparing and contrasting the low level imagery aquired with the remotely controlled plane with the images from the available remote sensing systems (both high and low resolution satellite and aerial). Is the low cost imagery a good surrogate for the more expensive forms of imagery? · And finally, looking at other potential applications for the use of technology As the research is ongoing, progress with the above investigations will be reported on at the conference. The platform and camera system will also be described.

How can we cokrige when the soil and ancillary data are not collocated?

R. Kerry and M.A. Oliver

Site-specific management of land requires detailed information on the within-site variation of several soil properties. Geostatistical methods can provide reliable estimates at unsampled locations, but they generally require more data than can be afforded by many surveys. If there are existing variograms of the properties they can be used to guide future sampling to ensure spatially dependent data for interpolation. The soil data are likely to be sparse, nevertheless. If the soil properties of interest are related to ancillary data, such as yield, digital information from aerial photographs or electrical conductivity, which are generally more intensive and less expensive, they could be used to improve prediction. The spatial patterns evident in ancillary and soil data suggest that they are related. Variogram analyses have confirmed that the data from aerial photographs and electrical conductivity have similar ranges to each other and to some of the more permanent soil properties. These relations suggest that such ancillary data should be suitable for improving the spatial predictions of the more sparse soil properties. The coregionalization between two properties measured at different intensities, but at collocated sites, can be described by the cross variogram, which can be modelled to provide the parameters for cokriging. The analysis with ancillary data is not straightforward because they are rarely collocated with the soil data; however, the difference between locations is usually small. Since the ancillary data are intensive and there is less unresolved spatial variation (as seen in the smaller nugget variances of their variograms) the aim is to relate these measurements in some way to those of the soil at the sampling locations. We examine several methods of obtaining data from an aerial photograph at the same locations as percentage clay, namely using the closest value, an average of all values within a 15-m radius, punctual kriging, and the pseudo cross variogram. A field on the Yattendon Estate, Berkshire, England was used for this study. The topsoil was sampled on a 30-m grid, and percentage clay was determined. An aerial photograph taken when the soil was bare in 1991 was scanned to create a ground pixel size of 3.4-m and was geo-corrected. Variograms were computed for the three wavebands from the photograph (green, red and blue) and for the clay content. The variogram for the blue waveband was the most similar to that for clay, and the different data sets on the 30-m grid for this waveband were used to compute the cross variograms with clay. The clay data (30-m grid) were sub-sampled to a 60-m grid, and the sampling points removed were retained for validation. The model parameters of the cross variograms were used together with the clay (60-m grid) and ancillary data (30-m grid and the full data for the blue waveband) to cokrige clay content at the validation sites. The clay data (60-m grid) were also used to predict at the validation sites by ordinary kriging using the variogram for the full data. The cokriged and kriged predictions at the validation points were compared with the original values and the mean squared errors (MSE) were computed. Finally, the pseudo cross variogram was computed using the non-collocated data (clay 30-m grid and blue waveband 3.4-m grid). The parameters of the pseudo cross variograms were then used to cokrige the 60-m clay data, and the MSE for the validation points was computed. The results for each method of obtaining collocated data and of interpolation will be discussed in detail, together with the reasons why certain methods out-perform others.

Site specific nitrogen fertilization recommendations based on simulation

K.C. Kersebaum, A. Giebel, K. Lorenz, H.I. Reuter, J. Schwarz and O. Wendroth

Site specific crop management provides a better efficiency of applied nutrients combined with lower emissions of agro-chemicals. The spatial and temporal coincidence of nutrient supply and the demand of the crops is especially important for nitrogen because a surplus can be easily leached into the groundwater. Nitrogen fertilizer recommendations for entire fields are usually based on measurements of soil mineral nitrogen content in early spring supported at later stages by measurements of the crop nitrogen status by optic sensors. Both methods are just snapshots of a present situation which do no enlighten neither the reason for an observed phenomena nor the probable future development. High spatial and temporal variability of soil mineral nitrogen within fields makes it doubtful that frequent and dense soil sampling might be a realistic approach under practical conditions for a site specific fertilizer application. Therefore methods are required to estimate easily the local nitrogen demand considering the spatial variability of soil nitrogen supply and crop yield potentials. Agricultural system models provide a tool to transfer the spatial heterogeneity of time stable soil and terrain attributes which have to be estimated once for a field into a temporal dynamic of the relevant state variables of the soil-crop nitrogen dynamics. We show examples of fields in different landscapes in Germany where intensive investigations and measurements of the spatial distribution of relevant soil characteristics, soil mineral nitrogen and grain yield of winter wheat were performed. On all fields soil data were measured with a nested grid sampling. Nitrogen dynamics and crop growth were simulated with the model HERMES using these spatially distributed data and results are compared to measured distributions of state variables like water content, soil mineral nitrogen and grain yield. It will be shown that grain yield formation is determined by various factors differing in their relevance from site to site and from year to year. The model can be used to derive fertilizer recommendations by subsequent model runs at different development stages combining actualized real weather data from an automatic weather station and a typical weather scenario of the specific site for a predictive calculation of the nitrogen deficiency. An example is shown which demonstrates the effect of different levels of information (e.g. different soil maps) on model performance and their consequences on model based fertilizer recommendations. Depending on the information density differences of 40 kg N/ha on field average occur. On another field the model was used in two different years (2000 and 2002) to give real time fertilizer recommendations during the growing season for 64 different field plots. In this fertilizer experiment the site specific fertilizer recommendation of the model is applied to 32 plots. On the rest of the field 4 different fertilization strategies were applied each to 8 plots: no nitrogen fertilization, the average of the model recommendation, the average altered by 30 % and a site specific strategy based on the average soil mineral nitrogen in early spring and measurements of an optic sensor (Hydro-N-sensor). Fertilization amounts, grain yields and residual soil mineral nitrogen of the different strategies are compared. In 2000 the model recommendation was on average 40 kg N/ha lower than the strategy based on soil and crop measurements without any reduction in grain yield. For 2002 results are not yet fully analyzed but will be available for the publication.

Low-Cost-Sensor for online measurements of soil properties

A. Kielhorn, N. Emeis, B. Ramler and D. Trautz

The Precision Farming concept depends on accurate site specific information in order to manage site variability. The financial, technical and temporal input to generate site specific information should not be prohibitive. At the moment low cost sensors for property measurements are in the centre of discussion, since soil parameters determine yield and crop growth significantly. Therefore during the interdisciplinary research project Intelligent Sensor Systems – ISYS at the University of Applied Sciences in Osnabrueck (Germany) a sensor system for online measurements of soil properties, soil moisture content and texture, was developed and tested by field conditions. To measure material moisture offline numerous methods are known (KUPFER 1997). Many of them are based on the measurement of the relative electrical permittivity of the medium. For online measurements only few appliances are commercially available and a limited number of experiments has been reported. Our online method uses the principle of a plate capacitor. The soil between the plates is the dielectric substrate. The tractor-mounted equipment is made of two parallel plates and an analog measuring device. The real conductivity and the capacity are determined by applying an alternating voltage of about 100 kHz. The latter enables the calculation of the relative permittivity of the substrate. The plates are pulled through the soil. By adjusting the penetration depth the measurement horizon is set. In contrast to commercially available systems not only conductivity or capacity alone are investigated but both of them simultaneously. The electrical capacity gives mainly an information about the soil moisture while the susceptability reflects additional factors of influence such as the soil texture. In laboratory and field test the developed tractor-mounted sensor system was optimised for dynamic measurements. The system is mounted on the three-point-linkage of the tractor and has a high mechanical stability. The electrical part of the system was built into a blackbox with the possibility of data-logging via DGPS-Bordterminal and following evaluation. Under common field condition of crop production, in a range of 15 to 30 percent volumetric soil moisture content, the system yields good results. Correlations between the soil moisture content and the capacity depends on soil types. In the latest development phase positive correlation between capacity and yield data were calculated. With the development of the described low cost sensor system a tool for online measurement of important grid properties is available. It has to be noted that no soil characteristic can be regarded isolated, but only in combination with other soil characteristics. Together with the measured information also offline spatial information like digitised soil maps or other digitally available soil examinations can be combined to get a more complete picture of the analysed location. These could be adjusted and combined during the passage over the field with online measured data, in order to receive improved information about the soil. Prospectively practical use for the introduced system could be for instance the online control of the depth and intensity of soil tillage or the online control of the drilling depth and seed rate. Kupfer, K.: Materialfeuchtemessung. Renningen-Malmsheim: Expert-Verlag, 1997.

GVIS - A hyperspectral imaging spectrometer

Peter Klotz

The growing importance of precision farming requires new data acquisition systems and techniques to obtain detailed information about the vegetation status. In 2001 the Institute for Geography of the University of Munich and the Ground Truth Center Oberbayern (GTCO) developed the 'Groundoperated Visible/Near-Infrared Imaging Spectrometer' (GVIS) with similar specifications as the airborne sensor 'AVIS' (see additional abstract). GVIS is a hyperspectral scanner, that allows the recording of continuous reflection spectra of the vegetation under observation. A custom made fibre-optic system consisting of 16 aligned lenses enables the vertical recording in a range of up to 12 m across driving direction with a spatial resolution of about 80 cm. A special foldable mounting holds the lenses and can be fixed to any type of tractor equipped with a standard 3-point lift system (front or rear) to measure the areas on the left and right side of the tractor-lane The collected light is transmitted to a spectrograph where it is split up into spectral units and afterwards screened on a camera CCD to be recorded. This system provides data in 64 channels with a spectral range from 400 to 900 nm and a spectral resolution of about 8nm. Simultaneous measurements of reference panels allow the calculation of the relative reflectance of the vegetation. The advantage of this procedure is a high independence from atmospheric influences and changes in the illumination, so that the system can also be used on days with inhomogeneous cloud coverage. After internal processing steps the data represents the reflection attitude of the scanned surface. It is generally known, that the reflectance of vegetation depends on a variety of influencing factors like chemical composition, growth status, nutrients supply and others. In contrast to common multi-spectral remote sensing systems like Landsat ETM, SPOT or IKONOS, where spectral information is obtained in a few broadband channels only, the hyperspectral GVIS acquires a continuous high-resolution spectrum. The collected data builds a starting basis for numerous accurate analysis of vegetation properties and conditions. Besides the standard vegetation indices like NDVI, SAVI or others, the hyperspectral approach provides possibilities to develop new interpretation and analysis procedures. A special attention was paid to the typical reflectance increase in the near infrared region, the so called Red Edge. During a AVIS campaign in 1999/2000 the Chlorophyll Absorption Integral (CAI) was developed. The CAI derives the chlorophyll content by measuring a specific area within the Red Edge region. Therefore it is an approach on the basis of a spectral envelope measurement and only possible with hyperspectral data. Since the CAI correlates with the Leaf Area Index (LAI), it is possible to easily obtain detailed LAI-maps for the fields under observation. In 2001 GVIS came into operation on sugar beet fields in western Germany. Measurements and field-tests proved the functionality of the new developed optic system and of the used calibration technique. LAI-maps were created and show accurately the large variability within the fields. A high correlation could be achieved by comparing the GVIS LAI-results from mid August with the sugar beet yield data from late November. The use of this correlation enables spatial high-resolution yield estimations for sugar beet fields. GVIS is a new approach to analyse different vegetation parameters. Depending on the required information suitable bands or indices can be derived from the hyperspectral data to visualise Leaf Area Index, chlorophyll and nitrogen content, NDVI and further indices. Due to the tractor-based measurement-system a flexible, accurate and fast data acquisition for any kind of vegetation can be realised by GVIS.

Standardization in data management to increase interoperability of spatial precision agriculture data

P. Korduan

In precision agriculture a lot of heterogeneous data sets are collected, e.g. administrative data, soil data, crop data, cadastral data, digital terrain models, image data and application maps. In practice there are many problems making use of the tremendous amount of data. Different data formats resulting from the different machines and applications as well as technical problems are to be named here. Hence the costs of data management is significant in precision agriculture. To solve this problems more automated processing routines, data- and workflows have to be established. Therefore we need interoperability between different software and data models. One first important step is the use of metadata to describe the data in spatial, semantical and temporal relation. But also this metadata must be an integrative part in the workflow. The most essential international metadata standard ISO 19115 is under development. This standard based on the Content Standard for Digital Geospatial Metadata (CSDGM) which is valid and in common use since 1998. Some public GIS products eg. ArcGIS, MapInfo, MapServer support the CSDGM and also FGDC metadata profiles that conform to the ISO 19115 metadata standard is under development. The CSDGM contain mechanism to extend it by additional elements. A collection of some aggreed extended elements build a new profile together with the basic minimum set of metadata collected to the specification of this standard. A profile is a document that describes the application of the standard to a specific user community. In this paper the author wishes to introduce such a profile for the precision agriculture information community. The fundament is based on the german research project "preagro" in the last four years. Especially the entity and attribute section was extended. Detailed descriptions about the semantical characteristics of kategories, data typs and sub-typs as well as attribute definition were inserted. The description includes a definition of the element, the reason for creating the extended element, the source, the type (integer, real, text, etc.), the domain, a short name, the parent and the children of the element. Additional to the semantically dependent characteristics, definitions of attributes for classes of geographic features in GIS tables were also implemented. Therewith not also metadata can be standardized but only the data set which was been described. The author will present concepts to develop meta information system by using conventions and standards. The meta information system of the preagro project based on internet technology and a relational database with an internal project dependent data model. The developed web-interface for data retrival and meta data management also allowes the export of the internal project meta data to CSDGM conformal data. The extensible markup laguage (XML) is used to store and transport metadata outsite of the project system. Furthermore CSDGM conformed XML metadata can be used to integrate precision agriculture data sets into datawarehouses and clearinghouses. By consequent using of XML representation of both metadata and data, using Geography Markup Language (GML), best possibilities for interoperability are given. To sum up, standardized metadata and data sets are useful to get a sustainable stock of data and support the development of data-, workflow management and finally help to save time and costs for data management in precision agriculture.

Strength of spatial structure affects mapping accuracy and performance of interpolation methods

A. Kravchenko

Effectiveness of precision agriculture depends on accurate and efficient mapping of soil properties. Number of soil samples, distances between sampling locations, and choice of interpolation procedure are among the factors that affect soil property mapping the most. The objectives of this study are to evaluate effect of the data variability as a whole and of the strength of spatial correlation in data distribution on (i) performance of grid soil sampling with different numbers of samples and (ii) performance of two most commonly used interpolation procedures, namely, ordinary point kriging and optimal inverse distance weighting (IDW). Soil properties with diverse variability, such as soil organic matter content (OM) with coefficient of variation (CV) of 12%, soil test K content (CV of 40%) and soil test P content (CV of 67%), sampled on a 16x16 grid with 30 m distance between grid points were used to simulate intense data sets with spatial structures of different strength. Strength of simulated spatial structures ranged from weak with nugget to sill ratio (N/S) of 0.6, to medium (N/S of 0.3) to strong (N/S of 0.1). From each simulated data set we selected (i) grid sample data sets of different size which were used for mapping soil properties and (ii) a series of 100 test data sets which were used for checking accuracy of the produced maps. The results indicated that, regardless of CV values, soil properties with strong spatial structure were mapped much more accurately than those with weak spatial structure. For the largest grid size (23x23) the average G values for the data with N/S ratio of 0.1 were around 60-70% indicating that the predictions of the test data sets based on kriging from 23x23 grid sampled data were 60-70% more accurate than the accuracy achieved by using a field average value. However, for the data with weaker spatial structure (N/S of 0.3) the average G values were equal to 40-45 %, and only to 12-18% for the data with N/S of 0.6. Kriging with known variogram parameters performed significantly better (P<0.01) than the IDW for most of the studied cases. The only data set size at which accuracy of kriging was not different from that of IDW for all soil properties and spatial structures was the data set with 529 data points, indicating that choice of interpolation procedure is inessential for large intensive data sets. However, kriging was much less accurate than the IDW when a reliable sample variogram could not be obtained due to either insufficient number of data points or too large a distance between the data points. Even when the distance between the grid data points exceeded the spatial correlation range of the studied data, IDW still was a valuable interpolation method, particularly for the soil properties with medium and strong spatial structure. The range of G values obtained from 100 test data sets in each scenario was relatively large. For example, G values for low CV with N/S of 0.3 and 36 grid samples ranged from -3.9% to 23.5 % leaving a lot of room for inconsistency in conclusions. Hence, at least several test data sets are needed for making decisive conclusions regarding either selection of an optimal sampling scheme or performance of an interpolation procedure.

An interactive tool to determine consistent within field patterns based on satellite imagery

F. Lahoche, J. Bergerou, L. Layrol and D. Lepoutre

Management zones offer great promises in terms of improved bottom lines for farmers, and decreased pressure to the environment. This paper outlines how a combination of wavelet and segmentation algorithms can be used to define meaningful management zones solely based on satellite imagery. The input dataset was composed of thirty-eight archived satellite images (Hyperion, SPOT, IRS and Landsat TM / ETM) acquired across 10 growing seasons over Blue Earth, USA. A simple stack of the various wavelet transformed images was then used to obtain a vector-valued image representative of the within-field variations over the last decade. A segmentation scheme using a combined classifier approach was then applied for an an efficient clustering of each pixel of interest into corresponding management zones. To determine which bands contained most of the information, we compared the segmented zones using all of the bands at our disposal with the segmented zones using combinations of these bands for 36 different fields. When either the near or mid infrared bands were removed from the database the resulting zones were considerably altered. A closer study of the hyperspectral Hyperion data (242 bands) confirmed the importance of the near and mid infrared bands in determining relevant management zones. On the opposite, the red, and especially the green, blue, and far infrared bands have all proved to be of relatively little importance to the overall segmentation process. A more carefully developped statistical analysis confirmed these conclusions. In terms of dates of image acquisition, a large amount of information was redundant within our database. As a consequence, consistent management zones were appearing with even a relatively limited number of archived images. However, five different acquisition dates appears to be a limit to avoid yearly-induced fluctuations in management zones. The segmentation algorithm was then applied on several fields over Blue Earth, USA and also over Baziege, France. Whenever possible, the result was presented to farmers for feedback. Farmers in our study area were confident that the segmentation of satellite images produced a meaningful picture of the within-field variability. In large part, the detected zones reflected problems and advantages related to within-field variations in topology and drainage. In addition, the segmentation algorithm proved to maintain its efficiency when used on other areas than the ones on which it was developed (i.e., France instead of US) despite a significant change in cropping patterns and fields size. Extensive ground-truth data was also acquired to check on the quality of our algorithm. Around 3000 acres of yield data were collected every year for 6 years over Blue Earth, and electromagnetic conductivity was measured with a Geonics EM38 for fields over Baziege, France. Zones derived from these different measurements based on an ISODATA classification algorithm were compared to zones derived from satellite images using our wavelet and clustering algorithm. This study confirmed the efficiency of our algorithm in identifying relevant within-field management zones. Based on this research, an interactive tool for management zones delineation is developed based on our segmentation scheme and web technologies, and will be presented for demonstration at the conference. This interface allows the farmer to determine management zones in real-time over any fields contained within the area of interest, the only condition being the existence of sufficient archived satellite imagery. Various factors (e.g. yield, elevation or ground sampling data) can be overlaid to help with interpretation of these zones. This research provided valuable insight as to which spectral bands should be used, and the type of algorithms necessary to extract meaningful management zones from satellite images. Using archived images appears a good way to rapidly gather information across several growing seasons without a large cost of investment for the farmer.

Adoption of precision agriculture in Slovenia and precision viticulture in the Meranovo vineyard managed by the university Maribor agricultural centre

Miran Lakota, Stanko Vrsic, Denis Stajnko and Janez Valdhuber

With 20,250 km2 of surface area and about 2 mio. inhabitants, Slovenia can be considered as one of Europes smaller countries. A rough division into the following four main landscape types can be made: Alpine, Panonian, Mediterranean and transitional. According to the latest available statistical data (SORS, 1996), slightly less than 43% of the national territory, i.e. 780,000 ha, is characterized as agricultural land. Only about a third of this land is used for cultivation. Covering just fewer than three-quarters of the agricultural land in use, grassland and permanent meadows predominate strongly within the structure of agricultural land. The size structure of Slovene farms is practically incomparable to that of EU farms. More than half of them are less than two hectares in size, and almost 70 per cent are less than five hectares. Coupled with farms of between six and ten hectares in size, those farms constitute the foundation of agricultural production in Slovenia, and cover as much as two-thirds of the agricultural area in Slovenia. In addition to an unfavorable structure, the relatively fragmented nature of holdings presents a large problem in Slovenia. It is interesting that this situation is intensifying despite various measures for a greater land concentration. Because of the above written facts, in the past there were only few applications in the field of precision agriculture in Slovenia. On the other hand a good basis for adopting of precision agriculture was already founded in the last ten years, when the Basic topographical maps, Topographic maps, Ariel survey, Ortho-photo and Digital terrain model maps were issued by The Surveying and Mapping Authority of the Ministry of the Environment, Spatial Planning and Energy. On the basis of available maps the first basically GIS data were collected at the beginning of 2001 in the field of precision viticulture. The variety of climatic and geological conditions in our wine-growing regions contributes to a diverse assortment of wines. In eastern Slovenia, with Panonian (continental) climatic conditions, there are about 9.000 ha of vineyards areas with relatively good yield. Data were collected in the Meranovo vineyard which is one of the most famous sites in Styria over Limbus at Maribor where Archduke Johann of the former ruling house of Austria set the founding stone of the modern Styrian wine-growing as early as in 1822. During 2002 we researched a time of grapevine ripening of different varieties Chardonnay, Riesling, Welschriesling and Sauvignon. Data were collected from nine different sites of vineyards and heights above sea level (top of the vineyards, bottom of the vineyards) to establish an optimal time of the harvest. The main monitoring parameters included content of sugars (oOe), total titratable acidity (g/l), wine acids (g/l) and pH. All samples were acquired manually and the locations of monitored parameters were stored with GPS receiver MARCH II with post processed data. Data of the sugar concentration of grapes were the basis for obtaining sugar maps. The obtained experiences, knowledge and GIS database including collected sugar concentration data, the type of vineyard, altitude, diseases maps and soil variability will form the basis for forecasting an optimal time of the harvest.

Experimental and analytical methods for studying within-field variation of crop responses to inputs

R.M. Lark and H.C. Wheeler

Precision agriculture hypothesizes that the optimum rate of inputs to a crop varies spatially within-field. Direct experimental evidence for this hypothesis is scarce, and is generally limited to comparisons between subregions of the field based on some prior classification, due to the practical and theoretical difficulties of designing appropriate experiments. A procedure is proposed for testing the hypothesis of precision agriculture for combinable crops. An input is applied according to a treatment map generated by a randomized block design. The crop is harvested using a yield monitor, the data are assumed to be a convolution of yield along the combine's path with an impulse response function which characterizes the smoothing effect of grain flow, threshing and separating on the rate of mass-flow. This smoothing is accounted for by exploiting mathematical properties of the convolution operation, namely that a linear relationship is invariant under the convolution. A spatial transformation of the input values determined by the experimental design, based on a preselected response function is applied. The resulting values can be used to estimate the parameters of the response function for the whole field or within a local window from the corresponding yield monitor data. The estimation is done by maximum likelihood to provide estimates of the parameters of the response function and their covariance matrix. The latter is necessary in order to obtain unbiased estimates of the local optimum rate of the input. A bootstrapping test is proposed for the null hypothesis that the spatial variation observed in parameters of the response function arises from random variation about a uniform underlying response function. This method was applied to cereal crops in three fields over two successive seasons. Using this analysis it was possible to show that in five of the six crops there was significant spatial variation in the nitrogen response. The local economically optimal nitrogen rates varied substantially from zero to over 250 kg/ha. In all fields the economic benefits of spatially variable application could be assessed. These varied between seasons and between fields, but in all cases most economic gains came from reducing nitrogen rates in less responsive parts of the field rather than from increasing rates in the more responsive parts. The analytical method was extended to compare the nitrogen response functions in subregions of one field defined from soil maps and from past yield data. While these alternative classifications accounted for significant variation in the response function they did not capture all of the important variations in the field. As well as estimating within-field response functions for single seasons, joint models were fitted to the two seasons results for each field. This revealed substantial seasonal variation in the response, indicating that season specific information on crop development will be necessary for effective within-field management of nitrogen. Radiometric measurements of the canopy of these experimental crops were also analyzed. This suggested that early-season measurements of the vegetation index may allow a local nitrogen rate to be prescribed so that nitrogen does not cause any local limitation on crop yield, but with rates varying by over 100 kg/ha within the field. Some of the implications of these findings for on-farm experimentation and decision support will be discussed.

An N-tier system architecture for distributed agronomic registration applications based on open source components

G. Lemoine and P. van der Voet

Modern agricultural management requires the frequent registration of a diverse set of location specific observations, ranging in complexity for simple crop status reports to detailed soil and plant sampling or disease diagnosis. Apart from their value as instant observables in a particular context (e.g. management control, fertilizer recommendation, plant protection), have these observables a considerable added value in expert models that operate at regional level (e.g. disease prediction, yield forecasts, food safety assessment). This is particularly true if they can be combined with regionalised variables, such as soil parameters, meteorological records and remote sensing imagery. Currently, though, there are few efforts aimed at facilitating the systematic collection of local and regional observables in such a manner that they can be easily made available to support expert applications whose output quality rely on extensive observation data. The PAC sector of the Joint Research Centre's MARS unit focuses its R&D on registration systems that have a particular relevance in the control of EU applications for crop income aid under the Common European Policy (CAP). Elaborate control systems are currently in place in most of the EU Member States. These systems, which are often based on a complete graphical identification system for agricultural parcels, offer significant potential for dual use in the agricultural users and research community as a geographical reference for many agronomic registration applications that have a spatio-temporal character. Such registration applications are not necessarily restricted to the provision of official reference data for institutional support (e.g. EU area income support) or legally required commitments (e.g. manure management regulation). They should increasingly be applicable in agronomic management in which regionalized information plays an important role, such as food quality management, disease control, environmental monitoring and logistics. Modern n-tier client/server computer architecture has become the state-of-the-art solution for facilitating distributed applications that rely on collaborative models of information exchange. To guarantee wide-spread use of such systems in practical applications, the client-side, i.e. the interface through which most end-users access the system, has to remain light-weight. At the same time, the components on the server must remain sufficiently generic, so that new application contexts that address newly conceived ideas can be rapidly assembled. The J2EE (Java 2 Enterprise Edition) platform is ideally suited to support modular n-tier system with a full range of functional components. Moreover, J2EE components can be generated as Open Source components, including back-end components, such as a MySQL database, or middleware, such as a Tomcat web-server or a JBOSS application server. Use of Open Source J2EE components allows for low cost rapid prototyping and fairly robust medium scale commercial application solutions, while leaving the possibility to migrate to robust industrial strength server solution within the programmer's realm. Open Source J2EE component use in combination with open data standards guarantees an open software solution that is easily maintainable and extendable. In the paper, we demonstrate an n-tier system built up with open source components (Java, Tomcat, MySQL) that allow sophisticated registration of agronomic parameters that drive an expert model for forecasting of infection risk of Phytophtera in potatoes. The system combines the use of satellite image based mapping with simple mouse and form-based information entry and localization for relevant data entry and retrieval. The system client side, which works on a standard Internet browser (JavaScript) was tested by a panel of farmers in 2001 and very well received. As a demonstration of its modularity, the system is planned to support mobile GPS-supported devices at the end of 2002. Future issues like accounting models and collaborative networks of servers will also be addressed.

Remote sensing for pastoral agriculture

H. Lilienthal, K. Betteridge, A. Gillingham, D. Costall and E. Schnug

The concept of Precision Agriculture (PA) is widely acknowledged in agricultural cropping systems. Pastoral farming, the predominant farming system in New Zealand, has yet to adopt the advanced features of PA, although farmers subconsciously use such concepts in day-to-day planning. Long term data from a New Zealand dairy trial site comprising 64 paddocks showed annual pasture production to be from 50% to 150% greater in the best compared to the poorest paddock. In another study within a 2 ha dairy pasture, three distinct zones of pasture growth were determined, from which different N responses were measured (Gillingham and Betteridge, 2001). Thus, it is clear that sufficient spatial variability exists within pastoral farms to test PA applications to optimise production, economic and environmental outcomes. In order to evaluate variability a digital elevation model (DEM) provides good basic information. Apart from general topography and geomorphology, relative elevation differences, the distribution of hills and depressions, slope and aspect are responsible for microclimatic variation, nutrient fluxes and water budget (Schroeder 1998). Using an elevation model general sources of variability can be accounted for. Remote sensing technologies allow the mapping of variation in pasture biomass production. Simple vegetation indices (e.g. Normalised Difference Vegetation Index, NDVI) explain only qualitative differences in biomass. Using more advanced algorithms like linear spectral unmixing (Adams et al. 1989) the quantity and quality of pasture components (e.g. proportions of fresh green pasture, dried-green pasture, dead pasture and bare soil) can be determined. The freely available New Zealand 20 m contour data are inaccurate at the within-farm extent and, when converted to a DEM, often wrongly classify slopes and aspects (compared to ground survey) because river and ridge breaklines are not known (Costall et al 2000, 2001). Stereography, using commercial aerial photos, can result in high resolution DEMs but at a high cost. Lake Taupo is a major tourist attraction located in the central North Island of New Zealand (E:175°50' S:38°45'). It is famous for its exceptional water quality but, during the period 1974-1999, water clarity has fallen from 12 to 10 m (Environment Waikato 2001) due in part to nitrate eutrophication from the 100 livestock farms in the Taupo catchment. This study compares methods of gathering baseline resource data (DEM, pasture mass and quality) to assist in the overall programme aimed at a 20% reduction in nitrate leaching from farms, through changed landuse. In this pilot-project, methods by which data describing resource variability can be gathered by remote sensing for creation of resource maps for use in site specific interpretation and management are evaluated. Specifically, two aspects are considered: 1. Use of satellite imagery to generate DEM (contour maps) at both the regional and within-farm extent, compared with DEMs generated from aerial photos, and 20 m contours created 40 years ago by less sophisticated means. 2. Evaluation of pasture mass and quality by using multi-spectral remote sensing data. Satellite imagery offers a cost effective way to retrieve information at a regional extent. In 1999 the NASA launched the earth observation satellite TERRA. Among other instruments on board, the Advanced Space-borne Thermal Emission and Reflection Radiometer (ASTER) records spectral measurements in 14 different wavebands. The geometric resolution is 15 m in the visible, 30 m in short-wave infrared and 90 m in the thermal infrared. Data from the ASTER sensor are used to generate a within-farm (2 500 ha Case Study farm) and a sub catchment-extent photogrametric elevation model from stereo-paired images (one looking in nadir, one in an off-nadir direction). The multi-spectral information was used to perform 'spectral unmixing' to derive quantitative information on pasture quality and biomass production. Reference spectra and pasture mass and quality are calibrated against ground truth samples.

Site-specific N fertilization based on remote sensing - is it necessary to take the yield potential into account?

A. Link and J. Jasper

Site-specific N fertilization is one of the main objectives in precision agriculture. This is due to the importance of nitrogen in plant nutrition, the costs of N fertilization and its environmental impacts. N fertilizer demand of a crop is often assessed by considering yield goals, soil nitrate content in spring and credits for increased N supply, e. g. due to previous grown legumes or manure application. Since these methods are not able to consider the year-to-year variability in N mineralization during the growth period plant analysis are used for detecting the actual N status of the crop in order to adjust the N fertilization. Traditional methods of plant analysis are not adapted for use in precision farming since getting data of high spatial resolution is very time-consuming and costly. In recent years it has been shown that using remote sensing data is a suitable way to close this gap. There is evidence that measurements of crop canopy reflectance, using spectral indices of particular wavelengths, can give reliable information about biomass and chlorophyll production and therefore the actual N uptake of a crop. This can be used for the calculation of the N fertilizer demand. Nevertheless, since the total N demand of the crop depends on its yield, it is widely assumed that N recommendations based on spectral reflection could be improved by taking the yield potential into account. This hypothesis was tested by using the data of 31 N response trials with cereals conducted in 1999 to 2001. For each trial the yield function was calculated and measurements with the Hydro N-Sensor were done at growth stage 31. Yield potential and optimum N rate were derived from the yield function and it was tested whether the optimum N rate is correlated with yield potential and N-Sensor values. A simulation was run where the yield functions of the 31 N response trials represent 31 equal sized areas within a heterogeneous field. This simulated field with known variability based on real data was used to investigate the suitability of three linear models to predict optimum N fertilization. The models are based on yield potential, N-Sensor value and a combination of yield potential and N-Sensor value respectively. The results of the N response trials show that the optimum N rate is not related to the yield potential but that there is a strong relation between N-Sensor values and optimum N rate. The higher the sensor values are, i. e. the higher the actual N uptake of the crop stand is, the lower the required N fertilizer rate to reach optimum yield. The results of the simulation confirm that there is no significant relationship between yield potential and optimum N rate. Optimising the N fertilization for the simulated field based on N-Sensor values indicates a highly significant relation between the spectral index and optimum N rate. This relation can be slightly improved by using the yield potential as additional information. It is concluded that strategies for site-specific N fertilization that are solely based on yield potential can not succeed. Measurements of the N uptake of the crop stand, e. g. with the Hydro N-Sensor, are applicable to the calculation of optimum N rates for site-specific fertilization. Using the yield potential as a supplementary parameter in addition to the actual N status may increase the quality of the prediction to a small extend but seems not to be important for practical use.

The influence of different spatial soil information levels on the performance of dynamic crop yield simulation

K. Lorenz and K.C. Kersebaum

Soil properties are an important factor determining crop yield formation. The heterogeneity of soils within fields are often the dominating origin of spatial yield variation. Therefore the availablity of spatial soil information is essential for yield estimation and to derive site specific management measures like fertilization especially when process oriented simulation models are applied. Usually soil data in Germany are available in the form of soil maps for all agricultural used areas coming from the agricultural land evaluation (Reichsbodenschätzung) started in the 30ies. Soil map units are attributed with generalized soil profile information which should represent the average soil properties within the unit. The question arises if such information is sufficient to explain the variation of crop yield and soil nutrient supply within the field. Therefore we compared model results of grain yield and residual mineral nitrogen content after harvest obtained from simulations with different soil information densities to measured spatial distributions of these state variables for different fields in Germany.

Beside the available standard soil map a more detailed soil map of the farms has been used for comparison. As a reference the soil properties required by the model HERMES were obtained by an intensive grid sampling. Depending on the underlying basic soil maps samples were taken in adopted standard grids with integrated nests to estimate small scale variability. So the total number of sampling points vary from field to field from 60 to 94. Model results based on these three levels are compared to measurements of yield, soil moisture and mineral nitrogen after harvest at the standard grid points.

Results of the grid sampling demonstrate that soil properties within soil map units can vary severely. We show examples that the agreement between simulated and measured spatial patterns increase if more detailed soil information are used. In some cases it is useful to separate different relevant soil properties from each other in separate coverages instead of linking them to a soil map unit. Differences also occur in the average yield estimations. The results show that the calculated residual soil mineral nitrogen after harvest is strongly related to the estimated yield. Therefore an underestimation of the average yields leads to a corresponding overestimation of residual mineral nitrogen contents. Deviations between simulations and observed values are higher when soil map information was used. Simulations using the grid data agreed well on average with yield observations as well as residual nitrogen contents in the root zone after harvest.

The differences in crop yield estimations has consequences when used to derive the crop demand for nitrogen fertilization. In our example differences in the average fertilizer recommendation of about ±20 kg N/ha are calculated. For single locations differences of 80 kg N/ha occurred.

Appropriate on-farm trial designs for precision farming

J. Lowenberg-DeBoer, Dayton Lambert and Rodolfo Bongiovanni

In the early 20th Century Fischer and others developed statistical methods that could reliably answer agronomic questions given the technology available at the time, including hand harvest of small plots and manual calculation of statistics. The random assignment of treatments to plots and blocking were methods of dealing with the spatial variability of soils and other factors. Later these designs were adapted to mechanical harvest, computerized analysis and larger plots. The cost per observation with this method was high. Relatively few farmers participated in on-farm trials because layout of the designs and collecting data interfered with cropping operations. Adaptation of the analysis to data with spatial structure has focused on eliminating spatial correlation by adjusting the data (e.g. nearest neighbor). Precision farming technology has dramatically changed this situation. Yield monitors and other sensors can provide many observations per hectare at a low cost. This change in technology has prompted a renewed farmer interest in on-farm trials, but most trials still use methods developed for early 20th century conditions. This paper summarizes an alternative approach and tests that approach using simulated trials based on soils and crop response data from Argentina. This paper will be of interest to anyone involved in designing on-farm trials and analyzing the resulting data, including agronomists, agricultural engineers, agricultural economists, and farm input suppliers. This alternative model starts from the spatial econometrics analyses (Anselin, 1988). This paper focuses on the spatial error model because it has proven most appropriate for a range of on-farm trial data. Formally, a spatial error model can be expressed as: $y = Xb + u$ with $u = dWu + e$ (1) where y is a vector (n by 1) of observations on the dependent variable, X an n by k matrix of observations on the explanatory variables, and u an error term that follows a spatial autoregressive (SAR) specification with autoregressive coefficient d. In the spatial autoregression, the vector of errors is expressed as a sum of a vector of innovation terms (e) and a so-called spatially lagged error, Wu. The latter is essentially a weighted average of errors in the neighboring locations. The selection of neighbors is carried out through the n by n spatial weights matrix W. The key point for on-farm trial design and analysis is that the model is mathematically identical for data from numerous strips and/or subplots within strips, and for data from a smaller number of blocks. For example, a classic strip design for a nitrogen rate trial might have three blocks with field lengths strips of five different rates of nitrogen. The blocking and the randomization of the rate strips within the blocks were used to deal with spatial variability. With the spatial error model, the same information could potentially be extracted from a simpler design with the five rates planted in five larger blocks. The spatial error model can potentially do this because it explicitly models the spatial error structure, instead of trying to eliminate the spatial correlation. This paper will test the potential of the large block design by a Monte Carlo simulation of nitrogen rate trials under alternative designs. The simulation is used to provide a preliminary assessment of the potential for this kind of design before it is field tested. The stochastic Monte Carlo simulation provides information on what results would be if a certain design were used over many weather years. The simulation will use the maize nitrogen response functions and soils data reported by Bongiovanni and Lowenberg-DeBoer (2000, 2001, 2002).

Approaches to changing land management options based on economic and environmental criteria

G. Lyle and M.T.F Wong

Issues facing the wheatbelt of WA are spatially variable financial field performance and a growing encroachment of salt in the landscape. The onset of salinity is due to change from native vegetation to annual cropping that uses less water than the initial vegetation. This hydrologic imbalance has resulted in a rising saline water table. A solution to salt encroachment is re-assignment of land use to perennial vegetation that would use more water and lower the ground water table. We developed an approach for identifying areas for alternative land use based on both spatially variable financial and environmental criteria. This study involved a typical two thousand hectare intensive cropping wheatbelt farm that was yield-mapped from 1998 to 2001 with an Agleader yield monitor. The on-farm crop rotation included wheat, lupins, canola and barley. The yield monitor data was pre-processed using a tool that removes yield spikes, checks the mapped yield against the paddock average physically weighed yield. The processed files were then mapped in ESRI ArcView. Yield maps and financial analysis showed that beside spatial variability, the annual gross revenue fluctuated due to seasonally variable yields, grains prices, input costs and the rotation phase. The farmer's margins were calculated from variable input costs derived from the farmer's records. Fixed infrastructure costs were estimated from regional economic data. Spatially variable income was calculated from grain maps and their corresponding grain prices. The analysis showed that within certain years all areas of the farm operated on positive margins due to favourable grain prices and this varied spatially across the farm. Water balance modelling based on soil type and spatially variable soil moisture retention properties showed that deep drainage varied spatially according to the water holding characteristics of the soil profile and the rooting depth. In the Mediterranean type environment of WA, wheat yield is limited by water availability and hence soil water-holding characteristics. This dependence of both drainage and wheat yield on water holding properties result in a convenient correlation between yield and drainage: low yielding areas tending to be on coarse textured, poor water holding soils and these areas corresponding to the areas of highest drainage risks. This correlation enables a performance-based decision on land use to be made. The performance-based land use decision took out of grains production a fraction of the recharge areas. A Digital Terrain Model and satellite imagery allowed us to identify saline discharge areas. A further criterion imposed on the land use decision included the proximity to these degraded saline areas. The area of land taken out of production was based on trade-off scenarios carried out in the raster-based IDRISI program which modelled loss of earnings offset against environmental gains derived from reduced drainage. The Dempster-Shafer model in IDRISI was used to provide an evidence-based approach for evaluating financial and environmental performance based on the opinion of the growers, scientists and other community stakeholders. This approach facilitated adoption of a relatively unpalatable decision of removing land out of current production by minimising adverse financial impact on growers while attempting to maximise the environmental impact.

Site-specific variable rate precision drilling of onions

Seamus Maguire, Richard Earl, David Smith, Paul Cripsey and Richard J. Godwin

The results from recent experimental fieldwork by a Bedfordshire onion grower in the UK suggest that by tailoring the onion seed rate to the water holding capacity of the soil, a more profitable crop results at harvest. This is because a significant increase in the amount of onions of a diameter suitable for sale is produced. However in order to implement this on a commercial scale, there is a need to automate the seed drilling system. The aim of this project was to investigate the feasibility and quantify the benefits of utilising an automated system capable of varying seed rates of a precision drill according to pre-drawn seed application plans. To achieve this, the methods used by precision farming for the variable application of inputs have been adopted. This involved the adaptation of the AGCO Fieldstar control system used on White Planters for use with a Stanhay Singulaire 780 single bed precision vacuum drill. Alterations to the machinery and control software were made to fulfil the system requirements. During spring 2000 the performance of the variable rate precision drill was tested in the laboratory and in the field. The laboratory tests consisted of a greased strip seed spacing test, and a seed disc speed accuracy test. In the field, a trial area was drilled in the centre of a commercial onion crop using a pre-determined seed application map. In addition, assessments of speed of response to required changes in seed population were conducted in a field at the Silsoe campus of Cranfield University. During all tests three forward speeds, 3.2, 4.0 and 4.8 km/h were used. The performance of the developed system was found to be satisfactory with a mean error in seed spacing between actual and required of 2.57% observed in the greased strip test, and 3.15% in the field test. The system was found to be capable of compensating for changes in planting speed such that seed population was unaffected, however, as speed was increased there was also increased variation in seed spacing. This variation in seed spacing observed needs to be reduced if the benefits of variable rate drilling are to be fully realised. At a forward speed of 4.0 km/h, the speed of response of the system to a change in desired seed population was found to be 2.26 seconds. When preparing seed application maps a minimum grid size of 10 m is recommended. The grower has adapted the White Planter control system for use on their Stanhay Dart precision drill. They have continued the work by planting the majority of their onion crop by the variable rate method. Harvest results have indicated an approximate 10% increase in the saleable yield of crop. They emphasis that in order to achieve the benefits of variable rate drilling the basics of soil preparation and machine settings need to be correct prior to drilling. The project has shown how precision farming methods used for the variable application of seed in cereal crops have been extended to include commercial horticultural crops.

Hay and forage yield measurement and mapping

Seamus Maguire, Richard J. Godwin and David Smith

Hay and forage crops are essential for the support of ruminant animals during winter conditions. High-density large square balers are used within one system of harvesting hay and forage crops. The objective of this project is to develop sensors and data processing algorithms required to record the yield of crops at the time of baling, both on a whole field and on a site-specific basis. The Massey Ferguson 187 baler includes a pre-compression chamber to form flakes of crop; these are then pushed into the main chamber and compressed into the final bale. If the bale weight and number of flakes per bale is known, then this can be correlated with the GPS position of were each flake was formed in the field to produce a yield map. The availability of straw in the Bedfordshire area of the UK provided opportunities for testing yield-mapping systems on the baler. The work to date has included weighing studies of bales produced from barley straw and the design and installation of a bale weighing system that uses the bale chute at the rear of the baler. The weights of barley straw bales from four fields provided proof that individual bale weights are required for a yield mapping system (standard deviations from the mean bale weight in four fields of 9, 21, 24 and 15 kg). For the weighing system, two cantilever beams measure the vertical force at the bale chute pivot points and two tension dynamometers measure tension force in the bale chute support chains. Preliminary field-testing of the weighing system with 12 wheat straw bales during the harvest of 2001 produced a mean relative error between indicated bale mass and actual bale mass of 0.1%, and a maximum error of 10.7%. This was after smoothing the signals with a moving average algorithm. Errors were attributed to machine noise and field undulations causing spurious oscillations of sensor signals and a data-logging rate that was to slow. A redesigned logging system provided somewhat better results from testing in barley straw during the harvest of 2002. Using data from 69 bales, mean relative error was 0.42% and the maximum error in predicting bale weight was 5.8%. An encoder (1000 pulses/revolution) was mounted onto the star wheel of the baler for testing in 2002. Using the encoder data for the barley straw bales the average pulse count per second was calculated and this was well correlated with Wet Mass Flow Rate ($R^2 = 0.99$). The method described for weighing bales on the bale chute appears to provide a good indication of bale weight; further data analysis methods may increase accuracy. Mass flow rate can be accurately determined from an encoder connected to the star wheel. However this method requires calibration with a number of bales for the crop under consideration. The next step is to correlate the bale weight data with GPS information. When the in-field position of each flake is known it will be possible to create a yield map for the baled field.

Ergonomic evaluation of lightbars for agricultural guidance systems

D.D. Mann, S.J. Young and C.S. Ima

Precision farming refers to the philosophy of efficient farm management. Inefficiencies occur in all farm operations; therefore, the precision of farming can be improved in many ways. The focus of my research has been the ergonomic improvement of guidance aids. A guidance aid is a device that provides guidance information to the driver rather than replacing the driver (i.e., it is not an automated guidance system). Consequently, it is critical that the ergonomics of such systems be considered so that the transmission of information from the system to the human driver is as efficient as possible. In North America, guidance aids using global positioning system (GPS) signals are becoming more common for operation of high clearance sprayers. Many of these GPS systems use a horizontal array of light emitting diodes (LEDs) to display the guidance error for the driver. In our research, we have determined that the GPS system is usually used in conjunction with external field cues, with the field cues being used most often, and therefore, of primary importance. Consequently, the lightbar is placed so that it does not interfere with the view of the field (i.e., it is visible only in the periphery of the driver's vision). Based on its small dimensions, it can be hypothesized that a typical lightbar cannot be readily detected by the driver in the periphery of his visual field due to the fact that visual acuity decreases linearly as the angle from the fovea increases. Rather, the driver's eyes must be focused on the lightbar, and consequently removed from the external field cues, when guidance information is to be obtained from the GPS system. It would be better if the driver could detect signals from the lightbar without moving his attention from the external field cues. To enable comparison of various lightbar designs in a controlled laboratory setting, a driving simulator was fabricated using a salvaged tractor cab. An experimental protocol has been developed that enables the driver's mental workload to be measured. Mental workload refers to the mental processing that is required to complete a given task. Previous research has shown that steering performance decreased when operators were required to share their visual attention between both the steering and rear-monitoring tasks. This result can be explained by the fact that two competing visual channels cannot be watched simultaneously. The driver overcomes this problem by scanning one or both of the visual channels; thereby increasing the mental workload. Mental workload can be measured using performance, physiological, and subjective measures. In this research, volunteer drivers complete driving sessions in which they must steer the simulator in response to a pre-programmed path that is displayed by the lightbar. Simultaneously, the drivers must monitor and respond to field cues located both ahead of and to the rear of the simulator. Steering error and reaction time are recorded. In addition to these performance measures, heart rate variability is recorded throughout the driving session. Finally, at the conclusion of each driving session, the drivers are required to complete a subjective questionnaire. Using the simulator and the experimental protocol, it has been determined that continuously-flashing, peripherally-located red lights annoy drivers. Large clusters of LEDs reduce lateral error compared with small LEDs. Blue and yellow LEDs tend to be more noticeable than red and green LEDs. Data is currently being analyzed to determine the most effective lightbar configuration. The data seems to suggest that there is room for improvement in the design of lightbars. Plans are currently underway to verify these findings in a field setting.

Soil properties and the productivity maps obtained by precision agriculture techniques

José Rafael Marques da Silva and Carlos Alexandre

The development of the Graphical Engineer in general and the Geographical Information System, Remote Sensing and the Global Positioning System in particular has lead, that in more recent times, some farmers stopped to treat agricultural parcel as uniform units and started to treat them as heterogeneous units, this type of strategy was called Precision Agriculture. This article shows, through a practical case that the graphical information it's extremely important to analyze the spatial variability of a crop productivity, in this particular case the Mays under center pivot irrigation. The main objective of this work was to relate soil parameters and soil productivity and see, if it's possible to elaborate a soil quality map from this type of relation. Soil parameters were obtained after collecting 200 soil monoliths. It was used a Differential Global Positioning System to geo-reference the instantaneous productivity, measured by a productivity sensor installed in a Mays combine, and to geo-reference 200 soil monoliths collected in the studied area. A Geographical Information System was used has a data storage system, a decision support system and a system responsible to elaborate all the thematic cartography. Geostatistics it was also used to analyze spatial variability of the main studied variables. This type of analysis should allows us to define the better strategies in how to allocate the different production factors, but the preliminary collected data showed us that this particular agricultural production system is a very complex one. Areas of low productivity can occur in places rather close to each other and the controlling factors for that low productivity can be very different from one area to another. We also could observe that there exists a few productivity controlling factors, but this controlling factors can only explain 20 to 50 % of the productivity variation. The main controlling factors were related with the topography, for instance, the slope, the upslope contributing area and the plan curvature. The topography of the studied area is undulated and therefore exists a great difference in soil depth and soil characteristics between the convexities and the concavities of the landscape. Soil depth is another important controlling factor, specially when it is analyzed in conjunction with the slope gradient, because shallow soils situated in regions with high slope gradients usually define low productivity areas. Soil depth in a concave planar curvature area can have more than one meter depth and in a convex planar curvature usually has less than thirty centimeters. Soil texture didn't show a great impact on soil productivity. Precision Agriculture have to answer to a certain practical questions if we want that this technique start to be used has a tool by the environmental agent, or should I say the farmer. In this particular study, more questions were raised up and not all the answers were satisfactory to interpret the spatial variability of the Mays productivity, nevertheless, a precious information was collected for future developments and we believe that the benefits of having this information are enormous for the farmer and for the environment has a whole in a close future. Key words: Soil Conservation, Soil Quality, Geographical Information System, Global Positioning System, Precision Farming.

Evaluation of precision farming techniques via multicriteria approaches

Fabrizio Mazzetto, Massimo Lazzari and Sacco Pasqualina

The assessment of Precision Farming Techniques (PFTs) - both from farm and institutional standpoints - is difficult to be obtained when is carried out by using conventional cost-benefit methods. This is because: a) the PFTs adoption imposes deep changes to the on-farm procedures for managing operations, which entails extensive re-training of all the workers involved; b) the introduction of Information Technologies generally takes place in several stages; c) the expected economic benefits can rarely be estimated in a unequivocal manner; d) it is difficult to evaluate the evolution of costs and benefits over time; e) the environmental effects of PFTs are difficult to be evaluated in economic terms; f) the adoption of these techniques is characterised by social impacts that affect the speed of the technologies introduction. Starting from the conventional cost-benefit analysis, a model taking into account the effects of the phenomena listed in the previous points 'a' to 'd' has been implemented. In addition, the model's results can be organized into a database to be used as input datasource for a further multi-criteria evaluation approach (MEA) based upon the computational procedures of the Concordance analysis. A MEA procedure enables decision makers in investigating a number of different alternatives (projects) in the light of multiple criteria and conflicting priorities, even including his subjective standpoints. Thus the overall software procedure even permits to consider the problems listed to the above 'e' and 'f' points. The methodology has been applied to study cases of italian farms. In particular, the introduction of PFTs into rice and wine farms have been taken into consideration. An example related to a Northern Italy 500 ha rice-farm will be presented and discussed in the paper. Five solutions referring to an increasing level of complexity in the application of PFTs will be here considered: 1. ZERO: current situation at the farm; 2. MIS: adoption of Management Information Systems limited to software procedures (use of simplified GIS packages for managing field data, use of advanced field databases; etc.); 3. advMIS: as previous point 2), but including the adoption of hardware devices for the automatic recording of operational data related to every field event (use of information field log-book); 4. PF: as previous point 3), but including a typical configuration of precision farming technologies (yield mapping and VRT devices); 5. advPF: as previous point 4), but even including the management of information supplied by remote sensing applications. The above solutions will be then evaluated on the basis of a set of 13 criteria. They were selected in order to take into consideration: · ECO: economic related factors (such as investment levels, expected benefits, payback periods, etc.); · ENV: environmental related factors (such as saving of chemicals, both pests and fertilisers); · WORK: work-quality related factors (such as being a status symbol, training work required per year, saving of scouting tasks, etc.). The interviewed farmer assigned to this study case the following priorities: ECO 60%; ENV 20%; WORK 20%. Under this decisional profile, all the PFTs solutions are refused. Anyhow, a sensitivity analysis on priority weights shows that only when the ENV aspects assume greater importance (40%) the alternative PF can be preferred to the conventional cultivation techniques. A relevant interest for advMIS and advPF solutions can be also achieved when WORK aspects tend to prevail (60%) together with a very low interest for ECO aspects (10%). By using a MEA approach a more general and realistic assessment of PFT can be obtained. This is particularly true in the European situation where the environmental aspects are continuously increasing their importance into the agricultural sector.

A potential role for soil electrical conductivity mapping in the site-specific management of grassland

S. McCormick, J.S. Bailey and C. Jordan

Introduction Soil electrical conductivity (EC) mapping is a simple inexpensive tool that farmers might use to rapidly characterise within-field differences in soil characteristics. Soil EC values can provide surrogate assessments of other soil properties, including extractable Mg, extractable Ca, soil moisture content and topsoil depth (Hartsock et al., 2000). Recently, it was found that variability in sward production in a grass silage field was linked to differences in net soil N mineralization, which in turn appear to have been linked with differences in topsoil depth (Bailey et al., 2001). The aim of the present study was to see if soil EC measurements could be used to predict differences in the main soil properties controlling within-field variability in grass production and thus have a useful role in the site-specific management of grassland. Materials and Methods A 50ha grassland site under 3-cut silage management, in Co. Down, was selected for the study. The site comprised 17 fields and was undulating in topography with a maximum height difference of 40m. Soil EC was mapped across the site in November 2001. A non-contact EC sensor, using electromagnetic induction (EMI) to measure EC to a depth of 0.5m, was drawn (using a vehicle equipped with a GPS) on a trolley along 25m-spaced sampling routes. EMI data were collected at 2m intervals and incorporated into Geographic Information System software. In February 2002, soil samples were collected across the site to a depth of 75mm using a 25m grid-sampling regime (333 points), and were analysed for pH, extractable P, K, Mg and sulphate-S. Soil depth measurements were also made at selected locations. Prior to 1st cut in May, grass strips (2m x 6m) were cut at 80 of the grid locations and sward dry matter (DM) and N yields determined; soil moisture measurements being made simultaneously using a Theta Probe. Ordinary kriging was used to estimate values for soil and crop variables at unsampled locations and maps were produced showing the spatial distributions of these variables. Linear regression analysis was used to investigate relationships between soil EC measurements and the different soil and crop variables. Results and discussion Soil EC measurements varied considerably across the site (from 0 to 23 mS m-1) and were positively correlated with soil water content (r2 = 0.36), which is a good indicator of soil texture and drainage characteristics. On half of the site furthermost away from the farmyard, the EC measurements were negatively correlated with soil depth (r2 = 0.32) and with sward DM (r2 = 0.21) and N (r2 = 0.54) yields; the latter being indicative of soil N- mineralization potential (plus fertiliser N efficiency). In contrast, on the fields closest to the farmyard and heavily manured, soil EC measurements were strongly and positively correlated with extractable Mg (r2 = 0.44) and extractable P (r2 = 0.38), but were not significantly correlated with sward N yield. Evidently, the high levels of Mg and P in the soil had masked the implicit relationship between soil EC and N-mineralization potential. It is concluded that soil EC mapping ought to be beneficial for identifying areas of grassland with shallow topsoil and low soil N-mineralization potential, and thus in need of additional fertiliser N. Knowledge of field management history, however, would also be needed to prevent erroneous interpretations of high soil EC measurements on heavily manured areas. References Bailey, J.S., Wang, K., Jordan, C.; Higgins, A. (2001). Chemosphere 42: 131-40. Hartsock, N.J; Mueller, T.G; Thomas, G.W; Barnhisel, R.I and Wells, K.L (2000). Soil Electrical Conductivity Variability. Proceedings of the Fifth International Conference on Precision Agriculture, Madison, USA. ASA-CSSA- SSSA (in press).

Facilitating technologies for delivering spatially variable insecticide applications in cotton

James M. McKinion, Johnie N. Jenkins, Jeffrey L. Willers and Sammy B. Turner

One of us, Dr. Jeffrey Willers, has development a new methodology for controlling cotton insect pests which when used properly has shown to save as much as 60% of the current cost of commercial insecticide applications. This new concept uses multispectral images of cotton fields processed to generate a normalized difference vegetation index (NDVI) image. The NDVI image emphasizes healthy, growing areas of a fields versus stressed, slow growing areas of a field. Using the NDVI image to direct field sampling and using the line-quadrate sampling methodology, a very accurate estimate of the insect population can be obtained while using from 40 to 60% less time than traditional cotton scouting. The strategy employed is to provide for early season control on insect populations while the population is very low with initial infestations. Research has shown that the cotton insects prefer areas of the field where the crop is growing best over areas of the field where the crop is either less developed or is under stress. The sampling methodology then calls for visiting areas of the field where the insects are likely to appear (high NDVI areas) and a few sites where the are not likely to appear (low NDVI sites) for verification. Once this is done, then the image is processed to generate a spatially variable insecticide application map where areas of infestation are designated to be sprayed and the non-infested areas are designated not to be sprayed. In traditional insect control, the entire field is sprayed when insects are found to be above economic threshold levels. The application map is then exported to a file that is specific to the controller on the spray machinery equipped with DGPS capability. This entire process from initial image acquisition to delivery of the application map to the spray controller and finally the actual application of the spatially variable insecticide control spray is very time critical. Ideally no more than 24 hours from the time of image acquisition until the actual spray is applied should elapse. Worst case still acceptable time elapse is 48 hours. If this time is exceeded, then the entire process must be repeated with new image acquisition because the biology will have changed too much for the spray to be completely reliable in insect population control. To address this need, a systems engineering approach was applied to the entire process. Ways were then explored to find ways to reduce the time element. Finally, for the southeastern US cotton belt it was found that the delivery and collection of information was the problem area. Since cotton production is conducted, by definition, in rural areas, a source of high speed connection to the Internet was pursued. Two satellite services were found which provided economically at least 600 Kbps downlink capability: Starband for residential service only and DirecWay serving both residential and commercial customers. DirecWay was selected and installed on two cooperating commercial cotton farms. To minimize time requirements for delivery of application maps to farm machinery and for collection of as-applied maps and field data, a wireless local area network was installed which provided high-speed, bidirectional data transfer anywhere on the farms with appropriately equipped PDA's, laptop computers, or farm equipment controllers. This paper will describe in detail the actual application of these technologies in a commercial cotton farm operation. The development of the automated process of converting multispectral images to spatially variable insecticide application maps while including site verification data will be discussed. The design and application of the wireless local area network with concomitant software for delivery and collection of information will also be discussed.

Temporal stability of tillage implement draft maps

N.B. McLaughlin, D.R. Lapen and E.G. Gregorich

Soil strength is known to affect crop growth, particularly in finer textured soils which are prone to compaction. Tillage implement draft is closely related to soil strength. Recent work has shown relationships between tillage implement draft and crop performance parameters. Consequently, spatial measurements of tillage implement draft have potential application in defining field management zones requiring remedial action or different treatments. Many modern farm tractors are already fitted with draft sensors as part of the integral electronic draft and depth control system. With a GPS and an appropriate data logger, spatial tillage implement draft data could be collected in conjunction with normal tillage operations. It is well known that crops respond to both soil and weather parameters, and that combine yield maps from the same field and different years often appear quite different. Patterns of variability in soil strength are evident in tillage implement draft maps, but it is not known if these patterns are consistent from year to year. Soil strength, and therefore tillage implement draft is affected both by soil water content and compaction resulting from wheel traffic, both of which may vary from year to year. This paper examines the consistency of the patterns observed in tillage implement draft maps over time. Field experiments were conducted near Winchester, Ontario, Canada on a clay loam soil. The field site had been in forage for 10 years prior to 1995. The field was mouldboard plowed in fall 1995, and a crop rotation experiment was established in 1996. Conventional tillage was performed with fall mouldboard plowing each year after harvest. The Agriculture and Agri-Food Canada instrumented research tractor was used to make spatial measurements of mouldboard plow draft and fuel consumption each year since 1995. The data were recorded at a logging frequency of 100 Hz resulting in a grid spacing of 13 mm in the direction of travel, and 1.2 m (plow width) perpendicular to the direction of travel. The data for each pass of the plow were plotted and examined for anomalies. Segments of data with obvious problems such as a plugged plow were removed. The spatial draft data were then filtered to remove the high frequency components resulting from implement vibration and brittle soil failure. High density data in the direction of travel were aggregated to create a mosaic with cell aspect ratio near unity. The data were then normalized. Data for each year were then divided into broad categories according to magnitude, and mapped. Soil water content relationships with draft were also evaluated to help interpretation of draft patterns. The maps visually showed some consistency in patterns of high and low draft among years although there were some distinct differences. One problem was that adjacent plow passes in a given year often had different mean draft, likely due to proximity to a wheel track, and ended up in different categories. This suggests that bidirectional filtering should be done, both in the direction of travel, and perpendicular to the direction of travel. The degree of stability was quantified by determining the percentage of cells in the same categories for each pair of years. Year to year variability was high when absolute draft data were used. Differences in soil water content, both during the growing season and at the time of plowing, likely contributed to this year to year variability. When the data were normalized, the draft maps were relatively stable over time for broad categories of draft magnitude. However, the stability broke down if either the cell dimensions were too small, or categories of draft magnitude were too finely divided.

Evaluation of environmental impacts of horticultural production systems

H. Mempel, H. Bauersachs and J. Meyer

During the last years, there has been an increasing public concern regarding food safety and quality as well as regarding the environmental risks and impacts associated with agricultural or horticultural production processes. In the future the growers are expected to report on the environmental impacts of their production systems and activities. While maintaining the productivity of their production systems, they simultaneously need to reduce environmental impacts. With the systems approach of life cycle assessment the environmental impact of a product or process can be evaluated. Such a systems analysis helps to identify major contributors to the environmental impact and to improve the environmental performance of agricultural or horticultural production. In this paper a simplified Life Cycle Assessment for two vegetable production systems with different production intensities (outdoor cropping and protected cultivation in a greenhouse) is presented. One of the first steps in such a Life Cycle Assessment is the Life Cycle Inventory Analysis (LCIA), where the relevant inputs and outputs have to be determined for all the different production steps. Unit process steps have to be defined for the whole production chain that represent the smallest portions of a product system for which data are collected. In the impact assessment phase, the detailed data collected in the inventory are assigned to the environmental impact categories. In this study the impact categories 'cumulative primary energy demand', 'global warming potential', 'acidification potential' and 'photo-oxidant formation potential' were considered initially. The potential environmental impacts associated with each unit process were calculated. For an evaluation of these production systems with different production intensities different functional units (production area, yield, volume of sales) have to be taken into account as well. In the presented open field production system the highest environmental impact is caused by field-farm transport processes in form of fuel consumption. With regard to material consumption, however, the process steps of fertilizing and plant protection cause the major contribution due to fertilizer and pesticide production. Additional demand for material by machinery and tractors are of minor importance. Therefore in the analysed system there is a great potential for increasing the ecological efficiency by optimising the transport system. In the greenhouse production system a large portion of the environmental impact is related to the greenhouse material itself and to the heating process. Accordingly, in this production system a high potential for increasing the ecological efficiency can be identified in increased productivity and energy saving cultivation strategies. The influences of different production strategies or different energy supply systems on the environmental impacts are shown. The systems analysis performed in this study allows for the identification of those process steps in horticultural production that are major contributors to the environmental impact and also includes judgements on the relative environmental benefit and disadvantages of different systems. Therefore the results can help the growers to identify weak points in the production chain and to improve the environmental performance of their production system. The input costs of the production process (labour, machinery, land, water, fertilisers pesticides and energy) are a significant portion of the overall farm operation costs. Thus a more efficient use of environmentally sensitive inputs can reduce both, the production costs for the farmer as well as the environmental risks or impacts. The detailed environmental system analysis of different production systems finally improves the knowledge about these systems and therefore supports the development of environmental policies and programs in horticulture or agriculture.

Remote mapping of crop water status for spatially variable irrigation scheduling

M. Meron, J. Tsipris and D. Charit

Spatial variability of crop water status and irrigation demands are combinations of the natural soil spatial variability, man and machinery induced variability of the crop and the water spreading systems uniformity. The difference within one irrigation section may be up to 100% between dry and wet extremes. In conventional irrigation scheduling, all the variability is accounted for in an efficiency factor. To ensure that the whole field will receive adequate watering, the irrigation intervals and efficiency factors are tailored to the driest part of the field, resulting large water losses. Scheduling to the mean stress of the field will result partial over and under watering, with corresponding yield reduction at the dry parts and water logging on the wet ones. Detailed mapping of the crop water status before irrigation, with quantification of the stress levels and the corresponding crop area, provides the necessary knowledge for decision on when and how much to irrigate. Where the spreading system enables variable rate application of the water, such as in center pivots or solid set micro irrigation, the crop water status map provides the blueprint for variable amount irrigation to match local crop water demand. Remote determination of crop water status by thermal methods relies on measurement of crop canopy temperatures and normalizing them to crop characteristics and environmental conditions. The Crop Water Stress Index (CWSI), as was formulated and applied originally, stumbled on two impediments: 1. Separation of relevant crop canopy temperatures from the leaf population and the soil background by the hand held or high altitude flown radiometers. 2. Proper definition and measurement of the well watered reference line from empirical measurements and the relation of them to meteorological data. The new approach presented here relies on better temperature separation and employment of Artificial Reference Surfaces (ARS) to normalize crop and environmental references. Low altitude, high-resolution thermal scan was shown earlier as instrumental to differentiate between crop canopy temperature populations and background soil temperatures. Pixel sizes over 0.1 m of cotton canopies distorted temperature separation. Introducing simultaneous remote temperature measurements of ARS, built from dark and bright colored, dry and wet materials, enabled calculation of well-watered reference temperatures from the energy balance equations, with the only additional input of global radiation. The new CWSI algorithm begins with statistical extraction of the canopy temperature, calculation of the theoretical maximum and minimum canopy temperatures under ambient conditions from ARS temperatures and global radiation and relating the extracted canopy temperature to the extremes as a fraction: CWSI 0 = well watered, 1 = fully stressed. After initial tests in Israel at 2000, the concept was tested on full field scale in the Griffith NSW are in Australia at 2001-2 season on cotton and vineyards. ADS SPECTRA-VIEW® system equipped with FLIR SC3000 thermal scanner was airborne on 32x24m footprints, at 32 m wide scan swaths in 40 m intervals over the test fields and the ARS. Ground meteorological data were collected from instrumented ARS. All data were time and geo-referenced and compiled into a single database. CWSI maps were generated using the SURFER mapping program. Results are presented from a 1500 ha cotton field, irrigated by furrow, lateral move and drip irrigation, and a 240 ha vineyard entirely drip irrigated. Crop water stress data compared well with irrigation history, and local agronomy data. Spatial variability of cotton water stress fitted former land leveling patterns, lateral move sprinkler flow variability and drip irrigation uniformities. Vineyard irrigation analysis revealed need for immediate changes in some of the fields. A sample irrigation scheduling table is presented, based on quantitative analysis of stress levels and stressed area.

Tree shape and foliage volume guided precision orchard sprayer - the PRECISPRAY FP5 project

M. Meron, J. Van de Zande, R. Van Zuydam, M. Shragai, J. Liberman, A. Hetzroni, P.G. Andersen, B. Heijne and E. Shimborsky

Precision Horticulture - implemented as tree specific management of orchards - foreseen to be the next step in application of Integrated Fruit Production and ISO 1400 for fruit quality. Such management systems will take orchard management much beyond simple harvest operations from the tree to the market with accompanying "paper trail" of crop history. Enabling variable rate application of agrochemicals, fertilizers, and water, supported by related site specific economic and horticultural information, monitoring of pest and disease infestation, plant stress, fruit growth, water and nutrient status, and other activities, will allow immense increase in production efficiency and economic returns. Reduction of farm chemical applications is one of the high priority goals of R & D in the EC for reasons of public health and the contamination of the environment. Precision spraying by matching spray volume to tree size and foliage density enables such reduction, even in the current state of pest control chemistry or biology, just by cutting the wasted chemicals in customary indiscriminate applications, aimed to ensure full cover of the foliage. The 5th Framework PRECISPRAY project was initiated to develop a segmented boom cross flow sprayer, where every segment will follow the outer contour lines of the tree and adjust airflow and spray volume to the foliage shape and volume in front of the outlet.. This mode of operation matches tightly the tree structure, applies large amounts at the wider and denser parts of the tree, reduces application at the thinner parts and closes the spray if no foliage is present, like in gaps between trees. The sprayer is guided by DGPS control, according to a tree shape and volume map (TPV) provided from aerial survey and photogrammetric processing. Photogrammetric mapping was chosen for tree contour delineation as earlier sound, light or electromagnetic echo methods (SONAR, LIDAR, RADAR) could not adequately trace tree contours, because the fuzzy nature of the foliage impeding sharp echo returns. Preliminary experiments showed that high resolution aerial photography enabled such mapping. The main components of the system consist of: 1.) A lightweight, inexpensive aerial photography system, providing ground referenced stereoscopic digital image sequences of the orchard. 2.) Digital stereoscopic program, processing the aerial imagery to provide tree shape and volume maps of the orchard and contour lines to follow by the sprayer outlets. 3.) Variable rate segmented vertical boom cross flow sprayer, capable of keeping the outlets in constant distance from the tree contour line and changing airflow and spray volume according to the foliage volume in front of each outlet. 4.) Sprayer guidance and control system which receives the spray order, directs the sprayer and returns actual execution details. 5.) Farm management GIS containing the infrastructure and physical properties of the orchard, including the TPV maps. It is the management component of the system where the pest management knowledge base is stored, field data is collected, infestation severity is evaluated, pest control decisions are made, spraying orders are issued and the feedback data from the sprayer is stored. The project is supported by research into parameter optimization of air and spray volumes according to variable foliage volumes and densities, and evaluation of the sprayers actual performance. Most of the components are already assembled and tested, and the system as a whole will be field tested in 2003. Detailed description of the components will be presented in the congress by their developers.

Geostatistical analysis of soil properties, landscape attributes, and corn quality

Yuxin Miao, Pierre C. Robert and David J. Mulla

Before precision agricultural technologies can be used to manage crops for both yield and quality, the spatial and temporal variability of crop quality parameters and their site-specific relationships to soil and landscape characteristics need to be clearly understood. Significant within field hybrid differences in corn (Zea mays L.) yield and quality have been reported, and their spatial variability were found to be related to soil properties, topography, electrical conductivity, and order 1 soil survey. These relationships were field-, and hybrid-specific. The objectives of this study are to characterize spatial structures of soil and corn quality parameters, compare within-field hybrid differences in spatial structures, and determine the feasibility of estimating spatial patterns in corn protein, oil and starch content using surrogate data and cokriging. Two sites from eastern Illinois were chosen for this study. Site 1 is a 32.8 ha no-till field and very flat, with a 4.57 m difference in relative elevation. It is composed of two principle soil mapping units: Flanagan silt loam in the west half, and Drummer silty clay loam in the east half of the field. Site 2 is a 8 ha conventional-till field with rolling topography. Five major soil mapping units compose this field: Drummer silty clay loam, Starks silt loam, Millbrook silt loam, Camden loam, and Fincastle silt loam. Two different hybrids (Pioneer 33G26 vs. 33Y18 in site 1, and Pioneer 34W67 vs. 34K77 in site 2) were planted side-by-side using the Pioneer split-planter comparison method in each field. P and K fertilizers were applied using variable rate technology (VRT), and N was applied uniformly at a rate of 181.4 kg/ha actual N in site 1 and 192.7 kg/ha in site 2. Corn samples for two hybrids were collected by hand before harvest on a regular grid of 47 x 60 m, and analyzed for corn oil, protein, and starch content using INFRATEC© 1229 NIR Grain Analyzer. Soil samples were taken from the same locations and analyzed for different properties. The fields were also surveyed for topography, electrical conductivity (EC), and soil at 1:7920 scale. Semivariograms and cross-semivariograms were developed for all properties. More intensively measured properties (yield, EC, and topography) were used to estimate less intensively measured properties (protein, oil and starch content). Preliminary analysis showed that soil properties displaying strong spatial dependence included soil test P, CEC, Ca, Mg, Mg% of base saturation, S, and Mn. Hybrid differences in corn quality spatial structures were found. Corn oil from 34W67 at site 2 showed moderate spatial dependence, with 74% of the total variance explained by spatial structure. Corn oil from the other three hybrids showed pure nugget effects. Corn protein from 33Y18 displayed strong spatial dependence, with a range of 81 meter, but corn protein from other hybrids all had moderate spatial dependence, and larger ranges. Corn starch from 34W67 at site 2 was strongly correlated in space, but starch from 34K77 showed pure nugget effect, and corn starch at site 1 showed moderate spatial dependence. The spatial structures were influenced by incidence of Diplodia, a corn ear disease. When samples affected by Diplodia were removed, corn oil from 33Y18 displayed moderate spatial dependence, but corn oil from 33G26 was still not spatially correlated. These results indicated that corn sampling intensities may need to vary with hybrids. Correlation analysis showed corn quality parameters at site 1 were significantly correlated with either EC or relative elevation, but not with yield. At site 2, corn oil from 34W67 was negatively correlated with EC, protein from 34K77 was significantly correlated with both relative elevation and EC, and starch from 34K77 was negatively correlated with relative elevation.

Teaching principles of variable-rate spraying in the lab

Gaines E. Miles and Daniel R. Ess

This presentation describes an apparatus which has been constructed for the purpose of demonstrating variable-rate spraying in a laboratory setting. Physically the device is capable of demonstrating the effects of boom height, droplet size, and nozzle tip selections. Electronically, the apparatus can demonstrate the pulse-width modulated nozzles, and follow a predetermined variable-rate application map. A data logger has been added to provide as applied mapping of the boom pressure. The apparatus consists of two mobile stands. The wet stand supports the spray boom, spray curtain, and collects the water sprayed from the two nozzles. The electronic stand contains a pump to recirculate the spray, the Case-Tyler AIM control module, the Raven SCS 440 spray controller, a Databoy data logger, and a laptop Personal Computer (PC) running Mid-Tech's Fieldware software. The flexible curtains on the wet stand permit access to the boom so students can adjust the height, change nozzle tips, and measure the nozzle flow rates with graduated cylinders. Students can also use water-sensitive paper to catch spray droplets and determine the relationships among boom pressure, type of tip, tip size, and the droplet size spectrum. With this hardware and software, students are able to plan a variable-rate spray application, then monitor the flow from the nozzle tips as the sprayer traverses the field as simulated by the software on the PC. By collecting the nozzle output, the discharge rate can be determined at various locations and compared to the desired application rate. Students also use this apparatus to learn how to calibrate a sensor. The boom pressure is adjusted by a switch on the AIM controller and the output of the electronic pressure sensor is measured by a digital multimeter. For each pressure setting, a gage mounted on the boom is read and the results recorded manually. The nozzle flow rates (volume per minute) are also measured with graduated cylinders and recorded for each pressure setting. Students record the data and use the statistical functions in Excel to determine the least-squares regression coefficients for the slope and intercept of a line through the data. The results are compared to the nozzle manufacturer's tables for new spray tips. Students can also use this apparatus to measure the performance of the pulse-width modulated control valve. With the AIM control module on, the desired boom pressure may be set to different levels. While this changes the droplet size spectrum, the flow should stay at the prescribed rate. This apparatus was constructed by seniors in the Agricultural Systems Management program at Purdue University as part of their capstone experience course. The substantial contributions by Case, Raven, and Mid-Tech are deeply appreciated. This apparatus will be used in academic courses on crop production equipment, technologies of precision agriculture, automatic controls, and in workshops for extension educators and farmers. This presentation will discuss the experiences of using this apparatus during the 2002-2003 academic year. The educational materials used in these sessions will also be documented.

Teaching the technologies of precision agriculture through experiential learning

G.E. Miles, D.R. Ess, R.M. Strickland and M.T. Morgan

This presentation will describe the content and experiences teaching an undergraduate, elective course in Technologies for Precision Agriculture at Purdue University. The primary objective of this course is to prepare agricultural systems management students for successful careers in information-intensive agriculture. The course is also taken by students in other disciplines such as agronomy and agricultural economics. The subjects include the technologies of global positioning systems, yield monitors, geographic information systems, site-specific data acquisition, remote sensing, variable-rate application, and economics. In order to help students bridge the gap between general understanding of the basic, fundamental principles and being able to use the technologies, handouts specific to Purdue have been prepared. These handouts provide step-by-step instructions and illustrations for how to use the electronic hardware and software available at Purdue. For some technologies, multiple sets of hardware/software are available, so students form teams to collect the data during class or lab. At other times, only a single set is available, so students reserve the equipment and do the work outside scheduled class/lab periods. Even though students may study and work in teams or groups, they are held accountable and evaluated individually. The class/lab activities lead the learner through all the steps required to connect the instruments, calibrate sensors, and acquire, transfer, and process data. The handouts are valuable resources for students during classes and labs, and are especially good for reviewing and preparing for the evaluation. The presentation will include a description of a variable-rate seeding simulator (VRSS) developed by seniors in Agricultural Systems Management curriculum as part of their senior, capstone experience. This laboratory apparatus is capable of performing all the functions of a variable-rate planter. It uses a planter row unit coupled to a hydraulic drive which is controlled by a Rawson Accu-Rate controller. An Ag Leader PF3000 monitor with GPS provides the seeding rate signal to the controller. A GPS simulator running on a laptop PC is used to provide position information while in the laboratory. A Dickey-john planter monitor and a radar sensor are used to provide seeding rate and speed information to the controller and PF3000. With this apparatus, students are able to build variable-rate seeding maps, then measure the seed drop for various locations across the simulated field, in order to calibrate the variable-rate controller. Students also use a variable-rate chemical application simulator (VRCAS) which is based on a Case Aim spray controller with pulse-width modulated nozzle tips, a Raven spray controller, and Fieldware application software from Mid-Tech. With this apparatus, students are able to plan a variable-rate application, then measure the sprayer performance as it simulates traversing the field. The prototype apparatus was also constructed by seniors in Agricultural Systems Management. Since the 2000 Fall semester, over 80 students have completed the 3-credit, junior-level course at Purdue. Student reviews of recent revisions to teaching methodologies are very positive. For each subject, students are required to complete a pre-test based on assigned reading materials; class and lab activities which provide hands-on (experiential learning) data collection, processing and analysis; and an in-lab unit evaluation. The presentation will include experiences in extension workshops and classes taught during the Fall of 2002 and the Spring of 2003.

Simultaneous identification of plant stresses and diseases in arable crops based on self-organising neural networks

Dimitrios Moshou, Cédric Bravo, Jon West, Alastair McCartney, Herman Ramon and Josse De Baerdemaeker

The objective of this research was to detect and recognise plant stress caused to disease infestation and to discriminate this type of stress from nutrient deficiency stress in field conditions using spectral reflectance information between 450 and 900nm. Yellow Rust (Puccinia striiformis) infected winter wheat plants were compared to nutrient stressed and healthy control winter wheat plants.

In-field hyperspectral reflectance images were taken with an imaging spectrograph mounted at spray boom height. Leaf recognition was possible using a NDVI threshold (normal differential vegetation index). A normalisation method based on reflectance and light intensity adjustments was applied. Feature selection was performed using stepwise waveband selection. For achieving a highly performant stress identification neural networks were introduced to refine discrimination. Simultaneously, the spectral data were analysed using Self-Organising Maps in order to visualise the spectral feature relationships.

Results and assessment of their significance

Winter wheat infected with Yellow Rust was successfully recognised from nutrient stressed and control plants. Spectra were taken at random plots around the field. The nutrient stressed plants were lacking one week of fertilisation, meanwhile flag leaves of the infected plants were not infestated by Yellow Rust and the underlying leaf showed 5% area infestation. The high spectral variability, caused by canopy architecture and different illumination levels, could be drastically reduced using light intensity normalization. Discrimination results were enhanced by the use of a spatial averaging window as wide as one plant. Using 5 wavebands of 20nm wide, it was possible to achieve more than 99% correct recognition of each of the stress conditions and the healthy plants.

Three wavebands were found to discriminate very well the normalised data when a spatial averaging window of 300 pixels was applied. These were the 725, 680 and 475nm +/- 10 nm wavebands. Two additional wavebands were added, since they are needed for leaf recognition using the NDVI estimation: 750 and 630nm +/- 10nm.

Nutrient stressed canopies show very clear spectral differences with control and diseased canopies. Diseased canopies however show high spectral variations, causing the necessity of spatial averaging. This method offers the opportunity of building a robust decision algorithm for on-line disease detection and spraying recommendation.

A general conclusion from observing the results is that the NDVI associated wavebands are not enough for performant disease and stress detection. The inclusion of the band centered at 725nm affects the recognition performance dramatically since the nitrogen stress recognition increases from 75% to 100%. Further the disease detection (among healthy and nitrogen stressed plants) increases from 84% to 97.4%. Additionally to these increases in performance, one can observe that the healthy plants recognition increases from 89% to 96%. Overall performance using 5 wavebands is more than 99%.

Development of dry bulk density map for a sandy loam field based on numerical-statistical hybrid compaction model

A.M. Mouazen, K. Dumont, K. Maertens and H. Ramon

The development of a soil compaction map, based on real-time measurement of the draught of a compaction sensor provides a quick view of positions of extremely compacted zones. A commercially available COSMOS/DesignSTAR 1.0 finite element program was used to perform a numerical simulation of soil cutting by a subsoiler used as a compaction sensor. One hundred twenty six finite element analyses were performed for various combinations among six gravimetric moisture contents; 0.03, 0.07, 0.10, 0.13, 0.17 and 0.22 kg kg-1, various dry bulk densities ranging from 1150 to 1820 kg m-3 and six depths; 0.10, 0.17, 0.22, 0.27, 0.32 and 0.37 m. A multiple linear regression analysis was performed on calculated sensor draught values obtained from the finite element analyses to establish a mathematical relationship for estimating the dry bulk density as a function of sensor draught, moisture content and cutting depth. This regression analysis was performed with S-PLUS, 4.5 statistical software at a confidence level of 0.95. The resulted numerical-statistical model is written as follows:

$$\rho_d = \sqrt[3]{\frac{D + 0.2136w - 73.9313d^2}{1673.4}}$$

where D is draught [kN], w is gravimetric moisture content [kg kg-1], d is cutting depth [m] and ρ_d is dry bulk density [kg m-3].

The measured draught of the subsoiler was utilised to draw the two-dimensional spatial variation in soil compaction of a sandy loam field (Arenic Cambisol). On the basis of Eq., dry bulk density indicating soil compaction was estimated as a function of the measured draught, cutting depth and moisture content. The ARCVIEW 3.1 GIS software was used to draw the field maps of measurement- and model-based dry bulk density, depth and moisture content. Since measurement-based dry bulk density and moisture content were measured in a 10*10 m grid, the maps for measurement- and model-based soil dry bulk density were developed from a 10*10 m grid data in order to uniform the resolution of these two maps. The map grid was interpolated using the inverse distance weighing (IDW) method. The interpolation grid size of all maps was of a radius of 25 m and a power of 2. The map cell size was 1 m2 with 50 rows and 137 columns.

Results showed that the measurement- and model-based dry bulk density ranged from 1343 to 1750 kg m-3 and from 1271 to 1523 kg m-3, respectively. The model-based dry bulk density underestimated the measurement-based dry bulk density by a mean error of 14%. A comparison of measurement-based dry bulk density map shown and model-based dry bulk density map indicates a similar tendency of spatial variation in soil compaction, particularly positions of extremely compacted zones.

This allows providing the farmer with a compaction map, which illustrates the extreme zones of soil compaction.

A correction factor of 14% in the calculated dry bulk density was incorporated into the developed numerical-statistical model, which improved the magnitude of the model-based predicted soil compaction. Furthermore, the real-time measurement of moisture content with better control of depth might be helpful to improve the magnitude of dry bulk density predicted and spatial distribution of soil compaction.

Student design contests promote hands-on learning and innovation breeding in precision agriculture

J. Müller, J. Smit, J.W. Hofstee and D. Goense

Introduction of recent scientific methods catalysed by Precision Agriculture research has proven to be an innovation motor in Agricultural Engineering. The effects are also beneficial in education: The high-tech approach is highly attractive for students addressing the IT skills that are developed as side effect of recreation activities amongst the upcoming e-generation. Meantime, most agricultural faculties respond to that arising demand by taking up courses in Precision Agriculture into their curriculum. Wageningen University is going a step ahead, activating the unique resource of universities: the creativity pool of students. Based on long-term positive experience with problem-oriented teaching, including self-educating teamwork on practical cases, an off-curriculum project offer was developed for Agricultural and Bioresource Engineering students. The general idea is to stimulate creative voluntarily and self-responsible teamwork during spare time. Tutors assign a challenging task and set a time limit. A fixed budget is offered to buy the required hardware. The student teams are free to choose any method to solve the task and observe time and monetary budget on own responsibility. The students have access to a workshop and assistance of professional staff for security reasons. Consulting resource persons is allowed and has turned out to inspirit the traditional universitas magistrorum et scholarium. The project results are presented to a public audience during open day of the University. For instance, one task was to collect images of a plant row by unmanned field scouting at a budget of 500 in three weeks time. The limited resources even challenged the students ambition and creativity. The solution was a remote controlled vehicle based on a second-hand toy car equipped with a CMOS camera transferring the video signal telemetrically to a receiver at the field end. This basic field scout was awarded in a competition of young scientists at the AgEng2002 conference in Budapest. Inspired by the positive resonance amongst students an inter-campus contest for field robots was initiated, inviting student teams of universities worldwide to contribute with own prototypes of autonomous vehicles, similar to the famous design contests of Engineering students at the Massachusetts Institute of Technology (MIT) in Cambridge, USA. Instead of developing new robot playgrounds every year as practised in the MIT contest, a challenging playground is provided by the natural degree of non-uniformity of an agricultural field. Navigating in the plant rows, turning at the headland and treating obstacles are program disciplines. Scouting capabilities are performed in freestyle and rated by an expert jury. In contrast to the MIT contest, the choice of components is not limited to a standard kit to allow unhampered creativity of the participants. In return, the jury assesses the hardware costs of the robot, based on the bill of material, admitting economy to an additional selection criterion. By accepting freestyle robots that are built at the home universities, the contribution of non-student resource persons cannot be ruled out. This is not an insufficiency but an increase of value, because identifying and motivating resource persons as well as fund-raising are skills, students also are earning from the project. Furthermore, the system is more promoting the spirit of university competition than individual competition, releasing additional resources and creative energy, thus being an innovation motor particularly for navigation and field scouting technologies but also for Precision Agriculture in general. Communication platform for the field robot event is an interactive web-site, created and maintained by Wageningen students of Agricultural and Bioresource Engineering: http://www.aenf.wau.nl/agrotechnologie/event.

Evaluation of fertilising and spreading strategies of fertiliser spreaders

A.T. Nieuwenhuizen, J.W. Hofstee, C. Lokhorst and J. Müller

Arable farming is faced with increasing legislation on the environment. Fertiliser is usually required for growing crops at a high quality. Therefore it is necessary to look for efficient fertilising strategies. Out of this need, split application strategies have been developed. A new development in precision farming is measuring the crop status in the tramlines during the growing season and adjusting the fertiliser to the local spatially variable crop requirements. Most spreaders spread much wider than the tramline distance and therefore affect the spread amount in the neighbour tramlines. Therefore it can be questioned whether information from these tramlines should be incorporated into the spreading of the present tramline. For this purpose a simulation program was developed to simulate location specific fertilisation taking into account crop data from neighbour tramlines too. The tramline distance in the virtual field was 24 meters; headlands and the sides of the field were not taken into account. Crop data was available per tramline with a distance of 1 meter between the data points and was obtained with a Hydro N sensor in a potato crop. Crop data was linearly interpolated between the tramlines to create a map with crop needs with a grid size of 1x1 meter. Four scenarios were used to analyse the effect of using crop information from neighbour tramlines on the final result. Scenario A spread a uniform amount on the field. Scenario B spread an amount that depended on the crop status in the current tramline. Scenario C was based on the crop status in the current and previous tramline and finally Scenario D took into account the crop information of the previous, the current and the next tramline. The set points for the spreader were calculated from the crop status. Scenario A was divided in two parts. The set point of Scenario A1 was a uniform amount of 50 kg N/ha, the set point of Scenario A2 was a uniform amount that corresponded with the overall average crop need of the field. Scenario B had a set point based on the crop status in the current tramline. Scenario C had a set point based on the average of the previous and the current tramline. The set point of Scenario D was based on the average of the crop status in the previous, current and next tramline. Subsequently a field was spread using two-dimensional spread patterns of a centrifugal and a pneumatic spreader with a grid size of 1x1 meter. The sum of squares of the difference between crop need and spread amount of fertiliser was used to judge the quality of spreading. These values were respectively 0.2522 (A1), 0.0704 (A2), 0.0069 (B centrifugal), 0.0096 (B pneumatic), 0.0225 (C centrifugal), 0.0149 (C pneumatic), 0.0227 (D centrifugal), and 0.0174 (D pneumatic). The results from the research showed that using crop information from neighbour tramlines did not result in a better-applied rate. Scenario B, only taking in account the crop status of the current tramline, gave the best approximation of the crop need. It appeared that the added value of the crop status of the neighbour tramlines was negative because of poor approximation of the crop requirement curve by the spread patterns based on using crop status information from neighbour tramlines to calculate set points. The spread patterns should have a shape that approximates the shape of the crop need, both in transversal and longitudinal direction.

The 'H method' – algorithms for automatic detection and elimination of defective data collected with yield monitoring systems

P.O. Noack, T. Muhr and M. Demmel

Yield monitoring systems have become an integral part of modern agriculture. Many new combine harvesters are shipped with yield monitoring systems being a standard component. They are extensively used to collect yield data in academic and industrial research projects. Yield maps created from this data often presents the sole input in precision farming processes as the creation of application maps. Yield sensors on combine harvesters are operating in harsh environments and are therefore producing measurements that contain defective data. Also, the change of external parameters influencing the grain flow (e.g. combine speed, lodged grain) will affect the quality of yield measurements. When creating yield maps with wide grid sizes (e.g. 50m) using geospatial methods (Kriging, Inverse Distance) healthy measurements will prevail over defective data resulting in reasonable local yield estimates. However, when diminishing the size of the area for which the local yield is to be estimated the weight of defective data gains influence on the result of the calculation. Yield data collected with different combine harvesters on one field have to be merged in order to create yield maps. The use of different yield mapping systems and/or insufficient or lacking calibration of the sensor devices influences the comparability of the data and therefore affects the quality of the yield maps created with the merged dataset. The preagro project (funded by the German Federal Ministry for Education and Research) is approaching the investigation of precision agriculture in a broad manner. Amongst other, yield data collected in eight different regions of Germany using three different types of yield mapping systems have been monitored and analysed over 4 years (1999 to 2002) and an algorithms have been developed in order to accommodate the demand for accurate yield maps with high resolutions. These algorithms called the 'H method' are based on the identification of combine tracks and the detection of their neighbourhood relations. Investigations have shown that mean yields on neighbouring tracks as well as yieldsums of neighbouring tracks show high correlations. This is due to the prevailing effect of healthy measurements over defective values described above. However, the method developed allows to derive local fits to yield measurements taken before and after distinct positions within the tracks. It also allows for the detection of local fits for the measurements in the neighbouring tracks. Comparing the local fits calculated for a track and those calculated for neighbouring tracks allows for the detection of deviations between the fits and the underlying yield measurements. Measurements can thereby be classified as being potentially erroneous and hence be excluded from further calculations or corrected by an estimation based on the measurements in the same track before and after the current position or even based on the measurements in the neighbouring tracks. The 'H method' is also designed to compare the trend of yield measurements on parallel and neighbouring tracks of two different combine harvesters operating on one field. This comparison may allow for the correction of poor or lacking calibration of one of the sensors. The 'H method' is currently integrated in a software product. Results comparing corrected and uncorrected yield maps will be presented.

Long term field study on the relative accuracy of different yield mapping systems installed on a single combine harvester

P.O. Noack, T. Muhr and M. Demmel

Yield mapping is becoming a common feature of combine harvesters nowadays. Many combine harvesters are shipped with yield mapping systems being part of the standard configuration. Yield data recorded with yield mapping systems is used by a wide variety of professionals in the agricultural field such as farmers, contractors, scientists and members of the agricultural industry. However, these users have different expectations concerning the accuracy and the resolution of yield maps. Over the past decade, numerous studies have been carried out aiming at investigating the accuracy of yield data measurements on combine harvesters. In order to accomplish this elaborated trials have been carried out in labatories and in the field. The results are often only based on the analysis of data from one field and one yield mapping system collected in one year. On the experimental farms Duernast and Scheyern (Technical University of Munich) a combine harvester has been equipped with two yield mapping systems (MF Fieldstar and LH 565) in 1993. Another combine harvester (Deutz Fahr 3670) in Gut Wittenfeld, Germany, has been fitted with two yield mapping systems (AgLeader PF 3000 and RDS Ceres) in 1999. In 2001 the monitors and sensors have been mounted on a MF72 that was already equipped with a FieldStar II yield mapping system. Analysing data collected with different yield mapping systems on a single combine harvester allows for the detection of contradictory yield measurements under practical conditions. Due to the lack of a reference system this method does not allow for the detection of absolute errors. However, the results may help to understand the influence of dynamics of grain flow within the machine and its influence on the yield data measurements gained with different sensor systems. The results may also be helpful when trying to generate yield maps from data collected with different yield mapping systems on one field. Yield data and positions from the two resp. three yield mapping systems have been imported into a database and then processed into gridded yield maps with different grid sizes using the Inverse Distance method. Yield maps generated from the different data sources were finally compared on a point-to-point basis (points with less than 1 m distance) and on a gridcell-to-gridcell basis. The results of the comparison of yield data measurements on a point-to-point basis indicate that single measurements are not appropiate for the estimation of yield. During the trials in Gut Wittenfeld in 2000 25 % percent of the yield data measurements from the yield sensors differed more than 25 %. Averaging the yield measurements over space using geostatistical methods (Inverse Distance) helped to increase the comapatability of the yield maps created from the yield data measurements. Increasing grid cell sizes had a positive influence on the comparability in terms of reducing the difference between the grid cell values. The above mentioned results were confirmed by the analysis of yield data collected in 1999 in Duernast : the correlation between the grid cell values of the two yield maps tremendously increased with increasing grid cells sizes (3m : 0.29; 12.5m : 0.68; 25 m : 0.87). Data from Gut Wittenfeld (2001 and 2002) collected with three different yield monitors will be analysed and results will be presented in detail.

A method for high accuracy geo-referencing of data from field operations

M. Noerremark and H.-W. Griepentrog

In this project we used a real time kinematic global position system (RTK GPS) to provide high accurate seeding data. We investigated deviations and errors of the RTK GPS when used in static and dynamic modes. We also investigated the reliability of the centimeter-level positioning over time. Furthermore we have investigated the accuracy of RTK GPS in eastern Denmark (55 40 N, 12 18 E). The project introduces a novel real time data aqusition system and post-processing algorithms for the positioning of seeds by merging RTK GPS data and pitch and roll data of the antenna triggered by individual seed drop detections. A conventional six row monopill seeder was retrofitted with a RTK GPS and a novel real time data aquisition system. Optical sensors were mounted directly below the seed metering discs and detected the seeds as they dropped into the furrow. The pitch and roll of the GPS antenna were measured with a digital 2-way inclinometer. The impulses from the optical sensors and the one pulse per second (1 PPS) impulse from the RTK GPS were logged by a DaqBook 120 connected to a 400 MHz laptop. The NMEA strings from the RTK GPS and from the inclinometer were logged at the 400 MHz laptop via serial ports. NMEA strings of RTK GPS and pitch/roll were logged 20 times and 10 times every second respectively. Data aquisition was done in DASYLab. All data were formatted and stored as text files. Each line of data included a hardware time of thousandths of a second The post-processing of a digital seed map was done in MatLab by procedures merging all data on basis of time syncronization and interpolation. The RTK GPS NMEA string consists of UTC time of position fix, UTM coordinates, height and other useful information of GPS positioning quality. To interpolate the time impulses from seed drop sensors to a position, the UTC time of the position fix needs to be synchronized with the hardware time. This time synchronization was done on basis of the 1 second impulse (1 PPS) generating hardware time for each impulse. With a known time lead from the 1 second impulse, the RTK GPS is generating a NMEA string with the UTC time for the corresponding 1 second impulse. It is then possible to make a table with corresponding UTC time and hardware time. The post-processing procedure also includes a control of GPS data and removal of GPS outliers. The measurement of pith and roll of the GPS antenna was used to correct the position at ground level. Next step was smoothing of GPS Northing and Easting using the time as independent variable. The smoothing was done by second order polynomial curve fitting. Heading calculations were then done by linear regression on smoothed GPS data. Estimation of seed drop position was done by linear interpolation using smoothed GPS data, hardware time as the variable and the heading vectors for geo-referencing each seed drop sensor relative to the GPS antenna positions. In this paper we present the results of the field tests which were conducted to check the accuracy of the estimated seed drop positions.

Experiences on site specific weed control in winter cereals

H. Nordmeyer, A. Zuk and A. Häusler

Patchy weed distribution is the normal situation on agricultural fields. This was demonstrated in numerous studies. Herbicide patch spraying resulted in reduction of herbicide use and in decrease of treated areas. By that, economical and ecological benefits are obvious. Patch spraying can be done by using different herbicides or no herbicides at all according to spatial weed distribution and economic thresholds. The use of variable dosages is an additional possibility in this concept. In this study, site specific weed control in winter wheat was done according to the weed mapping concept. Experiments on site specific weed control in winter cereals were carried out for 4 years in the northern part of Germany (region Braunschweig). All fields (size 9 to 18 ha) were in conventional agricultural use with winter barley, winter wheat and sugar beet in the crop rotation. The studies on site specific weed control were carried out only in winter cereals. Weed distribution and location was recorded by field walking and vehicle-based mapping using a Differential Global Positioning System (DGPS). Weed densities were counted or estimated in a irregular grid according to the tramlines. The weed mapping was done shortly before weed control and the results were given in plants/m2. Using geostatistical methods and kriging interpolation weed maps for single and grouped weed species were created. Weed species were grouped into monocotyledonous (MOCOT), dicotyledonous (DICOT) without Galium aparine (GALAP) and for weeds with extremly low economic threshold levels, i. e. Galium aparine. Only areas with weed densities above the threshold level were treated with herbicides. For each field threshold levels were established separately for site specific weed control. On many fields heterogenous occurrence of single and grouped weed species with densities above or below threshold levels were identified. Therefore, patchy herbicide application was possible. The most important weed in the MOCOT group was Apera spica-venti. Based on the weed distribution maps and according to threshold levels the herbicide application maps were created for MOCOT, DICOT and GALAP, separately. The results (1999 to 2002) showed the potential of herbicide reduction in winter cereals. Averaging the results over a period of 3 years (1999-2001), 16.5 to 55.3% were treated with herbicides for monocotyledonous weeds, 23.9 to 53.5% were determined in case of dicotyledonous weeds (without Galium aparine) and 25.5 to 66.7% were calculated for Galium aparine. Site specific weed control offers the possibility of saving herbicides. As a consequence of limited field travelling it may also contribute to a reduction of man hours and machine work. In the year 2000 the herbicide and machine costs (mean of 5 fields) could be reduced by 12.02 Euro/ha and 6.06 Euro/ha, respectively. The results presented on the spatial and temporal dynamics of weed seedling populations indicate a heterogeneous occurrence of weeds and a relatively high continuity of distribution patterns in time, even for different crops. The small variations of spatial distributions of total weed infestations throughout the investigation period are particularly promising for crop management. Subsequent investigations with greater spatial resolution and covering a longer period of time are required to verify these results. In all, the findings could contribute to an improvement of weed monitoring and therefore facilitate the integration of patchy weed control in the concept of precision agriculture. The paper describes the procedure, the tools and the experiences with site specific weed control in agricultural practice.

On political economy aspects of precision farming and promotion of conforming technologies by non-profit and profit-organization in agriculture

Ernst-August Nuppenau

The search for instruments to encourage the development of less polluting technologies and to promote the adoption of precision farming technologies has become a pivotal research topic in agricultural economics. Precision farming is considered as technological innovation that benefits farmers due to higher productivity and lower production costs and that preserves the environment due to less negative externalities and lower abatement costs. But, technology development needs incentives and policy design. Apparently, many attempts that work with the induced innovation hypothesis, in general, and precision farming, in special, have been suggested. Following the sequence of invention, proliferation, and adoption of techniques, mostly, the process of risky investment to invent technologies for environmental reason is not treated under economic consideration. Neither totally voluntary technology innovation nor regulation on technology development seem to be an appropriate route and commercial interest in innovation of precision techniques, to be paid by farmers, is not self evident. The paper investigates a model in which innovations towards precision technologies are costly. Innovations are undertaken by a large agricultural input providing company, such as a farm equipment or pesticide producing multi-national. The aim of precision techniques is to reduce side effects of inputs, i.e. the intention is to close the gap between application and impact of industrial inputs. The company shall have a quasi monopoly on a input market and conventional farmers are buying precision inputs for profit purpose; but, do not care about environmental impacts. In contrast, parts of agricultural land is occupied by non-profit-oriented farmers. These farmers restrict practises and industrial inputs to the benefit of the environment and their objective function is a mix of profits and environmental goals. A government supports these farmers, but also research for precision farming, i.e. both types farmers compete for public funds. Support for groups of farmers increases their competitiveness on land markets, determines market shares for technologies, and reduces externalities. The government maximises net benefits from reduced negative externalities, subject to taxpayers money, paying for precision farming research and compensation non-profit farmers. We will further show how farmers, that prefer voluntary programs on restriction in current technology, and farmers, that prefer better or precision technologies, interact, and what role an input industry plays and what happens to shares in land markets, respectively. In the past, the necessity of public funding for better technologies, as public good, has been frequently stressed; but, arguments becomes even more evident and complicated with negative externalities. In particular, we will focus on governments opportunities to co-finance better technologies via research instead of financing more environmental friendly agriculture via best practice, notably, as alternatives to be explored. However, a government has always the alternatives to support directly environmentally friendly farming, for instance organic farming, or indirectly conventional farming, for instance precision farming. Money spent in precision farming research may show higher ecological payoffs than organic farming. In a model exercise to provide arguments, parts of farm land are occupied by a non-profit-oriented farm community. This community can restrict agricultural practise and its objective function is a mix of profits and environmental quality of the landscape. Expansion reduces payoffs for precision farming in conventional farming. If a government wants to achieve its environmental objectives, it has to support the farm community and/or research departments for precision farming. Higher support for selected farm communities increases competitiveness on the land market and higher support for technical innovation reduces negative externalities: a trade-off. The question is: Where to direct the money and how much money should be raised? The paper is organised in four sections. First the interest groups and their interaction are presented. Second, objective functions are stated and, third, the interaction is modelled and, fourth, partial optimisations for interest groups are conducted. Fifth and finally, suggestions for application are discussed.

Spatial dependence of available phosphorus in grazed silvopastoral system using semivariogram models

Leonard Nwaigbo and G. Hudson

Semivariogrm models were used to study spatial dependence of soil available phosphorus (P) at tree scale in a grazed silvopastoral system site at Glensaugh, Northeast Scotland. Treatments were plots of about 0.8ha each planted with Hybrid larch (Larix europepis Henry) at 5m x 5m (HL400), 7m x 7m (HL200) and 10m x 10m (HL100) spacings on a predominantly rye grass (Lolium perenne) pasture which was grazed (Agricultural control, AC) pasture plots without trees. All treatments were replicated three times in a randomized complete block design. Soil samples (0-20cm depth) were collected from around two randomly selected but diagonally opposite trees separated by a distance of between 60-75m within each silvopastoral plot. The total number of soil samples collected from each plot was a function of tree planting distance (espacement) in the plot. From the replicate plots of HL400, HL200, HL100 and from the two hypothetical tree positions in the AC plots, soil samples were collected starting at 0.5m from the base of the tree and at 1m intervals thereafter in the 4 main compass directions. The limit of soil sample collection in each treatment plot is the mid point of the separation distance between the trees. Soil samples were analysed for available phosphorus using the ICP-MS procedures of the Macaulay Land Use Research Institute (MLURI), Aberdeen. The results from the analyses which were rationalized against the result of a known test soil and then corrected for moisture content were used for spatial correlation analyses. The semivariogram models were tested for the spatial distribution of soil available P at tree scale. Three geostatistical models were tested (a) Power model, whose variogram indicated a linear spatial gradient (b) Spherical model and (c) Exponential model. The considered resolution was for 8-lag distance (1 lag separation distance =1metre) more than half the maximum distance in any direction. Each model type was characterized in three ways (i) non-application of weight (ii) weighted by the number of pairs (sample pairs) and (iii) weighted by the number of pairs divided by the sampling distance. Spatial dependence of soil available P was observed in the Agricultural Control (AC) and HL100 treatments. While the AC showed isotropic (similar) spatial dependence, the HL100 had an based on least square technique and weighted by the number of samples pairs divided by the sampling distance gave the best fit of 82.4% and 67.5% for the AC and HL100 respectively. The estimated range of spatial dependence was between 5 and 6m in both treatments. There was no spatial dependence of available P observed in HL200 and HL400 and this may indicate a random distribution pattern of available P in these plots. It may also suggest that tree density might have influenced the spatial distribution pattern of soil available P in these grazed silvopastoral system plots. The results observed may have an effect on the subsequent need for application of fertilizer inputs. *This abstract is for the 4ECPA Conference **Corresponding author

Comparison of methods to extract correlations for canonical correlation analysis of cotton yields

S.J. Officer, R.J. Lascano, J. Booker and S. Maas

Development of efficient methods to identify the main factors generating yield patterns of a particular field is an important step to define land management zones for precision agriculture (PA). Product-moment correlations are often used, which provide a spatially averaged relationship between two variables over the area examined. However, the data-intensive and spatially referenced nature of PA data sets allows the spatial dependence of the correlations to be derived, which may provide a better estimate of the relationship between two variables. In particular the short-range error, or nugget variability, can be filtered out. In this study, the effect of deriving product-moment correlations from the raw data was compared with three methods of deriving correlations after geostatistical treatments of the data that filtered out nugget variability. Deep (0-900 mm) and shallow (0-300 mm) in-situ electrical conductivity (EC) values were collected from one pivot circle (400 m radius), under cotton (Gossypium hirsutum L.) in Texas. A yield monitor collected cotton lint values in 2000 and 2001. The EC and lint values were standardized to a Gaussian distribution and individually block kriged to a 20-m grid using a combination of a nugget and a spherical model. Correlations were generated from the kriged values of each grid node. A nugget model plus three nested spherical models (ranges: 131.0-m, 232.4-m, 350.1-m) were fitted to all the variograms and crossvariograms and a codispersion coefficient that filtered for nugget semi-variance was obtained from the sum of the b values. A structural correlation that filtered for the nugget semi-variance was generated using semivariance and cross-semivariance values at the maximum model range (350.6-m), less the nugget variance. All correlations between lint and EC were positive but the correlations where nugget variability was filtered out were higher than those derived from the raw data. The filtered correlations were all similar, indicating that the methods were comparable. Canonical correlation analysis (CCA) was applied to all sets of correlations to describe the lint yield patterns in terms of deep and shallow EC. All the CCA's formed one pair of canonical variates that had a relatively large and significant canonical correlation. For all CCA's, both cotton years contributed to the first crop variate, and were also highly correlated to that variate. For all CCA's, the contribution from deep EC to the first canonical EC variate was always relatively large and positive, reflecting a significant positive correlation between yield and deep EC in both years. Shallow EC had a negative contribution to the variate, despite a positive correlation to yield and to the canonical EC variate. The CCA utilized shallow EC as a suppressor variable to model a non-linear relationship between deep and shallow EC and yield. Yields were relatively low where both deep and shallow EC values were relatively low. Yields were only moderate where both deep and shallow EC values were relatively high in a depression, possibly because the wetter soils encouraged rank growth. Yields were relatively high in areas where deep EC values were relatively high, but the shallow EC values were in the medium range. The relatively high deep EC values indicated a finer textured layer present between 300-mm and 1-m, which could provide water storage under dry conditions. The combination of deep and shallow EC information in a multivariate analysis was concluded to be a more effective descriptor of the yield pattern than linear correlation alone. The use of the raw and the filtered correlations in CCA found similar outputs and interpretation indicating that, although the variability of the raw data was high, the use of the raw data was sufficient to identify the relationship between yield and EC. For a more accurate determination of the correlations, the calculation of filtered structural correlations was the most rapid method.

Exploring the spatial variation of take-all (*Gaeumannomyces graminis* var. *tritici*) for site-specific managment

M.A. Oliver; S.D. Heming, G. Gibson and N. Adams

The fungus *Gaeumannomyces graminis* var. *tritici* (Ggt), commonly known as Take-all, affects the roots of both wheat and barley. It causes damage to roots that limits crop growth and results in a loss of yield. The presence of take-all is evident in the discolouration and loss of roots, chlorotic plants, stunted growth and sometimes premature ripening leading to 'whiteheads'. These symptoms are not conclusive on their own and the plant roots should be examined for the infection. Interest in take-all is increasing because effective fungicides are becoming available, the proportion of winter wheat crops at risk has increased, and cereal production in a monoculture system is likely to increase in Europe. Much research has been done on the management of soil acidity and moisture, and plant nutrients in the precision agriculture context, but little on the management of diseases in plants. The implications for economy and environmental protection should also be considered with regard to pesticides, herbicides and fungicides. There is little knowledge available on the within-field spatial variation of take-all and how it might relate to features observed in the growing crop, selected soil properties and spectral information from remotely sensed imagery. The aims of the investigation were to determine whether take-all has a patchy distribution; to determine how crop attributes, spectral wavebands and properties of the soil relate to its spatial distribution; and to develop methods for site-specific disease management. We did an intensive survey on a 24-m grid, with additional samples along two transects, of a 6 ha area within a field with a second wheat (Triticum aestivum, cv (Soissons)) at growth stage 69 (late anthesis). At each sampling site three adjacent plants and their roots were removed. The plant roots were then examined in the laboratory and an assessment of damage by the fungus was made based on the take-all index which has a range of 0 to 100%. The soil was sampled at the same time to a depth of 20 cm by taking two cores on either side of the plants that had been removed; the cores were bulked and then analysed in the laboratory. Crop attributes were recorded in the field, and a high-resolution image (2-m pixels) with 25 spectral bands was taken after the soil and plants were sampled. Preliminary geostatistical analyses show that there is marked spatial variation in the incidence of take-all in this field that encompasses the full range of values in the take-all index from 0 to 100%. The range of spatial dependence for take-all and chlorosis is about 70 m and their kriged estimates show similar patterns of variation. The variogram ranges for leaf area index and plant height are about 90 m, and they have a similar spatial pattern. Of the soil properties examined so far, pH has the most similar range of spatial dependence and pattern of variation to take-all. The variation of calcium carbonate is somewhat less similar. Statistical and geostatistical analyses of soil particle size and plant nutrients, and of the spectral wavebands are being done and these results will be available for presentation and for the full text. The results available at present suggest that take-all could be managed in a precise way and that certain features of the plant and soil appear to be linked with its spatial variation.

Image processing performance assessment using crop weed competition models

Christine Onyango, John Marchant, Andrea Grundy, Kath Phelps and Richard Reader

Vegetable growers in the UK use over 550 tonnes of herbicides annually. As environmental pressures increase and the number of chemical weed controls declines, there is greater emphasis on improved targeting of pesticides and finding sustainable alternatives. A key component in implementing this strategy is the development of sensors capable of identifying targets. Since no appropriate sensor technology will ever give totally accurate discrimination between crop and weed, it is also important to know how degradation in the performance of the sensor affects crop yield. As far as the authors are aware there is no published work which attempts to relate sensor performance to yield. This paper describes an image processing algorithm for separating crop from weed and the consequent impact of misclassification as measured by a crop/weed competition model.

Several methods of automated weed detection have been proposed for both chemical and non-chemical systems. One example is that of active shape models (ASM), whereby an ASM is constructed for different weed species using a set of training images (Sogaard & Heisel, 2002). Early work by Marchant et al (1998) used monochrome, near-infra-red cameras to obtain high contrast between vegetation and soil. As the technological capabilities of image capture equipment improved, algorithms were developed which exploited other parts of the light spectrum. Manh et al (2001) developed deformable templates for weed leaf segmentation in colour images of crops and weeds. Hemming & Rath (2001) at IMAG in Holland, used colour and plant morphology to identify weeds under controlled illumination.

The crop weed competition model, Park et al (2001), simulates the light interception of individual plants by calculating crown zone area and the within-crown leaf area index from total plant dry weight. Crown zone area is defined as the smallest circle in plan view that encompasses all the leaves of an individual plant. In the growth model plants grow according to the solar radiation received, this is reduced when the leaf area exceeds the crown area because of self-shading. Analysis of plan view images of the crop allows the image processing algorithm to estimate crown zone area and to predict the estimated effect on crown zone area of altering the algorithm parameters.

The work presented here combines colour, morphology and knowledge about planting patterns to distinguish crop plants from weeds. As the crop growth stage advances, the algorithm is forced to trade improved crop recognition for reduced weed classification. Depending on the chosen method of weed removal, misclassification may result in inadvertent damage to the crop or even complete removal of crop plants and loss of yield. However incomplete removal of weeds might result in competition and subsequent yield reduction. Extensions to the model described in Park et al,(2001) by weed modellers at Horticulture Research International in the UK allow us to predict final crop yield after weed or crop removal. The model also allows us to investigate the impact on yield of misclassification in the presence of both aggressive and benign weed types. The competition model and the image analysis algorithm have been linked successfully to investigate a range of misclassification scenarios.

Hyperspectral remote sensing - A tool for the derivation of plant nitrogen and its spatial variability within maize and wheat canopies

Natascha Oppelt and Wolfram Mauser

Nitrogen is very important in plant nutrition because it is a key parameter for plant growth and development. Therefore, knowledge about the nitrogen status of crops as well as its spatial variability within a field provide valuable information for an optimised field management. The application of hyperspectral sensors enables identification of individual biochemical compounds of a plant or canopy as well as their quantitative analysis. Furthermore, the two-dimensional measurement enables the derivation of their spatial distribution. Nitrogen itself does not reflect or absorb, but a significantly large amount of nitrogen is bound to chlorophyll proteins in the plant leaves. Therefore, the nitrogen content can be derived indirectly via the chlorophyll content, which has characteristic absorption features in the visible wavelength region (400-700nm). A strong relationship between leaf chlorophyll and nitrogen content has been observed for several plant species including maize and wheat. In this paper, a cost-effective airborne imaging spectrometer used in an operational manner is presented. The Sensor AVIS (Airborne Visible / near Infrared imaging Spectrometer) was built at the Chair for Geography and Remote Sensing of the Ludwig-Maximilians-University Munich and operates in the 550-1000nm wavelength region. Using this system, a two-dimensional approach, the Chlorophyll Absorption Integral (CAI), is presented for the derivation of the chlorophyll and nitrogen status of maize and wheat. The CAI determines the chlorophyll content by measuring the area between a straight line joining two points of the red edge, which can be chosen interactively, and the curve of the red edge itself. The database was compiled during a multitemporal campaign flown in 1999 and 2000 in the Bavarian alpine foothills between Starnberger See and Ammersee. Three fields of maize (Zea mays) were chosen as test fields in 1999 and 2000 respectively. Six and three fields of winter wheat (Triticum aestivum) respectively form the sample size for wheat during this time period. The sensor was flown on board a Dornier Do-27 aircraft. Between April and September 1999 twelve data sets were flown. In 2000 nine data sets were gathered between May and August. First the data had to be system-corrected and wavelength calibrated. Then the data were atmospherically corrected and reflection calibrated. The resulting reflectances were analysed for the biochemical compounds of the canopies investigated. For the validation of this approach, weekly ground measurements of several plant parameters (plant height, biomass, leaf chlorophyll content, leaf nitrogen content) were carried out during the measuring periods. The CAI turned out to be a good predictor of the chlorophyll and nitrogen contents of the maize and wheat canopies investigated. The results depend strongly on a priori knowledge; even though significantly high coefficients of correlation can be derived for the whole sample sizes of wheat and maize, the coefficients of correlation are higher when the varieties are analysed individually. Ratios and indices are known to become saturated at chlorophyll contents above $1g/m^2$. With wheat, the CAI becomes saturated at chlorophyll and nitrogen contents above $2.5g/m^2$ and $8g/m^2$ respectively. No saturation effects can be observed with maize. Therefore, the CAI can be used for a wider range of chlorophyll contents than ratios and indices and provides better results for the chlorophyll and nitrogen derivation. Remotely sensed data provide the ability to retrieve the two-dimensional distribution of the parameters investigated. Therefore, CAI measurements derived from AVIS data can be used to monitor the spatial variability of the nitrogen content within a field.

Spatial variability of wine grape yield and quality in chilean vineyards

R. Ortega, A. Esser and O. Santibañez

During the last decade Chilean wine industry had undergone an important growth in terms of planted area and production. Chile's wine exports represent about 3.5 % of the world's total, with annual sales for approximately USD 600 millions. Chilean wine industry is focused on producing high quality wines with high efficiency and low environmental impact. Due to the nature of alluvial and granitic soils, where most of the vineyards are planted, it is expected that they present high spatial variability in terms of yield and quality. This means that, in any given vineyard, there would be areas producing high quality grapes that, on the average, would go unnoticed during harvest. With the objective of studying the spatial variability of wine grape yield and quality as well as soil chemical and physical properties, a study was performed during the 2001-2002 growing season. Two contrasting fields located in commercial vineyards were selected for study. The first field of 3.01 ha was a 10-yr old Carmenérè vineyard,located at the Maipo Valley (33°60; south latitude, 70°50 west longitude) on a alluvial soil. The second field was a 7.6-ha, 4-yr old Pinot Noir, on a granitic soil at the Leyda Valley (33°67; south latitude, 71°49 west longitude). Both vineyards were focused to premium wines. Through intensive grid sampling, with the help of differential GPS (DGPS), we determined the spatial variability of grape yield, grape quality (°Brix, titrable acidity, and pH) and soil chemical (electrical conductivity, pH, available N, Olsen-P, and extractable K) and physical (texture, water retention curve) properties. Sampling for yield and quality was performed two days before harvest, using systematic designs, with a intensity of 20 and 18 samples/ha, for the Carmenérè and Pinot Noir fields, respectively. Each sampling point corresponded to an individual plant, which was carefully selected to be representative of that particular location, and fully harvested and weighted at the field. A representative sample, of about 50 % of the harvested bunches, was taken to the laboratory for quality analyses. Sugar content was determined with a manual refractmeter, while pH and titrable acidity were determined by potentiometry. Soil sampling was performed using a systematic design with intensities from 2 to 8 samples/ha, depending on the field and type of analysis. Higher densities were used for chemical analyses and for the Carmenérè field. Each sample was composed of 6 sub samples taken to 30-cm depth within an area of 1-m radius. Georeferenced data was entered to a geographic information system (GIS) for analyzing and mapping the measured variables. Two methods of interpolation, inverse distance weighting (IDW) and kriging, were used. Variograms were estimated and adjusted in Vesper version 1.5. Cluster analyses using the algorithm fuzzy k-means were performed on soil properties to define uniform management zones. By defining specific ranges for sugar content (°Brix), pH and titrable acidity, areas of low, medium and high grape quality were defined in each field. With the actual prices for grapes of different qualities, the economical impact of zoning and differentially harvesting by quality was determined, comparing it to the whole field harvest. Results indicated that most of the evaluated variables showed a significant spatial variability which would justify site-specific management. Grape yields varied 8 to 10 fold, which is consistent with previous foreign studies. Spatial dependence for most measured variables was strong, with ranges varying from 20 to 30 m for grape yield and quality and from 50 to 200 m for soil properties. The economical impact of identifying high quality zones within the field was significant. These areas represented more than 20 % of the studied fields, while the benefits of identifying them reached about USD 200/ha, compared to uniform management. According to the obtained results Precision Viticulture has a big potential for achieving the objectives of the Chilean wine industry.

Spatial variability of the weed seed bank in an irrigated alluvial soil in chile

R. Ortega, J. Fogliatti and M. Kogan

The estimation of the soil seed bank is used to predict the potential weed problems in a given area. Usually, the first 0-2.5 cm soil depth concentrate most of the seeds that will germinate during the growing season. Under equilibrium conditions it is expected that the soil seed bank can be systematically sampled and spatially mapped in order to design site-specific management programs. The objective of this work was to study the factors that determine the size and composition of the seed bank, its spatial variability, and its relationship with the weeds present in a sweet corn crop. We spatially studied the soil seed bank to a depth of 2.5 cm, in terms of germination, viability, and dormancy per species. The study was performed in a 7.4-ha commercial field (34°41 south latitude, 70°58 west longitude), located in the Central Valley of Chile on a alluvial soil, during the 2001-2002 growing season. Before planting, soil samples, for seed bank analyses (0-0.025 m depth), and for chemical analyses (0-20 cm depth) were collected, with a sampling intensity of 4 samples/ha, using a systematic design, with the help of differential GPS (DGPS). The weight of the seed bank samples varied from 800 to 1450 g. Soil samples were air-dried, at room temperature, and protected from direct sunlight. Weed seeds were separated from the soil, washing the samples through sieves of different sizes (16, 32, and 60 mesh). Seed were dried at 30°C for 2 hours and left at room temperature for 24 hours. Seeds were manually separated by species, counted and put in a germination chamber at 22°C for 14 days. Germinated seeds were counted and the ones that did not germinate were tested for viability using the tetrazolium test. The inputs to the seed bank were estimated by determining the amount and species of seeds present in the irrigation water, in several opportunities during the growing season. At the field, sampled places were protected, with a 1-m2 plastic cover, before pre-planting herbicide application, to allow for germination of the seed bank on those places. At the stage of 8th leaf of the corn, we counted and identified emerged weeds at each sampling location. Immediately before harvest, weed dry matter, weed nutrient extraction, and weed composition was determined at each sampled point. Crop yield and corn nutrient extraction was determined in weeded and non-weeded locations. All georeferenced data was entered to a geographic information system (GIS) for management and analysis. Preliminary results indicated that only some weed species such as Chenopodium album and Digitaria sanguinalis showed spatial dependence under the sampling intensity used, which would allow adequately mapping them. However, higher intensities should be used for mapping all the species present. The total seed bank was, on the average, 270,600,000, varying from 4,815 to 73,800 seeds/m2. The input to the seed bank from irrigation water was 9 ± 2 seeds/m3, which meant $112,000 \pm 2,530$ seeds/ha during the growing season. Weed nitrogen extraction varied from 8 to 195 kg N/ha and negatively correlated with sweet corn yield. Results show that, making the necessary adjustments, sampling for weed seed bank estimation would be a good tool for site-specific weed management.

Application of simple technologies for a precision farming model in Mexican agriculture

Hipolito Ortiz-Laurel and Dietmar Rössel

Precision agriculture means the application of many new technologies to crop production, including global positioning systems (GPS), geographic information systems (GIS), yield monitors on harvesting equipment, and site-specific and variable-rate applications. Precision agriculture grew up in the United States and has entered in an era of rapid growth. The reasons are the extensive surface area planted for single crops, significant variations in soil and production, a technologically advanced range of machinery, and technical assistance and business capacity of high level. In Mexico, despite of initial interest and enthusiasm shown, it seems that such a vision of precision agriculture would be difficult to apply. Farmers showing great interest in precision agriculture are to some extent unwilling to accept it because of the cost and complexity. It looks unlikely that the American precision agriculture model for big farming enterprises becomes widespread in Mexico due to: a) The reduced average size and irregular shape of plots, b) Higher investment for the monitoring equipment and associated devices, c) Lack of training for using information technology and advanced electronics, d) Limited availability of specific software and e) Difficulty in explaining how to interpret production data. However, some type of simplified precision agriculture will have to be aimed at for Mexico, one which would still maintain the goals of curbing costs and protecting the environment, one in which the differentiated management of the entire set of production factors is exercised through direct intervention, one which would be applied to facilitate and ensure respect for the prevailing environmental norms. Although, reduction in pesticides and fertiliser used is a key element in the cost reductions stemming from precision farming, it is not in itself sufficient justification for its adoption. That at least is the case in purely business terms. However, on environmental terms there is a different panorama taking into account the reduction in the use of contaminant agents. Under this context, it is possible to develop a new agriculture for Mexico over the medium term as shown by the examples in this work, incorporating electronics and information technology with goals which are, in large part, identical to those of the American model through automation and by controlling machine functions in real time. Research is being undertaken to examine areas of implement control, which will lead to higher levels of accuracy being achieved. Soil tillage for seedbed preparation, weed management and crop fertilisation appear to offer the opportunity for the development and spread of this technology. Precise depth control in seedbed preparation conserves moisture, incorporates chemicals and creates a quality soil structure for seeding. A cultivator designed for precise and accurate seedbed tillage uses wheel sensors and electronic circuitry to level the frame and to maintain depth measurements. Weed control is a major constraint to integrated and organic farming systems. Mechanical precision weed control reduces herbicide use. A vision guided steerage tined implement, uses camera guidance to ensure accurate, mechanical inter-row weeding at high work rates. Patch weed control is a technique that uses optic sensors to detect the presence of weeds. Overall treatment rates can be reduced to the easy-to-kill targets. A differentiated distribution of nitrogen includes detection, through specific sensors of the chlorophyll content of the crop to be fertilized and in relation to this content, intervenes to regulate the spreader. This report ends with a cautious assessment of the prospects of precision farming in Mexico, for any numbers of reasons. The most weighty of these is the vast array of multidisciplinary know-how required for the application of these new techniques to our own farming situation, know-how that we do not yet fully possess. Keywords: Precision agriculture, information technology, advanced electronics, implement control, sensors.

Realisation of an „In-field Controller" for an Agricultural BUS-System based on open source program library "LBS-Lib"

R. Ostermeier, H. Auernhammer and M. Demmel

Introduction: Precision farming is using mapping systems ("map approach"), real-time sensor-actuator systems ("sensor approach") or combinations of both ("real time approach with map overlay"). Map based systems use GPS or other locator systems to establish a geographic basis for precision farming. They allow information to be gathered from various automatic and manual sources. Then a desired control map can be generated to guide such field operations as variable planting, irrigation, fertilization, or pesticide application. Real-time Systems do not require locators or mapping software and hardware. The relevant quantity is sensed and then an appropriate action is immediately taken. Fertilization with N-sensor based variable rate technology is an example for this approach. Both approaches hold system-dependly different disadvantages. Data from the past (mapping approach) describe static statuses and they therefore cannot be assigned at all the weather related variabilities within the growth period. The exclusive consideration of the current sensor data (sensor approach) disregards experience about fundamental growth and yield potential as well as enviromental constraints of the current site. "Real-time systems with map overlay" combine the sensor approach and the map approach and enable to overcome the deficits of each. Maps of yields, soil types and nutrients can be used together with real-time sensors for plant growth, soil moisture, and weed infestation to control field operations. These systems allow the optimization of the field operations. The realisation of this sophisticated approach requires an effective and future-oriented agricultural engineering implementation. Definition of an „In-field Controller" as core piece of a "Real-time system with map overlay": With the availability of Agricultural-BUS Systems (LBS by DIN 9684, ISOBUS by ISO 11783) for tractors and agricultural machinery powerful distributed electronic systems could be designed. With LBS a national standard is available since 1997 and a successful conclusion of the international standard ISOBUS is imminent. Besides company implementations at the Technical University Munich the conception and implementation of an Open Source program library „LBS-Lib" was developed which serves as a foundation stone of a really open and compatible electronic communication system based on LBS (DIN 9684) and ISOBUS (ISO 11783) (SPANGLER et al., 2001). Agricultural BUS-Systems allow smooth multiple-shift usage of tractor sensors, of GPS with permanent actual information about position and time, of data input and output units for management, as well as of the user terminal. By the integration of additional sensors further „online-available" information could be inserted in the system. Problems will occur by combining map based with sensor based systems which are not integrated on a proprietary way. A solution would be possible, if in addition to the „Task Controller" (processes only predefined tasks) a specificly defined service „In-field Controller" (controls in-field activities) would be available. As intermediary between „Task Controller" and application system it would: * be activated by the „Task Controller" and take over the setpoints, * receive or poll the actual position from the positioning service, * record measurement readings of the online-sensor(s), * integrate an expert system and derivate consequentially the „local setpoint". Documentation of local information is likewise provided. Therewith all electronically controllable application systems (seeding, fertilization, pest management) could also be utilized in a sensor based system with map overlay without any modification. Admittedly the integration of online sensors in application systems is neither defined nor standardised so far. Differentiated proposals for special services in such systems aren't known just as little as overlay algorithms for application maps and online sensor technology. If any this systems are implemented in an proprietary way. Aims and results: The presentation will inform on the analysis and the definition of the tasks and architecture of the "In-field Controller" (exemplified for a N-fertilization system) and of first steps of it's realisation based on the "LBS-Lib".

Decision support tool for extension and sales purposes to evaluate the economic effects of introducing precision farming technology to a farm

Susanne Otter-Nacke

Decision support tool for extension and sales purposes to evaluate the economic effects of introducing Precision Farming Technology to a farm. Farmers are continuously forced to find new strategies to reduce costs and maximize profits. Therefore, before investing in new technology it is vital to get a good picture about the economic effects of such a decision. When trying to evaluate whether the step into the information age technology will pay for a future user it is most important to employ relevant figures. The key questions for potential Precision Farming users are: - How much do I gain when using this technology? - How is the variability within my fields? - Which is the best way for me to use this technical opportunities provided by Precision Farming? General answers are hard to give, however, since meaningful figures on economic effects that could be generalized are virtually not available. Most reports on savings realized using Precision Farming Technology result from few observations rather than being based on a representative sample. Research into the economic effects of Precision Farming Technology is limited, however, and scientific results are difficult to obtain due to the particularities of the technology. On the other hand, all that counts for a given user is the economic consideration of his particular farming situation which may be realized using his numbers. In order to be able to put all the relevant information together making sure that no relevant aspect of the decision is missing, a scheme was developed which assembles the necessary figures allowing the user to compare investment costs required with potential profits under real farm conditions using the farmer's own inputs. Thus, are custom tailored decision results giving the user more confidence to find the right solution for himself out of a confusing number of possibilities. This calculation table defines different levels of Precision Farming which correspond to certain cost saving strategies. There are different options for the calculation: - use typical figures derived from reports of other users who operate their farm under comparable conditions - use own estimates based on the farmer's own long term experience. - use a combination of the above. In addition the scheme enables the farmer to create several scenarios to account for situations that are difficult to quantify at times. In this scheme figures may be varied, accounting for contrasting scenarios defining the complete range of economic conditions, thus helping to define the risk or potential profit of this investment. There are many different approaches to Precision Farming, such as: - Sowing date or density - Fertilisation - Crop protection - Liming - Drainage - Deep tillage Any of the above approaches stands for a certain economic benefit. The introduced scheme helps to find the most profitable option for a given farm by comparing different scenarios. It could be employed by consultants specializing in Precision Farming. In addition sales representatives may use this tool to demonstrate the economic reserves and potential of this technology to help farmers to profit from precision. Evaluation of the system with respect to solving the farmers practical questions when employing Precision Agriculture strategies on their farms is discussed.

Methods to define confidence intervals for kriged values: application on precision viticulture data

J.N. Paoli, B. Tisseyre, O. Strauss and J.M. Roger

Professionals who work in the field of agriculture and more specifically of viticulture will have access to an increasing number of information sources. The precision and the resolution of these data are heterogeneous, and the major problem is to combine them in order to define homogeneous zones into fields or to compute data which could be used by professionals [Bramley, 2001; Tisseyre, 2001]. In Precision Agriculture, geostatistics and kriging methods are mostly used to estimate the different data onto a common grid. Confidence intervals are essential to analyze kriged values. It is generally considered that kriging variance could be used to compute these intervals. However, this approach has to be discussed. This variance :

is based on the modeled variogram which is estimated from a variogram cloud. Usually, the model doesn't take into account inaccuracy of variogram estimation. Moreover, it is defined for all the field, and it ignores local variations of measurement variance.

depends on the distance between known values and estimated values

The originality of our work is to present methods to compute confidence intervals

without using kriging variance

by taking into account

the inaccuracy of variogram cloud,

or the measurement variance.

We worked on grape yield data, measured with an on-line sensor embarked on a grape harvesting machine. Our experimental vineyard is a piece of Syrah variety trained in Royat cordon. Stocks are 8 years old (density of 4000 stocks/ha), and trained with a height of 1.7 m. This piece of 1,2 ha is on the "Clape limestone massif". We used two different estimation methods on these data. To validate these methods,

the database were divided in a test base and a learning base

the estimation processes were performed on the learning base

error between the test base and the prediction given by the previous estimation was evaluated

correlation between errors and kriging variance was computed.

1st method: The aim of this method is to take into account inaccuracy of variogram estimation. It is an iterative method based on the repetition of the kriging procedure using different variogram models without modifying data. To estimate different variogram models, a variogram cloud was computed. Then, according to it, a parametric function was chosen and intervals were calculated for the parameters of this function (range, nugget effect…).

2nd method: To take into account measurements variance, it is possible to compute a local variogram on the neighborhood of each estimated point [Mac Bratney, 1999].

Nevertheless, another method could be defined to solve this problem. A single model could be considered as representative enough of all the parts of the field [this aspect will be developed in the full paper]. The estimation process could be tested statistically.

A "Bootstrap" method was used [Lecoutre, Tassi, 1987]. Different sub-samples were drawn randomly from the data and the Kriging procedure was repeated using these sub-samples.

In each estimated points, the different values generated were used to compute means and variances. On our study case, variances obtained by these methods are better correlated with errors than "kriging variance". The 2nd method (based on "Bootstrap") seems to be particularly efficient. Our future work is dedicated to the use of those estimated values to provide a segmentation of the field into homogeneous region. We think about using a region growing segmentation process that could use uncertainty measure into a fuzzy decision process base on rules.

Precision agriculture - economic aspects

Jan Pawlak

Implementation of precision farming has multiple effects both on inputs and results of the agricultural production and on environment. Its application is connected with a need of detection techniques (for instance image processing for weed and crop recognition, enabling to obtain information on the growth stage of the weeds). Many techniques of the detection can be used that can enable to study the spatial variability of soil, crop and weed conditions and consequently to determine the time and intensity of relevant treatments. The spatially variable inputs of fertilisers, pesticides and other materials can only be realised with using machines equipped with devices for automatic regulation of emissions. These instruments for automatic control make the prices of machines and their costs of operation to increase. On the other hand there is a chance to make the inputs of chemicals and other materials more efficient. Also the negative effects of the treatments on environment would be reduced. The advantages in the field of environment can be estimated by means of calculation of the cost of recovering fertiliser or pesticide not used by the crop and polluting soil and/or water. One approach is to estimate it as a coefficient of the cost of the chemical material. In a case of nitrogen fertiliser, professor Bresson of INRA estimates the value of this coefficient as 4 (times the cost of fertiliser). Thus, the precision agriculture enables not only the direct reduction of costs of chemicals, but also diminution of environment loses, that in case of nitrogen fertilisers are even more important. The difference between monetary value of inputs in a case of implementation of the precision farming system as compared to the traditional one can be estimated using the formula:

$$R = \sum_{k=1}^{n} (I_{k0} - I_{k1})P_k + (Ce_0 - Ce_1) - \sum_{m=1}^{k} (Cu_{m1} - Cu_{m0}$$

Where: R –changes in costs of crop production on a farm caused by implementation of the precision agriculture,

I_{k0} – inputs of k-ths types of fertilisers, pesticides and other materials on farm before implementation of the of the precision farming system, in convenient units,

I_{k1} – inputs of k-ths types of of fertilisers, pesticides and other materials on farm after implementation of the of the precision farming system, in convenient units,

I_{k1} – prices per unit of k-ths types of of fertilisers, pesticides and other materials,

Ce_0 – costs of environment deterioration, on farm before implementation of the of the precision farming system,

Cu_{m1} – operation costs of farm machines after implementation of the of the precision farming system,

Cu_{m0} – operation costs of farm machines before implementation of the precision farming system.

The implementation of precision farming effects also the value of crop production. Optimisation of production inputs does not always lead to the increase of production. Sometimes the necessity to improve the efficiency of inputs cause the production to decrease. This not always means the drop of the value of production, if you take into consideration the quality of obtained material. Particular streams of inputs can not be examined separately when making evaluation of economic efficiency of a precision farming system. There are important interrelations between particular streams of inputs as well as between the structure and intensity of inputs and the efficiency of an adapted production system. Therefore, the holistic approach is necessary. Only the complex and harmonised utilisation of the progress in different fields of the biological and the agricultural engineering connected with correct management system may bring the positive results for farms. The economic efficiency is a main criterion in evaluation of different solutions. This is very important under present low profitability of the agricultural production.

Potato yields and variation in tuber size measured by an optical sensor

H.H. Pedersen and P.A. Algerbo

A new sensor AgriCount has been tested on two potato harvesters in Sweden and in Denmark. The sensor measures the size of each tuber in two dimensions by use of a CCD camera. The pixel value of each tuber measured by the camera is converted to weight and size using calibration curves created from sensor values of a number of tubers of known size and weight. A one-row Juko potato harvester equipped with the sensor was used to harvest trail plots in 1999, 2000 and 2001 in Sweden. The harvester was also equipped with weight cells to validate the sensor values. Six varieties of potatoes were harvested. There was a significant difference in the pixel-weight and the pixel-size relation ship between the varieties. This means that specific calibration curves for each variety are needed. The average yield estimated by the sensor was between 1 and 10 % higher than the yield measured by the weight for the six varieties respectively. In the 2002 harvest the sensor has been used in Denmark in four fields to estimate the distribution of tuber sizes and the amount of tubers harvested. The sensor was mounted on at one-row Wuhlmaus harvester. It is of great value to the factory packing the potatoes to have an estimate of the size distribution of the tubers before they arrive as the potatoes are divided into classes according to purpose at the factory. After a specific calibration to each variety the sensor was able to estimate the distribution with few percent's error only. The total yield was overestimated by up to 15 %. All data from the sensor were stored on a PSION Workabout Computer together with positions from a DGPS receiver. Data were afterwards managed in a GIS program to produce yield maps and maps showing differences in tuber sizes across the fields. From both the Swedish and the Danish series of measurements it was found that the yield measurement were overestimated. The errors for one variety were however all at the same level. To compensate for this error the first load on a field can be weighed and the factor can be used more precisely to measure the yields in the rest of the field. The sensor requires that light from a lightbar can be measured by the camera disturbed by the tubers only. Other objects including soil and stones will be measured as well. Objects smaller or bigger than the normal pixel range for tubers are filtered out. The sensor must be well-covered to avoid disturbance from incoming sunlight. Moist conditions can cause dew on the camera lens. The sensor is found to be a valuable tool. It can be used to ease the work of potato trails that normally require much manpower weighing yields. In connection with trials it can as well provide useful values for the size distribution of tubers. In big-scale farming the sensor can be used to find the differences in yield and quality across the fields. This information can be used for future optimisations in the field. The sensor also gives the farmer information on the size distribution of the tubers he has in stock. This information is valuable when the potatoes are to be traded. Mounting the present version of the sensor on a harvester is not easy. AGEC AB, the Swedish company that has developed the sensor, is working on a new compact version that is easier to fit in the flow of potatoes on more different harvesters.

Adoption of precision farming in Denmark

Soren Marcus Pedersen, Spyros Fountas and Simon Blackmore

Precision farming and the GPS-system have been prevalent in Denmark for about 5-10 years. The early adopters started with yield monitoring and mapping and continued with variable rate application of lime and nitrogen. Innovators were the first to adopt precision farming practices and so far about 400 farmers have adopted some site-specific and GPS related technologies on their farms. Even though many farmers expect that precision farming can help them to gain economic and environmental benefits, they face many challenges when implementing the technology, interpreting the results and making management decisions. This study summarises the conclusions from two mail surveys, which were undertaken in 2001 and 2002 with the same group of Danish farmers. The first survey addressed adoption trends, profitability and environmental issues and the second survey focused on information sources and management issues. Use of equipment and software were also addressed in the last survey. The farmers in the two surveys were not chosen randomly they were producers who had already used some precision farming practices on their farms. To get a better understanding of the survey results we extended the analysis with focus group meetings and personal interviews. In total, 35 stakeholders have participated at focus group meetings and 28 have been interviewed in 2001 and 2002. The target group was precision farming stakeholders, including consultants, commercial companies, researchers, consumers and farmers to get an in-debt knowledge about obstacles for adopting precision farming technologies and the potential economic and technical benefits. The results of the surveys, focus groups and interviews were put together and analysed in a holistic perspective in order to point out parameters, which have a particular impact on the adoption of precision farming in Denmark. We also discuss the perceived advantages and disadvantages of precision farming in practical applications, focusing on information management and decision-making in a complex management system. Furthermore, adoption rates and trends as well as future opportunities and constraints are addressed. Preliminary results from the two surveys, focus group meetings and interviews show that farmers and experts in general are optimistic about the future perspectives for precision farming in Denmark, however it is difficult to verify the economic and environmental gains in practice. In particularly, it has been difficult to demonstrate continuous yield improvements from site-specific fertiliser application and the costs of gathering information for patch spraying is relatively costly, compared with the potential gains. For some farmers variable lime application has been a profitable practise. Moreover, compatibility between hardware and software, as well as user friendliness are for many producers a serious impediment for adoption. A number of experts and stakeholders seem to agree upon that it is difficult to conduct variable applications based on historic yields and soil maps. Given the uncertainty related to yearly temporal variability, real time canopy management is required, to conduct precise and variable applications of nutrients and chemicals. More focus should also be put on site-specific economic and environmental effects, better logistics when harvesting, bale handling and field planning. Moreover, data handling and interpretation is also regarded as important obstacles for adoption.

Probable distribution of soil moisture in a well-drained paleudalf as influenced by landscape characteristics

E.M. Pena-Yewtukhiw and J.H. Grove

Soil moisture levels and their distribution affect most of the production and environmental processes that occur in soil environments. One of the principal objectives of precision or site-specific agriculture is to delimit and manage the soil-based production system in accordance with both its potential and its limitations in order to maintain maximum economic/environmental benefits. Many advances have been made in the study of soil moisture, water movement, contaminant transport, runoff, erosion and other related processes and properties. However, new knowledge is necessary regarding the spatial distribution of soil moisture, and its temporal dynamics, at a landscape scale. Uncertainty arises from natural variation in seasonal climate, biological (plant) response to that climate, and economic forces in the grain marketplace. Soil moisture under rain fed conditions will be affected by different sources of uncertainty, including climate and processes like movement of water within the soil mass. These factors can be studied at different scales, and a few examples of the effect of landscape on soil moisture uncertainty exist. In the USA, largely under rain fed agriculture, the most important limiting factor in agricultural production is soil moisture. We hypothesize that landscape is an important factor affecting the temporally-spatially-probabilistic distribution of soil moisture, even in otherwise well-drained soil. The study area was located near Lexington, Kentucky, in a udic environment. The soil was Maury (fine, mixed mesic Typic Paleudalf) silt loam, characterized as a deep, well-drained upland soil developed in residuum of loess and Ordovician limestone. The landscape was gently rolling, with small ranges in elevation (maximum difference of 1.55m) and slope (between 0 and 9.7%) in a study area of 0.36 ha, where 38 sampling points were established. Intact cores were taken at each sampling point and both texture and bulk density were determined in 10-cm intervals, to a depth of 60 cm. Volumetric soil moisture was measured over the 0 to 60 cm depth, at 10 cm depth intervals, every two days between October 2001 and October 2002, using a calibrated portable Sentek capacitance probe. Annual crops were present in the study area from 15 April until 15 October, 2002. Field capacity, wilting point, plant available water, and root zone saturation were determined in situ from the temporal capacitance data. There were two pronounced hydrologic periods; wet-from October 2001 until May 2002, and dry-from May 2002 until October 2002. Although the soil in the study area was fairly uniform, we were able to measure differences in the frequency distribution for measured soil moisture at each sampling point within this area. The frequency appeared to be related to both landscape variables as well as the spatial distribution of soil properties (texture, bulk density). Vertical flow is predominant in this well-drained soil, but lateral flow was also observed, and this affected the ability of the soil to retain moisture across both the wet and dry periods within the study year. Uncertainty (risk) lies at the heart of grower decision making as regards crop selection for rain fed soils/fields. A stochastic analysis of the distribution of probable soil moisture may be an aide to grower decision making.

A variable rate pivot irrigation control system

C. Perry, S. Pocknee and O. Hansen

The past decade has seen a rapid increase in the availability of geospatial tools such as Geographic Information Systems (GIS), Remote Sensing (RS), and Global Positioning System (GPS). The use of this technology has been grouped under the umbrella of "Precision Agriculture"- a spatial management philosophy that espouses matching inputs to exact needs everywhere. Precision Agriculture (PA) advocates methods that allow farmers to alter input application rates relative to the needs of individual areas within fields. Technologies that facilitate such methods have come to be known collectively as Variable Rate Technologies (VRT). While many see the promise of precision agriculture, there are clearly some factors that complicate the real world application of site specific theories. One of the most quoted complicating factors is a farmer's inability to predict or control the weather. One could argue, however, that weather related risk has long been managed (at least partially) through the use of irrigation. In total, some 25 million hectares of US farmland are irrigated . While VRT systems exist commercially for inputs such as fertilizers, herbicides, insecticides, and seed, it is curious that no commercial VRT system exists to apply water. Logically, producers will adopt precision agriculture technologies only if they impact directly on those factors that have greatest influence on their production systems. For a large portion of the US, the single most important production factor is irrigation management. Although water is the major yield determiner, irrigation also has negative impacts. There is increasing concern that irrigation is contributing to excessive ground and surface water abuse because of inefficient water management. One of the ongoing challenges the irrigation industry faces is the optimization of water use so that neither yield nor the environment are negatively impacted. To do this, more efficient irrigation methodologies must be found. In the US, CP systems account for roughly one third of all irrigated acreage and this proportion has been steadily growing over the past decade. Although CP systems are relatively efficient water applicators, there are a variety of reasons why uniform CP systems result in sub-optimal irrigation efficiencies. All of these are the necessary result of applying a uniform application of water to an inherently non-uniform medium. The non-uniform nature of a field may result from any combination of the factors below. 1. irregularly shaped fields resulting in off-site application of water through: i. internal inclusion of non-cropped areas (such as ponds or wetlands) ii. external intrusion of non-cropped areas (such as roads or other fields) 2. overlapping CP systems resulting in double applications of water 3. multiple crops under the same CP system 4. variability in soil types 5. variability in topography The solution to sub-optimal application lies in matching the field non-uniformity with an appropriate non-uniform CP application. The technology to do this is known as Variable-Rate Irrigation (VRI). VRI (also called precision irrigation) has yet to gain a foothold in commercial agriculture. The existence of a substantial body of recent research indicates that there is increasing interest in VRI. CP systems are particularly well suited to VRI because of their current level of automation and large coverage with a single pipe. VRI systems supply water in rates relative to the needs of individual areas within fields. A VRI system was developed through collaboration between the FarmScan group (Perth, Western Australia) and the University of Georgia Precision Farming team. The VRI system cycles sprinkler on/off time and controls CP travel speed to vary application rates. Desktop PC software is used to define application maps which are loaded into theVRI controller. The VRI system uses GPS to determine pivot position/angle of the CP mainline. Results from VRI system performance testing indicated good correlation between actual and target application rates and also showed that sprinkler cycling on/off did not alter the CP uniformity.

Use of guidance systems in precision agriculture

Krister Persson and Hans Skovsgaard

Use of guidance systems in precision agriculture Objective The objective of the proposed paper is to show the performance of DGPS based guidance systems that help the driver to hold the correct distance between each single tramline in fields where no fixed tramlines have been established, as it is normal procedure in Europe. It will be discussed how guidance systems will enable the use of machines with different working widths in the same field and how they will make it possible to make a stop in the field and afterwards easily return to the point where the job had been stopped. The second objective is to make proposals for a test procedure for guidance systems in order to establish comparable results. Methods The work has been carried out at Research Centre Bygholm as a part of a research project concerning techniques related to precision agriculture – fertiliser distribution. The project concerns precision in fertiliser application and accuracy of GPS/DGPS receivers. The work, which has been carried out in relation to guidance systems, covers development and test of methods for determination of the precision of guidance systems. The proposed methods cover tests of guidance systems used in fields where the working directions are either north-south, east-west or diagonal. The reason for testing systems in different directions is that earlier tests have proved different accuracies of the GPS receivers in those directions. The proposed methods cover operations at different working widths, i.e. 3-36 m in steps of minimum 9 m, in order to show differences in accuracy depending on the desired working width. The method covers tests of different guidance methods – either in straight lines or in curves. The proposed methods are based on registration of actual positions in the field by use of RTK/GPS. The test results will give much information to characterise the accuracy of the guidance system. The pass accuracy expresses the deviation from the actual track (measured by the RTK receiver), the pass-to-pass accuracy expresses the distance between each track, and the pass distance expresses the distance to the base line that will be the first track in the field. Methods for sensitivity testing of guidance systems are being developed. The tests will show which deviations from "correct position" will be allowed by the systems before information is given to the driver on how to correct the steering. Results In the work carried out until now, a guidance system from OUTBACK has been used. This system is based on Omnistar DGPS signals, but prepared for the EGNOS. The system allows working widths from 1 m and up in steps of 0.1 m. The sensitivity of the guidance information may be adjusted in three levels. As reference, a Topcon RTK-DGPS was used with a base station within a distance of max. 2.5 km. Tests have been carried out in different directions and at different working widths. The results show a good pass-to-pass accuracy in relation to fertilisation carried out with centrifugal spreaders, which is the most common spreader type in Europe. The pass accuracy tracks are found to be acceptable, also in view of the used fertiliser distributors. In the case where full-width spreaders are used, the deviation seems to be too high, as the guidance system will accept deviations of ±0.4 m. This means that some areas will receive too much fertiliser and other areas too little.

Analysis and theoretical estimation of the errors of the agriculture crop production mapping

Cârdei Petru, Muraru Vergil, Gangu Vergil and Pirna Ion

The paper consist in finding the errors variation of measurement parameters as a function from the process commands errors in the precision agriculture. Improving the quality of agriculture production from the point of view of power consumptions, and of the environment protection by diminishing the amount of managed fertilizers there are developed the new managing methods based on precision agriculture. For obtaining these improvements, the treatments of cultivated fields should be correlated with the soil local conditions. This method requires a lot of information.The production maps give information on the soil state maps (soil state parameters). The main objective of this paper is to establish the relationships between the final error sent to the combine harvester performing the production mapping and the errors of the process parameters (displacement speed, working width, flow, transfer through the harvester, etc.) supposed to be known. For analysis and estimation of errors we are using the mathematical model from [1]. The minimization of mapping errors assure finding of information as close as possible to the real one.By information, in this stage, it is meant the data collection containing the values of the real crop production on the network of the mapping points. Any information obtained by mapping is expected to be different from the real one. The differences between the data obtained by mapping process and the real distribution of the specific crop production are estimated by the standard deviation and variance. The errors estimation may be done by theoretical way or by numerical simulation. The analysis and simulations was made using computer software MATHCAD 2002 and our programs written in BASIC The experimental estimation of the mapping errors are practically impossible, on one hand due to the necessity of destroying the real information if we wish to have knowledge on the specific crop production before mapping and on the other hand, due to the operation unrepeatability because after a mapping the harvest is processed, the repeatability in identical crop conditions being excluded. For these reasons it is significant to emphasize the importance of each type of error occurring during the mapping process. In the paper we will analyze the following errors: - errors in working speed; - errors in working width; - errors in the flow transfer through the combine harvester; - errors of stopping or/and unloading; - errors of GPS system; From the point of view of the errors origin, they are considered as being of two types: - determinist; - stochastic. We studied the dependence of the error of specific crop production from the above-mentioned errors, but we can try to find the dependence from the other factors, like as: the quantity and quality distribution of the specific crop production or the type of crop for which the mapping is performed. The most important parameters within the models offered in this study of this type are those characterizing the dependence between the input flow and the output one of the material into and out of the harvester. The errors of sowing precision are unconditionally registered in the mapping process of specific crop production. [1] R. Vansichen, J De Baerdemaeker, Processing of Signals and Dynamic Systems at the Production Continuous Measuring Performed on the Harvesting Combines, AgEng 92; June 1st; 4th, Uppsala, Sweden.

Determining a methodology for characterising within-field variability

F. Piraux, F. Gasc and D. Boisgontier

The process of precision agriculture can be divided into three phases. The first consist in the detection of within-field variability zones. The second phase includes the observation and the agronomic interpretation of the homogeneous zones. And finally, the third step of precision agriculture is the implementation of the decisions taken according to the interpretation of management zones. Too often, precision agriculture is just restricted to modulating inputs (the third step of the process), whereas the most important phases are the determination and the characterisation of the management zones within the field. Indeed, the definition of the homogeneity and the study of the variability of each parameter which characterise the within-field management zones, will give the clues for the right decisions in terms of what should be modulated, and how. This research was carried out to investigate the variability of each factor participating to the characterisation of homogeneous zones within a field. The overall aim of the study is to describe the variability within a field given by each factor, for different pedoclimatic environments in France. The first objective of this project is the characterisation of eight different pedoclimatic environments in France. Georeferenced data were collected for three years in eight fields, each belonging to one typical French cereal basin. The average area of these eight fields is 10 ha. The data collection has been as exhaustive as possible, including soil data, meteorological data, remotely sensed data, plant measurements data, and other assessments directly carried out in the fields (such as weeds, wheel tracks, ...). The density of the punctual sampling, ten points per hectare, was chosen in order to get the most accurate interpolations over the whole field for each variable. The description of the variability for each parameter and for each group of parameters is an important step in the process of interpretation of homogeneous zones in a field. Indeed, the variability will be interesting for modulation if it follows three rules. First, the variability should be high enough to be agronomically interesting; that is, values of a variable should range between two extrema enough far apart from each other, for the modulation recommendation to take this variation into account. The second rule is that the variability should be within an interesting range of values, from the agronomic point of view (a field, for example, composed of several types of very rich soils, will have variability, but all the values will be in the higher class and therefore the recommendation will be the same for the whole field). The third rule for variability is that it should be spatially structured, that is, if a value is high, for example, its neighbourhood values should be somehow high too. Nowadays, within-field variability is mostly characterised using factors requiring samplings (such as soil texture, depth, mineral residue,) which appears to be an expensive method, especially when accurate management zones are needed. Previous studies have shown that other parameters (such as yield or electrical resistivity mapping) could reveal a certain variability within a field. The final objective of this study is to derive a methodology that could be used for characterising any field within a given pedoclimatic environment, reducing the cost of the determination of management zones. This methodology, based on the relation between the variables involving samplings and other factors (yield, electrical resistivity) that characterise the variability, allows to give information, for a given type of modulation in a given environment, on : 1/. which factors are the most relevant for the within-field variability, 2/. which sampling density should be used for the best quality/price ratio possible for each of these parameters, and 3/. what other data (such as electrical resistivity, previous year yields, other farmer knowledge of the field) could be used to refine the homogeneous zones.

Selecting the optimum locations to apply soil moisture sensors and other soil investigations

M. Plöchl, H. Zimmermann, C. Luckhaus and H. Domsch

The survey of soil data, which are precise and representative for the whole survey area, is a critical issue for many agricultural applications. The variability of soil structure and therefore soil data measured in one field can be relatively high. However, in most cases a limited number of soil moisture sensors is applicable, to assess the data of one field. In the following a method is introduced which assists in the selection of the optimum sites, i.e. which are representative for the surveyed area. This theoretical concept will be applied on soil moisture sensors and other soil investigations necessary to improve irrigation scheduling and to control/monitor nutrient leaching. Based on rapidly surveyed electromagnetic induction (ECa) data an algorithm is searched able to select measurement sites that are spatially representative of the entire survey area as well as suitable from a statistical viewpoint. The ECa data are measured once by a single overpass of the field by an EM38 attached to a tractor or any other vehicle. In order to minimise the disturbance and compaction of the soil the ECa data are surveyed along the usual work trails. Hence, the coverage of the whole field is obtained from Kriging interpolation routine. After this routine the data are transformed by a principal component analysis and a second-order central composite response surface design to form a pseudo response surface (PRS) design. The PRS design finally is used to represent calibration sites, spatially and statistical representative and well suited for point measurements and model applications. In the case of this study online and continuous measured soil moisture data are required to model optimum irrigation scheduling on large-scale agricultural areas. This irrigation optimisation is achieved by integrating a soil-vegetation-atmosphere transfer (SVAT) model, soil moisture measurements and remote sensed data of plant growth. The benefits, expected from this irrigation management, are an improved water use efficiency as well as a reduced energy consumption. Further it is envisaged to couple the plant growth model with nutrient requirements with the aim of a controlled fertilisation concept. As the soil moisture sensors applied are based on the Time Domain Reflectometry (TDR) principles they can be also used to monitor ion activities in the soil. Hence, the reduction of wasteful leaching of nutrients might be a benefit of the system envisaged, as well. Continuous online soil moisture measurements by several soil moisture probes are used to facilitate the accurate estimation of model parameters and validate the model performance. Remotely sensed plant growth data are surveyed by hand held or mounted radiometers, which perform measurements in the blue, green, red, near IR and thermal IR range of radiation. The growth data are used to validate the system during runtime and to trigger the model, if necessary. In addition the system uses meteorological data, monitored with a close-by weather station recording solar radiation, air temperature, wind velocity, air humidity, and precipitation. This allows for the spatial modelling of plant growth, plant development, soil moisture, and local water balance in order to calculate the up-to-date and near future crop water requirements. The system currently is in the state of testing for practical use by a pilot study (field experiments with potato and corn) in cooperation with companies for sensor technique, data transfer, and irrigation technique.

End tower position accuracies required for variable rate center pivot irrigation

S. Pocknee, C.D. Perry and C. Kvien

Center pivot (CP) irrigation systems cover some 8.5 million hectares of cropland in the USA. Recent developments have shown that it is technically feasible to retrofit these systems to enable variable rate irrigation (VRI). Given the importance of water to any cropping system and the ubiquity of CP systems it seems likely that there will be strong demand for VRI. The immaturity of VRI technology, however, leaves many practical questions yet to be answered. One unknown is the end tower position accuracy (ETPA) required to provide adequately-precise in-field applications. On current systems, it is the position of the end tower that is used to calculate positions for each of the individual sprinkler nozzles. If relatively low accuracy positioning devices (such as uncorrected global positioning system (GPS) receivers, or wireless network positioning services) could be used it would substantially lower cost and reliability implementation barriers. Because the end tower of a CP is necessarily constrained geographically by it's fixed relationship to the stationary pivot point, rotational measurements can be shown to be more precise than the absolute GPS measurements they are calculated from. Additionally, the limited speed of rotation allows the possibility of increased accuracy through repeated measurements at any given CP position. The authors devised a computer model relating ETPA, CP length, CP travel speed, CP nozzle configuration, and CP curvature to the in-field application accuracies. Environmental considerations such as wind speed & direction were not considered. To test this model two VRI enabled CP's were equipped with three separate GPS receivers. These receivers were a non-differentially corrected GPS, a Wide Area Augmentation System (WAAS) enabled DGPS, and a precision Real Time Kinematic (RTK) GPS. The CP's were 180m and 275m in length and were respectively equipped with drop-nozzles and overhead sprinklers. Errors in GPS and DGPS measurements were calculated relative to the RTK-GPS. Dynamic accuracy results seen were similar to those calculated from short term static tests with the GPS and DGPS receivers. By applying dynamic position filters it was possible to increase the rotational accuracy observed although it was not possible to remove temporally-correlated GPS errors. The observed results verified the model predictions and showed that it is possible to predict expected in-field accuracies using pivot characteristics and a measurement of ETPA. Because acceptable in-field accuracy is a function of the management task at hand, it is difficult to generalize about the suitability of any given positioning method. To accommodate this the authors proposed a number of possible VRI management goals with increasingly more stringent spatial accuracy requirements and compared the suitability of each positioning system with them. Stand-alone GPS was shown to be sufficiently accurate for differential application to compact zones larger than one hectare in size. Differential application on thin, linear zones (such as ditches, waterways, and roads could only be accurately achieved with DGPS or RTK-GPS. However, while positional accuracies allowed management of these features it was seen that the geometry of the feature relative to the minimum fixed application grid of the VRI system (a function of nozzle spacing and control zone grouping) was often the limiting factor.

Maximising profit from orchard product tracking techniques

J.P. Praat, A.F. Bollen and A. Mowat

The supply chain for perishable products is complex where product variability, supply and demand pressures and poor information flows result in imperfect knowledge for managers at all levels. However, demands are increasing for information on product origin, production inputs, environment impacts and quality. The data does generally exist but not usually in an easily accessible form. A key point of differentiation between competing supply chains is their ability to transfer information seamlessly throughout the chain. Frequently, information required by the postharvest and/or marketing components of the supply chain must be sourced directly from the producer. Increasingly, the information sought from the producer needs to be based on complex data collected under a wide range of field conditions. This paper reports on techniques that can add value to the information required of the producer for all chain participants. Techniques can be applied to improve the quality of information at one level of the supply chain, which is pertinent to other levels of the supply chain. These techniques include geographic information systems, handheld data capture devices, remote sensing equipment and database management at different levels in the supply chain. Also, the information generated from these techniques can be used to optimise physical and financial aspects of horticultural production systems, in addition to the provision of auditable product traceability. This aspect may help to motivate improvements to current information systems. While there are exciting opportunities for further progress, horticultural production and its supply chain has challenging demands for these technologies, not the least of which is the question how will it work in practice?. Case studies from the pipfruit and kiwifruit industries explore the implications, and identify current limitation, for researchers, developers and users of the various technologies in New Zealand. Temporal and spatial variability of product quality (fruit size, dry matter and brix), yield and profitability have been analysed for several seasons. Relationships between canopy architecture, slope, aspect, elevation, soil conductivity, soil quality air temperature dynamics and crop performance are explored. The profitability of orchard blocks relates to product yield and product price. Systems have been developed to discretise product information in order to identify opportunities for improving crop quality and yield. Technologies such as barcodes, radio frequency tags, GPS and electronic compass to track harvest equipment have been used to trace product from the orchard to the packhouse. The intimate linkage between the orchard harvest site and the packhouse is seen as the major issue with hand harvested horticultural products. The systems described here can integrate with regional systems for predicting and managing national crop volumes and quality. Of particular interest to New Zealand growers is the ability of these systems to provide tracebility back to origin of individual bins filled during harvest. Product variability makes it difficult for growers to produce consistently produce of sufficient quality to attract these premiums. Using spatial information systems has not only improved the reliability for product tracing but also assisted with orchard management.

Use of airborne gamma radiometric data for soil property and crop biomass assessment, northern dryland agricultural region, Western Australia

G. Pracilio, M.L. Adams and K.R.J. Smettem

The west Australian agricultural landscape is variable in soil and crop yield. Airborne radiometric data potentially provides an improvement in the assessment of such variation at a more detailed scale than current methods. In particular soil texture, soil fertility and crop biomass variation were investigated. Paddock or farm scale variations of the former two variables are not readily available. Radiometric data (potassium, thorium, uranium, total count) was acquired at farm and paddock scale to potentially provide such detail (100m line spacing and 20m flying height). A simple exploratory analysis of linear relationships between radiometric and soil/crop properties was investigated. Spatial structure from semi-variogram plots determined the potential for mapping soil properties. Three study areas in the northern wheat belt region near Yuna, Western Australia were chosen. In general, linear relationships were more likely in decreasing order across the study sites at Summerset, Nolba north and Nolba south. Geology and soil texture were two important factors in understanding this difference. The primary radiometric source within Nolba south was freshly weathered granitic gneiss, hindering strong relationships with soil property or fertility. Nolba north and Summerset occur in a weathered sedimentary terrain, where primary minerals were not likely and correlation with soil texture more likely. However correlation coefficients differed greatly between these two sedimentary study sites. The lack of correlation in Nolba north was attributed to dominance of sand and a major fault at the study site edge. Within Summerset, up to 68% of the log clay content variation was explained by log thorium. Since organic carbon is also related to clay content, strong relationships with total count were also evident ($r2 = 0.67$). Such relationships could be used to predict these soil properties over the Summerset farm, particularly since the spatial structure between these data sets were similar. Spatial prediction of percent gravel was not feasible as there was insufficient gravel content at greater than 10% to develop relationships with any confidence in Summerset. Colwell potassium was the most related of all the soil fertility data investigated with radiometric data across Nolba and Summerset farms, with r2 of 0.74 and 0.62, respectively. The Nolba farm relationship is localised to the geological boundary conditions of the site. Spatial prediction of potassium would be feasible since matching spatial structure was evident between the laboratory and radiometric based potassium. Principal component analysis highlighted the need for further analysis of potential relationships of nitrogen with conductivity, terrain and thorium. Despite the generally weak correlation between soil properties and radiometric data across Nolba north, radiometric data explained up to 69% of crop biomass variance of a paddock. This was comparable with Summerset, where relationships explained up to 74% of the biomass variance. These are based on maximum r2, where biomass and radiometric relationships vary according to season. In Nolba north, lateritic gravel relative to the sands, was a driver in higher crop biomass. In Summerset, potassium was identified as a positive factor. Where sand or granitic parent material dominate paddocks, less biomass variation was explained by radiometric data (at up to 38 and 15%, respectively). In sedimentary dominated terrain with some soil texture variation, radiometrics can be used to develop simple linear relationships to predict % clay, potassium and biomass (in some seasons and paddocks). The matching spatial structure between soil sampled and airborne radiometric based measurements indicates that such soil property variables could be represented spatially at a scale not currently available.

Optimizing soil organic matter supply considering soil heterogeneity

M. Puhlmann and U. Franko

A crucial part of precision farming is the identification of significant indicators which are related to crop yield variability. Another requirement is that the indicators should be usable to modify crop management strategies, for example to reduce nitrogen fluxes into groundwater. It is known that different soil properties (clay content, soil organic matter, soil moisture, etc.) or the nitrogen fertilizer application can be key factors for yield variation. These key factors can vary from one field to another or even within a single field. This paper will discuss a possibility for the optimization of organic matter supply in order to control nitrogen cycling for heterogeneous fields. Yield mapping on an approximately 73 ha large agricultural field (high intensity farming) which is located in the Chernozem area of Saxony-Anhalt (Querfurter Platte, Central Germany) has shown remarkable yield differences for winter barley from 3 to 10.8 t ha^{-1} (mean 7.5 t ha^{-1}). Investigated soil parameters were content of total carbon (Ct) and nitrogen (Nt), content of hot water soluble C and N (Chwl, Nhwl; easily available soil organic C and N compounds), clay-, silt- and sand content, soil moisture, NO_3-N and NH_4-N. These parameters are necessary for an assessment of N-cycling and N-leaching. Soil samples were taken in two depths 0 to 30 cm (Ct, Nt, Chwl, Nhwl, NO_3-N, NH_4-N, soil moisture, clay-, silt- and sand content) and 30 to 60 cm (Ct, Nt, Chwl, Nhwl, soil moisture, clay-, silt- and sand content) on an 150 x 150 m grid on the field after harvest. Additionally the conductivity of the soil was determined continuously on field scale by means of electromagnetic mapping (EM38). First results from statistical analysis show that the soil parameters Ct and Nt are closely related to crop yield. Moreover, the yield is related to silt and clay content, soil moisture and the electromagnetic conductivity (descending order) determined immediately after harvest. The electromagnetic conductivity was also related to silt and clay content. For each point on the 150 x 150 m grid model simulations were carried out to estimate the actual risk of N-leaching into groundwater. For the simulations the CANDY model (Carbon and Nitrogen Dynamics) was used. CANDY has been developed and validated for the considered side conditions (Franko et al. 1995; Franko et al. 1997). In the model not only organic and inorganic fertilisers but also the diffuse nitrogen input from the atmosphere, which is estimated to be about 60 kg N ha-1 a-1 in the region, were considered. In addition the management strategy in the model simulations was changed. The aim of the strategy change was a reduction of the N-surplus in the soil after harvest. From the Static Experiment Bad Lauchstädt the optimum rate of ca. 0.8 t ha-1 a-1 for soil organic matter turnover has been identified as a goal for sustainable agriculture. With a management scenario that keeps the carbon turnover near this rate one could expect an utilisation of atmospheric nitrogen and a reduction of N-surplus in the soil after harvest. Both (actual versus optimum carbon supply) scenarios will be compared with special consideration of soil heterogeneity effects. Franko, U., Oelschlägel, B., Schenk, S. ,1995. Simulation of temperature-, water- and nitrogen dynamics using the model CANDY. Ecological Modelling, 81: 213-222. Franko, U., Crocker, G.J., Grace, P.R., Klír, J., Körschens, M., Poulton, P.R., Richter, D.D. 1997. Simulating trends in soil organic carbon in long-term experiments using the CANDY model. Geoderma, 81 (1-2): 109-120.

GIS-supported simulation of the economic risk of different site-specific nitrogen fertilization strategies

Antje Reh

At present site-specific fertilization strategies are discussed as a possibe increase of profitability in agriculture. Beside anthropogene influences spatial differences in soil fertility lead to crop yield variation within a single field. For crop yield level causal relations between soil, climate and crop management are characteristic. In order achieve economic and also ecological advantages, crop input intensity must be adapted to the spatial yield potential within a field. Due to the uncertainty of data farmers are running the risk either to waste resources (nitrogen) or to put insufficient fertilizer quantity to a crop and thereby to abstain from possible profits. For the determination of the economical optimal input rate of nitrogen according to spatial yield potential within a single field, simulation techniques like plant growth models and Monte Carlo methods were evaluated.

For the spatial demarcation of yield variability Geographic Informations Systems (GIS) were used, which offer complex data processing tools like supervised classification of different information sources. After the vectorisation of classified zones vector polygones were created describing areas of different productive levels (low, medium and high yield potential). These poligones served as a basis for subsequent simulation calculations and provided information about the size and location of each management unit within the analyzed field.

The modelling of site-specific yield response to nitrogen required the estimation of equations for production functions, which should consider spatially different soil conditions as well as the influence of annually changing climatic conditions. So far only very few empirical data was existing to derive production functions from conducted yield response trials for individual management areas. In order to compense this data lack causal plant growth models were used for the estimation of stochastic production functions. The simulation model CERES-IM was able to evaluate the impact of spatially different N-inputs on plant growth and yield of each management unit under different climatic conditions.

The evaluation of the economic risc of different site-specific nitrogen input strategies should take into account the uncertainty of the principle independent variables like climate and product price. By defining the uncertain variables as probability distributions the computer model was able to simulate this uncertainty appropriately. For simplyfying modelling it was assumed that only nitrogen should be considered as a possible yield limiting crop input.

The results of simulation calculations provided cumulative frequency distributions for each nitrogen strategy. For decision making farmers can compare and evaluate each strategy in accordance to their individual preference for risk. The results of this research indicated that GIS-supported simulation techniques are reliable decision-making tools for risk analysis of site-specific crop input strategies.

Site-specific land use as demonstrated by planning variable seeding rates

E. Reining, R. Roth and J. Kühn

One of the first steps in cropping is the establishment of crop stands that are in correspondence with the yield potential of the site, the previous crop, the target yield, the time of seeding and the quality of seed and seedbed. Until nowadays, a uniform seed rate for the entire field is used in most cases. Nevertheless these fields often show large variations in plant and stand development in spite of uniform management because site properties vary within the field. The adaptation of the seeding rate to the in-field variability in yield potential offers the possibility to reduce the use of resources or to increase yields. Prerequisites for applying site-specific seeding rates are: - distinct variability in site quality - sufficient and reliable information on the cropping site for assessing yield potentials for the different parts of the field - adequate technology for realizing the necessary application maps (seed rates planned and actually applied) - exact information on the wheat cultivar to be used - technology for establishment of yield maps as a basis for controlling the results of the application. Within the framework of the research project pre agro (Management system for precision agriculture to increase the efficiency of farming and promote its environmental compatibility) a GIS-based decision support system was devised for the differential seeding of winter wheat. A fundamental prerequisite for planning site-specific seeding rates is reliable and spatially explicit information on the yield potential of the site and its in-field variation. This information can be derived or compiled e.g. from yield maps of preceding years, detailed soil maps, aerial photographs, digital terrain maps or results of crop growth models. Since the availability of such information cannot in every case be taken for granted, the present module offers the possibility to assesses the yield potential of the site by using information on soil parameters (maps of the German Soil Rating, RBS) and local climate data (annual precipitation). Further adaptation to the actual conditions are possible by accounting for the date of seeding, the previous crop, and, if necessary, the terrain inclination. In addition, the farmer is enabled to apply corrections to the average yield resulting from his experience from preceding years. The expected yield and information on the wheat cultivar are the basis for the calculation of target stand densities (ears m^{-2} and plants m^{-2}) on the different field parts. At this stage of the calculation, corrections are applied for the quality of the seedbed, germination capacity of the seeds, plant emergence rate, and plant losses over winter. The result is the seeding rate in number of grains per m^2. By using the thousand kernel weight this value is converted to the seeding mass (kg ha^{-1}) for the field and the field parts. The results of the calculation are used to produce application maps. Within the framework the pre agro project differential seeding rates were applied on a variety of fields in different regions of Germany. The results of the applications were controlled by spot checks of plant density, ear density and sample yield. Results so far show that site-specific seeding rates can be successfully applied. In some cases, however, site information is not sufficient to account for observed differences in plant stand density or yield.

Optimisation of oblique-view remote measurement of crop N-uptake under changing irradiance conditions

Stefan Reusch

In the recent years several spectral indices have been successfully related to crop biomass, chlorophyll production and nitrogen uptake. Especially the knowledge of the actual site-specific N-uptake is regarded to be essential for variable fertilizer application. However, most of the data have been acquired at nadir view and under optimum irradiance conditions, i. e. around solar noon and/or at clear skies. This may be sufficient for aerial and satellite imaging, but will definitely be problematic if ground-based routine measurements are considered. In those cases an oblique viewing geometry is much more practical as it avoids long booms to carry the spectral sensor out of the shadowing area of the vehicle. To reliably measure the N-uptake at an oblique view all over the day under changing irradiance conditions, two steps are taken: Firstly, a special viewing geometry is designed to suppress solar azimuth effects on the reflectance spectrum. Secondly, the spectral index is optimised to be as insensitive as possible to remaining solar zenith and cloudiness effects and at the same time as sensitive as possible to N-uptake. Typically, at an oblique view strong effects of the relative azimutal angle between the solar and the viewing direction are observed. These effects can be well visualized and quantified using the PROSPECT/SAIL canopy reflectance model. They were found to significantly interfere in the determination of the N-uptake out of the reflectance spectrum. However, simulations on the canopy reflectance model also show that these effects can be drastically reduced if the oblique measurements are made in two or more azimutal directions symmetrically and simultaneously. It is concluded that with four optical inputs (and an angle of 90° between each two of them) solar azimuth effects can almost be removed completely. Though by this azimutal effects can be removed very well, the reflectance spectra are still affected by the solar zenith angle due to the non-lambertian reflectance of all canopies. For example, it has been observed that the infrared reflectance (around 800 nm) of a canopy may more than double from noon to evening when the sun reaches the horizon. If spectral indices (e. g. spectral ratios) are used instead of single wavelengths, these effects can be reduced. However, it has been found that well-known indices like the infrared-to-red ratio or the NDVI often do not perform very well in that respect. To find better indices, two sets of reflectance spectra have been acquired: One dataset (referred as signal) contains reflectance data from canopies with different N-uptakes, while a second dataset (referred as noise) contains data that have been taken continuously on the same spot of an average canopy across the whole growing period. Spectral indices are then calculated from the reflectance spectra and linearly related to the N-uptake using the signal dataset. The same spectral index is calculated from the noise dataset and transformed into N-uptake using the linear regression derived before. The performance of a spectral index is then characterised by the daily standard deviation of the N-uptake from the noise dataset. The less this standard deviation, the less sensitive the spectral index is to irradiance changes and the better it can be used for remote measurement of N-uptake. Various spectral indices have been ranked according to this procedure. It has been found that even with simple but well-chosen two-wavelength reflectance ratios much better results can be achieved than with the standard infrared-to-red indices. In addition, it turned out that for different crop types different spectral indices are optimal.

MOSAIC: crop yield observation - Can landform units improve our understanding of yield variability

H.I. Reuter, O. Wendroth, K.C. Kersebaum and J. Schwarz

In fields, crop yields can vary strongly. Still, farmers typically manage fields as one single unit, applying the same management practice across a field, although it consists of different textural classes, different landform units, as well as different soil nutrient contents. In order to manage fields according to the demand of plants, one has to understand, how and why yield variability occurs. This knowledge allows the implementation of management decision making. Research in North America and in Europe has shown that a small fraction of a given field exhibits a temporally stable crop yield pattern. Crop yield patterns are related to specific landforms (LF) and might be helpful to explain yield variability at the field scale. For example: crop yield at footslope positions decreases during a wet year, whereas during a dry year, yield at footslope position may increase. A Digital Elevation Model (DEM) describes a landscape or a field site in terms of topographic features. A relief analysis tool was developed within a Geographical Information System (GIS), which uses DEMs to classify different landform units within a field site. A DEM was obtained by laser scanning with a spatial resolution of 1m on a field site in Lüttewitz in south-eastern Germany, Federal State of Saxony. The published routines by ({Pennock, Anderson, et al. 1994 369 /id}) were implemented into the GIS Arc Info. The four components slope, profile curvature, planform curvature and flow accumulation area were used to segment the landscape into up to eleven LF for a spatial resolution of 10 m. The four major LFs Shoulder, Backslope, Footslope and Level positions were classified. Further classification was performed for convergent, planar and divergent areas inside the major LF. This resulted in a map containing the LF elements which can be provided for the field site. Local small size LFs can occur due to I) errors in the dataset provided, or II) the representation of local small-scale topographic features. This misclassified LF increases the difficulty to interpret crop yield. These misclassified areas, which were smaller than a given area threshold, were replaced by an iterative process by using a local filter. Yield properties in a 20ha field site were measured for spring barley in 1999, and for winter rye in 2000 and 2001. Grain yield was obtained using hand harvesting at 192 plots of 0.5-m^2 size. Additionally, yield mapping was performed with a Claas Combine using a CEBIS system. In general, less than 50% of the area showed consistent spatial yield ranking from one year to the following year, i.e. crop yield grown at the field site showed variable behaviour depending on weather conditions. In 1999 yield at level positions was highest with 0.2 t/ha above field average. In contrast, during the wet year of 2001 yield at shoulder position was 0.7 t/ha above the field average. Spike analysis showed that yield per spike was homogeneously distributed over all LF units and years, whereas Thousand kernel mass was not. Therefore the differences in yield were related to the number of spikes per area as well as to the number of kernels per spike. Both yield properties are indicators of stress during development. The impact of soil moisture and aeration at different LF units on yield development will be discussed. Landform analysis proved to be helpful for the understanding of yield and yield components during three years of on farm research. Conclusions based on the spatial distribution of yield components related to landforms for variable nitrogen applications will be drawn.

Improving the information of the yield maps

A. Ribeiro, M. García-Alegre, L. Navarrete and C. Fernandez-Quintanilla

The foundation of precision agriculture rely on the ability to continuously monitor yield during the harvest operation. Precision agriculture will be beneficial only if the information provided increases the certainty of a decision. Yield monitoring combined with the ability to establish a geographic reference for these data allows the producer to construct yields maps and track field performance from year to year. Yield maps provide a rich information to analyze the sources of variation in crop yield and there are many examples which show this variation related to other site attributes. A major difficulty in the yield maps interpretation is the ambiguity in the meaning, that provokes the assignation of yield variation to a wide range of possible causes. The present work aims to help in the analysis of the yield maps from the weed presence perspective. The application allows to acquire georeferenced short videos, accomplishing a sampling plan, simultaneously to the harvest job. The sampling plan is previously defined by the user that can apply different strategies, such as grid or random. Then in the field, the sampling points are detected by a DGPS receiver and the short videos acquired in the sampling point are stored with their associated position. The PC recorded videos can be later visualized by the expert and/or the farmer to jointly estimate the weed abundance. In this stage videos appear as small icons in a map and users can click on it for visualization purposes. Once the video has been selected, its sequences are continually repeated until user decides to stop them. The video size forces this repetitive operation. The weed abundance estimation can be numerically ranged by the expert and displayed on the yield map. Finally, the expert can assess the relation between weed abundance and a lower yield. This management improves the design of an effective and selective plan for weed treatment. The implemented application does not need any control or surveillance from the harvester driver and videos are automatically acquiring in the processing system (a laptop PC). The developed system has three main components: a digital video camera, a Pentium PC and a GPS receiver with differential correction. The digital video camera has been located over the cutterbar looking ahead; in front of the harvester. The camera is a SONY DCR-PC110E with 1,070,000 pixels CCD and 520 lines horizontal resolution. The Personal Computer is a Fujitsu Stylistic 2300SM pen tablet, a high-performance pen based computer that is designed to support, in this case, Windows 98. It is specially designed to work on outdoor environment, with Colour Transflective (CTF) SVGA LCD display measuring 8.4. The video camera and PC are interconnected trough a IEEE1394 device that has a transfer rate up to 400 Mbits/sec. At last, the Global Positioning System (GPS) is a 3100LR12 from OmniSTAR a Fugo Group Company; a 12-channel GPS receiver and a L-Band differential receiver, both housed within a single unit. Position accuracy is typically less than one metre and the receiver is suitable for both ground and air applications. The DGPS receiver is supplied with a combined GPS and OmniSTAR antenna for signal reception. The connection to computer is via a standard RS-232C port.

The critical challenge of learning precision agriculture new skills: Grower learning groups

Pierre C. Robert and Christopher J. Iremonger

One of the most important impact of precision agriculture (PA) -site-specific management- has been to create the need for spatial detailed information about soil, landscape, and crop characteristics. Early adopters within the U.S. Midwest have significantly benefited from the creation of farm information systems in many ways, including unexpected ones (e.g., banking, crop insurance, land value, and food safety). Now that a greater number of growers are adopting some aspects of PA, there is a very critical need for helping them in many ways including providing unbiased information about the new PA technologies and practices; starting to use PA progressively and efficiently; developing on-farm simple but correct experiments to define optimum site- specific practices; and, perhaps more importantly, process, manage, and use efficiently all the data collected over the years related to soil properties, seeds, fertilizers, pests control, and yields. This is an overwhelming challenge for most growers. The Precision Agriculture Center (PAC), University of Minnesota, has been testing and adapting an outreach model based on Grower Learning Groups (GLGs). One of the main characteristic of this project is our emphasis on empowering growers and their advisors to conduct their own experiments, build local databases, and create local recommendations. METHODS The first year, the PAC has sponsored three GLGs dedicated to sharing knowledge of precision agriculture practices and conducting on-farm experiments. The GLGs represent distinct regions defined by soil/landscape, climate, and cropping system (agroecoregion). Growers identified limiting factors for crop yield, quality or water quality management and then run experiments to test new management approaches. Extension agents, local consultants and agribusiness companies can participate in activities. PAC outreach scientists: - help select yearly goals, - help design on-farm field experiments, - help field soil/landscape/plant characterizations, - assist with data analysis and management. A data warehousing system with web - access for GLG members is constructed. - organize timely topic discussion/training and provide equipment for hands-on training. EXPECTED RESULTS - GLG members are more involved in learning new skills because they are in control; - GLG members gain more efficiently new skills by doing; - GLG members learn realistic protocols for on-farm research and data aggregation; - Each GLG member benefits from ALL member experiments; - A much larger number of on-farm field experiments without disturbing too much farm operations; - The adoption of changes facilitated from group experience sharing; - A better knowledge of limiting factors for crop quality, yield and water quality within distinct landscapes; - Participants profit from a friendly GLC data warehousing website with processed field information. CHALLENGES - Growers may not want to share all data with their neighbors and competitors; - Many growers will be interested in understanding what products or rates perform best on their fields, while researchers want to understand why the products perform differently; - GLGs may founder without committed project champions at the local level; - GLG members may not follow adopted protocols because of time constraint, weather related problems, temporary equipment or personnel resource difficulties; - Environmental quality experiments may be of limited interests or of concerns. FIRST YEAR RESULTS Summary of the first year activities, results, and problems. CONCLUSIONS Concluding remarks on what has been learned by GLG members and PAC outreach coordinators, and decisions for the second year.

Use of precision farming technologies in cotton production as influenced by producer and farm characteristics

Roland K. Roberts, Burton C. English, James A. Larson and Rebecca L. Cochran

Cotton is a high-value, high-input crop with potential for profitable precision farming and reductions in input use. Our objective was to determine the factors that influence cotton farmers to use precision soil testing and variable rate fertilizer and lime application. Identifying these factors could help policymakers encourage cotton precision farming if warranted. Let Us be the expected benefit from gathering site-specific information necessary to make the variable rate (VRT) versus uniform rate technology (URT) decision and let Uv|s and Uu|s be the respective expected benefits from using VRT and URT given that site-specific information was gathered. Further, let Uw be the expected benefit from gathering whole-field information. Given Us*=Us–Uw and Uv*=Uv|s–Uu|s, a farmer who maximizes expected benefits: Gathers site-specific information and uses VRT when
1) Us*>0 and Uv*>0;
Gathers site-specific information and uses URT when
2) Us*>0 and Uv*<0; and
Gathers whole-field information and uses URT when
3) Us*<0.
Gathering whole-field information and using VRT is not possible because the site-specific information necessary for VRT use was not gathered. These decisions are sequential and independent. By choosing to gather site-specific information, the farmer is self-selected into the sample of farmers who can choose between VRT and URT. This property implies the use of methods that account for sample selection. The unobservable latent variables, Us* and Uv*, are assumed to be random functions of observable vectors of exogenous variables Zs and Zv,
(4) Us*=ZsGs+Es, Uv*=ZvGv+Ev,
where Gs and Gv are vectors of unknown parameters and Es and Ev are random error terms. Although Us* and Uv* are not observed, a farmer's decisions can be characterized as observable zero-one variables,
(5) Is=1 if Us*>0, Is=0 otherwise,
(6) Iv=1 if Uv*>0 and Is=1, Iv=0 otherwise.
The probabilities of occurrence for the alternatives in equation 1 through 3 can be written in terms of the variables given in equations 5 and 6,
(7) Pr(Is=1; Iv=1)=F2(ZsGs, ZvGv, r),
(8) Pr(Is=1; Iv=0)=F(ZsGs)–Pr(Is=1; Iv=1),
(9) Pr(Is=0)=1–F(ZsGs),
where F2 and F are cumulative distribution functions for the standard bivariate normal and standard normal distributions, respectively, and r is the correlation between Es and Ev. If r is not zero, equations 5 and 6 form a system of equations that can be estimated as a bivariate probit model; otherwise, these equations can be estimated as separate binomial probit models. Data were collected from a mail survey of cotton producers in Alabama, Florida, Georgia, Mississippi, North Carolina, and Tennessee conducted in 2001. Of 5,976 cotton producers, 1,031 (17%) responded. The number of usable responses was reduced to 773 because of missing data. Of these 773 respondents, 136 (18%) said they used grid and/or management zone soil sampling and 76 of those respondents (56%) said they used VRT for fertilizer and/or lime application. Equations 5 and 6 were estimated with binomial probit models because r was not significantly different from zero. Farm size, attendance at college, age of the farmer, land quality, the farmer's perception about the importance of cotton precision farming five

years in the future, and the farmer's perception about whether precision farming was profitable on their farm were significant in determining whether a farmer used precision soil testing. Computer use for farm management, owned versus rented land, and the farmer's perception about the cost of a cotton yield monitoring system versus the actual cost of the system were not significant variables in explaining the use of precision soil testing. Use of VRT, given precision soil testing, was significantly explained by the farmer's age, land quality, and the farmer's perception about the cost of a cotton yield monitoring system versus the actual cost of the system. Other variables mentioned above were not significant in explaining application of fertilizer and/or lime using VRT.

Data management for transborder-farming

M. Rothmund and H. Auernhammer

Introduction and purpose: Transborder-farming as a virtual land consolidation is one possible way to improve the economical situation of farmers in small structured farming regions. But cultivating transborder-fields across existing property borders requires continuous documentation of all in-field working processes, if an exact allocation of yield and effort is needed afterwards. Only automatic data acquisition systems in tractors and other maschines guarantee high quality of acquired information. As a consequence of the data amount a management software for data handling and analysing is needed. Within the preagro research project in the section micro-precisions-farming a data management system for transborder-farming should been developed wich integrates online data acquisition on-the-go and data post processing at the PC. Methods: For data acquisition a system developed at the Munich University of Technology was used. The system stores GPS-Position together with important technical attributes during fieldwork each second. Furthermore yield measurement systems for harvesters were used which were partly available on the market and partly developed or adapted for the project. So the yield could also be allocated site specifically by GPS data. For data evaluation and analysing a completely new system had to be generated, in which special requests for transborder-farming were incorporated. The developed software is based on a Microsoft-Access database. The user interface is generated by Access-forms and programm sequence runs by Access-macros and VBA-modules. The System was installed and tested in Zeilitzheim, a small village in Lower Franconia in Germany. In cooperation with seven local farmers three transborder-fields, each about 7 hectares in size, were farmed together. Results: For two seasons all working processes on the three transborder-fields were documented automatically. Starting from a database concept two years ago, the data management software has been developted step by step till all important functions for manage transborder-farming data were included. These functions are: - import of different raw data types into a common database - raw data checking for plausibilty - allocation of each data set to a deposited area element (for instance a field) by GPS data - summarizing raw data to jobs by grouping according to certain values - storing job-related data sets for fast access on the one hand an job-linked original data sets on the other hand for using in GIS applications or analysing tools - automatic detection of transborder actions and accordingly job merging - correction of result data for validation in transborder-farming - generating invoices for the farmers Conclusions: Quality of results from a data management system depends on quality of input data. Technical problems at field work can result in a total data loss and, then as a consequence, there is no base for validation of field work. So the reliability of the data acquisition system should be improved further on. For the case of a technical breakdown during data acquisition a possibility of manual job data input was created in the software. But there is no way of a real work data based correction for validation with each farmer after cultivating a transborder-field, when there is no online documetantion. The developed system of on-the-go data acquisition and post processing allows data management for transborder-farming without notable time effort. So the economic advantages for farmers cooperating in a transborder farming system can be fully absorbed. The developed software concept should be reproduced by using an advanced programming language to make it independent from Microsft Access platform and to give it a higher performance.

Use of Electrical Resistivity (ER) maps derived from the ARP system to build up agronomic advices for precision agriculture

Dominique Rouiller, Jean-Marie Larcher and Michel Dabas

A considerable pedological variability, an intensive cereal production, environmental problems and the size of the farms (over 200ha) make the center of France a favorable region for the development of Precision Agriculture (PA). Owing to these characteristics, the agronomic department of the cooperative EPIS-CENTRE (E-C) offers its active members a service to implement PA. The offer consists of providing farmers with maps of intrafield agronomic advice on seeding, fertilization and agrochemical applications. To achieve this goal, the agricultural engineer must have a good knowledge of the delimitation of the management zones and their agronomic potential. For us, the difficulty was to define the areas of intrafield homogeneous plots with an accuracy compatible with PA.

In the past, engineers were encouraging farmers to establish yield variability maps over a minimum period of three years to characterize heterogeneity and define homogeneous zones through a year-basis comparison. We now consider that yield variability is not the most suitable parameter with which to define homogeneous zones, as a result of too many factors, including climatic and agronomic, errors of method or measurement…

Another method involves soil mapping. The spatial knowledge of soil conditions for PA, according to recent test and simulation using geostatistical models for different plots in France, should be acquired with a density in the order of 10 meters minimum in both directions. For economical reasons, this density is not compatible with traditional measurements like those performed with auger borings. Therefore, we decided to use a non-direct and faster method: the electrical profiling method. From the prototype developed in France by CNRS (MuCEP, CNRS and Paris VI patent), GEOCARTA has developed a commercial automated system for measuring ER. The ARP01 system, towed by an ATV, enables a real-time navigation and checking of three maps obtained for three depths of investigations. A specific resistivimeter and interface with a rugged computer was developed. Geo-codification of the data is carried out in real-time with a dGPS and a radar. The spatial resolution along the profiles can be as low as 10cm whatever the speed of motion. Maximum speed of acquisition is 6m/s but could be enhanced.

This system is now applied on a wide scale (thousand hectares) in France in 2002 by E-C. Measurements are done by the staff of E-C and data, send via the Internet to GEOCARTA, are processed in a few hours and the maps produced send back to E-C. The data, transformed into "pseudo-pedological units" are used in a GIS for the implementation of a precise pedological map and the positioning of soil measurements (auger borings or pits). These maps, as well as ones showing the clay, stone, chemical content are used to derive the management zones for the surveyed plots.

Agronomic models usually used by the engineers of the agronomic department of the cooperative, like for example: "EPICLES" model for Potassium and Phosphorus fertilization, "RAMSES" or "SCAN" methods for Nitrogen fertilization and "SEPALE" for fungicide protection, are involved in building up agronomic advice to farmers who had already implemented PA on their farms. For the campaign 2002-2003, farmers were given maps for seeding (rape, wheat, barley…). The seeding model was built up with the GIS ARCVIEW, in collaboration with the French agronomic institute ITCF. These maps were provided and used by the farmers owning a seed-sowing machine driven by an inboard computer and dGPS. The same process will be carried out for N.P.K. fertilization during winter and spring 2003.

Spatial variation of N fertilizer response in the southeastern USA Coastal Plain

E.J. Sadler, C.R. Camp, D.E. Evans and J.A. Millen

Within the southeastern USA Coastal Plain, spatial variation in grain yield poses a challenge to farm managers who desire to optimize crop management. Low native fertility and low organic matter content of shallow, sandy soils further exacerbate N management. These problems complicate site-specific management for maize. This paper outlines the results from two experiments examining irrigation and N response functions for maize. The experiments discussed here were conducted using the Site-Specific Center Pivot Irrigation Facility at the USDA-ARS Coastal Plains Soil, Water, and Plant Research Center in Florence, SC. These pivots are capable of independently irrigating and fertigating areas as small as 9 x 9 m. Both are modified commercial pivots, 140 m long. Additional manifolds, nozzles, valves, and controls were added to achieve this capability. Under center pivot 2, during 1999-2001, we examined irrigation and N fertilizer response functions for maize grown on twelve soil map units in a typical field, with emphasis on soil variation and irrigation response. Experimental design was a randomized complete block (RCB) with blocks placed within map units, plus incomplete blocks where map units were too small. Two N rates (135 and 225 kg/ha) were imposed for each of 4 irrigation levels (0%, 50%, 100%, and 150% of normal). In sum, there were 396 plots arranged within the 6-ha field. Earlier analysis of variance results had shown the N treatment to be insignificant in 1999 and 2000, but significant in 2001. Concurrently, under center pivot 1, we examined the same responses on a more uniform soil, predominantly one map unit, with the emphasis on N response. This design was a 4-block, replicated RCB design. There were 3 irrigation levels (0%, 75%, and 150% of normal irrigation) and 4 N rates (50%, 75%, 100%, and 125% of recommended rates, which were 135 for rainfed and 225 for irrigated culture). The area of this 144-plot experiment was much smaller than the one under CP2. Analysis of variance had indicated that the mean yields for the lowest N rate were significantly less than those for the other rates. To extend the classical statistical analysis to account for spatial variation, we calculated N response at all locations where the highest irrigation rates had been applied. N response was calculated as $R = (Y_{hiN} - Y_{loN}) / (N_{hi} - N_{lo})$, where Y is yield in kg/ha, N is the N rate in kg/ha, and hi and lo indicate the specific high and low N treatment used. Where the CP1 experiment had several N rates from which to choose, CP2 was limited to 135 and 225 kg/ha. Therefore, the evaluation under CP1 was limited to the closest match to those two N rates. Under CP2, values of R ranged both positive and negative in all years, from -47 to 39 kg/kg in 1999, rom -22 to 60 kg/kg in 2000, and from -30 to 50 kg/kg in 2001. Negative values suggested that N was not the yield-limiting factor. Irrigation response curves indicate that, even with irrigation at 1.5x the expected full amount, there were areas in the field in which water was the limiting factor. Interestingly, these areas of negative N response and insufficient irrigation did not correspond well. Further, in 2001, the year in which the N treatment was significant, and which had a well-distributed rainfall pattern (and record high producer yields), the spatial pattern in the N response surface was just as striking as in the other years. Spatial variation under CP1, on the smaller area with more- uniform soil, was slightly less, but not negligible. Spatial variation in the R response surface showed a distinct pattern that was much more informative than the analysis of variance results. The challenge presented to researchers is how to incorporate this information into site-specific crop management recommendations.

Possibilities of LASER-induced chlorophyll fluorescence measurements in wheat

Josef Schächtl, Franz-Xaver Maidl, Georg Huber and Ewald Sticksel

The heterogeneous nitrogen conditions of plant canopies should be considered when making a nitrogen application in precision agriculture. With this task sensors can help the farmer by doing an online-detection of the nitrogen status of crops. Passive sensors like the N-Sensor of Hydro-Agri have been developed for a practical appliance in the last years. Nevertheless, passive sensors have some problems like the detection of a mixed signal of plant and soil reflectance. They also depend on the reflection of irradiance. External factors like sunlight and cloudiness have an impact on the reflection signature. Active sensors like the LASER-sensor have an own excitation light and therefore the advantage of being independent of weather conditions. But active sensors are still in the stage of development. We tested the possibilities of a LASER-sensor for detecting the nitrogen status of crops in field trials of winter wheat. The trials were carried out in Southern Bavaria with the factors variety, nitrogen fertilisation and site. Five varieties with different chlorophyll content and properties were tested with four nitrogen treatments on the site Dürnast. The factor site was tested with two different soil types in Thalhausen, one variety and five nitrogen treatments. Plant samples were gathered at characteristic growth stages (EC 30, EC 32, EC 37, EC 49, EC 65, EC 92). Each time the total aboveground biomass was cut out and dried. The nitrogen content was examined according to Kjeldahl. The LASER-sensor was used at the same time for ascertaining which information the sensor values provide according to growth stage and variety. Additional measurements were made with a passive sensor, a two-channel spectrometer. Thus the different measurement techniques of active and passive sensors could be compared. The LASER-sensor was a hand-held device constructed by the DLR in Oberpfaffenhofen and the firm Fritzmeier. It was carried at a defined position above the ground and along the rows of the plots. The excitation source was a red-light LASER, and high-sensitive optical components detected the intensity of the fluorescence light at the wavelengths of 690 nm and 730 nm The red-light LASER excites the photosystem emitting fluorescent light. The fluorescence emission and the pigment absorption bands are overlapping around 690 nm. Therefore the fluorescence light is selective reabsorbed by chlorophyll during its way through a leaf. The reabsorption of the fluorescence light increases with increasing chlorophyll content. So the detected fluorescence light decreases mainly at 690 nm. The vegetation index ratio was calculated as proportion of the detected intensities at 690 nm and 730 nm. It is known from literature that the ratio decreases with increasing chlorophyll content. Our results suggested this negative correlation between ratio and nitrogen content of the plant. We were able to distinguish statistical significance between the different nitrogen treatments at early developmental stages (EC 30 and EC 32) with the LASER-sensor. At this time the spectrometer showed the well-known problems of passive sensors due to the detection of a mixed signal of plant and soil reflectance. The varieties influenced the fluorescence signature due to their different chlorophyll content. The site had an effect on the fluorescence spectra. Therefore the regression between nitrogen content and ratio was different for the two examined sites. An all-time valid calibration of the sensor cannot be done. We ascertained that the LASER-sensor works under different environmental conditions with additional measurements. The different nitrogen conditions between variably fertilised plots were detected under daylight conditions and during night. We found no influence of waterdrops on the surface of leafs on the fluorescence signal. The sensor can be used under nearly all environmental conditions. The passive sensor could not detect the reflectance spectra of plots without irradiance during night.

Field-scale validation of a tractor based crop scanner to determine biomass and nitrogen uptake of winter wheat

U. Schmidhalter, S. Jungert, C. Bredemeier, R. Gutser, R. Manhart, B. Mistele and G. Gerl

Optimising nitrogen inputs is an essential requirement for high yields and decreasing nitrogen losses to the environment. Heterogeneous fields require a targeted, site-specific input of nitrogen. The potential of a tractor based crop scanner to detect differences in biomass, nitrogen content and uptake of wheat was tested in field experiments in 2001 and 2002. The experiments were conducted with the goal to spatially match destructive ground-truth measurements of biomass and nitrogen content with reflectance measurements. To match the area scanned by the sensor two different approaches were used. In the first year the scanner was mounted lateral on the backside of the tractor at a height of 1 m above the plant canopy which allowed to reduce the scanned area to 2 m^2 and in the second year the scanner was mounted in 2.2 m above the crop stand and the area scanned was between 18 to 10 m^2. The sensor used has four optical inputs with 90° azimutal angle between them and an average view zenith angle of 64° each. Light is collected from four inputs and optically averaged through a four-split light fibre. This arrangement allows to take an average measurement from four ellipsoids located around the tractor practically independent of solar azimuth direction. In the first year, the measurements were done on an area reflecting two ellipsoids, whereas in the second year they were made on the sensed area. To our knowledge no such measurements have been reported. Spectral measurements were conducted at 5 wavelengths (550, 670, 700, 740 and 780 nm ± 5 nm) and the following spectral indices were calculated: REIP; SAVI; NDVI; G/R, IR/G, IR/G. In the first year biomass was cut manually and in the second year destructive harvests were done with a green forage chopper with 1.5 m cutting width equipped with a weighing unit. Dry and fresh weight of biomass was determined and the shoot nitrogen content measured. Randomised treatments were used with seeding rate (250, 450 and 650 grains/m^2) and nitrogen dose (100, 135, 170 and 205 kg N ha^{-1}) as variables in 2001 and nitrogen dose (90, 130, 170 and 210 kg) as variable in 2002. Spectral measurements were conducted at BBCH 32, 55 and 73-75. Measurements during milk ripeness were conducted to predict the final yield. Close relationships between spectral measurements and biomass, nitrogen content and uptake were determined in 2002 and a reasonable agreement with biomass and yield was observed in 2001. Crop stands in the different plots were much more homogeneous in 2002 and the comparative measurements were located on exactly the same positions. This was not the case for the 2001 measurements. In 2001, goodness of linear fits (R2-values) between IR/R measurements and biomass was 0.51, whereas final yield correlated to REIP measurements at BBCH 32 and 55 with 0.62 and 0.42. In 2002 at BBCH 32, the goodness of non-linear fits between biomass, nitrogen content and uptake to IR/R, IR/G, REIP was as follows (linear fits are given in brackets): 0.82, 0.59, 0.85 (0.70, 0.65, 0.73); 0,83, 0,80, 0,92 (0.75, 0,82, 0.86); 0.83, 0.82, 0.94 (0.77, 0.78, 0.84). For BBCH 55 in 2002 the goodness of non-linear fits between biomass, nitrogen content and uptake to IR/R, IR/G, REIP was as follows (linear fits are indicated in brackets): 0.74, 0.50, 0.74 (0.69, 0.41, 0.55); 0,69, 0,72, 0,82 (0.65, 0,65, 0.68); 0,65, 0,78, 0,82 (0.62, 0.69, 0.69). All relationships reported were highly significant. The results indicate that biomass and nitrogen uptake can reliably be detected with a tractor-based non-contacting sensor. Further applications are investigated with other crops.

A method for advanced soil modeling with digital elevation models and geophysical data

F. Schmidt, R. Gebbers and E. Lück

Both soil and terrain exert a large influence on small-scale site differences. Soil and terrain control the lateral and vertical movement of water and thus the pattern of soil moisture content. Soil and terrain usually interact in the process of soil formation. Terrain can be represented using a Digital Elevation Model (DEM) and its derivatives such as contours, slope, aspect and topographical indices. A complete cover of soil properties pattern can be derived with electrical conductivity mapping (ECa). Both types of data, however, can not explain all aspects of field heterogeneity. Whereas ECa maps show an integral summary signal down to 1.5-2 m soil depth, DEMs provide explanations for the lateral movement of water on the surface and thus the transport of fine particles. For three test sites representing various climatic conditions in Germany, ECa data were evaluated in combination with terrain attributes. The most commonly used terrain attribute to describe the soil water potential is Beven and Kirkby's topographic wetness index ln(As/tan). High values of the topographic wetness index (TWI) are found in converging, flat terrain. Low values are typical for steep, diverging areas. Previous results showed a good correlation (r^2 up to 0.81) between soil moisture measurements and TWI for specific dates, locations and soil depths. However, due to variations in soil texture, the TWI does not predict the real long-term soil moisture pattern on all fields. In this study, the TWI was combined with ECa maps using overlay functions. The crucial idea behind this approach is to use the relative information about vertical permeability of soils assessed from the ECa signal for the correction of the lateral flow pattern. A relation of hydraulic and electrical conductivity is assumed for this approach: ECa ~ 1/ksat. The ECa signal is known to rise with clay content, ksat is rising with sand content. Problems result from layered soils, e.g. thin clay layers in sandy soils and should be treated carefully. Limitations of the concept will be discussed. Three distinct cases could be identified when comparing ECa and TWI pattern: 1) high ECa coinciding with high TWI, 2) low ECa coinciding with high TWI and 3) no clear pattern at all. Case 1) could be confirmed on all test sites as wet spots on the field. Low ECa usually indicates lighter texture, lower cation exchange capacity and lower organic content. Thus, on spots where low ECa values coincide with linear patterns of high TWI, run-off is assumed to alter soil conditions in the long run, i.e. by erosion of clay particles and deposition of coarse material respectively. This results in lower soil moisture for most times of the season. These areas can be detected with the combination of ECa maps and TWI. Such results were found on fields with a clear drainage pattern shown by the TWI. Overlay operations are beneficial for case 2) in order to incorporate areas of high potential vertical losses to the calculated lateral drainage pattern. An additional advantage of the method is that ECa maps and DEM can be captured at the same time. However, a precise GPS is needed. Quality requirements for DEM generation will be presented. Laser scanning is discussed as an alternative method, which might be a cheaper data source when provided by mapping agencies in the near future. As a result, the common interpretation of the TWI pattern should be adjusted. Since calculated flow lines (high soil moisture potential) do not correspond with long-term saturation but high intensity of transport processes, the TWI should be interpreted more carefully in terms of the terrain-soil interactions. The simultaneous analysis of ECa and TWI maps helps to understand the dominating processes on an agricultural plot and can lead to improved site maps.

Imaging spectroscopy for grassland management

A.G.T. Schut, J.J.M.H. Ketelaars and C. Lokhorst

The potential of an imaging spectroscopy system was explored as tool for grassland management. The system detects reflection in image-lines in the wavelength range from 405-1650 nm with three sensors, from 1.3 m above the soil surface in a regular sampling pattern. The spectral resolution varied between 5-13 nm, and spatial resolution between 0.28-1.45 mm2 per pixel. Aspects of grassland management considered were growth monitoring and assessments of clover content, feeding value and sward production capacity and nitrogen and drought stress detection. Experiments were conducted with mini-swards grown in containers. Under a rain shelter in 1999 and 2000, an experiment was conducted with 36 mini-swards were sward damage varied, from undamaged to severely damaged due to drought or removal of circular patches from the sward. In two experiments (2000 and 2001), nitrogen application varied (0, 30, 60, 90 and 120 kg N ha^{-1} harvest^{-1}). The drought stress experiment was conducted in a climate chamber and there was a control with high N supply and two drought stress treatments with low and high N supply. Drought stress treatments received no additional water during growth. During the experiments, 42 image-lines per mini-sward were recorded 2-4 times per week. Image-lines were classified to separate pixels containing soil, dead material and green leaves. These classes were subdivided into reflection intensity classes. Ground cover (GC), reflection intensity, image-line texture, spatial heterogeneity and patterns, and spectral characteristics of green leaves were quantified. The index of reflection intensity (IRI) quantified the distribution of green pixels over intensity classes. Spatial heterogeneity was quantified with the spatial standard deviation of ground cover (GC-SSD), and image-line texture and spatial patterns with wavelet entropy (WE). Spectral characteristics were quantified with shifts of spectra at various edges. Partial least squares (PLS) models combining spectral and spatial information were calibrated and validation on two separate data-sets from the sward damage and N experiment in 2000, to predict DM yield, feeding quality and nutrient content. As result of the design of the experimental system, reflection intensity varied with sward height and leaf angle. Therefore, IRI quantified vertical canopy geometry. GC was linearly related to light interception under a cloudy sky and logistically under a clear sky (R2adj = 0.87-0.95). The relations differed significantly between dense and open swards. Growth was accurately monitored with GC and IRI. Seasonal means of GC and IRI were strongly (R2adj = 0.77-0.93) related to seasonal dry matter yield (SDM) and light interception capacity. There was a clear relation between seasonal means of GC-SSD and spatial variability of tiller density (R2adj=0.69). Seasonal means of GC-SSD differentiated dense from damaged swards. GC-SSD and WE of GC were related to SDM. The WE of image-line texture differentiated clover from grass swards, where mixtures had intermediate values. Drought stress was detected in an early stage, when DM content of leaves was below 20%, with shifts at the edges of water absorption features. The combination of inhibited or reversed shifts at the green and red edge was strongly related (R2=0.95) to yield reduction relative to high N supply. The PLS models accurately predicted DM yield, nutritive value and nutrient contents with mean prediction errors of 235-268 kg DM ha^{-1}, 0.96-1.68 % DM, 5.8-6.5 g kg^{-1} DM ash, 8.4-10.4 g kg^{-1} DM crude fiber and 0.24-0.34 % N. The predictions of P, K, S, Mg, Na and Fe were robust in both experiments, in contrast Zn, Mn and Ca. It is concluded that imaging spectroscopy can provide valuable information for grassland management on dairy farms.

Draft force mapping as a tool to map spatial variability of soil properties

B. Schutte and H.D. Kutzbach

Site specific farming in precision agriculture requires knowledge about spatial differences and properties of soil in order to make management decisions for sowing, fertilizing etc.. Current practice is analysing soil samples, taken at several points in a field, to win the knowledge about the variable soil physical properties. Due to the large amount of work in collecting and analysing the soil samples in some cases mapping of electrical conductivity is taken to describe the variability of soil within a field. However, these data has to be analysed subsequently and statistical calculations are necessary in order to get area wide soil property maps. Maps off electrical conductivity have to be critical analysed because the measured parameters depend on a lot of soil properties like soil type, water content, salinity or soil compactions. The recent development in precision agriculture has shown the benefit of automatically techniques during anyway necessary working processes such as yield mapping on combine harvesters. A promising way to collect data about local soil properties is the position-related measurement of draft forces during conventional tillage operations as cultivating or ploughing. Especially in view of modern tractors which are already equipped with sensors to determine draft forces, mapping of draft forces in connection with a (D)GPS-system might be a low-cost, time-saving and exact way to win information on spatial variability of soil properties. Former experimental investigations have shown the potential of mapping draft and/or PTO–forces values during the conventional tillage operations. The measured data is influenced by lots of parameters. Once there is the tillage tool itself, than different soil parameters like moisture, compaction, soil type and humus matter and process determined influences like speed, working depth and width. At the Hohenheim Institute of Agricultural Engineering these measurements were carried out by the use of an 6 component measurement frame between tractor an implement. Six load cells measure all three components of forces and torques between tractor and implement. In combination with high accuracy position data of an RTK GPS unit it is possible to map the spatial variation of draft force and PTO - force over a plot. A new far-reaching measurement system allows logging of the engine and ground speed, working depth/width, implement draft forces (draft measurement frame), implement draft forces (lower link draft sensors), fuel consumption and positioning data during normal field work. Measurements can be done with various implements like ploughs, cultivators or rotor harrows. In addition the influence of soil moisture and soil compaction on the draft values is being examined. The test fields are selected to determine the practical value of draft maps for site-specific managements systems. A comparison between other ways (mapping electrical conductivity, soil samples) to get soil information in site specific management is accomplished in the test fields. With regard to the implementation of draft measurements on standard tractors and implements in practice the suitability of the lower link draft sensors compared to the measurement frame is tested. To enable the application of mapping draft forces in practical use under various conditions and with different implements/tractors algorithms on base of the experimental data are drawn up.

Three years results with site-specific nitrogen fertilization

J. Schwarz, C. Kersebaum, H. Reuter and O. Wendroth

A field experiment with different nitrogen fertilizer strategies has been established in Lüttewitz, Federal State of Saxony, Germany. The project was a joint research project between the Institute of Agricultural Engineering Bornim (Potsdam), the Center for Agricultural Landscape and Land Use Research (Müncheberg), Südzucker AG (Mannheim/Ochsenfurt), Amazonen Werke (Hasbergen), and Agrocom (Bielefeld). The aim of this study was to identify the effect of different nitrogen strategies on yield, soil and plant response as a basis for subsequent recommendation for site-specific nitrogen fertilization recommendations. Beyond that the model HERMES (crop growth, soil and crop nitrogen dynamics) should be calibrated for new field crops and specific field conditions. The experiment took place within the years 2000 and 2002 in Lüttewitz / Federal State of Saxony in the South-East of Germany. Some facts about the experimental site. The average annual precipitation is 660 millimeters, the average annual temperature is 8 degree Celsius. In Lüttewitz the ground is very hilly. The average German soil fertility index in Lüttewitz is about 70. The grid size of the investigated plots is 54 meters by 54 meters on 64 different plots; the whole field covers an area of about 30 hectares. The crops grown were winter wheat in the years 2000 and 2002, and winter rape in the year 2001 to make investigations to calibrate the parameters for the model HERMES. In the years 2000 and 2002, five different nitrogen fertilizer treatments were applied: zero nitrogen, a variant with the first application based on soil analysis and the Hydro N Sensor for the second and third nitrogen application; and three strategies based on the nitrogen model HERMES. The fertilizers were applied on following three dates: 06. April, 27. April, 16. May in 2000, and 19. March, 24. April, 23. May in 2002. In the year 2001 only two different fertilizer treatments were applied. A zero fertilizer strategy and a homogeneous fertilizer application. This results from the calibration for the model HERMES. The fertilizers were applied on the 16. March and 18. April. In the year 2002 only one of the HERMES strategies was slightly changed. Before each fertilizer application soil and crop samples for nitrogen analysis were taken. Previous to the winter wheat harvest in 2000 and 2002 crop samples were taken for quality analysis, e.g. measuring of protein content. The field was harvested with a combine harvester which included a yield measurement system. In the year 2000 only the zero fertilizer strategy had a statistically significant yield difference from the others, but the range was very wide. The rang for the zero lasts from 2.0 tons per hectare up to 8.7 tons per hectare. The zero fertiliser strategy was found to result in a significantly lower protein content compared to the other four strategies. The protein content was 11.5 %, whereas for the other four strategies, it was about 14%. The mineral nitrogen content for the zero strategy decreases over the sampling period. For all other strategy no uniform trend is obvious. In the year 2001 only the zero strategy and a homogeneous strategy were applied. For the year 2002 the results are not yet available, but will be available for the full paper.

Measurement of Normalised Difference Vegetation Index (NDVI) and crop height to characterise a winter wheat canopy

I.M. Scotford and P.C.H. Miller

Cereal crop production requires a series of inputs e.g. fungicides and fertilisers which are applied to the crop, ideally in the right place, at the right time and at the correct dose rate. Crop canopy characteristics are one of the main factors influencing these inputs decisions. Hence, there would be obvious benefits if the crop canopy could be remotely sensed and linked to the agronomic decision making process. The aim of this research was to assess the usefulness of a tractor mounted radiometer (NDVI) and ultrasonic sensor (crop height) to characterise a winter wheat canopy.

To establish a range of different crop canopies, 3 varieties of winter wheat (Claire, Consort and Riband) were planted, each variety, at low and high seed rates (150 and 250 kg ha-1 respectively). The plots were drilled on 19 October 2001 on field with a heavy clay soil, typical of that used commercially for growing of winter wheat in the United Kingdom. The whole plot area was treated uniformly in terms of weed control, fungicides and fertilisers. The aim being to eliminate any variability associated with weeds, disease and fertiliser deficiencies.

A measuring system including both NDVI and ultrasonic height sensors was designed, built and fitted to the 3-point linkage of a tractor. Using this system all of the plots were scanned at approximately weekly intervals between March and August 2002. Each plot was scanned for a 10 s period during which time approximately 20 readings of NDVI and crop height were recorded. During the scanning period the forward speed of the tractor was 0.22 m s-1, therefore during a 10 s scan a linear distance just over 2 m was covered. For each plot a single value of NDVI and crop height was calculated by taking the average of the scans taken during the 10 s period.

For all 3 varieties, especially at the high seed rate, the quickest increase in NDVI occurred until early to mid May, when the crop was at growth stage (GS) 31/32 (early stem elongation). Following this period NDVI values increased more slowly peaking late May, corresponding to about GS 45 (mid booting), before starting to decrease. This generally represents a typical canopy expansion and death curve of winter wheat. However, it should be noted that maximum green area index (GAI) of winter wheat usually occurs at GS 59 (ear completely emerged) whereas the NDVI values generally peak at GS 45 (mid booting). In addition the greatest increase in GAI, from 2 to 6, occurs between GS 31 (first node detectable) and GS 39 (flag leaf all visible), during this period NDVI values only increase steadily. This evidence suggests that NDVI measurements are most useful up to GS 31 and of limited use once the crop has passed this stage.

The height of the crop, as measured with the ultrasonic sensor, does not start to increase until end of April when the crop is at GS 30 (stem elongation). It then increases linearly reaching its maximum mid June, corresponding to GS 59 (ear completely emerged) and maximum GAI. This evidence suggests that crop height measurement is of limited use before GS 30, but is useful up to and beyond GS 59.

1. NDVI is useful for monitoring the crop up to GS 31/32.
2. The ultrasonic height sensor is useful for monitoring the crop from GS 30 to beyond GS 59.
3. Combination of the output from the two complementary sensors is likely to provide a better estimation of canopy characteristics than either sensor in isolation.

Estimation of GAI, GS and density of winter wheat using Normalised Difference Vegetation Index (NDVI) and crop height measurements

I.M. Scotford and P.C.H. Miller

Research has shown that the Normalised Difference Vegetation Index (NDVI) is most useful for monitoring winter wheat up to growth stage (GS) 31/32. Whereas the crop height, as measured using ultrasonic sensors can be used to monitor the crop from GS 30 to beyond GS 59. It is hypothesised that by combining the output of these two sensing methods better characterisation of the canopy will be obtained. This paper details preliminary methods used to estimate green area index (GAI), growth stage (GS) and crop density.

Between March and August 2002 a tractor mounted scanning system was used to obtain weekly measurements of NDVI and crop height from 3 varieties of winter wheat (Claire, Consort and Riband) each planted at low and high seed rates (150 and 250 kg ha-1). Each measurement of NDVI and crop height was the average of approximately 20 readings taken as the tractor covered a distance of 2 m over a 10 s period. The plots were treated uniformly to eliminate variability associated with weeds, disease and fertiliser deficiencies.

NDVI and crop height values were plotted against time . From this graph it is clear to see that NDVI values start reaching their maximum from GS 31 whereas height values reach their maximum at GS 59. This clearly indicates the limitations of these two sensors used in isolation. However if the summation of the two values (NDVI and crop height) for each variety are plotted we can see that this combined value is linear until GS 59 is reached. Similarly it is reported (HGCA, The Wheat Growth Guide, 1998) that GAI also increases in a linear fashion during this period. No measurements of GAI were made during this experiment, but the evidence suggests that the summation of the NDVI and crop height values will be highly correlated with GAI.

Further analysis of the crop height data was conducted to determine its ability to estimate the crop density. Using the 20 readings obtained for each plot the Coefficient of Variation (CoV) was calculated for each plot using:

$$CoV = \frac{\sigma}{\overline{X}} \times 100$$

where $\overline{X}$ = mean value; and σ = standard deviation; of the 20 readings obtained for each plot. When the CoV values are plotted against time for the high and low seed rate plots for each variety it can be seen that the CoV is generally higher for the low rate plots, indicating the lower tiller densities associated with these plots. This method shows promise to estimate both plant and tiller density but further experiments are required verify this approach.

This preliminary study has indicated the potential for this technique to estimate GAI, important GS and density of a winter wheat crop, but the results in this report are based a limited number of experiments conducted during one growing season. Further work is required to establish the statistical validity of these preliminary conclusions.

Spatial detection of topsoil properties using hyperspectral sensing

T. Selige, L. Nätscher and U. Schmidhalter

Topsoil characteristics influence markedly crop development and yield and are related to erosion, sealing, infiltration, mineralisation, heat and water balance. Frequently significant heterogeneity across fields can be found in topsoil characteristics which cause differences in crop germination, nutrient and water uptake and thus influence crop growth. Topsoil heterogeneity has therefore implications on site specific management. For optimising crop growth and to sustain soil fertility especially soil tillage, seed bed preparation, fertilization and herbicide applications require adaptation to topsoil properties. But there is still a serious lack of data about physicochemical topsoil characteristics for a precise and site specific adaptation of harmonized management procedures. In this paper we present results from a research study aimed to locate, identify and quantify topsoil properties using hyperspectral remote sensing, field-spectroscopy and laboratory near-infrared spectroscopy. This should lead towards an optimal spectral analysis tool for quantitative determination of soil texture and organic matter from single spectral signatures. This would allow to map soil texture and organic matter simultaneously by remote or proximal sensing. The study area is located in Sachsen-Anhalt, Germany. The research is based on 72 topsoil samples representing the spectrum of soil texture and organic matter in the study area and most of Central European cropland. Soils were analysed for Corg and Nt using dry combustion by elemental-analyser. $CaCO_3$ content and pH-value were measured. Particle size distribution was analysed using sieve analysis for sand and coarse silt fractions and pipette analysis for silt and clay. Dithionite and oxalate soluble iron oxides were determined. Organic matter composition was analysed e.g. separating aliphatic and aromatic compounds. The spectral response of soil samples was measured in the lab using a FT-IR-Spectrometer (780-2850 nm). Bare topsoil samples were measured outdoor using a GER 3700 field-spectrometer (330-2500 nm). The airborne Hymap sensor was used for recording hyperspectral images of representative bare soils. The mathematical and statistical analysis covers spectral feature extraction techniques as principle component regression (PCR), partial least square regression (PLSR) and wavelets to extract spectral features from the different signature sources correlating with topsoil variables. A strong spectral response was found for the clay content. The relation is independent of the organic matter content (Corg, Nt) but was linked up with iron oxides which were closely correlated with clay ($r=0.94***$). Only wavebands 2300 nm showed a significant regression with the clay content ($R^2=0.78***$), whereas the reflection of ultra-violet and visible wavebands were depending on the amount of organic matter ($R^2=0.68***$). The scattering of data was then compared with the visible soil color. From this it became obvious, that the type of organic matter plays a significant role in the spectral absorption. This seems to be caused by the genesis of organic matter due to site specific transformation processes. The ongoing analysis using response models based on PCR, PLSR and wavelets shows so far clearly the advantage of PLSR over PCR. But the most promising results can be found with wavelet procedures. The two most important topsoil parameters for precision farming applications, organic matter and clay content, are correlating with spectral properties. It is necessary to consider wavebands 2300 nm for these applications. Organic matter can not be described only by the amount of Corg and Nt in relation to spectral properties. The quality of organic matter has to be considered. The effects of soil variables on soil properties will be discussed regarding management practice requirements related to soil aggregation, compaction, erosion, sealing and other risks but also germination, nutrient uptake and early crop growth. This paper will also discuss the chances and limits to apply spectral proximal or remote sensing for high resolution topsoil mapping in precision agriculture.

Site-specific crop response to temporal trend of soil variability determined by the real-time soil spectrophotometer

S. Shibusawa, C. Hache, A. Sasao, I. Made Anom S.W. and S. Hirako

Site-specific crop response to the soil variability has been an important issue to develop variable-rate technology. Our project group has already developed a real-time soil spectrophotometer to determine the detailed spatial variability of soil parameters. Next issue is to apply the soil maps in estimating spatial variability of crop response. Objectives of the work were to determine the temporal changes in spatial variability of soil parameters of a small field (155 m long and 30 m wide) using the real-time soil spectrophotometer, and to determine the site specific response of crop to the soil variability. Material and Method: A small field used for the experiment on the experimental farm of Tokyo University of Agriculture and Technology, had been divided into eight plots with three different types of soil management (fertilizer, fertilizer-manure and manure) for more than ten years. The soil type of the field was 20% clay, 28% silt, and 52% sand. Two-year three-crop rotation: wheat-corn-wheat-bean had also been performed more than 10 years. The real-time soil spectrophotometer mounted in a tractor with a RTK-GPS collected visible (VIS) to near-infrared (NIR) reflectance of the soil. The reflectance collected covered a range of 400 to 950 nm with a resolution of 3.6 nm and a range of 950 to 1700 nm with a resolution of 6 nm, respectively. The sensor traveled at a speed of 30 cm/s at 15 cm depth. Soil reflectance was collected on three segments in the field at a scanning interval of 1 m in longitudinal direction and 5 m spacing in lateral direction, for a total of 360 scanning points. To calibrate the soil reflectance, about 75 samples were taken every 5m at the same depth and location as the corresponding sensor's scanning point. These samples were analyzed in laboratory for moisture content (MC %), nitrate nitrogen (NO3-N), soil organic matter content (SOM %), total carbon (TC %), total nitrogen (TN %), C/N, EC and pH. The 75 soil samples were divided in 2 data sets, where 50 samples were used for calibration and 25 for validation of the prediction model. Both data sets had similar average and standard deviation values. For calibration procedure, the moving average smoothing method, Kubelka-Munk transformation method, multiplicative scatter correction, and stepwise multilinear regression analysis were performed. Data of plant length, SPAD and number of green leaves per plant were taken every week at fixed measurement points over the growing period up to harvest in the field. Using Kriging method, interpolated maps were created; spatial variability was present in both soil and crop parameters. Results and Discussion Temporal trends of soil parameters were evaluated during the two crop seasons in 2000 to 2001, and corn plant response was traced in the following season of 2001. The spatial trend maps of moisture content, SOM content, NO$_3$-N, pH and EC were evaluated, and the temporal stability was also evaluated for the respective soil parameters. The level of stability was determined here as stable (CV < 5 %), relatively stable (CV between 5 % and 10 %), and unstable (CV>10%) in convenience, and then a management unit map was presented. For crop parameters, correlation among crop parameters varied in time. Highest and stable values for crop parameters were observed in fertilizer-manure treatment. SOM, TN and TC condition just before sawing appeared to have strong effect in the development of corn in this field. In this experiment, corn plants responded well to the three types of treatments and some parts of the field showed high or low stability, while in other areas higher temporal variability was present. The previous trends of soil parameter variation partly affected the following crop response. More detailed discussion will be presented.

Digital tree shape mapping and its applications

E. Shimborsky

It is observed in [2] that precise mapping of tree location and canopy volumes (TPV maps) is of major importance in applying precision horticulture methods in orchards. Basic methodologies and preliminary results of an automated process of digital TPV map generation from aerial scans were presented in [1]. The methodologies presented therein were based on applying image-processing techniques for automatic matching of tree patterns extracted from aerial stereo pairs and generating tree elevation points by using an automated photogrammetric triangulation scheme. The resulting three-dimensional data sequence has to be processed in order to generate a three-dimensional geometric surface representing actual tree shapes. In the current work methodologies of digital tree shape modeling and their application to precision horticulture will be discussed, and applications and results of these methods regarding the PRECISPARY project (see [2]) will be presented. TPV mapping is a process in which digital models of orchard trees are constructed from aerial scans and embedded in the orchard data management system. TPV maps can be updated frequently by conducting digital aerial surveys. Digital image scans are processed in order to conduct a photogrammetric aerial triangulation for setting up a geometric model out of which orchard trees elevation points can be referenced throw a global coordinate system. In the next stage tree elevation points are extracted automatically by conducting an automated image matching process. The resulting stream of three-dimensional tree elevation points are then processed for setting up a three-dimensional surface model representing actual tree shapes. In the final processing stage tree shape layers are embedded as three-dimensional spatial layers of an orchard GIS management system. TPV layers containing digital tree shapes facilitate the visual presentation and spatial analysis of a virtual three-dimensional mockup orchard. Using the virtual mockup orchard, precision horticulture utilities can be enhanced. Combining original imagery data from aerial scans and the generated three-dimensional tree shapes, virtual orchard synthetic views can be generated for monitoring three-dimensional spatial data. In precision horticulture applications TPV maps can be processed and analyzed in order to simulate agricultural application devices driving through a virtual orchard. In the PRECISPRAY project ([1]) the ground application device consists of a variable rate segmented vertical boom sprayer that has to maneuver within the orchard 's tree rows. This sprayer device is capable of keeping its outlets in a constant distance from the tree contour lines. The three-dimensional simulation system accepts the sprayer's nominal routes and generates nominal distance contours for each boom segment by analyzing the sprayer's nominal position with respect to the orchard tree rows. Distance contours are constructed by measuring the horizontal distance of boom segments nominal position to the adjacent neighborhood of tree shapes. These nominal distance contours are the key elements for calculating the sprayer's boom segment outlet positions and for controlling outlet airflow and spraying volume deposits. Finally, results of a "drive through" simulation of the PRECISPRAY boom sprayer in a ground-testing orchard will be presented. 1. SHIMBORSKY E., MERON M. (2001). Automatic acquisition of tree shapes and foliage volume maps of tree orchards, 3rd ECPA, Montpellier, France. 2. Meron M. Tree shape and foliage volume guided precision orchard sprayer - the PRECISPRAY FP5 project, submitted to the 4th ECPA.

Sensor based nitrogen fertilization in winter wheat levelling out spatial variability in protein content

P. Skjødt, P.M. Hansen, R.N. Jørgensen, N.E. Nielsen

Precision farming has developed tremendously the latest years and especially site-specific nitrogen (N) application has been in focus. The basis of site-specific N application is that the farm manager can get access to spatial data from the field and use the data in a relaying algorithm or model. Easy and cheap access to global positioning systems has created the foundation for various kind of different spatial data, which can be measured by the farm manager or by contractors. At present, measurements of yield, electromagnetic induction (EMI), topography, and canopy reflectance have all been considered and used in algorithms either alone or in combination. However, the models developed optimising the use of nitrogen spatially have often proved to be inadequate, showing only a slight improvement in yield and grain quality. A strip experiment was established in winter wheat at Risø National Laboratory, Denmark in growth season 2000/2001 on a 14 ha field. The experimental field was divided into two macro plots, A and B, respectively. Macro plot A included four different nitrogen strategies each with six strips giving 24 strips in total. Macro plot B was used as an additional control experiment, where two of the four strategies were used with 2 times 10 strips. A site-specific N application strategy in winter wheat based on canopy reflectance data was developed (S4). The aim was to test the variable nitrogen application strategy with regard to maintaining yield and improving grain quality in winter wheat based solely on remotely sensed data. The idea was to spatially distribute the available nitrogen according to a three-split strategy. This implies the use of a site-specific application at the second and third nitrogen application using on average 66 % and 34 % of the remaining nitrogen after the first base fertilization, respectively. The intention in the first site-specific nitrogen application was to use normalized difference vegetation index (NDVI) as a measure of biomass to level out the spatial differences in biomass by supplying more nitrogen to areas with low biomass and less nitrogen to areas with high biomass. At the second variable nitrogen rate application, both NDVI and red edge inflection point (REIP) were used to describe the spatial variation of biomass and chlorophyll density. The strategy was to level out the spatial differences in expected final grain protein concentration by supplying more nitrogen to areas with high biomass and relatively low chlorophyll density and less nitrogen to area with low biomass and relatively high chlorophyll density. The nitrogen strategy has been tested against a uniform nitrogen application (S2) and another spatial N strategy developed by Kemira Agro OY, Finland (S3). The spatial analyses of the results show that protein yields and nitrogen uptake rates increased by 10 %, while the grain yield was unaffected compared to a variable nitrogen rate application strategy developed by Kemira Agro Finland and a uniform application. Hence, the data revealed that it is possible to develop a spatial nitrogen strategy and attain higher protein yields by using support from an integrated system of remote sensing of canopy reflection and differential global positioning system (DGPS) in relation to this field growth conditions and the Danish legal guidelines. No economical net return was obtained by using a three-split site-specific nitrogen application compared with a two-split uniform nitrogen application. Keywords: yield, protein, remote sensing, canopy reflectance, nitrogen application strategy, variable rate, winter wheat

Correcting the effects of field of view and cloud in spectral measurements of crops in the field

M.D. Steven

Spectral measurements of crops offer valuable information on the state of the canopy for precision agriculture. Current satellites may provide useful strategic information on within-field variability for future planning, but data acquisition is not always sufficiently reliable for tactical monitoring, where the information is used to guide management operations on a specific crop. Tractor-mounted sensors may offer an effective means of providing this information because they can operate in all weathers, the measurements may be made when the tractor is in the field for other purposes and the information is readily (and exclusively) accessible to the farmer without the need to wait for processing by a third party. However any operational system must account for instrumental characteristics and for variations introduced by the weather. This study compared spectroradiometer measurements (Licor LI-1800) made with different configurations over field plots of winter-wheat. Ten experimental plots of a Nitrogen trial at Terrington St Clements (52.8°N, 0.3°E) were selected for this study in order to provide a wide range of canopy covers on a single date. Pairs of measurements were made with a wide and a narrow instantaneous-field-of-view (IFOV) and narrow IFOV measurements were also made with the crop canopy shaded to simulate the effect of cloud cover, as well as in direct sunlight. Four repeat sets of measurements were made at evenly spaced locations in each plot. Two reference measurements were made for each set of four target measurements. The whole sequence was conducted as rapidly as possible to minimise changes of illumination with time. Large and consistent differences in spectral reflectance were found with measurement configuration, particularly in the near-infrared: measurements made with a hemispheric IFOV were up to 60% higher on average than those with a narrow IFOV and measurements in shade were up to 40% higher than in bright sunlight. In spite of these large differences, neither IFOV nor shading had any significant effect on the ability of the normalised difference vegetation index (NDVI) to measure canopy density. The slope of NDVI versus canopy cover was shallower with the wide IFOV and the shaded measurements, compared with the narrow IFOV sunlit measurements, but the values converged at high cover density and the coefficients of determination were 0.95, 0.92 and 0.94 respectively. According to the law of reciprocity, the problems of field of view and of shading are essentially equivalent. A simple theoretical model was developed that treated the signal from the canopy as a linear mixture of soil, foliage and shade components. The proportions of these components sampled with different instrument configurations or with the canopy shaded vary systematically. Normalising by an idealised spectrum of pure foliage (derived from the measurements) allowed the spectral dependence on leaf area to be separated from differences in the mixture coefficients, which are related only to viewing and illumination angles. Fitted coefficients of the model were applied in an attempt to reconstruct the narrow IFOV sunlit spectra from wide IFOV or shaded data. Spectra reconstructed from shade measurements were reasonably accurate, with root mean squared errors of about 10-15%. Reconstructions from wide IFOV were somewhat less accurate (15-25%), but more problematically they generated unrealistic spectral responses at low leaf area index. However, while the ability to correct for shade is essential to account for cloud cover variations during field surveys, the effect of IFOV is less critical and can be accounted for by recalibration of spectral relationships with canopy parameters. These results suggest that it should be possible to obtain accurate and consistent spectral information from tractor-mounted sensors under operational conditions.

Automatic operation planning for GPS-guided machinery

A. Stoll

Agricultural machines have to perform their tasks quickly, precisely, economically and reliably. These requirements are increasing permanently. Electronic control facilities shall fulfil these requirements. On the one hand, they are necessary to perform site-specific field operation. On the other hand, they can help to utilise the machine's capacity better than at manual use. Mainly steering, power train and implements are considered to be electronically controlled. Automatic control of at least one of these functional units is state-of-the-art. Even fully automated machines have already been brought up. Different problems came along with this development. It is not sufficient that a machine can carry out functions automatically. There is also the demand of a suitable operation plan describing the target values of the automatic functions. Suitable planning systems for the driving functions such as steering and speed are still lacking. Manual route planning tools can not satisfy in order to efficiently create operation plans. Only a limited number of automatic route planning systems is known. However, they can only handle simply shaped fields. Numerous research projects have considered route planning for service robots acting in indoor environments. However, these methods can only be applied conditionally to agricultural machines. Due to these problems the Institute of Agricultural Engineering at the Hohenheim University started a project dealing with operation planning methods for fully automated agricultural machines. A first result was an automatic path planning system for swath harvest operations of a GPS-guided forage harvester with automatic steering. The planning methods could be extended to field covering tasks (e.g. mowing). The guidance paths provide optimised target values for the driving route, the driving speed and the implements. In order to calculate guidance paths for an automated machine automatically different kinds of a-priori data have to be processed. The data can be divided into machine specific, field specific data and into the description of the operation strategy. Field specific data contain the field's border and obstacles inside the field. Neighbouring areas of the field are also considered. Attributes are attached to fields, neighbouring areas and obstacles in order to describe these objects adequately for automatic guidance. The fields can be divided into operational zones, in transient and non-transient turning areas (e.g. green corridors or a harvested neighbouring field) and in inhibited zones (e.g. street, obstacle, etc). Machine specific data contain the technological restrictions such as geometry, steering and power train properties of the machine as well as the setting data of the implements. The operation strategy defines the optimisation criteria. Examples for these criteria are shortest driving distance, high area capacity, reduction of repeated passes or reduction of curved passes during operation. The considered field is divided into headland and main operation area in a first step. The headland can be extended to neighbouring areas which are not inhibited. This allows to move turnings outwards the field. Consequently soil compaction on the headland and crossing of harvested crop is reduced. The guidance paths are calculated by dividing the main operation area into suitable partial areas. The driving direction is determined for each partial area. Priorities are assigned for the harvest order of the partial areas according to the spatial arrangement of the partial areas. According to the machine kinematics different kinds of turnings in forward and backward motion are possible. They are evaluated with respect to turning time and turning distance. The planning program selects the optimal turning manoeuvre for each change of the track. Therefore, the overall motion plan is optimised. Target values for the speed and implements are integrated into the plan, too. The automatic planning process results in an operation plan for the automatically guided machine.

A low cost platform for obtaining remote sensed imagery

T. Stombaugh, A. Simpson, J. Jacobs and T. Mueller

Effective management of agricultural crops is critical. Optimization of inputs, yield and quality is becoming of greater importance to all farmers. Researchers have estimated that losses due to insects, diseases, weeds, fertility, water problems, and other factors account for as much as 20 billion dollars annually in the United States alone. To reduce these losses, farmers are relying increasingly on diagnostics and subsequent recommendations from crop scouts or by diagnosing problems within the fields themselves. This crop scouting is slow, laborious, expensive, and often inaccurate due to small sample size and the limited training of personnel. Limited crop scouting efforts often result in unnecessary application of resources over large areas, improper timing, misplaced applications, or unnecessary replications. Remote sensing has shown promise as a tool to enhance crop scouting efforts. Remote sensed images can be used to identify areas within a field that may be experiencing some type of stress or pest pressure. Unfortunately, it is expensive to obtain remote sensed images through traditional satellite or aircraft services. Additionally, there is often a significant delay between the time when the image was acquired and when it is available to the person who will be making the management decision. New and innovative technologies are needed to reduce the costs involved in the characterization of small remote sites, and to obtain those images in a timely fashion. Unmanned Aerial Vehicle's (UAV's) have been proposed as an alternative to current remote sensing methods. UAV's provide a new and innovative strategy for the collection of remote sensing data because they can efficiently image large areas of croplands quickly, in remote locations at a fraction of the cost of conventional image capturing techniques (i.e. Satellites and Aircraft). This technology will be of great importance as precision agriculture starts relying more on precise diagnostics. The goal of this project was to demonstrate the use of inexpensive Unmanned Aerial Vehicles (UAV's) as a tool for capturing scientific quality remote sensed imagery for precision agriculture. Several design and performance parameters were identified for the vehicle. The system had to be easily transportable in common vehicles. Once delivered to the location, the air-borne system had to be easily assembled, hand-launched, and operated in a variety of weather conditions. Upon retrieval, a mechanism had to be in place that would allow immediate retrieval of digital images. There are two approaches to platform design were explored in this project. First, the RS equipment was installed into an acceptable commercially available remote controlled aircraft. Second, a completely new aircraft platform was designed for the specific purpose of carrying the RS equipment. The commercially available platform was less expensive, but the custom aircraft had better performance parameters. Both approaches are discussed in this paper. Several different options were also explored for the remote sensing equipment. The goal to develop an inexpensive platform as well as weight restrictions for the airborn platform precluded the use of specialized multispectral cameras. The equipment that was used consisted of CCD-based digital still cameras with appropriate light filtering to obtain images in the desired wavelengths.

A new N sensor concept is ready for agricultural use

Peter Thoren and Doreen Blesse

Variable-rate N fertilization is an important tool for fertilizer reduction, improvement of economic yield, and environment protection. A new sensor generation for site-specific determination of nitrogen requirement of the plants during fertilization has been developed, tested, and made ready for application during the last years. The underlying optical measurement principle is based upon the detection of laser-induced chlorophyll fluorescence and is therefore an active one which does not rely on sunlight as light source. The sensor, called Laser-N-Detector, determines the nitrogen content of the plants via measuring the chlorophyll density. Along with accompanying determination of the crop density a N fertilizing strategy can be calculated for online variable-rate fertilization. The Laser-N-Detector was developed by Planto in cooperation with AMAZONE as a leading supplier of fertilizer spreader equipment . It therefore fits into the fertilizing system comprising the fertilizer spreader with variable-rate technology, electronics, tractor terminal, and sensor as a means to adjust the application rates for the left and right side separately according to the simultaneously measured nitrogen need of the plants. The sensors are mounted at the tractor roof and monitor the crop to the left and right over a distance of several meters. The measurement of chlorophyll fluorescence is a research tool used by biologists for plant stress detection. Biologists also discovered the correlation of the ratio of two fluorescence wavelengths (F690/F730) with leaf chlorophyll density. The Laser-N-Detector has been developed by transferring this knowledge to agricultural application by using modern optical and electronic technologies. A pulsed laser beam is scanned across the crop canopy. The resulting fluorescence radiation is measured several thousand times per second. Fluorescence signals can therefore be related to individual plant leaves while soil does not contribute to the signals. Statistical analysis of fluorescence data delivers individual as well as collective nitrogen status of the scanned area. Furthermore plant density is determined as a by-product. The advantages of this new sensor generation are - all-day usability, even in dusk and into the night - measurement is independent of leaf color / surface structure - only plant tissue contributes to measured signal - plant density can be determined directly Since 2000 the Laser-N-Detector has been tested in pot and field trials in cooperation with scientific partners. Chemical and optical analyses of plant samples as well as yield determinations and raw protein measurements have been performed in order to calibrate the sensor. The correlation between different N supply levels and the chlorophyll density has been proved. Important insights into the chlorophyll distribution over plant height levels have been achieved. Field trials on different sites with the same parcels of different N application levels have been carried out. Measurements with the Laser-N-Detector have been done in conjunction with common single plant measurement methods for N. The results showed the suitability of the Laser-N-Detector for site-specific fertilization and provided calibration data. As an example the Laser-N-Detector has been tested on a farm site with 64 large parcels (52x52m) with different N levels. Furthermore there were heterogeneities because of slopes and accordingly different water supplies. The collected sensor data have been evaluated and interpolated with dedicated GIS software. Different N levels as well as heterogeneities caused by topology could be shown accompanied by differences in plant density. In 2003 some Laser-N-Detector prototypes will prove their suitability and efficiency in practical farm use.

Development of a commercial vision guided inter-row hoe: achievements, problems and future directions

Nick Tillett, Tony Hague, Philip Garford and Peter Watts

This paper outlines key technical aspects of research into computer vision for inter-row hoe guidance before going on to describe commercial developments and future research directions. The system developed at Silsoe Research Institute (SRI) applies a shadow invariant transformation to images which are then passed through a digital band pass filter to extract underlying crop row structure. The camera looks down at 45 degrees viewing several crop rows over approximately 3m. The advantage of the band pass filter is its tolerance to missing crop and ability to filter out weeds that do not match crop row structure. A Kalman filter is used to fuse vision observations with implement motion information, providing a robust estimate of camera lateral offset and heading with respect to crop rows. The recursive nature of the Kalman filter ensures that previous information is fully utilised in successive estimates, making the system robust to short periods (approximately 4m) in which vision observations are poor or non-existent. A simple on/off hydraulic control moves a side shift mechanism left if lateral error is right and vice-versa. To prevent valve chatter a 9mm dead band is included. The hoe blades and the camera are both mounted on the side-shifting frame. Evaluation under typical field conditions gave standard deviations in lateral error of between 9 and 12mm at speeds of up to 10kph. On this basis one would not expect peak error to exceed 3cm. In 2001 Garford Farm Machinery in conjunction with Robydome Electronics and SRI developed a commercial guidance system. These developments have been based on the original concept, with special attention to mechanical robustness and cost. For example PC hardware has been used and the single board computer uses flash memory. Attention has also been paid to the user interface based on a high quality colour LCD display. This provides a live video image, system performance status and a simple menu driven operator set up procedure. Whilst the original research was conducted on sugar beet and cereals (22cm row spacing), commercial uses include carrots, parsnips, onions, leeks, brasicas, field beans and pumpkins. Some customers have been growing organically, but most are conventional growers wishing to reduce herbicide inputs. Principle motivations are pressure from their customers (multiple retailers), a shortage of effective herbicides and an opportunity to reduce chemical costs. The advantage of vision guidance over traditional manual guidance is said by users to be increased precision, higher speed and reduced driver workload. Farmer reaction to vision guidance has been good with many prepared to experiment with the new technology. For example at least one customer is also using a guided toolbar purchased primarily for inter-row cultivation to apply other inputs in a precision-banded way. It has also become clear that despite efforts to minimise costs the 11,000 price for a guided toolframe over and above the cultivation part makes the system best suited to wider implements operated by larger enterprises. However, we hope that groups of growers or contractors will take up the technology, thus sharing the costs over a wider area. Up to now hoeing has been limited to one drill/transplanter width due to the difficulty of matching bouts with sufficient accuracy. Automatic guidance has the potential to break that barrier, as it would be possible to track multiple bouts from a single tractor using one camera per bout. Our current work is exploring ways in which the information from multiple bouts can be best combined to provide very high work rate systems. Reliability, automatic error recovery and ergonomic user information are particular challenges in such a complex system. We will also be looking to apply the same techniques to precision band spraying.

Data fusion for winegrape yield variability analysis

B. Tisseyre, J.N. Paoli, C. Mazzoni and N. Ardoin

The recent commercial availability of yield monitoring equipement for winegrape harvesters presents grape and wine producers with opportunities to tailor production of both grapes and wine according to expectations of vineyard performance and desired goals in term of both yield, quality and the environment (Bramley, 2001 ; Tisseyre et al. 2001). Within-vineyard variability might be driven by different factors (diseases, fertilizer deficiencies, soil texture, microclimate, soil water availability, etc.). Among all these factors, the knowledge of problems due to soil water availability constitutes a significant goal for several reasons:

- they are likely to be consistent over the years (vineyard irrigation is not allowed in France),
- they determine the quality level of the grapes and the wine (and thus its level of payment). Zones of high water availibility lies to an excessive plant vigor, significant yield, delays of maturity, dilution of sugars and other compounds like colouring compounds and flavours,
- once identified and located, it is easy to propose a site specific strategy to manage water availibility and the resulting excessive plant vigor (putting vine row under grass, reduction of the number of buds during winter pruning, fertilisers reduction).

This work proposes a method which allows, starting from the grape yield map of the year and complementary observations:

- to diagnose, identify and locate zones of high yield where high water availability (and high vigor) is involved,
- to determine if the observed phenomena is consistent since the vine plantation,
- to define a sampling strategy for complementary measurements necessar to explain the phenomena and to propose management strategy,

The originality of the method lies in the fact that data fusion :

- is not based on interpolated maps,
- does not require arbitrary classes of values but involves the measured data as well as their variability,
- is based on a test of the robustness of classification (zoning),

The cartographic representation is only made to validate the spatial result of the classification process.

Method

1)   choice of the parameter to measure

In addition to the grape yield (monitoring), several indices were assessed on the vines and the soil. These choices were made according to vine ecophysiologists and soil scientists [the full paper will explain the expert reasonning involved in this choice) :

- the topography of the piece (distribution of soil moisture),
- soil heterogeneity estimated by resistivimetry (Douzals 2000),
- the vine vigor estimated by the diameter of the trunk, (McFarlane et al. 1997) [the full paper will specify methodology used to assess this measurement. It will specify the possibility of automation of such a measure in the precision viticulture context]

2)   clustering

- Sampling intensity was not the same for all the parameters. A scaling method was applied to the data (mean of k near neihboorgs). [full paper will describe and discuss this methodological choice, nevertheless, results showed that this method was convenient in our context (Molin et al., 2001)]. The result of our scaling is a n*p table where n is the number of samples (located in the vineyard) and p the number of assessed factors.

- Clustering was performed with a non-supervised method : hierarchical classification (Ward's method)([full paper will justify the choice of this method compared to others : clustering around moving centers, k-means algorithm (Shatar et al. 2001, Saporta 1990)]. Coordinates of the points were not introduced in the classification process. After the clustering, points belong to three to five classes depending on the vineyard.
- The spatial relevance of the classification is validated by a simple mapping of the classes (cubic interpolation). The process is relevant if the number of within vineyard zones is the same as the number of class. [full paper will discrite a simple indice which allow to estimate the spatial relevance of the classification].

3)   Robustness and Interpretation

A discriminant analysis was then performed in order to notice factors which are strongly involved in the classification process (factors which explain the more the relevant zoning)(figure 3.b). A cross validation was then performed in order to test the robustness (prediction error) of the classification process.

Results

This method was performed on three different sites in 2002.

For one site, it highlighted that within vineyard yield variability was due to soil water availability and topography. Thanks to vigor measurements, it was possible to say that this phenomena was consistent since the vine plantation. It was possible to define a strategy with the viticulturist to decrease the yield (and increase the grape quality).

For a second site, our method highlighted that within vineyard yield variability was due to soil heterogeneity. Vigor measurements showed that this phenomena was annual and not consistant over the years. Further investigations will show the kind of problem encountered.

For a third site, results showed that yield variability was not due to soil water diponibility.

Visualisation methods in precision agriculture

Jonas Tornberg, Mats Söderström, Susanne Van Raalte and Johan Martinsson

Background and objectives Traditionally, data in precision agriculture are collected using field point sampling methods or by the use of various sensors in different applications. Often a considerable number of data layers are recorded for each field, and this amount of layers will increase with time. The large amount of data requires that the user can store and organise the data efficiently in a database or at least in an intelligent data file structure, and some commercial tools for this purpose has been developed. Precision farming data are normally interpolated and displayed in two-dimensional maps. Most precision farming practioneers agrees upon the importance of using appropriate methods to create such maps. So far, not so many publications deal with appropriate methods to visualise the data. The intensive spatial (and over the years temporal) data collection in precision farming makes it interesting to use a number of other mapping and visualisation methods. Geographic information systems (GIS) are in general used for the management and analyses of precision farming data in two-dimensions, but videos, animations, 2.5 and 3D GIS models as well as true virtual reality models are possible options when we want to present and facilitate interpretation of complex data that vary in space and time. The intention of this paper is to exemplify how such methods can be used to visualise and model data in a precision agricultural context. Applications From extracted data, collected at several test farms in southwestern Sweden, we exemplify and discuss the use of various visualisation and modelling methods, for example virtual reality models of a test farm as well as a single field, video clips and animations of spatio-temporal data, and soil mapping in 3D. 1. Virtual reality (VR) may be defined as a computer-based representation of a space in which users can move their viewpoint freely in real time. A VR-model may be a very detailed and realistic model of the real reality, but also very simple and without any finer details. The level of detail is accomplished through a combination of photographs and spatial objects. Simple VR-models may be created directly from vector data in some geographic information systems. Through the Virtual Reality Modeling Language, VR-models can be presented and used on the Internet without expensive software. In this study we have created three different types of VR-models, a simple model of a test farm with test plots, a model for presenting soil maps to farmers and a modelling technique to display crop growth and yield maps. 2. Precision farming data are typically spatio-temporal and animation is an excellent technique to introduce the temporal component of spatial data. Maps from multiple measurements of within-field crop growth during a season combined with meteorological data, may for example be animated as a cartoon to efficiently visualise crop development in the field in response to weather changes. But animations may also be non-temporal to display, for example, relationships of spatial attributes. The animation technique is exemplified by a video clip from repeated Hydro N-Sensor scannings, an animation of the effect on soil pH of precision liming and a fly through/fly-over model from aerial photographs, digital elevation models and yield maps. 3. Even if it is not common in precision farming, soil-sampling data may be available from several depths in some fields. In general, it possible to generate perspective views in most GIS, but it is more difficult to visualise true three-dimensional GIS-data. In this case we have tried to present three-dimensional soil data for an agricultural field by combining different types of simpler presentation techniques such as various diagrams, animations and VR-models.

Application of multivariate adaptive regression splines in precision agriculture

K. Turpin, D.R. Lapen, E.G. Gregorich, G.C. Topp, N.B. McLaughlin, W.E. Curnoe and M. Robin

With modern technology, it is possible to collect large amounts of spatial soil and crop parameter data in fields. Successful implementation of site specific management depends to a large extent on correct interpretation of the data to make appropriate decisions on both defining management zones, and determining site specific inputs. Interpretation difficulties increase when there are large numbers of measured parameters, each with a significant contribution, and each interacting with each other variably over space and time. In addition, many parameters have inherent measurement errors, and exhibit small scale variability, both of which can appear as noise in the data. This paper examines the potential of data mining techniques to help, i) identify soil factors potentially important in the interpretation of corn plant establishment and yield during a cooler-wetter season (2000), and ii) identify critical variable value thresholds at which environmental correlations change or activate. Field experiments were carried out on a clay loam field near Winchester, Ontario, Canada (45 03 N, 75 21 W). Crop rotations were established in 1996 on 14 m wide by 300 m long field plots with rotations typical of local livestock farming (corn-corn-alfalfa-alfalfa), and cash crop farming (corn-corn-soybean-wheat). Spatial soil and crop parameter measurements taken on a 10 m sample spacing, included total residual spring soil N, bulk density, texture, volumetric water content, soil cone penetrometer resistance, plant establishment, and final yield. Elevation and surface water catchment variables were derived from fine resolution GPS information. Multivariate adaptive regression spline (MARS) techniques were used to uncover dominant soil physical and chemical parameters that potentially impacted stand populations and final yield. Although MARS generates predictive functions, the objective was to use these functions and variable interaction surfaces to, i) identify variable thresholds where variable interactions/correlations changed or were effectively activated, ii) identify important variance minimizing environmental variables with respect to yield and corn establishment, and iii) interpret environmental correlations in the context of identifying crop risk areas in the field. MARS fits piecewise linear regression splines to the data and automatically calculates the optimal transition points or knots subject to user imposed constraints. A knot is where the behaviour of the function changes. In many cases, one parameter may have a dominant effect over a certain range, but relinquishes dominance to another parameter when outside of that range. MARS identifies these ranges for large, complex data sets. The analysis showed that crop establishment was negatively related to soil water content (0-0.15m depth increment) and cone penetration resistance (0-0.15m depth increment). Soil water content, penetration resistance, total spring residual soil nitrogen, and elevation were the most important factors contributing to corn yield with soil water content and penetrometer resistance having a negative correlation, and spring residual soil N and elevation having a positive correlation. Breakpoints or knots occurred at volumetric soil water contents of 28.4% vol.; wetter soil areas having a predominantly negative affect on yield and soil N. The results were in line with expectations as these soils are poorly drained, and are prone to compaction problems. Although the field was relatively flat, subtle differences in elevation had an effect on soil water contents.

Simultaneous assessment of cotton yield monitors

G. Vellidis, C.D. Perry, G. Rains, D.L. Thomas, N. Wells and C.K. Kvien

The most essential component of precision farming is the yield monitor – a sensor – or group of sensors – installed on harvesting equipment that dynamically measure spatial yield variability. Yield maps, which are produced using data from yield monitors, are extremely useful in providing a visual image to clearly show the variability of yield across a field. In response to the demand for a reliable and accurate cotton yield monitor, several monitors have recently become commercially available. One of the first questions asked by potential users is "What is the accuracy of the system?" The trick is to understand how accuracy is defined. Instantaneous accuracy is the accuracy of each yield data point (practically impossible to measure), load accuracy or load error is the accuracy over a basket load of cotton, and field accuracy or field error is the accuracy over an entire field. Field accuracy is usually the smallest number of the three because over an entire field, measurement errors average themselves out. To provide users with unbiased evaluations, we assessed the AgLeader, AgriPlan, FarmScan, and Micro-Trak cotton yield monitors in the southeastern United States for five harvest seasons between 1997 and 2001. Each year, three or four yield monitors were mounted on a cotton harvester and used during harvest of several farmer-owned and managed fields. The accuracy of each sensor was tested by comparing load accuracy. Yield maps from each yield monitor were also produced with the respective software packages and compared. Feature comparisons of each monitor were included. Between 1997 and 2000, the Micro-Trak and Agri-Plan yield monitoring systems were directly compared in several fields. Agri-Plan errors fluctuated around zero while the Micro-Track was biased towards underestimating loads and typically had error measurements exceeding ±10%. Without question, on a load-by-load basis, Agri-Plan was more accurate. The Micro-Trak exhibited significant operational problems. It occasionally experienced problems with one or more of the eight emitter-receiver pairs on the sensors reading continuously high. This was caused either by a system malfunction, or more frequently, by repeatedly recording cotton lint which was caught on the sensor housing. Yield monitor manufacturers recommend that their systems be calibrated whenever field conditions change (variety, irrigated vs nonirrigated, defoliation quality, yield, etc.). Most users find it difficult to calibrate systems several times during the season because of the time required for calibration and difficulty in locating certified scales in close proximity to the fields. During the 2000 harvest season, we evaluated 3 commercial yield monitors, Agleader, Farmscan (manufactured in Australia) and the Agriplan, when used under real-world conditions. The results were disappointing. None of the systems were consistently accurate throughout the season. This was probably aggravated by variety changes and possibly moisture content changes in the crop. In general, AgLeader tended to over predict, FarmScan tended to under predict, and Agri-Plan's response was mixed. Our experience indicates that frequent calibration is necessary to ensure accuracy. Each of the yield monitoring systems we assessed have something to offer the grower interested in creating yield maps. All the systems are capable of producing an adequate yield map provided the system is properly calibrated, operated, and maintained. The issue appears to be how much calibration and maintenance is required for good performance. All potential users should carefully research prospective cotton yield monitoring systems before purchasing for the following attributes: quality of the product, user-friendliness, ease of installation, GPS requirements, availability and responsiveness of technical support, skill level required of the picker operator, and time available for downloading data files.

Development of a sensor for continuous soil resistance measurement

R. Verschoore, T. Seps, J. Vangeyte and J. Pieters

Soil resistance can be measured in several ways. The most wide-spread and well documented one is the vertical penetrometer where the force needed to fore a cone into the soil is measured. The result is expressed as a pressure (Pa). Although this method has been specifically developed for civil engineering purposes, it is also used as a reference for "soil compaction". A second type of systems for determining soil resistance measures the force needed to pull a nearly horizontal flat or curved plate trough the soil at a depth of about 50 mm. In this case, shear forces are measured, so this method is rather appropriate for traction purposes.

In this study, a third type, namely a horizontal penetrometer, was investigated. This method uses a cone or other object, which is pulled horizontally through the ground by a tractor, while the penetration force is measured. Horizontal penetrometers allow the soil resistance to be measured on-the-go in a continuous and relatively quick way.

A first horizontal penetrometer studied was analogous to the classic vertical one, it means that the force needed to penetrate with a cone the soil in a horizontal trajectory is measured continuously and gives an indication for the soil resistance. This method, however, was not used because of its horizontal levelling stability problems.

The second penetrometer investigated used a vertical smooth blade, equipped with strain gauges at several heights, pulled through the soil. Combining the results of the measured strains the soil compression profile was determined. Sensitivity analyses showed that measuring the cutting forces would result in poor resolutions for the upper soil layers, so a third system was developed.

This third penetrometer also used a vertical smooth blade pulled through the soil, but now the blade was equipped with 50 mm high "wings", symmetrically mounted on the blade, as indicated in the figure. The bending moment on the wings was measured and used as a measure of soil resistance. Six couples of wings were mounted, so the total depth of the measurements was 300 mm. The raw data of a measurement gave a time/distance history of the bending of the different wings at the six different depths. After treatment of the data, these values could be transformed into soil resistances in the six layers.

A series of laboratory tests was carried out in a soil bin to study the behaviour of the measuring system and to study the correlation between the results of the horizontal penetrometer and a classical vertical penetrometer. Because of the large variances in the results, however, only a poor correlation was obtained.

Proximal sensing of soil reflectance to characterise field soil organic carbon

Raphael A. Viscarra Rossel, Christian Walter and Youssef Fouad

Soil colour is an important visible characteristic of soils that has long been used for soil identification and qualitative determinations of soil characteristics (e.g. Webster & Butler, 1976). The reason may be that various soil properties exhibit spectral response in the visible range of the electromagnetic spectrum. Some of these properties have been shown to have good correlations with measurements of soil colour. For example, there is a large volume of work that reports relationships between soil colour and soil organic carbon (SOC) (Ben-Dor et al., 1997; Lindbo et al., 1998). The dark colour of soils has generally been associated with high organic matter contents and high inherent soilfertility. In classification systems, soils with dark surface horizons are often separated from other soils at the highest categorical level, determining their importance as a medium for plant growth (Schultze et al., 1993). The main objectives of this work were to (i.) establish relationships between soil colour (using RGB, CIE (Commmission Internationale de l'Eclairage, 1978) and Munsell colour systems) and soil organic carbon (SOC), (ii.) determine whether quantitative measurements of soil colour could be used to predict SOC content of agricultural soil and (iii.) to compare predictions to spectrometric measurements made using a visible range spectrometer. Soil colour was determined using different systems for representing colour space: qualitatively using the Munsell Soil Colour Charts and quantitatively using RGB (red, green, blue) tristimulus values from soil images acquired using a digital camera. The RGB pixel data was extracted out of each image and the median R, G, and B pixel values used in the analysis. The RGB values were converted to CIE XYZ tristimulus and their resulting CIEL*a*b* (CIELAB) and CIEL*u*v (CIELUV) transforms. To establish relationships between SOC and soil colour, forty-three different soils were collected from various locations across Brittany, France. Visual measurements of Munsell soil colour and digital images of these soil samples were acquired in the laboratory under 'ideal' lighting conditions. The soil was also analysed for SOC using conventional laboratory analysis. Statistical relationships between Munsell value units, corrected RGB image-intensities, CIEL*a*b* and CIEL*u*v coordinates and SOC were derived for predictions of SOC content of field soil. The visible spectra (400 nm – 700 nm) of these soils were also measured using a spectrometer and partial least-squares (PLS) regression (Martens & Naes, 1989) implemented on the spectra for predictions of SOC. Field soil was sampled from two different fields with different mean levels of SOC and transported to the laboratory for image acquisition, spectral analysis and SOC determinations. Predictions were validated against chemical analyses and statistics that relate the accuracy of predictions to their precision and bias were used to quantify their quality. Quantitative soil colour measurements using either RGB image-intensity values (corrected for light) or CIELAB / CIELUV colour coordinates showed good response for SOC. Predictions of field SOC (using simple RGB/CIE vs SOC relationships) were good (RMSE 0.34 for field 1 and 0.36 dag/kg for field 2). In fact they were better than the more complex spectrometric measurements (using the visible range spectrometer and PLS regression) (RMSE 0.36 dag/kg for field 1 and 0.54 dag/kg for field 2). Using an appropriate calibration model, accurate predictions of field SOC using digital soil images are possible. The digital technique provides a quick, cheap and easy way to collect useful, quantitative soil data. Not yet better but more efficient than conventional analysis when spatially dense data are needed.

Site-specific primary soil tillage and cultivation

H.-H. Voßhenrich and C. Sommer

Today, crop production methods have to be cost saving and soil protecting. Therefore, the aim of sustainable agricultural soil use is to structure production methods in such a way that the least amount of energy is used and the lowest costs maintained at the same time achieving high yields with good product quality without affecting the stability of the ecosystem involved and/or neighbouring ecosystems. Soil protecting production methods and management are significant aspects in sustainable agriculture. The objectives of the research- and development sub-project Site-specific Soil Tillage and Cultivation in the framework of preagro (supported by the Federal Ministry of Education and Research (BMBF), Berlin) is to optimise site-specific primary soil tillage and cultivation to make a contribution towards cost reduction and avoidance of harmful soil compaction and soil erosion, as well. Part of the project was to prepare a management system for a DGPS-supported soil tillage that takes into account the local soil specifications within a field. The first implement of its kind world-wide was built which allows tillage working depth of a chisel plough to be varied during the ride. The prototype (in co-operation with Amazone-Werke, Hude, Germany), a 3 m wide combination of a highly variable pre-loosening tool, a pto-driven rotary cultivator and a V-ring roller was designed. Meanwhile, a second prototype (in co-operation with Amazone-Werke, Hude, Germany), a drawn tillage implement with a working width of 6 m, has been specially built. The rules for site-specific tillage state that soil needs to be loosened when unfavourable growth conditions for plant roots (compaction, lack of oxygen) occur. This is often the case at sandy locations with a tendency towards compaction, on soils with a poor structure or heavily soil water-influenced soil types, such as gley and pseudo-gley. The algorithm developed uses different information levels (German soil inventory, soil type maps, conductivity measurements, humus content drainage and others) for the derivation of tillage depths for different field sub-areas. Site-differentiated tillage was tested (Farm Täger-Farny, Groß Twülpstedt, Germany). For comparability's sake, the fields were divided up into plots with conventional and differentiated tillage. Plant evaluation at several monitoring points showed slightly different rates of field emergence which, however, must be attributed to the individual soil conditions. Even the intentional omission of deep tillage did not have a negative effect. The energy requirements for superficial (8-10 cm) and deep (25 cm) tillage were determined for the above described first prototype. For this purpose, a strip experiment with four measurement rides was carried out on a field which was still traffickable at that time. The measurements show that traction forces, fuel demands and working time can be halved by switching from deep to superficial tillage. Under practical conditions the extent of the savings largely depends on the relative share of superficial tillage within a site. Locations with a small percentage of sites which need loosening this is often the case in areas with a pronounced relief offer the largest savings potential with regard to energy consumption and technical requirements. Conventionally, these sites are usually tilled deeply. The paper will present results from farm field experiments. The conclusion is that site-specific conservational soil tillage can successfully be realised. Less primary soil tillage intensity at selected parts of the field by means of algorithm provide equal plant development and yields compared with the conventionally tilled field. Related to the field heterogeneity, fuel demands and working time can be reduced up to 40-60 %. Therefore, economical advantages are obviously. First results will show that additionally further aspects of sustainable crop production are on the way. So: site-specific soil tillage and cultivation will realise economical and environmental goals, as well.

Preliminary practices of precision agriculture and priorities to promote technological innovation in P.R.China

Maohua Wang

At the crossing of centuries, China's agricultural development has changed from simply seeking high yields to seeking a combination of yield, quality, and efficiency. The major constraints to agricultural development have evolved from mere resource constraints to combined constraints from resources, market demand, benefit and environment. The progress of science and technology has become the main driving force of agriculture and rural economic development. Joining the WTO will allow encourage making use of both domestic and foreign production essential factors to improve the levels of equipment and management in agricultural modernization, speed up the transformation of traditional agro-industries to modern agriculture, improve the quality of agricultural products, and increase the efficiency of agricultural production system. The traditional concept to manage agricultural systems is gradually changing. The government has identified biotechnology and information technology as two leading technologies to guide the agricultural science and technology revolution in the new millennium. IT is defined as a mean by which we acquire, record, process, apply, and communicate information. It embraces data (the facts), knowledge (interpretation of observations), applications, and technology to support production system management decision and control. The topic on precision agriculture (PA) has drawn a strong interest in China since past 3-4 years. It is regarded as an information revolution in agriculture and agricultural system in the future. As a new concept to manage resources and production system with information technology, precision agriculture strategy and technology are not only suitable to large-scale farming, but also could be effectively extended to improve small-scale farming and conventional agriculture reconstruction in developing countries with appropriate technologies through capture of data at an appropriate scale, interpretation and analysis of that data, and implementation of a management resources at an appropriate scale and time as well as with appropriate tools. The highest possible accuracy is not always needed. Reliability, cost, technical support, and education to diverse conditions are quite important. The end users are key factors. The concept of precision agriculture needs to be in the mind of farmers, not only in used machines. This requires dissemination of the concepts and training of extension service staff. Concepts and infrastructure for this have to be developed and implemented with the support from government. Recent surveys are showing that the infrastructure of PA services is developing and the adoption by producers in some practices continues to increase. Better and more complex models to deal with real-world problems will be needed. More intelligent machines and processes are becoming reachable based on analysis of human behavior. This paper will highlight the preliminary approaches and practice of precision agriculture in some areas in China in the last 3 years. The first results of testing and technological innovation based on the domestic conditions will be analyzed. Investigating technical integration mode based on real conditions and extension of the corrective understanding would be quite important. Some priorities in favour of technological innovation and adoption of precision agriculture are proposed to meet the urgent needs for the development. Strategic problems discussion would be beneficial for extension and possible adoption of PA concept in less developed countries.

Differential GPS Positioning over the Internet for precision agriculture

Georg Weber, Harald Gebhard, Matthias Gröschel and Denise Dettmering

Due to the increased capacities of the Internet, applications which transfer continuous data-streams by IP-packages, such as Internet Radio, have become well-established services. Compared to these applications, the bandwidth required for the transfer of real-time GPS positioning data is relatively small. As a consequence, the Federal Agency for Cartography and Geodesy (BKG) developed a new technique using the open Internet for the real-time collection and exchange of GNSS data, as well as for broadcasting derived products. A major purpose of these activities is the dissemination of Differential GPS corrections (DGPS) in support of real-time GIS and navigation for precision agriculture

This paper describes the http-based technique for streaming DGPS data or other kinds of GNSS data to mobile receivers via the Internet. It allows multiple PDA, Laptop, or GPS-Receiver connections via Mobile IP-Networks like GSM, GPRS, EDGE, or UMTS. The technique establishes a format called Networked Transport of RTCM via Internet Protocol (Ntrip), because of its main application being the dissemination of corrections in the popular RTCM-104 streaming format. Sufficient precision is obtained if correction data are not older than a few seconds. The RTCM standard is used worldwide. Most, if not all DGPS receivers accept it.

Ntrip's system components, disseminating RTCM data, are
- NtripSources, which generate DGPS data-streams at a specific location,
- NtripServers, which transfer the data from one or multiple sources in Ntrip format,
- NtripCaster, the major stream-splitting system component,
- NtripClients, receiving data of desired sources from the caster.

The data streaming is accomplished by TCP/IP protocol stack. Several attempts, based on plain Serial-to-TCP conversion of streaming data on the reference-side (server) and TCP-to-Serial re-conversion on the rover-side (client), have shown the suitability of the TCP/IP protocol for streaming data to mobile IP clients. A number of public/private institutions, directly or indirectly supporting precision agriculture, show a distinguished interest in providing DGPS or RTK-GPS data over the Internet. Among these are the
- Sub-commission for Europe (EUREF) of the International Association of Geodesy. EUREF maintains a GPS Permanent Network (EPN) encompassing about 160 continental-wide distributed reference-stations. Current interests are focussed on the dissemination of DGPS-corrections for GIS and navigation.
- International GPS Service (IGS) of the International Association of Geodesy. The IGS intends to exchange GPS data for the purpose of real-time orbit determination. Precise orbit information is one of the prerequisites for Real Time Kinematic GPS (RTK-GPS).
- European Space Agency (ESA) with its SISNeT, a system disseminating RTCA-corrections from Europe's geo-stationary EGNOS satellite for navigation purposes.
- German State Survey Administrations with their Satellite Positioning Service SAPOS. Its main objective is to provide a nation-wide public DGPS/RTK service.

ASCOS company, a commercial DGPS/RTK service provider with the intention to cover Germany.

This paper focuses on implementation and availability aspects of Ntrip-based DGPS services. As an example, the current status of the EUREF-IP project (IP for Internet Protocol) is being described. It aims to stream RTCM corrections over the Internet and GSM/GPRS cellular phone networks. Server, client, and caster software characteristics are being outlined. Problem areas like latency and IP-Multicast are being discussed. The achievable accuracy will be demonstrated based on most recent experiences. Conclusions are then being drawn concerning the potential of DGPS over the Internet for establishing services in various fields of precision agriculture.

Developement of decision rules for site specific N-fertilization by the application of data mining-techniques

Georg Weigert, Peter Wagner and Hubert Linseisen

For the realisation of the potential economical benefits of Precision Farming one critical task still remains: the development of a system that is able to collect site-specific data efficiently and to use this data for site-specific management decisions. This paper proposes a competitive way to generate decision rules for site specific N-fertilization by the use of data mining-methods within a Management Information System (MIS). The concept of the MIS has been developed within the project "Information System for Site-Specific Management" founded by the German Research Foundation (DFG) and enables various forms of data analysis. Essential spatial information is gained by the use of mobile process technique installed on the machinery. This technology allows to record automatically spatial point data for both the (on-field) operation itself and - by the use of sensors – various data about plant vitality and soil properties.

These data are stored regardless of different management aims in a stationary database. The 'data warehouse' is completed by another database for non-spatial data such as climatologically data. Data processing is done by Geo Information Systems (GIS). The following functions of the MIS are necessary to get consistent and valid datasets:

- Data preparation
- Building surfaces by various methods of interpolation
- Variable rastering resp. variable grid sizes, esp. in tramline direction and partial working-width size
- Aggregation of data onto sites

These aggregated datasets are essential for the here proposed method for developing N-fertilization rules by ex-post data analysis. For this task a simplified N-cost adjusted Output-function is used. Its components are the site specific yield (Y), the N-fertilization-rate (f_N) as the only variable input factor and the corresponding factor prices (p_Y, p_N). Assuming that there exists a site specific profit function as a result of a specific underlying yield function the goal is to maximise financial gain (G) by site-specific optimised fertilization rates (f_N).

$$G = p_Y * Y(x_1,x_2,..,z_1,z_2,..,f_N) - p_N * f_N \text{ à max!}$$

The underlying yield function (Y) is determined and characterised by the variable input factor (f_N), spatial factors ($x_1,x_2,..$) and non-spatial factors ($z_1,z_2,..$). Spatial factors promising to have significant influence on the yield function are gained within the MIS by spectroscopic, electromagnetic conductivity, yield and draught-force sensors. Similarly non-spatial data such as rainfall, temperature, etc. up to the planned date of the application are assumed to play a decisive role

The idea is to predict the site specific yield by an artificial neural network (ANN). The yield is assumed to be (linear or non-linear) depending on the fertilization rate and spatial and non-spatial factors. The ANN can be trained with the above mentioned datasets gained by the MIS.

With a trained ANN it is possible to get the profit-optimising fertilization rate (f_N) for a given site by the simulation of yields for various fertilization rates. By the simulation of this operation for various site-specific factor-values it is possible to get datasets consisting of site specific factor values and the profit-optimising fertilization rates.

The last task is to extract decision rules from this datasets. For this task data mining methods are planned to be applied to get rules by the use of classification techniques. As the output of the models are susceptible to various sources of error, experts are interviewed to prove results. By the use of MIS and mobile process technique it is possible to get the requisite amount of plot data to train the ANN and to prove gained decision rules in field trials. Furthermore the paper tries to illustrate and discuss the critical factors for the success of the here described method.

Soil protecting mechanisation and precision farming

Peter Weisskopf, Claus Sommer and Eugen Kramer

During the last decades, technological development has lead to efficient, powerful and heavy agricultural machines. Subsequently the weight of tractors, self-propelled machines and transport vehicles increased inevitably, thereby intensifying the risk of subsoil compaction, especially under wet soil conditions. Control and prevention of this risk are a matter of concern for the farmers – regarding the protection of soil fertility as their fundamental means of production as well as in order to reduce follow-up expenditures in field work (money and time needed for soil tillage). On the other hand there is a public interest in the preservation of soils and its important functions, which is put more and more into legal regulations (e.g. the German Bundesbodenschutzgesetz). Due to concern for physical soil protection, threshold values for machinery parameters (e.g. maximum allowable wheel load) have been put forward, which were derived from wheeling experiments. However, these propositions result in fears that agricultural mechanisation would have to be reconsidered with unforeseeable consequences for its design and cost. In order to offer practicable solutions, it is useful

- to analyse the causes and processes leading to damages to soil structure;
- to derive cultivation technologies contributing to precautionary solutions and to avoid or reduce risks.

A successful risk assessment is ensured with the fulfilment of three requirements:

1. describing the pressures and stresses which are transferred by the wheels of a vehicle in a soil, at least for a stationary situation;
2. describing the stress distribution in a soil depending on soil conditions;
3. estimating the change of soil structure due to vertical stresses, especially concerning its vital functions regarding plant production, living space and hydrological processes.

Based on this information it will be possible to show limits of soil strength which a mechanisation respecting soil structure has to accomplish. This can be done by means of decision schemes which do no longer depend on general threshold values (e.g. for axle loads), but rather on threshold limits in soil strength / soil stress-diagrams. The risk of damage to soil structure can be reduced by several measures, the most important being agricultural engineering, cultivation techniques and field traffic management. In a precision farming strategy to promote sustainable soil use and management, three engineering approaches can contribute to site and situation-specific cultivation practices: 1. Planning a mechanisation respecting the site-typical limitations of soil strength. The general risk for damages to soil structure due to a certain vehicle on a given site can be assessed for long-term average conditions by means of a decision scheme (e.g. for the selection procedure to buy or hire a vehicle). 2. Adaptation of vehicles regarding to prevailing soil conditions. Based on continually measured and interpreted soil and vehicle data determining the risk of soil compaction, properties of a vehicle working in the field are actively influenced. This way the risk of damages to soil structure can be controlled depending on vehicle load and changing soil conditions. Controlling systems for wheel load distribution and tyre inflation pressure offer interesting possibilities to adapt vehicle parameters actively to current soil strengths. 3. Design and construction of vehicles respecting the limitations of soil structure. Engineering approaches relevant to soil protection such as a maximum allowable wheel load with a given soil strength, an optimum distribution of contact areas or enough space for the adjustment of tyres are examples where a more precise assessment can be achieved. This allows for a more specific optimisation of vehicle properties regarding to limitations of soil structural stability. Moreover, heavy off-road vehicles can be equipped with active control systems in order to influence directly and instantly the risk of soil compaction.

GPS receiver accuracy test - dynamic and static for best comparison of results

C. Weltzien, P.O. Noack and K. Persson

Objective GPS receivers have been and are being widely tested. The biggest drawback on most tests and measures of accuracy is the lack of comparability or the lack of applicability for agricultural purposes. Between geo-konzept and DLG a setup has been developed to test static as well as dynamic, relative and absolute accuracy of positioning of D/GPS receivers common for agricultural application. The main goal of these tests is to produce meaningful, comparable and reproducible measures of performance in order to define different classes of GPS receivers for agricultural proposes. Methods Tests for both static and dynamic measurements with a duration of 24 hours each have been carried out in three repetitions with 2 to 4 weeks time between the repetitions. The endurance test setup from the DLG test centre for agricultural machinery has been adapted for the purpose of the dynamic GPS receiver test. This setup is an automatic self propelled test rig travelling along a round course at a user defined speed, pivoted in the centre of the circle (called 'round-about' in the latter). The round-about has a diameter of 44 m, four reference points along the circumference have been surveyed according to the local landsurvey system (Lagestatus 489, Hessisches Landesvermessungsamt, HKVV) with cm precision using a Trimble 4000ssi kinematic 2 receiver GPS system. The reference positions along the track are postprocessed for each full second (UTC) using the exact time when passing the known reference points recorded with the Trimble 4000ssi. The travelling speed of the round-about is constant. The reference positions for each full second and the data from the D/GPS receivers in the test is synchronised using the UTC time from the NMEA string ($GPGGA). For both static and dynamic measurements the absolute accuracy (deviation from the reference point) as well as the relative accuracy (deviation from the mean) are calculated. The dynamic setup is very hoarse (particularly fastidiously) for the use of GPS receivers as the edge of the canopy (porch roof at 4m height) of the nearby testhall touches the circumference of the round about like a tangent. Thus on each pass of the round-about the receivers are suspect to shading of the canopy. This is an intended strain on the receiver performance in order to emphasis differences in the ability to cope with agricultural environment. Results All numeric results can only be true for the time and date of the test as well as the location of Groß-Umstadt as accuracy of GPS measurements is always a function of several parameters including the exact location the time, the date and more. For this reason a system of classification rather than a numerical measure of accuracy has been developed to judge the different receivers. Four classes of receivers will be distinguished as presented before by others (Buick 2002) which include all agricultural applications. The results of these 24h tests are also compared to data from in-field testing gathered in the preagro research project in several locations throughout Germany and Denmark. It can be shown that the test condition at the DLG are hoarse but realistic and that the shading is a useful tool to emphasise differences in receiver performance which under optimal environmental circumstances wouldn't show. The results of this test shall help to decide upon the best equipment for each purpose. For those users who need receivers for mapping the absolute accuracy from the test results will be the right measure. Others who use the GPS for guidance purposes need the relative positioning to be as accurate as possible. It's very likely to be the majority of users who don't want to spend a lot of money on a GPS system.They can use the test results to learn about the restrictions as well as the abilities of a low-cost systems.

Technical solutions for variable rate fertilisation

Cornelia Weltzien and Krister Persson

The objective of the proposed paper is to show how the choice of spreader and working width influence the possibilities of performing variable rate fertilisation. It will be discussed in which way high heterogeneity of fertiliser requirement in the field reduce the possibilities of applying correct fertiliser dosage. The second objective is to show how the spread patterns are influenced by application rate changes which again affects the working quality during variable rate fertilisation. The importance of optimisation (adaption) of the spread pattern during spreading will be shown. The importance of the best possible setting of the machine before starting the application work and the correct adjustment during work will be discussed. The discussion will be based on measurements of 2 dimensional fertiliser distribution patterns from different fertiliser distributors. Methods In a research project carried out in collaboration between DLG/preagro and DIAS the distribution quality of fertiliser spreaders for variable rate application have been tested with granular fertiliser covering the German and the Danish market. The work that has been carried out covers spreading with 12, (20), 24 and 36 metres working width and different spreaders. The spatial resolution of typical requirements was determined according to an application map using the KemiraLoris Software. Tests have been executed in order to establish the 2 dimensional fertiliser distribution pattern at application rates of 30, 60, 90 and 120 kg N/ha. The Spreader settings have been adjusted according to manufacturers instructions for spreader setup given in handbooks or on the internet. The 2 dimensional spread patterns have been established by measuring the transversal distribution pattern at steps of 2 m offset in longitudinal direction. Dependent of the test set-up and the spreader type 5-20 tests have been carried out per spreader, fertiliser and application rate. Based on the measured spread patterns the final fertiliser distribution in the field has been calculated by simulation of the distributor moving up and down in the field in adjacent parallel tramlines at wanted working width. Results The results show that: - Wide working widths are limiting the possibilities for variable rate fertilisation when requirements call for high variation in short lateral distances or variations within limited areas of the field. - Fertiliser distributors producing a wide and overlapping distribution pattern in longitudinal direction introduce limitations to addressing heterogeneity with small spatial resolution and fast reaction time requirements in longitudinal direction. - Full width spreaders with sections of 10-12 metres which are equipped with individually controlled metering systems to apply different rates at each section will be required to address demands at a small scale resolution for fast and high changes in application rates. Calculations show that the average application error in an example field - that is to be applied with variable rates - changes with a factor 2 when going up in working width from 12 metres to 36 metres. This dramatic change in average application error is mostly caused by the change from an in principle triangular spread pattern at 12 meter working width to a spread pattern of rather trapezium form at 36 metre working width. It can also be shown that the average application error may be reduced by 30% by changing from a centrifugal spreader to a full width spreader at e.g. 20 metre working width. The lower average application error obtained through the full width spreader is due to reduced spreading width in longitudinal direction. Simulations of total fertiliser distribution from a full width spreader having 2 or more individual adjustable sections will show an even lower average application error.

MOSAIC: crop yield prediction – Compiling several years' soil and remote sensing information

O. Wendroth, K.C. Kersebaum, H.I. Reuter, A. Giebel, N. Wypler, M. Heisig and J. Schwarz

Although farmers' fields are managed homogeneously, grain yields vary considerably within the same field. Moreover, in a particular field, spatial yield patterns are unstable for different years even for the same crop. The question remains unsolved so far, where to expect high or low yields within a respective field site, and how to decide upon distribution of spatially varying amounts of fertilizer, especially nitrogen, and pesticides. On several field sites on a farm in Lüttewitz in South-Eastern Germany, Federal State of Saxony, soil properties such as textural properties, soil organic matter, nitrogen, phosphorous, potassium content, and land surface elevation obtained from a laser scan, were analysed in spatial grid designs with different spatial and temporal resolutions (see: Poster on the MOSAIC-Project). Spatial grain yield distributions were monitored with a yield mapping system on the harvester combine during harvest. At one or two specific times during the growing season (spring time), normalized differential vegetation index (NDVI) was computed from color infra-red aerial photographs. For recent years, crop nitrogen status was determined on-the-go with a crop sensor (Hydro-N). For previous years, where no sensor data were available, crop nitrogen status was obtained from simulations with the uncalibrated nitrogen dynamics and crop growth model HERMES. Spatial yield distributions for different fields, crops, and years were predicted using multivariate auto regressive models. Transition coefficients were obtained in state-space analysis. In this analytical method, auto regression coefficients are estimated for describing a spatial or temporal series. The matrix of auto regression coefficients is optimised within an iterative scheme, i.e., the Kalman Filter, while - unlike in ordinary regression or auto regression analysis - measurement and model uncertainty are accounted for in a filtering scheme which consists of an updating and smoothing part. In order to compare the log-likelihood prediction quality for different state vectors, and for reasons of numerical stability during the optimisation, data were normalized. Transition coefficients varied for different spatial resolutions, and for different auto and cross correlation structures. Coefficients derived from state-space analysis were successfully applied for pure predictions, hence without stochastic filtering (updating and smoothing) in the respective year. Different soil, crop and remote sensing state variables yield thereby different prediction quality. From this result we conclude that in cases where laborious soil monitoring campaigns cannot be accomplished, on-site remotely sensed information (NDVI, crop nitrogen status, and land surface elevation) taken during spring time combined with autoregressive state-space models supports grain yield prediction. In our investigated fields, the quality of prediction based on remotely sensed information was not lower compared to that obtained from soil based information (texture, soil organic matter). It can be observed that for different years and a particular spatial resolution, transition coefficients between yield and NDVI observed during the respective growing season remained relatively stable. A further critical examination of auto regression coefficients derived from state-space models for several years' data shows whether stability of coefficients over different years is sufficient to support real predictions based on spring time observations of NDVI and crop nitrogen status.

Development of produciton level management zones for nitrogen fertilization

D.G. Westfall, R. Khosla, K. Fleming, B. Kock, A. Hornung, B. Gangloff, R. Reich, T. Shaver and D. Heermann

Grid soil sampling has been used to develop nutrient concentration isoquant maps to guide users of precision fertilizer application since the inception of precision agriculture. However, the cost associated with collection and analysis of soil samples at a density that accurately define soil spatial dependency is generally prohibitive, especially for most agronomic crops. Also, most commercial users of grid sampling do not test their data for spatial dependency. If the data are not spatially dependent, users are not justified in interpolating the data set for generating fertilizer prescription maps. As a result of these limitations, much interest is now being directed to the use of "management zones≅ (MZ) to allow implementation of precision agriculture. The objective of our research is to evaluate four methods of development of production level MZ for precision N fertilizer management as compared to grid soil sampling. Each of the methods uses a unique set of soils, yield, and/or remotely sensed data. The level of technological expertise required to develop the MZ ranges from very simple that most producers with limited computer skills could use to techniques that could only be used by trained crop consultants that are computer and data management literate. The data layers being evaluated, from least to most technical, to develop production level MZ are: 1) Bare-soil panchromatic aerial photography, topography and the producer=s past production experience of the field; 2) Apparent soil electrical conductivity (EC_a) as measured by the Veris® E.C. cart; 3) Bare-soil panchromatic aerial photography, topography, soil OM content, CEC, texture and previous year=s yield map; 4) Bare soil image (Red, Green, Blue, and NIR) to guide cluster based soil sampling, followed by complex multi-variate spatial statistical analysis to generate a MZ surface map. We classify MZ into high, medium and low production potential. The crop production and soil parameters defined by these four MZ delineation methods are being compared to each other as well as the standard grid soil sampling method. Nitrogen recommendations are based on an algorithm that includes soil organic matter, residual soil nitrate, and yield goal. Our hypothesis is that the use data layers of Astable≅ soil parameters will result in the development of production level MZ that are stable for several years. Composite soil samples can then be taken within each production level MZ each year to develop fertilizer N recommendations for each MZ, thus requiring only three composite soil samples each year after MZ development. All four MZ delineation method were evaluated in 2001 and 2002 in two to three farmers= fields under center pivot and furrow irrigated corn (*Zea Mays* L.) production in eastern Colorado, USA. Our results from 1999 and 2000 with Methods 1 and 2 showed that homogenous sub-regions regions in the field could be identified that had similar soil properties and yield. At one location only two MZ were identified with Method 1. However, when Method 2 was used, it was found that the areas of high productivity were included in the medium productivity management zones identified in Method 1. Scale of resolution in Method 1 did not allow identification of the high productivity MZ. The data from 2001 and 2002 will also be discussed. The MZ approach appears to be a viable method of implementing precision agriculture. However, an easy, accurate method of delineating MZ must be developed. This is our project goal.

On-farm field experiments for precision agriculture

B.M. Whelan, A.B. McBratney, A. Stein and J. Cupitt

The philosophy of Precision Agriculture (PA) embodies an information-gathering and decision-making process. At the farm field scale, once a decision has been made to move towards evaluation of a variable-rate treatment program, two main options present as feasible at this stage. The simplest option is to gather spatial data to construct a simple production budget for the input of interest. This process is possible for many nutrients that are applied as fertilisers, but usually requires a number of agronomic assumptions. It is a much less suitable process for evaluating amelioration treatments such as soil physical (e.g. deep ripping) or chemical (lime and gypsum) amendments. Here it is necessary to measure a response to each treatment and structured field experimentation becomes important. Structured field experimentation is also necessary if knowledge of whole-field and within-field input-response functions (e.g. to fertiliser and seed rates) is required for economically and environmentally optimising production. Variable-rate technology allows the setting down of sophisticated field experiments in farmers' fields by the farmers themselves to acquire this knowledge and yield-monitoring technology allows the measurement of response. However, the spatial design of such 'on-farm' experiments are in their infancy. Where the object is to produce a local moving-window response function, and the total production from the experimental field is not crucial, systematic designs such as a modified draught board or an egg-box design have been used. These designs are highly invasive, in the sense that they may impact heavily on a farmer's production and therefore may be very expensive. More practical, less-invasive, designs are needed. For fields divided up into management zones, an efficient design would seem to be the fleck design where randomised block experimentation is done with spatial constraints and economic considerations. The economic consideration being that one does not want to penalise the grower's expected profit by using sub-optimal application rates over much of the field. Most of the field can have a uniform treatment which the grower considers best practice. Data from all of the field can be used in the analysis. The proper objective function and design for these experiments have yet to be developed, but an approach homologous with the use of spatial simulated annealing for spatial sampling seems the most obvious one. Here we make a first tentative attempt at such a design. The objective function is divided into two parts. First, we select designs that meet the economic criterion of x% penalty on expected profit. The consensus (in Australia) seems to be that x should be no larger than around two and a half. From those candidate designs we optimise the position of the plots in some biometric (e.g., D or A optimality) sense. We give an example, for a 70 hectare field in Eastern Australia Finally, results from more simplistic nitrogen response experiments are used to highlight need for repeating the experiments with the same treatments in the same locations for a number of seasons. Relatively flat response functions have been observed in a number of seasons where water is limiting crop growth. Significant soil mineralisation of nitrogen is also believed to be partially obscuring the response to applied N. With new experimental designs and existing technology, agricultural science may be able to help farmers remove the soil nutrient buffer that has operated in the name of risk management and operate the nutrient system closer to the optimum for production and the environment.

Determination and evaluation of the optimal site specific seed rate

M. Wiesehoff and K. Köller

Until now scientific research in the area of Precision Agriculture mainly generated exten-sive field data about spatial variability such as digital soil maps, aerial photographs, maps of the electrical conductivity or yield maps. But very often the generated maps describe the situation just in a certain moment. Some methods, e.g. digital terrain models show longevity and contain valuable information, but are often very expensive and not for practical use. Therefore site-specific information of previous years results in an application map with so-called management units. These management units result from the different conditions con-cerning location, cultivation system and short-term parameters. Based on this mapped infor-mation application maps for various site-specific variable rate applications can be derived. Sowing of small grains is a field operation that has not been very intensively investigated in terms of precision agriculture. Anyhow, research results show a reasonable potential of yield increase, weed suppression as well as savings of seed, pesticides and fertilizer by applying optimum site-specific seed rates. Meantime, seed drills equipped with a variable metering system via electrical or hydraulic transmission adjustments and DGPS navigation are avail-able enabling map-based sowing. To generate such application maps on a PC with GIS-programs a decision support system for map-based sowing has to be developed. The use of decision support systems enables to convert the generated information into useful application maps. In classical decision support systems the user is asked and must enter facts, which depend on an individual problem. Therefore, the ability of the user has an important influence on the results. Strategies without large aid of the farmer to derive maps of optimum seed rate are still missing. Therefore, objective of this research is data mining concerning the interference of seed spacing and various influence parameters such as topography, microclimate, soil type, soil moisture or cultivation. The results of the survey are organized in a data base to develope algorithms as backbone of a decision support system. In the structure of the database the pa-rameters and criteria concerning the variation of the seed rate are determined from the litera-ture. As the seed rate is often not the main point of investigations, also literature with different emphasis is evaluated. Results from literature must be standardized, classified and feed to the database to derive algorithms. As source of supply agricultural universities and available da-tabases come into consideration. The algorithms are derived from scientific knowledge and must be constantly adapted to the current conditions, all economic and ecological terms should be considered. To evaluate the practical value of the decision support system, additional field trials have been realized. At different locations in Germany the seed rate was varied in three different levels to investigate the impact on yield. Independent of the local conditions three different seed rates (conventional and 20%, more and less) were applied in a field strip experimental design. The locations are showing different conditions in terms of relief, i.e. the variation of depressions, slopes and hilltops as well as the soil type. In some cases maps of electrical conductivity are available, which will be compared with yield maps. Aside from the field emergence, the tiller-ing and the yield structure in summer (ear/m2), kernel/ear and thousand kernel mass has been determined.

Pulse radar systems for yield measurements in harvesting machines

K. Wild, S. Ruhland and S. Haedicke

Yield monitoring is a basic prerequisite of precision farming. Different systems were developed for combine harvesters and are commercially available now. Therefore, the yield of grain can be monitored, but for other harvesting materials suitable systems are still missing. Hence, the objectives of research work was to develop and test new systems which can close this gap. Radar systems are becoming more and more popular and due to increasing production lots their prices are going down. Based on promising results of pulse radar systems for measuring crop density and soil moisture, such devices were investigated on their potential for yield measurements. For this task a test stand was developed in which three different pulse radar sensors could be placed in parallel. Besides distance measurements pulse radar systems can also determine the intensity of the reflected radiation. A main difference between the systems was the working frequency (5.8 GHz, 10 GHz, and 26 GHz). The sensors were mounted on top of the test stand so the microwaves were radiated downwards on a table, where the material which should be measured was placed on. Below the table, there was a metal plate which reflected the radiation, so the microwaves passed through the material two times before coming back to the sensor. Therefore, the measurements were carried out in pseudo transmission mode. The reflected radiation was detected and the difference in intensity of the emitted and received radiation was calculated. This setup with the metal plate was necessary, because the radiation is also reflected on top of the material and by the table. In order to determine which proportion of the radiation is reflected by the metal plate all distances between the sensor and reflection areas were determined in the measured signals. Due to the know distance between the sensor and the metal plate the portion of the signal curve reflected by the metal plate could be separated and used as wanted signal. For the investigations grass, hay, corn silage, oats and wheat were employed. The material was placed on the table of the test stand. After the readings from the three sensors were taken the next layer with material was placed on the table. This procedure was carried until no reflection from the metal plate could be detected any longer. First, for all investigated materials calibration curves were determined. This could be done with a very high accuracy. Depending on the measured substance and the employed sensor the calculated coefficients of determination ranged from 0.96 to 0.99. For the sensor with a working frequency of 5.8 GHz measuring oats and wheat the highest coefficients were registered. The tests with the sensor based on a working frequency of 10 GHz showed no significant correlation between the mass of the material and the intensity of the reflected signal. The reason for this is probably the analysis of the signals which is different to the algorithms which are used for the other two sensors. However, they could not be changed and therefore this sensor was not tested any longer. With the further measurements comparisons between the measured masses and the actual masses of the investigated materials were carried out. Calculated regression lines showed values of 0.90 up to 0.99 for the coefficients of determination. The detected measurement deviations ranged from 0.1 % up to 10 % and above. The magnitude of the errors depended on the sensor's working frequency and on the investigated material. Also, with larger mass respectively with taller heights of the pile (when the intensity of the reflected signal was very low) larger deviations were registered. In summary, pulse radar systems show a high potential for yield detection. Therefore, the next step will be a sensor modification so they fulfil the individual prerequisites for installation and usage in harvesting machines.

The application of electromagnetic induction mapping for target soil surveying

Stefan Wolharn

Detailed soil maps are the basis for interpretation of precision farming results as illustrated in yield maps. The yield map details how crop yield varies spatially within a field but the amount of yield depends on plant growth, which is related to soil type weather conditions, crop type and farming process. Surveying soil which contains layering and texture is carried out by invasive sampling of the soil to a depth of about one metre using a conventional screw auger, a window sampler or physical excavation of trial pits might be to expensive when using a 100 m grid spacing or smaller. Therefore the application of geophysical methods involving continues galvanic resistivity or non-invasive conductivity mapping as a basis for target soil sampling appear to be a more cost effective possibility. This paper describes the use of electromagnetic induction to investigate soil variation, in the context of precision farming. An electromagnetic induction (EMI) sensor EM38 manufactured by Geonics company of Canada was integrated into agricultural operations using a specially constructed sledge, in order to measure within-field variation in soil conductivity. Data were recorded in vertical and horizontal dipole mode from fields located at Silsoe, Bedfordshire (United Kingdom). Within these maps influences of various factors on EMI observations were found: 1. Absolute calibration of the Geonics EM 38 sensor is not possible using the current recommended procedure, however, patterns of relative variation are consistent. The response function of the instrument is not linear with the penetration depth. Therefore the relative contribution of different soil layers may be important. 2. The most relevant factor influencing conductivity readings is the total water content. Therefore: a) Local variations in water content caused by topography can caused variations in conductivity unrelated to soil variation b) Compaction or blocked drainage within the field causes higher conductivity readings c) When agricultural operations such as sub soiling increase the local soil infiltration rate, these zones exhibit a higher conductivity after rainfall 3. Historical artefacts such as old field boundaries are visible in the conductivity map. Subsequently research of three fields were undertaken by correlating EMI observation to crop yield from harvest 1999 and 2000. The coefficient of determination ($R2$) between EMI readings and crop yield varied between 0.003 and 0.28. Therefore, additional information is needed to explain yield variation. In one of these fields the soil conductivity map in vertical dipole mode is used for targeted soil survey. The cause for different EMI values are investigated by estimating trail pits within the field. Also it will be verified if the same EMI value at different location represent the same soil properties. In order to classified pit locations the German soil estimating law (Bodenschaetzungsgesetz) is applied to identify differences in the natural yield potential. In general the relationship of increasing conductivity from lighter to heavier soil series (sand to clay) in uniform layered profiles has been confirmed. However, two significant anomalous are identified. In the locations with similar low conductivity e.g. 3-4.5 mS/m, the natural yield potential ranged between Sl3D35 and lS3D50. This is due to the nature of layered soils and the present of different material at depth. In the high conductive range, 30-70mS/m, the same soil type LT4D 55-60 is present. At these locations, the different water content caused by the topography within the field (hydraulic potential), increases the EMI values in lower areas. It can be concluded, that conductivity maps are no direct indicator for natural yield potential. However, the EMI technology is a useful basis for target soil survey. Further development of this technology is necessary which enable to identify differences in soil layers.

Model for land use decisions based on analysis of yield and soil property maps and remote sensing

Mike Wong and Greg Lyle

Our recent history of yield mapping in Western Australia since 1996 reveals that wheat yield typically varies spatially between 0.4 to 4.0 t/ha within fields. These fields are normally about 100 ha. By applying economic analysis we have shown that some parts of the field are consistently operating at a loss. This variability occurs in a farming system characterised by the replacement of perennial native vegetation with annual winter crops. The inadequate water use of the annual crops compared with the native vegetation is responsible for rising saline ground water table. Water balance modelling suggests that about 50% of the wheatbelt must be reassigned to an alternative perennial land use in order to have a useful impact on water use and salinity. The scale of this land use change is massive and needs to be based on a sound and transparent decision-making process. The Dempster-Shafer Weight-of- Evidence model offers a rigorous methodology for assigning land use based on independent lines of evidence. We used this model in a case study at Three Springs to identify areas suitable for cropping and those that should be reassigned to perennial vegetation. The model used maps of historical gross margins derived from yield maps, soil property, modelled drainage values, soil type, remotely sensed biomass and proximally sensed gamma-ray emission and soil electrical conductivity as evidence. These layers of evidence relate to profits, environmental impact and the causes of yield variability. As the spatial pattern of yield variability varies from year to year and as there is no clear-cut distinction between suitable and unsuitable cropping areas based on soil property maps and other spatial data, the evidence maps were converted to fuzzy sets. These fuzzy sets include expert knowledge and hard data evidence to define the degree to which areas are suitable for cropping. Although the concept of fuzzy sets and Dempster Schafer modelling is somewhat new in agriculture, it is increasingly clear that building more intelligent modelling systems is a prerequisite for land allocation decisions. The expert in conjunction with the farmer and/or agronomist decides for each line of evidence where our understanding lies about the relationship between the evidence and the hypothesis. Additional lines of evidence can be added to include other discipline leaders such as entomologists, pathologists, weed scientists etc. By focusing on profits and environmental outcomes, the model has the potential to facilitate the adoption of land use change based on the combined contributions of the grower and discipline leaders. The work at Three Springs showed that the areas that were consistently operating at a loss coincided with areas of inherently infertile soil due to coarse sandy texture and hence poor water holding properties and nutrient status. These areas can be identified by gamma-ray spectrometry due to emission of 40K from clay minerals and by mapping soil electrical conductivity using EM38. Water balance modelling showed that these sandy areas were the most leaky in the field. Based on the evidence provided, the grower agreed that it would be uneconomic to improve the poor performing area of his field in this particular instance. He plans to grow native trees adapted to the poor performing areas. This is a win-win decision for him economically and for the environment.

Modeling rice growth using hyperspectral reflectance data

Chwen-Ming Yang and Rong-Kuen Chen

Field experiments were conducted at the experimental farm of Taiwan Agricultural Research Institute, Wufeng (24°45'N, 120°54'E, elevation of 60 m), Taiwan to study the seasonal changes of rice reflectance spectra and the approaches to modeling rice growth from reflectance data. Ground-based remotely sensed high-resolution canopy reflectance spectra and the corresponding growth traits of rice plants (Oryza sativa L. cv. Tainung 67) were measured periodically during the first and the second cropping seasons in 2000-2002. Reflectance spectra were taken by a field-portable spectroradiometer (model GER-2600, Geophysical and Environmental Research Corp., Millbrook, NY), and growth traits of fresh weights of leaves and aboveground, dry weights of leaves and aboveground, plant height and leaf area index were collected on the same days of spectral measurements. Spectral indices of RED/NIR ratio, GREEN/NIR ratio, RED/GREEN ratio and NDVI were calculated, where GREEN was reflectance at green light maximum, RED was reflectance at red light minimum, and NIR was reflectance at near-infrared peak. It showed that curve of spectral reflectance of rice canopy was similar to that of green plants, reflectance was lower in the ultraviolet and the visible wavebands, higher in the near infrared, and varied considerably in the short-wave and middle infrared during the growth. Within the visible light, the maximum reflectance was located in green region and the minimum reflectance was somewhere in violet or red region. It also indicated that the seasonal patterns of the examined spectral indices and growth traits were curvilinear during rice growth in both cropping seasons. Generally growth traits reached the climax near heading when the vegetative growth climbed to the climax, and decreased after this stage when the plants grown toward maturity. Seasonal changes in spectral indices and growth traits differed between cropping seasons implying the existence of growth and development effects as well as environmental impact. As there was a significant morphological transformation from vegetative to reproduction growth, it was found that modeling between spectral indices and growth traits can be improved by dividing growing period into the pre-heading and post-heading phases. Some of the selected spectral indices were shown to have significant relationships with growth traits, linearly or nonlinearly. The correlation may be further compared on a relative time scale by using the unit time normalization technique. Spectral indices were normalized with the relative growing days of the specific cropping season on the unit time baseline, therefore the differences in growth between cropping seasons were compared on the real time baseline. In addition to the spectral index approach, models established by the multiple linear regression (MLR) analysis may also be used for estimating rice growth. The best-2, -3 or more variables MLR models can be determined with a choice of user-dependent coefficient of determination, which decided the number of variable to be added to the models. The multiple regression models provided flexibility in choosing the individual narrow bands and improved the relationships to estimate plant growth. Also, the combining spectral reflectance from two, three or more wavebands by the mathematical formula into single numbers exhibited a greater sensitivity to plant vegetation variation and provided a better estimation.

Economic payback from GPS guidance systems

I.J. Yule and R. Buick

Non-adopters of precision agriculture have often given the reason as being that the technology is not proven in terms of producing a consistent economic return for investment. While this is to some extent true it ignores the facts of the situation. Craighead and Yule (2002) demonstrated improved output of NZ$75 to $100 from mixed cropping situations using yield mapping, soil sampling and variable rate in Canterbury New Zealand. Other studies are sighted which show a range of returns. GPS is an enabling technology, it assists us in terms of making sure fertiliser is being spread from the correct position, it does not guarantee that the crop will not be subsequently attacked by disease or be subject to weed infestation later in the growing cycle. Precision agricultural technologies can help raise the overall level of performance, but if one aspect of the farming system is weak then all the potential gains can be easily lost. The non-adopters conveniently forget that the same applies to their system, uniform fertiliser rates over the whole farm do not give uniform results and changes in pesticide treatment do not give automatic improvements. They are just as site- and time-specific as those quoted in precision agriculture trials, it is just that by not measuring the performance they remain blissfully unaware of the true performance and the true variability in performance. In general, precision agricultural technologies are tools to the farmer there is not one secret recipe for success on every farm or with every farmer or for every farm management system. GPS guidance is one area of precision agriculture that is proving to provide very consistent and easily demonstrated payback. Guidance system payback centers around gains in operational productivity and efficiency. New advances in automated guidance are finding new ways to achieve improvements in agronomic productivity also. Yule and Buick(2002) described a second generation of GPS equipment which was capable of delivering direct economic benefits as well as providing a platform for further developments in automation and agro-mechatronics. This generation includes such options as affordable DGPS (submeter and decimetre accuracy) and RTK (Real Time Kinematic for centimetre accuracy) GPS guidance systems. Due to the easily demonstrated return on investment, commercial uptake of these systems is advancing at a faster rate than more agronomically-focused applications such as yield mapping. Benefits of GPS guidance that translate to payback and direct economic benefits include reduced overlap and skip in fertilizer and crop protection chemicals, reduced fuel use, reduced labor costs, extended hours of operation and improved productivity at critical times of year (e.g., planting) and in some management systems improved crop productiion. The case for adoption would appear to be compelling and evidence from North America would suggest an early payback for such a major investment. The paper gives examples of economic payback calculations as well as a generalised spreadsheet, which could be used to estimate economic return. The paper draws on a number of previous studies to identify the advantages of guidance as well as an assessment of their financial impact. The economic opportunities created by overcoming some of the difficulties in operating large field machines is also explored. References: Craighead, M. Yule, I.J. Variability, Crop Rotation & Cultural Practises- The NZ Experience, In: Precision Agriculture in Australasia, Proceedings of the 6th Annual Symposium on Australian Research & Application. University of Sydney, August 2002. Yule, I.J. Buick, R. Precision Agriculture: Are we being one-dimensional? In: Precision Agriculture in Australasia, Proceedings of the 6th Annual Symposium on Australian Research & Application. University of Sydney, August 2002

Interpreting soil electro-magnetic survey results

I.J. Yule, C.B. Hedley, C.E. Eastwood, J. Dando, R.I. Murray and P.R. Stephens

Apparent Electrical Conductivity (ECa) mapping using a Geonics EM38® sensor is seen as a rapid and non-destructive means of detecting soil textural variability in the surface 1 - 1.2 m layer of soil. This has many uses and is now being exploited commercially, in New Zealand the largest demand has come from the wine industry, when either establishing a vineyard, carrying out capital improvements or attempting to improve management. Further studies have been completed on land used for vegetable production and arable cropping either dryland or under irrigation. A number of studies have been completed to help explain what the ECa measurement actually relates to, Doolittle et al. 1994 stated that in non-saline conditions it was primarily related to texture, moisture and cation exchange capacity, with the possibility of being used to estimate other more difficult to measure soil parameters. The problem when mapping is that clearly all of these factors are varying in all positions making up the mapped area. Some variation will be related to the topography and soil formation and erosion processes, other changes will be artifacts of agricultural (crop rotation and production methods) or cultural practices, such as roadways or traffic lanes. All of this makes interpretation of ECa values difficult. A study was initiated to examine the usefulness of the EM38 in determining patterns of soil variability in an area that had been subject to intensive soil sampling. The soils of the study area are mapped in six phases of the Kairanga silt loam (Shephard 1992); they are classified as Gleyed Fluvial Recent soils from the NZ Soil Classification, (Hewitt 1998). Two ECa maps were produced of the 12 ha pastoral-cropping site in two successive years. In the second year, four replicate sites in each soil-mapping unit were sampled to approximately 1m depths for soil moisture and bulk density determination. Soil chemistry was also measured through the profile, this included pH, C, Total N, NO3, NH4, P, Ca, Mg, K, Na and CEC. Results indicated the soil parameter most strongly related to ECa was percentage clay content, (weighted to a profile average, $r^2 = 0.72$). The ECa also had a positive relationship with volumetric water content ($r^2 = 0.43$). Management effects such as surface compaction through grazing and contrasting fertiliser applications were also noted. The EM38 has proved to be an extremely useful management tool in identifying soil variability. Work in this study would appear to confirm earlier work that soil texture in non-saline conditions is the main driver for changes in soil ECa. The work has been successfully developed into a commercial service provided by the NZCPA and is used in conjunction with a Trimble Ag214, RTKDGPS to give accurate topographical maps along with soil ECa. This has allowed the use of 3D graphics to model surface topography that is useful in explaining the data and soil formation processes. References Shepherd TG (1992) Soils of the Pasture-Cropping Systems Unit, Aorangi Research Station, Manawatu District. DSIR Land Resources Scientific Report No.30. Doolittle JA, Sudduth KA, Kitchen NR and SJ Indorante (1994) Estimating depths to claypans using electromagnetic induction methods. Journal of Soil and Water Conservation 49(6):572-575. Hewitt AE (1998) New Zealand Soil Classification (Landcare Research Science Series No.1) 133pp.

Ecological effects of site-specific weed control - weed distribution and occurrence of springtails (*Collembola*) in the soil

A. Zuk and H. Nordmeyer

Site specific weed control has economical and ecological benefits. This was showed by experiences on site specific weed control in agricultural practice. The herbicide use could be reduced. The aim of this work is to show possible benefits of site-specific weed control on the agro-ecosystem, especially on the occurrence of springtails (Collembola (Insecta)) in the soil. The studies were carried out within a farm (size 450 ha, main crops sugar beet, winter wheat and winter barley) near Helmstedt, Northern Germany. The farm was in conventional agricultural use. Since 1999 an area of 106 ha (8 fields) were under site-specific weed control. For weed monitoring, a weed mapping procedure using the Differential Global Positioning System (DGPS) was chosen. The weed distribution was recorded by field walking in a grid with spacings of 25 m x 36 m according to the tramlines or vehicle based mapping. Weed species were counted in a 0.1 m^2 area at every grid point. Weed maps were created and patch spraying was done according to threshold values. Areas below thresholds remained unsprayed. The results showed that site specific weed control offers the possibility of saving herbicides. For the zoological studies a 17 ha field of the farm was selected. 53 grid-points (total 168) with different soil characteristics were chosen for soil sampling. It was an field area of partly high and low weed densities. The soil samples were taken with a special soil auger (diameter of 5 cm, 30 cm length). The soil cores were seperated into 3 segments of 10 cm so that depth profiles of springtail species could be recorded. The samples were processed in an extractor for microarthropodes. Each core was put into a small bucket with a fine screen-wire on the bottom. By increasing the soil surface temperature from 20 °C to 60 °C, continously, the springtails were collected in vessels, filled with alcohol. Three soil samplings were carried out during the vegetation period. 1. Three days before herbicide use (for claryfing the occurence of the springtails on this field before spraying) 2. 14 days after herbicide use (effects of of herbicides and weeds on springtails) 3. 42 days after herbicide use (long-term effects of herbicides and weeds on springtails). The results indicate a relation between the presence of springtail species in unsprayed areas in comparison to sprayed areas of the field (e.g. Folsomia spec., Onychiurus spec.) Some phytophagous species profit by weed species like Matricaria chamomilla, Stellaria media and Veronica hederifolia. For example the springtail species Folsomia candida, known for its sensitivity against chemical influences, react quite clearly on the herbicide application, so usually it could be found on unsprayed parts of the field. An increase of more rare species was found in unsprayed areas and an increase in abundance of particular species, mostly omnivorous species like Onychiurus armatus, could be found in sprayed areas. The concept of site specific weed control with unsprayed areas has a potential of many ecological benefits. Effects on springtails on agricultural fields could be found in our investigations. This raise the valence of site specific weed control.

Full poster papers for ECPA

Development of a canopy density adjusted segmented cross-flow orchard sprayer equipped with a canopy contour guidance system (PreciSpray)

V.T.J.M. Achten[1], R.P van Zuydam[1], J.C. van de Zande[1] and P.G. Andersen[2]
[1]*Institute of Agricultural and Environmental Engineering (IMAG), P.O. Box 43, 6700 AA Wageningen, The Netherlands*
[2]*Hardi International, Taastrup, Denmark*
Vincent.Achten@wur.nl

Abstract

In order to control an orchard sprayer based on a digital tree placement and -volume map a sprayer and control system have been developed. The mechanics and the control system of the prototype cross-flow orchard sprayer are described. A master sprayer controller links the sprayers momentary position to an application map in its memory and controls the sprayer output accordingly using section micro controllers. The communication between the controllers is based on CAN. The sprayer functionality and safety features were tested on a test ground in a test orchard with artificial trees. After several tests and optimizations the sprayer software operated to full satisfaction.

Keywords: precision horticulture, orchard sprayer, aerial survey, GPS, guidance

Introduction

The PreciSpray (EU-QLRT-1999-1630) project was initiated as a part of the precision horticulture concept to reduce pesticide use by tree shape and -volume specific precise application of agrochemicals. Matching spray volume and -direction to tree size and -shape can reduce operational costs and environmental pollution. Actuated segmented boom sprayers (manually or by the use of IR sensors) have shown reductions in agrochemical use of 30% and more. A prototype variable volume precision orchard sprayer that can be guided by a tree foliage shape and -volume map has been developed for this project.

Tree maps that are obtained by a 3D-aerial survey method, which is converted from defence applications, are input to a plant protection and orchard management GIS database. This GIS database is also capable to store monitored infestation and utilise plant protection management and decision support systems. A sprayer control system for the prototype sprayer had to be developed. This system should control spray amount and direction according to the GIS data and automatically record the realised spray application while spraying. This information is fed back into the orchard GIS for registration and analysis purposes.

Materials and methods

A one-sided prototype cross-flow sprayer, equipped with 5 segments of 0.50 m height on top of each other has been developed. The segments are moveable perpendicular to the driving direction by means of electrically actuated side-shifts. Each section has an air outlet and three nozzles that can be switched on/off individually. The air speed of each outlet can be adjusted in four steps.

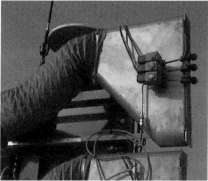

Figure 1. The prototype one-sided cross-flow orchard sprayer and a close-up of a section.

To control the sprayer outputs according to the sprayer position within the orchard a cascade control system with an overall sprayer controller and three section controllers was developed. The section micro controllers are responsible for the control of the sprayer outputs (section position, spray liquid and air amount) and are programmed with several safety features. Whenever a fault occurs all spray liquid nozzles are closed and the sprayer sections are retracted.

The overall sprayer controller is responsible for controlling the sprayer's overall behaviour in the orchard. The controller uses the sprayer position, measured by RTK-dGPS, the actual sprayer settings and the digital orchard map to determine the new settings for the sprayer outputs. These settings are transmitted over an ISO 11783 compatible CAN-bus system to the section micro controllers. The overall sprayer controller also monitored the momentary sprayer performance during spraying. In the development stage this information was used to analyse and optimise the sprayer's performance.

Results

During development tests were performed to ensure the reliability of the section micro-controllers and the sprayer controller. A test track and a small orchard with artificial trees have been created for this purpose. The sprayer was driven over the test track and the sprayer had to adjust its outputs according to the digital map of the (virtual) orchard. On-line recorded data was used to analyse the sprayer's performance and was the basis for modifications in the software. During the tests the software was tested for functionality and reliability by deliberately simulate broken wiring, faulty GPS receipt and power failures. After running several tests and numerous small modifications the sprayer software operated to full satisfaction.

Conclusions

The design of the CAN-based control system and its implementation resulted in a safe, reliable and robust sprayer control system. The sprayer control system was tested and able to control the sprayer according to local orchard conditions, based on a digital orchard map (canopy density and –volume).

Determining spatial variability of yield and reflectance of a cotton crop in the Guadalquivir Valley

J. Agüera[1], M. Pérez[1], J. Gil[1], A. Madueño[2], P. Zarco[3] and G. Blanco[1]
[1]Univ. of Córdoba, Agricultural Engineering Dpt., Menendez Pidal s/n 14004, Spain
[2]Univ. of Sevilla, Escuela Universitaria de Ingeniería Técnica Agrícola, Spain
[3]Univ. of Valladolid, Escuela Técnica Superior de Ingenierías Agrarias, Spain
jaguera@uco.es

Abstract

This work shows the first results of the calibration and yield mapping by using an Agleader PF3000 cotton yield monitor in Andalusia, Spain. Also, a methodology which enable the economical generation of hyperspectral imagery at post-processing from ground-based remote sensor data is proposed, in order to study the relationship between spectral vegetation indices and crop yields. Even though theses results are still preliminary, they make possible to anticipate that both equipments provide correct and accuracy data which are of a great interest for the study of the cotton yield spatial variability.

Keywords: yield monitor, yield mapping, hyperspectral imagery, cotton, remote sensor.

Introduction

The cotton crop occupies in Spain a surface of about 90.000 ha, 95% of them are located in Andalusia, along the Guadalquivir Valley. This crop requires a high amount and variety of chemical, as well as an intense tillage and irrigation water which involve a considerable economical and environmental cost. Wilkerson et al, (1994) developed an optical-attenuation-based sensor to measure cotton flow that has been significantly improved since then (Moody et al, 2000) reaching to be marketed as the Agleader Cotton Yield Monitor. On the other hand, remote sensing can be very useful in collecting data from the vegetal canopy (Koller & Upadhayaya, 2001) and ground-based remote sensors may be used (Haberland et al, 2001). This work shows the first results from the setup of a cotton yield monitor and proposes a methodology for spectral data collection from a ground-based remote sensing system. Theses are the first steps of a larger project in progress that started during the 2002 season which aim is to evaluate the potential application of precision agriculture in the cotton crop under our Mediterranean conditions.

Material and methods

A cotton yield monitor Agleader PF3000 was mounted on a John Deere 9965 harvester. It uses optical cotton flow sensor in each of the 4 ducts toward the basket. A Trimble AgGPS 132 receiver with Omnistar corrections was connected enabling the acquisition of geo-referenced yield data with submetric precision. In a first test, a 6,8 ha field was harvested dividing the production in 6 loads that were independently weighed by scales in order to compare with PF3000 measurements. No GPS data were logged in this test. In a later test, a second field of 8,6 ha was harvested in second pick-up, logging both yield and GPS data. Also, a two-channel Ocean Optics spectrometer was adapted in order to collect spectral data in the field along with the same AgGPS receiver previously described. Both instruments were simultaneously linked to a laptop to enable the acquisition of hyperspectral radiance, irradiance and GPS data every second as the equipment was moved over the row of the crop.

Variability of hyperspectral reflectance and vegetation indices calculated in the 400-950 nm spectral region from the vegetation canopy were mapped along 14 rows of 150 m and 19 m in between as a preliminary test.

Results

Result and percent of error of the yield monitor calibration test are shown in table 1 where a pondered average error of 1,56% is deduced. Figure 1 shows the yield map from the second test with differences up to 10 times in yield within the second field. Spatial variability of NDVI within a part of the third field is shown in the figure 2.

Table 1. Calibration test of the yield monitor.

Yield Monitor (kg)	Scales (kg)	Error (%)
1729	1670	3.50
1545	1560	0.96
1783	1800	0.94
1715	1760	2.55
4993	4930	1.27
6029	5800	3.79

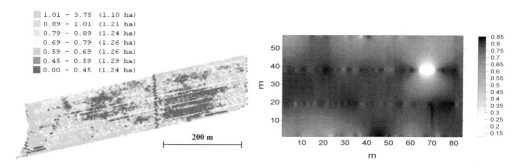

Figure 1. Yield map of the second field.

Figure 2. Spatial variability of NDVI within field 3.

Conclusions

Even though theses results are still preliminary, they make possible to anticipate that both equipments provide correct and accuracy data which are of a great interest for the study of the cotton yield spatial variability.

References

Haberland, J., Colaizzi, P., Kostrzewski, M, Waller, P., Choi, C., Eaton, F., Barnes, E. and Clarke, T. 2001. Ground-based remote sensing system mounted on a linear move irrigation machine. ASAE Paper N° 01-1145. St. Joseph, Mich.

Koller, M. and Upadhyaya, S. 2001. Relationship between a soil adjusted vegetation index and processing tomato yield. ASAE Paper N° 01-1143. St. Joseph, Mich.

Moody, F. H., Wilkerson, J. B., Hart, W. E., Goodwin, J. E. and Funk, P. A. 2000. Non-intrusive flow rate sensor for harvester and gin application. Proceeding of the Beltwide Cotton Conference. Memphis, Tenn.: National Cotton Coun. of America. pp 410-415.

Wilkerson, J. B., Kirby, J. S., Hart, W. E. and Womac, A. R. 1994. Real-time cotton flow sensor. ASAE Paper N° 94-1054. St. Joseph, Mich.

Acknowledgements

This work is being funded by the project AGL2001-2436 of the Spanish Ministry of Science and Technology and the European Union by FEDER funds.

Implementation of a geographical information system for precision agriculture in selected sugarcane areas of Holguin province, Cuba

Ricardo Alvarez, Sara Interian and Jose J. Rodriguez
National Center for Laboratory Animal Production, CENPALAB, Finca Tirabeque, Bejucal, Havana, Cuba
geomatic@cenpalab.inf.cu

Introduction

The projected development of PA in sugarcane agriculture, Cuba's main economic crop, is justified by centuries of experience, existing infrastructure and inminent restructuring of the industry. The fact that large continuous areas are dedicated to this crop is a premise for efficient application of PA technologies. The establishment of a national Differential Geographical Positioning System (DGPS) service is already underway, with 2 base stations installed in Havana and Holguin provinces that transmit a differential correction signal. This allows positioning errors of less than 1m, essential for PA technologies. The Electronics and Automation Group of CENPALAB has converted several agricultural machinery and implements to hydraulic and microprocessor-controlled systems, with DGPS receivers installed, to enable them for site-specific variable-rate application as well as yield mapping in the case of the combine harvester.

Geographical information system for precision agriculture

The present work is part of the first experience in Cuba on the application of Geomatics in Precision Agriculture (PA), it's design, development and implementation, as well as integration with other advanced technologies such as remote sensing, satellite positioning, vehicle tracking, crop yield mapping and others.

The design of a GIS for PA has to take into account the integration of differing technologies that is typical of PA systems: data processing, remote sensing, satellite positioning, automation of agricultural equipment, among others. The database includes geographical, socioeconomic and environmental aspects as well those related to PA technology itself, such as soil characteristics, soil and crop treatments (fertilization, herbicide spraying, etc), crop yield and others which are referenced to a spatial grid of pre-determined size; it becomes, in effect, a knowledgebase. Real-time data input by radio from microcontroller-based units on tractors, harvesters and other agricultural equipment, as well as from sugarcane transports, goes to a vehicle-tracking system which is interfaced with the GIS database.

Apart from the more traditional advantages and uses of a GIS, there are many new possibilities for it's application in the sugarcane PA:

- Storage of soil analysis data based on precise grid location (DGPS positioning).
- Generation of grid-based soil fertility maps.
- Precise grid-based fertilization recommendations for automated machinery.
- Generation of grid-based yield maps from harvester data (see Figure 1).
- Yield, costs and efficiency comparison between different harvests.
- Highlighting of "problem" grids for further analysis.
- Calculation of optimal vehicle trajectories and fieldwork assignments.
- "Just-in-time" resource planning (fuel, fertilizers, workforce, availability of tractors, harvesters or agricultural implements, etc.).

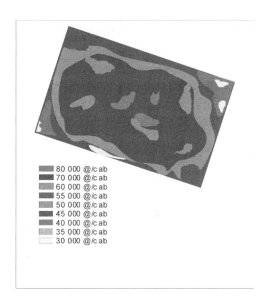

Figure 1. Yield map of sample sugarcane field from PA GIS.

The detection of relative height of weeds as a main determinant of an early competition index estimate in the field

L. Assémat and M. Chapron
INRA, UMR Biologie et Gestion des Adventices, BP 86510, 21065 Dijon-Cedex, France &
ETIS ENSEA, CNRS 2235, 6 Avenue du Ponceau, 95014 Cergy, France.

Abstract

We estimate future prospects of stereoscopic imaging in order to characterize the 3D structure of a weed-crop cover for both weed species identification and competition modelling.

Keywords: stereovision, plant height, light efficiency, competition index.

Introduction

Height has been constantly suggested as one of the key parameter for driving light competition and most competition models have to posses a module that simulate expected height growth in relation to time and/or development stage of species. Whether height differences between species were the result of emergence differences, whether only species growth characteristics and canopy architecture explain these differences, almost no remote sensing at any scale in the field of this variable has been attempted. However, small initial differences at these early stages - where control decisions have to be undertaken - are known to explain a bigger part of variability in the outcome of competition. In the context of Precision Agriculture, where localized control of weeds is the aim, a better evaluation of local competitive pressure should be devised.

Materials and method

Our demonstration is in 2 parts. Image analysis of color photography of weed cover at seedling stage and simulation of light competition between species.
First, we use a methodology recently developed (Chapron *et al.*, 2nd European Conference on Precision Agriculture, 1999) in order to analyse stereo images of experimental plots of weed seedlings. In this method, an initial classification of pixels of stereo images in two classes (vegetation, soil) is performed from the red, green, blue and near-infrared images using the Principal Component Analysis. Then, an adaptive cross-correlation is carried out on the contour pixels of leaves in the left and right images. This provides the disparity and correlation coefficient images. The correlation coefficient represents the reliability of the stereo-matching and the disparity gives the height through the calibration parameters which model the stereovision system. The intrinsic and extrinsic calibration parameters model the projections of the 3D world on 2D image planes. The heights inside the contours of leaves are obtained by a diffusion technique based upon a partial differential equation. The results of this stereovision method are compared with the true values of measured heights of leaves.
Second, a 3D simulation model for radiative transfer balance in a multi-species row structure is constructed (from Sinoquet & Bonhomme, Agric. For.Meteorol., 1992) in order to estimate what is the importance of the height information of plants on the light efficiency (part of incident light intercepted) of each species. Competitive pressure at this local scale is then estimated by the ratio of actual light efficiency over light efficiency of the same species when all (or part of) other species have been removed.

Results

We present the results for experimental plots measuring 48 x 84 cm. We identify possible sources of errors for all steps from picture shot in the field to the computation of competition indices. Finally, we try to estimate what would be the importance of plant height knowledge in comparison with the identification and estimation of plant cover of the species only, for a correct estimation of potential weed competition at this local scale in the field. For this, we use path coefficients in a causal diagram where plants width and height, distance between maize rows and weed density are the measured variables, and light efficiencies and competitive index are the explained variables.

Conclusions

The discussion concerns how much effort we should continue to put in a 3D description of weed-crop cover, relative to a 2D description much simpler, yet difficult. One conclusion will be that height can also be a good criteria for species identification where all other attributes (color, texture, shape, ..) have failed. The spatial nature of weed competition concerns the horizontal patchy distribution of weeds in the field, and the vertical distribution of species hierarchy : both control the strength of competition with the crop and ultimately the economic and environmental benefits of any future precision weed management.

Use of geographic information systems to evaluate the plausibility of weed maps

M. Backes and L. Plümer
Institute for Cartography and Geoinformation, University of Bonn, Meckenheimer Allee 172, 53115 Bonn
backes@ikg.uni-bonn.de

Abstract

Even though the knowledge about weed spatial distribution and weed biology has increased in the last decades, the production of convincing weed maps is still very difficult. In this paper the weaknesses of weed maps are visualised by comparing those created with different levels of precision in sampling from the same area of interest. By using non-simulated data for this comparison, an assessment of the errors made by sampling strategy and counting weeds will be demonstrated.

Keywords: weeds, weed map, spatial distribution, uncertainty, Fumaria officinalis

Introduction

With the objective of generating continuous maps in a GIS (Geographic Information System), discrete sampling data is used in Precision Farming. These maps are the basis for many measures in crop production. In the case of e.g. soil- or yield maps, calibrated and established systems and methods are available. Albeit for the generation of weed maps such systems exist, there are doubts according to the plausibility of the maps produced by the above described systems. This phenomenon is caused by the large variance and the patchy appearance of weeds in a field. Furthermore the sampling of weeds is much more time consuming, challenging and error-prone as the assessment of yield or soil properties. The basis for the analysis presented was a maize field of about 2 hectares on which sampling was done manually, counting weeds in a quadrate of 0.25 m^2 size. Simultaneously the sampling was done with a camera system (approx. 0.16 m^2), used successfully in other studies (Gerhards et al, 2002).

Plausibility assessment of weed maps

The results of the manually sampled area (grid size of 7.5 x 15m) were compared with data resulting from the camera system of the same area, having had a grid size of about 2 x 3 metres. Data could be imported into a GIS in order to achieve the purpose of generating weed maps by interpolation (linear triangulation interpolation) and point-shapefiles (GIS specific data) for further computing. Due to the larger number of samples achieved by the camera system, an algorithm for scanning the point-shapefiles generated could be used to extract the nearest neighbours in the camera related points to the manually sampled points. As an outcome, there was a second weed map of same grid size (7.5 x 15m), with the data basis for weed infestation from the camera sampling inside. Five additional weed maps of the mentioned maize field could be generated out of the camera-derived data in the same way, representing the non-sampled areas of the manually sampled grid. This could be implemented by shifting the starting points for sampling by 1.5 metres in the GIS. Figure 1 shows the results of shifting in case of *Fumaria officinalis* L. in the mentioned field. It appears to be obvious that there is a discrepancy between the presented weed maps. For weeds with uniform appearance this behaviour of weed maps was observed previously (Cousens et al, 2002). A statistical evaluation could be realized to compare the resulting weed maps and create a prediction towards the plausibility of weed maps by cross-correlation.

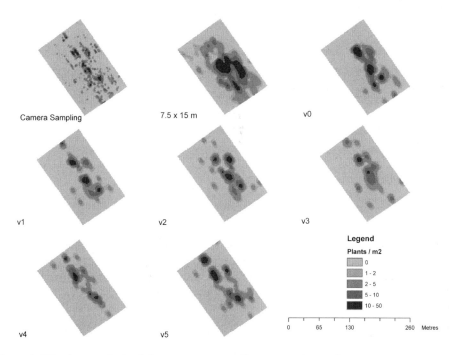

Figure 1. Weed maps derived from camera sampling, manual sampling (7.5 x 15m) and five shifted starting points (each shifted by 1.5m) for sampling (s1-s5). s0 is convergent to the manual sampling map with weed infestation ratio from the camera sampling. The weed mapped is *Fumaria officinalis* L.

Conclusion

The results reveal that there are several potential weed maps produced for the same area with a GIS. The accuracy and plausibility of the maps produced is not completely known. The risks of such maps are very often embezzled or not realised. Evaluating the most effective and adequate ways for sampling weeds in any further discussion on weed maps, and the decisions derived from sparse sampling strategies is clearly necessary.

References

Cousens, R.D., Brown, R.W., McBratney, A.B., Whelean, B., Moerkerk, M. 2002. Sampling strategy is important for producing weed maps: a case study using kriging. Weed Science 50 542-546.
Gerhards, R., Sökefeld, M., Nabout, A., Therburg, R.-D., Kühbauch, W. 2002. Online weed control using digital image analysis. Journal of Plant Diseases and Protection, Special Issue XVIII : 421-427.

Estimating oil content based on oil palm fruit color

S.K. Balasundram, P.C. Robert and P.H. Gowda[1]
Precision Agriculture Center, University of Minnesota, St. Paul, MN 55108, USA
Department of Soil, Water and Climate, University of Minnesota, St. Paul, MN 55108, USA
sivakb@soils.umn.edu

Abstract

Harvesting oil palm based on the minimum ripeness criterion is a challenge. This study aimed at developing an alternative harvesting approach that could potentially complement site-specific oil palm management in South Sumatra, Indonesia. Two regression models that describe the relationship between oil content and surface fruit color were developed. Both models explained about 80% of the variation in fruit color and gave a predictive accuracy of about 77%.

Keywords: oil palm, surface color, total oil, image analysis, regression modeling

Introduction

Harvesting of oil palm (*Elaeis guineensis*) fresh fruit bunches (FFB) begins 36 months after planting and stretches for a period of 18-22 years at 10-14 day intervals. At present, FFB ripeness is ascertained based on the number or percentage of fruit detached from the bunch. Theoretically, it would take several hundred detached fruits to indicate that a bunch has reached maximum oil content (Corley and Law, 2001). Manual collection of detached fruits is costly but highly necessary because they influence oil extraction rates. For this reason, commercial plantations harvest FFBs based on the minimum ripeness criterion (MRC), which usually ranges from 1-5 detached fruit(s) per FFB (Ariffin et al., 1991; Teo and Yeang, 1992). However, the use of MRC can be questionable due to physical problems such as entrapment of fruits within the tree trunk and feeding of detached fruits by pests. This study is part of a site-specific oil palm management effort to optimize oil palm yield and FFB quality. A preliminary step toward optimizing FFB quality so as to complement site-specific strategies is to understand the relationship between fruit quality and its quantifiable physical characteristics such as color and shape. The objective of this study was to explore the relationship between fruit quality, parameterized as percent total oil, and its exterior surface color distribution.

Materials and methods

This study was conducted at a commercial oil palm plantation located in South Sumatra, Indonesia. Eighty fruit samples were collected from random fields bearing 5 year-old palms. Unbruised fruits were obtained consistently from the central periphery of healthy harvested FFBs. Each fruit sample comprised 2 individual fruits. Samples were photographed using a Canon Powershot 600 digital camera (0.5 million effective pixels, 7 mm lens) under room temperature (27°C) with illumination sourced from a white 34-watt fluorescent tube (length: 1.22 m, diameter: 0.46 m). To overcome the glaring effect on the fruit image as a result of direct exposure to light, a diffused lighting setup was created using a cardboard box laced with white paper sheets in the interior and positioned 3.5 m below the fluorescent tube. Each sample was photographed against a white background to obtain front and rear surface views at a constant distance of 25 cm between lens and the object. The photographed samples were analyzed for total oil content using soxhlet extraction techniques. Fruit images were sequentially processed using a suite of computer software. Image rectification to establish the zone of interest for each fruit was performed in *Adobe Photoshop*. Isodata and Maximum likelihood algorithms were employed to classify the rectified images using *MultiSpec*, an image processing software. Classified images were re-classed in *ArcView* to quantify the distribution (%) of black, red, orange and yellow. Relationship between color components and total oil content was explored using multiple regression analysis. Model selection was made using a stepwise backward elimination approach.

Results and discussion

Two regression models were developed. The regression of total oil on surface color distribution was explained by the following mean functions:

Model I: E (Total Oil | Black, Yellow) = 88.08 – 0.52*Black + 1.30*log(Yellow)

Model II: E (Total Oil | Red, Yellow) = 36.84 + 0.63*Red + 1.52*log(Yellow)

Essentially, total oil is inversely related to black but proportional to red and yellow. Both models explain about 80% of the variation in fruit color. Figures 1(a-b) illustrates the relationship between observed and predicted total oil content from both models, respectively. The R^2 values obtained for models I and II were 0.78 and 0.76, respectively, indicating that both models seem to predict total oil content with fair accuracy.

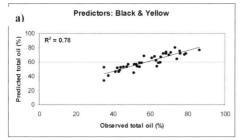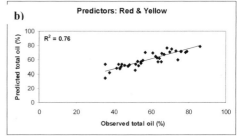

Figure 1. Relationship between observed and predicted total oil content: a)Model I, b)Model II.

Conclusion

Relationship between total oil and surface color distribution of oil palm fruits was quantified using an empirical modeling approach. This finding offers the prospect for development of a sensor-based system that can estimate oil yield accurately and rapidly to facilitate FFB quality determination and yield mapping for site-specific management.

References

Ariffin, A.A., R.M. Som, M. Banjari, W.Z.W. Omar and C.K. Woo. 1991. Ripeness standard: Any sign of loose fruit and with one loose fruit per bunch as the minimum standard. Paper presented at Int. Oil Palm Conf. PORIM-ISP, Kuala Lumpur, Malaysia. 31 p.

Corley, R.H.V. and I.H. Law. 2001. Ripening, harvesting and oil extraction. The Planter, 77: 507-524.

Teo, H.T. and T.S. Yeang. 1992. Harvesting standards and quality control for higher productivity in oil palm. The Planter, 68: 249-255.

Using high resolution satellite imagery to detect spatial variability in corn

B.J. Beesley[1], G. Carbone[1], E.J. Sadler[2], D.E. Evans[2], J.A. Millen[2] and C.R. Camp[2]
[1]Department of Geography, University of South Carolina, Columbia, SC, 29208, USA
[2]Coastal Plains Soil, Water and Plant Research and Education Center, USDA-ARS, Florence, SC, USA
greg.carbone@sc.edu

Abstract

Remote sensing of biophysical variables plays a potentially important role in precision agriculture. This study assesses the use of high-resolution satellite imagery for detecting nitrogen and water stress in a corn field. Vegetation indices computed from 4-meter IKONOS imagery. A simple ratio of NIR to red reflectance varied significantly across classes defined by two nitrogen and four irrigation treatments. Yield also varied significantly as a function of the same treatments. However, the relationship between vegetation indices and yield did not show strong spatial correlation across the field, indicating that other spatially varying phenomena (probably related to soil properties) complicated the relationship.

Keywords: corn, remote sensing, spatial variability.

Introduction

Improved knowledge about spatially-varying crop response to irrigation and fertilizer application could improve field management and productivity. Previous work has demonstrated the utility of remote sensing for detecting intra-field variability in corn response to environmental variables (Osborne et al., 2002, Scharf et al., 2002, Shannahan et al., 2001). This study evaluates the use of high-resolution satellite imagery to detect differences in corn response to varying irrigation and nitrogen application. It complements a series of multi-year experiments investigating how spatial differences in soil type, irrigation, and fertilizer application interact to affect corn yield in a U.S. Atlantic Coastal Plain field (Sadler et al., 2002).

Materials and methods

Four irrigation treatments and two nitrogen treatments were used on twelve soil mapping units. Vegetation indices were computed from 4-meter multispectral IKONOS satellite imagery collected on 3 July 2000. We examined these indices with respect to the eight possible irrigation/fertilizer treatment classes and twelve soil types to determine whether spectral signals were significantly different as a function of the amount of irrigation or fertilizer application. We also examined the relationship between vegetation indices and yield.

Results and discussion

We found that a simple infra-red to red reflectance ratio showed significant higher greenness values for corn with a nitrogen application rate of 225 kg/ha N versus 134 kg/ha. Additionally, greenness was higher for corn irrigated at 50% and 100% of the rate necessary to keep soil water constant, compared with rain fed corn. However, we found diminishing improvement at higher irrigation levels. There was no significant difference in greenness between sample plots with an irrigation application of 150% of the rate required to maintain constant soil water versus 100% of the rate required to keep soil water constant. The specific relationships between greenness and irrigation or nitrogen application also varied as a function of soil type, reflecting the interacting response of moisture-holding capacity with nitrogen and irrigation application. Corn yield also varied in response to nitrogen and water treatments across the field. However, the relationship between yield and vegetation indices was relatively poor.

Conclusions

Both yield and vegetation indices responded to spatially varying rates of nitrogen and fertilizer inputs. These relationships were complicated, however, by varying soil moisture holding capacity across the field. Equally significant were variations *within* soil units. Closer examination of in-situ crop conditions will be required to incorporate remote sensing data into management decisions more effectively (Robert, 2002).

References

Osborne, S.L., Schepers, J.S., Francis, D.D., and Schlemmer, M.R. 2002. Use of spectral radiance to estimate in-season biomass and grain yield in nitrogen- and water-stressed corn. Crop Science 42 (1) 165-171
Robert, P.C. 2002. Precision agriculture: a challenge for crop nutrition management, Plant and Soil 247 (1) 143-149.
Sadler, E.J., Camp, C.R., Evans, D.E., and Millen, J.A. 2002. Spatial variation of corn response to irrigation. Transactions of the ASAE 45 (6) 1869-1881 247 (1) 143-149.
Scharf, P.C., Schmidt, J.P., Kitchen, N.R., Sudduth, K.A., Hong, S.Y., Lory, J.A., and Davis, J.G. 2002. Remote sensing for nitrogen management. Journal of Soil and Water Conservation. 57(6) 518-524.
Shanahan, J.F., Schepers, J.S., Francis, D.D., Varvel, G.E., Wilhelm, W.W., Tringe, J.M., Schlemmer, M.R., and Major, D.J. 2001. Use of remote-sensing imagery to estimate corn grain yield. Agronomy Journal 93(3) 583-589.

Acknowledgements

This research was supported by the National Aeronautics and Space Administration, MTPE (Grant No. OA99073, W-19, 080), and the U.S. Department of Agriculture, NRICGP (98-35106-6837).

Beet seedlings automatic counting system by artificial vision

Bernard Benet[1], Thierry Humbert[1] and Didier Guiraud[2]
[1]Cemagref: 24 Avenue des Landais - 63172 Aubière Cedex - France
[2]Institut Technique de la Betterave: 45, rue de Naples - 75008 Paris – France
bernard.benet@cemagref.fr

Abstract

An automatic counting system of sugar beet seedlings on three lines in parallel was developped. The objective is to determine on different microparcelles the beet population, the distance between the various seedlings and the leaf area of each one, for each line. Two types of sensors are used for this machine: the artificial vision for the positionning of sheet of beet on the lines and two position sensors to know the position of seedlings on the ground and the information on the orientation of the machine. This system is guided manually by an operator. The software developped uses information of the sensors to realize the counting, in real time.

Keywords: sugar beet seedling, real time image processing, position sensor, cartography

Introduction

Each year, the Technical Institute of Beet (ITB) carries out many experiments on various varieties of beet, in relation to the seed-producers ones. The percentage of lifting of seeds is a significant element for the selection of the species. Thus, each year the ITB technicians spend a lot of time to count the seedlings throughout course of the process of germination. This counting, currently realized manually, represents a relatively tiresome and repetitive task. ITB thus linked a collaboration with the Cemagref to build an automatic system of counting on line of the beet seedlings. An automatic counting system of sugar beet seedlings on three lines in parallel was carried out. This machine determines on a piece given the beet population, the distance between the various seedlings and the leaf area of each one, on three lines, in parallel. Two types of sensors are used: the artificial vision for the positionning of sheet of beet on the lines and two position sensors to know the position of seedling on the ground and have an information on the orientation of the machine. This system is guided manually by an operator. The displacement speed on the lines is variable and limited to 4 km/h. This system allows to detect in real time the seedlings on the beet lines.

Materials and method

Concerning the system of artificial vision, three color video cameras are used to view the three lines and to detect the beet seedlings. High frequency linear spots are used to light the scene. The system is entirely closed to obtain a constant lighting, independant of external light and to avoid the problems of shadow. It works in real time conditions. So the image processing algorithms used must be fast. A study relating to the discrimination of the color of the beet sheets compared to the colors of the ground, in the three-dimensional RGB color space, was done. The objective was to determine an envelope of color points in order to separate the colors from the beet sheets of those of the various grounds. These ones can have various colors: white, black, clear and dark brown, gray, with various noises (stones, grass.).
The position encoders make it possible to be located on the ground, by giving the position in real time of the machine on the lines. Thus, we know the distance between each beet seedling

361

and their leaf area, on the lines. Also, this information of position on the ground is used not to count several times the same seedling and not to forget some.

The software uses information of both sensors (vision and position sensor) to count the beet seedlings. A first algorithm, in real time, is used to realize a first counting and to record geometrical and colorimetric information about the beet seedlings on the three beet lines. A second algorithm, in differed time uses this information relating to the profiles of the lines, to refine the counting of the seedlings. A mechanical prototype of counting of the beet seedlings was carried out.

| A beet seedling | A beet line | Prototype of beet counting |

Results

The system must be able to detect seedlings corresponding at a first stage (2 small sheets) up to a more advanced stage (4 large sheets or more). The results obtained showed that the counting system can detect with an optimal output all the beet seedlings whose width of the sheets is higher than 5 millimeters and spaced at least 30 millimeters. The seedlings of width ranging between 2 and 5 millimeters are detected with an output of 90 %. The detection of the very small seedlings and the accuracy relating to the position of the seedlings on the lines depends of the characteristics of the position sensors and guidance on the ground which can be more or less fatty and be more or less embossed. The standard deviation of the position error obtained for the detected seedlings is 5 millimeters. Concerning the estimation of the leaf area projected for each seedling, the standard deviation of the error is 10 mm^2.

At the end of one counting operation on differents microparcelles, the user obtains data files which represent a cartography of beet lines. By making regular measurements of counting, the operator can realized statistical calculations with the data files to compare the various varieties of sugar beet.

Conclusion

The future works consist in improving the results of detection of the very small seedling beet, with width inferior to 5 millimeters, and to apply this work for others products (maize,…).

Site specific calibration of a wheat simulation model using sensor and yield data

U. Böttcher[1], H. Kage[1] and Y. Reckleben[2]
[1]Institut für Pflanzenbau und Pflanzenzüchtung, Christian-Albrechts-Universität Kiel, Hermann-Rodewald-Str. 6, D-24118 Kiel
[2]Institut für Landwirtschaftliche Verfahrenstechnik, Christian-Albrechts-Universität Kiel, Max-Eyth-Str. 6, D-24118 Kiel
boettcher@pflanzenbau.uni-kiel.de

Abstract

First results of a project are presented were a plant growth model for wheat derived from CERES-wheat is used for the site specific prediction of yield potential and nitrogen demand. The model has been adjusted site specifically using estimates of crop parameters derived from spectral sensors and by using actual weather data. The aim of the project is to achieve a better yield prediction and therefore more precisely adapted fertilizing strategies.

Introduction

In contrast to the huge progress made in the technology for site specific crop husbandry during the last years, convincing concepts for the automated decision support are still lacking. Dynamic plant growth models have been suggested as a tool for the site specific prediction of yield potential, nitrogen demand and fertilisation rates (Paz et al. 1998). However, the often limited accuracy of these models may seriously reduce their applicability for this purpose especially when information about the application site is scarce. The usage of sensor data for model calibration and initialisation has been shown to improve the prediction accuracy of crop growth models (Guérif and Duke. 2000). The aim of the presented work, therefore, is to develop a framework which allows the combination of crop growth models, sensor and yield monitoring data in order to allow more precise fertilisation strategies and to validate it against data from field experiments.

Material and methods

On a practical field of 15 ha at the experimental farm Hohenschulen near Kiel winter wheat was grown. The field was fertilized according to usual practice except for two strips which were left unfertilized. Canopy characteristics were determined using the Hydro Agri N-Sensor several times during the vegetation period in 2002. Along transects dry matter and nitrogen content of the different parts of the plant were measured destructively. Soil water content and mineral nitrogen were determined from soil samples. At harvest time, site specific yield and nitrogen content were recorded using a combine based yield monitor and NIR system.
Yield, grain nitrogen content, sensor and available soil data were used for a site specific parameterization and calibration of a winter wheat simulation model based upon modules derived from CERES Wheat. The whole model is implemented within the HUME modelling environment (Kage and Stützel, 1999) , which supports parameter estimation techniques .

Results and discussion

There were distinctive differences in yield and sensor data reflecting the pronounced relief and soil variability in the region (Ostholsteinisches Hügelland). Yield of winter wheat was

363

between 60 and 130 dt/ha. First calculations indicate that after site specific calibration the model may be able to reproduce much of this variability as indicated by modelled data for transects of the field.

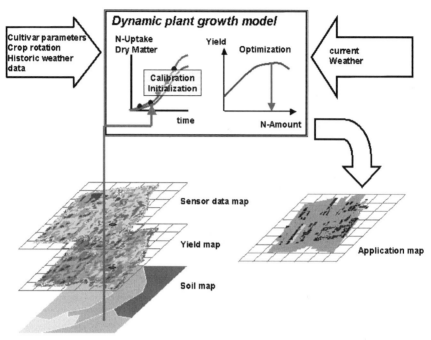

Figure 1. Concept of the model as a decision support tool for a site specific fertilizing strategy.

After further validation the model may be able to calculate site and year specific yield potentials and may therefore help to derive optimised fertilizing strategies.

Further input data for the site specific parameterisation of the model may also include yield maps from previous years as well as cultivar specific parameters (Figure 1).

References

Guérif, M.O., and C.L. Duke. 2000. Adjustment procedures of a crop model to the site specific characteristics of soil and crop using remote sensing data assimilation. Agriculture Ecosystems and Environment. 81(1):57-69.

Kage, H. and H. Stützel 1999. An object oriented component library for generic modular modelling of dynamic systems. In: "Modelling cropping systems". (C. S. M. Donatelli, F. Villalobos, J. M. Villar, ed.), pp. 299-300, Lleida. June 1999. European Society of Agronomy.

Paz, J.O., Batchelor, W.D., Colvin, T.S., Logsdon, S.D., Kaspar, T.C., and Karlen D.L. 1998. Calibration of a crop growth model to predict spatial yield variability. Transactions of the ASAE 41(5): 1527-1534.

Assessing the potential for site specific irrigation using CERES-MAIZE crop model

R. Braga[1] and B. Basso[2]
[1]Esc. Sup. Agrária de Elvas, Inst. Polt. de Portalegre, Apartado 254 - 7351 Elvas, Portugal
[2]Dept. of Agro-forestry and Environmental Sciences, Univ. of Basilicata, Potenza , Italy
ricardo_braga@esaelvas.pt

Abstract

Site-specific Irrigation (SSI) is a relatively new Precision Agriculture domain (Nijbroek et al. 2003; Sadler et al., 2002). Before investment is made, potential benefits should be assessed. Crop models are excellent tools for predicting crop performance and evaluate the gains of SSI. We have studied a 100 ha corn field located in Elvas, Portugal over 14 years. At the water cost of 0.017 €/m3 (without application costs), SSI brought savings on average of € 597 (€ 238 - €1107) for the whole 100 ha. This resulted in average gains of about 6 €/ha (€ 2.4 – € 11.1). Considering the costs of the new Site Specific irrigation equipment, the investment in SSI for this particular field is not beneficial.

Keywords: Irrigation, Crop models, CERES-MAIZE, Corn, economic benefits

Introduction

Site-specific Irrigation (SSI) technology is a relatively new Precision Agriculture domain. It addresses the application of different irrigation rates across a field according to different crops and/or different soil types. SSI, as all precision agriculture technologies, is intuitively appealing. Nevertheless, before investment is made, one should always assess the potential benefits. Crop models are tools that simulate the behavior of crops according to different soil types, genotypes, weather and management. They are excellent for predicting crop performance in real or futures environments.
We have used CERES-MAIZE crop model to study the gains from SSI on a 100 ha corn field located in Elvas, Portugal.

Material and methods

The field was divided into 40 areas of 2.5 ha each with similar soil water behavior. Each area was characterized using Field Capacity and Permanent Wilting Point and the resulting plant available soil water (PASW).
SSI was simulated as the irrigation strategy that would give no stress to each of the 40 areas. This was simulated using automated irrigation whenever soil water content dropped below 50% of plant available soil water at 50 cm depth. The flat irrigation amount strategy was simulated using the automated irrigation strategy for the site with less plant available soil water (site 32). This assured that all sites had no water deficit under this irrigation strategy. Instead this caused excess irrigation on most sites.
Both irrigation strategies (SSI and flat amount) were simulated for 14 years (1988/2001). The economic benefits were calculated assuming a water price of 0.017 €/m3 (without application costs) from a local farmer.

Results

The field average plant available soil water (PASW) was 170 mm with standard deviation of 26 mm. Soil depth varied from 120 to 150 cm, leading to average values of PASW in mm/m of 127 with standard deviation of 11.

Average yield across years was 11508.3 kg/ha (8711 – 12996 kg/ha). Since it was assured that no stress occurred in both irrigation strategies, and that excess of irrigation in flat amount strategy caused no stress, grain yield across sites had no significant variation. On the contrary, excess irrigation in flat amount strategy varied across sites and years. The field average excess of irrigation varied from 14 mm to 65 mm across years with an average of 35 mm (14000, 65000 and 35000 m3 for the whole 100 ha). These amount represented 2.8%, 12.9% and 6.1% of the amount of water applied (498mm, 686mm and 579 mm). Figure 1 shows the irrigation amounts across sites for four representative years.

At the water cost of 0.017 €/m3 (without application costs), SSI brought savings on average of € 597 (€ 238 - € 1107) for the whole 100 ha. This resulted in average gains of about 6 €/ha (€ 2.4 – € 11.1).

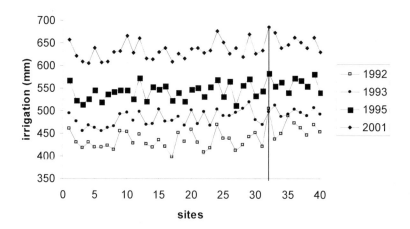

Figure 1. Irrigation amounts across sites for four representative years.

Discussion and conclusions

Considering the costs of the new Site Specific irrigation equipment, the investment in SSI for this particular field is not beneficial. No environmental gains were considered. Any judgment on the profitability of a precision agriculture technology is field and farm specific.

References

Nijbroek, R; Hoogenboom, G; Jones, JW (2003). Optimizing Irrigation Management for a spatially variable field. Agr. Syst. 76: 359-377

Sadler, EJ; Camp, CR; Evans, DE; Millen, JA (2002). Spatial variation of corn response to irrigation. Trans. of ASAE 45: 1869-1881.

Geostatistical analysis of soil chemical properties and chickpea yield in a reduced tillage experiment in southern Spain

C. Bravo[1], J.V. Giráldez[1], P. Gónzalez[2] and R. Ordóñez[2]
[1]*Enviromental Engineering, Dept of Agronomy, University of Córdoba, Apdo. 3048,14080 Córdoba, Spain*
[2]*Dept. of Soils and Irrigation, CIFA. Apdo 3092, 14080 Córdoba, Spain.*
g12brmec@uco.es

Abstract

This paper examines the spatial variation in crop yield and soil fertility indicators for two different systems: conventional tillage (CT), and no tillage (NT). using geostatistics. Semivariograms are described with exponential and spherical models.soil fertility indicators and chickpea yield semivariograms varied slightly according to tillage systems. There is a weak relationship between yield and soil fertility.the range of spatial correlation for yield, shown by the semivariograms in CT was similar to that of the soil chemical properties such as, organic matter (OM), Ca and Mg contents. In NT, yield semivariograms were much better correlated with OM, P, and Ca contents

Keywords: soil fertility indicators , geostatistics, spatial variability, tillage, crop yield maps

Introduction

Traditionally fertilisers, chemical and other crops inputs have been applied without taking into account spatial variation in field characteristics. The evaluation of the spatial patterns of soil properties, and particularly, the fertility is a critical parameter for the establishment of crop management systems(Frogbrook et al, 2002). The objectives of this study are: (i) The characterization the spatial variability of some soil fertility indicator as well as the chickpea yield in plots under long term conventional tillage (CT) and no tillage (NT) treatments and (ii) the determination of the relationship between the spatial variability of crop yield and soil properties.

Materials and methods

A long term experimental started in 1982,(Giráldez and González ,1995), was used for the study. Tillage treatments were CT and NT. Soil samples from the 0-6 cm depth were collected from each plot of the tillage treatment, applying a square sampling grid with a 3 m unit cell. The soil was analysed for organic matter (OM), and available. phosphorus (P), potassium (K), calcium (Ca), and magnesium (Mg). Crop yield was measured from each plot by harvesting within a square frame of 1 m^2. The soil properties and crop yield data were analysed using geostatistics. (Isaaks and Srivastava, 1989).

Results

The model parameters are given in table 1. Semivariograms were described with an exponential and spherical models .Contents of OM, P, K, Ca, Mg and CEC in CT were spatially correlated to smaller distances of 6 m, whereas in NT the soil fertility indicators were spatially correlated to a distance < 9 m. The ranges of spatial correlation of chipckpea yield varied from 3.50 m for CT to 6 m for NT.

Table 1. Parameters of the semivariograms under conventional, CT, and no tillage, NT.

N: 102	Conventional Tillage				No tillage			
	Models	Nugget	Sill	Range (m)	Models	Nugget	Sill	Range (m)
OM %	S	0.003	0.009	6.09	E	0.003	0.047	8.96
P , mg.kg^{-1}	E	0.003	0.039	6.00	E	2.310	23.43	6.00
K , μg.kg^{-1}	E	0.100	2.000	6.30	E	1.000	3.400	6.00
Ca, g.kg^{-1}	E	0.174	1.102	6.00	S	0.120	0.960	4.20
Mg, mg.kg^{-1}	E	0.000	1.232	6.00	E	0.280	2.099	5.04
CEC, mol$_c$.kg^{-1}	E	0.180	0.630	4.32	E	0.068	0.980	5.88
Yield, Mg. m^{-2}	S	0.000	38.00	3.57	E	18.60	93.00	6.00

S: Spherical; E: Exponential

Discussion

This study shows that, although crop management systems were different, the spatial distribution and spatial dependence level of some soil fertility indicators and chickpea yield were similar. On the other hand, although this NT plots has been managed long term without major disturbances, the surface may have experienced some slight changes due to imperfections of the drilling machine. The results indicate that concentration of soil nutrients were generally high enough to maintain a normal crop growth. Therefore factors others than soil nutrients, were causing the variation in yield, such as the incidence of weeds, namely broomrape which persisted in the soil many years even the last after herbicide application. The low nugget/total variance ratio and small ranges values for the majority of soil properties and yield suggest the patchiness of the distribution, which could be attributed to an effect of sampling. It has been hypothesized that strongly spatial dependent soil variables may be controlled by intrinsic variations in soil characteristic (Cambardella et al, 1994).

Conclusions

Results suggested that agricultural management slightly affects soil properties and crop yield. Others factors such as the incidence of weeds might have more influence on the variation in yield The soil properties with the strongest relation to the spatial variation in crop yield were OM, Ca and Mg in CT, and OM, P, and Ca in NT.The range of correlation of soil properties changes between 4.50 m and 9 m, while for crop yield varies between 3.50 m and 6 m.

References

Cambardella C., Moorman, T.B.; NovaK, J.M., Parkin,T.B.; Karlen, D.K.; Turco, R.F. and Konopl,a A. E 1994. Field-scale variability of soil properties in central Iowa soils. Soil. Sci. Soc. Am. J. 58:1501-1511.

Frogbrook, K.L., Oliver, M.A., Salahi, M. and Ellis, R.H. 2002 Exploring the spatial relations between cereal yield and soil chemical properties and the implications for sampling. Soil Use Manag. 18: 1-9.

Giráldez, J.V and P. González 1995. No Tillage in clay soils under mediterranean climate: physical aspects. In: Proccedings of the WorkShop. Giessen, Germany Vol I: 111-117 pp.

Isaaks E.H. and Srivastava R. M. 1989. *An introduction to Applied Geostatistics,* Oxford University Press.New York:

Fungal disease evolution on winter wheat monitored by multifiber input spectrograph

Cédric Bravo[1], Dimitrios Moshou[1], Dimitri Lemaire[2], Timo Hyvärinen[3], Henri Maraite[2], Josse De Beardemaeker[1] and Herman Ramon[1]
[1]KULeuven, Lab. for Agro-Machinery, Kasteelpark Arenberg 30, 3001 Heverlee, Belgium
[2]UCL, Fac. Sc. Agronomiques, Pl. Croix du Sud 2, bte 2, 1348 Louvain-la-Neuve, Belgium
[3]SPECIM Oy Ltd, P.O. Box 110, Teknologiantie 9, Fin-90571 Oulu, Finland
cedric.bravo@agr.kuleuven.ac.be

Abstract

As fungal infestations on winter wheat have been proven to be automatically detectable by high-tech optical research instruments (Bravo et al., 2003, Moshou et al., 2003, West et al. 2003), the need of simplifying the instrumentation towards a practicale low-cost optical eye has risen. In this paper a first step has been investigated on the feasiblility of detecting and recognising two types of fungal infestation by using a multifiber input spectrograph as a versatile and more practical version of the imaging line spectrograph. This investigation was carried out on greenhouse plots.

Experiment

16 trays were sown in greenhouse with winter wheat (cultivar Beaufort) on November 15th 2001 and immediately placed outside for low temperature treatments. They were then brought back on 18/01/02 to an unheated greenhouse. Inoculations were made on March 19/03/02 at growth stage 33. The first pycnidia (Septoria) were expected between April 4th 2002 (latent period calculated for susceptible cultivar) or April 6th (resistant cultivar) and were detected on April the 5th. The first urediospores were expected on April 1st 2002 and were effectively detected on this date.

Methodology

Spectra were acquired outdoors by an imaging spectrograph on which the objective was replaced by four optical fibres. Three of the fiber optics were placed at 75cm height pointing downwards on the plants meanwhile one optic was used as illumination reference. It was pointing upwards, and covered by a down well irradiance diffraction device (figure 1a). Measurements were caried out from April 5th until April 17th 2002.

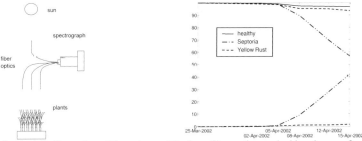

Figure 1. (a) On the left: illustration of the set-up: while three fiber optics are looking down, a fourth one looks to the illumination. (b) on the right: the higher 3 curves illustrate the averaged green area develoment of the plant, while the 2 lower curves represent the average percentage of disease progression on the plants.

A normalisation method based on reflectance and light intensity adjustments was applied. A new method for discrimant waveband selection was introduced. Five 20nm wide wavebands were enough to recognise the diseases.

Results and assessment of their significance

Expert observations pointed out that symptoms started to be clear around April 9th 2002 (*fig. 1b*). In parallel, clear spectral disease patterns were detected on April 10th 2002, though (*fig. 2*) small spectral differences were already visible on April 8th 2002. Disease development was further on detectable. Due to disease presence a shift towards higher reflectance was observed in the 600 to 700nm region, meanwhile a shift towards lower reflectance was noticed in the 750 to 800nm region.

While Septoria seemed better detectable than Yellow Rust, both diseases were clearly separable starting from April 10th 2002. Neural networks were introduced for discriminating the diseases. A Multilayer perceptron with 5 inputs, one hidden layer and 3 outputs could recognise the 3 treatments with an initial error of 12% on April 10th. This discrimination error fell down to 3% on April 17th 2002.

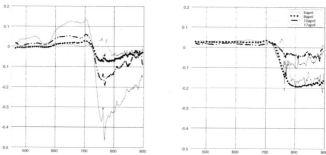

Figure 2. both graphs represent the spectral difference in the 450-900nm region along the measurement dates (a) On the left: illustration of the spectral differences between Septoria infected reflectance and healthy plants (b) idem but with respect to Yellow Rust

Conclusion

The proposed method offers the opportunity of building a robust decision algorithm for on-line disease detection and site-specific spraying recommendation. The innovative equipment will be brought on the field and tested further on. As disease infestation patterns evolve, the spectral approach could be used as input into disease evolution models, which are integrated into spraying decision algorithms.

References

C. Bravo, D. Moshou, J. West, A. McCartney, H. Ramon, "Detailed Spectral Reflection Information for Early Disease Detection in Wheat Fields", Journal of Biosystems Engineering, 2003, Vol 84 (2), 137-145.

D. Moshou, C. Bravo, J. West, A. McCartney, H. Ramon, "Intelligent plant disease detection at an early stage of development based on Neural Networks", Journal of Computers and Electronics in Agriculture (COMPAG), accepted 9/2002.

J. West, C. Bravo, R. Oberti, D.Lemaire, D. Moshou and A. McCartney. The potential of optical canopy measurement for targeted control of field crop diseases. Annual Review of Phytopathology, accepted 01/2003.

Prediction of Soil Organic Matter (SOM) and clay content from reflectance data of different spectral resolutions

N.H. Broge
Danish Institute of Agricultural Sciences, Department of Agroecology, Research Centre Foulum, P.O. Box 50, DK-8830 Tjele

Abstract

Soil organic matter (SOM) and clay content of a soil characterised as a coarse sandy loam was modelled using hyperspectral reflectance data acquired with a spectrometer. The partial least squares (PLS) regression method was applied, and the results validated using cross validation.

Keywords: organic matter, clay content, spectral reflectance, partial least squares regression

Introduction

Knowledge of plant growth potential is important for farm managers seeking to optimise agricultural production at minimal economical and environmental costs. Hence, soil textural parameters, in particular the abundance of clay and organic matter, play an important role in the complex water and nutrient exchange processes occurring in the root zone. Soil organic matter (SOM) content and soil moisture content, which is related to clay content under field conditions, are both known to have a strong influence on soil reflectance. The objective of the study was to evaluate and compare the information content between high spectral resolution data and four broad bands characterising common satellite sensors, with respect to estimation of SOM and clay content of the plough layer.

Materials and methods

Electric conductivity of a field characterised as a coarse sandy loam soil or an Orthic Acrisol according to the FAO/UNESCO legend was mapped using a mobilized version of the EM38 instrument (Nehmdal and Greve, 2001). Five zones representing the full range of observed soil electric conductivity within the field were identified. Five soil samples were collected at random within each of these zones giving a total of 25 samples, which were subjected to textural analysis in the laboratory. Each sample represents an average of 5 core samples collected within a radius of 1 meter. In addition, six samples of the topsoil were collected (for reflectance measurements in the laboratory) at the location of each of the averaged textural samples, giving a total of 150 topsoil samples.

Spectral reflectances were obtained for each topsoil sample in the laboratory using a GER-2600 spectrometer (Figure 1). The spectral reflectance was measured with the samples still in the sample rings in order to maintain the natural textural distribution at the sample surface. All spectra were resampled to a spectral resolution of 2 nm in the 400-1000 nm region and 10 nm in the 1000-2500 nm region, resembling the actual spectral resolution of the spectrometer. An additional data set was constructed by spectral resampling to match the configuration of the sensor onboard the new commercial satellite, QuickBird. After completion of these measurements, all topsoil samples were dried at 60 °C for 96 hours, and spectral reflectance was then measured again using the same protocol. The Partial Least Squares (PLS) calibration models were generated to assess the predictability of clay and SOM content from the processed reflectance data for both wet and dry soil. Each model was optimised with respect to outlier detection and number of latent variables to be included. The models were validated using segmented cross-validation. Useless or unreliable spectral bands were eliminated as part of the cross-validation procedure, using the jack-knife approach.

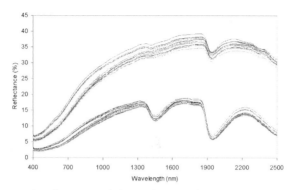

Figure 1. Mean spectral reflectance of dry and wet soil samples for each sample location. Spectra of dry samples are characterised by high reflectance whereas wet samples are characterised by low reflectance.

Results and discussion

The results summarised in Table 2 show that clay content is best estimated from the high resolution spectral reflectance (FSR) of dry soil samples, whereas SOM content may be predicted with the same accuracy from wet and dry soil reflectance respectively. Reducing spectral resolution to 4 bands resembling the QuickBird instrument (QB) only marginally decreased the accuracy of the SOM content prediction models in terms of the root mean square error of prediction (RMSEP), i.e. 5% for wet soil and 2% for dry soil. However, the accuracy of the clay content prediction models was significantly decreased (32% for wet soil and 51% for dry soil). The results also show that clay content predictions based on spectral reflectance data are more robust than SOM predictions (Relative RMSEP).

Table 2. PLS-1 model statistics associated with predicted versus modelled clay and SOM contents based on the spectral reflectance of wet and dry soil using full spectral resolution (FSR) and the four band QuickBird configuration (QB).

| | SOM (%) | | | | Clay (%) | | | |
| | Wet | | Dry | | Wet | | Dry | |
	FSR	QB	FSR	QB	FSR	QB	FSR	QB
Slope	0.54	0.48	0.56	0.50	0.85	0.76	0.89	0.77
Offset	1.75	1.98	1.73	1.92	1.25	2.05	0.97	1.97
Correlation	0.70	0.65	0.69	0.68	0.93	0.87	0.95	0.88
RMSEP	0.34	0.36	0.34	0.35	0.44	0.58	0.37	0.56
RRMSEP (%)	8.7	9.3	8.8	9.0	5.1	6.7	4.3	6.6

References

Nehmdal,H. and Greve,M.H. (2001), Using soil electrical conductivity measurements for delineating management zones on highly variable soils in Denmark. *Proceedings of the 3rd European conference on precision agriculture, June 18-20. Montpellier, France.* (G.Grenier and S.Blackmore, Eds.), pp. 461-466.

Estimating soil properties combining reflectance, topographical and conductivity data

N.H. Broge[1]; M.H. Greve[1] and H. Nehmdahl[2]
[1]*Danish Institute of Agricultural Science (DIAS), Department of Agroecology, Research Centre Foulum, P.O. Box 50, DK-8830 Tjele.*
[2]*ConTerra ApS; Agro Business Park, P.O. Box 10, DK-8830 Tjele.*

Abstract

Partial Least Squares (PLS) calibration models have been applied to identify variables retrieved from digital elevation models (DEM), soil electrical conductivity (SEC) measurements and reflectance measurements describing the spatial variation of soil clay, soil organic matter (SOM), extractable phosphorus and grain yield. Results indicate that the use of multiple data layers could enhances management zone delineation using clustering techniques.

Keywords: organic matter, soil properties, spectral reflectance, soil conductivity, wetness index.

Introduction

For some years measurements of SEC has been used for delineating management zones on variable soil in Denmark. Highly correlated to soil texture and especially soil clay content, SEC maps are still not easily transformed into maps reflecting the yield potential at field level. SEC measurements do not provide information on the spatial variation of SOM in mineral soils in DK typically ranging from 0.5 to 3.5 % SOM. Water retention characteristics and plant nutrient status will often be described inadequately and consequently the delineation of management zones solely based on SEC data therefore questionable. Topographical data and spectral reflectance data has been used to describe the spatial variation of SOM but usually high quality data is expensive and difficult to retrieve. Recently though, high quality digital elevation models have been made accessible in DK along with airborne reflectance measurements sparking interest among researchers and farmers to access their potential.

Materials and methods

The 10 ha test field is located on a young moraine landscape with covered with loamy and clay till, glaciofluvial sand and peat. Elevation ranges from 54 to 67 m. The natural drainage is good except for a small peat basin in the central part. Soil data was collected in a 20 x 20 m grid at soil depth of 0.25 cm and 25-75 cm. The clay content varies from 3 to 24 %. Generally the clay content for sub soil samples is slightly higher than for topsoil samples. Data from the peat basin and the edges of the field were not included in the study and SOM values vary between 1.3 and 3.3 % over a total of 206 sampling points. SEC data were collected in a 5 x 10 m grid using a Geonics EM38DD sensor simultaneously measuring SEC for soil depths of app. 0-75 cm (EM38h) and 0-150 cm (EM38v). The sensor was towed behind an ATV and data from the EM38DD were merged with elevation data from a RTK-GPS using the correction signal provided nation wide via the Trimble GPS-net. A mapping system simultaneously recording EM38h, EM38v and precise elevation data has been introduced in DK by the Danish Cooperate Farm Supply (DLG). Maps of topographical wetness index (WI), slope and aspect were calculated from elevation data. Spectral reflectance data obtained at visible (456nm = ρb, 554nm = ρg and 660nm = ρr) and near-infrared (806nm = ρbNIR) wavelengths were acquired using an airborne digital camera system designed and developed at DIAS. The images were geo-rectified and converted to reflectance using ground panels with known position and reflectance, re-sampled to a spatial resolution of 0.5m and smoothed using a 7 by 7 low-pass filter. Yield data from 1993 (dry growing season) and 1994 (wet growing season) are included in the analysis. Interpolated values (10m grid) for topographical, reflectance and SEC data were correlated with soil and yield data.

Results and discussion

The correlation matrix (Table 1) and results from PLS analyses (Table 2) indicate the importance of incorporating topographical and reflectance data in models used to delineate management zones.

Both tables do indicate that topographical and reflectance data improves the prediction of the spatial variation of SOM with a mean prediction error of 0.24 % (Table 2). For moraine soils in DK varying water retention characteristics determines the yield potential. Combining SEC data which reflects soil clay content with topographical/reflectance data should therefore improve the prediction of the grain yield reflecting the yield potential especially following dry growing seasons. A slightly higher correlation coefficient (Table 2) for the dry 1993 season as compared to the wet 1994 season supports this assumption. It has to be noted that the correlation coefficient between yield and WI in the PLS analysis is positive in the dry 1993 season and negative in the wet 1994 season.

With the content of extractable P which is closely linked to SOM content, a better prediction of the spatial variation of P is expected, as indicated in the results, when including reflectance and topographical data in the PLS-analysis.

Although differences in soil clay content for top- and sub soil in the test field are minor, results indicate that EM38v data yields a better description of the variation in sub soil clay content than EM38h data, which is consistent with theory.

Table 1. Correlation coefficients between descriptive variables and soil/yield data respectively. Bold figures: significant correlations at p < 0.05 (n=206).

Parameter	Elevation	Aspect	Slope	WI	EM38h	EM38v	ρb	ρg	ρNIR	ρr
SOM %	-0.12	0.00	-0-07	*0.45*	-0.06	-0.08	*-0.23*	*-0.28*	*-0.29*	*-0.34*
Clay (0-25cm) %	*-0.27*	-0.04	*0.31*	0.26	*0.79*	*0.77*	0.22	*0.19*	-0.02	*0.32*
Clay (25-75cm) %	-0.09	0.02	*0.34*	0.27	*0.75*	*0.77*	0.28	*0.29*	0.09	*0.32*
P mg/kg	-0.02	0.05	*-0.26*	*-0.43*	*-0.57*	*-0.52*	*-0.27*	*-0.24*	-0.05	*-0.23*
Yield 1993 ton/ha	-0.06	0.06	0.12	*0.46*	*0.51*	*0.49*	0.11	0.03	*-0.17*	0.01
Yield 1994 ton/ha	0.05	0.00	0.03	-0.03	*0.38*	*0.34*	*0.27*	*0.19*	0.06	0.10

Table 2. Descriptive variables identified using PLS models, correlation coefficients and the root mean square error of prediction associated with predicted vs. measured plots (RMSEP).

Parameter	Elevation	Aspect	Slope	WI	EM38h	EM38v	ρb	ρNIR	ρr	r	RMSEP
SOM %	*			*					*	0.57	0.24
Clay (0-25cm) %	*			*	*				*	0.79	1.02
Clay (25-75cm) %				*		*			*	0.77	2.23
P mg/kg				*	*		*			0.66	4.00
Yield 1993 ton/ha		*		*	*			*		0.63	0.86
Yield 1994 ton/ha	*	*	*	*	*		*			0.57	0.45

Towards the quantification of the relationship between patch sprayer performance and weed distribution

J.P. Carroll and N.M. Holden
Department of Agricultural and Food Engineering, University College Dublin, Earlsfort Terrace, Dublin 2, Ireland.
john.carroll@ucd.ie

Abstract

This paper outlines the development of a method for modelling patch sprayer performance in the field. For a specified control time the model produces an output file that represents spray application as true positive (correct application to target), false positive (application to non-target), true negative (no application to non-target) and false negative (no application to target). The quality of application can be summarised using a Spray Quality Index (based on the Tanimoto Similarity Criterion), which can be related to a description of a weed distribution map. It was concluded that spatially discontinuous weed patterns (i.e many small patches) result in less accurate patch spraying, and that with more control over the boom (i.e. divided into smaller segments) it is possible to have more accurate spraying. The methodology will allow improved patch sprayer design and better matching of machine to field conditions.

Keywords: Patch spraying, distance transform, weed pattern, spray quality.

Introduction

Targeted application of herbicides to weed patches has the potential to significantly reduce herbicide use, which has both economic and environmental advantages (Lutman *et al.*, 1998). Spatially selective herbicide treatments are based on visually created weed maps that are used to control the sprayer. These maps show the spatial distribution of the weed populations. The weed map is used in conjunction with a geographic information system, a positioning system and the control system of the sprayer to produce a spatially variable application system. The simplest such system is patch spraying using one rate, which is either on or off. The challenge is to design suitable equipment that actually sprays weed patches at the correct application rate in a reliable manner. Sprayer performance is related to two factors: the spatial pattern of weeds being targeted and the control time of the sprayer. The spatial pattern dictates how often the sprayer has to be switched off or on, and the control time dictates the time it takes for the chemical output to reach a desired application rate.

Materials and methods

The sprayer/weed interaction is being modelled using a computer simulation. The program starts by importing a weed map in bitmap form. The model then performs a virtual patch spraying operation. The sprayer moves in the common sprayer pattern along parallel tramlines across the field and also completes a fixed headland operation around the perimeter. There may be some overlap but this is inherent in any spraying operation. User modification of boom length (vary the length of the sprayer boom to simulate the different sizes of sprayers available on the market), boom segmentation (facilitates the division of the boom into individually controllable segments) and sprayer speed (relates to the control time of the sprayer) is permitted. Once the program has run, it exports data to a spreadsheet. Processing of the data then calculates number of correctly sprayed pixels (true positive), incorrectly sprayed (false positive), correctly not sprayed (true negative) and incorrectly not sprayed (false negative). Tests to date have assumed no control time delay in the sprayer, as there are no graduations of spray rate (100% or nothing). Spray quality can be calculated as

$$SQI = \frac{(T.P. + F.N.) - F.N.}{(T.P. + F.N.) + F.P.}$$

Results and discussion

Table 1. Sample of results for a 12m boom.

Field	True Positive (%)		False Positive (%)		True Negative (%)		False Negative (%)		SQI	
	6*2m	2*6m	6*2m	2*6m	6*2m	2*6m	6*2m	2*6m	6*2m	2*6m
A	77	77	4	14	19	9	0	0	0.95	0.83
B	25	25	4	18	71	57	0	0	0.88	0.6
C	68	68	7	20	25	12	0	0	0.91	0.86

The results for a small sample of processed fields for a 12m boom in 2 different segmentations (i.e. 6x2 m segments and 2x6 m segments) are shown in Table 1. Even for a system with no control delay both the field and the boom segmentation have a major bearing on the SQI. For example comparing fields A and B. Field A has a much higher weed concentration as can be seen from the true positives columns. The weed patches are more uniform and spatially aggregated than in field B as can be seen by the higher SQIs at both boom segmentations. The dramatic increase in the SQI for field B from the 2x6 m to the 6x2 m segmentation is due to the large amount of small weed patches in this field, which are well detected at the finer spray resolution. The effect of aggregation of weed patches into larger areas is to permit more accurate chemical application. This method will be used to examine control time effects and to specify design requirements for patch sprayers. The same principle can be used for designing other variable rate field equipment.

Conclusion

It was concluded that the more spatially variable the weed pattern (i.e. more small patches spread unevenly throughout the field) the less accurate the patch spraying system will be. It was also shown that when there is more control of the boom (i.e. divided up into smaller controllable segments) it is possible to have a much more accurate spraying system provided that spray application response rates are adequate. The method can readily be extended to account for control time.

References

Balena F. 1999. Programming Visual Basic 6.0. *Microsoft Press International.*
Lutman, P.J.W., Rew, L.J., Cussans, G.W., Miller, P.C.H., Paice, M.E.R., & Stafford, J.E. 1998. Development of a 'Patch Spraying' system to control weeds in winter wheat. *HGCA Project Report No.* 158

Acknowledgements

I would like to thank the University College Dublin Faculty of Engineering and Enterprise Ireland for their scholarship awards.

A new optical signature of nitrogen deficiency in wheat useful for decision support in precision agriculture

A. Cartelat[1], Y. Goulas[1], C. Lelarge[2], A. Barbottin[3], M.-H. Jeuffroy[3] and Z.G. Cerovic[1]

[1]*Equipe Photosynthèse et Télédétection, LURE/CNRS, Bât 203, BP34, 91898 Orsay, France*
[2]*IBP, Bât 630, Centre Universitaire Paris-Sud, 91405 Orsay, France*
[3]*UMR d'Agronomie, INRA INA-PG, BP01, 78850 Thiverval-Grignon, France*
Zoran.Cerovic@lure.u-psud.fr

Abstract

Nitrogen (N) deficiency induces a decrease in leaf chlorophyll content (Chl) and an increase in leaf polyphenol content (Phen). Using dedicated optical devices, we followed these two variables during the growth season of winter wheat. We found gradients of Phen and Chl along the leaves, the variables increasing from the ligula towards the leaf tip. The Chl/Phen ratio could discriminate among crops grown at different levels of N-fertilization despite the presence of these gradients. So, this ratio that can be assessed remotely by chlorophyll fluorescence, is a potential decision support signature for site-specific N-fertilization.

Keyword: wheat, leaf chlorophyll, leaf polyphenol, autofluorescence, nitrogen fertilization

Introduction

It is well known that N-deficiency induces a decrease in leaf Chl. Furthermore, the carbon-nitrogen balance hypothesis predicts that it should also induce a concomitant increase in leaf Phen (Lambers et al, 1998). On the one hand, Chl can be assessed from its absorption or reflectance features in the red and near infra-red. On the other hand, the screening of chlorophyll fluorescence excitation in the UV by polyphenols can be used for their quantification (Ounis et al, 2000) knowing that they are mainly accumulated in the epidermis. We, therefore, designed and tested a new optical signature of wheat N-status based on both leaf Phen and Chl content. In Monocotyledons, the age of the cells increases along the leaf, due to its basal growth, so older leaf segments accumulate more Chl and Phen than younger ones (Meyer et al, 2003). These gradients can blur the effect of N on Chl and Phen at the leaf and canopy levels. Here we show that at the leaf level, the new signature, Chl/Phen, can discriminate crops grown at different N-fertilization rates.

Materials and methods

Field experiments were conducted on winter wheat (*Triticum aestivum* L.), cv. Isengrain and Récital, grown on a clay loam (INRA, Grignon, 48.9°N, 1.9°E). The total amounts of added N ranged from 0 to 260 kg ha^{-1}, applied in three fractions between tillering and earring. Crops were fully protected against weeds and pests, and irrigated when necessary. We used the Minolta SPAD-502 for Chl and the Dualex for Phen measurements. Dualex uses chlorophyll fluorescence to measure the UV-absorbance of leaf epidermis, and therefore, Phen (Goulas et al, 2001). These two leaf-clip devices were used in 2001 and 2002, on a weekly basis, to follow Phen and Chl in the upper leaves of the canopy. Methanolic extracts of 0.25-cm^{-2} discs of leaves were used to calibrate the SPAD in chlorophyll units (µg cm^{-2}). Dualex values from both the adaxial and abaxial side of the leaf were summed in order to assess the total Phen.

Results

For both cultivars, and in both years, the average Chl of the upper leaves increased, and the average Phen decreased, with the amount of N applied to the field. N-deficiency also reduced the length and the width of the leaves. The Chl and Phen content increased along the leaves, starting from the ligula (fig. 1). This was observed at any stage of development. Taking all this into account, in 2002, we measured Chl and Phen only in the middle of fully developed upper leaves. At earring, the Chl/Phen ratio could discriminate crops grown at 140 and 260 kg N ha^{-1} (n = 30, P < 0.01), whose final yield and protein content were different.

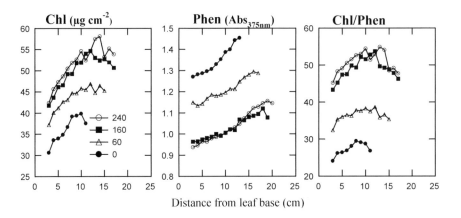

Figure 1. Chl, Phen and the Chl/Phen ratio along the leaves at earring in 2001 (cv. Isengrain). The amount of N applied to the field is indicated in kg ha^{-1}.

Discussion

The combined signature, Chl/Phen, alleviates partially the problem of gradients along the leaves, which otherwise would blur the signal at the canopy level. Chl itself is a good indicator of the N-status of the crop, but Phen can be more specific because Chl is sensitive to S-deficiency whereas Phen is not (Samson et al, 2000). Phen might also help to separate the different stresses a crop endures during growth, like water-stress, that can change the leaf Chl content. As we have shown previously that both Chl and Phen can be assessed remotely using chlorophyll fluorescence (Ounis et al, 2000), we can propose to use Chl/Phen as an improved optical decision support for site-specific wheat N-fertilization.

References

Goulas Y., Cerovic Z.G. and Moya I. 2001. Measuring device for the assessment of light absorption characteristics of biological tissue samples, associated measuring procedure, and application in the field of plant analysis and medicine, FR 0112551 28 September 2001 (France).

Lambers H., Chapin F.S. III and Pons T.L. 1998 Plant Physiological Ecology, New York Berlin, Springer, 540 pp.

Meyer S., Cartelat A., Moya I. and Cerovic Z.G. 2003. UV-induced blue-green and far-red fluorescence along wheat leaves: a potential signature of leaf ageing. Journal of Experimental Botany 54 757-769.

Ounis A., Cerovic Z.G., Moya I. and Briantais J.-M. 2001. Dual excitation FLIDAR for the estimation of epidermal UV absorption in leaves and canopies. Remote Sensing and Environment 76 33-48.

Samson G., Tremblay N., Dudelzak A.E., Babichenko S.M., Dextraze L. and Wollring J. 2000. Nutrient stress of corn plants: early detection and discrimination using a compact multiwavelength fluorescent lidar. In: Proceedings of the 20th EARSeL Symposium, Dresden, Germany, edited by R. Reuter, pp 214-223.

A methodology to define soil units in support to precision agriculture

A. Castrignanò[1], M. Maiorana[1], M. Pisante[2], A.V. Vonella[1], F. Fornaro[1] and G. Fecondo[2]
[1]Istituto Sperimentale Agronomico, Bari, Italy
[2]COTIR, Vasto (PE), Italy
annamaria.castrignano@tin.it

Abstract

This paper presents a new methodology to derive basic soil units for precision agriculture. Geostatistical and non parametric clustering techniques were applied to a set of soil properties The obtained results produced the subdivision of the field into five soil fertility classes.

Keywords: soil variability, geostatistics, clustering, non parametric density approach.

Introduction

In the precision agriculture soil tillage and crop management are adjusted to local soil properties and crop requirements, which requires a detailed knowledge of soil within-field spatial variability. Research suggests dividing a field into a series of homogeneous zones on the basis of soil characteristics, which can be assumed as a basis for varying agronomic inputs. The main purpose of the paper is to propose a new methodology to reduce soil within-field variability to a restrict set of clusters in support to precision agriculture.

Materials and methods

The research was carried out in the experimental farm of the Institute, Foggia (41° 27' lat. N, 3° 04' long. E, south Italy) and the soil is classified as Typic, chromoxerert, fine, Mesic (Soil Taxonomy-USDA). On a 2-ha field, split in four 0.5-ha plots, the most relevant soil properties (N, P, K, pH, limestone content, organic carbon, sand, clay and silt) were determined on a 10 gy 20-m grid at two depths (0-20 cm and 20-40 cm). The spatial structure was investigated by variography and the variables were interpolated by kriging and cokriging techniques (Goovaerts, 1997). To divide the fields into a number of soil clusters or classes, a non-parametric clustering method, which makes use of density estimation (Scott, 1992), was used. In order to obtain contiguous clusters the spatial coordinates were added to the interpolated values of the variables as inputs to the clustering procedure. All the geostatistical analyses were performed using ISATIS – Geovariances software, whereas the non parametric density approach was implemented in the MODECLUS procedure of the SAS/STAT software.

Results

From the statistical analysis (not reported) it results that there are no clear differences between the two soil depths, because soil was continuously ploughed to 40-50 cm depth over time. The variography analysis showed that most of variables, with the exception of P and N, looked well spatially structured up to an average distance (range) of 40-50 m, related to the size of the agronomic plot. Since there are no significative differences between the two soil depths, the geostatistical techniques were applied only to the surface variables: cokriging separately to the two groups of variables: sand, silt, clay for the first group and total carbonate, total N, available P and exchangeable K for the second one, because the variables within each group showed a significative spatial correlation; kriging to pH, because it was not correlated with the other variables. As an example, we report the estimated maps for clay and total carbonate in figure 1 for the presence of relevant spatial patterns.

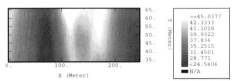

Figure 1. Cokriged maps of clay content (on the left) and total N (on the right).

The clay map (figure 1 left) shows that the field can be divided roughly into four zones: the central and the left ones are characterised by the highest values. A different behaviour is showed by total N (figure 1 right) with an increasing trend along the W-E direction. The application of the clustering approach produced the subdivision of the field into 5 distinct classes. In table 1 there are reported the basic statistics of the five clusters, whereas figure 2 shows the spatial partition of the field into the five homogeneous zones.

Table 1. Mean and standard deviation of each cluster.

Variable	cluster 1		cluster 2		cluster 3		cluster 4		cluster 5	
	Mean	Std	Mean	Std	Mean	Std	Mean	Std	Mean	Std
Clay	37.8	2.6	39.5	3.1	42.4	1.1	30.9	2.5	28.3	2.1
Silt	45.4	2.1	44.6	2.2	42.2	1.3	52.1	1.8	52.4	1.4
Sand	16.8	0.6	15.9	1.3	15.5	0.5	17.0	1.1	19.3	0.8
Total carbonate	37.9	7.3	89.2	10.6	87.4	5.7	87.0	12.5	56.0	9.4
Total C	13.1	0.6	15.1	0.7	15.2	0.2	15.6	0.3	14.7	0.4
Exchangeable K	1306.3	56.3	1500.2	82.1	1601.8	74.3	1522.4	20.2	1471.5	29.3
Total N	1.5	0.1	1.7	0.1	1.7	0.0	1.7	0.1	1.5	0.1
Available P	36.2	2.4	28.4	5.7	30.5	3.5	30.4	3.9	29.3	3.4
PH	8.2	0.0	8.2	0.0	8.2	0.0	8.3	0.0	8.2	0.0

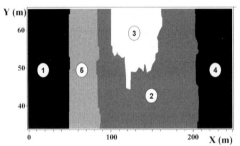

Figure 2. Spatial distribution of the five clusters.

The multivariate clustering has essentially reproduced the spatial patterns of the clay, with two wide central clusters characterised by the highest levels of physical and chemical fertility.

Conclusions

The proposed approach has then resulted to be a quick and effective method to identify different potential soil units as a basis for the variable-rate application of inputs.

References

Goovaerts, P., 1997. Geostatistics for Natural Resources Evaluation. Oxford Univ. Press, New York, 512 pp.
Scott, D.W., 1992. Multivariate Density Estimation: Theory, Practice, and Visualization, New York: Jhon Wiley & Sons, Inc.

An information tool for the management of animal waste at farm and regional level

E. Cavallo and I. Albertin
Institute for Agriculture and Earthmoving Machinery. Italian National Research Council. Strada delle Cacce 73. 10135 Torino. Italy.
e.cavallo@imamoter.cnr.it

Abstract

A prototype of the GIS database has been developed in order helps agricultural technicians to optimise the management of pigs' slurry, reducing the impact on the environment.
ArcView software has been used to build and manage the data of the GIS database.
The GIS database is made up of 4 types of information: geo-referenced thematic maps, geo-referenced data regarding the farms and their fields, data specific to each individual farm, and the environmental regulation criteria to be satisfied.
The GIS database can manage geo-referenced graphical queries on a multi-layer map of the area considered or alphanumeric queries on farm records.

Keywords: ArcView, environment, GIS, pigs, slurry

Introduction

Italian legislation concerning animal waste management, as in most European countries, is becoming increasingly stringent.
Pig farmers that produce slurry as waste are considered responsible for the pollution of soil and water in large areas of Italy. These farms, that for economic reasons, are becoming larger and are tending to be concentrated in specific areas, spread their slurry on the soil thus creating a real threat to the environment.
In order to reduce the risk of pollution, pig farms have to comply with a number of regionally-established regulations. These regard a list of aspects that cover the whole farm. For example, the number of pigs and their "weight" with relation to the surface area of the farm, the physical characteristics of the soil and the distance at which the slurry is spread, the existence and type of treatment plants for slurry, the average nutrient and pollutant content of the slurry, and planning restrictions. Farmers are required to report all of these aspects in official documents and analyses submitted to the public authority in charge of regulating the area. In general, pig farm associations prepare and conserve the above papers. Over the years a large database, on paper, of all these characteristics has been set up. Any queries are fairly difficult because the information is contained in several different documents.
In cooperation with the pig producer association, a project has been launched to integrate existing data with thematic geo-referenced maps of the area where farms are located, thus making all of the information available in a GIS database. Queries on the GIS database will be directly referenced to maps or farm records.

Method

The prototype of the GIS database has been developed in a small municipality in the north-west of Italy, where specialized pig farms are very popular. The surface area of the municipality is 842 ha, and it includes 25 pig farms with a total of 22,300 animals.
ArcView software has been used to build and manage the data of the GIS database.

The GIS database is made up of 4 types of information: geo-referenced thematic maps, geo-referenced data regarding the farms and their fields, data specific to each individual farm, and the environmental regulation criteria to be satisfied.

The geo-referenced digital maps used for the area of the municipality in question are the land registry map (1:2000 in dwg format), the regional vector map (1:10,000 in E00 format), the regional land-use map (1:100,000 in E00 format) and regional limitation of use map (1:100,000 in E00 format).

Specific geo-referenced data, such as the location of the farm and the location of the analysed samples of soil, have been collected through surveys.

Alphanumeric data concerning the farms, physical characteristics of field soils and the content of nutrients and pollutants are associated to the geo-referenced points. For example the name and address of the farm, the number of animals, the farm's surface area, the storage capacity of slurry, the type of flooring, the type of housing, and the sand, clay and loam content of soils, and the N-P-K and heavy metal content of slurry.

The environmental regulation parameters are generally in the form of a ratio between numerical data.

Results

The result of this project is a GIS database, that can manage geo-referenced graphical queries on a multi-layer map of the municipality or alphanumeric queries on farm records.

Technicians of the pig farm association are currently testing the GIS database. It runs on a portable pc, has a very friendly interface and can be easily updated "on-site".

The prototype of the GIS database is a tool that helps technicians optimise the resources of the farm, thus reducing the impact on the environment. It is made possible by cross-referencing GIS database information.

For example the "weight" of pigs that a farm can breed depends on the quantity of slurry (t) that can be spread on soil (ha); this in turn depends on the sand content (%) of soil and on the nutrient content (NPK) of the slurry, considering the crops cultivated in the fields (from the land registry map), the limitation of use imposed by regulations (from the thematic map), the slope of the fields and the distance from river or wells (from the thematic map).

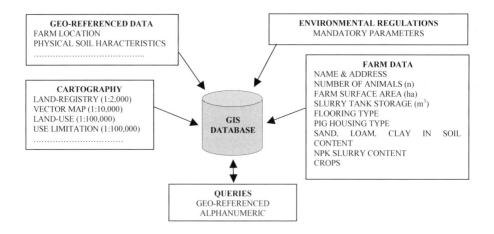

Analysis of reflectance spectra and spectral characteristics for rice grown in different cropping seasons

Rong-Kuen Chen and Chwen-Ming Yang
Taiwan Agricultural Research Institute (TARI), Department of Agronomy, 189 Chung-Cheng Road, Wufeng, Taichung Hsien 413, Taiwan
cmyang@wufeng.tari.gov.tw

Abstract

Ground-based reflectance spectra of hyper-spectral resolution from field-grown rice canopy were acquired from 4 cropping seasons in 2000-2002. Reflectance spectrum was shown a typical green plants-like pattern with reflectance lower in UV (<8%) and VIS (<15%) and change considerably in NIR, SWIR and MIR during growth. In VIS the maximum reflectance was in green region and the minimum was in violet, blue or red. Seasonal variability of spectra in First crops was greater than that in Second crops as growth performance. Some spectral characteristics were selected for a dynamic comparison within and between crops.

Keywords: reflectance spectrum, spectral characteristics, rice, high resolution, variability.

Introduction

Rice is the most important food crop occupying the largest cultivated area with the highest production value in Taiwan. By incorporating remote sensing techniques with proper management practices, it is expected to reduce environmental pollution and production cost while maintaining stabilized yield and quality in growing rice. Prior to practical applications, however, spectral data of ground truth and spectral-biophysical relationships for rice crop should be well established. This study collected high-resolution reflectance spectra of rice canopy to analyze the temporal variations of reflectance spectra and the selected spectral characteristics for evaluating their biophysico-physiological meanings along rice growth.

Materials and methods

Reflectance spectra (350-2500 nm) of hyper-spectral resolution from ground-based remotely sensed field-grown rice canopy were acquired at Taiwan Agricultural Research Institute Experimental Farm (24°45'N, 120°54'E, elevation of 60 m) during First and Second cropping seasons in 2000-2002. The spectral data were taken by a portable spectroradiometer, and seasonal changes in reflectance spectra and 11 selected spectral characteristics were analyzed.

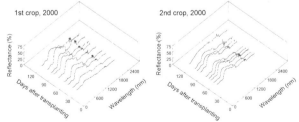

Figure 1. Three-dimensional dynamic variations of canopy reflectance spectra from rice plants grown in the first cropping season of 2000.

Results and discussion

Morphological modifications and color changes are the external appearance commonly observed to most plants in nature. The symptoms reflect the processes of growth, development and age of plant cells and tissues, and result in an alteration in optical property of plant populations and thus a dynamic variation in canopy reflectance spectrum. The three-dimensional variations in rice reflectance spectra across First and Second cropping seasons were plotted in Fig.1, confirming that spectral curves changed with time and spectral reflectance varied with wavebands. Rice reflectance spectrum was a typical green plants-like pattern and similar in every season. Generally seasonal variability of reflectance spectrum in First crops was greater than that in Second crops, indicating the influence of environmental effects. Reflectance was lower in UV (350-400nm, <8%) and VIS (400-740nm, <15%), and higher in NIR (740-1300nm, 16-54%), SWIR (1300-1800nm, 8-31%) and MIR (1800-2500nm, 4-24%). MIR was sensitive to environment and plant water content, and hence was not suggested used in monitoring rice growth. Reflectance within VIS was higher in green light and lower in two ends of visible region echoing photosynthetic activity changed during the growth (Blackburn, 1998; Yang and Lee, 2001). Reflectance change in NIR followed the covering percentage and aging of rice plant as that indicated by Gausman (1982). 11 selected spectral characteristics changed curvilinearly during the growing periods, a compromise mixture under the influence of growth and environment. Among them, seasonal variability in RED/NIR ratio, NDVI and red edge slope may be reasonably modeled and estimated.

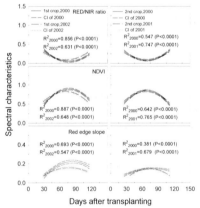

Figure 2. Seasonal variations of different spectral indices calculated from canopy reflectance spectra of rice plants grown in the first and the second cropping seasons of 2000. CI was the 95% confidence interval.

References

Blackburn, G.A. 1998. Spectral indices for estimating photosynthetic pigment concentrations: a test using senescent tree leaves. Intl. J. Remote Sens. 19(4):657-675.

Gausman, H. W. 1982. Visible light reflectance, transmittance, and absorptance of differently pigmented cotton leaves. Remote Sens. Environ. 13:233-238.

Yang, C.M. and Y.J. Lee. 2001. Seasonal changes of chlorophyll content in field-grown rice crop and their relationships with growth. Proc. Natl. Sci. Counc. ROC(B) 25:233-238.

Platform experiments with a wide-swath, variable-rate, granular-material applicator

T. Chosa, Y. Shibata and M. Omine
NARC Hokuriku, 1-2-1 Inada Joetsu Niigata, 943-0193, Japan
tchosa@affrc.go.jp

Abstract

This paper presents platform experiments that examined a wide-swath, variable-rate, granular-material applicator. In an indoor platform experiment, uniform cross-sectional distribution of granular material was observed, with a time lag from signal output to actual application. The coefficient of variation for the cross-sectional distribution was less than 15%. The time lag from signal output to actual application was 0.75 - 1.25 seconds. The platform experiment simulated variable rata application using GPS signals.

Keywords: VRT, granular-material applicator, map-based, GPS, platform experiment

Introduction

Variable-rate application of fertilizer is essential for introducing precision farming to paddies. Map-based variable-rate technology for paddies has already been developed (Chosa et al., 2000, 2003). However, it is difficult to evaluate its performance in the field, since it is difficult to collect all the material that is spread during field operations. This paper presents a platform system for testing a wide-swath, variable-rate, granular-material applicator, and the results of its performance in platform experiments.

Materials and methods

The variable-rate granular-material applicator consists of a hopper, boom, ventilation unit, four rotary applicators, and four DC motors. Each applicator can spread fertilizer, agricultural chemicals, or seed at different rates over one fourth (2.5 m) of the total width covered (10 m). The material is blown from the rotary applicators to the edge of the boom. A 0 to 5-V analog signal controls the application rate. The signal is calculated from GPS and application map data to realize variable-rate application. The experimental platform for wide swath applicators was developed by the Bio-oriented Technology Research Advancement Institution (BRAIN) and consists of sampling boxes, electronic scales, and a control unit. Each sampling box is 300 x 1,000 mm. The sampling boxes are placed across the width of the swath of the applicator, and are connected by electronic scales. They collect and weigh all the material spread. The control unit monitors the measured weights. The minimum-monitoring interval is 0.25 seconds. The application rate is continuously measured to calculate the variation in the monitored weights.

The first experiment examined the cross-sectional distribution of granular material, and the second examined the time lag from signal output to actual application. A computer and DA converter were used to adjust the control signal and control time of the variable-rate applicator. The third experiment simulated map-based variable-rate application using GPS signals.

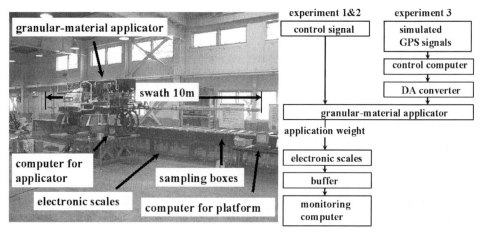

Figure 1. Photograph (left) and signal flow diagram (right) of the indoor platform experiments.

Results and discussions

In the first experiment, several kinds of fertilizer, agricultural chemicals, and rice seed were used as experimental materials. A uniform distribution was observed with every material. The distributions of rice seed and fertilizer were especially fine; the coefficients of variation were less than 15%. In the second experiment, the time lag from signal output to actual application at the center of the swath was less than 0.75 seconds, while that to application at the edges was 0.75 - 1.25 seconds. These time lags have to be taken into account to realize more accurate control. In the third experiment, the measured application rate and simulated position information were used to generate a post-application map that was similar to the application map, demonstrating successful variable-rate application.

Conclusions

The performance of the variable-rate application system that we developed was examined using platform experiments. The results of each experiment were used to improve the variable rate technology, and the improved system was used for actual fertilizer application.

References

Chosa, T., Kobayashi, K., Omine, M., 2000. Site specific variable rate application by automatic driving vehicle, Journal of Japanese Society of Agricultural Machinery, 62(5), 123-125.
Chosa, T., Shibata, Y., Omine, M., Kobayashi, K., Toriyama, K., Sasaki, R. 2003. Map-based variable control system for granule applicator, Journal of Japanese Society of Agricultural Machinery, 65(3) (in press)

Acknowledgements

The authors would especially like to thank Dr. S. Miyahara (BRAIN) for taking care of the experimental platform that was used for the wide swath applicator.

Precision viticulture in Portugal: The beginning of a process

L.A. Conceição, J.P. Mendes, R.P. Braga, S. Dias, and F. Mondragão Rodrigues
Esc. Sup. Agrária de Elvas, Inst. Polt. de Portalegre, Apartado 254 - 7351 Elvas, Portugal
ricardo_braga@esaelvas.pt

Abstract

To begin exploring the spatial variability of a vineyard we measured grape weight and sugar content in a 10 ha farm field located 3 km from Estremoz, Alentejo, Portugal. We also measured soil related properties. Average yield was 6446.5 kg/ha with a standard deviation of 2506.0 kg/ha. Average Brix value was 21.4 with a standard deviation of 0.9. There was a significant effect of slope direction (N and S) on yield (N = 8353.13, S = 5175.33 kg/ha) and Brix (N = 20.91, S = 21.68) value (p=0.047 and 0.003, respectively). No significant correlation was found between yield and Brix values.

Keywords: Viticulture, Slope Direction, Yield, Quality, Soil Properties

Introduction

Viticulture is a large agricultural business in Portugal. Alentejo is one of the largest wine producing regions in the Country. There is considerable interest in all technologies that can further increase quantity as well as quality of the product. We foresee a large potential for precision agriculture technologies to be applied and achieve such objectives. Two specific strategies are envisioned. On one side, it could help reduce costs and environmental damage by allowing for a more spatially precise application of nutrients and pesticides. On the other side, by having the knowledge of the spatial variability of sugar content (and other wine quality related traits) one could sub-divide the overall production into lots of different wine quality potential.
The objective of this preliminary work was to measure the amount of variability found in a typical vineyard and try to explain it using topographic and soil properties. We did not want to obtain any type of geo-statistically valuable maps at this stage.

Material and methods

We sampled the field to obtain two types of variables: plant variables (yield and brix values) and soil variables (pH, OM, Cond, P, K, Ca, Mg, Sand, Silt, Clay, BulkD, FC, PWP, PASW). Brix values was considered to be the quality related variable of yield (1 Baumé ≈ 0.55 Brix). Both variables were measured along a N-S transect. Plant variables were originally measured on 10 locations: 6 facing South and 4 facing North. Soil variables were measured on two depths (0-30 cm and 30-60 cm) and 4 locations: 2 facing South and 2 facing North. The plant variables were averaged to 4 locations in order to allow for comparison with the soil variables. The sampling to place in September. We performed ANOVA analysis using slope direction as independent variable (factor) and considered any dependent variable to be significant at 0.05. The soil dependent variables were analyzed as 0 - 30 and 30 – 60 values and also as 0 – 60 averages. All statistical analysis were performed with STATISTICA software (StatSoft, Inc., 1995).

Results

Average yield was 6446.5 kg/ha with a standard deviation of 2506.0 kg/ha. Average Brix value was 21.4 with a standard deviation of 0.9. There was a significant effect of slope direction (N and S) on yield (N = 8353.13, S = 5175.33 kg/ha) and Brix (N = 20.91, S = 21.68) value (p=0.047 and 0.003, respectively). No significant correlation was found between yield and Brix values (BRIX = 21.978 – 0.001 YIELD, r= -0.2716; p= 0.247).

Table 1 shows the average values for different soil variables according to slope direction. Slope direction South represented areas of lower yield and higher Brix whereas slope direction North represented areas of higher yield and lower Brix values. Soil variables related significantly related to slope direction were electrical conductivity, organic matter, Ca and Plant available soil moisture.

Table 1. Probability and values of soil variables for different depths and slope direction. Significant relationships are in bold.

	0-30			30-60			0-60		
	p	S	N	p	S	N	p	S	N
pH water	0,217	7.6	5.8	0,110	7.8	5.7	0,158	7.7	5.7
pH KCL	0,101	6.7	4.8	0,067	6.8	4.6	0,080	6.7	4.7
Cond mS/cm	**0,001**	**0.060**	**0.205**	0,423	0.060	0.070	**0,009**	**0.060**	**0.138**
OM %	**0,038**	**0.7**	**0.95**	0,553	0.7	0.8	0,089	0.70	0.88
P2O5 ppm	0,547	56	110	0,435	39	28	0,606	47	69
K2O ppm	0,499	214	337	0,632	99	56	0,648	156	216
Ca ppm	0,059	2360	780	**0,041**	**2346**	**761**	**0,050**	**2353**	**770**
Mg ppm	0,394	247	159	0,157	420	155	0,175	333	157
SAND %	0,536	44.8	49.1	0,714	46.2	48.7	0,620	45.5	48.9
SILT %	0,361	30.0	32.9	0,277	28.2	32.5	0,311	29.1	32.7
CLAY %	0,171	25.4	18.0	0,173	25.6	18.8	0,172	25.5	18.4
DAP	0,423	1.60	1.66	0,423	1.60	1.66	0,423	1.60	1.66
FC %	0,423	0.350	0.334	0,423	0.350	0.334	0,423	0.350	0.334
PWP %	0,851	0.159	0.155	0,633	0.156	0.149	0,748	0.157	0.152
PASW mm/m	**0,001**	**191**	**179**	0,133	194	185	**0,024**	**193**	**182**

Discussion and conclusions

In the Northern hemisphere, slopes facing South have in general higher levels of radiation and therefore are warmer and drier than slopes facing North. This has impacted yield and quality in an expected way: slopes facing South had lower yields with higher Brix whereas slopes facing North had higher yield with lower Brix values. Plant Available Soil Moisture was significantly related to slope direction and therefore to yield. This makes sense considering that the studied vineyard was irrigated under regulated stress. This study shows room vineyard management improvements with precision viticulture.

References

StatSoft, Inc. (1995). STATISTICA for Windows. Tulsa, OK: StatSoft, Inc., 2300 East 14th Street, Tulsa, OK, WEB: http://www.statsoftinc.com.

Acknowledgements

We want to acknowledge Eng° José Castro Duarte for allowing us perform this study at Quinta da Esperança, Estremoz.

Automation concepts for the variable rate fertilizer applicator for tree farming

S. Cugati[1], W. Miller[2] and J. Schueller[1]
[1]University of Florida, Agricultural and Biological Engineering, Gainesville, FL 32611, USA
[2]University of Florida, Citrus Research and Education Center, Lake Alfred, FL 33850, USA
cugati@ufl.edu

Abstract

A broad overview of various steps involved in developing an autonomous variable rate granular fertilizer applicator (AVRA) capable of generating optimum applications according to the spatially variable needs of tree crops is presented. An investigation on the concepts and components of an autonomous variable rate granular fertilizer applicator is made. A choice of alternatives for each subsystem is stated. Various overall system integration concepts are presented and evaluated.

Keywords: automation, variable rate, citrus, sensing, fertilizer applicator.

Introduction

Citrus growers in Florida are currently considering precision agriculture applications in order to maximize the profitability through such technologies as optimizing the use of fertilizers. This variability in fertilizer application at any point can be achieved by the variable rate fertilizer application. Spatial variability implies different needs, often from tree to tree. By implementing variable rate application of fertilizer, the grower is not only contributing to the economic prosperity, but is also an environmental benefit through reduction of problems such as leaching of chemical fertilizers leading to contamination of ground water.

Concepts

In order to automate the entire process of fertilizer application to a tree in a field, it is not possible to work only on the data that is acquired previously from the GIS layer. It becomes necessary to check for the validation of this data in real time and make the decisions. The quantity of fertilizer to be deposited at any point is dependent on the *spatial variables* such as, soil properties, nutrient content, pH value, crop growth, disease, pest, tree age and vigour; *temporal variables* such as seasonal changes and growth changes; and *predictive variables* such as yield, weather.

The automated fertilizer applicator is comprised of three modules: the *input module*, the *decision module* and the *output module* as shown in Figure 1. The input module primarily constitutes the GIS data and the data from the real-time sensing. The parameters for all the variables that affect the quantity and the spread pattern of the fertilizer are determined in this module. The parameters for real time sensing include the tree canopy size, position of the equipment (using GPS), bearing and direction from path guidance system, wind direction and speed. The real time data acquisition systems can either be mounted on the applicator or it can be mounted on a separate vehicle independent and moving in front of the applicator. The communication between the independent real-time data acquisition system can be accomplished by wireless technology such as Bluetooth. This type of arrangement permits buffering of the real-time data.

The decision module consists of a mathematical model and control system that is developed based on the above-mentioned variables to calculate the optimum quantity of fertilizer and the spread pattern required to maximize the yield.

The output module comprises of the actuators and the equipment that regulates the application of the fertilizer in the optimum spread pattern as chosen by the decision module. The spreading equipment can be a spinner disc spreader or a pneumatic spreader.

Once the complete system is assembled, it has to be evaluated as per the existing and proposed ASAE and European standards. A testing procedure is to be prepared to evaluate the performance of this applicator under normal working conditions and its response to the variation of the parameters should be studied.

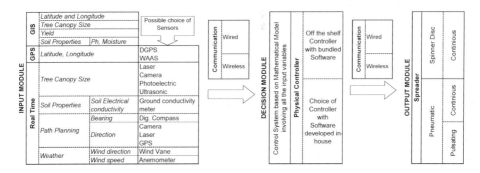

Figure 1. Schematic of the subsystems comprising the AVRA

Conclusion

A wide array of choices can be made for different subsystems of this applicator. For example, machine vision, photoelectric, ultrasonic or laser-based sensor can accomplish the measurement of tree canopy size, which is an input for predicting the yield and also to determine the treatment spray quantities. The choice of the sensor is also influenced by its feasibility to be used for other applications such as spray applications and path planning. Each of the choice has to be evaluated for both its advantages and limitations. The choices available for the modules and the evaluation procedures are listed.

References

D.L. Needham, S.D. Reed, M.L. Stone, J.B. Solie, K.W. Freeman, W.R. Raun. 2002. Development of a robust precision fertilizer application system utilizing real-time, ground-based optical sensors and fluid application control. 2002 ASAE Annual International Meeting, paper number 021180.

R.L. Parish. 2002. Broadcast spreader pattern sensitivity to impeller/spout height and PTO speed. Applied Engineering in Agriculture, Vol 18(3): 297 – 299. American Society of Agricultural Engineers.

S.D. Tumbo, M. Salyani, J.D. Whitney, T.A. Wheaton, W.M. Miller. 2001. Laser, ultrasonic and manual measurements of citrus tree canopy volume. 2001 ASAE International Meeting, paper number 01-011068.

Y. Zhou, L. Tian, L. Tang. 2002. Aided guidance system based on GPS and optical encoders for smart sprayer. 2002 ASAE Annual International Meeting, paper number 001018.

Precision agriculture in Cuba: Precedents and perspectives

Miguel Esquivel[1], José Julio Rodriguez[2], Britaldo Hernández[3], Farnia Fernández[3], Enrique Ponce[3] and Lázaro Quintana[3]
[1]*Projects Division,* [2]*Computer Laboratory,* [3]*Automation and Control Laboratory, National Center for Laboratory Animal Production, CENPALAB, Finca Tirabeque, Bejucal, Havana, Cuba*
esquivel@cenpalab.inf.cu

It has been widely recognized that Precision Agriculture (PA) is one of the top emerging technologies in the present century. Even when it still has not been established as common practice, there are several countries that are evaluating it and Cuba is not an exception. For a country where agriculture is one of the main economic activities, it represents the possibility to improve the efficiency of agricultural production, as well as to reduce it's impact on the environment.

There are several precedents that make it possible to establish PA in Cuba. During the last decades Cuba has achieved a considerable level of development in technologies such as remote sensing, geographic information systems and automatic control of agricultural machinery. Due to the specific characteristics of the socioeconomic system existing in Cuba, the strategy to develop PA can be based on a national outlook instead of applied to individual producers; several large investments (DGPS base stations, agriculture machinery automation, etc.) can be optimized for the whole country.

Specialists have recognized that there are three main groups of challenges for the adoption of PA, namely socioeconomic, agricultural and technological.

The socioeconomic challenges are related to the cost of new services or machinery, and the abilities required of farmers. There are some services that already exist, such as soil nutrient mapping. The Institute of Sugarcane Research (INICA) provides the nationwide fertilization diagnostics service. Soil maps exists for all sugarcane-planted area in the country. Diagnostic is made by traditional sampling methods, and Geographical Information Systems (GIS) are used to compile soil fertilizer prescriptions. A study is presently under way with the intention to reduce the cost of soil studies by several alternatives, which combines the use of automatic soil sample machines and evaluation of several direct measurements sensors for soil and plants, as well as indirect methods such as NIR spectrometry, etc.

Instead of importing new machinery, reconversion of existing machinery to automatic control in order to PA-ready is carried out when possible. A practical example is the FT350 mechanical fertilizing machine. About 1200 such machines exist in the country and they do not support variable rate application. A prototype has been modified for PA use where all mechanical systems have been replaced with hydraulics including a hydro-motor, a regulatory hydraulic valve with electro-hydraulic control and a sensor to control the rotation speed of the fertilizer distributor. The system is controlled by a microcontroller (AGROMATIC). When the mechanical system is replaced, there is enough space to include a third fertilizer deposit, all three with independent rate control, making it possible to apply simultaneously three different kinds of fertilizer (e.g. N, P, K) with variable doses.

The need for highly qualified human resources is not a problem in Cuba, since there is a large number of technicians available, not only in the agricultural sciences, but also in electronics and computer applications. An important achievement has been the creation of the University of Informatics Sciences (UCI) near Havana. It presently has around 2000 students from all over the country, and 25 labs with more that 500 modern PCs, among other facilities. Total amount of students will be around 10000. A total of eight faculties will carry out teaching and research on the fields of geomatics, GIS, image processing, automatic control, among others related to PA. A revolutionary concept is the existence of a Technological Park within the UCI, where more that 800 students are involved in enterprise incubator projects. The development of PA technology for potatoes and related crops is just one of the projects under proposal to be developed at the UCI Technological Park.

Most of the agricultural challenges are common to any country, such as the lack of prescriptions for PA or the inefficient use of agricultural information. The sugarcane fertilizing service provided by INICA, has already established prescriptions depending on soil characteristics and sugarcane variety. In the same way that sensing methods are getting cheaper and more accurate, and automatic machinery ready for PA is able to act with higher precision, the knowledge base that supports the agricultural prescriptions will need to be more accurate, too. Multifactorial studies to determine the response of sugarcane varieties to discrete changes in fertilization levels must be undertaken.

Technological challenges are related to a lack of PA-ready machinery, software for GIS or data processing and the availability of DGPS signal. As previously explained, the reconversion of existing machinery is the main alternative, but in some cases such as the potatoes PA system, some PA-ready machinery has been imported.

The use of open source GIS software is being strongly discussed in Cuba. Many servers are switching from Windows to the Linux operating system, and the community of Linux users is growing fast, as is the availability of software developers for this open source OS. A national network of DGPS beacons is being created, which will offer DGPS correction signal all over the country, and will guarantee the necessary precision for PA applications.

Some of the results of the PA program in Cuba are already finding their way to other countries. The automatic control system for sugarcane harvesters has been tested successfully in Brazil, and the possibility to add sugarcane yield mapping is already at the prototype stage.

From the organizational point of view, several steps have already been taken to facilitate a rapid transition to PA. A National Program for the development of PA has been created and is coordinated by the Ministry of Science, Technology and Environment. CENPALAB is currently the Program Head.

The Ministry of Sugarcane Industry has merged the three previous entities it had for information technology, automation and communications into a single entity called TEICO, which, together with CENPALAB, is developing PA technology for sugarcane.

Use of reflectance measurements to evaluate water status in wheat

Z.X. Fan, S. Graeff and W. Claupein
Institute of Crop Production and Grassland Research, University of Hohenheim, Fruwirthstr.23, 70599 Stuttgart, Germany
fzhongxue@yahoo.com

Abstract

A greenhouse study was conducted to determine specific wavelengths ranges responsive to water stress of wheat plants. Reflectance of wheat leaves, from plants grown under five different water treatments (65%-25% FC), was determined from the 4th to 6th leaf stage. Reflectance measurements were performed at the 4th leaf in the visible and infrared spectra with a digital camera under controlled light conditions. Reflectance patterns at 510-780 nm, 516-1300 nm and 540-1300 nm were found most responsive at a water content $\langle$ 75%. The results indicate that reflectance measurements may serve as a rapid, non-destructive approach to identify water stress.

Keywords: reflectance measurements, wheat, water stress.

Introduction

Available water is, in addition to nitrogen, the most important yield-determining factor for crops. An evaluation of water content of crops for site-specific crop management, by use of reflectance measurements, could improve existing irrigation systems. Hunt et al. (1987) pointed out a possibility to estimate relative leaf water content by reflectance at some specific wavelengths in the range of near infrared. Carter et al. (1989) found that visible wavelengths correlated best with the total water potential of loblolly pine needles. Thus, the objectives of this study were a) to evaluate reflectance changes of wheat leaves at different water stress levels, b) to determine important wavelength ranges in the visible and infrared spectra responsive to water stress, and c) to evaluate a possible correlation between leaf reflectance and plant water content.

Materials and methods

A pot experiment was conducted in a greenhouse with summer wheat (*Triticum aestivum* L.) cv. Thasos. Five different water levels were chosen based on the field capacity (FC) of the soil and designed as 65% (Control), 52%, 39%, 33%, and 26% FC with three replications. The plants were allowed to grow for 30 d under identical conditions. Afterwards, all pots were weighed and watered each day according to the designed water levels. Reflectance measurements were carried out at the 4th leaf, 9, 15, 21 and 24 d after starting of water stress. Measurements were performed with a digital, light-sensitive, high-spatial resolution camera (S1 PRO, Leica, Germany) in conjunction with a constant light source (HMI 21 W/D~10 Wm^{-2}, Sachtler, Germany) of total daylight spectra. The spectrum was split into various wavelength ranges using long-pass filters, active at wavelengths longer than 380 nm, 490 nm, 510 nm, 516 nm, 540 nm, and 600 nm. Scans were carried out with the software SILVERFAST V. 4.1.4 and analyzed with ADOBE$^{®}$ Photoshop 5.0 in the L*a*b*-color system. Wheat plants were harvested immediately after scanning for determination of plant water content. Analysis of variance was performed on all crop and reflectance data using Tukey-test of SAS 6.12 on a basis of 5% likelihood.

Results and discussion

The reflectance parameter b* increased in the visible and near-infrared wavelength range with increasing water stress level. Significant changes of the parameter b* were obtained at a water content below 75% (Figure1). There was a negative linear relationship between parameter b* and total plant water content. Correlation coefficients differed from 0.80 to 0.83 (data not shown). The best correlation was obtained in the wavelength range 510-780 nm with $R^2=0.96$ (Figure 2). Woolley (1971) and Bowman (1989) have also found that reflectance tends to increase in the 400-1300 nm region when water is lost from a leaf. The reason for reflectance changes in this region has been inferred as the changing of the internal structure of the leaf besides of water losing. This might also be the reason for the determined reflectance changes in this study and explain the close correlation between reflectance changes and plant water content. Further studies have to be carried out to determine the involved physiological changes in detail.

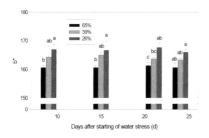

Figure 1. Change of reflectance parameter b* of 4th leaf in the wavelength range 510-780 nm.

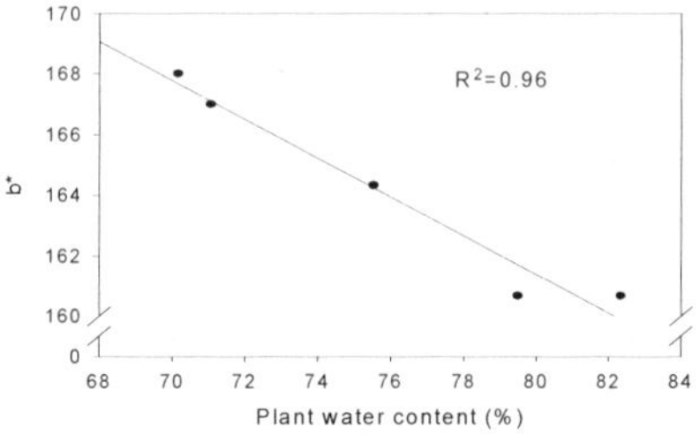

Figure 2. Correlation between plant water content and reflectance parameter b* of 4th leaf in the wavelength range 510-780 nm.

Conclusions

The results of this study indicate that reflectance measurements may provide a powerful tool for the evaluation of the water status of wheat plants. Further investigations are needed to discriminate leaf reflectance patterns of water stress from other plant stresses.

References

Bowman, W.D. 1989. The relationship between leaf water status, gas exchange, and spectral reflectance in cotton leaves. Remote Sensing of Environment 30(3) 249-255.
Carter, G.A., Paliwal, K., Pathre, U., Green, T.H., Mitchell, R.J., and Gjerstad, D.H. 1989. Effect of competition and leaf age on visible and infrared reflectance in pine foliage. Plant Cell Environment 12(3) 309-315.
Hunt, E.R., Jr., Rock, B.N., and Nobel, P.S. 1987. Measurement of leaf relative water content by infrared reflectance. Remote Sensing of Environment 22(3) 429-435.
Woolley, J.T. 1971. Reflectance and transmittance of light by leaves. Plant Physiology 47(5) 656-662.

Acknowledgement

The financial support by DAAD-K.C. Wong Fellowship (A0103969) to Zhongxue Fan is gratefully acknowledged.

Feasibility of GPS-based vehicle navigation aid

A. Fekete[1], I. Foldesi[2] and L. Kovacs[2]
[1]SzIE University, Physics & Control Department, 1118 Budapest, Hungary
[2]Hungarian Institute of Agricultural Engineering, 2101 Godollo, Hungary
h10835fek@ella,hu

Abstract

There is a definite need for a navigation aid or rather a steering aid that could reduce the error of the working width with machines of wide working width, such as fertilizers, sprayers, etc. A dGPS and an appropriate display can be used as a navigation aid. Experiments were performed with such a system. The results show that relative to the conventional solutions a favorable steering accuracy and an acceptable error of the working width can be obtained by such a technique and procedure.

Keywords: GPS, navigation, parallel swathing, error

Introduction

GPS is a principal tool for site-specific farming however it can be used as an agricultural vehicle navigation aid, or rather as a steering aid for agricultural machines in different field operations as well. There is a definite navigation and steering problem with machines of wide working width such as with fertilizers, sprayers, some tillage implements, etc. This problem is the overlapping and/or having a gap between the two covered or cultivated strips next to each other. The application of a dGPS can be a solution of the problem. The dGPS can be used for determining the distance between a definite run and the previous one. Therefore the dGPS can and the display can be used for displaying the proposed steering correction when following a straight line determined by two points. Furthermore a GPS can be fitted with a special purpose display to be used as an on-line navigation and/or steering aid. There are displays for helping the driver in following a special leading line along the field and indicating the proposed steering wheel corrections.
The objective of the work reported herein was to determine the feasibility of such a dGPS-based navigation or steering aid for parallel swathing of tractor and implement combinations along the field. Furthermore the objective was to determine the accuracy of the navigation and steering aid, the error in the distance between a designated straight line to be followed at a definite distance and the result line.

Materials and methods

Experiments have been performed under different field and soil conditions along approximately flat fields without considerable elevations. A straight line was determined along the edge of the test field to be the leading line and it was stored in the memory of the on-board computer to which the dGPS was connected. The test runs were done with a four wheel drive tractor and sprayer combination. The set point of the distance was 13.0 m. The purpose of the driver was to follow the leading straight line according to the indications displayed by the navigation aid display of the dGPS. This display indicated the direction of the proposed steering correction therefore it continuously helped the driver to follow the leading straight line determined earlier. The result of the steering was determined by the distance between the wheel ruts of the two subsequent runs. Each rut was characterized by the centerline of the rut. The length of the test runs was 800 m and the interval between the measurements along the rut was 20.0 m. Therefore the goodness of the steering was evaluated by the deviation of the distance between the two subsequent center lines. Histograms of the distances between the subsequent wheel rut lines were determined and analyzed.

Results and discussion

The dGPS-based vehicle navigation aid is a multi purpose device. It can be used for location, to determine the coordinates of a machine within the field, to measure and display working width and to designate a straight line by two coordinates. With our experiments the dGPS-based navigation aid was used to determine the set point for position of the tractor and implement combination during the headland turning and help the driver to find the spot that is in a definite distance relative to the former run. During the operation, such as spraying, fertilizing, etc. the display of the system works as a steering aid to help the driver to follow the straight line.

The results of several test runs showed that the average distance between two subsequent rut lines was 12.84 m instead of the 13.0 m set point. Therefore the average error was 0.16 m that is an acceptable value for the majority of field operations, such as for fertilizing, spraying, etc. The standard deviation of the rut line was 0.45 m and the average deviation was 0.36 m. The accuracy of the navigation and steering is dependent on the abilities of the driver as well. The 0.16 m average error, the value of the deviation between the two subsequent runs is definitely smaller than a deviation that can be obtained with conventional methods and without a dGPS-based vehicle navigation or steering aid.

Conclusions

A dGPS-based navigation aid was found to be suitable for positioning the tractor and implement combination during headland turning and especially for the adjustment of the working width and for the designation of the straight line to be followed while working. Consequently the average error between two subsequent runs was found to be 0.16 cm that is an acceptable value for the steering. Such an error can be determined both in the overlapping and in the gap between the working widths joining to each other.

References

Fekete, A., Foldesi, I., Kovacs, L. 2001. Automatic control aspects of precision farming. Bulletin of Hungarian Institute of Agricultural Engineering, Godollo, Vol. XL. 34.

Fekete, A., Foldesi, I., Kovacs, L. 2003. Report on the advancement on the development of automatic control aspects of site specific farming. SzIE University report, Budapest, 30.

Acknowledgements

The work reported herein was sponsored by the National Research and Development Program no. 00303/2001.

Yield and soil variability

A. Fekete[1], I. Foldesi[2] and L. Kovacs[2]
[1]SzIE University, Physics & Control Department, 1118 Budapest, Hungary
[2]Hungarian Institute of Agricultural Engineering, 2101 Godollo, Hungary
h10835fek@ella.hu

Abstract

The objective of the work reported herein was to determine relationship between soil nutrient content characteristics and yield for three experimental fields. Furthermore the objective was to determine soil fertility by a definite transfer characteristic. The variations in the soil nutrient content and in the yield were analyzed for the experimental fields. The frequency function of the nutrient content was determined. Soil fertility coefficient was introduced to describe an element of the soil-crop system for a definite management zone.

Keywords: nutrient content, yield, variability, fertility coefficient

Introduction

There is a definite variability in the soil characteristics across the field. This variability is of high importance for appropriate use of variable application rate control with sprayers and fertilizers. Therefore is a need for appropriate characteristics to describe the variability. The objective of the work reported herein was to determine relationship between soil nutrient content characteristics and yield for three experimental fields. Furthermore the objective was to determine soil fertility by a definite transfer characteristic.

Materials and methods

The experimental fields were relatively flat and of loamy soil. The total area of three fields was approximately 300 ha. These fields were divided into quasi-homogenous management zones for soil sampling. The border lines of these management zones were determined on the basis of soil type and yield maps. The principle of sampling was to take two samples from the soil each management zone along two diagonal straight lines of the zone. Approximately 20 sample parts were mixed and composed a single sample of 500 g where the sampling depth was 0 to 30 cm. The soil nutrient content characteristics determined were as follows: nitrate-N content, organic matter content, P-content and K-content. The expected value, the standard deviation and the coefficient of variation were determined for each nutrient content and field. The nutrient content range was divided into ten categories. The same characteristics were determined for the yield as well.

Results and discussion

The soil-crop system of the management zones were assumed as elementary systems. Therefore each management zone was taken as a soil-crop element that has a transfer characteristic. This characteristic was determined as the ratio of the output and input parameter of the element. Soil nutrient content was taken into account as input variable and the yield as output variable. Frequency functions of the soil nutrients were determined and used to describe the variations across the fields. The coefficient of variation was found to have a relatively high value for the experimental fields (Figure 1).

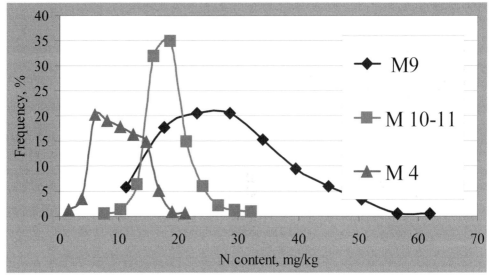

Figure 1. Frequency function of soil N-nitrate content for the experimental fields.

Soil fertility coefficient was introduced to describe the soil-crop system by the transfer characteristic of the system that is the ratio of the yield to a definite soil nutrient content property, such as N-nitrate, P, K, etc. This transfer characteristic is a principal characteristic of a definite element, or rather as the fertility coefficient of the soil-crop element for the definite management zone. Consequently the soil-crop elements as management zones can be characterized by the soil fertility coefficient.

Conclusions

Frequency function was determined for the experimental fields and the soil fertility coefficient was introduced to describe an element of the soil-crop system for a definite management zone. The soil fertility coefficient can be used for the better characterization of soil fertility of a definite management zone and therefore the use of this coefficient can encourage variable application rate strategies. Furthermore multivariate correlations among the soil nutrient characteristics should be determined and new soil nutrient characteristics should be introduced for the development of a multipurpose expert system.

References

Fekete, A., Foldesi, I., Kovacs, L. 2002. Variations in the nutrient content within the fields. International Conference of AgEng, Budapest, 1-7.
Fekete, A., Foldesi, I., Kovacs, L. 2003. Report on the advancement on the development of automatic control aspects of site specific farming. SzIE University report, Budapest, 30.

Acknowledgements

This work was sponsored by the National Research and Development Program (00303/2001).

Precision agriculture system for sugarcane in Cuba

Farnia Fernández[1], Britaldo Hernandez[1], Enrique Ponce[1], Lazaro Quintana[1], Miguel Esquivel[2] and Jose Julio Rodriguez[3]
[1]Automation and Control Laboratory, [2]Projects Division, [3]Computer Laboratory, National Center for Laboratory Animal Production, CENPALAB, Finca Tirabeque, Bejucal, Havana, Cuba
briti@cenpalab.inf.cu

Abstract

Sugarcane fields represent more that 50% of the total agricultural land in Cuba. In 2001 1 100 000 ha of sugarcane were harvested. It is also one of the main cash crops. The design of a precision agriculture (PA) system for sugarcane in Cuba was conceived in five main steps. First, a national Differential Geographical Positioning System service was established, on the basis of several base stations that transmit a differential correction signal. So far, two base stations have been installed on Havana and Holguín provinces. The second step was to study the costs structure of sugarcane production. This allowed us to determine the technical and economic factors of highest incidence, these being soil preparation, sowing, fertilizing, weed control and harvest. On this basis, it was decided to install a DGPS receptor on a tractor, combined with a microcontroller system (AGROMATIC). This tractor could be hooked up to several soil preparation implements, as well as to a fertilizer and herbicide sprayer. Both the fertilizer and sprayer were previously transformed from mechanical to hydraulic systems, and further automated, giving the possibility to be controlled by AGROMATIC. The third step includes the establishment of communication protocols and data transfer between all components, as well as the necessary management software. The fourth step is the implementation of a completely reorganized agricultural management system. Finally, step five is a pilot test of the whole system in a field of 130 ha, with a set of tractor, fertilizer, sprayer and sugarcane harvester, to demonstrate that the technological system created allows the management of sugarcane production with a precision agriculture system.

Introduction

Sugarcane cultivation is the most important agricultural activity in Cuba from the economical point of view. The need to increase the efficiency in sugarcane yields makes it necesarry to introduce modern technologies. The pesent work outlines the Precision Agriculture (PA) System for sugarcane cultivation in Cuba, based on the existing realities and possibilities.

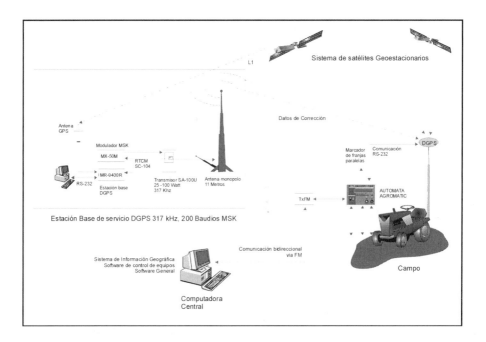

General diagram of precision agriculture technology employed.

A DGPS base station is used, located aproximately 30 km. from the sugarcane field, which transmits differential correction signals to the DGPS receiver installed in the tractor, where the rest of the electronical system is. In this manner, one tractor is able to work with different agricultural implements such as fertilizers, sprayers, and soil preparation implements. The heart of this system is a microcontroller-based unit, designed for this purpose, which we named AGROMATIC.

The other components of the PA system are:

1- Yield monitor designed and installed in a cuban KTP 3S sugarcane harvester.

2- Transformation of the FT 350 Fertilizer for variable-rate application.

3- Adaptation of a herbicide sprayer for variable-rate application.

4- Creation of a sofware for controlling agricultural implements.

5- Creation of a Geographical Information System.

Contribution to the design of an autopilot for tractors: considerations on the set-up of the system

A. Formato[1], S. Faugno and G. Paolillo[1]
[1]Department of Agricultural Engineering and Agronomy –University of Naples- Italy
formato@unina.it

Abstract

An economic compact equipment and easily to use by the operators has been realized. It allows to check the direction of a tractor with an error of the order of the decimetres. It is formed by a rudder with two pressure sensors connected to the hydraulic power steering of a tractor. Such system can be also installed as "retrofit" on different types of tractors and it can be used or in way "stand alone" or in combination with other systems of autonomous guide.

Keywords: mechanical actuator, directional control, servomechanism

Introduction

One of the demands very felt in agriculture it is that to redeem and to assist the tractor driver during the long and monotonous field operations, this with the purpose to improve the safety and the job conditions. Nowadays many devices of automatic guide are available, they are generally used in aeronautic, naval and transports sectors. Such systems have been used also in agricultural machineries with satisfactory results, even if they are very expensive and they require for their use, high-levels of competence and this blocks their ampler diffusion. Insofar we decided to study a device that satisfies the aforesaid demands.

Materials and methods

To achieve the appointed purpose any servomechanisms that make reference to soil guides have been considered. They can eventually be integrated also in systems of automatic guide with GPS devices. Insofar a mechanical actuator has been installed on the power steering system of a tractor, and it is droved by a signal that is originated from pressure sensors located on the rudder set on the soil (figure 1).

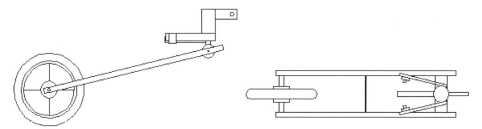

Figure1. Directional rudder.

The rudder is located in the back part of the tractor and it has two pressure sensors on both sides, indeed if there is a change of direction, a pressure difference between the two faces is generated, such signal is detected by an hydraulic system that acts on the power steering, which corrects the direction of the tractor wheels until to annul the pressure difference on the two sides of the rudder and, therefore, to annul the deviation. For a first implementation of the used system, the directional signal has also been compared with that furnished by a compass of fluxgate type and the two complementary signals have been registered and constantly compared by an electronic system. The compass has been located in advanced position in respect of the tractor on a wood plinth set aloft away from the magnetic masses of the tractor. Also the compass sends the signal to the hydraulic power steering, and therefore also it performs corrections of the direction. The compass has a sensibility of 5 degrees, and this value causes not acceptable maximum errors during the soil workmanship, while the signal furnished by the rudder results very more precise. The strength agent on the two faces of the rudder can be modelled with a second order equation:

$$F = m\frac{d^2x}{dt^2} + c\frac{dx}{dt} + kx \tag{1}$$

where F is the applied force to be measured and that acts on the pressure transducers. The damping can be adjusted to obtain fast responses without oscillatory behaviour. The used sensor is of piezoresistive type with low histeresis value, and it is able to evaluate strength until 50 N. One important specification of the sensors is the minimum thrust that can be measured.

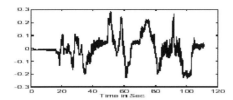

Figure 2. Raw force.

The problems of greater importance in the operation of the considered servomechanism concern the rapidity of answer, the stability of the infinite loop, the sensibility to the variations of load and the troubles overlapped to the signal, the required power to the alimentation, the maximum levels of signal that can present, the operation under exorbitant conditions from limits of linearity.

Discussion and conclusions

The error values obtained with the rudder and with the compass have been evaluated and compared. The system has given good results for variations of direction in the range of 10-15 deg. The data results values obtained with the system of control used with those theoretical imposed (straight line) have been compared, noticing the maximum error of 10-15 cm. The equipment has given satisfactory results in soils in which there were not many stones or elements very hard to induce numerous deviations. Insofar such system results reliable, not expensive and can be also applied on existing agricultural machines as retrofit.

References

Balsari,P. 1999 Actual available technologies for precision agricolture. Proceedings of conv.AIIA Grugliasco (TO)-Italy-22-23 giugno.

Blackmore B.S. 1996. An information system for precision farming. Proceedings of the conference on pests and diseases Brighton, U.K.

Morgan M. D. Ess 1997 The precision farming guide for agriculturist. An agricultural primes series created by Deere&Company JohnDeere publishing Illinois USA.

Scheding, E. M. Nebot , M. Stevens and H. Durrant-Whyte, J. Roberts and P.Corke, Cunningham and B. Cook, "Experiments in Autonomous Underground Guidance", IEEE Conference on Robotics and Automation, 1997, pp. 1898-1903.

Durrant Whyte H., "An autonomous guided vehicle for cargo handling applications", Int. Journal of Robotics Research, 15(5), pp. 407-441, 1996.

An economic assessment of site-specific and homogenous nitrate fertilization management in consideration of environmental restrictions

Markus Gandorfer, Andreas Meyer-Aurich and Alois Heissenhuber
Chair of Agricultural Economics and Farm Management; Technische Universität München/ Weihenstephan
markus.gandorfer@wzw.tum.de

Abstract

In recent studies only little or no economic benefits and economic potentials of site specific nitrogen fertilization management techniques could be found. Therefore, the approach of this simulation is to demonstrate the relationship between ecological purposes and their economic consequences. It can be shown that with variable rate application of nitrogen fertilizer environmental restrictions (e.g. nitrate concentration in seepage water) can be reached with lower abatement costs.

Keywords: marginal abatement costs, nitrogen leaching, environmental restrictions.

Introduction

Ecological benefits of variable nitrogen application in contrast to uniform cultivation are commonly calculated at profit max N-rate. In this study, the economic superiority of precision farming techniques in contrast to common cultivation techniques in case of applying legislative environmental restrictions is to be demonstrated.

The **hypothesis** is: "A specific threshold (e.g. 50 ppm NO_3/l seepage water) can be reached at lower costs with variable rate application than with uniform treatment".

Material and methods

For simulation purposes, a model field with two specific yield/management zones is created. The main characteristic to differentiate these management zones is the average field capacity of rooted soil layers (high yield zone: 338mm; low yield zone: 207 mm). For each management zone quadratic yield response functions to nitrogen ($y = a + b\,N + c\,N^2$) for winter wheat have been estimated (fig. 1). Experimental data (2002) from Freising-Thalhausen, Bavaria were used. Potential nitrate leaching into groundwater is treated as a function of soil type (field capacity), climate and nitrogen balance. Potential nitrate in seepage water (NO_3 i.S.) can be calculated according to equation 1 (OSTHEIM, 2000):

$$NO_3 \text{ i.S.} = ((Nbil * AF) / S) * 4{,}43 * 100 \tag{1}$$

(Nbil: nitrogen balance (kg N/ha*a), AF: leaching factor S: seepage water (mm/a))
Seepage water (S) is estimated for every single day according to RIESS (1993):

$$S = N\text{-Etpot} + V \tag{2}$$

(S = seepage water, N = rainfall, ET_{akt} = actual evapotranspiration, V = change in water storage) The meteorological input parameters are daily rainfall, average air temperature and vapor pressure.

To compare both systems (site specific nitrogen application and common cultivation), returns above fertilizer costs (€/ha) (using site specific response functions) are optimized as the objective function of a linear optimization (MS EXCEL SOLVER). Potential leaching water has been considered as constraint.

Results

The site specific response functions indicate a strong heterogeneity of the model field (fig.1). Response functions have been estimated with Space.Stat (econometric model: Ordinary Least Square), but no significant spatial dependence could be found. The marginal abatement cost function of VRT shows a smaller slope, therefore it can be concluded that with precision farming environmental targets can be reached with lower abatement costs than with uniform application techniques (fig. 2). Results also show that uniform rate application has great potentials to reduce groundwater N pollution without having large negative economic effects. The price of reducing 28 ppm NO₃/l seepage water is only 5 €/ha with uniform rate application (fig. 2) in that particular case.

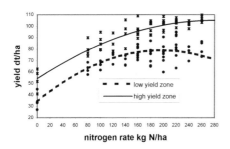 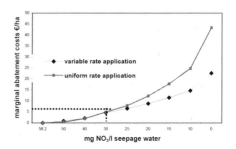

Figure1. Nitrogen response functions for winter wheat using experimental data from Thalhausen 2002

Figure 2. marginal abatement costs for potential nitrate concentration in seepage water

Discussion and conclusions

Potential nitrate concentration in seepage water has proved to be a sensitive indicator to compare variable and constant rate techniques. On the basis of this indicator has been shown that environmental restrictions can be reached with lower abatement costs with precision farming techniques. The effect of lower marginal abatement costs with VRT counts only on a very restrictive threshold of 25 ppm NO₃/l seepage water (fig 2). It can be concluded that uniform cultivation systems have also great potentials to reduce nitrogen leaching into groundwater if the objective is not only to maximize profit.

References

Ostheim, K-W. (2000): Prüfung der ökologischen Vorzüglichkeit einer GPS-gestützten teilflächenspezifischen Landbewirtschaftung. Landwirtschaftsverlag GmbH, Münster-Hiltrup.

Rieß, F. (1993) : Untersuchungen zur Nitratauswaschung nach mineralischer und organischer Düngung von Ackerland und Grünland mittels der Saugkerzenmethode. Diss., TU-München-Weihenstephan.

Acknowledgement

"This project is funded by the "Deutsche Forschungsgemeinschaft-DFG"

MOSAIC: Spatial representativity of mineral soil nitrogen monitoring

A. Giebel[1], O. Wendroth[2], H.I. Reuter[2], K.C. Kersebaum[3], J. Schwarz[1]

[1]*ATB, Institut for of Agricultural Engineering, Max-Eyth-Allee 100, 14469 Potsdam, Germany*
[2]*ZALF, Institute for Soil Landscape Research, Eberswalder Str.84, 15374 Muencheberg, Germany*
[3]*ZALF, Insitute for Landscape Systems Analysis, Eberswalder Str.84, 15374 Muencheberg, Germany*
agiebel@atb-potsdam.de; owendroth@zalf.de

Abstract

Spatial variability pattern of mineral soil nitrogen is known to differ within the same field site in different years. To achieve a better understanding of spatial variation structure and association between different years was the main objective of this study. Spatial variation of mineral soil nitrogen N_{min} after harvest was examined in three successive years. An upper limit of sampling cell size was calculated. Standardized variograms and covariograms were compared between different years. In order to reduce the sampling efforts cokriging was used with common standardized models for estimation of spatial variability of N_{min} in successive years.

Keywords: mineral soil nitrogen, spatial variability, cokriging

Introduction

Mineral soil nitrogen is an important state variable reflecting mineralization status at the time of sampling. It provides information on soil nitrogen available for crop uptake or leaching within a limited period of time. Therefore, nitrogen fertilization recommendations have been widely based on mineral soil nitrogen content. A remaining question is how the spatial distribution of mineral soil nitrogen in the field can be determined and how it is related between different years. Moreover, is it possible to estimate it with reduced amount of samples? The objective of our study was also to determine the range of spatial representativity of mineral soil nitrogen and to apply a common model for the estimation of spatial variability in successive years.

Material and methods

The sampling area was a 19 ha field of the Südzucker AG farm in Lüttewitz, southeastern Germany, Saxonia. The soil is a Luvisol. From 1998 to 2001, soil samples were taken at 192 locations (depth 0-60 cm). Mineral soil nitrogen N_{min} was analysed after harvest. The sampling scheme was a regular grid with a separation distance of 27 m. At five locations nested samples were taken. Geostatistical methods were used to determine the spatial distribution of N_{min}. The upper limit of cell size is calculated as mean correlation distance (MCD), proposed as a practical sample size based on soil variability by Han et al. (1994). For the estimation of N_{min} in successive years we used a common standardized variogram and covariogram model with all values from the first year, 50% (8 samples ha^{-1}) and 25% values (4 samples ha^{-1}), respectively, of the estimated years. The MAD (median absolute deviation) between measured and estimated values yielded the precision of estimations. The difference between classes (4 percentiles) of measured and estimated sampled grid cells was used to judge whether the values were similar or not.

Results and discussion

In the first two years average N_{min} values were between 1.3 and 1.52 mg (100g soil)$^{-1}$ with standard deviations between 0.49 and 0.57 mg (100g soil)$^{-1}$. In the following two years N_{min} concentration was only 50-70% of that in the previous years and the standard deviations were lower. However, coefficients of variance were similar between all years (34-41%).

The spatial variability of N_{min} differed with respect to the magnitude of the sill (data variance) and the spatial correlation length (range). In those years when the magnitude of semivariance was large, the range was large as well, and vice versa. Approximately 50% of the sample variance (C) was spatially structured in each year regardless of the magnitude of the variance. The other 50% (C_0) were caused by the short scale variability of N_{min}, sampling and analytical errors. The MCD was between 9 and 16m. Hence, standardized semivariograms and covariograms were calculated and for both, common models were derived (Table 1).

The precision of estimation (MAD) ranged between 0.184 and 0.346 mg (100g soil)$^{-1}$. Especially with extreme values present, the absolute deviation was high. If the N_{min} content was grouped into 4 classes approximately about 85% (when 8 samples ha^{-1}) or 75% (when 4 samples ha^{-1}) of estimated values are in the same or successive class like analysed values (Table 2).

Table 1. Standardized variogram parameters of spatial variability of N_{min} ; common variogram and covariogram for the estimation of N_{min} in successive years.

Year	Grown	Data Variance (mg/100g soil)^2	Standardized variograms spherical model		
			C	C_0	Range
1998	Winterwheat	0.208	0.620	0.548	80 m
1999	Summer barley	0.345	0.523	0.523	92 m
2000	Winter ray	0.055	0.473	0.552	52 m
2001	Winter ray	0.097	0.424	0.369	37 m

Common Variogramm	⟹ C_0=0.50	C=0.5	Range=80m
Common Covariogramm	⟹ C_0=0.05	C=0.45	Range=80m

Table 2. Absolut frequency of class differences and MAD between measured and estimated N_{min}.

Class difference	Estimated with					
	8 samples ha^{-1}			4 samples ha^{-1}		
	1999	2000	2001	1999	2000	2001
Absolut frequency %						
0 classes*	56.8	52.6	64.9	46.9	34.4	51.1
1 classes	36.5	32.3	24.1	37.5	42.7	25.8
2 classes	5.7	13.0	8.9	13.5	15.6	17.4
3 classes	1.0	2.1	2.1	2.1	7.3	5.8
MAD mg (100g soil)$^{-1}$	0.324	0.184	0.194	0.346	0.200	0.206

* with input samples

Conclusion

N_{min} content after harvest showed similar spatial variability structure regardless of the magnitude of the variance in different years. The years were different in magnitude of variance and range. The calculated upper cell size limit for sampling was 9 to 16 meters. Cokriging based on standardized N_{min}, a unique variogram and a covariogram exhibited promising results for the prediction of variability patterns under reduced sampling density.

Acknowledgements

The Support of this project from German Research Foundation (DFG, Bonn), Südzucker AG and Agrocom is gratefully acknowledged.

References

Han, S., Hummel, J.W., Georing, C.E., Cahn, M.D. 1994. Cell Size Selection for Site-Specific Crop Management, Transactions of the American Society Agriculture Engineers, 37:19-26.

Delimiting intra-field zones from yield and soil maps

J.M. Gilliot, A. Jullien, P. Huet and J. Michelin
UMR EGC, INA P-G INRA, 78850 Thiverval-Grignon BP 01, France
Jean-Marc.Gilliot@grignon.inra.fr

Abstract

Delimiting intra-field zones is an important step in Precision Agriculture, that may lead to the creation of management zones. Yield variations, associated with terrain conditions, are due to time stable factors (soil) and time variable factors (climatic). When trying to analyse this variability from yield maps, several years have to be considered. A method, combining several years of yield maps in a fuzzy logic approach, tries to map stable and unstable intra-field zones. Zones are compared with aerial photography and with soil electrical resistivity map.

Keywords: yield map, soil map, intra-field zones, fuzzy logic, electrical resistivity

Introduction

Various approaches have been proposed for field segmentation from yield data. Methods combining data from several years have been introduced to isolate the effect of terrain conditions (Blackmore, 1999).

Materials and methods

The study was conducted on a 15 ha field in the north of France (Grignon, Yvelines). The "pont cailloux" field presents various topographic situations and soil compositions. With altitude ranges between 90 to 125m, and maximum slope of 20°. Calcareous soils in the north of the field, chalk soils in the south and presence of few clay soils and colluvial soils. Five years of yield maps were used : 1998 (wheat), 1999 (winter barley), 2000 (corn), 2001 (wheat), 2002 (wheat). To be compared, yield data have been normalised (SNV : Standard Normal Variable) (1).

Figure 1. Field localisation.

$$SNV = \frac{yield - \overline{yield}}{\sigma} \quad (1)$$

1998 (wheat)   1999 (winter barley)   2000 (corn)   2001 (wheat)   2002 (wheat)
Figure 2. Normalised yield Maps (SNV).

Yield data have been interpolated (nearest neighbourhood) on a 2.5m resolution grid, with the ArcGis GIS system, a spatial smoothing filtering was applied on the resulting map.
Membership functions (MF), based on fuzzy logic, were used to define classes corresponding to low, medium or high yield level, as proposed by Panneton (Panneton, 2002). Decision rules

411

were elaborated to combine data of different years and to define stable / unstable zones and high / low level zones.

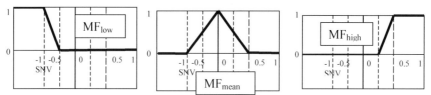

Figure 3. Membership functions of the three considered classes.

Results

The utilised method permit to produce simplified maps, from a yield maps serie, taking into account yield stability over years. Resulting maps will be compared with aerial photography and with soil electrical resistivity map. High correlation has been found between black and white aerial photography grey level and soil electrical resistivity (figure 5)

Figure 4. A 12 meter resolution resulting map.

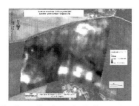
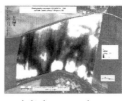
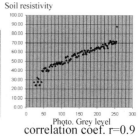

soil electrical resistivity map (0-50cm)   aerial photography   correlation coef. r=0.9

Figure 5. Comparison between resistivity and aerial photography.

Conclusions

To obtain more realistic intra-field zones for field management, a better knowledge of the spatial heterogeneities is necessary. Combining several years of data could help to understand the importance of these heterogeneities on the crop in various conditions.

References

Blackmore, S., Godwin, RJ, Taylor, J.C., Cosser, N.D., Wood, G.A. 1999. Understanding variability in four fields in the United Kingdom. In Proceedings of the fourth international conference on precision agriculture 3-18.

Panneton, B., Brouillard, M. and Piekutowski, T. 2002. Gestion de series pluriannuelles de cartes de rendement. Canadian biosystems engineering vol. 44 7.23-7.29.

Acknowledgements

This work has been supported by the agriculture precision program of the french institute for Agronomy research : INRA (Institut National de la Recherche Agronomique).

Assessment of 10 years commercial nutrient maping

S.J. Griffin
SOYL Ltd, Red Shute Mill, Hermitage, Berks, RG18 9QU, U.K.
simong@soyl.co.uk

SOYL Ltd are a precision farming company that was established in 1993 to provide nutrient mapping services for P, K, Mg and pH. SOYL also make GPS variable application fertiliser plans based on these maps. To date over 200,000 hectares of agricultural land throughout the U.K. has been mapped. Approximately 60% of this land has had Phosphorus and Potassium fertiliser applied on the basis of the nutrient maps. After initial soil sampling there is a period of 4-6 years of variable application of P and K. After this fields are remapped. This involves sampling at the same locations to monitor nutrient levels. Over 20,000 hectares have been resampled in this way.

This means there is now available a large body of data that can be examined to assess the benefits of commercial P & K mapping services. Commercial nutrient mapping is typically based on samples taken every 100 metres (one per hectare). This approach is borne out of a necessity to keep costs to an affordable level. However it has attracted much criticism as being beyond the range of spatial dependence and not suitable for producing interpolated maps.

Therefore the first question to be addressed is has soil sampling at this intensity produced useful nutrient maps?

Sampling data taken from the farms containing 100 soil sample points or more, is used. The data from each farm is used to construct a variogram. Each farm has data for P and K analysed. For each variogram the analysis includes the range of the variogram, the Model used and nugget variance. The data was analysed using Vesper software, produced by the Australian Centre for Precision Agriculture.

Phosphorus

Farm	Ha	Range (m)	Model	Nugget	Soil Type
Essex	253	342	Spherical	99	Clay
Sussex	283	550	Circular	28	Sandy Loam
Norfolk	165	463	PSpherical	25	Chalky Till
Borders	214	380	Circular	35	Variable
Berkshire	190	299	Spherical	81	Clay&Flints
Hereford	100	481	Circular	13	Silty Clay BEarth
Oxon	170	464	Circular	19	Rendzina

Potassium

Farm	Ha	Range (m)	Model	Nugget	Soil Type
Essex	253	286	Exponential	1130	Clay
Sussex	283	313	Circular	442	Sandy Loam
Norfolk	165	417	Circular	970	Chalky Till
Borders	214	493	Spherical	521	Variable
Berkshire	190	350	Spherical	1039	Clay&Flints
Hereford	100	352	Circular	1019	Silty Clay BEarth
Oxon	170	443	PSpherical	147	Rendzina

The range for both P and K is above 100 metres and therefore the sampling interval intensity of 1 per hectare on the **farm scale** may be justified.

The next question is does variable application of fertiliser work?

The 20,000+ hectares of remapping allows an assessment of the success of the variable application of P & K fertiliser over the five year period between initial sampling and resampling.

The objective of any P & K policy is to build nutrient poor areas to a point at which they are no longer deficient and limiting to optimum yield. In the U.K there is an index system used for nutrient management. The optimum index for Cereals is Index 2. If nutrient levels are above that which is necessary for optimum yield (index 3+), then fertiliser rates can be reduced and nutrient levels run down to a maintenance level (index 2). The calculation of these build up and run down rates are vital to a successful P & K policy. If the variable application of fertiliser has been a success then the range of nutrient values in should have narrowed. A survey of 3,750 ha showed this to be the case, low levels were increasing but high levels did not seem to be falling as quickly, suggesting many of the fields were buffered in some way.

% of all samples in each index range before and after variable fertiliser application

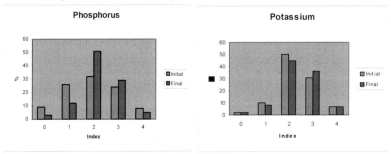

Average within field variation before and after variable application of P and K

Nutrient	Initial Range (mg/l)	Final Range (mg/l)	Initial sd	Final sd
Phosphorus	27.8	21	7.5	5.9
Potassium	165	123	45.9	35.3

These figures show that within field P and K variation within field has reduced after variable application of fertiliser based on nutrient mapping

References

Minasny, B. McBratney,A.B. and Whelan, B.M. 2002. VESPER version 1.5, Australian Centre for Precision Agriculture, McMillan Building A05, The University of Sydney, NSW 2006

Spatial variability of mineral nitrogen content in topsoil and subsoil

J. Haberle, M. Kroulík, P. Svoboda, J. Krejčová and D. Cerhanová
Research Institute of Crop Production, Prague-Ruzyně, Czech Republic

Abstract

The content of soil mineral nitrogen (N_{min}) in a soil profile down to 60 cm and 90 cm was determined in 19-ha experimental field in three years. We found similar variability of N_{min} content in top and subsoil but the relation between N_{min} content in the zones was mostly poor. Observation of apparent nitrate N depletion confirmed that wheat utilizes N supply from the depth at least 90 cm that corresponds with observed rooting depth. Geostatistical analysis showed a poor spatial dependence between N_{min} in sampling points.

Keywords: available nitrogen, spatial variability, subsoil, rooting depth, winter wheat, oats

Introduction

Accounting not only for spatial variability of mineral nitrogen in crop production but also for variable distribution of N in a soil profile has several promises. Besides optimizing N doses with aim of maximum yield and/or maximum nitrogen use effectiveness, a good quality of products is often decisive for a high profit. For example the supply of nitrate leached to a deep subsoil and utilized to the end of growth may negatively affect the quality of sugar beet or malt barley, and positively the quality of wheat grain. Further, decreasing the risk of nitrate leaching by reducing the amount of residual nitrogen at autumn became priority.

Materials and methods

Soil mineral nitrogen content N_{min} (NH_4-N, NO_3-N) was determined in 19-ha experimental field in Prague-Ruzyně, Czech Republic (Lipavský et al., 2002). Winter wheat was grown in the field in years 2000 and 2001, oats in 2002. Soil from sixty locations (45 in autumn 2000) was sampled from a topsoil layer 0-30 cm and a shallow subsoil layer 30-60 cm at spring 2001 and 2002 and at autumn 2000 and 2002. Deep subsoil layer 60-90 cm and changes of N_{min} during crop growth at twelve selected locations were monitored to estimate the distribution of apparent depletion of N from the soil profile. Root length distribution and maximal rooting depth of the crops were determined in four locations as an indicator of potential depth of nitrogen depletion. The locations were selected to represent different soil conditions. Obtained data were evaluated by standard statistical and geostatistical methods.

Results

The variability of nitrate N in top soil and subsoil (30-60 cm) ranged between 19-39 % and 27-37 %, resp. It was mostly the same or slightly greater in subsoil than in topsoil. The variability of ammonium N in top soil and subsoil ranged between 17-60 % (the second highest being 40 %) and 21-25 %, resp. It was similar at top- and subsoil, except for autumn 2000.

There was a positive relation between nitrate content in top and subsoil in all sampling terms (r= 0.65, 0.27, 0.62 and 0.40, resp.). We found a weak or none such a relation in ammonium nitrogen content (r=0.23, 0.14, 0.07, 0.03) and in relation of ammonium N to nitrate N in the same layer (r= -0.27 to +0.31; mostly between -0.11 to +0.10). The content of N_{min} and nitrate in top soil correlated strongly with total N_{min} in layer 0-60 cm in 2000, but poorly in other experimental years. Roots of both cereals penetrated to the depth of at least 90 cm. The apparent depletion of readily available nitrate N from top and subsoil is shown by progressive decrease from the start of fast growth of winter wheat on 24.4.2001.

Spherical model of variogram was used for approximation of obtained data. The relation between values of mineral N and values which were predicted by cross validation was poor (R^2 = 0.02 to 0.15). Pearson's test of linearity showed similar results (r = -0.17 to +0.37).

Table 1. The content of nitrate N (in mg N/1 kg dry soil).

Term of sampling	Depth (cm)		
	0-30	30-60	60-90
24th April	4.4	4.7	7.9
9th May	2.3	1.6	7.3
13th June	2.8	1.8	2.2
21st November (tilled)	8.7	3.1	1.7

Discussion

Our results on the spatial variability of subsoil N_{min} and of apparent depletion of nitrate N from a soil profile suggest that the factor may contribute to variability of growth of the crops. Experiments in near fields proved that wheat utilizes N supply down to at least 90 cm and even deeper layers are depleted if plant demand is not covered by (fertilized) top layer (Svoboda et al., 2000). However, geostatistical analysis showed that predicting N_{min} content for construction of application maps poses a problem. Plant itself is the best indicator of nitrogen availability supposing the effect of other growth limiting factors can be reliably estimated. N status of a crop reflects nitrogen supply of layers that were already occupied and it may not be a good predictor of N supply for the rest of growth period. Evidently, extensive sampling of subsoil for N is impracticable and unprofitable. One possibility is to combine mineral nitrogen sampling at selected reference points with knowledge of empirical relation of soil N content at the rest of a field. Mathematical models (Haberle& Svoboda, 2000) offer methods to account for interacting factors which affect distribution and availability of nitrogen in a soil and its utilization for yield. Especially understanding interaction of N and soil water (Haberle et al., 2002) may improve interpretation of yield variability in a field.

References

Haberle, J., Svoboda, P.2000 The simulation of nitrogen depletion in winter wheat crop. In: 3rd International Crop Science Congress, Hamburg, p. 152.

Haberle, J., Svoboda, P. and Trčková, M. 2002 Available supply of soil water in relation to development of root system of wheat. In: Ecophysiology of plant stress. Proceedings of the 5th International Conference, Nitra, pp.158-159.

Lipavský, J., Slejška, A., Kroulík, M.,Haberle J. and Baierová, V. 2002 Spatial variability of nutrients content in soil and winter wheat plants. In: Book of Proceedings, VII ESA Congress, Cordoba, Spain, pp. 639-640.

Svoboda, P., Haberle, J. and Krejčová, J. 2000. The relation between nitrogen supply in rooted soil volume and plant nitrogen status. In: 3rd International Crop Science Congress, Hamburg, p. 166.

Acknowledgements

The project was supported by MZe ČR research plan M01-01-01.

416

Strategy to identify soil parameters influencing spatial variability of soybean yield in a small field in Japan

C. Hache, S. Shibusawa, A. Sasao and K. Sakai
Tokyo University of Agriculture and Technology, Agricultural and Environmental Systems Engineering, Fuchu-Shi, Saiwai-Cho 3-5-8, 183-8509 Tokyo, Japan
sshibu@cc.tuat.ac.jp

Abstract

Soil parameters influencing soybean yield were identified after dividing the field according to productivity in the different areas as well as combining partial least square (PLS) and multilinear regression (MLR) analyses.

Keywords: soil, production areas, spatial variability, PLS, multilinear regression.

Introduction

"Obtaining the maximum production potential of a particular crop depends on the environment during the growing season and the skill of the producer in identifying and eliminating or minimizing those factors that reduce yield potential" (Havlin *et al.*, 1999). To identify those factors several approaches are being applied in precision agriculture.

Materials and methods

This research was conducted in a small field at the university. The field presented a slope from West to East, favoring the movement of water towards East (Figure 1a). The experimental design corresponded to eight plots receiving two types of management (*i.e.* chemical-fertilizer and manure). In June 2002 soil reflectance was collected every 1m with Real-Time Soil Spectrophotometer (Shibusawa *et al.*, 2000), as well as soil sampling every 5m for laboratory analysis. Using reflectance results for moisture content (MC) and soil humic fraction, and laboratory results for several other parameters, soil spatial variability was understood. Later, soybean was planted and forty measurement points were positioned in the field (Figure 1b), matching with the lines where soil experiment was conducted. At the end of the season, yield corresponding to $1m^2$ around each measurement point was collected.

Results

Highest yield surpassed lowest production in 460%. A yield map was created and classified according to Central Limit Theorem (Figure 1c). The map clearly shows high, average and low production areas, regardless of experimental design, where high yield corresponded to higher elevation zone. Soil samples and reflectance locations matching with yield measurement points were selected to generate a database, creating three datasets and introducing their respective models: a) whole field (n=40), b) high yield area (n=25), c) low yield area (n=15). Following, PLS and MLR analyses were applied to identify soil parameters influencing production. Figure 2 presents the algorithm used for the analysis. Both methods presented similar results, indicating that increments in MC diminished productivity. Sulfate caused a positive effect in both whole field and high yield area. Nitrate negatively influenced production in low yield area. Negative effect of nitrate could be caused by its conversion to nitrite due to excess of MC in the shallower zone.

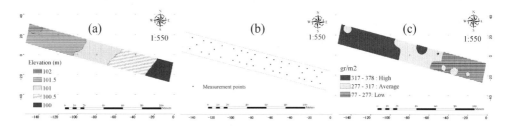

Figure 1. (a) elevation levels in experimental field; (b) locations of yield measurement points; (c) yield spatial variability map.

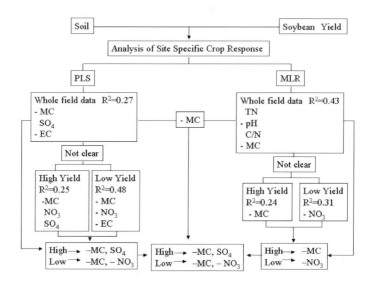

Figure 2. Algorithm used to analyze and identify soil parameters influencing yield.

Conclusions

Introduction of different datasets according to production areas, as well as the combination of multivariate analyses as PLS and MLR, allowed the identification of particular factors affecting specific zones in this field. For further research, physical and other chemical properties of the soil should also be analyzed.

References

Havlin, J., Beaton, J., Tisdale, S., Nelson, W. 1999. Soil Fertility and Fertilizers. 6th Edition. Prentice Hall, Inc. NJ 07458. USA.pp.6.

Shibusawa, S., I Made S.W., Sato, H. Sasao, A. 2000. On-Line Real-Time Soil Spectrophotometer. Presented at 6th International Conference on Precision Agriculture, July 16-19th. Minnesota, USA.

Spatial and temporal variability in the recovery of alfalfa following drought

D. Hancock[1] and M. Collins[2]

[1]*University of Kentucky, Department of Biosystems and Agricultural Engineering, 119 C.E. Barnhart Bldg., Lexington, KY 40546, USA*
[2]*University of Kentucky, Department of Agronomy, N-222 Ag. Science North, Lexington, KY 40546, USA*
dhancock@bae.uky.edu

Abstract

Alfalfa (*Medicago sativa* L.) is valuable in forage-based livestock systems in the Midwestern United States due in part to its well-documented drought tolerance. However, recent severe summer droughts in the region have caused alfalfa to respond variably within some fields. A 4 year old alfalfa stand exhibited extreme short-scale variability in productivity in areas as small as 5 to 10 m as the result of drought during 2002 followed by ample moisture supply. Samples were used to characterize yield and shoot diameter/weight, number, and height. Results indicate that visible expression of drought symptoms may not portend the productivity of the plant upon recovery for a given location.

Keywords: alfalfa, spatial variability, drought tolerance

Introduction

In addition to its high quality and productivity, the well-documented drought tolerance of alfalfa (*Medicago sativa* L.) has been of critical importance to many livestock systems in the Midwestern United States. However, recent severe summer droughts in the region have caused alfalfa productivity during the life of the crop to respond differently in various areas within a given field. This has raised concern regarding the persistence of alfalfa in these areas as affected by this prolonged moisture stress. Spatial and temporal variability in alfalfa productivity and persistence has been well documented relative to various parameters such as phosphorus, potassium, soil pH, weed competition, seedbed preparation/method, soil type, fall dormancy, snow depth, and grazing intensity (Cosgrove & Collins, 2003; Kallenbach et al., 2002; Leep et al., 2000). The spatial variability in alfalfa's response during and following drought has not been evaluated. The ability to georeference those areas within a field that respond differently to drought stress could lead to site-specific management strategies. However, this is contingent on finding a significant temporal and spatial relationship in alfalfa's response to prolonged moisture stress.

Materials and methods

A 4 year old alfalfa stand on the University of Kentucky's Maine Chance Research Farm (84° 29' 16.54" W, 38° 6' 13.54" N), was severely affected by drought from June through mid-September 2002. Extreme short-scale variability in productivity was observed in areas as small as 5 to 10 m. These spatial patterns were documented near the end of the drought period in August by mapping severely and mildly affected areas. Ample moisture supply followed the moisture stress, beginning in mid-September. Four locations within the 10 ha field that exhibited high levels of short-scale variability were randomly selected. On 18 November, four shoot samples were collected from the severely and mildly affected areas at each location, using a 0.1 m quadrat. Samples were used to characterize yield and shoot diameter, weight, number, and height. In each area, crop moisture was determined using a representative grab sample by drying to a constant weight at 60°C in a forced air dryer (72 hr). All areas within a location were at the pre-bud stage of maturity. Shoots from a quadrat sample were sorted by length into three categories: less than 15 cm, 15-30 cm, and greater than 30 cm. Within each category, total weight was assessed, shoots were counted, and 5 stems were randomly selected for stem diameter measurements using a micrometer and measured to the nearest 0.1 mm.

Results

Drought response category by location within field interaction effects were absent (P>0.05) for the parameters measured. Areas showing extreme drought response exhibited significantly (P<0.05) increased numbers (15.81 vs. 12.94 stems 0.01 m^{-2}, respectively) and weight (0.482 g vs. 0.277 g 0.01 m^{-2}, respectively) of stems less than 15 cm tall. For the other parameters of yield, stem diameter, and stem numbers within height categories, drought had no significant (P>0.05) effect. Location effects were more pronounced.

Discussion

These preliminary data indicate that within a given area, severity of drought stress expression does not affect yield, shoot diameter or shoot height. In drought affected areas, shorter stems (less than 15 cm) were of higher number and weight and contributed more to total yield than taller stems (≥ 15 cm). Variability in yield and yield components between locations superceded effects of drought. A prolonged moisture stress extending through the summer and into the fall may have longer term effects due to insufficient carbohydrate reserve buildup prior to winter dormancy. Therefore, additional measurements will be made during spring green-up to evaluate temporal effects of such prolonged drought stress.

Conclusion

Visible expression of drought symptoms may not portend the productivity of the plant upon recovery for a given location. These data suggest that spatial variability within an alfalfa field due to reasons other than moisture stress may be confused with drought effects. Longer term effects of prolonged moisture stress may still exist. Measurements during spring green-up will help to further understand the temporal effects of drought stress in alfalfa.

References

Cosgrove, D.R. and M. Collins. 2003. Forage establishment. In R.F. Barnes, C. J. Nelson, M. Collins, and K.J. Moore (eds.) Forages, Vol. I. An Introduction to Grassland Agriculture. pgs 239-261. Iowa State Press. Ames, IA.

Kallenbach, R.L., C.J. Nelson, and J.H. Coutts. 2002. Yield, quality, and persistence of grazing- and hay-type alfalfa under three harvest frequencies. Agron. J. 94 1094-1103.

Leep, R.H., P. Jeranyama, D. Warncke, D. Pennington, M. McNabney, and R. Brook. 2000. The effect of variability in soil pH and available potassium on growth of alfalfa in Michigan fields. In. P.C. Robert et al. (ed.) Proc. 5th International Conference on Precision Agriculture. ASA Misc. Publ., ASA, CSSA, and SSSA, Madison, WI.

Electromagnetic classification of cultivated soil in southern Finland

P. Hänninen[1], L. Alakukku[2], A. Jaakkola[3], M. Jakonen[4], S. Penttinen[5] and A. Ristolainen
[1]Geological Survey of Finland, P.O. Box 96, 02151 Espoo, Finland, pekka.hanninen@gsf.fi
[2]MTT Agrifood Research Finland, Soils and Environment, E-House, 31600 Jokioinen, Finland, antti.ristolainen@mtt.fi, laura.alakukku@mtt.fi
[3]University of Helsinki, Dep. of applied Chemistry and Microbiology, P.O. Box 27, 00014 University of Helsinki, Finland, antti.jaakkola@helsinki.fi
[4]Lahti Polytechnic, P.O. Box 214, 15101 Lahti, Finland, mjakonen@lpt.fi
[5]Geological Survey of Finland, P.O. Box 1237, 70211 Kuopio, Finland, sari.penttinen@gsf.fi

Abstract

In many studies EM38 (Geonics Ltd., Canada) has been used to classify cultivated soils. However, if EM38 measurements have been done 0.7 m above soil surface, the response integral is bigger from the subsurface (0.3 - 1.5 m) than the tillage layer (0 - 0.3 m). In this study the resistivity fork (GTK, Finland) has been used to measure the electrical conductivity of topsoil (about 90% of its response integral is from the tillage layer). The dielectric coefficient i.e. the soil moisture of the topsoil has been measured by percometer (Adek Ltd., Estonia). The main result of this study is that the soil moisture and the electrical conductivity have temporal variation, but the electrical conductivity of subsurface has only slightly dependent on time. Another result is that the snow thaw has the largest influence to soil moisture in northern countries. The study will be continued in order to develop methods for soil classification for precision farming purposes.

Keywords: spatiotemporal variation, electrical conductivity, soil moisture, cultivated soils

Introduction

The determination and control of the variation of soil properties affecting yield and yield quality is essential for precision agriculture. Electrical conductivity (σ_a) and dielectric coefficient (ε_r) depend on soil mineral composition, water content and soil solute electrical conductivity (Ward 1990). The aim of this study is that in many studies EM38 (Geonics Ltd., Canada) has been used to classify cultivated soils. However the response volume of EM38 reaches up to 1.5 m of depth. If EM38 measurements have been done 0.7 m above soil surface, the response integral is bigger from the subsurface (0.3 - 1.5 m) than the tillage layer (0 - 0.3 m).

Material and methods

Effects of soil physical properties are studied in Southern Finland on a clay soil (Vertic Cambisol).The electrical conductivity σ_a measured with resistivity fork (GTK, Finland), whose four electrode-Wenner system with 16-cm-spacing of the 15-cm-long electrodes (Puranen et al. 1999). Main effect (about 90%) of its response integral is from the tillage layer. Electromagnetic measurements of soil moisture are based on the dependence of dielectric coefficient on the volumetric water content (Topp et al. 1980, Hänninen 1997, Penttinen 2000). The dielectric coefficient of the topsoil has been measured by percometer (Adek Ltd., Estonia).

Results and discussion

The main result of this study is that the soil moisture (Figure 1.) and the electrical conductivity (Figure 1.) have temporal variation, but the electrical conductivity of subsurface has only slightly dependent on time. By using together EM38 and resistivity fork a two layer model of subsurface can be used to find out sectors where the topsoil's conductivity is anomalous compared to its basement.

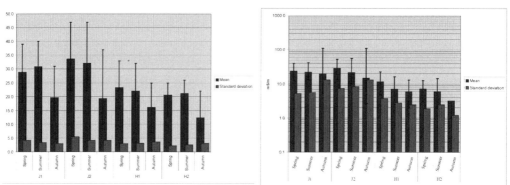

Figure 1. Range (bars), mean, standard deviation of dielectric coefficient (left figure) and electrical conductivity of resistivity fork (right figure) in different measurement time and study sites (J1 and J2 =Jokioinen 1 and 2, H1and H2 =Hausjärvi 1 and 2).

Another result is that the snow thaw has the largest influence to soil moisture in northern countries (snow cover during the winter). Measurements, done less than one month after the snow thaw, will classify the water holding capacity of the topsoil. After the snow thaw time the soil moisture does not vary, but the electrical conductivity change, which means that the ionic concentration of top soil change.

Conclusion

The study will be continued in order to develop methods for soil classification for precision farming purposes.

References

Hänninen, P. 1997. Dielectric coefficient surveying for overburden classification. Geolocical Survey of Finland, *Bulletin* 396, 72 p.
Parkhomenko, E.I. 1967. Electrical properties of rocks. Transl. And ed. By G.V. Keller. Plenum Press, New York, 314 p.
Puranen, R., Sulkanen, K., Nissinen, R. & Simelius, P. 1999. Ominaisvastusluotainet ja vastustalikot. Geologian tutkimuskeskus (in Finnish), Q15/27.4/99/2, 8p.
Penttinen, S. 2000. Electrical and hydraulic classification of forest till soils in Central Lapland, Finland. Geological Survey of Finland, *Bulletin* 398, 88p.
Pernu, T. 1991. Model and field studies of direct current resistivity measurements with the combined (half-Schlumberger) array AMM, MNB. Acta Universitatis Ouluensis. Ser A, Sci. Rev. Nat. 221, 71p.

Topp, G.C., Davis, J.L. & Annan, A.P. 1980. Electromagnetic determination of soil water content: Measurements in coaxial transmission lines. Water Resources Research, 16(3), pp 574-582.

Ward, S.H 1990. Resistivity and induced polarization methods. In: Investigations in Geophysics 5: Geotechnical and environmental geophysics. Ed. S.H. Ward, Soc. Expl. Geoph., Tulsa, pp. 191-218.

Acknowledgements

This project was funded by Finnish Government Ministry of Agriculture and Forestry.

Management zones delineation using soil electrical conductivity and topographic attributes

B. Hanquet[1], M.F. Destain[1] and M. Frankinet[2]
[1]Department of Agricultural Engineering, Gembloux Agricultural University, Passage des Déportés 2, B-5030 Gembloux, Belgium.
[2]Département Production Végétale, Centre de Recherches Agronomiques de Gembloux, Rue du Bordia 4, B-5030 Gembloux, Belgium.
hanquet.b@fsagx.ac.be

Abstract

The usefulness of soil electrical conductivity measurements (by means of electro-magnetic induction) was investigated. Several field campaigns allowed to confirm the sensitivity of such measurements to clay content in the particular soil conditions of the experiment, namely deep silty soils in Belgium. Moreover, this study permitted to test a methodology for homogenous zones delineation based on the clustering of soil electrical conductivity and topographical data.

Keywords: Management zones; Soil electrical conductivity; Cluster analysis

Introduction

Within the context of precision agriculture, the knowledge of spatial variability of soil's physical and chemical properties is essential as a decision support information for cultural operations modulation. The measurement of soil's apparent electrical conductivity (ECa) is a potential way to determine this variability. Indeed, ECa is known to be dependent of properties such as clay content, cationic exchange capacity, water content, salinity and porosity.

The purpose of this study was to asses the capability of electromagnetic induction sensing of apparent electrical conductivity for the characterisation of within-field variability of soil's physical and chemical properties. Furthermore, we evaluated the possibility of delineating management zones by means of a combination of ECa and topographic characteristics, following a method inspired by FRAISSE et al. (2001).

Materials and methods

The study was conducted on an experimental field of 7 ha located in Sauvenière (Belgium). ECa measurements were performed in May '01, September '01, April '02 and September '02 using a Geonics EM38 sensor. The ECa data acquisition was coupled with a 12-channel DGPS, allowing to import the data set in a GIS (ArcView), in order to create accurate electrical conductivity maps. Topographic attributes of the field were computed on the basis of the DGPS survey.

In order to determine homogenous management zones, an unsupervised classification (cluster analysis) was applied to ECa (Sept. '01 measurements) and elevation data. This allowed to divide the plot in several classes (from 2 to 6). On the other hand, chemico-physical properties of the experimental field were determined in February '01, by means of an intensive soil sampling (0.3 m depth, 16 points per hectare). The following measurements were made: exchangeable elements content (P, K, Mg, Ca), textural analysis (clay, silt and sand content), organic matter content and pH. These observations were then put into the classes determined before and the variance of each soil parameter was computed for each class. The total intra-zone variance for each parameter was defined as the average variance for each class weighted by the corresponding areas.

Results and discussion

The correlation coefficients between ECa data and soil chemico-physical properties were computed. Highest correlation was found with clay content (stable in time; r=0.8), and associated elements : K content (r=0.8) and Ca content (r=0.7). Correlation between water content and ECa was low, but this may be due to the low variability of this parameter, both spatially and temporally. The evolution of the variance of each parameter with the number of divisions was studied. For example, when the field is divided in 3 zones, the decrease of the total intra-zone variance is, on average for the 9 parameters, of 35%. For the clay content this decrease is about 56%. The figure represents the clay content map and the management zones map (3 zones). The comparison of these maps prove the relevance of such method for characterisation of within field soil variability.

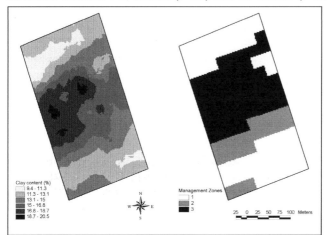

Conclusions

Considering the results obtained on the experimental field, it appears that the delineation of homogenous zones based on the measurement of ECa and on topographic attributes of field can be useful. Indeed, the knowledge of such zones may serve: 1) to limit the amount of soil analysis to do inside the field, and to precisely know where to take the samples; 2) to create application maps for the modulation of certain cultural operations (by means of phytotechnical modelisation).

References

Fraisse, C.W., Sudduth, K.A., Kitchen, N.R. [2001]. Delineation of site-specific management zones by unsupervised classification of topographic attributes and soil electrical conductivity. Transactions of the ASAE 44 (1), 155-166.
Sudduth, K.A., Drummond, S.T., Kitchen, N.R. [2001]. Accuracy issues in electromagnetic induction sensing of soil electrical conductivity for precision agriculture. Computers and Electronics in Agriculture 31, 239-264.

Acknowledgements

The Belgian Ministry of Small Trade and Agriculture is gratefully acknowledged for its financial support to this research.

A study on potato yield sensor and monitoring system for potato harvester

Yoshiyuki Hara, Hideyuki Takenaka and Kenji Sekiguchi
Hokkaido Central Agric. Experiment Station, Naganuma-cho Hokkaido, 069-1301 Japan
haraysyk@agri.pref.hokkaido.jp, takenaka@agri.pref.hokkaido.jp

Keywords: Optical Yield Sensor, Potato Harvester, Precision Farming

Introduction

Potatoes are important and well-established crop suitable for cool climate of Hokkaido Island that is located in northern part of Japan. In 2000, Hokkaido produced 2,160,000 tons of potatoes from about 60,000ha field. Conventionally, most of the potatoes are harvested with 1-row pulled-type potato harvester. The small size potatoes, clods, stones, greenery potatoes and haulms have to be discriminated during harvesting by several workers at the selection conveyer of the harvester to enhance evaluation in the market.

Method

Yield mapping is an important part of precision farming. We advanced development of the optical yield sensor and yield monitoring system on the potato harvester, which can measure potato weight during harvesting. The advantage of optical line sensor is easiness to mount on the potato harvester, and to be resistant to vibration and noise, and to be inexpensive in comparison with impact sensor or camera sensor. The yield sensor consists of a pair of line sensors, control box and AD-converter (Figure 1). Each line sensor has 32-photo diodes lined at even pitch of 2cm. Length of line sensor is 62cm and distance of each sensor is 40cm. The sensor was mounted at the end of the final selection conveyer, and measured the weight of potatoes, which were falling into the tank.

When potatoes pass between the optical sensors, potatoes shut off several optical beams. Shut-off length of a potato is depending on the potato size and the dropping speed. The number of optical beams shut-off by potato is counted in some sampling period and that is integrated. Sampling period controlled by PC (Figure 2). Integrated number of the optical beams shut-off by dropping potatoes has relation to potatoes weight, so can extrapolate potato weight. We tested under the following conditions. Drop height of potatoes from final selection conveyer is 15cm, and sampling period is from 5ms to 10ms and photo diodes pitch is 2cm and 1cm. The position of the potato harvester in the field is measured with RTK-GPS (RT-2, NovAtel).

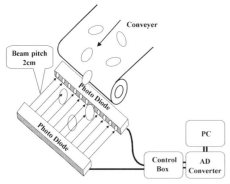

Fig1 Optical yield sensor and measuring system

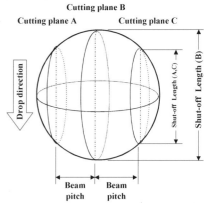

Fig2 Measuring Model with Optical yield sensor

Result and discussion

We tested the yield sensing system using different diameter of sphere before field examination. Number of optical beams shut-off had a linear relation to a diameter of sphere. The ratio of optical beams and diameter of sphere depend on the beam pitch and sampling period (Figure 3).

Different sizes of potatoes, which weight was from 70g to 140g/pieces and 8 verities of potatoes are tested. At each test 20kg potatoes were examined. The estimated accuracy of the yield sensor using a variety (May Queen, Elliptical shape) was less than 10% in the RMS error (Figure 4). As the result of field test, the integrated number of the optical beams shut-off and potato weight ratio differed depending on the potato varieties, size distribution and dropping posture of the potatoes. However an accuracy of measurement was increased by narrowing beam pitch from 2cm to 1cm and by shortening sampling period.

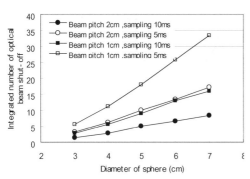

Fig 3 Relationship between integrated number of optical beams shut-off and sphere size

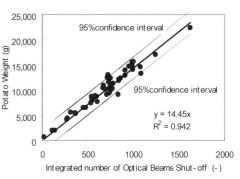

Fig 4 Relationship between integrated number of optical beams shut-off and potato weight

428

Derivation of soil texture and soil water content from electromagnetic induction measurements

K. Heil and U. Schmidhalter
Technical University Munich, Chair of Plant Nutrition, Germany
kheil@wzw.tum.de

Abstract

Surveying soils can be reduced by measuring the apparent soil electrical conductivity and converting the readings into textural parameters as well as soil water contents. The relationship between EC_a and clay, silt, sand as well as water content at field capacity was investigated in representative soils of the 150 ha agroecosystem research study site Scheyern. Field wide distribution of these parameters were obtained with a sequence of calibrations. The spatial variability of soil texture and soil water as well as their boundaries could well be described.

Keywords: apparent soil conductivity, EM 38, topography, soil pattern

Introduction

Site-specific agriculture requires detailed information about the spatial soil heterogeneity of fields. Earlier investigations have shown that the apparent electrical conductivity (EC_a) measured with EM38 represents the influence of several factors, including volumetric soil water content, gravimetric content of soil texture and further soil conditions like conductivity of the pore water, cation exchange capacity and organic matter. However, on the field/farm level simplified field wide derivations to predict soil texture and water content are not sufficiently developed. This study describes an approach to calculate these parameters with a sequence of multi-linear regressions.

Materials and methods

The study was carried out on the experimental farm at Scheyern, located in the tertiary hill side, a hilly landscape where tertiary sediments are partly covered by Pleistocene loess. The study included 133 sites with a broad range in soil texture. For each site soil texture and water content were determined and weighted according to the characteristic EM 38 response with depth.

Results and discussion

Field wide distribution of soil texture and water content at field capacity were obtained with the following steps: (i) Serial conductivity measurements with EM38 within a distance of 20 m; (ii) soil sampling on selected points and determination of the contents of clay, silt, sand and water (iii); classification of the data set according to different criterions (empirical derivation and statistical evaluation) e.g. content of soil texture and soil water, cultivation, topography, geology; (iv) derivation of soil texture by means of apparent electrical conductivity (EC_a) and topography (altitude, inclination, exposition) for every cluster with the purpose to obtain field wide contents of clay, silt and sand; (v) calculation of the water content by means of the apparent electrical conductivity, topography parameters and soil texture for every cluster with the aim to get field wide contents of soil water. The best results for the derivation of clay, silt and sand content were achieved with a segmentation of the data set into five different groups: Miocene freshwater molasses and Pleistocene loess each with integrated and organic farming, and gleyic soils. The multi-linear regressions show adjusted R^2 between 0,62 and 0,86 in the vertical as well as in the horizontal mode (Tab. 1). Such a good representation of sand and silt content was not reported until now in the literature. This is due to the segmentation of the data set and further due to the inclusion of topography parameters in the equations. The greatest influence on EC_a exerted the clay content at field capacity, but silt and sand should not be neglected, particularly if the target variable is the water content.

Table 1. Adjusted R^2 between soil texture and the independent variables EC_a and topography parameters for the clusters Miocene freshwater molasse and Pleistocene loess each with integrated and organic farming, and gleyic soils at Scheyern.

Dependent variable: clay, silt, sand [g*g-1]
Independent variables: EC_a, topography

Cultivation	Mode	Geology: Pleistocene loess			Miocene freshwater molasse		
		Clay	Silt	Sand	Clay	Silt	Sand
Integrated	vertical	0,71 ***	0,71 ***	0,72***	0,72 ***	0,63 ***	0,77 ***
	horizontal	0,68 ***	0,82 ***	0,74 ***	0,79 **	0,62 *	0,76 **
		n = 34			n = 24		
Organic	vertical	0,73 ***	0,76 ***	0,72 ***	0,73***	0,70 ***	0,66 ***
	horizontal	0,72 ***	0,82 ***	0,76**	0,78***	0,64***	0,86 ***
		n = 14			n = 54		
Gleyic soils	vertical	0,67*	0,72*	0,79*			
	horizontal	0,79*	0,96**	0,83*			
		n = 7					

By means of discriminant analysis the partitioning into the geological substrates was calculated. The classification results show a predicted group membership of 88% (Pleistocene loess) and 83% (Miocene freshwater molasse), respectively, with a significant canonical correlation coefficient of 0,7. In contrast to the simulation of the soil texture, the estimation of the water content at field capacity requires only a segmentation into the groups hydromorphic and terrestrial soils. In both clusters, soil texture predominantly influences the soil water content. Adjusted R^2 of 0,99 (V-, H-mode) were achieved in the terrestrial soils and of 0,95 (V-mode) and 0,89 (H-mode), respectively, in the water influenced soils.

Conclusion

Based on the delineation of more homogeneous entities the procedure described displays an effective method to derive the spatial variability of soil texture and soil water as well as their boundaries.

References

Schmidhalter U., A. Zintel und E. Neudecker, 2001. Calibration of electromagnetic induction measurements to survey the spatial variability of soils. In "Proceedings of the 3rd European Conference on Precision Agriculture". Montpellier. (Eds. G. Grenier and S. Blackmore), 479-484.

Acknowledgement

This study is a part of the agroecological research network "Forschungsverbund Agrarökosysteme München" (FAM), which is financially supported by the German Federal Ministry of Education and Research (BMBF 0339370).

Spatial estimation of available water capacity by pedo-transferfunctions based on German soil rating data and electrical conductivity maps

Ruprecht Herbst and Jürgen Lamp
C.A. University Kiel, AG Soil Informatic, Olshausenstraße 40, 24118 Kiel
rupherbst@aol.com, jlamp@soils.uni-kiel.de

Keywords: soil water, pedo-transferfunctions, soil rating, electrical conductivity

Introduction

The distribution of the available water capacity (AWC) is one of the most important soil functions for predicting yield variation of agricultural fields, especially in rain-limited landscapes. The AWC can be estimated by pedo-transferfunctions (PTF) as specified in the German manual for soil surveys (AG BODEN, 1994). This PTF needs field data of texture, humus and bulk density from all soil horizons until the efficient rooting depth (RDe) which by itself depends on texture and density classes. These parameters can be assessed pointwise by the SoilRover soil augerings. But variogramm analyses on seven preagro fields have shown that 50% of the geo-structured semi-variance (between sill and nugget) of most topsoil attributes occurs at median distances <50m, and of subsoil functions, like AWC, at <35 m (HERBST, 2002). Experiences gained in very different landscapes of Germany reveal that the quality of areal soil texture assessments depends more on the availability and quality of pre-information than on sampling density (LAMP et. al., 2000). Therefore, soil-designating areal pre-informations are necessary to predict the PTF parameters and the AWC, efficiently. Especially, old, but extensive German soil rating (SR) and geo-electrical conductivity data can help to predict texture information for the PTF in order to estimate the AWC.

Pre-information sources

Data German Soil Ratings (SR): Primarily, they serve taxation purposes and are consistantly available – partly still in analog (1:2000 maps, pit coding booklets), partly now in digital form - over all agricultural land of the entire "German Reich". SRs have three information layers, the 50 m borings, the class polygons with according class symbols, and soil layer descriptions of class-designating pits. Unfortunately, the 50m grid boring data are not available anywhere and are laborious to digitize. The class polygons and symbols include three data, the depth-weigted main soil texture (9 classes), the development status of soils (7 codes) and the geologic formation (5 classes). The third information layer are the designating pits. Their survey depth is usually < 8 dm with 2-4 different layers for which the layer depth, texture, humus, hydromorphie classes are specifed. These old profile codings have to be updated and translated into modern soil terms.

Geo-electrical soil conductivity (EC): The areal survey of ECs by the Geonics EM38-sonde within tramlines is an efficient technique to get fast and high resoluted, but indirect informations about the average clay contents of soil profiles. But with decreasing order, also soil moisture, temperature and bulk density have impacts on EC signals. The EM38 can be used in two different modi, the vertical an the horizontal (ECv, ECh), for which different sensivity curves and mean response depths (ECv > ECh) are given by the Geonics manual.

Results

Texture correlations of Soil Rating (SR) classes: Data about the development status and the geology of soils proved to be not very useful for predicting the AWC, but the main texture classes are more profitable. Correlations were evaluated at point positions by comparing field assessed textures and derived clay contents of augered soil profiles with SR main texture classes. In heterogeneous landscapes with a strong texture differentiation (e.g. sandy topsoils on loamy/clayey subsoils or on dystric histosols) the correlations are quite well ($r \sim 0.6$), but in less structured soilscapes the prediction power is rather limited ($r \sim 0.2$). A general problem is that SR has not directly assessed the silt fraction of soils which is highly relevant for water available pore volumes.

Spatial extendability of SR designating pits: This is question of high importance for all point-to-area predictions and answered by median distances of variograms. Analyses on preagro fields in Germany show that the AWC varies about ten times smaller than the class polygons of the SR (HERBST, 2002). Further problems are silty soils because the SR didn't know this texture class and, with a few exceptions, the missing of Colluvia.

Texture correlations of EC-values: Clay content estimation by ECs may be hampered by co- or cross-varying water contents of soil profiles. Therefore, by taking parallel auger core probes, analysing their water contents and evaluating water corrected EC data, the correlation coefficients (r) between the mean clay content of profiles (averaged from 0 to 15dm) could be raised from 0.6 to 0.75, but for well structured soilscapes only. Very sandy soil zones as well as loamy to clayey textures (>35 % clay) in the subsoil can be detected with a high (relative) repeatability, but absolute EC values are difficult to get. It is not possible to detect soil texture stratification by differentiate evaluation of ECv and ECh values.

Combination of SR and EC pre-informations: An important point in the investigations is that neither the SR nor the EC-measurement can be used for predicting especially the silt contents and the AWC very convincingly. Both sources have to validated by soil borings and locally adapted in different soil regions. It should be considered that the AWC cannot explain alone yield variation inside of fields, especially in not rain-limited areas with distincted relief. In these areas, the lateral water as well as the capillary rise of the ground water must be modelled.

References

AG Boden. 1994. Bodenkundliche Kartieranleitung, Schweizerbart'sche Verlags-buchhandlung, Stuttgart, 4. Auflage, 331 S.

Herbst, R. 2002. Bodenschätzung, geoelektrische Sondierung und pedostatistische Modellierungen als Basis von digitalen Hof-Bodenkarten im Präzisen Landbau. In: Schriftenreihe für Pflanzenernährung und Bodenkunde, Universität Kiel, 60.

Lamp, J., Herbst, R., Reimer, G. 2000. Preciser and efficient soil surveys as basis for application maps in Precision Agrculture. In: Precision Agriculture: Third European Conference on Precision Agriculture, edited by J. V. Stafford, BIOS Scientific Publishers Ltd, Oxford, UK, pp. 49-52.

Sugarcane yield mapping from the harvester biomass input flux

Britaldo Hernandez[1], Farnia Fernandez[1], Enrique Ponce[1], Lazaro Quintana[1], Miguel Esquivel[2] and Jose Julio Rodriguez[3]
[1]Automation and Control Laboratory, [2]Projects Division, [3]Computer Laboratory
National Center for Laboratory Animal Production, CENPALAB, Finca Tirabeque, Bejucal, Havana, Cuba
briti@cenpalab.inf.cu

Abstract

It is widely recognized that yield mapping is the most important step towards the establishment of a precision agriculture system. Yield monitors have been mainly developed for grain crops. Up to now, there are no yield monitors for sugarcane harvesters. From 1996 to 1999 several studies were carried out on the automation of the Cuban KTP sugarcane combine harvester. Two control loops were established. The first is related to the ground speed regulation, measuring biomass input per time unit. The second loop controls the height of the cutting discs, on the basis of hydraulic pressure. Both loops are controlled by a microcontroller system developed specifically for this application, called AGROMATIC. Using as a reference the automatic translation loop, biomass input was measured from the height variability of the harvester main roller feeder. A DGPS receiver was connected to AGROMATIC, giving the possibility to acquire exact positioning data simultaneously with biomass measurements. To verify the results, several tests were carried out in fields with diverse estimated yields and different sugarcane varieties. A high correlation was obtained between weight of sugarcane harvested and its distribution along the field, with data registered by AGROMATIC. The proposed system can be used as a reliable source for sugarcane yield mapping. In the next sugarcane harvest season it will be tested again in four harvesters on 700 ha.

Introduction

Yield Mapping is one of most important parts of a Precision Agriculture (PA) System. There is no possibility of creating a yield map with the current sugarcane harvesters, and this fact limits the implementation of PA in this crop. This work describes the creation of a yield map with a cuban KTP 3S sugarcane harvester. A necessary previous work was the automation of the harvester's ground speed and cutting-disk height control, which was carried out by the authors of the present work.

Bellow, we show the general diagram of the system used for the creation of the yield map.

Legend:
1- Main roller feeder
2- Linear displacement transmitter
3- Microcontroller
4- Flash Memory
5- DGPS Receiver

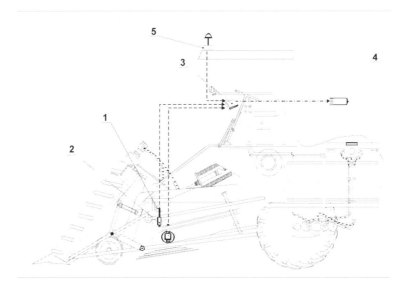

The Microcontroller (3) receives, in real time, the geodetic coordinates and horizontal speed of the harvester from the DGPS receiver (5), through an RS-232 port on the AGROMATIC. Simultaneously, it receives the position of the roller feeder (1) through the linear displacement transmitter (2), which is, in effect, a reference of the biomass input that enters the feeding system of the harvester. This information is recorded onto a flash memory (4) , that is later sent to a computer where the proccessing is carried out.

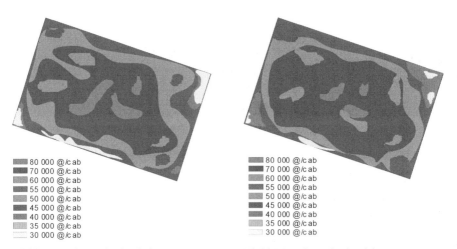

■ 80 000 @/cab	■ 80 000 @/cab
■ 70 000 @/cab	■ 70 000 @/cab
60 000 @/cab	60 000 @/cab
■ 55 000 @/cab	■ 55 000 @/cab
50 000 @/cab	50 000 @/cab
■ 45 000 @/cab	■ 45 000 @/cab
■ 40 000 @/cab	■ 40 000 @/cab
35 000 @/cab	35 000 @/cab
30 000 @/cab	30 000 @/cab

Yield Mapping obtained in a sugarcane caballería (13.4 hectares), weighing sugarcane fields of 100 m².

Yield Mapping obtained in a sugarcane caballería with the yield monitor of our work .

On-line methods for determining lag time and mix uniformity in direct injection systems for the site-specific application of herbicides

P. Hlobeň, M. Sökefeld and P. Schulze Lammers
Institute of Agricultural Engineering (IfL), University of Bonn, Nussallee 5, 53115 Bonn, Germany
phloben@uni-bonn.de

Abstract

Two optical methods and one conductance method were developed for the on-line measuring of mixture concentration. These methods were, used in a laboratory model of an injection sprayer system for immediate determining lag time and mix uniformity.

Keywords: plant protection, sprayers, direct injection, herbicides, lag time.

Introduction

Site-specific herbicide application based on information about weed distribution requires the use of sprayers with an integrated direct injection system (DIS). In injection sprayers, herbicides and carrier are kept separately. According to the indications on the weed treatment map, the herbicides are metered into the carrier and mixed immediately before reaching the nozzles. There are two crucial factors of current DISs, which influence the accuracy of application. The first one is lag time, i.e., the time it takes for the mixed solution to flow from the injection point to the spray nozzles. The second one is the occurrence of non-uniform mixtures in the boom. To make it possible to control these factors, accurate on-line methods for measuring mixture concentration must be developed.

Materials and methods

The first method (Fig.1a), uses a CCD camera which takes monochromatic images of a coloured mixture flowing through an in-line sight glass cell in order to determine the degree of mix uniformity. The mixture is a solution of black colorant (E161) in water. The concentration of the mixture is then determined according to the mean value of the grey level of each image. The second method is based on sensing the transmittance of light by means of a photodiode. The LED emits its green light (emission band 565 nm), through the coloured mixture (E161+water), towards a receiving photodiode (Fig.1b). The light intensity (the voltage produced by the photodiode) is directly related to the concentration of the solution.

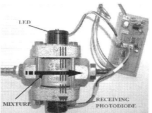

Figure1. a) CCD camera - static test, b) Transmittance method, c) Conductance method– measuring cell.

The third method is based on sensing the electrical conductivity of a common salt solution which flows through between two electrodes with a distance between them 1.2 mm, in a measuring cell with a constant C = 0.1. A suitable electronic circuit provides the output voltage, which indicates mixture concentration (Fig.1c).

Results and discussion

For the CCD camera method, only static trials were performed. These resulted in calibration curve of the device which is based on the average values of samples of known concentration (c) ranging from 0 to1.35 mg/ml of E161. There is an exponential relation between the grey levels of monochromatic images obtained from the camera and the mixture concentrations.
For the second optical method, based on transmittance of light, there is a similar relation. For this method, dynamic trials were performed with a constant of mixture flow of Q = 14 l/min in the sight-glass cell. Ranging from 0 to 2.8 μg/ml, the concentrations of the colorant were lower than in the CCD camera method. As is evident in the calibration curve (Fig. 2), the relation between the output voltage from photodiode U_D [mV] and the mixture concentration is an exponential function.
The results from the conductance method show also evince an exponential relation between electric resistance, which depends on the concentrations of salt mixture ranging from 0 to 0,5 mg/ml, and output voltage U_c [V]. The calibration curve (Fig. 3) has the same decreasing course as the curve obtained by the transmittance method. The flow through the measuring cell was maintained at a constant level of 14 l/min. The temperature of the mixture during the trials was kept constant at about 20°C.

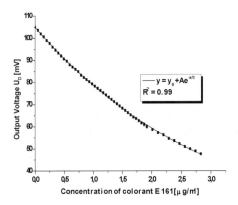

Figure 2. Calibration curve of the transmittance sensor.

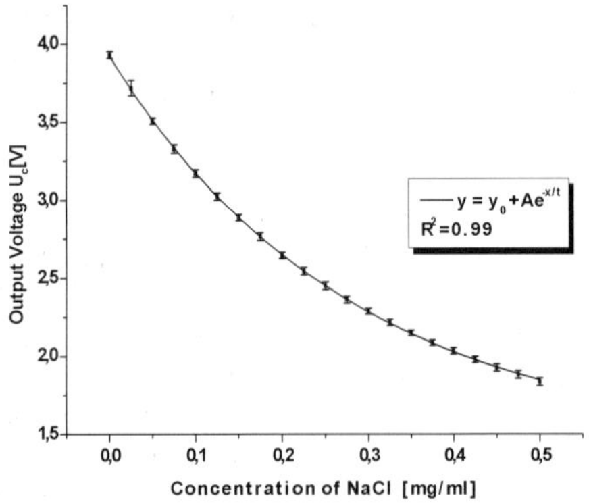

Figure 3. Calibration curve of the conductance sensor.

Conclusions

The results from the static and dynamic trials prove the practicability of these methods. There are significant functional relations between mixture concentration and output signal for each of these methods. The in-line measuring cells can be installed in any place throughout the length of the spray boom. Thus, they can be used to determine on-line time lag as well as mix uniformity. There is an influence of mixture temperature on outgoing voltage U_c. If the temperature of the mixture varies during measuring with the conductance sensor, a temperature compensation must be performed.

A method for optimising nitrogen fertilisation of wheat within a field based on a crop model approach

V. Houlès[1], M. Guérif[2], B. Mary[1], J.M. Machet[1] and D. Makowski[3]
[1]INRA Unité d'Agronomie de Laon-Reims-Mons, Rue F. Christ, 02007 Laon, France
[2]INRA Unité CSE, Site Agroparc, Domaine Saint-Paul, 84914 Avignon Cedex 9, France
[3]INRA Unité D'Agronomie INRA INA P-G, B.P. 01, 78850 Thiverval-Grignon, France
houles@laon.inra.fr

Abstract

Variable rate application of nitrogen fertiliser in the frame of precision agriculture should allow to adapt the nitrogen dose to the local requirement of the crop, in order to optimise crop production and minimise environmental hazards. One way for achieving these objectives consists in using a crop model that takes into account soil properties, climatic and N management scenarios to predict crop yield, grain protein content and environmental variables, such as nitrogen leaching estimate or a nitrogen balance. This study intends to test several criteria based on these variables to determine an optimal fertiliser rate.

Keywords: crop model, optimal nitrogen rate, decision rule, gross margin, N balance

Introduction

This study aims at comparing different criteria to determine an optimal nitrogen rate at several points in a heterogeneous field. This approach is based on a crop model simulating the impact of different climatic and technical scenarios on agronomic and environmental variables.

Materials and methods

The crop model STICS (Brisson et al, 1998) was used to simulate C, N and water balance on a heterogeneous field (10 ha) located near Laon, northern France (Guérif et al, 2001). The simulations were implemented on 81 points distributed on a regular grid. The soil model inputs were either measured or assessed through pedotransfer functions established previously (Houlès et al, 2002). 300 simulations were performed on each point: 30 different climates (corresponding to the past 30 years) were crossed with 10 N management scenarios corresponding to 4 dates of application and differing by the third supply. The range of N fertiliser rates tested was 180 to 270 kg ha[-1]. The method used to determine the optimal third N rate considers both economic and environmental outputs and is based on hypothetical and not actual rules. We calculated for each of the points i a gross margin $C_i(d_i)$ and a simple nitrogen balance $B_i(d_i)$ between N input and output:

$$C_i(d_i) = p.Y_i(d_i) - a.d_i \qquad (€ \ ha^{-1}) \qquad (1)$$

where d_i is the total N rate (kg.ha[-1]), p the price of wheat grain (€.t[-1]), $Y_i(d_i)$ the grain yield (t.ha[-1]), and a the cost of the fertiliser (€.kg[-1]) and:

$$B_i(d_i) = d_i - 10.Y_i(d_i).N^g_i(d_i) \qquad (kg \ ha^{-1}) \qquad (2)$$

where $N^g_i(d_i)$ is the grain nitrogen content (%).
The values of $C_i(d_i)$ and $B_i(d_i)$ used were the mean of the 30 values obtained for the 30 climatic series. Based on these variables, we proposed two criteria to determine an optimal total nitrogen dose d_i^* for each point i such as:

$C_i(d_i^*) = \max[C_i(d_i)]$ and $B_i(d_i^*) < T_B$ for a given i ('Local threshold')   (3)

$\Sigma_i[C_i(d_i^*)] = \max[\Sigma_i\{C_i(d_i)\}]$ and $1/n.\Sigma_i[B_i(d_i^*)] < T_B$ ('Global threshold') (4)

where n is the number of points i (n=81) and T_B is a threshold for which different values were tested from 20 to 100 kg ha^{-1}. The value of p was 100 € t^{-1} and of a was 0.45 € kg^{-1}.

Results and discussion

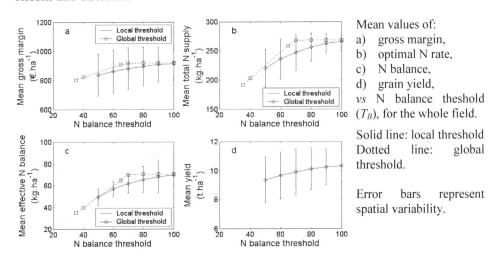

Mean values of:
a) gross margin,
b) optimal N rate,
c) N balance,
d) grain yield,
vs N balance theshold (T_B), for the whole field.

Solid line: local threshold
Dotted line: global threshold.

Error bars represent spatial variability.

The spatial variability is more important than the variability due to T_B. The interest of spatial varying N rate is greater when T_B is smaller (b). The optimisation was not possible when T_B was lower than 50 kg ha^{-1} in the case of 'local' approach and than 35 kg ha^{-1} for 'global' approach because these thresholds could not be reached in the range of total N rates tested. This shows that the 'local threshold' criterion is more severe than the 'local' one. Moreover it leads to recommend lower N rate and consequently to obtain lower gross margin and N balance, even if the difference concerning this last output is smaller. In conclusion, these results show that (i) a constraint on nitrogen balance can decrease farmer's gross margin and (ii) reasoning fertilising within a field is more requiring than at the whole field scale, which is a logical result. Precision agriculture's goal of matching supply up to local requirement doesn't maximise the farmer's income that is calculated at the field scale. Another approach would consist in including an environmental tax ('eco-conditionality') in the gross margin.

References

Brisson, N. et al., 1998. STICS : a generic model for the simulation of crops and their water and nitrogen balances. Agronomie 18 311-346.

Guérif, M., et al., 2001. Designing a field experiment for assessing soil and crop spatial variability and defining site-specific management strategies. In: Proceedings of the 3rd European Conference on Precision Agriculture, edited by G. Grenier & S. Blackmore, © Agro Montpellier, France, pp. 677-680.

Houlès, V., Nicoullaud, B., Beaudoin, N., Mary, B., 2002. Sensitivity of a crop model to pedotransfer functions at the field scale. In: Proceedings of the 7th congress of the European Society of Agronomy, edited by F.J. Villalobos & L. Testi, © Junta de Andalucía, Spain, pp. 631-632.

Spectroscopic detection of maize canopies as a requirement for variable fertilisation regimes on heterogeneous sites

G. Huber, F.-X. Maidl and A. Schmid
TU-München, Chair of Agronomy and Plant Breeding, 85350 Weihenstephan, Germany
ghuber@wzw.tum.de

Abstract

The late nitrogen demand of maize offers the possibility of substantial N fertiliser savings. Sensor controlled fertilisation regimes can be a vital approach to achieve this aim. Conventional spectroscopy methods usually cannot provide reliable information about agronomic parameters of row crops at early development stages. It could be found that the accuracy of the early assessment can be improved by orientating the sensor directly above the plant row and limiting the view area. For every tested variety a universal regression equation could be found, which described the correlation between nitrogen uptake and spectrometer readings from five leaf stage to tasseling.

Keywords: spectroscopy, maize, fertilisation

Introduction

The vegetative growth and the main nitrogen uptake of maize occurs in the summer months, when the N mineralisation from soil reaches maximum values. Hence maize is able to use the mineralised nitrogen with high efficiency. This crop can meet a high portion of its N demand by the supply from soil. Due to changing soil quality in heterogeneous fields the minerali-sation can feature a wide spatial variation. An estimation of the nutrition status of maize, that could serve as a basis for the fixation of N doses as needed, must appear during early growth when the ground is not yet fully covered. Conventional spectroscopy methods, that integrate a mixed signal from canopy and ground reflection, usually cannot provide reliable information about agronomic parameters of row crops at early development stages. The aim of the present examination was the optimisation of the spectroscopic assessment of maize under field scale conditions, regarding the usefulness for sensor-based fertilisation systems.

Methods

Field trials were conducted on a heterogeneous site in the Tertiary Hills of Upper Bavaria, Germany. In plot trials the factor "variety" was combined with the factor "nitrogen fertilisation". In order to create a wide range of diverse canopy types, four varieties with a distinct variation in terms of leaf colour and leaf orientation were chosen. The applied nitrogen amount ranged between 0 and 170 kg N ha^{-1}. Intermediate harvests were made at six dates during the growing season. Dry matter yield and nitrogen concentration of the biomass samples was detected. At physiological maturity corn and foliage were separately harvested and analysed. We periodically detected the test plots, using a two-channel reflection spectrometer. The downwards view angle for the detection of the canopy reflection was 25°. The irradiance was simultaneously recorded with a view angle of 180°. The spectral resolution was 3.2 nm between 360 and 1050 nm. We carried the hand-held sensor system directly above the planting rows at a distance of approximately ten centimetres.

Results and discussion

The reflection signature, recorded by spectrometer, was transformed into the vegetation index "Red Edge Inflection Point (REIP)" (Guyot & Baret, 1988). The REIP reading was closely correlated with the nitrogen uptake in above ground biomass investigated from intermediate harvests. A specific calibration for every variety improved the quality of the biomass estimation. In the range of maximum N uptake rates the REIP readings showed a saturation, that was the more pronounced the earlier in the season the measurement took place. (Figure 1) The coefficient of determination of the power function regression was always higher than the coefficient of the linear regression. For every variety a universal regression equation could be found, that describes the correlation in the whole period from three leaf stage to tasseling. (Table. 1).

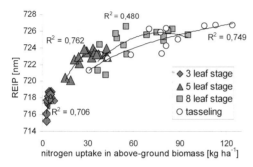

Figure 1: correlation between N uptake and REIP reading, variety LG 32.26

Table 1. Coefficient of determination of the regression between N uptake and REIP reading.

	Banguy	LG 32.26	Magister	Major
3 leaf stage	0.62	0.71	-	-
5 leaf stage	0.75	0.76	0.76	0.86
8 leaf stage	0.85	0.48	0.82	0.74
tasseling	0.66	0.75	0.68	0.83
pooled across all dates	0.88	0.79	0.63	0.83

The accuracy of the earliest measurements (only examined for two cultivars) was only few lower than at later dates. In our trials we carried the sensor closely above the plant row. Consequently the downwards channel mostly collected light from canopy reflection. From the five leaf stage the REIP reading could be used for a reliable assessment of the nitrogen supply of all tested varieties.

Conclusions

Spectroscopic methods can be suitable for non-destructive detection of the nutrition status of row crops such as maize. The vegetation index REIP can describe the biomass parameters with one regression equation pooled across several development stages of the plants. The accuracy of early estimations can be successfully improved. The impairment from soil reflection can be reduced by orientating the sensor directly towards the plant rows and limiting the field of vision in a narrow view angle.

References

Guyot, G. and Baret, F. 1988. Utilisation de la haute resolution spectrale pour suivre l' état des couverts vegetaux. In: Proceedings of the 4th Colloquium on Spectral Signatures of Objects in Remote Sensing. ESA SP-287 279-286.

Acknowledgements

The project was funded within the Forschungsverbund Agrarökosysteme München (FAM) by the German Federal Ministry of Education and Research.

Assessment of the site heterogeneity of the soil workability indicators

J. Hůla[1], P. Kovaříček[1], V. Mayer[1] and A. Rybka[2]

[1]*Research Institute of Agricultural Engineering, Drnovská 507, 161 01 Praque 6 – Ruzyně*
[2]*Czech University of Agriculture in Prague*
kovaricek.hula@bon.cz

Abstract

Before the ploughing the penetration resistance in topsoil was measured as basic for evaluation of its site heterogeneity. During ploughing was recorded the tractor trajectory and tensile resistance of one ploughing organ was registered, ploughing depth and operational width of plough were measured. The site heterogeneity of ploughing specific resistance was evaluated and differences of ploughing quality on plot separate parts were assessed. During following analysis was confirmed, that ploughing specific resistance and penetration resistance found out in the topsoil profile before ploughing are not mutually substitutive characteristics.

Keywords: site heterogeneity; soil workability indicators

Introduction

In current systems of the soil tillage there is not sufficiently considered the site irregularity of the soil characteristics (Scarlet et al., 1997). Often consequence of soil tillage on large plots regardless the site heterogeneity of soil workability is irregular quality of soil tillage. Other unfavourable consequence can also be extensive energy consumption of soil tillage operations. Sirjacobs et Destain (2000) present possibilities of machine intensive effect on soil controlling on basic of site heterogeneity evaluation of soil workability indicators. The scope of the measuring realized during ploughing after harvesting of the winter wheat was to record and assess the site heterogeneity of the cone index, specific ploughing resistance and ploughing quality indicators and by use of the measuring results on other sites to determine conclusions on the site tillage heterogeneity effect onto quality of the working operations.

Materials and methods

Measuring was carried-out in 1999 on plot after winter wheat harvesting. Soil conditions: loamy soil, soil moisture suitable for soil tillage. Before ploughing was measured the penetration resistance in topsoil in beared points in network 30x30 m. The trajectory of tractor movement with plough was recorded during ploughing including synchronous registration of dynamometer data with computer continually evaluating the tensile resistance of central ploughing body of the three-blade plough. The ploughing depth was measured by the standard method and real working width of the plough was found out. The result value was record of the specific ploughing resistance. Average set operational speed was 7,4 km.h^{-1}, ploughing adjusted depth was 0,25 m. Measuring was conducted on part of plot of acreage of 5,6 ha.

Results

Despite the plough adjustment and drive manner did not change during the measuring on the chosen plot, the course of the specific ploughing resistance has varied significantly. In the graph in figure 1 is expressed frequency of measured values of the specific ploughing resistance on the plot during ploughing to depth of 0,25 m. In fig. 2 are presented values of ploughing specific resistance measured on partial parts of the plot (in rows a-g). Diesel consumption during ploughing on that plot was within wide range from 12,8 to 21,3 l.ha[-1]. The ploughing quality has varied as well within a wide range. The average clods ratio after ploughing on some parts of the plot with a good soil workability was 9,8 % of mass (clods size above 100 mm), meanwhile plots with worse soil workability the number of these big clods was considerately higher: 52,8 % of mass in average. The following assessment was focused to finding of eventual dependence of the specific ploughing resistance onto the penetration resistance recorded on the identical points of the plot before ploughing. No statistically decisive model of the regressive function was found out during that assessment. The realized measurements have confirmed results of the measuring and evaluation of the previous years-ploughing specific resistance and penetration resistance found out in the topsoil profile before ploughing are not mutually substitutive characteristics.

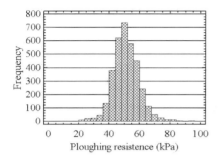

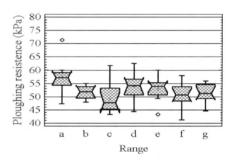

Figure 1. Frequency of measured values of the specific ploughing resistance on the plot.

Figure 2.The selected sets comparion-specific ploughing resistance in rows a-g.

Conclusions

The big heterogeneity indicators of the soil workability on the plots resulted unfavourably mainly at deeper soil cultivation connected with the cultivated soil layer turning, i.e. during ploughing and at the following soil tillage after the ploughing. The unfavourable consequence then is the conditions irregularity for the crops stand laying-out. Partial solution of that problem is utilization of mineralised soil cultivation for cereals and other crops.

During substitution of the ploughing by shallow loosening the local differences in the soil workability and their unfavourable consequences for crops cover establishment will result in considerably smaller scale in comparison with the conventional soil tillage with ploughing. Measuring within the period 2000-2002 has confirmed that hypothesis.

References

Scarlett, A. J., Lowe, J. C., Semple, D. A., 1997: Precision tillage: in field, real time control of seedbed quality. In: Proceedings of the Fierst European Conference on Precision Agriculture. Vol. II. Technology. BIOS Scientific Publishers Ltd. SCI, 503-510.

Sirjacobs, D., Destain, M. F., 2000: A soil mechanical resistance sensor for on-line application in precision agriculture. In: AGENG Warwick, Paper Number 00-PA-021.

Acknowledgements

Results presented in this paper were obtained during solution of the project MZEM05 9901 (Ministry of Agriculture Czech Republic).

Measurement of sunlight-induced chlorophyll fluorescence in the daily course under ambient conditions

C. Idelberger and E. Sticksel*Department for Agronomy and Plant Breeding, TUM-Weihenstephan, Germany*
claudia.i@wzw.tum.de

Abstract

The sunlight-induced chlorophyll fluorescence and the photochemical reflectance index (PRI) can provide information about the photosynthetic rate of a crop under ambient conditions. This remotelessly acquired information could serve as a new tool in Precision Farming.

Keywords: sunlight-induced chlorophyll fluorescence, PRI, photosynthetic rate, spectrometer

Introduction

Absorbed photosynthetically active radiation is dissipated by plants as chemical work, heat and fluorescence light. With decreasing efficiency of photosynthesis an increasing part of absorbed energy is emitted as fluorescence light. According to Maier et al. (1999) chlorophyll fluorescence can be measured under ambient conditions. Another protection mechanism against excess irradiation is the xantophyll cycle, which can be described with the PRI (Gamon et al., 1995). An active xantophyll-cycle is measurable via decreasing PRI values.

Material and methods

A hand-held spectrometer (Tec5) with a resolution of 0.8 nm was used to sequentially measuring irradiance and reflection. Sunlight-induced chlorophyll fluorescence and PRI were computed from the reflexion signature. Photosynthetic rate was measured as a reference with a LI-COR 6400. Two types of experiments were conducted: A) Dim-out experiment: The plant canopy was covered for approx. 20 minutes to ensure that plants were dark-adapted. Fluorescence, PRI, and photosynthetic rate were measured immediately with the removal of the cover. B) Continuous measurements: ca. every 20 min.

Results and discussion

Dim-out experiment: After removing the cover photosynthetic rate and xantophyll cycle slowly increased (Figure 1) while fluorescence light decreased, thus indicating an increasing efficiency of photosynthesis.
Continuous measurements: In the morning fluorescence decreased steadily while photosynthetic rate increased (Figure 2). From ca. 9.15 a.m. on decreasing PRI values and an increasing fluorescence were noted. Obviously irradiance plus temperature exceeded optimum values so that the plant activated protection mechanims.

Conclusions

Sunlight-induced chlorophyll fluorescence and xantophyll-cycle proved to be closely correlated with the photosynthetic rate. The next step is to verify these results under practical conditions in order to develop a new application in precision agriculture.

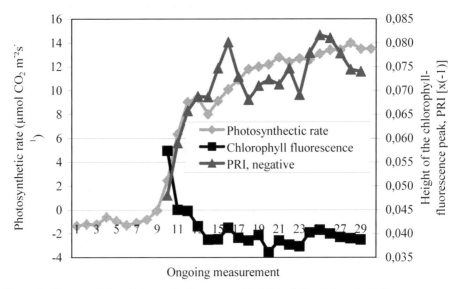

Figure 1. Course of the photosynthetic rate and height of the chlorophyll fluorescence peak (Dim-out experiment) with *Galinsoga*. Box removal between the 10th and 11th measurement.

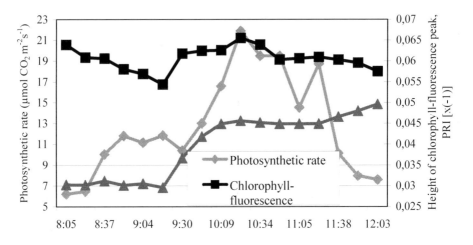

Figure 2. Continuous course of chlorophyll fluorescence and photosynthetic rate with rhubarb.

References

Gamon, J.A., Roberts, D.A. and Green, R.O. 1995: Evaluation of the photochemical reflectance index in AVIRIS imagery. AVIRIS Proceedings, JPL Publication 95-1.

Maier, S. W., Günther, K. P., Lüdeker, W. and Dahn, H.G. 1999: A new method for remote sensing of vegetation status. ALPS.

Acknowledgements

Funded by the Bayerisches Staatsministerium für Landesentwicklung und Umweltfragen.

Evaluating within-field spatial variability in soil properties and biomass using an early season multispectral satellite image

Anne Jacobsen[1] and Mats Söderström[2]
[1]*AgroVäst Livsmedel AB, Swedish Agricultural University, Box 234, S-532 23 Skara, Sweden*
[2]*The Swedish Farmers Supply and Crop Marketing Association, S - 531 87 Lidköping, Sweden*
anne.jacobsen@agrovast.se

Abstract

The investigation was performed on five test fields (25-50 ha in size) in south-western Sweden. Principal component analysis was performed on IRS 1D LISS3 multispectral data from 6th May 2000. Principal component regressions using PC scores, measures of biomass equivalence (Hydro N-sensor registrations) and measures of clay and organic matter (laboratory estimates) were performed to evaluate the explained variance. The investigation indicated i) that PC1 and PC2 may be used to separate the reflectance from clay and organic matter under bare soil conditions and ii) that PC1 and PC2 may be used to separate crop and soil reflectance of sown fields.

Keywords: multispectral data, Principal Component Analysis, regression modelling, soil reflectance, biomass variation.

Introduction

The spectral reflectance from perennial fields early in the season is either a 'pure' reflectance from bare soil influenced by the amount of clay and organic matter (OM) or a spectral mixing of soil and crop reflectance (fields with spring or winter crops). It is the objective of the study to investigate i) whether a simple multispectral three band image (IRS 1D LISS3) may be used to describe within field variability of soil properties from bare fields and ii) whether it may be used to separate the spectral signal from crop and soil from sown fields.

Materials and methods

The investigation was performed on two fields with winter crops, two fields with spring crops and one field with bare soil. Experimental data included IRS 1D LISS3 spectral data in green, red and near infrared registered 6th May 2000, biomass estimates (biomass$_{HNS}$) registered by the Hydro N-sensor 5th, 11th and 22nd May 2000 and measures of the fractional content of clay and OM registered year 2000. Principal component analysis was performed for the fields using the IRS 1D LISS3 spectral data. Linear regression modelling was performed as Principal component regression (PCR) (Esbensen et al, 1998) using PC scores, measures of biomass$_{HNS}$, and content of clay and organic matter to evaluate the explained variance.

Results

The PCR analysis was performed as a set of multiple regressions. The result of the PCR is shown in summary in Table 1.

Tabel 1. Results of the PCR analysis. Numbers indicates explained variance (r^2). 05, 11 & 22 under Biomass$_{NHS}$ is the measurement date in May.

	Bare soil		Spring crop Field 1 & Field 2				Winter crop Field 1 & Field 2			
	PC1	PC2	PC1	PC2	PC1	PC2	PC1	PC2	PC1	PC2
Clay & OM	0.25	0.12	0.67	0.00*	0.02*	0.37	0.14	0.04*	0.09*	0.38
Biomass$_{NHS}$										
05					0.00*	0.49	N/A	N/A	0.49	0.04*
11			0.05*	0.37					0.66	0.02*
22					0.00*	0.20				

* pr > F > 0.01

Discussion

Bare Field: PC1 and PC2 are both significant in explaining clay and OM variance. A single lineær regression of clay and OM on PC scores, show that the explained variance was additive along PC1 whereas PC2 explained variance in OM only. Spring crop – Field 1: PC1 explains the variance in soil properties, PC2 explains the variance in biomass. The PCA has separated the spectral signal from soil and vegetation. For this spring crop, soil reflectance is the most dominant spectral component on the field. Spring crop – field 2: PC1 do not explain the variance in any of the parameters investigated. The explained variance along PC2 is additive for soil properties and biomass. The PCA has not succeeded in separating the spectral signal for soil and vegetation. There is a very varied local relief in the field and it is likely that this has the most important influence on the spectral signal. Winter crop – field 1: Soil properties is explained along PC1- there is no biomass data available. Winter crop – field 2: PC1 explains the variance in biomass, PC 2 explains the variance in soil properties. The PCA has separated the spectral signal from soil and vegetation. For this winter crop, vegetation is the most dominant spectral component on the field.

Conclusions

In summary, the investigation indicated i) that the spectral signal from clay and organic matter may be separated using PCA of bare soil and ii) that the spectral signal from crop and soil may be separated using PCA on sown fields. The dominant underlying structure of the spectral data for one field with spring crop was soil properties, whereas the dominant underlying structure of the spectral data for one field with winter crop was biomass. PCA extracts the most dominant feature along PC1, the next most dominant feature along PC2 and so on. The variation in the underlying spectral structure for spring and winter crop, respectively, has been extracted successfully using this method.

References

Esbensen, K., S. Schönkopf, T. Midtgaard, D. Guyot. 1998: Multivariate Analysis in Practice. Camo ASA, Norway, 3rd edition. ISBN No. 82-993330-1-6.

Acknowledgements

Funding: VL stiftelsen, Sweden and Swedish Farmers' Foundation for Agricultural Research.

The requirements and prospects of precision agriculture in sugar cane – a case example in Mauritius

I. Jhoty and L.J.C. Autrey
Mauritius Sugar Industry Research Institute, Reduit, Mauritius

Abstract

Precision Agriculture, an innovative agricultural management tool suited to a wide range of crops, aims at controlling crop production and achieving optimal financial and managerial benefits. In the application of precision agriculture to sugar cane, farm planning and other requirements are essential, as it is the case in Mauritius. These requirements and some results of yield monitoring and soil electromagnetic conductivity mapping in 2001 and 2002 are briefly discussed.

Keywords: mechanization, farm planning, yield monitoring, soil conductivity.

Introduction

The sugar cane crop is an appropriate agro-system for the application of precision agriculture in as much as a yield monitor has been specifically developed for it. Mauritius, a subtropical volcanic island, cultivates sugar cane on about 41% of its area in diverse topography, soils and climatic conditions. Farm planning techniques and other measures have been implemented to allow crop mechanization for timeliness of cost-effective work operations. These are valuable preconditions for the application of precision agriculture, which has recently been initiated. Some results of the application of precision agriculture are described while prospects and further work are also examined.

Farm planning requirements

A prerequisite to mechanization of sugar cane production in Mauritius is that land should be suitable and planned for the purpose (Jhoty et al, 2002a). Because of soils rockiness and areas criss-crossed by drains and watercourses, farm planning techniques such as field derocking and complete removal of surface rocks, land cut and fill, redesign of drains and roads, and establishment of longer cane rows are introduced. The efforts of farm planning are intensified and proving useful to the application of precision agriculture.

Yield monitoring

In 2001, the first cane yield monitoring revealed that yield variability within fields was significant (Jhoty et al, 2002b). Differences in irrigation water efficiency, soil types and tiller density accounted for yield variability. Soil electromagnetic conductivity surveys with EM38 showed a good relationship between yield variability and soil conductivity, i.e zones of higher yields related to zones of higher conductivity values. In 2002, yield variability patterns within fields were similar to those observed in 2001. In the one field monitored in 2001 and 2002, zones of high and low yields occurred in nearly the same geographical positions (Figure1), indicating an opportunity for variability management.

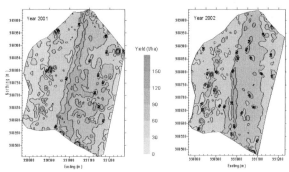

Figure 1. Yield variability within the same field in two successive years.

Repeated EM 38 measurements in an irrigated field demonstrated the reliability of the instrument in recording zones of consistent conductivity values. In the harsh working conditions of the harvester, frequent monitor calibration, breakage of leads and cables and dust infiltrating the system should be addressed under constant technical supervision to ensure success.

Prospect and future work

Results of yield monitoring and soil electromagnetic conductivity surveys in 2001 and 2002 are encouraging and can effectively lead to variability management through control of input applications and other adjustments in management strategies. Further work will continue on determining the relationship between yield variability and other factors such as topography, soil properties and cane growth.

Conclusion

Farm planning conditions are essential to implement mechanization and precision agriculture in sugar cane. Constant supervision is required to solve yield monitoring related problems. The conditions described for Mauritius can serve as examples in other countries where soils and land characteristics are similar.

Acknowledgement

The authors thank Mr Siram Ramasamy, of the Land Resources Dept. of the MSIRI, for the report on problems encountered with yield monitoring.

References

Jhoty, I., Ramasamy, S., Baggonauth, D. and Tulloo, P. K. T. 2002a. Suitability of cane land for mechanization. Occasional Report No. 29, Mauritius Sugar Industry Research Institute, Reduit, Mauritius, August 2002.
Jhoty, I., Ramasamy, S., Blackmore, S. and Autrey, L. J. C. 2002b. Yield variability in sugar cane. In: Proc. 6th International Conference on Precision Agriculture, 14-17 July, 2002, Minneapolis. In press.

Which context makes site-specific N management best valuable? Effects of climate, crop rotations, spatial variability, and economic factors

A. Jullien[1], R. Roche[1], B. Gabrielle[1], M.H. Jeuffroy[2] and P. Huet[1]
[1]Unité Mixte de Recherche INRA-INA P-G Environnement et Grandes Cultures, 78850 Thiverval-Grignon
[2]Unité Mixte de Recherche INRA-INA P-G Agronomie, 78850 Thiverval-Grignon
jullien@grignon.inra.fr

Abstract

The benefits of site-specific N management depend on many factors including climate, crop rotations, spatial variability, and economic parameters. Here, we used the CERES and Azodyn crop models to investigate variable-rate N management for a wheat crop in the Northern France. It appears that the potential benefits of this technique are greatly offset by inter-annual climate variability. Simulations with crop rotations show that risks of N leaching should be considered over the whole rotation, and that the structure of the heterogeneity also influences economic results.

Keywords: N management, crop model, climate, crop rotation, economic factors

Introduction

The benefits of site-specific N management depends on many factors including climate, crop rotations, spatial variability, and economic parameters. Is it possible to evaluate this value a priori from soil characteristics using crop models (CERES, Azodyn)?

Materials and methods:

A field experiment was conducted with wheat in 1998-1999 at the experimental farm of the Institut National Agronomique de Paris-Grignon, Thiverval-Grignon, France (48,9 N, 1,9 E). The field is 15ha in size and contains 3 homogeneous soil zones: a flat plateau made up of calcareous loamy soil, a sloping zone of shallow sandy calcareous soil, and a lowland with a thick loamy soil (colluvium). The crop models CERES-wheat (Ritchie and Otter, 1985), as modified from its original version ((Gabrielle, Huet *et al.* 2001)) and Azodyn (Jeuffroy and Recous 1999) were validated for each of the three 3 homogeneous soil zones. Models were subsequently used for site-specific simulations. For each zone, potential yields were calculated using a 15-year series of weather data using CERES. Optimal amounts and dates of N fertilization were then calculated with Azodyn (Figure 1). Both models were used to simulate yield, quality (Azodyn) and leaching (CERES). The economic gross margin was calculated in euro/ha as: GM = yield * 10 - fertiliser amount * 0.6). Thus, site-specific N management could be evaluated following on environmental and economic criteria.

Results

Models simulations were in accordance with yield observations for each zone. They were thus used to simulate the effect of the above-cited factors. Results of site-specific fertilisation are presented for 1999 in Table 1. It appears that climate variations strongly reduce site-specific N gains when calculated over 15 years (Economic margin gain variation (in euro/ha, N-site specific – conventional): min = -16.8 max = 16.8 mean = 3.5).

Table 1. Effect of site-specific N-fertilisation.

		Lowland	Sloping zone	Plateau	Field	GM
N-site-specific	N amount (kg/ha)	190	140	220	183	737
	Ceres yield (qtx/ha)	89	70	95	85	
	Grain N content (%)	10.4	9.5	9.9	9.9	
	Mineral N at harvest (kg/ha)	40	24	39	34	
Conventional	N amount (kg/ha)	184	184	184	184	774
	Ceres yield (qtx/ha)	88	74	87	83	
	Grain N content (%)	10.2	10.7	9.4	10.1	
	Mineral N at harvest (kg/ha)	40	25	39	35	
Difference in straw N content (N-site-specific - conventional) kg/ha		-1	-	14	9	-1

Heterogeneity structure also influenced economic results (Figure 2): the worst situation corresponds to large number of homogeneous small zones. From a time point of view, simulations with crop rotations show that the risks of N leaching only appear under the wheat crop after rapeseed, which indicates that variable rate management may be more valuable with the latter crop (Figure 3). This technique should thus be devised for the whole rotation. Anyway, results also indict that in a situation of low N price (intensive agriculture), economic gains for N modulation may always be low unless pollution risks are taxed.

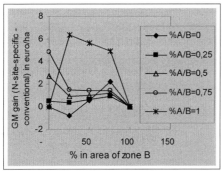

Figure 2. Effect of heterogeneity structure on site-specific management gain

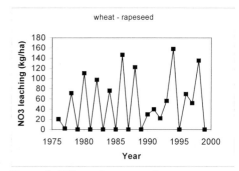

Figure 3. Effect of crop rotation on site-specific management gain

Conclusion

This investigation presents interesting results to characterize potential value of N modulation in a specific situation. It is difficult to conclude on site-specific management value it is dependent on many external factors. It may be interesting to further work on virtual fields to draw a typology of interesting situations for precision agriculture, combining different factors of variation. To better investigate the economic value and effect of economic complex, a combined use of crop model and economic model is currently on-going.

References

Gabrielle, B., P. Huet, et al. (2001). Prediction of wheat crop yield map using post-anthesis radiometrical data. Third ECPA, Montpellier, France.

Jeuffroy, M. H. and S. Recous (1999). "Azodyn: a simple model simulating the date of nitrogen deficiency for decision support in wheat fertilization." European Journal of Agronomy 10(2): 129-144.

Ecological and economic assessment of sensor-based site-specific nitrogen fertilisation

W.-D. Kalk, R. Schlauderer, U. Völker and D. Ehlert
ATB-Potsdam, Max-Eyth-Allee 100, D-14469 Potsdam, Germany
wkalk@atb-potsdam.de

Abstract

Based on positive results in development of a mechanical sensor (pendulum-meter) for indirect measurement of grown plant mass in standing grain populations, the late nitrogen fertilisation and since 2002 also the second N-application in farm-scale strip trials for winter wheat on altogether four fields was arranged to indicate the ecological and economic effects of site-specific fertilising. The summarised results show significant advantages of variable fertilisation for the N-surplus in the fertilising variants on all fields, for the N-efficiency on two fields. No disadvantages with regard to yield, grain quality and contribution margin were ascertained as a result of precision fertilising.

Keywords: on-farm-trials, site-specific nitrogen fertilising, assessment, N-efficiency

Introduction

One of the basic questions for a successful introduction of site-specific nitrogen fertilising is, if the changed management leads to improved ecological effects and particularly to economical benefits for the farmers. On-farm-trials were conducted in the last three years in Brandenburg state to study the potential of ecological and economic effects of site-specific fertilising.

Methods

The study presents new methodology approaches to illustrate and assess these effects. Significant tests were used to proof statistical differences of the results. The objective is a low cost solution for the praxis. This affords a real time measurement and direct application of adapted nitrogen fertiliser rates. Based on positive results in development of a new mechanical sensor (pendulum-meter) for indirect measurement of grown plant mass in standing grain populations, the late nitrogen fertilisation in farm-scale strip trials for winter wheat was site-specific varied in the first years, since 2002 also the second nitrogen application rate.

The **ecological assessment** of nitrogen fertiliser application is based on the requirement that no superfluous fertiliser is applied particularly in field areas with low stand development, in order to reduce nitrate leaching into surface and ground waters. The calculation of site-specific nitrogen surplus was used as indicator for the assessment of nitrogen loss potentials. Nitrogen efficiency was calculated as indicator for the nitrogen utilisation. To calculate precision balances the nitrogen quantities were allocated to 6 respectively 7 classes.

For the **economic assessment** it was differed between a) the annual costs for precision agriculture equipment, b) the benefits of reduced nitrogen fertiliser input, c) the benefit of changed/increased yield level and d) protein content of the grain.

Results and discussion

In figure 1 an example of ecological assessment is shown. The calculation of the late nitrogen application with 65 kg N/ha was the starting point for assessing the precision fertilising. Assuming pragmatically a targeted nitrogen surplus of below 50 kg N/ha for all sites in Germany (Frede and Dabbert, 1998), the average nitrogen surplus of about 100 kg/ha is still too high.

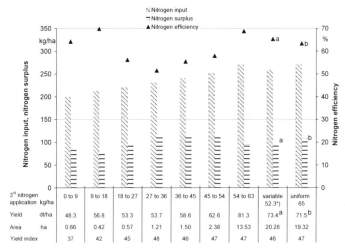

Figure 1. Nitrogen efficiency of site-specific fertilising in winter wheat.
(*) mean value; a b significantly different values)

With a mean of 65 %, the nitrogen efficiency at this site was lower as the value targeted for this location of more than 70 %. The classification chosen depended to the specific field characteristics and has to be interpreted accordingly. Further examples showed different results and will be presented. Additionally the impact of varying the second nitrogen application rate will be shown. The summarised results of the significance test show significant advantages of variable fertilisation for the N-surplus in the fertilising variants on all fields. As regards nitrogen efficiency, only two fields showed significant advantages of variable fertilising (see figure 1). Advantages of variable fertilising are particularly evident on plots with low plant growth. It also was studied, if variable fertiliser rates, so saving of fertiliser, do have an impact on the general yield level, grain quality or gross margin. The statistical tests showed, that no disadvantages in this respects were ascertained as a result of precision fertilising. An estimation of the economic potential showed advantages of site-specific cropping. The saving of nitrogen fertiliser amounted 10.1 kg/ha (4,23 EUR/ha) and additional yield of 0.7 dt/ha (7.49 EUR/ha), in total 11.73 EUR/ha. The fixed cost of the special equipment was estimated to be 1000 EUR/a. Under this conditions, the break even is achieved with a cropping area of 85 ha..

Conclusions

The ecological and economical assessment method shows interesting results but has to be developed further step-by-step in order to validate the assessment of the plant production impact of precision N-fertilising.

References

Frede, H.-G.; Dabbert, S. [HRSG.] (1998): Handbuch zum Gewässerschutz in der Landwirtschaft. Ecomed Verlagsgesellschaft Landberg.

Soil analysis and pretreatment for calibration data of the real-time soil spectrophotometer

Y. Kato, S. Shibusawa and A. Sasao
Tokyo University of Agriculture and Technology, Fuchu, Tokyo 183-8509, Japan
yuco@cc.tuat.ac.jp

Abstract

The real-time soil spectrophotometer collects a large amount of soil data from the field to describe detailed soil parameter maps. A tough work is to get the ground-truth data for calibration by soil sample analysis in the lab because of time-consuming. Three pretreatments for soil analysis; raw soil, oven-dried soil and air-dried soil, were tested in order to accelerate the soil analysis process. Results showed high correlation in EC value between different treatments. There were similar patterns on EC maps using respective values for calibration.

Keywords: real-time sensing, calibration, soil parameter, mapping, sampling strategy

Introduction

A real-time soil sensing system requires four important elements; reservation of reliable calibration curves, collection of soil reflectance, correct soil mapping technique and soil information service. We have been developed the real-time soil spectrophotometer with an EC electrode, a soil compaction meter and a GPS. Focusing is how to increase accuracy of calibration using soil reflectance and soil analysis data on pH and EC for time saving purpose.

Materials and methods

A 0.3 ha upland and a 1.2 ha paddy fields were provided for the experiment. The real-time soil spectrophotometer collected underground soil reflectance by 1m spacing, and soil samples were also collected every 5m or 10m at the same locations and depths as data scanned. To investigate an available soil treatment for calibration purpose, three different treatments; wet soil in-situ, air-dried soil in room temperature, and oven-dried soil under 110°C-24h, were tested and compared. Values for pH and EC were analyzed with a potable ion meter (D-24, HORIBA).

Results

Results showed that correlation between pH values of wet and air-dried soil treatments scored 0.70 in R^2, while $R^2=0.73$ was obtained for air-dried and oven-dried soil treatments, and $R^2=0.49$ for wet and oven-dried soil treatments. High correlation between EC values of wet and air-dried soil treatments showed $R^2=0.95$, while $R^2=0.92$ for air-dried and oven-dried soil treatments, and $R^2-0.91$ for wet and oven-dried soil treatments in Figure 1. A similar pattern of spatial distribution of EC appeared on maps using the calibration data by three different soil treatments.

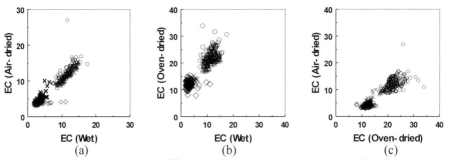

Figure 1. Comparison between different soil pretreatments about EC values; (a) Wet and air-dried soil, (b) Wet and oven-dried soil, (c) Oven-dried and air-dried soil.

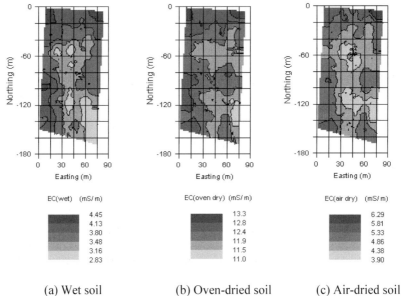

(a) Wet soil (b) Oven-dried soil (c) Air-dried soil

Figure 2. EC maps using value of each pretreatment soil for calibration.

Conclusion

Wet and air-dried soil treatments were available, but oven-dried soil treatment was not available, as the values were as double as that by wet soil or air-dried soil treatment. In terms of time saving, moreover, wet soil treatment will be candidate for a real-time soil sensing system.

References

Shibusawa, S., S. Hirako, 2001. Real-Time Soil Spectrophotometer for Precision Farming. Journal of the Spectroscopic Research. 50(6): p. 251-260 (in Japanese)

An optimal design of grain flow transducer for instantaneous measurements for applications on combine harvester

W. Keska
Poznan University of Technology, Poland
keska@sol.put.poznan.pl

Abstract

The general aim of the presented work was to improve the design of a transducer for continuous measuring of the grain flow transported to the grain tank of the combine harvester. This transducer is the key part of the system for grain yield mapping. The improved transducer consists of a sensing plate suspended on the two parallel circular plate springs, which are a part of a viscotic damper, and opto-electronic sensor for measuring the displacement of the springs. The optimal values of construction parameters of the mechanical parts of the transducer were obtained by way of computer simulation.

Keywords: yield mapping, combine harvester, grain flow, measuring system

Introduction

The maps of local yield play a very important role in precision agriculture. The maps are constructed on the basis of values of instantaneous flow rate of grain stream fed into the grain tank of a combine harvester. There are numerous methods used for measuring the grain flow rate. One widely applied method applies an impact type transducer. Which measures the force generated by the grain stream. The stream hits a sensing plate in a sensor, causing the displacement of the spring on which the plate is suspended. The displacement is then translated into proportional electrical signal. This method is considered to be one of the best methods. However, the dynamic behavior of the mechanical part of the transducer may introduce serious measuring errors, especially near resonance. Since on one hand the grain stream acting upon the sensing plate id of pulse type, and on the other hand, the transducer frame is subjected to intensive vibrations of the combine harvester. Therefore, the practical question is to select the values of spring stiffness and damping coefficient. So, as to obtain reliable measurements, is advisable to investigate the mathematical model of the mechanical part of the transducer.

Materials and methods

The mechanical part of the transducer was modeled as a two body system, with masses connected by springs and damping elements. The first body represents the sensing plate, the second – the transducer frame. This system is excited by the vibrations acting on the transducer frame, and by the force impulses, acting on the sensor plate. The special computer procedure written in Object Pascal, enables us to investigate the transducer's response, represented by displacement of the sensing plate. To obtain the time history of the force acting upon the sensing plate of transducer, the mathematical model of the grain motion in combine harvester we built. The formulation of this model was based on the results of photo registrations of the grain flow in the release chamber of a grain conveyor. This conveyor, obtained from a serially produced combine, was investigated on a laboratory stand (Keska & Jankowiak 2002). To identify the vibration excitation, the time history of vibration acceleration of the conveyor frame in the direction of the main axis of the transducer, was registered, using HBM DMC+ measuring system.

Results

We carried out the numerous simulations with this model, changing values of construction parameters of modeled transducer. Figure 1 shows, for example, the time response of the mechanical system with intensive damping. The bold curve represents the registered response of the force measured by the transducer, while the thin curve represents the force values simulated force acting on the sensing plate.

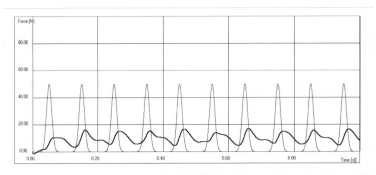

Figure 1. Time curve of the sensor response

The response of the sensor shows lower fluctuations then the exciting force, nevertheless the average values can be equal. This mechanical low pass filtering effect can play positive role in measuring system, because any short period fluctuations of the measured force (shorter then s), ought to be averaged in the electronic part of measuring system. The pulse type excitation originating from grain stream has a period ca. 0.1 s. So, it must be averaged. Other simulations has pointed out, that by some values of spring stiffness and weak damping multiplying of the amplitude of the response takes place, due to resonance effect. This effect diminishes the accuracy of the measurements. On the other hand, too high damping rate causes delayed and fuzzy sensor response. It was observed too, that nonlinear damping may cause significant shift of the signal average.

Conclusions

The optimal selection of the damping and stiffness coefficient of the suspension of the sensing plate can improve the accuracy of measurements. This can be achieved by applying the transducer of special design. This transducer consists of two circular plate springs, and stiff internal wall between them, forming two flat chambers, connected by the narrow channel of controlled size. The space of those chambers is filled with a damping fluid. The damping ratio of such system can be easily controlled and optimized. For translating the displacement of suspended parts of the transducer, into electrical signal, the opto- electronic sensor of displacement was applied.

References

Kęska W. Jankowiak S. 2002. Badania symulacyjne oraz empiryczne naporu dynamicznego strugi ziarna na płaską płytę przetwornika pomiarowego do pomiaru lokalnej wysokości plonu na kombajnie zbożowym. (The simulation and empirical investigation of the dynamic pressure of grain stream on a sensing plate of the yield sensor for measuring of momentary yield distribution, mounted on combine harvester). Journal of Research and Applications in Agricultural Engineering, vol 42,. PIMR Poznan.

Sociology of farm mechanisation of district Rohtak in Haryana state (India)

R.S. Khatry
Research Scientist KVK Rohtak, Haryana, India

Abstract

The kind of social system human being adept reflects the kind of work relationship from which they earn their livelihood. Haryana state is still a predominantly agrarian society. Majority of its population (69%) live in rural areas and more than 60% of them earn their livelihood wholly or partially, directly or indirectly from farming.
Mechanisation of farm and off farm work in Haryana followed with the initiation of green revolution. Green revolution brought extra income and saving which made it possible for farmers to think of purchasing tractors & other implements for timely completion of there farm operations. Institutional finance helped the farmers to a greater extent for speedy mechanisation of their farm.
With the passage of time, share of purchased inputs gradually increased, but production and income remained more or less static, resulting in reduced savings. To sustain farm mechanisation, farmers started relying on finance – institutional finance if available and private finance if institutional finance is not available.
Progression of green revolution and breaking of family structure are positively co-related in Haryana. More than 6o% family have gone nuclear with the advent of twenty first century. Apart from it elders have lost their pre-eminence in decision making process.
Smaller land holdings, more than required availability of tractors ,increasing cost of cultivation is turning possession of tractor an un-viable enterprise. All these factors are not for an healthy social and economic order and require corrective steps.

Keywords :sociology, farm mechanisation ,development

Introduction

Human spirit always yearns and strive for fulfillment of his self in all facets of life : physically, mentally, morally ,materially, which is popularly called - development. Food is one such item which is required to sustain his body and spirit to continue his endeavor for perfection. In his journey of development ,human being have devised certain tools to handle the process of greater food-grains production-in modern language we call it mechanization. But sometime, certain things go beyond an optimum limit and does not fit into the whole scheme of things and require correction .This situation{over mechanization} has arrived in this part of the earth and present paper is part of the effort to look into the issue and to suggest some remedial measures.
Various studies even in eighties have indicated that pattern of mechanization needs correction .Nandal et al. in their study on" impact of farm mechanization on farm productivity, income and employment in Haryana" have concluded that there is no significant difference regarding cropping pattern and cropping intensity between tractor operated farm and animal operated farms .But in spite ot the reservation expressed by a number of specialists about the advisability of mechanization in a labour surplus country , mechanization continued unabated.

Materials and methods

Two villages kharawar and kharak jatan were taken purposively , one well connected from the district HQ. And national capital via rail and road and other one a little bit of remote area though all villages in the state are well connected by all weather-metalled road. Then a sample of thirty respondent farmers from each village and total of sixty respondent farmers were taken randomly, half tractor owners and half non tractor owners, to collect the primary data by survey method . Considering the nature of the subject, very rigorous statistical model is not used, rather simple tabular analysis was done and these simple facts were used to analyse and understand the complicated human decision making process.

Results and discussion : With the increased population and division in family, holdings per family, are getting smaller and smaller. Majority of the farmers in the state and in the study area are having less than one hectare of land holding. But Ploughing have gone totally mechanized. 6.6 percent farmer in khark jatan and less than one percent farmer in kharawar are having bullocks but they too got plough their field with tractor in exchange of their bullocks doing some other light activities on the farms of tractor owners.

Family composition of the respondent farm family is : 68.3% reported nuclear family living husband wife and their unmarried children and 31.7% are living in a joint family having more than two generation together . 75% respondent reported the consultation of the elders while taking social decisions like marriage etc. but while taking economic decisions 60% reported consultation not necessary though legally elders are the owners of the whole property.

WE found no significant difference regarding cropping pattern, cropping intensity and yield of tractor owners and non tractor owners. In human beings decision making process is very complicated one. In economics we assume human being as a rational creature. But he is not always that much calculating and his decisions are taken to balance the economic rationality and his social ego. When it was asked from the respondent whether they consider purchasing a tractor a social issue or an economic one . Though 81.6%agreed that it is an economic issue but hundred percent agreed that it increased their social standing as well. 80% respondent farmers got finance to purchase tractor and 20% purchased it from their own sources. Out of these loanees 60%are facing problem in repayment of their loan as they are not generating sufficient income and savings to pay for the loan. They are facing difficulty to hire out their tractor for off farm operations as there are number of tractors available competing for the available work.

In village kharawar 83.3%farmers reported sale of milk and 20%from village khark jatan to augment their income . All the sample farmers from both the categories and from both the villages reported preference for the job outside farming .All the tractor owners got 85%paddy transplantation and harvesting done from migrated hired labour and 33.3% non owner farmer got 40% work done from hired labour.

Conclusion

Mechanisation of agriculture is a fact not an option. But unbearable cost of mechanisation ,decreasing income and savings and finding alternative work avenues for the displaced work force, are some of the issues which needs urgent attention of the thinkers planners and administrators . If we fail in resolving these issues rationally and to the satisfaction of all affected population our all developmental achievements will be of no avail. It shall be harmful for our social harmony .Sustainable development needs stable social order. And discontentment anywhere is a potential threat to a just and civil order everywhere.

References:Nandal,D.S. Kadian,R.S.Singh,V.K.and khatry,R.S.1986 "Impact of farm mechanisation on farm productivity, income and employment in Haryana" Research bulletin no 18 by department of agricultural economics Haryana Agricultural University Hisar India.

Software development for vision-based crop remote sensing

Yunseop Kim and John F. Reid
John Deere Technology Center, One John Deere Place, Moline, Illinois 61265 USA
KimJames@JohnDeere.com and ReidJohnF@JohnDeere.com

Abstract

Assessment of crop health condition from vision sensor has been of interest worldwide. However, limited work has been done for developing software for real-time image sensing and processing of crop remote sensing. Therefore, a study was conducted to develop software to acquire and process crop image for real-time assessment of crop characteristics.

Hardware and software configuration of the system were illustrated in Figure 1 and 2, respectively. A crop image from the camera is captured by a frame grabber and signals from other sensors are collected in analog and/or digital format through a data acquisition I/O card. Camera parameters such as gain and exposure are controlled by a computer in each individual channel though a serial communication. Each image is recorded and associated with GPS position through a serial communication. The serial communication is also used for reference chlorophyll readings from a SPAD meter.

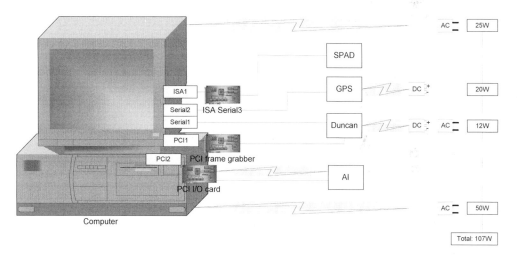

Figure 1. Hardware configuration of crop sensing system.

Figure 2. Software configuration of crop sensing system.

In many cases, researchers develop own algorithms for signal processing, image analysis, control, and need to integrate them into built-in image analysis functions provided by a commercial image software. The purpose of this paper is to show how to utilize commercial image software and integrated custom-developed algorithm into commercial software. A case study was applied on multi-spectral remote sensing.

Consideration of the agro-ecological variability of soil and population parameters in the rationale of measures for precision farming

D. Köppen and B. Eichler

University of Rostock, Department of Agroecology, J. v. Liebig-Weg 6, D-18051 Rostock
bettina.eichler@auf.uni-rostock.de

Abstract

The variability of important parameters of soil fertility and yields were quantified field by field evaluating 23 production experiments from different sites for the calculation of coefficients of variance (s%), their effect on the target values for plot specific fertilization is discussed. These values were rendered more precisely by linking them to the specific variability, under consideration of the half confidential interval (K%). This adjusted target values allow agro-technical measures and reduce the probability of economical and ecological wrong decisions.

Keywords: fertilization, target values, soil management

Introduction

The use of soil analysis data for the calculation of fertilization measures is one main point for precision farming. But the success of measures depends mainly on the precision of these data. Precision farming requires well caused target values as well as actual values which are mainly estimated by evaluating samples. A method was developed for precision of target values under consideration of the variability of actual values.

Materials and methods

The variability of important parameters of soil fertility and yields were quantified field by field evaluating 23 production experiments (experiments without interference = VoE) from different sites for the calculation of coefficients of variance (s%), their effect on the target values for plot specific fertilization is discussed. Practical examples are used to elucidate the methodology.

Results and discussion

These values were rendered more precisely by linking them to the specific variability, under consideration of the half confidential interval (K%) (formulas 1 and 2).
The variability of the actual values strongly influences the essential sample size. P and K showed a low precision in both locations, with s% lying between 17.7 and 62.0 for P and 17.0 and 64.9 for K. When s% = 17.7 and K% = 3.3 the necessary sample size (n_1) for a precise estimate of phosphorus as a parameter of soil fertility (BFK) is n = 114. As this sample size is not feasible in practice, the target value (P = 5.6 mg P 100 g^{-1} soil) must be adjusted accordingly. There is a reduction of precision of soil parameters in the following descending order: pH value, organic matter, Mg, B, Zn, Cu, N_{min}, P, K, Mn. For population parameter the descending order was found as follows: corn weight, amount of corns per ear, yield, density of ears.

formula 1: Calculation of the half confidence interval (K%) and adjusted target value(BFK$_{sp}$) for phosphorus

$$K\% = t_{(\alpha,\ FG)}\ s\%\ n^{-\frac{1}{2}}$$

example from table 1, row 4: $\quad K\% = 1.98_{(5\%,\ 108)} * 49.0 * 109^{-\frac{1}{2}} = 9.29\%$

$BFK_{sp} = BFK_s - (K\%\ BFK_s)\ 100^{-1}$ (algorithm)

example from table 1, row 4: $\quad BFK_{sp} = 5.6 - (9.29 * 5.6)\ 100^{-1} = 5.08$ mg P

K%	= half confidence interval
s%	= coefficient of variation for the parameters (Köppen, 1993)
n	= sample number of VoE
BFK$_{sp}$	= parameter of soil fertility (adjusted target value, mg 100g^{-1} soil)
BFK$_s$	= parameter of soil fertility (target value, mg 100g^{-1} soil)

t-distribution with a significance level $\alpha = 0.05$ for a two-sided questioning and n-1 degree of freedom (FG).

formula 2: Calculation of the half confidence interval (K = mg P 100g^{-1} soil) and adjusted target value for phosphorus

example from table 1   s % = 62,00   K % = 47,66
(row 2)   content class C = 5.6 – 8.0 mg P 100 g^{-1} soil

algorithm from formula 1 $BFK_{sp} = BFK_s - (K\%\ BFK_s)\ 100^{-1}$

		BFK$_s$	K	BFK$_{sp}$
-	upper threshold value + K:	= 8.0 + 3.81	= 11.8	
-	upper threshold value - K:=	8.0 – 3.81	= 4.2	
-	lower threshold value + K:	= 5.6 + 2.67	= 8.3	
-	lower threshold value - K:=	5.6 – 2.67	= 2.9	

Conclusion

The precision of soil analysis data is an important factor for precision soil management. The exactness can be improved by increasing the sample number. But due to the high effort in practise this possibility often cannot applied. Adjusted target values allow agro-technical measures and reduce the probability of economical and ecological wrong decisions.
The developed algorithm pays attention to the specific variation coefficient, and is practicable for already finished samplings too.

Reference

Köppen, D., 1993: Agrochemische Bodenfruchtbarkeitskennziffern zur agrarökologischen Beurteilung von Bodennutzungssystemen. VDLUFA-Schriftenreihe 36, 186 S.

Relations between soil apparent electrical conductivity, available N, pH and soil moisture on a morainic soil under growing barley (*Hordeum vulgare* L.) in SE Norway

A. Korsaeth

The Norwegian Crop Research Institute, Apelsvoll Research Centre, N-2849 Kapp, Norway
audun.korsaeth@planteforsk.no

Abstract

Soil apparent electrical conductivity (EC_a), available N, pH and soil moisture were measured on five occasions during the growth of spring barley on morainic loam in SE Norway. EC_a was measured using a magnetic dipole soil conductivity meter (EM38, Geonics Ltd., Canada). The device was operated manually in both horizontal (EM_h) and vertical (EM_v) modes. The EM_h measurements which gave the overall best results, accounted for 46%, 39%, 30% and 27% of the variation in levels of pH, nitrate-N, moisture and ammonium-N, respectively. Combining the variables with soil moisture in a two-predictor regression model, increased adjusted R^2 in all cases. The strongest relationship was found between EM_h, nitrate-N and soil moisture (adjusted R^2=0.65).

Keywords: ammonium, morainic loam, nitrate, pH, soil apparent electrical conductivity

Introduction

Dense datasets are required to describe within-field variation precisely. Sensor techniques appear to be promising to reduce labour, time and costs compared with traditional methods of sampling and analysis. Soil apparent electrical conductivity (EC_a) has been found to be well correlated with a range of soil properties, such as salinity, topsoil thickness and clay content (Sudduth et al. 2001). EC_a has also been reported to be sensitive to available soil N (Eigenberg et al. 2002). In this paper we present relationships found between EC_a and available N, pH and soil moisture during the growth of spring barley on morainic loam.

Material and methods

Measurements were conducted in a 160 m long field trial, established in barley (*Hordeum vulgare* L.) in 2002 at Kise Research Station (60°46'N, 10°48'E, 130 m asl). The trial had 20 replicate blocks containing five N-level treatments (0, 60, 90, 120 and 150 kg N ha^{-1}, given as calcium ammonium nitrate). Each plot was 1.5 x 8 m. Soil samples were taken from all five treatments in three selected blocks at 0-15 cm depth, shortly before fertilizing/sowing (10.05.02) and then at fortnightly intervals until the beginning of July (23.05, 06.06, 20.06 and 04.07). Analyses comprised nitrate-N, ammonium-N, pH and moisture. At each sampling, EC_a was measured in the same plots, using a magnetic dipole soil conductivity meter (EM38, Geonics Ltd., Canada). The device was operated manually in both horizontal (EM_h) and vertical (EM_v) modes. Linear regression models were used to analyse the data.

Results

Linear regressions showed that both EM_h and EM_v correlated well with the measured variables. EM_h was a better predictor than EM_v (higher adjusted R^2) for available N and pH, but not for soil moisture (table 1). Combining the variables with soil moisture in a two-predictor regression model, increased adjusted R^2 in all cases. The strongest relationship was found between EM_h, nitrate-N and soil moisture (table 1).

Table 1. Relationships found between soil apparent electrical conductivity (mS/m) measured both horizontally (EM_h) and vertically (EM_v) and available N (kg/ha), pH and soil moisture (Water, g $100g^{-1}$).

Dependent variable (Y)	Predictor 1 (X_1)	Predictor 2 (X_2)	Equation	Adjusted R^2 (%)	p-value
EM_h	NO_3-N	-	$Y=61.6+0.42\,X_1$	38.5	<0.001
	NH_4-N	-	$Y=65.9+0.14\,X_1$	26.5	<0.001
	pH	-	$Y=886-124\,X_1$	45.7	<0.001
	Water	-	$Y=35.0+1.14\,X_1$	30.3	<0.001
	NO_3-N	Water	$Y=26.0+0.39\,X_1+1.04\,X_2$	64.6	<0.001
	NH_4-N	Water	$Y=26.2+0.14\,X_1+1.14\,X_2$	57.6	<0.001
	pH	Water	$Y=715-100\,X_1+0.41\,X_2$	47.4	<0.001
EM_v	NO_3-N	-	$Y=74.9+0.20\,X_1$	17.2	<0.001
	NH_4-N	-	$Y=76.8+0.67\,X_1$	12.2	0.001
	pH	-	$Y=628-83.8\,X_1$	41.8	<0.001
	Water	-	$Y=47.0+0.98\,X_1$	46.1	<0.001
	NO_3-N	Water	$Y=42.9+0.18\,X_1+0.94\,X_2$	60.0	<0.001
	NH_4-N	Water	$Y=42.6+0.67\,X_1+0.98\,X_2$	59.1	<0.001
	pH	Water	$Y=360-46.2\,X_1+0.65\,X_2$	53.1	<0.001

Discussion

It seems reasonable that EM_h was a better predictor than EM_v for available N and pH measured in the topsoil, since the relative contribution to the instrument reading from the 0-15 cm layer is larger for EM_h than for EM_v (Sudduth et al., 2001). The correlation between topsoil and subsoil moisture was probably stronger than that between topsoil and subsoil pH and available N content. Moreover, the relative contribution from soil moisture to EC_a would probably increase with depth, due to the presence of a larger proportion of soluble salts from fertilizer in the topsoil. This may explain why soil moisture in the upper 15 cm could be better predicted from instrument readings which were affected more by the layer below 15 cm (EM_v).

Conclusion

Measurement of soil apparent electrical conductivity appears to be a promising method for time and cost effective monitoring of available N during plant growth. The method must nevertheless be calibrated to take into account variations in soil moisture and pH.

References

Eigenberg, R.A., Doran, J.W., Nienaber, J.A., Ferguson, R.B. and Woodbury, B.L. 2002. Electrical conductivity monitoring of soil condition and available N with animal manure and a cover crop. Agriculture, Ecosystem and Environment 88 183-193.

Sudduth, K.A., Drummond, S.T. and Kitchen, N.R, 2001. Accuracy issues in electromagnetic induction sensing of soil electrical conductivity for precision agriculture. Computers and Electronics in Agriculture 31 239-264.

The effect of various parameters on mowing machine material feed rate measurement

F. Kumhála, V. Prošek and J. Mašek
Czech University of Agriculture in Prague, Technical Faculty, Department of Agricultural Machines, Kamýcká 129, 165 21 Prague 6 – Suchdol, Czech Republic
kumhala@tf.czu.cz

Abstract

The effects of various materials and conditions (parameters) changes and their impact on mowing machine material feed rate measurement accuracy were measured under laboratory conditions. The influence of crop variety, crop maturity and moisture, and intensity of conditioning were tested. The impact of the changing parameters in two measurement methods (based on torque-meter and/or on impact plate) was studied. Eight files of torque-meter and/or impact plate measurement were obtained during our experiments. Statistical analysis was used for data evaluation. Two-sample comparisons were used for torque-meter measurement. It is evident from that evaluation that changing crop variety, crop maturity and intensity of conditioning can have statistically significant influence on the measurement based on torque-meter. For impact plate measurement the analysis of variance was used. It was find out that it is not possible statistically determine the influence of tested factors on our measurement. Considering these results the material feed rate measurement based on impact plate is better from practical point of view.

Keywords: mowing machines, feed rate measurement, accuracy

Introduction

Feed rate measurement techniques for mowing machines equipped with conditioner were tested by Kumhala et al. (2001). Results were obtained with mowing machine material feed rate measured by torque-meter placed on a conditioner shaft and by means of curved impact plate mounted on the machine's material output. The carried out measurements proved a very good linear relationship between the conditioner's power input, output frequency of the apparatus measuring impact force by means of the impact plate, and material feed rate through the mowing machine. The calculated coefficients of correlation were about 0,95.

The main aim of the research was to find out effects of various materials and conditions (parameters) and their impact on accuracy of mowing machine material feed rate measurement under laboratory conditions. The influence of crop variety, crop maturity and moisture, and intensity of conditioning were measured. The impact of changing parameters in both measurement methods (torque-meter and impact plate) was studied.

Materials and methods

A mowing ŽTR 216 H machine (Agrostroj Pelhřimov Company, Czech Republic) was used for the measurements. The mowing machine was equipped with an electronic measuring unit developed in our laboratory. The mowing machine's conditioner shaft was supplied with a torque-meter based on resistant strain gauges and with a RPM optical sensor measuring number of conditioner shaft revolutions. Besides the torque-meter, the mowing machine was equipped with a curved impact plate mounted at the exit of the machine. The material ejected from the mowing machine conditioner struck the impact plate. The force created from the change in direction was measured in this way. The curved impact plate was equipped with four elastic members with strain gauge bridges for force measurement. Two types of material were used for our measurement: alfalfa with grass mixture and grass from natural meadow. These crops were harvested on the same landholding but at different crop maturity and moisture content. Moisture content varied from 82 % to 72 % in alfalfa with grass mixture and from 77 % to 74 % in grass. Two levels of conditioning intensity were set up and tested during the measurement with each different material.

Results and discussion

The Statgraphics®Plus computer software was used for data evaluation. Eight files of torque-meter measurement were obtained during our experiments. Since there was found out a statistically significant difference amongst standard deviations at 95 % confidence level, it was not possible to use analysis of variance. It was decided to use two-sample comparison for that reason. Every file of measured data was compared with each other. It results from that comparison that out of 28 tested pairs of files in 12 cases exists a statistically significant difference between the means of two samples at the 95 % confidence level. It is clear from some comparisons than the change of intensity of conditioning itself or the change of crop maturity itself or the change of crop variety itself can make statistically significant difference between measured files. All tested factors can have statistically significant influence on the measurement based on torque-meter for that reason.

Eight files of impact plate measurement were obtained during our experiments as well. Statistically significant difference amongst the standard deviation of tested files was not indicated. It was possible to use analysis of variance for data evaluation in that case. The results of analysis of variance for these files are in Table 1.

Since the P-value of the F-test is in this case greater to 0,05, there is not statistically significant difference between the means of tested files at 95 % confidence level. It follows from the analysis that it is not possible to find the mean significantly different from others. It is not possible to determine statistically significant influence of tested factors changes on impact plate measurement from our experiments.

Table 1. Analysis of variance results for files obtained during impact plate measurement.

Source	Sum of Squares	Df	Mean Square	F-Ratio	P-value
Between groups	13669.6	7	1952.81	0.61	0.7441
Within groups	244620.0	83	3188.19		
Total (Corr.)	278290.0	90			

Conclusion

As it follows from the measurement carried out, the measurement method based on using impact plate can be less sensitive to material and intensity of conditioning changes. Another advantage for the impact plate utilization can be its simplicity and lower costs in comparison with the torque-meter. It could be possible to recommend an infield measurement arrangement to obtain relationships under the real field conditions.

References

Kumhála, F., Kroulík, M., Heřmánek, P., Prošek, V.: Yield Mapping of Forage Harvested by Mowing Machines. Tagung Landtechnik 2001 (Conference Agricultural Engineering 2001) VDI-MEG Düsseldorf, VDI Verlag GmbH, 2001, Germany, p. 267-272.

A agro-ecosystem simulation model for precision agriculture

M. Ladányi, M. Gaál, L. Horváth, L. Hufnagel, A. Révész and É. Erdélyi
Szent István University, Budapest, Villányi út 29. 1118, Hungary
mladanyi@kee.hu

Abstract

In this paper five different models, as five modules of a complex agro-ecosystem are investigated. The water and nutrient flow in soil is simulated by the *nutrient-in-soil model* while the biomass changes according to the seasonal weather aspects (handled by the *activity evaluator model*), the nutrient content of soil and the biotic interactions amongst the other terms of the food web are simulated by the *food web population dynamical model* that is constructed for a piece of homogeneous field. The numbers of individuals of the different populations in every phenological phase are given by the *phenology model*. The *food web model* is extended for inhomogeneous field by the *spatial extension model*. Finally, as an additional module, an application of the above models for *multivariate state-plane*, is given. The modules built in the system are closely connected to each other as they can utilize each other's outputs, nevertheless, they work separately, too.

Keywords: food web model, activity function, spatio-temporal simulation, multivariate state-planes

Introduction

Applying informatics and electronics, agricultural production adjusted to a special region became more and more urgent to be realized. By 'precision and sustainable agriculture' we mean not only a new production method but also a complex system that handles biologic, technological and economic factors and joins the natural circumstances flexibly which aims the optimisation of proficiency and environmental protection together with the forecast of risk, damages and profit.

Materials and methods

To simulate the interactions a discrete difference equation system with daily scale is used. As in the literature there are plenty of excellent models describing certain parts of processes quite exactly, our aim was to create a model that describes the whole interaction process in order to be able to apply it also in case detailed data are missing and to extend it in case more complex data are available. The models were created in such way that they can work separately. The outputs of both the *activity function evaluator model* and the *nutrient-in-soil model* can be applied by the *food web population dynamical model*. The output of the *food web population dynamical model* can be built in the *nutrient-in-soil model*, the *phenology model* and the *spatial extension model*. The *multivariate state-plane module* can be based both on the *food web* and the *spatial models* and it can be completed with data related to the environment.

Results

An agro-ecosystem is directed mainly by the interactions among the populations living in the agro-ecosystem together. Several indirect or hidden types of interactions that can not be expressed as different kinds of material flow such as competition, the indirect interactions that can be derived from the escape from or the defence against a common predator as well as the so-called 'top down and bottom up regulations' are involved in our *food web population dynamical model* for the biomass change. The general equation of the *food web model* is based on three elements: the first one is to express the activity of individuals depending on temperature, the second one is to describe the effect of the quality and the quantity of the nutrient available for the populations and the third one is to display the effect of the predators. The model can be based both on a simplified *nutrient-in-soil model*, that describes the water and nutrient flow in soil, and the *activity function evaluator model*, that expresses the effect of temperature.

Applying the above models it is obvious to ask how the number of individuals can be derived from a given amount of biomass. More exactly, if the phenological phases of the population together with their biological properties are known, how to define the number of the individuals in the different phases at a given point of time. The sketched problem aimed to be solved by the so-called *phenology model*.

The above methods have been worked out for a homogeneous piece of field. For web systems, however, which contain migratory animal populations, too, the models should be extended for inhomogeneous field as well. Therefore, the examined field was divided to parcels and considered to be separately homogeneous. Based on the hypothesis of abundance and the one of uniformity the *spatial extension model* describes how the quantity of the elements of the food chain changes on the different parcels in time. There are individuals of two kinds of pests represented in the model, the first one can arrive from the neighbouring parcels, only, the second one can attack from any parcel.

The above mentioned spatio-temporal simulation models offered new possibilities to use the multivariate state-planes.

For short- and long-term simulated results with daily average temperature and precipitation data see the poster. Right there you can find a case study, that simulates the temporal changes of the different parcels in an inhomogeneous field.

Discussion

Based on the case studies, our systems seem to be suitable to solve the problems mentioned in the introduction, so the systems are still under thorough and extensive validation, development and generalization.

References

Horváth, L., Gaál, M.,- Hufnagel, L., (in press): Modelling of Spatio-temporal Patterns of Eco-systems in Agricultural Fields. In: EFITA 2003, Debrecen, Hungary

Ladányi, M., Hufnagel, L., (in press): A Phenology Model Nested in an Eco-system Model for Agro-ecological Processes. In: EFITA 2003, Debrecen, Hungary.

Ladányi, M., Erdélyi, É., Révész, A., (in press): An Ecosystem Model to Simulate Agro-ecological Processes. In: EFITA 2003, Debrecen, Hungary.

Acknowledgements

We would like to express our thanks to Professor Harnos, Zs. for supporting our work at Szent István University, Department of Mathematics and Informatics.

Using NDVI as a support to sugar cane yield estimates

R.A.C. Lamparelli, J.V. Rocha, H.M. Machado and J. Zullo Jr.
Unicamp, Cidade Universitária, Campinas - São Paulo - CEP:13083-970, Brasil
jansle@agr.unicamp.br;rubens@cpa.unicamp.br;jurandir@cpa.unicamp.br;machado@agr.unicamp.br

Abstract

Production estimates of sugar cane do not take into account the spatial variability of biomass. This study aims to demonstrate the viability of using remote sensing to improve yield estimates. Multitemporal analysis of NDVI calculated from 5 Landsat 7 /ETM+ images was carried out in order to monitor the crop vegetative cycle.This methodology showed an increase in accuracy of close to 50% in the estimates. In order to evaluate if the NDVI corresponded with biomass values classified as low, regular and high, a sample of biomass for each level was taken and the results showed a good correlation between biomass and NDVI values.

Keywords: sugar cane, yield estimate, Landsat 7, NDVI

The yield estimate process in sugar cane mills in Brazil is normally carried out without considering the spatial variability of the plantation within the fields. Remote Sensing (RS) has two characteristics that can improve this process: the temporal availability and the sinotic view (Diker et al., 1998; Pellegrino, 2001). Within this context this study aimed at demonstrating the viability of using RS data to improve the yield estimate process carried out by sugar mills.

Methodology

Five Landsat 7/ETM+ images of different dates representing the vegetative cycle of sugar cane were used in the study. NDVI was calculated from the imagery, with the dates matching with those used by the mill to estimate biomass using the traditional process of field observation. The NDVI images were reclassified in 3 different classes, using unsupervised digital classification, representing the biomass quantity (low, medium, high). The resulting maps were used by the field scouters to re-evaluate their estimates, now with exact location for field observation and biomass estimates. Using the new estimates for each class o NDVI values, a new production estimate for the fields was calculated using the sum of the areas of each class multiplied their respective estimates.

Results

Figure 1a shows NDVI image of one of the observed areas, where it is demonstrated the NDVI differences within the fields, corresponding to biomass production spatial variability measured on field (Area 1). Figures 1b and 1c show the location from where biomass samples were taken in one of the areas studied (Area 3) as well as the classified image with 3 classes of NDVI. The yield values for each class in Figure 1c were multiplied by the class area and integrated in order to obtain the field total production, shown in Table 1, for the 3 test areas.

Figure 1. a) sample points (correlation between NDVI and biomass); b) biomass differences; c) estimates from the samples based on the unsupervised classification of NDVI.

Table 1. Comparison of the total production (in kilos) in test areas using traditional field scout methods (3) and field scout based on NDVI classified image (1).

Area	NDVI estimate (1)	Total Harvested (2)	Mill estimate (3)	$(2-1)/2$ %	$(2-3)/2$ %	Improvement in Accuracy %
Area 1	12567,00	12011,91	13792,80	- 4.62	-14.8	24.5
Area 2	10515,70	10833,50	11400,40	2.93	- 5.23	43.9
Area 3	5226.70	4813.09	5501.32	- 8.6	- 14.22	39.5

Discussion

The results showed that there was an agreement between NDVI variability and biomass data colected at field level. That correlation allowed to generate new adjusted estimates based on the NDVI varibility and classification. This methodology has shown a gain in accuracy between 20% and 50% in the final estimates as showing the data in Table 1.

Conclusions

It is possible to map differences in sugar cane biomass production using NDVI derived from Landsat7/ETM+ imagery.These maps can help field scouting, locating sites for more accurate field observation and yield estimates.There is a considerable improvement (20% to 50%) in accuracy of sugar cane production estimates using NDVI-derived maps.

Acknowledgements

The authors acknowledge FAPESP – São Paulo State Research Fund, Brazil, for funding this study (Project 99/07071-1 – Remote Sensing Applied to Precision Agriculture in Sugar Cane) as well as COSAN S.A., for providing the reference data.

References

Diker, K.; Bausch, W.C.; Podmore, T.H. 1998. GIS Mapping of Plant Parameters and Yield Potential Estimate by Remote Sensing. ASAE Meeting. Orlando/Flórida. July 12-16. Presentation Paper n.983143.
Pellegrino, G. P. Utilização de dados espectrais do satélite NOAA14/AVHRR como fonte de dados para modelos de estimativa da fitomassa da cana-de-açúcar (Sugar cane biomass models estimate using NOAA14/AVHRR spectral data). 2001. 114p. Tese. Feagri/Unicamp-SP. Brasil.

High resolution satellite images for land use classification

R. Laudien, G. Bareth and R. Doluschitz
Department of Agricultural Economics, Division: Agricultural Informatics and Farm Management, University of Hohenheim, Stuttgart, Germany
laudien@uni-hohenheim.de

Abstract

This paper presents the approaches and methods of generating land use classifications with high resolution QuickBird data, using the software packages ERDAS® Imagine and eCognition. A comparison between these two classification approaches is discussed.

Keywords: Land use classification, object/pixel oriented analysis, QuickBird

Introduction

The aim of the project "Application of remote sensing and GIS for sugar beet companies" is to create a user-friendly management information system. One objective in this project is to use high resolution QuickBird satellite data to detect sugar beet fields which are located in the area of interest. The very high spatial resolution of 61cm in PAN mode and 2.44m in multispectral mode delivers very accurate data for remote sensing analysis. Especially the near infrared band (760 to 900nm) is important for agricultural and land use classification applications.

Materials and methods

A QuickBird image of October 2nd, 2002 was used to generate a land use classification for the area of interest. As a pixel oriented analysis software, ERDAS® Imagine 8.6 was used to create a supervised land use classification. For each crop, training samples were digitized to enable the supervised classification. These training samples were collected in field campaigns and GPS measurement results were added to the dataset. For pixel based classification, the maximum likelihood decision rule was used (ERDAS® LLC, 2002, 234). eCognition analysis software is knowledge as well as object based. It was used to point out the difference between the pixel based approach and the object oriented one. The first step in eCognition is to create polygon objects out of the satellite scene. This is carried out by segmentation, in order to extract image objects in the first step. In the second step, the classification of the objects is done by using the same training samples and the nearest neighbourhood analysis (Baats et al. 2001, 4-48ff). The results of the two different classification approaches and the training samples were used for evaluation.

Results

Figure 1 shows a QuickBird subset image of the study area (2030x1200m). It was taken on October 2nd, 2002. Since QuickBird is a very high resolution satellite, details and differences of every field could be pointed out.

Figure 1. MS QuickBird subset image of the study area (left); Land use: ERDAS® Imagine 8.6 result (middle) and eCognition 2.1 result (right).

Different land use classes could be displayed (Figure 1, middle, right). The classification result of ERDAS® Imagine is characterized in heterogeneous fields, which contain many different classes. The overall impression regarding quality of the classification is less then the one of eCognition, were the fields are clearly separated form each other by visible borders.

Discussion

Problems of pixel based classification application, i.e. computing performance, result in using object oriented image analysis software. Blaschke and Strobl (2001) as well as Hoffmann and Van der Vegt (2001) described that, especially in high-resolution images, it is very likely that neighbouring pixels belong to the same land cover class as the pixel under consideration. This fact is also demonstrated in Figure 1 (middle and right). With per-pixel classification methods, pre-processing is required to avoid "salt-and-pepper" effects and reduce the number of objects in the scene (Bauer & Steinnocher, 2001, 27). The eCognition software generates more homogeneous classification results than these of the pixel based one. Field borders could be identified a lot easier and the well known mixed pixel phenomenon is irrelevant due to the fact that this approach classifies whole segments (Baats et al., 2001, 1-1).

References

Baats, M., Benz, U., Dehghani, S., Heynen, M. Höltje, A., Hofmann, P., Lingenfelder, I., Mimler, M., Sohlbach, M., Weber, M. and Willhauck, G. 2001. eCognition User Guide. München, Definiens Imaging GmbH, Germany.
Bauer, T and Steinnocher, K. 2001. Per-parcel land use classification in urban areas applying a rule-based technique. GeoBIT/GIS 6 24-27.
Blaschke, T. and Strobl, J. 2001. What's wrong with pixels? Some recent developments interfacing remote sensing and GIS. GeoBIT/GIS 6 12-17.
ERDAS® LLC 2002. ERDAS Field Guide, Sixth Edition. Atlanta/Georgia, USA.
Hoffmann, A and Van der Vegt, J. 2001. New Sensor systems and new Classification Methods: Laser-and Digital Camera-data meet object-oriented strategies. GeoBIT/GIS 6 18-23.

Acknowledgment

This study was carried out in cooperation between a sugar beet company in southern Germany, and the University of Hohenheim, Germany.

Variable rate fertilization of nitrogen in direct seeded rice field

H.J. Lee, J.H. Chung, J.H. Seo, S.H. Lee, A. Chun and C.H. Yi
Department of Agronomy, Seoul National University, Suwon, Korea
hojinlee@snu.ac.kr

Abstract

Decreasing variation of yield and growth is a main effect of variable rate fertilization. In direct seeded rice field, using variable rate nitrogen fertilization by soil testing was effective in decreasing yield variation and soil pollution.

Keywords: variable rate fertilization, direct seeded rice, maps, nitrogen fertilizer.

Introduction

Variable rate fertilization is a developing technique that can increase the efficiency of fertilizer and can decrease the environmental pollution. This experiment was conducted to verify the effect of variable rate nitrogen fertilization in rice paddy field.

Materials and methods

Rice cultivar was *Oryza sativa* L. cv Nampyeongbyeo, a late maturing Japonica. We divided a rice field to control and VRT (variable rate treatment) plots. The control plot was applied with conventional rate of N fertilizer. While, VRT plot was applied with variable rate of nitrogen that based on nitrogen fertilizer recommendation. Soil samples of each plot were collected before puddling to make the map of nitrogen fertilizer recommendation for variable rate fertilization. Grid sampling was taken from each of the 9.8m × 8.9m plots in each 107.6m × 35.6m fields. We collected the data; plant height, number of tillers, dry weight, LAI (leaf area index), chlorophyll meter values and yield at heading stage from each plot of each field. Furthermore, we took soil water samples by irrometer to determine the soil pollution. Geostatistical analysis and mappings were performed by GS+ software.

Results

Average amount of nitrogen fertilizer reduced in VRT plot was 18kg/ha less than in control. Most of growth values in VRT plot were higher than in control. Yield variations were 1.96t/ha to 6.43t/ha in control plot, but 5.87t/ha to 8.73t/ha in VRT plot. Average yields were 4.94t/ha and 6.60t/ha in control and VRT plot respectively (Figure 1.).

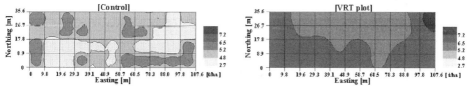

Figure 1. Kriged maps of direct seeded rice yield.

Conclusion

In VRT plot, variation of yield and growth parameters was minimized than control. Variation of soil water nitrate concentration and average soil water nitrate concentration in VRT plot was decreased than control.

References

J.E. Sawyer. 1994. Concepts of variable rate technology with considerations for fertilizer application. J. Prod. Agric. 7: 195-201.

C.S. Lee. 1998. Recommendation of fertilizer amounts for crops based on soil testing, symposium for soil testing and fertilizer: Jeju national university: 139-151.

J.H. Seo, H. J. Lee, Y. S. Jung and S. H. Lee. 2002. Soil and yield mapping and nutrient recommendation for precision agriculture in rice paddy. J. Korean. Society of Precision Agriculture 1: 51-60

Acknowledgements

This study was supported by Technology Development Program for Agridulture and Forestry, Ministry of Agriculture and Forestry, Republic of Korea.

Thermography as a tool for the forecasting of fungal infections of wheat

J.-H. Lenthe, E.-C. Oerke and H.-W. Dehne
University of Bonn, Institute for Plant Disease, Nussallee 9, 53115 Bonn

Abstract

Thermography as a technique for monitoring the micro-climate within wheat field has been tested. Special interest is focused on the detection of liquid water on wheat leaves and the recording of the process of canopy drying. Recordings of the microclimate whithin wheat plots showed potential for the differentiation of disease risk zones. Identification of leaf wetness was attempted in preliminary experiments. These informations are to be used for a site-specific forecast of climatic conditions conductive to an infection of wheat by the fungus *Puccinia recondita* causing brown rust.

Keywords: thermography, microclimate, leafwetness, site-specific, forecast

Introduction

For the initiation of germination uredospores of *Puccinia recondita* causing leaf rust of wheat demand contact with waterdrops for at least 4 hours (Kluge et al. 1999). Growth stage, plant habitus and weather data such as precipitation, air temperature, soil temperature, relative humidity (RH) and windspeed are relevant parameters for the incidence and duration of water on leaves within the crop. In regard of a site-specific fungicide application information on the heterogeneity of micro-climatic conditions resulting in site-specific differences in the epidemic development of fungal diseases are needed. Portable weather stations permit the monitoring of these parameters, however, they are limited in regard of planar resolution, depending on the number and sensitivity of sensors. Thermography permits the monitoring of surface temperatures with high thermal and spatial resolution (Schuster & Kolobrodov, 2000) This method is going to be evaluated for its potential in monitoring the microclimate of wheat crops in order to forecast the incidence of leaf diseases.

Materials and methods

Seven weather stations with sensors for the measurement of air temperature, relative humidity, soil temperature and leaf wetness were distributed over three plots (45 x 45m) within a wheat field in the summer 2002, recording from 16.5.02 until 4.7.02.
Preliminary experiments were carried out with the thermography system Varioscan 3021 (Jenoptik, Jena). The drying of wet wheat canopies was simulated in the greenhouse in 0,8 x 1,2m plots at 16 °C and 70% RH and 18 °C and 30% RH. The focus angle was 50° off the horizontal.

Results

Measurements in the field with common equipment have shown variation in leaf wetness occurrence ranging from 18 to 29% of the recording period of 1196 h, resulting in 11 to 18 periods of wetness exceeding the minimal required leaf wetness duration of 4 hours for uredospores of *Puccinia recondita*. Within one plot the greatest range in number of wetness periods was 11 to 17, equivalent to 226 to 364 hours of wetness. Within a distance of 5 m differences of up to 30% of summed wetness duration were detected. The longest recorded wetness period lasted 92 hours. The RH in this period never fell below 50% even in the afternoon when commonly values around 30% are measured.

Thermography of a wheat canopy in greenhouse scale showed the maximum temperature difference between the wet and dry region (including a soil proportion) of 1,8 °K at high RH and difference of up to 4,5 °K at low RH.

Discussion

The use of common sensors for leaf moisture showed a considerable heterogeneity in leaf wetness duration within a wheatfield. However, a coverage that would allow interpolation for a complete coverage of a plot and correlation of microclimate with disease occurrence will require a much closer net of sensors.

First attempts to detect humidity within a canopy by thermography showed the cooling effect of vaporizing water. However, the influence of the air to leaf water vapor pressure saturation deficit has to be considered. At high RH the vapor pressure saturation deficit of the surrounding air is low and water on leaves will vaporize at a lower rate than at low RH. Therefore, the cooling effect will be smaller. Since rain and dew in nature coincide with high RH, there is a possibility that the effect of canopy cooling due to vaporisation could be rather small. Furthermore, under field conditions there is no dry reference region that can be compared to a wet canopy. Since canopy temperature by itself cannot tell wether the canopy is wet or dry a reference is necessary. One approach may be the modeling of a theoretical dry-canopy-temperature through air temperature. Another approach may be the interpretation of patterns of different canopy temperatures within the canopy.

Conclusions

Within a crop canopy considerable differences in wetness condition at points 5 to 100m apart can be found. This holds potential for a good correlation of local microclimate and the occurrence of brown rust in wheat.

These first insights into thermography have to be confirmed under field conditions because simulation of defined climatic conditions on a greenhouse scale puts surpassing demands on technical equipment.

References

Kluge, E., Enzian, S., Gutsche, V. 1999. Atlas der potentiellen Befallsgefährdung durch wichtige Schadorganismen im Ackerbau Deutschlands. BBA für Land-u. Forstwirtschaft Berlin u. Braunschweig, ISBN 3-930037-01-7.

Schuster, N., Kolobrodov, V. 2000. Infrarotthermographie, Wiley-Vch, Berlin 2000, ISBN 352740130X.

Acknowledgements

This project is part of a graduate research and training program funded by DFG (Deutsche Forschungsgemeinschaft).

We thank InfraTec GmbH, Dresden for their support of the purchase of the thermography system Varioscan 3021.

Visualization of downy mildew development in cucumber using thermography

M. Lindenthal[1], E.-C. Oerke[1], U. Steiner[1], M. Blanke[2] and H.-W. Dehne[1]

[1]*Institut for Plant Diseases, University of Bonn, Nußallee 9, 53115 Bonn, Germany*
[2]*Department of Horticulture, University of Bonn, Auf dem Hügel 6, 53115 Bonn, Germany*
miriam.lindenthal@gmx.de

Abstract

The development of *Pseudoperonospora cubensis* causing downy mildew in cucumber was investigated using non-destructive thermography. Early infection stages and expansion of fungal growth in the leaves were analysed in thermographic images. Leaf temperature was strongly correlated with transpiration rate. To differentiate stomatal from transcuticular transpiration due to damaged tissue the water loss from healthy and infected leaf areas was measured under light and dark conditions.

Keywords: infrared thermography, leaf temperature, fungal infection, transpiration rate.

Introduction

Plant pathogens damaging leaf tissue are known to have an influence on the transpiration rate of plants (Gould et al, 1996) and therefore on the leaf temperature. In this study the effect of a fungal infection on the leaf temperature was investigated by infrared thermography, using cucumber and *P. cubensis* as model species. Additionally the stomatal conductance which is supposed to affect leaf temperature was considered (Chaerle et al, 1999).

Materials and methods

Two weeks-old cucumber leaves were inoculated with 2.5 ml sporangia suspension of *P. cubensis* containing $50*10^4$ zoosporangia per millilitre. High-resolution thermographic images from healthy and infected tissues were obtained using an infrared scanning camera (VARIOSCAN 3201 ST, Jenoptic Laser, Jena, Germany) with a spectral sensitivity from 8 to 12 μm and a geometric resolution of 1.5 mrad at the 3rd, 5th and 7th days after inoculation. Simultaneously transpiration rate and stomatal conductance were measured with a portable porometer type CIRAS-1 with automated gas mixing and a Parkinson leaf chamber type PLC-B (PPSystems, Hitchin, UK).

Results

An initial thermal response of cucumber leaf tissue to *P. cubensis* was observed one day before visual symptoms caused by *P. cubensis* appeared; thermal spots developing typical symptoms of downy mildew during later stages of pathogen development were colder up to 0.8 °C than the surrounding non-infected plant tissue. The colder areas expanded into neighbouring tissue; simultaneously the tissue in the centre of the spots became necrotic and temperature increased above the level of healthy leaves. Finally areas with visible necrotic cell death had a temperature about 1 °C higher than non-infected tissue.

The supposed negative correlation between leaf temperature and transpiration rate could be ascertained (Figure 1).

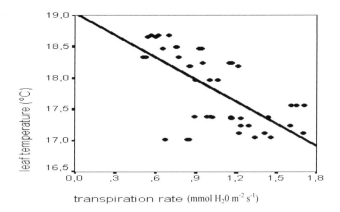

Figure 1. Correlation between transpiration rate and leaf temperature of cucumber leaves infected by *Pseudoperonospora cubensis* (p < 0.01; r = -0.7).

Discussion

The thermal effect at early stage of infection results from enhanced transcuticular transpiration probably due to ruptured leaf tissue by *P. cubensis*. As a result of the further spread of the fungal mycelium into plant tissue at an advanced state of infection proximate cell regions were also effected and showed higher transpiration rates, consequently the front of lower temperature expanded. Subsequently the temperature of the tissue increased due to interrupted water supply. Ultimately the high temperature of necrotic spots results from the desiccation of dying tissue due to the lack of natural cooling from transpiration (Chaerle et al, 2001).

Conclusions

A non-destructive pre-symptomatic detection of fungal diseases and the observation of disease progress at plant level seems to be possible by thermography. However, for the interpretation of thermographic images the influence of the infection stage of pathogens on transpiration rate and temperature of plant leaves has to be taken into account.

References

Chaerle, L., De Boever, F., Van Montagu, M., Van Der Straeten, D. 2001. Thermographic visualization of cell death in tobacco and *Arabidopsis*. Plant, Cell and Environment 24 15-25.
Chaerle, L., Van Caeneghem, W., Messens, E., Lambers, H., Van Montagu, M., Van Der Straeten, D. 1999. Presymptomatic visualization of plant-virus interactions by thermography. Nature Biotechnology 17 813-816.
Gould, A., Aldrich, J., Andersen, P. 1996. Influence of pecan scabon gas exchange and chlorophyll content of pecan leaves. Plant Disease 80 317-321.

Acknowledgements

This research was supported by grants from the German Research association (DFG).

Using reflectance measurements to evaluate water stress and take-all disease (*Gaeumannomyces graminis* var. *tritici*) of *Triticum aestivum* L.

J. Link, S. Graeff and W. Claupein
Institute of crop production and grassland research (340), University of Hohenheim, Fruwirthstr. 23, 70599 Stuttgart, Germany
johannalink@web.de

Abstract

Two greenhouse studies were carried out to evaluate water stress and take-all disease by use of reflectance measurements. Leaf scans were performed on the youngest, fully developed leaf with a digital LEICA S1 Pro camera. Data were analyzed within the L*a*b*-color system. Water stress led to significant reflectance changes, which differed from changes due to take-all disease in selected wavelength ranges (e.g. 510 – 780 nm). Reflectance changes as a result of water stress were attributed to changes in leaf pigments and anatomical leaf structure, whereas reflectance changes, because of take-all disease may be a result of additional nutrient deficiencies.

Keywords: reflectance, take-all, water stress, wheat.

Introduction

Water stress, due to drought, is one of the most limiting factors in crop production. Take-all disease is a widely occurring parasite of cereals. Under field conditions, it is difficult to discriminate between these two stress factors, because visible symptoms are nearly similar. In order to decide on suitable treatments for these stress factors, non-destructive remote sensing methods might be of advantage. Thus, the objectives of this study were i) to evaluate reflectance changes throughout the visible and near-infrared spectra due to water stress and take-all disease, ii) to select wavelength ranges responsive to water stress and take-all disease, and iii) to assess a possible discrimination of water stress and take-all disease based on stress specific reflectance patterns.

Materials and methods

Two greenhouse studies were conducted to investigate the use of reflectance measurements for crop water status assessment (experiment WS) and take-all infestation (experiment TA) of winter wheat genotypes. Plants used in experiment WS were germinated in soil over 22 d and afterwards cultivated in nutrient solution. Water stress was induced after 36 d by adding Polyethylene-Glycol (PEG) in different concentration levels (0, 36, 72 and 144 g PEG l^{-1}) to the nutrient solution. Plants used in the experiment TA were grown in Mitscherlich pots (5 kg). The soil was a loamy sand with a pH of 7.4. Take-all disease was induced by adding inoculated rye seeds to the soil at the sowing date. The inoculum was placed in a layer 5 cm below the wheat seeds in rates of 0, 5, 10, 15 and 20 g inoculum g^{-1} soil. Leaf scans were performed on the youngest, fully developed leaf with a digital LEICA S1 Pro camera under controlled light conditions. Measurements were conducted each day over a period of 10 d in experiment WS, and once a week in experiment TA over a period of 5 w. The employment of different longpass filters allowed the investigation of single wavelength ranges throughout the spectra of 380 – 1300 nm. Leaf scans were analyzed with the Software ADOBE® Photoshop 6.0 in the L*a*b*-color system by splitting the scans into a* parameters (green/red percentage

481

of a color) and b* parameters (blue/yellow percentage of a color). Analysis of variance (ANOVA) was performed on all crop and reflectance data using the general procedures of Sigma Stat 2.0 (Jandel Scientific, San Rafael, CA). Tukey tests were carried out for comparison of means.

Results und discussion

Water stress led to significant reflectance changes of wheat leaves. As water stress increased, the b* parameter reached higher values, both in the visible and near-infrared spectra. In the visible domain reflectance is influenced by leaf pigments such as chlorophyll and in the near-infrared domain it is affected by anatomical structure of leaves (Guyot, 1990). Two days after the induction of water stress changes in reflectance within the wavelength range of 510 - 780 nm were detected. Correlation between the water content of the youngest, fully developed leaves and the b*-values increased during the experiment (data not shown). Within all wavelength ranges the a*-parameter was not responsive to water stress (Fig.1). Take-all disease affected reflectance, both in the visible and near-infrared spectra, too. The change in reflectance in some wavelength ranges seems to be similar as under water stress (data not shown). But in contrast to water stress, both a* and b* parameters were affected, as shown for the wavelength range of 510 – 780 nm (Fig. 2). These differences may be caused by additional nutrient deficiencies, e.g. manganese as a consequence of take-all disease. Manganese deficiencies influence cell expansion (Abott, 1967) and ingredients. Further investigations are necessary to evaluate the physiological parameters responsible for the reflectance changes.

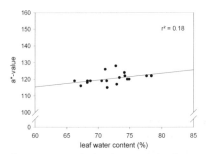 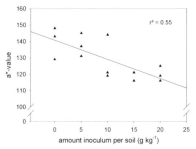

Figure 1. Correlation between leaf water content (%) and the a*-values within the wavelength range of 510 – 780 nm.

Figure 2. Correlation between amount inoculum per soil (g kg^{-1}) and the a*-values within the wavelength range of 510 – 780 nm.

Conclusions

The results indicate that reflectance measurements in selected wavelength ranges (e.g. 510 – 780 nm) may provide a powerful tool for the specific identification of water stress and take-all disease in wheat plants. Further investigation to detect respective reasons are necessary.

References

Abbott, A. J.. 1967. Physiological effects of micronutrient deficiencies in isolated roots of *Lycopersicon esculentum*. New Phytol.. 66. 419-437

Guyot, G.. 1990. Optical properties of vegetation canopies. In: Applications of remote sensing in agriculture. Steven, M. D. & J. A. Clark (eds.). Butterworths. 19-43.

Processing and mapping of yield data from combine harvesters

Gang Liu[1], Yichun Xing[2] and Limei Zhang[2]

[1]*Research Center for Precision Agriculture, Key Laboratory of Modern Precision Agriculture System Integration, China Agricultural University P.O.Box No.63, No.17 Qinghua Donglu Road, Haidian District, Beijing, P.R. China, 100083*

[2]*College of Mechanical and Electric Engineering, Agricultural University of Hebei, Baoding, Hebei, P.R. China, 071001*

Abstract

The understanding of spatial variability of crop yield is fundamental to the development of precision agriculture practices. The purpose of the research described here is: (1) to pre-process the yield data from combine harvesters and eliminate as many erroneous yield data as possible without changing the actual fluctuations in the yield, (2) to assess the suitable spatial interpolation techniques to create the higher accuracy crop yield map. The research constituted of a field of approximately 23 hectare in Beijing Precision Agriculture Demo Farm, in which crop yield were measured by the CASE 2366 combine. Some error data were eliminated using the statistic method (Interval Estimation), and the original crop yield data sets were modified and transferred by the Moving Weighted Average Method. Three interpolation methods such as Inverse-Distance-Weight (IDW), Spline and Kriging were compared and analyzed by MSE. The best interpolation methods were finally determined. The precise yield spatial distribution map was constructed by GIS software ArcView.

Keywords: Precision agriculture; IDW; Kriging; GIS

Introduction

Precision agriculture has been advanced rapidly in recent years by the production of yield maps which are created by continuously "weighing" the flow of crop as it is harvested from the field by the combine harvester, and recording their position by reference to a satellite based global positioning system (GPS). Complexities of throughput lag (grain residence time in the harvester), actual cutting width, vehicle attitude, load, speed and other factors have all contributed to yield errors of magnitude and location thus limiting the accuracy of combine harvester yield measurements. With the growing use of yield mapping systems in combine harvesters, more attention is being paid to questions of the quality of the yield data, problems connected with further processing of these data and problems of the suitable spatial interpolation techniques. The precise yield spatial distribution maps are finally constructed.

Materials and methods

The experimental field is located at Beijing Precision Agriculture Demonstration Farm. The field surveyed is approximately 23 hectare and. Crop yield data including the grain yield, the grain moisture and theirs corresponding position were collected by a CASE 2366 combine fitted a grain mass flow monitor, a grain moisture monitor and a DGPS.

The basic idea of point data pre-processing involved in identifying extremely high and extremely low values using the statistic method (Interval Estimation) and transferring original crop yield data sets to a rational data sets by the moving weighted average method. Firstly, the descriptive statistics of a wide variety of yield data set suggested that all yield data were normally distributed therefore; the mean values (μ) and standard deviations (σ) of the yield data were used to define the value areas (Juerschik, 1999). The yield data outside mean yield ±2σ were removed in term of 95% reliability. Secondly, the mean of all data of the area in every selected grid is present to every point in the grid through the moving weighted average method. Every two neighboring grids cover the same strip as 6m (the width of the cutter), and points in that strip has their weight as 2, others 1. The grids with grid widths of 10m, 20m, 30m, 40m and 50m were examined. The programming language, VC++6.0 was applied to process the calculation and analysis.

With the GIS ArcView, three interpolation methods (IDW, spline, kriging) were compared and analyzed by MSE (Mean Square Error). MSE is calculated by kriging with spherical, exponential, gauss models, spline with the degree of 2, IDW with the distance power 1, 2 and 3. The best interpolation methods were finally determined. The crop yield spatial distribution maps are constructed using the best interpolation methods.

Results and discussion

The average yield of the total field was approx. 5.1 t/ha, and the points with values below 2.13 t/ha and points with values of above 6.8 t/ha were deleted (mean ±2σ). It is obvious that these points concerned places at which the combine harvester passed the headland and the combine harvester passed the core area of the field without the whole width of the cutterbar. The effect of various grid sizes (10m×10m, 20m×20m, 30m×30m, 40m×40m, 50m×50m) on correlation analysis indicates that the grid size of 30m×30m above which the correlations are similar would be appropriate.

The results comparing among three interpolation methods show that Kriging and IDW are collectively the best, and Kriging is the best with spherical, IDW is the best with the distance power of 2. According to above results, the crop yield spatial information distribution maps were made with the GIS ArcView.

References

P. Juerschik, A. Giebel. (1999). Processing of point data from combine harvesters for precision farming. In: Proc. of the 2 th European Conference on Precision Agriculture. Ed. John V.Stafford. Silsoe Research Institute, UK. pp 297-307

Liu Gang, Kuang Jishuang☐The study on field soil sampling strategies and interpolation techniques, session 6:information technology of agriculture, ICAST, 2001, pp331-336

Acknowledgement

The research project is funded by the Project 863—"Research and demonstration of Precision Agriculture Technology System" (2001AA245011) and "Research of Precise Irrigated Technique-Research and Development of Laser Leveling System" (2002AA6Z3041).

Estimation of soil properties by electrical and electromagnetic methods

E. Lück[1], U. Spangenberg[1], M. Dubnitzki[1], R. Gebbers[2] and J. Rühlmann[3]

[1]Institut of geosciences, University of Potsdam, Germany
[2]Centre for Agricultural Landscape and Land Use Research, Müncheberg, Germany
[3]Institute of Vegetable an Ornamental Crops, Großbeeren, Germany
elueck@geo.uni-potsdam.de

Abstract

The aim of electrical field measurements in agriculture is to get indications about the spatial distribution of soil properties like texture, soil moisture, N-content, and compactness. Because soil electrical conductivity (EC) can be influenced by various soil properties at the same time it is absolutely necessary to establish a correlation model on the bases of soil sampling and laboratory analysis. This model can than be used to interpret the EC maps from field measurements. But this interpretations has to be done carefully. The strength of the correlation between EC readings and a single soil parameter depends on the observed range of the parameter as well as on the correlation with other parameters. To identify the influence of the soil parameters more precisely, laboratory experiments are necessary. By systematically varying one parameter at a time we can observe the bivariate relationship between EC and the soil parameter and establish a regression function. Comparing the regression functions of different parameters we can derive the proportion of influence each parameter has on the EC readings.

Keywords: electrical conductivity, lab data, correlation, regression functions, Kassow

Introduction

Geophysics provides several methods for estimation of electrical soil properties. Electrical conductivity can be measured by direct current (DC) method, by electromagnetic method (EM) or by application of devices with a capacitive coupling. Each method has advantages and disadvantages. Worldwide, EM and DC are most common. We preferred the EM method because of the compactness of the instrument we used (EM38 by GEONICS).
Fieldwork was carried out in a ground moraine area in north east Germany (Kassow farm). In addition, laboratory experiments were conducted measuring EC of soil sample under defined conditions. Whereas most field measurements were carried out in a single frequency mode, in laboratory we also study the frequency behaviour, which can deliver auxiliary information.

Materials and methods

The fields of the farm Kassow are arranged along the river Warnow and cover an area of 18 km in north south direction and 8 km in east west direction. The landscape shows high small scale variability. This is, because the last ice age was only 10,000 years ago and relief as well as soil texture is predominantly influenced by glacial deposits. About 1800 ha were surveyed by conductivity mapping during the last three year. Field size ranges from 50 to 70 ha. From each field, usually 5 soil samples were collected for calibration. At some sites, samples were taken within a very short spacing (down to 5 m). The soil samples were analyzed for soil texture (grain size distribution), water and organic matter content. Additional, the N-content was estimated for some probes. The estimation of correlation coefficients between EC and other soil properties was done for several spatial scales (scale of the whole farm, field scale and scale of sub fields). The field data were collected over three years during several campaigns. That means, that seasonable conditions influences also the conductivity values. The influence of different soil temperature can be corrected easily, because all considered soil types have nearly the same temperature cures. We calibrated all conductivities to a temperature of 25°C by the formula:

$EC_{25} / EC = 0.256 + 1.29 \exp - [(T-13.8)/21.1]$, which is the result from lab data collected for a loamy sand. To improve the correlation between EC and soil texture we considered also the differences in soil moisture. The water content varied between 5 and 15%. Lab tests have shown, that soil moisture has an extreme influence on the electrical resistivities especially for dry soil samples (water content smaller than 5%). It becomes smaller for higher water contents. That's why the correction of differences in moisture doesn't improve the correlation significantly.

Results and discussion

The figure shows the correlation between the electrical resistivities (the inverse value of conductivity) and soil parameters of the whole farm. We know, that clay content increases conductivity values as well as organic matter does it. In Kassow we find both – regions with higher clay content and areas of peat. That's why we can improve the correlation, if we consider the cation exchange capacity (CEC) instead of pure clay content. For further improving we have chosen a modified expression for the CEC. We have not only taken into account the content of clay and organic matter, we have added also some portions of silt. In several studies we found out, that correlations became stronger on this way. Nevertheless the correlation coefficient R^2 is about 0.45. Estimating electrical resistivity in the field we integrate over a soil volume, which is not homogeneous and which should be described by a couple of properties.

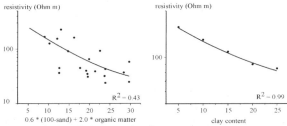

Figure. Correlation between electrical resistivity and soil texture.
Left: field data from Kassow, Right: data from laboratory (constant soil moisture of 15%)

Conclusion

Laboratory analysis gives us a better understanding of the influences on electrical conductivity. These influences can be modelled by regression functions. Under field conditions, these functions have to be calibrated for every site specificly. For future use of electrical properties in practice soil electric conductivity should be considered as a complex value described by amplitude and phase shift. On this way we can improve the unambiguousness

Acknowledgements

This project was funded by 'Deutsche Bundesstiftung Umwelt'.

References

E. Lück, M. Eisenreich, H. Domsch (2003). Innovative Kartiermethoden für die teilflächenspezifische Landwirtschaft, Innovative Methods for Precision Agriculture, Selbstverlag der Arbeitsgruppe Stoffdynamik in Geosystemen, Heft 7, ISSN 0949-4731, 155 S.

In-line quality measurements on forage maize

Koen Maertens, Piet Reyns and Josse De Baerdemaeker
KULeuven, Laboratory for Agro-Machinery and Processing, Kasteelpark Arenberg 30, B3001
Leuven, Belgium
koen.maertens@agr.kuleuven.ac.be

Abstract

During the selection of new forage maize varieties, breeders try to come to a better
digestibility of the plants. During their breeding program, thousands of plots are harvested
and evaluated for yield and quality. The samples, taken from each plot, have to be dried and
ground to allow NIR analysis in the laboratory. An on-line measurement system on the
harvester would strongly reduce the time and cost of sampling, drying, grounding and
laboratory NIR measurements.

The spout of a two rows forage harvester is used for in-line testing. Flowing initially in the
middle of the chute, the material is impacting the wear plate in the curved plate. At this place,
the flow has the highest density and is flowing at the outer side of the chute. This is the
optimal place to measure reflectance spectra. A square hole is cut in the wear plate to mount a
square quartz window and a ZEISS Corona 45 NIR 1.7 sensor is installed. The acquisition
and storage of the spectra was programmed in *Labview® 5.1*.

Two sets of samples are measured. The first set consists of 640 samples from one field for
which labNIRS reference measurements are available as well as oven determinations for the
dry matter content. Spectral measurements are also done on a second set of 243 samples from
a second field with different hybrids.

Selection of the test plot spectra

The signal obtained from one test plot varies strongly in time. At the beginning and end of the
flow, spectra are present of the transitional stage. In this transitional stage, the signal increases
steadily till it reaches the first local maximum. The spectra between the first and last local
maximum are selected as representative for the crop on the test plot.

Processing

The selected spectra are averaged in time, and converted to absorbance values using time-
interpolated spectra of black and white standards. The linear interpolation (in time) is
executed between the nearest standard's spectra taken before and after the sample
measurement. After wavelength interpolation, the spectra are imported in the *GRAMS32*
program for Partial Least Squares (PLS) calibration. Enough reference measurements were
available for the labNIRs parameters and the dry matter content. Therefore, those datasets
were split into a calibration and validation set. The accuracy of the calibration is expressed as
the Standard Error of Prediction (SEP), and determined by predicting the calibration
parameters on the validation set of spectra.

Results

The standard error of prediction is quite large for the most important labNIRS parameters
when compared to the range of the reference values of the calibration set (table 1). Dividing
the standard deviation (SD) of the calibration set by the error, the ratio does not exceed the

1.5 limit for any of the parameters. The worst results are obtained for the cell wall content, which can not be predicted with the calculated calibration.

Table 3. Results of Corona measurements on spout for the labNIRS referene values and the dry matter content. The last two columns reflect the calibration results of the laboratory NIRS instrument. SEP = Standard Error of Prediction, SD = Standard Deviation of the reference dataset, SECV = Standard Error of Cross Validation.

Crop Parameter	Corona calibration with labNIRS values			labNIRS instrument calibration with wet chemistry values	
	#factors	SEP	SD/SEP	SECV	SD/SECV
Ash content	7	0.43	1.23	0.47	2.32
Starch content	7	2.81	1.05	1.71	6.2
Crude Protein content	9	0.46	1.16	0.45	2.67
Soluble Sugar content	9	0.86	1.16	1.23	3.98
Crude Fibre content	2	1.3	0.99	1.01	4.16
Acid Detergent Lignine	3	0.45	1.04	0.43	2.37
Cell Wall content	6	1.95	0.96	1.87	3.32
Enzym. Organ. Matter Digest.	6	2.35	1.03	1.87	3.26

References

P. Reyns. 2002. Continuous Measurement of Grain and Forage Quality during Harvest. PhD Thesis. Department of Agro-Engineering and Economics, KULeuven, 229 pp.

A precision farming application in cotton in the small farms of Greece

Ath. Markinos[1], L. Toulios[2], D. Pateras[2], G. Zerva[2] and A.T. Gemtos[1]
[1]Agricultural Machinery Laboratory, University of Thessaly, Fytoko, N. Ionia, 38446, Geece
[2]NAGREF/ISM, Theofrastou 1, Larissa 41335, Geece
gemtos@agr.uth.gr

Abstract

A two years study of cotton fields in central Greece showed large infield variability in many factors like yield and soil physical and chemical properties. Even in small fields there were distinctive zones of equal yield. Yield is affected straightforward by a set of systematic factors relative to the soil properties, which specify the electrical conductivity value at every field point. The present study correlates yield and electrical conductivity maps and aims at finding out ways to apply a more precise management of cotton fields.

Keywords: cotton, maps, yield, soil EC, variability.

Introduction

Cotton is the most important Greek arable crop. It covers more than 400,000 ha that is nearly the 50% of the irrigated land. It is cultivated in small farms with average area 5.6 ha. Present work reports PF application in cotton and is a part of a project applying Precision Farming techniques in small scale farms in Greece (Gemtos et al, 2002).

A considerable yield spatial variability in any field is the basic criteria to define the beneficial use of management according to PF. Cotton yield mapping application in central Greece for two consecutive years shows that in each field can be observed distinctive zones of equal yield (Markinos et al, 2002). These differences in yield are caused mainly by systematic soil factors and by random events. The influence of soil properties factors in yield can be estimated with appropriate grid sampling that requires a lot of work and high cost. The zoning of the field based on soil properties variability and yield could lead to a cheaper and more reliable soil quality variability assessment by taking soil samples corresponding to distinctive zones. Actually, these field areas are management zones that express a relative homogenous combination of yield limiting factors and can be drawn using soil electrical conductivity (EC_a) measurements on the go.

Materials and methods

Cotton yield mapping performed for 2001 and 2002 harvesting periods in the same fields (overall about 80ha) located at Myrina in Karditsa prefecture, Central Greece. A yield monitor made by Farmscan® was used. At the same fields, soil electrical conductivity (EC_a) mapping was performed in early spring of 2002 before planting, using the direct contact method at depths 0-30 cm and 0-90 cm. A Veris® 3100 machine was used that consisted of a trailed vehicle with 6 rolling electrode discs. Collected data were processed to produce the yield and EC maps using ArcGIS® software. The kriging method was used for data interpolation.

Results

Soil EC mapping results for a field 4.3 ha can be seen at fig. 1 (depth 0-30 cm) and fig. 2 (depth 0-90 cm). The 1st picking yield maps can be seen at fig. 3 and 4 (periods 2001 and 02).

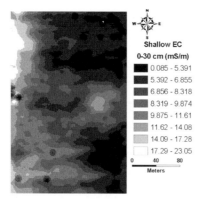

Figure 1. Soil EC at depth 0-30 cm.

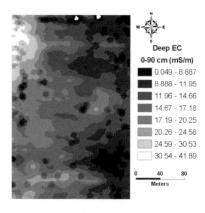

Figure 2. Soil EC at depth 0-90 cm.

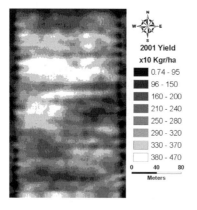

Figure 3. 1st picking yield map for 2001.

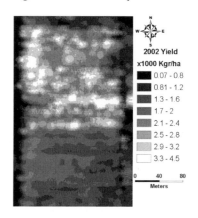

Figure 4. 1st picking yield map for 2002.

Discussion - Conclusions

The maps show that the fields have a high variability in yield and soil EC values. Distinctive zones can be defined in each field. The correlation coefficient R^2 between shallow EC and 2001 yield map equals to 0.38, between shallow EC and 2002 equals to 0.14. The coefficient between the two EC maps equals to 0.89, and between the two yield maps equals to 0.45. The two years under study represented two extreme weather periods. The 2001 was a very dry and hot year while 2002 a very wet. This can explain in some extent the low correlation between yields and the different correlation between yield and soil EC. It is also quite clear that many other factor affect yield and the under way analysis of the soil properties could reveal them. Meanwhile, the EC and yield mapping can permit the definition effective management zones.

References

Gemtos, T., Fountas, S., Blackmore, B.S., Greipentrog, H.W. 2002. Precision farming experience in Europe and the Greek potential. In: *Proceedings of the 1st HAICTA Conference*, Athens, 6-7 June 2002: 45-55.

Markinos, A., Gemtos, T., Toulios, L., Pateras, D., Zerva, G. 2002. Yield Mapping of Cotton crop in Greece. In: *Proceedings of the 1st HAICTA Conf.*, Athens, 6-7 June 2002: 56-62.

AGROffice: WinGIS based applications for precision farming

Walter H. Mayer
PROGIS Software AG, Postgasse 6, A-9500 Villach, Austria
office@progis.com

Abstract

The success of PROGIS AGROffice is based on the sales of technology but also on the

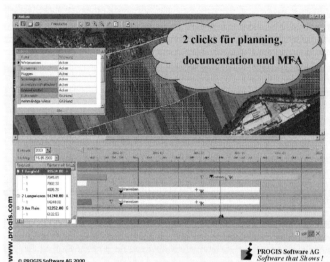

development of a complete new business- and partnership model. Figure 1 shows the application AGROffice as the integration of maps, field calculation, planing of cultivations and first time the PC GIS time management for easy visualising of all the activities. A complex planing step as there is cultivation planing AND cost- and earning-calculation for all fields AND EU subsidy paper (INVEKOS agreement) are reduced to 2 clicks or 1 drag and drop.

Following software solutions will be installed at partners locations together with a GIS solution and an ortho image:

- Documentation (= protocol of used seeds, fertilizers and pesticides per field (a must within EU from 2005) also in the version "geo-traceability".
- Cost-/revenue calculation per field based on local standard values for cultures, cost elements, to do´s, used tools, equipment etc. (all standard data can be modified), regional data can be developed.
- EU-subsidy calculation is done automatically and can be transferred to public bodies (incl. geo-data if necessary) on request.

Based on the same technology the logistic within the agriculture and forestry complex can be build up as a service for farmers and/or the industry. The logistic models have been used for sugar beet harvesting in cooperation with the German Bundesmaschinenring and the agro-giant Südzucker since years.

For further partners out of the segment „soil analyse", based on the same GIS-platform, a fertilizer calculation and recommendation application can be linked. More applications have been developed towards the segment „Precision Farming", integrating field data, machine data, GPS- and sensor technology etc.

Further applications are available for following market segments:

- Forestry: forest inventory / forest taxation, management planning etc.
- Smart community management: with construction admin, internet etc.
- Planing within rural areas

Figure 3: Logistics means a central GIS for planing, mobile GIS systems installed on roughed PC´s" or PDA´s, communication with SMS; after detailed planing at the headquarter the harvester is informed via SMS, returns acceptance and transmits the location (GPS-data) and the weight (machine-sensor-data) of the products (in the sample left sugar beet) to the HQ, independent if it is sugar beet, potatoes, round logs or also any packed elements.

Keywords: GIS, GEO-Traceability, logistics, precision farming, forestry

Introduction

Since 1994 PROGIS is successfully evolving the development of a "European" GIS system. Already 1998 PROGIS has been listed belonging to the „top ten" worldwide (GIS World 11/98). The development towards "easy to use" GIS-solutions for end users, five steps for a successful result have been necessary:

- Development of our base product WinGIS, an object oriented PC-GIS;
- Focus towards customer oriented applications for vertical market segments as there are agriculture, forestry, communities, pipeline-management, the rural area
- Integration of other geomatic technologies as there are GPS, dGPS, ortho images (air- or space born), sensors for meteorology-, soil- and water-data as well as sensors for machine-data to visualise these data spatial, communication technologies as SMS, long waves etc. as well as embedding of geo-data within internet;
- Cooperation with public or semi-public data suppliers (in Austria e.g. BEV (Bundesamt für Eich- u. Vermessungswesen) as well as the famous cartography company Freytag und Berndt);
- Analyse all the needs and possibilities of public-private-partnerships with the target to develop a business- and cooperation model for the market segment "The rural area".

Reference

Agriculture: Bundesverband der deutschen Maschinenringe, Agrarbezirksbehörde – National Government Lower Austria, Agrargemeinschaft Weißbriach, Landwirtschaftliche Landesschulen, Universität für Bodenkultur Wien, ARGE Geobauern.

Forestry: Ministry of Forestry in Bosnia and Serbia, Forest Directorate Prince DI Alfred Liechtentein, Von Pezold'sche Land & Forstwirtschaft Gusterheim, Waldwirtschaftsgemeinschaft Mostviertel West.

Community: Korneuburg, Mönchen-gladbach, Seehafen Rostock Verwaltungsgesellschaft mbH. PipeGIS:gasPetronas Res & Scientific, Malaysia, Transpetrol a.s., Slovakei, Gas of Portugal, Portugal

GeoINFOtainment: Koch Media, Wien, Freytag & Berndt,

Weed monitoring by infrared camera and panoramic annular lens

P.Á. Mesterházi, M. Neményi and S. Maniak
University of West-Hungary, Institute of Agricultural, Food and Environmental Engineering,
H-9200 Mosonmagyaróvár, Vár 2, Hungary
nemenyim@mtk.nyme.hu

Abstract

The complexity of the field conditions and the morphological variability of the plants still make the weed identification complicated (Manh et al., 2001). The authors review their optical sensor based weed monitoring system operating with CCD and infrared cameras, and a special optical device with a viewing angle of 360° to eliminate the limitation of the optical instruments.

Keywords: weed monitoring, machine vision, on-line image processing, infrared camera, Panoramic Annular Lens (PAL)

Introduction

The weed control based on machine vision and spectroscopy is a very popular but complex research field with special limiting factors. In this way, to compensate the difficulties more and more complicated systems operating with many parameters are necessary. However, more and more new experimental tools came in to light, they haven't been yet put into the practice. In this article the authors review their optical sensor based weed monitoring system.

Materials and methods

A possibility to significantly increase the operation speed of any plant recognition system is to divide only the soil and the plant parts from each other. The collected information is still satisfactory for weed observation between the rows (in case of row crops), or even stubble analysis can be used as a starting point for preventive weed control. Our purpose was to build up a quick practice-ready system, which can be the foundation of a VRA weed-control unit for both map and sensor based applications. Therefore, positioning is requested. For this purpose both of the DGPS receivers (CSI Wireless, DGPS MAX,) can be applied, which are available at the institute as parts of the RDS and the Agrocom ACT yield monitoring systems. For image capturing a CCD camera is applied, the images are digitised on-line by means of a PCI Frame Grabber card (WinTV Go, Hauppauge) installed into a portable computer (KP-5212T/A). On the portable computer self-developed software is running to capture and process video images, which are stored in a database. To get the weed density, each pixel in the blue colour component of the captured image is scanned and compared with a threshold. The ratio of the number of pixels, which are lying under the threshold and the total number of pixels result in the weed density in percent. Moreover, the software described above can export data to different Geographic Information Systems (GIS) (Neményi et al., 2003; Mesterházi et al., 2002) in order to build a weed density map (figure 1.). Over the CCD's capability, the infrared wavelengths may provide us more advantages. Due to the infrared camera (ThermCAM PM 675, FLIR) owned by the institute and the altered algorithm the limitation concerning to the spectral properties seems to be solved. For increasing the view angle a special imaging device, called Humanoid Machine Vision System (HMVS) was applied. It consists of two main parts: an imaging block such as the Panoramic Annular Lens (PAL) that renders omni directional panoramic view, and a so-called foveal lens (Greguss, 2002).

Results and discussion

The weed monitoring system built by the Institute of Agricultural, Food and Environmental Engineering proved to suffice its purpose. It is capable of discriminate the soil and weed parts in the CCD captured image with high confidence, and gives the value of weed (plant) density in on-line mode. The operating speed is limited only by the working speed of the carrying machine and the field conditions. The image captured in infrared range increases further the accuracy of plant and soil characterisation, and seems promising even for weed identification, pest and disease detection. The integration of the PAL device into the weed monitoring system was successful; the scanned area could have been significantly enlarged in this way. The self-developed MATLAB application proved to be adaptable for its planned task (figure 1.). As a next step, our goal is to get the proper position information for each pixel of the PAL image, and to build on-line software for it, based on our experiences with the AMTLAB application.

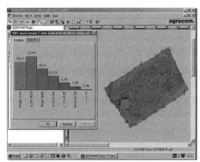 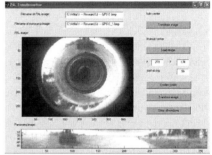

Figure1. The weed density map, and the transformation of the PAL image.

Conclusions

Due to the technical development, there is an increasing chance to develop a system for weed identification, which can recognise more weeds with higher accuracy than ever before. However, be it remembered that these systems must entirely fit to the agricultural practice. Consequently, can the recognition be even perfect, if other parameters do not make possible to adopt it into the practice.

References

Greguss, P., 2002. Centric-minded imaging and GPS. RAAB'02, Balatonfüred, Hungary, 30 Jun. – 2. July 2002.

Manh, AG., Rabatel, G., Assamat, L., Aldon, MJ. 2001. Weed leaf image segmentation by deformatle templates. Journal of Agricultural Engineering Research, 80 (2): 139-146 OCT 2001.

Mesterházi, P. Á, Neményi, M., Kacz, K., Stépán, Zs. 2002. Data transfer among precision farming systems. ASAE Annual International Meeting/CIGR World Congress, 2002. 07.28-31, Chicago, Illionis, USA, cd (021047)

Neményi, M., Mesterházi, P.Á., Pecze, Zs., Stépán Zs. 2003. The role of GIS és GPS in precision farming. Computers and Electronics in Agriculture. (in print).

Analysis of optimal site specific fertilisation in maize from a strip trial

A. Meyer-Aurich[1], G. Huber[2] and F.-X. Maidl[2]
[1]TU-München, Chair of Agricultural Economics, 85350 Freising-Weihenstephan, Germany
[2]TU-München, Chair of Plant Production, 85350 Freising-Weihenstephan, Germany
meyer-aurich@wzw.tum.de

Abstract

This study presents research results from a on-farm strip trial in which optimal site specific fertilisation rates for maize were investigated. It has been found, that yield response to nitrogen fertiliser input varies substantially in different yield response areas. While the economic advantage of site specific fertilisation is quite small, the environmental effects of site specific fertilisation are more significant. However, the positive effects of site specific fertilisation only occur, if the right fertilisation strategy are selected for the different yield response areas. A wrong decision on the site specific management can result in economic and environmental drawbacks compared to homogenous fertilisation.

Keywords: site specific fertilisation, maize, strip trial

Introduction

The expectations on economic and environmental potentials of site specific fertilisation are high. However, there is evidence, that economic benefits of site specific fertilisation does not always hold its promises. Up to now only few studies prove environmental benefits of site specific fertilisation strategies in maize (e.g. Thrikawala et al., 1999). The environmental effects of site specific fertilisation are mainly caused by reduced fertiliser input. However, these effects can mostly only be stated with modelling approaches (e.g.: Kersebaum & Lorenz, 2002). The presented investigations shall provide empirical evidence for the economic and environmental advantages of site specific fertilisation in maize.

Materials and methods

To investigate the economic and environmental potential of site specific fertilisation strategies in maize, on-farm strip trials were conducted on fields of the research station of the Research Network on Agro-ecosystems Munich from 1999 to 2001. This paper presents results from a trial, which was conducted on the plot "A20" in 2001. According to yield observations within the field a high and low yield response area has been identified. Within the trial four different fertiliser intensities (0, 120, 170, 220 kg N/ha) were tested within 8 strips over the high and low yield response areas of the field. The strips were divided into two sub-strips which all were fertilised homogenous. Within the strips maize yield was obtained from miniplots with nine repetitions for each fertiliser intensity in the high and the low yield response area.
Response functions were calculated on the basis of the energy yield of the maize. Because of the experimental design of the trial the calculated response functions were tested against spatial autocorrelation with the Software SpaceStat®.

Results

The regression analysis shows, that yield response to nitrogen differs significantly in the two yield response areas. Spatial autocorrelation can be neglected if the two yield response areas are considered as spatial regimes. The response functions demonstrate, that site specific nitrogen application rates for this field could have provided higher gross margins. In this case the benefits would mainly originate from the cost savings from reduced fertiliser inputs in the high yield response area. Environmental benefits would result from less nitrogen input of about 25 kg/ha N. However, the savings of about 16 €/ha would have to cover information and application technology to make variable nitrogen rate application profitable for this plot.

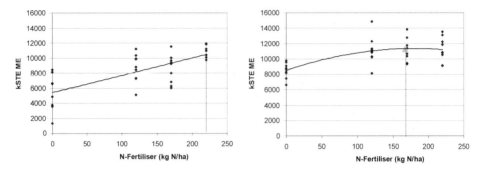

Figure 1. Response function in the high and low yield response area (economical optimum at dotted line).

Discussion and conclusions

This results state that if the management is optimised, variable nitrogen application provide some environmental benefits with a small economic advantage. However, for this example the benefits would be achieved with a high fertilisation in the area with low yield response and a reduced fertilisation in the high yield area. This shows, that the economic and environmental benefits of site specific fertilisation cannot be achieved with simplified rules, like high fertilisation in high yield response area and low fertilisation in low yield response area. In contrast, a wrong decision on the site specific management can result in economic and environmental drawbacks.

References

Kersebaum, K.-C. and Lorenz, K. 2002. Lokaler N-Austrag. In Werner, A. and Jarfe, A. Precision Agriculture – Herausforderung an integrative Forschung , Entwicklung und Anwendung in der Praxis. KTBL-Sonderveröffentlichung 38, Darmstadt.
Thrikawala, S., Weersink, A., Kachanoski, G., and Fox, G. (1999): Economic Feasibility of variable-rate technology for nitrogen on corn. American Journal of Agricultural Economics, 81, 914-927.

Acknowledgements

This research was funded within the Forschungsverbund Agrarökosysteme München (FAM) by the German Federal Ministry of Education and Research.

Spectral detection of nitrogen status, biomass and yield of field-grown maize plants

B. Mistele, C. Bredemeier, R. Gutser and U. Schmidhalter
Technical University Munich, Department of Plant Sciences, Chair of Plant Nutrition
mistele@wzw.tum.de

Abstract

On-the go measurements of biomass and nitrogen uptake allow to optimise fertilising strategies in heterogeneous fields. We evaluated the potential to non-destructively detect biomass and nitrogen uptake in maize with a tractor-based sensor. The measurements were conducted in 1.5 m wide strips on both sides of the tractor. To create a range of biomass and differences in nitrogen status, five N-fertilizing and three seeding rates were applied. Measurements were conducted seven times between 5-leaf-stage and ripeness. Nitrogen-uptake could remarkably well be detected with $R^2 = 0,85$. The results show a good correlation of about $R^2 = 0,80$ between REIP (red edge reflection point) and the final yield for all measurements.

Keywords: nitrogen uptake, proximal sensing, yield estimation

Introduction

Precision farming can increase yield, reduce fertilizer input and pollution by avoiding partial over-fertilization. Therefore the nitrogen status should be estimated, the biomass and yield be determined. Spectral measurements are a suitable tool to estimate these parameter. We examined how well they can be estimated in maize with a tractor based field spectrometer.

Materials and methods

The experiment was carried out in the south-east of Germany in 2002. The experimental field was about 4 ha in size and contained 52 plots which were 15 m wide and about 40 m long. The sowing rate was varied at 6, 10, and 14 plants per m² and the N-rate consisted of 5 treatments (25, 70, 120, 170 and 220 kg/ha). 25 kg N were applied at sowing, the remainder was applied at May 5 as N-fertilizer stabilised with a nitrification inhibitor.

Figure 1. Field sensor in front of the tractor.

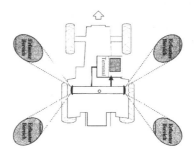

Figure 2. Sensing area with four ellipsoids, each about 2 m² in size.

The spectral measurements were conducted with a two channel spectrometer at five wavelengths using a modified Hydro-N Sensor (Fig. 1). It consists of a four fibre optics to measure the canopy reflection (Fig. 2) and a second spectrometer to measure the radiance. Measurements were done at 550, 670, 700, 740 and 780 nm and three spectral indices were calculated, namely red edge inflection point (REIP), infrared green ratio (G/IR) and red infrared ratio (700/780). The measurements were performed at 2 m above the canopy. Spectral indices were compared with destructive harvests of biomass, N-content and yield. Measurements were conducted seven times between the 5-leaf-stage and ripeness. Biomass was destructively measured exactly in the same area as the measurements were done in June 20 (5-leaf-stage), July 1 (6-leaf-stage), July 22 (8-leaf-stage) and the harvest was at September 16.

Results and discussion

N-concentration was moderately well determined (Fig. 3). High coefficients of determination for quadratic relationships were obtained between biomass, yield, N-uptake, and reflection indices (Figures 4-6).

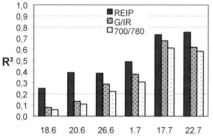

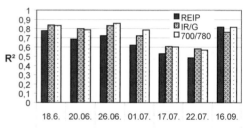

Figure 3. Coefficients of determination between concentration and reflection indices.

Figure 4. Coefficients of determination between biomass and reflection indices.

Measurements were probably influenced by soil reflectance at the 5-leaf-stage and by flowering at the 8-leaf-stage. Yield estimations with REIP were hardly influenced. N-uptake was consistently very well predicted.

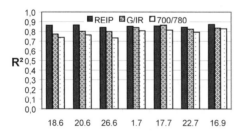

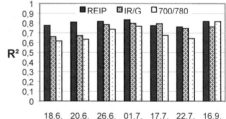

Figure 5. Coefficients of determination between N-uptake and reflection indices.

Figure 6. Coefficients of determination between yield and reflection indices.

Conclusion

The results show that nitrogen status, nitrogen uptake and yield can be detected already at an early stage by spectral measurements. The results suggest that nitrogen fertilisers could potentially more specifically and rationally be applied to maize crops.

498

A mobile imaging spectroscopy system as tool for crop characterization in agriculture

G.J. Molema[1], J. Meuleman[1], J.G. Kornet[1], A.G.T. Schut[2] and J.J.M.H. Ketelaars[2]
[1]Institute for Agricultural and Environmental Engineering (IMAG), P.O. Box 43, 6700 AA Wageningen, The Netherlands
[2]Plant Research International (PRI), P.O. Box 16, 6700 AA Wageningen, The Netherlands
geertjan.molema@wur.nl, tom.schut@wur.nl

Keywords: imaging spectroscopy, imspector, agriculture, reflection, crop characterization.

Abstract

Imaging spectroscopy combines the potential of digital imaging and spectroscopy. IMAG and PRI aimed to develop a new high resolution sensing technique for non-destructive measurement of quality parameters in agricultural crops, starting with grass canopies. Recently an experimental stationary prototype was developed and tested as tool for grass sward characterisation. A new mobile concept is under development for on-line measurement in the field, using a self-propelled GPS equipped vehicle.

Introduction

On-line imaging spectroscopy is a relatively new sensing technique in agriculture. Imaging spectroscopy combines the potential of digital imaging and spectroscopy. The potential of high spatial and spectral resolution, combined with hyperspectral imaging spectroscopy, for rapid and objective characterization of quality parameters in agricultural produce is enormous. For sustainable agricultural practices detailed information about crop condition and other relevant crop properties are necessary. IMAG and PRI aimed to develop a new high resolution sensing technique for non-destructive measurement of quality parameters in agricultural crops, starting with grass canopies.

Materials and methods

Recently a first experimental stationary prototype was developed and tested as tool for grass sward characterisation (Schut et al., 2002). The second version is under development: a new mobile concept for on-line measurement in the field, using a self-propelled GPS equipped vehicle (Figure 1). The measuring system consists of the imaging spectrometers V9 (450-900 nm) and N17 (900-1650 nm) and a user specified 3 CCD camera. In both imaging spectrometers an 80 μm slit is used, resulting in a spectral resolution of 5-13 nm (depending on wavelength). The spatial resolution is 0.16 mm^2 in the range of 450-900 nm and 1.45 mm^2 for 900-1650 nm. The high spatial resolution enables measuring reflections at leaf level. Both imaging spectrometer systems produce on line calibrated reflection curves, since a 50% reflection standard is in the images. The V9 is equipped with a 12 bits digital B/W CCD camera, the N17 with a 12 bits digital InGaAs camera. The 3 CCD digital area scan camera measures in narrow bands at 600, 710 and 800 nm. Typical half-band width ranges from 7.4 to 10.5 nm. The system is equipped with two types of Xenon flash lights (line and ring). All measurements are done simultaneously.

Figure 1. Artist impression of the mobile concept.

Results and conclusions

Combination of the imspector and area scan technology offers new perspectives. Canopy structure can be characterized in horizontal (ground cover, heterogeneity) and vertical components (height of crop, leaf angle). Spectral characteristics can be obtained at leaf level (Figure 2). For practice this implies new ways to quantify and qualify variables like ground coverage, biomass, ratio grass/clover, nutrient and drought stress, nutritive value, nutrient contents (*e.g.* N, P, K), canopy geometry, etc.

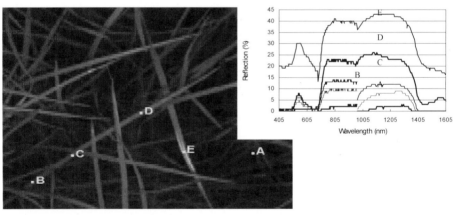

Figure 2. Grass sward reflection at points A-E.

References

Schut A.G.T., Ketelaars, J.J.M.H., Meuleman, J., Kornet, J.G. and Lokhorst, C. 2002. Novel Imaging Spectroscopy for Grass Sward Characterization. Biosystems Engineering **82** (2) 131-141.

Development of a soil electrical resistance device using a no-till planter and its coulter disks as sensors

J.P. Molin and L.A.A. Menegatti
Agricultural Machinery, Dept. of Rural Engineering, ESALQ, Un. Of São Paulo. Av Pádua dias, 11, 13418-900 Piracicaba – Brazil.
jpmolin@esalq.usp.br

Abstract

This work deals with the development of a soil electrical resistance (ER) sensor dedicated to no-till systems. Two isolated coulters from a no-till planter, a data logger and an electric circuit compose the system. When the coulters are in the soil, the current measured by the circuit is proportional to the soil ER. Results indicate that the sensor is sensitive to the soil variation on ER. Soil ER information is useful on indicating trends and on trying to understand soil behavior but more study is necessary to provide added knowledge about soil physical and chemical parameters related to ER.

Keywords: Spatial variability, yield factors, soil electrical conductivity.

Introduction

Precision farming is a powerful tool to recognise and manage spatial variability within fields. For grain crops, the yield mapping procedure is widely available for agriculturists, but the complete knowledge about yield spatial variation is a challenge for researchers and farmers. Several factors affect the yield response such as soil fertility, physical conditions and water availability. ER of the soil is influenced by cation exchange capacity, soil texture, moisture, bulk density and salinity and might be used to identify management zones within fields (McBRIDE et al., 1990; Clark et al., 2000). The objective of this work was to develop a soil ER sensor for no-till conditions and use this information to support management decisions.

Materials and methods

The system was composed by two isolated coulters from the no-till planter, a data logger and an electric circuit supplied with 36 V DC and stabilized at 24 V. The difference of potential was applied to the isolated coulters (electrodes). The current was measured by the circuit as a difference of potential across a known resistance of 19.9 ohms. The data logger read the voltage and stored it in a 0.5 Hz frequency and also recorded time. A cut-off sensor associated to the hydraulic system of the planter informed the data logger if the coulters were in contact with the soil or not. The positioning system was composed by a GPS with satellite differential correction and the position was stored in a palmtop recording also time. The data from the data logger and the positioning data were linked in a spreadsheet using the time column as identifier and a filtering process was performed to remove inconsistent data and outliers. A semivariance analysis was performed to the data to characterize its spatial dependence and the data were interpolated to produce the soil ER map. The test was run in an area of 17 ha where soil sampling, yield data and a recent soil cone index survey were available. A correlation matrix among soil ER and other soils factors was built trying to understand the reasons of variability on soil ER.

Results and discussion

At the total 7751 soil ER measured points were collected, resulting in a data density of 456 points.ha^{-1}. Table 1 shows the results of the semivariance analysis and the parameters for the adjusted semivariogram. Figure 1 shows the map of soil ER using krigging as interpolator.

Table 1. Parameters from semivariance analysis for soil ER.

Parameter	Model	Nugget	Sill	C	Range	Cross Validation
Soil ER	EXP	24300	71090	46790	23.73	0.901

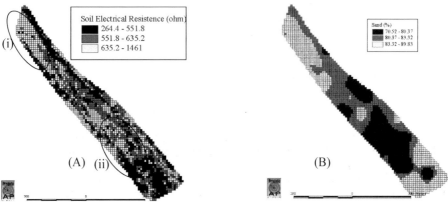

Figure 1. Soil ER map (A) and sand content map (B).

Soil ER data is useful on determining management zones, such as the high resistance zone shown on Figure 1 (i). This high resistance zone indicated in the map is probably caused by the sand content of the soil in that region, which varies between 83 and 89%. The zone of low resistance (ii) is probably caused by the high soil compaction expressed by the cone index found in that region. The correlation between soil ER and clay and soil ER and sand was - 0.12 and 0.16 respectively, indicating that the higher the clay content, the lower the resistance. Also results for base saturation, cation exchange capacity and cone index at 10, 15 and 20 cm depth, resulted in negative correlations, indicating coherent results.

Conclusions

The results indicate that the sensor is sensitive to the soil variation on ER. Soil ER information is useful on indicating trends and on trying to understand soil behavior but more study is necessary to provide added knowledge about soil physical and chemical parameters related to ER.

References

Clark, R.L., Chen, F., Kissel, D.E., Adkins, W. Mapping soil hardpans with the penetrometer and electrical conductivity. Proc. of the 5th International Conference on Precision Agriculture, Minneapolis. Ed. P.C. Robert. CD-Rom. 2000.

Mcbride, R.A., Gordon, A.M., Shrive, S.C. Estimating forest soil quality from terrain measurements of apparent electrical conductivity. Soil Science Society of American Journal, Madison, v.54, p.290-293. 1990.

Electrical conductivity by induction and its correlation with soil factors and crop yield in southern Brazil

J.P. Molin[1], L.M. Gimenez[2], V. Pauletti[2], U. Schmidhalter[3], J. Hammer[3] and J.C. Araújo
[1]*Department of Rural Engineering, ESALQ/University of São Paulo, Brazil*
[2]*ABC Foundation, Brazil*
[3]*Inst. of Plant Nutrition, Technische Universitaet Muenchen, Germany*
jpmolin@esalq.usp.br

Abstract

This work reports the experience done in a 19ha no-till field in Southern Brazil, aiming to correlate soil electrical conductivity (EC) measured by a induction sensor without contact and soil chemical fertility properties, soil texture, altitude, soil water content and crops yield before and after the measurement of EC.

Keywords: soil electrical conductivity, spatial variability, soil properties.

Introduction

The use of soil EC obtained by contact or induction sensors (Rhoades & Corwin, 1984) has been intensively used in some countries as a diagnostic technique for soil characteristics and their spatial variability. As EC is related to some physical and chemical properties of soil, it has potential on helping to define different management zones within a field (Domsch & Giebel, 2001). This is the first survey on induced soil EC in distroferric ferralsoils done in Southern Brazil between 2000 and 2002, aiming to correlate soil EC to soil and crops.

Material and methods

The experimental area is a 19ha no-till field. EC was obtained using an induction sensor without contact, based on a regular grid of 8.0 samples.ha^{-1} in the horizontal and vertical mode, corresponding to depths of 0.0 to 0.3m and 0.0 to 1.5m, respectively. At the same time soil texture and water content samples were collected at 0.0-0.2 and 0.2-0.4m depth on a regular grid of 2.0 samples.ha^{-1} in April 2001. Chemical fertility samples were collected at 2.8 samples.ha^{-1} and 0.0-0.1m depth, also on a regular grid, in late September 2001, before planting soybeans for the next season. Yield maps and altitude were obtained with yield monitor; corn was harvested in April 2001 and soybean in March 2002. A geostatistical analysis was performed defining the parameters for each component and kriging interpolations were done by using a dedicated GIS. A correlation matrix was generated from the surfaces, giving indication of tendencies and behavior of the different components values.

Results and discussion

The geostatistical analysis resulted in ranges above 150m and good cross validation, except for bases saturation (V%). The correlation matrix is shown on Table 1. Accentuated slopes and consequent soil erosion caused by intensive rains during the summer are characteristics of agricultural fields in Southern Brazil. No-till practices are being successfully used in the region for controlling it. The field presented slopes of up to 15% and it has been under no-till for 17 years, giving indications that before it intensive water erosion removed parts of the topsoil. The yield variability was intense on both, corn and soybeans, with correlation of 0.38, indicating that both crops are not well related in terms of spatial variability. Levels of EC at the 0.0 to 1.5m depth were significantly higher than at the 0.0 to 0.3m depth and the correlation between both was -0.28. Correlations with the major part of factors was opposite between both indicating that the soil profile changes significantly between the two layers where the sensor is measuring EC. As no other variable was monitored under 0.40m, the meaning of correlations between EC at 0.0 to 1.5m and those variables only has meaning if considered the effect of each variable on EC at its specific depth. Also, the correlation of clay and water content for EC between 0.0 to 0.3m depth was around 0.40, except for clay at 0.0-0.2m where it was only 0.26. Other indications were CEC and sum of bases (SB), giving positive correlations of 0.26 and 0.24, respectively, for EC at 0.0 to 0.3m. Strong and expected positive correlations were observed between clay and water content and also between clay and organic matter and negative between clay and pH. The sensor presented difficulties on operation, related to frequent necessary calibrations, being affected by local temperature changes. An additional survey on EC based on direct flow measurement could provide answers on the variabilities and increase the operational accuracy. One concern of using EC information on local soils is the iron content that may affect the results on indicating soil characteristics for field variability understanding and it will require more investigation.

Table 1. Correlation matrix generated from the surfaces of 10m cells, giving indication of tendencies and behavior of the different variables.

Variables	Soy.	Corn	CEC	V%	SB	OM	pH	Alt.	Moist.	Sand	Clay	Moist.	Sand	Clay	EC
			Depth 0.0-0.1m						Depth 0.2-0.4m			Depth 0.0-0.2m			0.0-0.3m
EC 0.0 -1.5m	0.19	-0.01	-0.19	0.40	-0.02	-0.3	0.36	0.19	-0.32	0.27	-0.36	-0.35	0.30	-0.30	-0.28
EC 0.0 - 0.3m	0.03	0.13	0.26	-0.11	0.24	0.22	-0.21	-0.45	0.40	-0.32	0.41	0.41	-0.30	0.26	
Clay 0.0 - 0.2m	-0.34	0.10	0.88	-0.56	0.67	0.94	-0.86	-0.17	0.95	-0.94	0.88	0.97	-0.99		
Sand 0.0 - 0.2m	0.36	-0.08	-0.86	0.54	-0.66	-0.93	0.85	0.18	-0.96	0.95	-0.87	-0.98			
Moist. 0.0 - 0.2m	-0.29	0.14	0.85	-0.52	0.66	0.90	-0.83	-0.21	0.97	-0.93	0.90				
Clay 0.2 - 0.4m	-0.20	0.20	0.79	-0.46	0.62	0.81	-0.74	-0.24	0.87	-0.76					
Sand 0.2 - 0.4m	0.32	-0.08	-0.79	0.51	-0.60	-0.87	0.80	0.17	-0.96						
Moist 0.2 - 0.4m	-0.26	0.16	0.82	-0.53	0.63	0.89	-0.83	-0.18							
Altitude	0.11	0.00	-0.23	-0.09	-0.25	-0.22	-0.05								
pH	0.30	-0.03	-0.72	0.84	-0.38	-0.86									
OM	-0.33	0.10	0.91	-0.56	0.69										
SB	-0.03	0.17	0.91	0.12											
V%	0.34	0.06	-0.31												
CEC	-0.18	0.15													
Corn	0.38														

Conclusions

Correlation analysis of EC with soil chemical fertility properties, soil texture, altitude, soil water content and crops yield resulted in important indications of sources of variability suggesting that EC may be a good measure on rapid soil diagnostic under this conditions.

References

Domsch, H; Giebel, A; Electrical conductivity of soils typical for the state of Bradenburg in Germany. In: European Conference on Precision Agriculture, 3., 2001, Montpellier. Proceedings of the 3rd European Conf. on Precision Agriculture, edited by G. Grenier and S. Blackmore, Ecole Nationale Supériure Agronomique, Montpellier 2001. CD.

Rhoades, J. D; Corwin, D. L. Measurment of inverted electrical conductivity profiles using electromagnetic induction. Soil Science Society of American Journal, Madison, v.44, p.288-291. 1984.

Development of liquid fertilizer applicator for precision agriculture in paddy field

E. Morimoto[1], M. Suguri[2], Y. Nii[2] and M. Umeda[2]
[1]Lab. of Environmental and Agricultural Systems Engineering Tokyo University of Agriculture and Technology, Tokyo, Japan
[2]Lab. of Field Robotics Kyoto University, Kyoto, Japan
eiji@elam.kais.kyoto-u.ac.jp

Abstract

The purpose of this research is to develop a liquid fertilizer applicator in order to sophisticate a conservative precision agriculture of rice production. A waterwheel (r=60mm) and micro-computer were applied for a flow sensor of inlet water. The coefficient of correlation between the flow volume of water and rotation speed of waterwheel was R^2=0.93, discharging time and total flow of liquid fertilizer was R^2=0.99 at a laboratory test.

Keywords: Animal waste, Liquid fertilizer, Micro-computer, Waterwheel, Solenoid valve

Introduction

Optimum fertilizer application is a key technology in Precision Agriculture (PA), and avoiding soil erosion is an important role for environmental friendly agriculture. In order to solve those problems, a liquid fertilizer applicator was developed as shown in Figure 1. Using animal waste for agriculture is environmental friendly compare to the chemical fertilizer. The goal of this research is to keep a stable ratio between inlet water and amount of fertilizer injection.

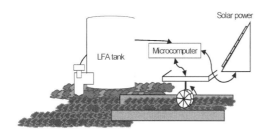

Figure 1. (a) Basic concept of LFA .

Figure 1. (b) Tank.

Figure 1. (c) Solenoid valve.

Figure 1. (d) Waterwheel.

Material and method

Basic concept of liquid fertilizer application (LFA) system consisted of 2 parts. First, a waterwheel was applied for measuring inlet water flow at channel. The amount of liquid fertilizer was controlled by using microcomputer, which could count the number of waterwheel rotation. Secondly, a solenoid valve was applied for discharging outlet. Figure 2 shows a relationship between flow volume of inlet water and the speed of waterwheel rotation. As a result, the speed of waterwheel rotation was strong enough (R^2=0.92) to adopt for flow volume estimator. In order to distribute fertilizer uniformly, Marriott method was applied. Figure 4 shows a relationship between discharge time and rate, the coefficient was R^2=0.99, thus, time required of 10 times rotation of waterwheel was applied for the time determination of discharge opening. Finally, the discharging time t (s) could be given by E.q.1.

$$t = \frac{0.19 \sum_{i=1}^{i=10} Q_{k_i} \cdot T_i + 2.5}{AQ_e} \quad (1)$$

where T_i is time required per one rotation (s).
A is mix proportion (water / fertilizer),
The N density of fertilizer was 0.3%, thus, the mixing ratio of water and fertilizer was expected 40:1 to 50:1 in this research.

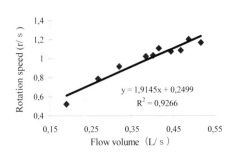

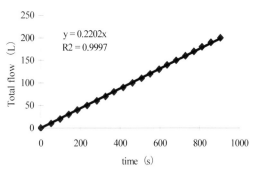

Figure 2. Relationship between flow volume and rotation speed of waterwheel.

Figure 3. Relationship between discharging time and total flow of fertilizer from discharging opening.

Result and discussion

Figure4 shows a relationship between flow of inlet water and mixing ratio of inlet water and liquid fertilizer. The mixing ratio was 40, even the different flow condition. In general, the flow of inlet range was between 0.25 and 0.45 (L/s), so that this applicator might be covered whole range in a practical use.

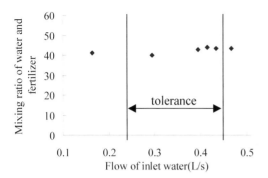

Figure 4. Relationship between flow of inlet water and mixing ratio of water and fertilizer.

Conclusion

The liquid fertilizer applicator was developed. The stabilization fertilizer distribution was discussed. The coefficient of flow volume and water wheel rotation was $R^2=0.92$, discharging time and total flow was $R^2=0.99$.

Comparison of 3 remote sensing sensors and 2 methods performances in the retrieval of biophysical variables on a wheat field

S. Moulin, R. Zurita-Milla, M. Guérif, V. Houlès and F. Baret

INRA/CSE, Bât. Climat - Site Agroparc - Domaine St-Paul, 84 914 Avignon cédex 9 , France
Sophie.Moulin@avignon.inra.fr

Abstract

Through a precision farming project, an intensive ground and airborne measurements campaign was performed. To address the estimation of biophysical variables related to the plant nitrogen status, 3 remote sensing sensors were utilized. The retrieval of variables was performed either via empirical relationships or through physical modeling. The drawbacks and advantages of different sensors/methods were discussed.

Keywords: remote sensing, wheat, reflectance, chlorophyll content

Introduction

In the frame of a precision farming project (Guérif, 2001), the nitrogen status of a wheat crop, and especially the characterization of field spatial heterogeneity was studied. Remote sensing measurements can help with building indicators based on the biophysical variables of the crop as green leaf area index (gLAI) and leaf chlorophyll content (Cab). The performances of 3 sensors and 2 methods in the determination of biophysical variables were compared.

Materials and methods

An experiment was conducted at Laon (North of France) on 2 winter wheat fields. Three optical domain sensors were used to overflow the fields with a 1 to 20m spatial resolution. They provided 4 images in 2000 (CASI), 4 in 2001 (SPOT), and 5 in 2002 (XYBION). SPOT satellite sensors acquired data in 3 or 4 broad bands. CASI radiometer provided hyperspectral measurements, and XYBION instrument is characterized by 6 bands. Data pre-processing was performed to account for atmospheric effects. Two different methods were implemented to estimate gLAI and Cab depending on the data quality and the sensor characteristics. The first method consists in using empirical relations between XYBION digital numbers and biophysical variables. The choice of indices was performed by minimizing the effects of atmosphere and soil background (see Zurita et al, 2003, this issue). A method based on the inversion of radiative transfer models (PROSAIL Jacquemoud and Baret, 1990; Verhoef, 1984) was tested for SPOT and CASI reflectances by accounting for the wheat structure, leaf biochemical composition and soil background. Prior information on the investigated variables was introduced in the inversion procedure to avoid for local minimum solutions.

Results

XYBION – empirical relationships : Figure 1a shows that when using NDVI (reduced central variable), a strong linear relationship ($r^2 = 0.8$) with LAI was obtained. LAI was estimated with a 0.5 m^2/m^2 rmse. Weak relationships were found between vegetation indices and Cab, although the estimation was improved when working with Cab integrated on the canopy.
SPOT – physically based models : The sensor, characterized by 3 broad wavebands, is not appropriated to estimate leaf chlorophyll content. However, LAI can be estimated through

inversion of radiative transfer models. Results (Fig. 1b) show the LAI was estimated with a 0.4 m²/m² rmse despite a weak correlation.

CASI – physically based models: Thanks to several thin bands, the CASI hyperspectral sensor was used to retrieve most of the radiative transfer inputs (10 parameters) and in particular the leaf and plant chlorophyll contents. Figure 1c displays a good correlation ($r^2 > 0.8$) and an estimation with a 10% error for LAI, Cab and plant chlorophyll content (LAIxCab).

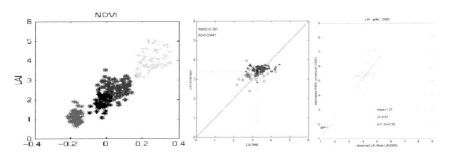

Figure 1. Comparison of retrieved gLAI estimates with field measurements : retrieval from (a) XYBION (5 dates), (b) SPOT (2 dates), (c) CASI (4 dates).

Discussion and conclusion

gLAI and Cab were estimated with a 10% accuracy. The weakness of correlation coefficients for a given date is related to the sampling strategy. The pixel surface is bigger than the ground sampled surface (1m²). Although SPOT calibration is robust, due to the sensor spectral characteristics, few parameters were estimated. Retrieval of biophysical variables using XYBION through empirical relationship was satisfactory, though the relations must be fitted for given conditions. The use of an hyperspectral sensor is the more promising method as one can estimate several variables and in particular the plant chlorophyll content that is a powerful variable in the plant nitrogen status estimation.

References

Guérif M, N. Beaudoin, C. Durr, J.M. Machet, B. Mary, D. Michot, S. Moulin, B. Nicoullaud, G.Richard, 2001, Designing a field experiment for assessing soil and crop spatial variability and defining site specific management strategies. *Third European Conference on Precision Farming*, 18-20 June 2001 – Montpellier, France, 677-682.

Jacquemoud S. and Baret F., 1990, PROSPECT : a model of leaf properties, *Remote Sens. Environ.*, 34:75-91.

Verhoef, W., 1984, Light scattering by leaf layers with application to canopy reflectance modeling : the SAIL model. Remote Sens. Environ., 16:125-141.

Zurita Milla R., Moulin S., Zago M., Guérif M., and Houlès V., Retrieving winter wheat LAI and Cab from Xybion airborne data, 2003, *European Conference on Precision Agriculture*, 15-19 June 2003, Berlin, Germany, this issue.

Acknowledgements

The authors thank Astrium for providing CASI data, M. Zago for developing a XYBION data pre-processing software, E. Pohl, N. Sanchez and M. Launay for data pre-processing.

A solution for interoperability of precision agriculture applications using web services

E. Murakami[1], A.M. Saraiva[1], C.E. Cugnasca[1], L.C.M. Ribeiro[1] and J.P. Molin[2]
[1]*Agricultural Automation Laboratory, Computer and Digital Systems Engineering Department, Polytechnic School of the University of São Paulo, São Paulo, Brazil*
[2]*Luiz de Queiroz School of Agriculture – University of Sao Paulo, Piracicaba, Brazil*
edson.murakami@poli.usp.br

Abstract

Experience has shown that the adoption of Precision Agriculture (PA) has been constrained because, although a vast amount of data has been collected, there is still a lack of applications to interpret those data, and the ones that exist have interoperability problems (Sorensen et al, 2002; Saraiva, Massola & Cugnasca, 1998). This paper presents a solution for interoperability among PA applications using Web services technology (Stal, 2002). A prototype for filtering yield monitor data was developed using the proposed solution.

Keywords: Precision Agriculture, Interoperability and Web Services.

Introduction

This article presents a solution for interoperability of applications for PA using Web services. Web service is core technology that is being used in the infrastructure proposed by Murakami et al. (2002) for development of component-based applications for PA. The architecture of this infrastructure is organized in layers: the first layer corresponds to the presentation-oriented Web applications build with Struts framework, from Apache Software Foundation; the second layer corresponds to the components with basic functionalities for PA build upon eMOSAICo framework, an extension of the MOSAICo (Object Model for Open Field Information Systems for Precision Agriculture) (Saraiva, Massola & Cugnasca, 1998); the third layer corresponds to "wrappers" responsible for the interoperability of the components of an application for PA. Components use basically Web services and EJB (Fingar, 2000). The interoperability solution proposal is presented in the next section using as reference the prototype for filtering yield monitor data.

Materials and methods

The prototype uses two interoperable components implemented as Web services: FilteringWS and MapperWS. FilteringWS is an eMOSAICo component that supplies the functionality of filtering yield monitor data (Molin & Menegatti, 2002). This service receives text files with raw data and returns the filtered data. Some representation formats can be accepted, for which XML descriptors where written. Before filtering, text files are submitted to the MapperWS component for the conversion of the data to an internal representation format (Patel, 2002) and returned to the FilteringWS. After filtering the data, the file is submitted again to the MapperWS to reformat the data to the initial format.

To supply a friendly interface to the user a presentation-oriented Web application was constructed with Struts framework, which communicates with the FilteringWS in asynchronous form, through a message queue (Filtering queue), to guarantee the availability of the application in case of manipulation of great amounts of data. Figure 1 depicts the application for filtering yield monitor data.

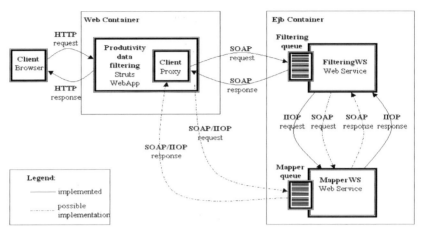

Figure 1. Information flow for the application of filtering yield monitor data.

Conclusions

The components FilteringWS and MapperWS use XML-based technology. Therefore, it is possible to access them from Client Proxies written in any language. The idea is to use the power of object oriented middleware to guarantee the interoperability between homogeneous applications (EJB/CORBA) and Web services to integrate heterogeneous applications (Web Services/SOAP). Although Web services technology is still maturing, due to its interoperability capabilities it becomes sufficiently interesting as a solution for global integration of reusable services, particularly for PA which is naturally multidisciplinary and frequently distributed in terms of information sources.

References

Fingar, P. 2000. Component-Based Frameworks for E-Commerce. Communications of the ACM. Vol. 43, No. 10.

Molin, J.P.; Menegatti, L.A A. 2002. Methodology for identification, characterization and removal of errors on yield maps. In: 2002 ASAE Annual International Meeting /CIGR XVTH World Congress, 2002, Chicago. St. Joseph: ASAE.

Murakami, E., Ribeiro, L.C.M., Cugnasca, C.E., Saraiva, A.M. 2002. An Infrastructure for Development of Information Systems for Precision Agriculture. 6th International Conference on Precision Agriculture and Others Precision Resources Management. Minneapolis, MN, USA. (in press).

Patel, S. 2002. Navigate data with the Mapper framework. [Online]. Available in http://www.javaworld.com/javaworld/jw-04-2002/jw-0426-mapper.html. (Accessible in june 2002).

Saraiva, A.M.; Massola, A.M.A.; Cugnasca, C.E. 1998. An object model for field information systems. In: Proceedings of the 4th International Conference on Precision Agriculture, Saint Paul, 1998. Minneapolis: ASA/CSSA/SSSA. Part B. pp.1355-1366.

Sorensen, C.G., Fountas, S., Pedersen, H.H., Blackmore, S. 2002. Information Sources for Decision Making on Precision Farming. 6th International Conference on Precision Agriculture and Others Precision Resources Management. Minneapolis, MN, USA. (in press).

Stal, M. 2002. Web Services: Beyond Component-Based Computing. Communications of the ACM. Vol. 45. No. 10. pp 71-76.

Role of bio-production robot in precision farming

N. Kondo[1] and S. Shibusawa[2]
[1]Department of Technology Development, Ishii Industry Co., Ltd., 422, Tomihisa, Matsuyama 791-8034, Japan
[2]Faculty of Agriculture, Tokyo University of Agriculture and Technology, 3-5-8 Saiwai-Cho, Fuchu, Tokyo 183-8509, Japan
kondo-n@ishii-ind.co.jp

Abstract

The world has been worrying about the quality of food supply because of recent problems with fruits, vegetables, meats, and other processed foods: food poisoning by bacteria, illegal unregistered agricultural chemicals, camouflage products, BSE, and etc. To solve these problems, a food traceability system which can show the food history during production, distribution and consumption, desirable. Bio-production robots have substituted for human labor and contributed to raise marketing value of product and to produce uniform products. Recent experience now shows these robots can also be useful and necessary to add precise information to data relating to field operations and thus can be valuable contributors to precision farming systems.

Keywords: bio-production robot, traceability, information added product, information oriented field

Introduction

Many bio-production robots have been researched and some of them were developed for practical use in Japan: a spraying robot, a cutting sticking robot, a grafting robot, a transplanting robot, and a grading robot. Those robots substituted labor and workers so human beings can be released from heavy, dangerous, or monotonous operations. In addition, the robots were effective to the increase market value of product, to produce uniform products, to make hygienic/aseptic production conditions and to give successors a hope for economic sustainability of small high value farm operations.

We have now found we can add another important role to bio-production robots, more precisely as a complement to precision farming, namely records of bio-production operations and of product information. It is obviously difficult for farmers to leave memo of all their works in fields. But it is possible for the robots to precisely record their operations in databases and to utilize the recorded information for the next operation or to accumulate information for farmer's decision support. In this report, this new role of bio-production robots will be discussed.

Robots and information

It is easy to record field ID, farmer's ID, and weather information surrounding field: temperature, humidity, radiation, and precipitation if sensors to measure those items are installed and these should be accumulated in a database for a decision support system as some of the most important information from robots or sensors. Operations were classified into four categories: filed management, seedling production & crop management, harvesting, and grading operations. Table 1 shows information obtained from robots or sensors for those operations. Some of the information and the robots in this table, are still under development for practical use and are not automatically obtained yet: P and K from soil sensor, harvesting robot, autonomous tractor, and residual chemicals (They are indicated in Italics). Bold letter items are already ready to go.

Table 1. Information from bio-production robots.

Operation	Robot or sensor	Information
Field management	**Soil sensor** *Autonomous tractor*	**N,** *P, K,* **EC, pH, moisture content, SOM soil temperature, firmness**
Seedling production & crop management	**Planting robot Grafting robot Spraying robot**	**plant variety, operation records: irrigation, chemical spray, fertilization**
Harvesting	*Harvesting robot* **Machine vision**	*crop No., positions of harvested production harvested time and date*
Grading	**Grading robot NIR sensor, X-ray Machine vision**	**size, color, shape, disease, inside injury sugar and acid contents,** *residual chemical*

Discussions

Below are few examples. In the area of agricultural chemicals, a spraying robot can spray the minimum chemicals based on information about crop damage and can store information about when, where and what kinds of chemicals were sprayed in computer databases. If the spray information is known and usable for the next operation, authors consider it possible to check residual chemicals in agricultural products in real time when the products are graded and to easily calculate total amount of chemicals lost in the surrounding environment. A fertilizing robot can control fertilizing operation according to variable information from the soil sensor (Shibusawa, 2003) and can calculate total amount of fertilizers. The combination of the robots and field soil sensor produce the concept of "information oriented" field.

A grading robot can accumulate data extracted from product images and the information can be used for farming next year with the operation records in the field. A grading robot for peach, pear, and apple fruit was commercialized last year in Japan by Ishii Industry Co., Ltd. (Kondo, 2003). The robot had a pair of manipulators for fruit picking from containers and a grading manipulator with 12 color TV cameras to measure size, color, shape, disease and insect injury. This robot produces the "information added" products, that can become "value added" products in the market.

References

Kondo, N. 2003. Fruit grading robot. In Proceedings of the 2003 IEEE/ASME International Conference on Advanced Intelligent Mechatronics (AIM 2003), Kobe, Japan.
Shibusawa, S. 2003. On-line Real Time Soil Sensor. In Proceedings of the 2003 IEEE/ASME International Conference on Advanced Intelligent Mechatronics (AIM 2003), Kobe, Japan.

Analysis of spatial variability for yield and soil chemical properties in a large scale transplanted rice field

Nguyen Tuan Anh[1], Kyung-Jin Choi[2], Jin-Chul Shin[2], Chung-Kuen Lee[2] and Byun-Woo Lee[1]
[1]*School of Plant Science, Seoul National University, Republic of Korea*
[2]*National Crop Experimental Station, RDA, Republic of Korea*

Abstract

A large-scale transplanted paddy rice field was evaluated for the spatial variability of rice yield and soil chemical properties. The coefficient of variation (CV) for yield was 11%. CV for soil variables varied from 2.21% in pH to 19% in CEC at transplanting and harvest. Soil variables with the highest correlation to grain yield was CEC (r =0.71) at transplanting and harvest. The stepwise regression procedure was applied to identify the soil variable with a significant influence on spatial variability of rice yield. Among the soil variables at transplanting, CEC, OM, available SiO_2, Ntot, Ca, Mg and K were stayed in the multiple regression model at significant probability level of 0.05, explaining as much as 76% of yield variability. CEC, OM, Mg, available P_2O_5, K and Ntot. at harvest were selected and explained about 80% of the variability of yield. CEC was the most important soil variable explaining about 51% of yield spatial variability.

Keywords: Spatial variation, soil, rice, grain yield, transplanted paddy field

Introduction

Spatial variability of rice yield in a transplanted paddy field results from inhomogeneity of management practice and soil properties and their complex interaction. This research therefore was carried out to investigate the causal factors of yield spatial variability in transplanting rice culture in Korea. Precise management considering this spatially variable condition is required to maximize rice production while minimizing use and leaching of fertilizes.

Material and method

A transplanted paddy field of 60m×110m (divided into 66 plots with 10m×10m) located in Suwon, Korea was used for the research. Sixty-six soil samples at transplanting and harvest were collected for the analysis of soil chemical properties. Grain yield for each plot was also measured. Correlation and stepwise regression analysis were applied to identify soil variables that significantly influenced the spatial variability of grain yield.

Results

The descriptive statistic for spatial variation analysis of rice yield and soil variables were presented in Table 1. Most of soil variables except for EC showed significant correlations with grain yield (Table 2). The largest correlation was obtained for CEC followed by OM, Ca, Ntot, etc.
Stepwise regression procedure with forward selection was applied to identify the soil variables with a significant effect on grain yield. The followings are the equations at transplanting (1) and harvest (2). Selected soil variables are written in increasing order of partial R-square.

Table 1. Descriptive statistics of soil chemical properties at transplanting and harvest in transplanted rice field (P_2O_5=Available P_2O_5, SiO_2=Available SiO_2).

Properties	Mean	STD	Min	Max	CV	Mean	STD	Min	Max	CV
	---------------------At transplanting-----------------					----------------------At harvest-----------------				
P_2O_5 (mg/kg)	71.8	9.62	52.1	94.5	13.4	68.0	6.77	53.3	87.7	9.96
Ntot. (%)	0.14	0.01	0.11	0.16	7.14	0.14	0.01	0.11	0.18	7.14
OM (g/kg)	25.0	2.44	19.2	31.3	9.76	24.14	2.27	19.8	29.2	9.40
SiO_2 (mg/kg)	100.3	10.8	80.1	122.4	10.7	99.08	5.97	83.6	109.9	6.03
Na ($cmol^+$/kg)	0.51	0.03	0.45	0.62	5.8	0.51	0.03	0.45	0.57	5.88
Mg ($cmol^+$/kg)	1.61	0.11	1.32	1.87	6.83	1.59	0.08	1.36	1.79	5.03
K ($cmol^+$/kg)	0.59	0.05	0.46	0.69	8.47	0.58	0.04	0.48	0.66	6.90
Ca ($cmol^+$/kg)	5.00	0.51	3.84	6.23	10.2	5.10	0.44	4.04	6.09	8.63
CEC $cmol^+$/kg)	7.90	1.55	5.61	11.4	19.6	7.90	1.55	5.61	11.4	19.6
EC (dSm^{-1})	0.47	0.06	0.35	0.62	12.7	0.36	0.05	0.26	0.45	13.8
pH	5.88	0.13	5.62	6.17	2.21	5.52	0.16	5.21	5.90	2.90
Yield (kg/10a)	569.1	65.9	402.5	723.3	11.5	569.1	65.9	402.5	723.3	11.5

Table 2. Correlation coefficient of rice yield with soil variables at seeding (lower diagonal) and harvest (upper diagonal) of transplanted field ([*]Significant at the 0.05 probability level, [**]Significant at the 0.001 probability level).

	P_2O_5	Ntot	OM	SiO_2	Na	Mg	K	Ca	CEC	EC	pH	Yield
P_2O_5	1	0.15	0.39**	0.14	0.32*	0.27*	0.06	0.41**	0.33*	-0.09	0.39**	0.51**
Ntot.	0.19	1	0.37*	0.32*	0.28*	0.22	0.21	0.11	0.43**	-0.16	0.19	0.52**
OM	0.11	0.26*	1	0.43**	0.51**	0.29*	0.26*	0.29*	0.40**	-0.02	0.20	0.62**
SiO_2	0.27*	0.22*	0.38**	1	0.20	0.41**	0.41**	0.30*	0.47**	-0.33	0.06	0.57**
Na	0.06	0.30*	0.25*	0.28*	1	0.40**	0.16	0.46**	0.41**	0.41	0.36**	0.47**
Mg	-0.02	0.19	0.37**	0.42**	0.34*	1	0.31**	0.64**	0.34**	-0.16	0.23*	0.57**
K	0.15	0.31*	0.39**	0.30**	0.19	0.22*	1	0.19	0.41**	-0.19	0.21	0.50**
Ca	0.38**	0.25*	0.44**	0.38**	0.22*	0.31*	0.27*	1	0.47**	0.03	0.44**	0.57**
CEC	0.33*	0.44**	0.52**	0.37**	0.24*	0.21	0.44**	0.44**	1	-0.10	0.18	0.71**
EC	0.13	-0.13	0.15	0.007	0.16	0.12	0.17	-0.03	0.12	1	-0.00	-0.19
pH	0.50**	0.06	0.31*	0.39**	0.07	0.18	0.18	0.32*	0.36*	0.05	1	0.24*
Yield	0.36**	0.48**	0.66**	0.59**	0.28*	0.48**	0.54**	0.58**	0.71**	0.06	0.43**	1

$$Yield = -231.9 + 12.4CEC + 1.3SiO_2 + 5.8OM + 642TN + 20.1Ca + 82.3Mg + 170K \quad r^2_{adj} = 0.76 \quad (1)$$
$$Yield = -382.5 + 13.4CEC + 6.5OM + 184Mg + 2.0P_2O_5 + 266K + 704TN \quad r^2_{adj} = 0.77 \quad (2)$$

CEC was a most important soil variable explaining more than 50% of the spatial variability of yield and other selected variables explained about 27% of it. CEC was also reported to be a most influential soil variable causing rice yield variability in Spain by Casanova (1999).

Conclusion

Many soil chemical properties at transplanting and harvest showed a spatial variability in transplanted paddy field in Korea. However, CEC, organic matter and total nitrogen are considered as the most important soil variables, which have significantly influenced spatial variability of rice yield.

Reference

D. Casanova, J. Goudriaan and etc. 1999. Yield gap analysis in relation to soil properties in direct-seeded flooded rice. Geodernma 91 (1999) 191-216.

Spatial variation of yield and soil chemical properties in a large scale direct-seeded rice field

Nguyen Tuan Anh[1], Kyung-Jin Choi[2], Jin-Chul Shin[2], Chung-Kuen Lee[2] and Byun-Woo Lee[1]
[1]*School of Plant Science, Seoul National University, Republic of Korea*
[2]*National Crop Experimental Station, RDA, Republic of Korea*

Abstract

A large scale direct-seeded paddy rice field was evaluated for the spatial variability of yield and soil chemical properties. The coefficient of variation (CV) for yield was 21.6%. CV for soil variables varied from 4.24% in pH to 20.1% in CEC at seeding and from 3.74 in pH to 20.1% in CEC. Soil variables with the highest correlation to grain yield were CEC ($r=0.73$) at seeding and harvest. The stepwise regression procedure was applied to identify the soil variables with a significant influence on spatial variation of rice yield. Among the soil variables at seeding, CEC, Ca, available P_2O_5, and Na were stayed in the multiple regression model at the significant probability level of 0.05, explaining as much as 77% of yield variability. Ca, K, OM, Ntot, CEC, Mg, and SiO_2 at harvest were selected and explained about 81% of variability of yield. CEC was the most important soil variable explaining about 53% of yield spatial variability.

Keywords: Spatial variation, soil, rice, grain yield, direct-seeded paddy field

Introduction

Spatial variability of rice yield in a direct-seeded paddy rice field results from inhomogeneity of management practice and soil properties and their complex interaction. This research therefore was carried out to investigate the causal factors of yield spatial variability in direct-seeded rice culture in Korea. Precise management considering this spatially variable condition is required to maximize rice production while minimizing use and leaching of fertilizes.

Material and method

A direct-seeded paddy rice field of 59m×110m (divided into 66 plots with 10m×10m) located in Suwon, Korea was used for the research. Sixty-five soil samples at seeding and harvest were collected for the analysis of soil chemical properties. Grain yield for each plot was also measured. Correlation and stepwise regression analysis were applied to identify soil variables that significantly influenced the spatial variability of grain yield

Results

The descriptive statistics for spatial variation analysis of rice yield and soil variables were presented in Table 1. Most of soil variables except for EC showed significant correlations with grain yield (Table 2). The largest correlation was obtained for CEC followed by Ca, SiO2, OM etc.
Stepwise regression procedure with forward selection was applied to identify the soil variables with a significant effect on grain yield. The followings are the equations at seeding (1) and harvest (2). Selected soil variables are written in increasing order of partial R-square.

517

Table 1. Descriptive statistics of soil chemical properties at seeding and harvest in direct-seeded paddy field (P_2O_5=Available P_2O_5, SiO_2=Available SiO_2).

Properties	Mean	STD	Min	Max	CV	Mean	STD	Min	Max	CV
	--------------------At seeding---------------------					--------------------At harvest-----------------				
P_2O_5 (mg/kg)	70.7	5.47	55.7	84.0	7.24	64.2	11.6	44.4	98.9	18.0
Ntot. (%)	0.14	0.01	0.11	0.16	7.14	0.13	0.01	0.10	0.16	7.69
OM (g/kg)	24.2	2.73	18.6	29.3	11.2	23.0	2.36	17.8	28.3	10.2
SiO_2 (mg/kg)	95.4	8.06	78.4	119.9	8.45	92.2	6.85	74.8	104.9	7.43
Na ($cmol^+$/kg)	0.52	0.05	0.38	0.64	9.62	0.52	0.03	0.40	0.60	5.77
Mg ($cmol^+$/kg)	1.60	0.10	1.31	1.95	6.25	1.59	0.15	1.20	1.90	9.43
K ($cmol^+$/kg)	0.57	0.07	0.41	0.73	12.2	0.56	0.07	0.39	0.69	12.5
Ca ($cmol^+$/kg)	4.97	0.53	3.46	5.89	10.6	4.69	0.54	3.25	5.86	11.5
CEC $cmol^+$/kg)	7.78	1.57	5.37	11.2	20.1	7.78	1.57	5.37	11.2	20.1
EC (dSm^{-1})	0.51	0.03	0.41	0.60	5.8	0.43	0.03	0.36	0.50	6.98
pH	5.89	0.25	5.09	6.46	4.24	5.61	0.21	5.18	6.08	3.74
Yield (kg/10a)	508.2	109.5	150.9	669.5	21.5	508.2	109.5	150.9	669.5	21.5

Table 2. Correlation coefficient of rice yield with soil variables at seeding (lower diagonal) and harvest (upper diagonal) of direct-seeded field ([*]Significant at the 0.05 probability level, [**]Significant at the 0.001 probability level).

	P_2O_5	Ntot	OM	SiO_2	Na	Mg	K	Ca	CEC	EC	pH	Yield
P_2O_5	1	0.51**	0.39**	0.47**	0.27*	0.42**	0.39**	0.40**	0.53**	0.07	0.03	0.52**
Ntot.	0.66**	1	0.42**	0.45**	0.49**	0.50**	0.52**	0.55**	0.41**	-0.27	0.07	0.66**
OM	0.47**	0.51**	1	0.44**	0.34**	0.56*	0.45**	0.60**	0.67**	-0.23	0.16	0.71**
SiO_2	0.42**	0.43**	0.56**	1	0.33*	0.33*	0.52**	0.50**	0.45**	-0.23	-0.10	0.63**
Na	0.37*	0.42**	0.44**	0.55**	1	0.26*	0.39**	0.23*	0.25**	-0.25	0.13	0.46**
Mg	0.52**	0.53**	0.25*	0.49**	0.49**	1	0.44**	0.63**	0.45**	-0.25	0.41**	0.69**
K	0.43**	0.59**	0.45**	0.43**	0.36*	0.39**	1	0.50**	0.50**	-0.32	0.10	0.68**
Ca	0.53**	0.52**	0.56**	0.62**	0.42**	0.43**	0.49**	1	0.63**	-0.21	0.39**	0.74**
CEC	0.46**	0.44**	0.68**	0.70**	0.45**	0.37*	0.40**	0.53**	1	-0.21	0.13	0.73**
EC	-0.16	-0.17	-0.28	-0.38	-0.33	-0.16	-0.08	-0.44	-0.41	1	-0.06	-0.41
pH	0.36*	0.30*	0.10	0.21	0.11	0.44**	0.28*	0.24*	0.20	0.21	1	0.08
Yield	0.62**	0.59**	0.67**	0.69**	0.69**	0.47**	0.55**	0.68**	0.73**	-0.50	0.10	1

$$Yield = -618.4 + 24.4\ CEC + 739Na + 49.4Ca + 4.23P_2O_5\ r^2_{adj}=0.77 \qquad (1)$$
$$Yield = -686.8 + 18.6Ca + 263K + 6.32OM + 1278TN + 16.2CEC + 157Mg + 2.78SiO_2\ r^2_{adj}=0.81 \quad (2)$$

CEC was a most important soil variable explaining more than 53% of the spatial variability of yield and other selected variables explained about 23% of it. CEC was also reported to be a most influential soil variable causing rice yield variability in Spain (Casanova, 1999).

Conclusion

Many soil chemical properties at seeding and harvest showed a spatial variability in direct-seeded paddy field in Korea. However, CEC, Ca, organic matter and available P are considered as the most important soil variables, which have significantly influenced spatial variability of rice yield.

Reference

D. Casanova, J. Goudriaan and etc. 1999. Yield gap analysis in relation to soil properties in direct-seeded flooded rice. Geodernma 91 (1999) 191-216.

GeoInformation-Technology: demanding more attention within the field of agricultural information management and quality assurance

O. Nölle

Institute for Agricultural and Forestal Informatics (afi) at the University of Münster, Robert-Kochstrasse 27, 48149 Münster, Germany
nolleo@afi.uni-muenster.de

Abstract

GeoInformation-(GI-) technology and especially the use of GeoInformationSystems (GIS) has always played a certain role within agriculture, not only in the field of precision agriculture. However, the documentation demands on agriculture are rising and regulations are already existing or are on the way to be established, which are forcing the use of this technology to enhance agricultural documentation in general. Beside this legal pressure, technical aspects like easy access to geoinformation (e.g. field-unit cadastres) via geo-services within geodata-infrastructures will certainly have its effects on the broader use of this technology in agricultural information management and quality assurance and corresponding IT-solutions.

Keywords: geoinformation, GIS, geodata-infrastructures, field-unit cadastre, IACS

Introduction, background

Over the last 15 years GeoInformation(GI-)-technology [often reduced just to the use of GeoinformationSystems (GIS), but also implying aspects like geodata- and mapservers, remote sensing, mobile geo-applications, GPS] has been worldwide established as the technological approach to work on digital spatial data processing in every field of application. Agriculture is without no doubt a spatial discipline, influencing its environment and getting influenced by it as well. So, agriculture is causing a lot spatial questions and hence is predestinated for the use of GI-technology. And this technology, especially the use of GIS has already a long history within agriculture, e.g. in precision agriculture. But time is changing with regard to the importance of this technology in the agricultural application domain. From the authors point of view GI-technology is moving from being just one technology in agriculture, to become one of the top ranking information technologies in it. The background of this evolution is manifold. One core reason are emerging or already existing legal regulations, basically coming from the European Union, like European Commission regulation 1593/2000, describing the new demands to the Integrated Administration and Control System (IACS) for agricultural subsidiaries. This regulation for example makes the use of GI-technology in IACS (which is of course one of the central activities of the agricultural administration) obligatory from the beginning of 2005 on. Apart of integrating this technology into IACS, one further important activity is the building of precise Land Parcel Information Systems (LPIS), which are in their most precise shape field-unit cadastres, based on regulation 1593/2000. Especially the exhaustive availability of this data is said to be one of the essential aspects when talking about the integration of GI-technology as a key-technology in agriculture, because the field-unit is always in the focus of agricultural activities.

Another reason for the rising importance of GI-technology in agriculture is the generally increasing availability of basic (e.g. digital orthophotos, topographic maps, digital elevation models) and specific (e.g. field-unit cadastre, soil maps) geodata relevant for agricultural purposes. But it is not only the simple availability of geodata, it is especially the near future perspective to get easy access to geodata via geo-services within so called geodata-infrastructures based on international standards like the specifications of the OpenGIS Consortium (OGC) and ISO that makes GI-technology more attractive for agricultural documentation objectives and topics like quality assurance, risk management and tracking and tracing then ever before.

Implications for agricultural documentation and quality assurance

Keeping in mind that the pressure to do precise agricultural documentation in general is rising, exhaustive LPIS are built within the framework of the new demands to IACS and geodata-infrastructures are erected to guarantee easy access to geoinformation, the author sees the clear chance for GI-technology to become a nearby standard-technology within the fields of agricultural documentation and quality assurance, where the geo-referenced documentation of agricultural activities is seen as a basis for topics like a sustainable agricultural quality management, especially in plant production. Access to field-unit geometries will offer the possibility to very precisely describe where a certain plant-production has taken place under which spatial conditions. From the farmers point of view this geo-referenced kind of agricultural documentation offers a new way of describing the quality of their products. The food producers will get the opportunity to ensure access to the origin of their products and to very quickly react on any kind of incident along their production chain in terms of tracking and tracing. From the consumers position there will be also the chance to examine the sources of the products they are consuming.

Conclusions

GI-technology has clearly the potential to become a turn-key-technique within agricultural documentation and quality assurance. In front of this background, research has to be done especially within the field of tracking and tracing and the integration of geodata-infrastructure-approaches into already existing agricultural business processes. A good starting point may be the objectives of the 6th European Framework Research Programme, priority 5, food quality and safety.

References

Nölle, O. 2003. Pushing GIS in agriculture: european commission regulation 1593/2000! To be published in: Proceedings of the twenty-third annual ESRI International Conference (July 7-11, 2003, San Diego, California, USA).

Nölle, O. 2003. GeoInformation-technology: a key-technology within the framework of quality- and risk-management and environmental control in agriculture! To be published in: Proceedings of the 2nd International European Seminar on Quality Assurance, Risk Management and Environmental Control in Agriculture and Food Supply Networks (May 14-16, 2003, Bonn, Germany).

Nölle, O. 2002. Geodata-infrastructures: the key to agricultural geo-business applications. In: Proceedings of the ESRI International User Conference, 8th – 12th July 2002, San Diego, California, USA.

Nölle, O. 2001. The central geodataserver of the department of agriculture for Westfalen-Lippe (Northrine-Westfalia, Germany) as the core aspect of the departments GIS-integrating strategy. In: Proceedings of the Third International Conference on Geospatial Information in Agriculture and Forestry, 5th – 7th November 2001, Denver, Colorado, USA.

Measuring feeding quality and dry matter yield in a lay sward using spectroscopic methods

Anna Nyberg
Department of Agricultural Research Skara, P.O. Box 234, SE-532 23 Skara, Sweden
anna.nyberg@jvsk.slu.se

Abstract

To facilitate the registration of dry matter (DM) and feeding qualities within a lay field reflectance measurements were performed in pilot-study using a portable NIR-instrument, FieldSpec Pro FR (ASDI, Boulder CO, USA). Plots in different field trials were measured before 2nd and 3rd cut in 2002. From each plot dry matter yield (DM) was obtained and samples were analysed for crude protein (CP), neutral detergent fibre (NDF) and metabolizable energy (ME). Evaluations were done by partial least square regression (PLS) in UNSCRAMBLER 7.8 (CAMO PROCESS AS, Norway). For 2nd and 3rd cut together the r^2-values were CP = 0.54, ME = 0.57, NDF = 0.41 and DM = 0.28. Predictions for 3rd cut were better than for the 2nd cut due to the larger differences in range for CP, ME and NDF. No highly correlated results but promising tendencies could be seen. More investigations are necessary to develop the technique to measure in the sward.

Keywords: spectroscopic reflectance, ley, grassland, NDF, crude protein

Introduction

Variations in dry matter (DM) and feeding qualities within a field have been studied in lays in previous investigations (Tuvesson, 1993, Nyberg& Lindén 2002). To facilitate the registration of these variations a tool for measuring grass yield and grass quality in the field is of highest interest.

Materials and methods

Reflectance measurements were performed in pilot-study using a portable NIR-instrument FieldSpec Pro FR (ASDI, Boulder CO, USA). Four plots at four different field trials were measured before 2nd and 3rd cut in 2002. All sites had four different grass and clover mixtures. In each plot 40 measurements were made and averaged into a mean spectrum. Measurements and sampling at 2nd cut were done at 10 July 2002 and for the 3rd cut, measurement were made 23 and 26 August 2002 with sampling 4 or 2 days after. At each cut dry matter yield (DM) was measured. Samples from each plot were analysed for crude protein (CP), neutral detergent fibre (NDF) and metabolizable energy (ME). Evaluations were done by partial least square regression (PLS) in UNSCRAMBLER 7.8 (CAMO PROCESS AS, Norway).

Results and discussion

Results from 2nd and 3rd cut are presented in Table 1. The wavelengths used in the PLS-models were 354-1350nm, 1460-1738nm and 2070-2318nm. There were largest variations in the predictions for ME. Prediction of NDF was almost equally in both harvests. There were over all better predictions for the 3rd cut than for the 2nd cut. This could be due to the differences in range for CP, ME and NDF. Lowest correlations were obtained for DM. There were difficulties to get good correlations in this small dataset. As shown there can be large variations between different cuts. For DM there is a problem in measuring close to the cut.

Table 1. PLS were made to predict crude protein (CP), metabolizable energy (ME), neutral detergent fibre (NDF) and dry matter yield (DM) in 2nd and 3rd cut 2002. Results shown as r^2-values, RMSEP, RER and range.

Analysis	Cut	Number of Plots	r^2	RMSEP	RER	Range
CP	2nd	12	0.49	17.3	4.34	101-179
	3rd	16	0.71	18.6	6.56	69-191
	$2^{nd} + 3^{rd}$	28	0.54	20.6	5.91	69-191
ME	2nd	12	0.16	0.5	3.19	9.8-11.3
	3rd	16	0.82	0.4	7.99	8.3-11.6
	$2^{nd} + 3^{rd}$	28	0.57	0.5	6.22	8.3-11.6
NDF	2nd	12	0.49	34.4	4.85	431-598
	3rd	16	0.59	40.0	5.03	413-614
	$2^{nd} + 3^{rd}$	28	0.41	43.0	4.67	413-614
DM	2nd	12	0.20	737.0	2.86	2450-4560
	3rd	16	0.02	451.5	1.61	3224-3950
	$2^{nd} + 3^{rd}$	28	0.28	583.0	3.62	2450-4560

RER = range error ratio
RMSEP = residual mean square error

Conclusion

It is difficult to make predictions with a small data set like this but promising tendencies in predicting feeding quality could be seen. More investigations are necessary to develop the technique to measure in the sward.

References

Nyberg, A. and Lindén, B., 2002. Within-field variations in forage yield and quality of a grass dominated ley in southwest Sweden, 1999-2001, Dep.of Agricultural Research Skara, Series B Crops and soils. Report 9. In Swedish with English summary.
Tuvesson, M. 1993 On variation in crop yield between and within grassland fields. Swedish J. agric. Res. 23:15-19.

Acknowledgement

This pilot-study was financed by Skaraborgsläns Nötkreaturförsökringsbolags Stiftelse in collaboration with Precision Agriculture Sweden (POS).

ISOBUS Task Controller with GSM data transfer

Matthias Pallmer and Denis Leuthold
WTK-Elektronik GmbH, D- 01844 Neustadt/Sa., Germany
info@wtk-elektronik.de

Abstract

The compatibility of tractors and agricultural implements of different manufacturers is provided by international standardization with the ISO 11783 (called ISOBUS). The lowest expense for Precision Farming equipment is given, if all necessary functions are integrated in the ISOBUS „Virtual Terminal". The tractor terminal *„field-operator 205"* belongs to this class of tractor bord computers. An additional wireless module can transfer Precision Farming data to a Geographic Information System (GIS). Although this system has a very wide function volume it is designed in order to optimizing device costs.

Keywords: ISOBUS, tractor terminal, task controller, wireless datatransfer

Introduction

In the near future the ISOBUS standard will become the base for Precision Farming. The functions are provided by a ISOBUS Task Controller. Together with the navigation module (NMEA 2000 standard) it fulfills all the necessary data processing to output setpoints and log process values in accordance to the geographic position.

Technical solution of whole system

To optimize costs of the Precision Farming equipment in the ISOBUS era it is profitable to use a tractor terminal, that includes as much as possible functions. The WTK solution (see figure 1) consits of the terminal *„field-operator 205"*, the navigation monitor, the wireless module, multifuncional handle, DGPS receiver, various implement electronic control units (ECU) and implement identifiers (IMI®). The terminal performs the following functions: ISOBUS Virtual Terminal for monitoring and controlling of implement functions, ISOBUS Task Controller, (partial) ISOBUS Tractor ECU, DGPS decoder for direct connected NMEA 0183 receiver, navigation module with external monitor (lightbar) for parallel swathing and driver software for external wireless module. To have all these ISOBUS functionality in one device minimizes costs for the Precision Farming equipment.

Task Controller with wiresless data transfer

The Task Controller function consists of the following:
- task data management with working time and area measurement
- output setpoints according to geographical position (grid data)
- process data recording with georeference
- coordination of the work of combined implements
- data transfer to office host PC

The normal Precision Farming data interface for the user is the flash memory card. But an additional wireless module can transfer an amount of Precision Farming files or small data packets to a far away information centre, where the data are processed, archived and visualized. The further extension is data transfer via GPRS regarding to reduce the costs and the transmission time.

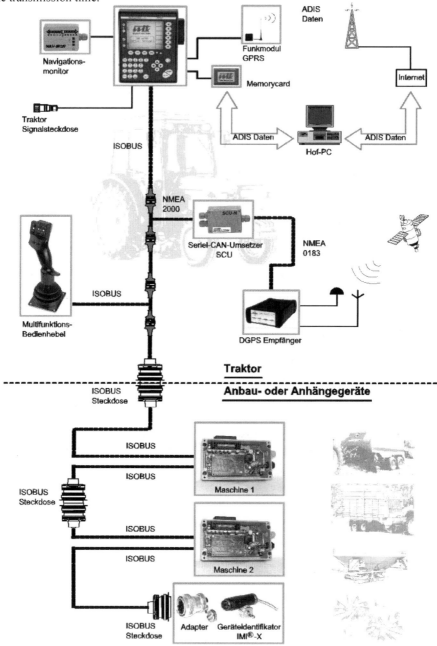

Figure1. Cost optimized equipment for precision farming in ISOBUS era.

A simulation of the economic and environmental impact of variable rate nitrogen application

S.M. Pedersen[1] and J.L. Pedersen[2]
[1]Danish Research Institute of Food Economics, Rolighedsvej 25, 1958 Frederiksberg C, DK
[2]The Technical University of Denmark, IPL, Produktionstorvet, bygning 423, 2800 Kgs. Lyngby,DK
Marcus@foi.dk

Abstract

This analysis shows that there is some potential yield benefit from site-specific application of nitrogen in winter wheat, barley and rape seed based on static knowledge about the soil structure and soil conductivity. A presumption is that some spatial variability of the soil structure occurs on the field. However, findings from this study also indicate that changing climatic conditions have a significant impact on yield response from variable nitrogen application. The study also establishes that site-specific N-application seems to have a positive but also small impact on nitrate leaching in cereals.

Keywords: Variable rate N-application, yield response, simulation model, nitrate leaching.

Introduction

The economic viability of variable rate nitrogen application has not yet been demonstrated. Specific tools are costly and the economic benefits are uncertain (Godwin et al, 2002, Bullock 1998, Swinton and Lowenberg-Deboer 1998). Moreover, the environmental impact remains unclear (Stafford, 2000). Variable rate application of nitrogen is aimed at yield improvements, better crop quality, reduction of nitrate leaching and input savings. Several studies have compared the yield potential and environmental impact of site-specific application of nitrogen in cereals but the uncertainty caused by climatic changes (temperature, radiation and precipitation) has often been neglected. The objective of this paper was to analyse the dynamic economic and environmental impact of site-specific application of nitrogen in common Danish crops.

Materials and methods

The calculations were based on the FASSET-model developed by the Danish Institute of Agricultural Science and The Danish Research Institute of Food Economics. FASSET is a dynamic simulation model that consists of a crop model and a soil model (Jacobsen et. al 1998). The model simulates the total production output on the field over a period of 8 years with either uniform or variable application. Each field was divided into 4 sub-fields, which varies from sandy soils to clay loam. From each scenario, the model estimates nitrate leaching and evaporation under different soil type and weather conditions. Yield response is thereby a function of soil type, precipitation, radiation, temperature, application of nitrogen and previous crop rotations. The model can manage crop rotations with barley, wheat, and rape seed.

Results

The analysis shows that there is some potential average yield benefit (8 years) from site-specific application of nitrogen in winter wheat, barley and rape seed (about 5-80 kg/ha) based on static knowledge about the soil structure and soil conductivity. A presumption is that some spatial variability of the soil structure occurs on the field. However, findings from this study also indicate that changing climatic conditions have a significant impact on yield response from variable nitrogen application and nitrate leaching in cereals. To optimise yields and economic benefit from site-specific application, the application strategy must depend on the interrelated marginal productivity of nitrogen between the different soil types on the field. In this matter, one application strategy one year is not necessarily appropriate the following year. Moreover the relatively small potential yield and economic benefits might not be enough to cover the cost of equipment and electronic devices for variable rate treatment. Given the climatic changes over time it might, for many farmers, be necessary to combine previous yield and soil data with temporal weather data and real time canopy management. Application strategies, which are solely based on historic information about yield variability and soil conductivity will probably not enable the farmer to establish an economic viable production from precision farming.

Conclusions

This simulation model shows that there is a small potential yield benefit from site-specific application (based on soil type and soil texture mapping) in winter wheat, barley and rape seed compared with uniform application. A presumption is that some spatial variability of the soil conditions occurs on the field. However the climatic conditions and precipitation have a significant impact on the potential yield from year to year which may overrule the yield potential from variable application. To optimise the economic benefit from site-specific application, the application strategy may include real time canopy management rather that historic data.

References

Bullock D.S. 1998. The Economics of Precision Farming: A Primer for Agronomists Designing Experiments. Precision Agriculture '99, Proceedings of the 2nd European Conference on Precision Agriculture, Ed J.V. Stafford. BIOS Scientific Publishers Ltd, pp 937-946.

Goodwin R.J., Earl R. Taylor J.C., Wood G. A., Bradley R.I., Wlesh J.P., Richards T, Blackmore B.S., Carver M.J., Knight S., Welti B.U. 2002: Precision Farming of cereals crops, A five-year experiment to develop management guidelines, HGCA, project report no. 267.

Jacobsen B. H., Petersen B. M., Berntsen J, Boye C, Sørensen C. G., Søgaard H. T. and Hansen J P 1998: An Integrated Economic and Environmental Farm Simulation Model (FASSET), Danish Institute of Agricultural and Fisheries Economics, report no. 102.

Stafford J. V. 2000: Implementing Precision Agriculture in the 21st Century, J. agric. Engng. Res. (2000) 267-275.

Swinton S.M. and Lowenberg-DeBoer 1998: Evaluating the Profitability of Site-Specific Farming, Journal of Production Agriculture, Volume 11 no 4. Oct.-Dec 1998 p. 439-446

Acknowledgement

This study is partly financed by the Danish Ministry of Food, Agriculture and Fisheries and the Danish Research Institute of Food Economics. It is also an integral part of a PhD-study at the Technical University of Denmark, focusing on technology assessment of precision farming practices in Denmark.

Algorithms for discrimination of weeds in sugar beet fields using digital image processing

Isabelle Philipp[1], Henning Nordmeyer[2] and Thomas Rath[1]

[1]*Institute of Horticultural and Agricultural Engineering. University of Hannover, Herrenhäuser Straße 2, D-30419 Hannover, Germany*
[2]*Institute for Weed Research, Federal Biological Research Centre for Agriculture and Forestry, Messeweg 11/12, D-38104 Braunschweig, Germany*
philipp@itg.uni-hannover.de, rath@itg.uni-hannover.de, h.nordmeyer@bba.de

Abstract

Experiments were carried out to develop algorithms for automatic detection and discrimination of monocotyledonous and dicotyledonous weeds in sugar beet by use of digital image processing. The comparison of weed distribution based on manual weed mapping and automatic weed mapping by digital camera showed, that the automatic method was able to detect the same patches of dicotyledonous weeds. The detection of monocotyledonous weeds by digital image processing still has to be improved. In general it could be shown, that an automatic weed mapping is realizable.

Keywords: automatic weed mapping, image processing, plant discrimination

Introduction

Herbicide reduction is becoming more and more important in agriculture. Instead of spraying an entire field, only infested parts should be sprayed, to have ecological and economical benefits (NORDMEYER et al. 1997). Hence, information about weed distribution is necessary. Manual weed mapping by field walking is very time-consuming and expensive in agricultural practice. Therefore, the implementation of an automatic weed mapping system and a spatial differentiated herbicide application is desirable.
The aim of this work was to develop and test an automatic weed mapping system in sugar beets and to prove its applicability.

Materials and methods

Experiments were implemented in 2001 and 2002 to develop an automatic system for detection and discrimination of monocotyledonous and dicotyledonous weeds in sugar beet. The experiments were carried out on two (9 ha and 14 ha) sugar beet fields on a farm near Helmstedt (Domäne St. Ludgeri), northern Germany.
First, manual weed mapping was carried out. Weed densities and species of weeds within the fields were counted manually at grid points (area 0,1 m²) to create a weed distribution map of each field. Grid spacing of 25 m x 36 m was used.
Second, colour images of the same grid points were taken by a digital photo camera (Olympus Camedia 3030 Zoom). The images were taken from a height of 25 cm with a resolution of 3.2 megapixels, without flash. Each image shows an area of about 34 cm x 26 cm. An automatic binarisation, based on the gray scale histogram, was implemented after transforming the colour images into the $i3_{new}$ colour channel. It is a specially adjusted colour channel to detect plants (green colour) in colour images (see PHILIPP and RATH 2002). By calculating and combining selected parameters a segmentation algorithm has been developed to discriminate monocotyledonous and dicotyledonous plants in the images. For all data

acquisitions - the manual and the automatic weed mapping - a Global Positioning System (GPS) was used to correlate the images with the field positions.

Results and discussion

Based on the GPS data and the mapping results, weed distribution maps for each field were created by use of a Geographic Information System (GIS). Weed maps based on automatic and manual mapping were compared (figure 1). The results show the applicability of the algorithms. Both methods detect the same weed patches. Although the automatic mapping leads to a lower density of dicotyledonous weeds, especially if the actual weed density is high (>100 plants/m²). This problem results from the overlapping of leaves in high density patches. For the detection of monocotyledonous plants the algorithm still needs to be improved to reduce the percentage of misclassification.

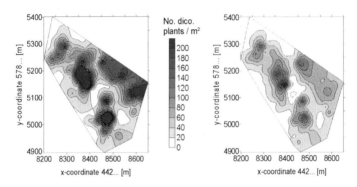

Figure 1. Distribution maps of dicotyledonous plants in sugar beet; based on manual mapping (left); based on automatic mapping by digital camera (right).

Conclusion

In general, it can be shown, that an automatic weed mapping is feasible. Although the developed algorithms need to be improved, to reduce the percentage of misclassification and to bring the system to an online-stage. In future the farmer will be able to implement weed maps in his spraying concept to realize a spatial variable herbicide application.

References

Nordmeyer, H., A. Häusler and P. Niemann (1997): Patchy weed control as an approach in precision farming. Proceedings 1. European Conference on Precision Agriculture, 307-314.

Philipp, I. and T. Rath (2002): Improving plant discrimination in image processing by use of different colour space transformations. Computers and Electronics in Agriculture, 35, 1-15.

Spectral characteristics of wheat in dry agriculture

A. Pimstein[1], M. Raz[1], A. Karnieli[1] and D.J. Bonfil[2]
[1] The Remote Sensing Laboratory, J. Blaustein Inst. for Desert Research, Ben Gurion Univ. of The Negev, Sede-Boker Campus 84990, Israel
[2] Agricultural Research Organization, Gilat Research Center 85280, Israel
pimstein@bgumail.bgu.ac.il

Abstract

The spectral responses of wheat to water and nitrogen stress at the canopy level are a basic aspect for the use of satellite images in the assessment of crop condition. This research looks to characterize wheat canopy using a spectroradiometer with ranging from 0.35 to 2.4 μm, and relate the spectra to different physiological parameters. Several indices were analyzed in different crop generated conditions, finding out good correlations with biomass, water and nitrogen content.

Keywords: spectral indices, biomass, water content, nitrogen content, reflectance

Introduction

Chlorophyll strongly absorbs energy in the spectral bands centered at about 0.45 and 0.65 μm. In the range from about 0.7 to 1.3 μm, the high reflectance of a plant leaf is a result of its internal structure. In longer wavelengths, water affects the spectral response of plants in the so-called *water absorption bands*, where there are deep drops of the reflectance at the 1.4, 1.9 and 2.7 μm bands (Lillesand & Kiefer, 2000). This is the starting point, from which several studies have found out correlations between different kinds and intensity of stresses with the spectral responses of plants (Peñuelas et al., 1993; Gitelson & Merzlyak, 1996). The purpose of this study is to validate the performance of different spectral indices at the canopy level of wheat crops in dryland agriculture, in order to identify mainly early water and nitrogen stresses.

Materials and methods

Wheat experiments with different fertilization programs, sowing densities and soil managements were established at the Gilat's Research Center in order to prompt different canopy characteristics and plant stresses. Experiment plots were placed under irrigated and rain-fed conditions.

Canopy radiance, of wheat plants 30 to 60 days after emergence, was detected with a 0.35 to 2.5 μm spectroradiometer fitted with 25° field of view optics (ASD's FieldSpec Pro FR flagship model) and expressed as spectral reflectance after periodic standardization by radiance of a level white standard (spectralon Labsphere Inc. panel). Data were collected in a nadir orientation at 1.4 m from the ground, around solar noon, under clear sky conditions.

Biomass, water and nitrogen content were determined for the vegetation that had been collected from the exact area where the reflectance measurements were done.

Different indices for determining biomass, nitrogen, chlorophyll and water content, based on the VIS, NIR, and MIR bands, were evaluated with respect to laboratory measurements. (Malthus et al., 1993; Peñuelas et al., 1993; Haboudane et al., 2002; Sims & Gamon, 2003).

Results and discussion

Figure 1 illustrates several spectral indices that have high correlations with the biomass, water, and nitrogen content.

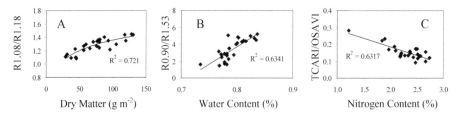

Figure 1. Correlations between different indices and dry matter, water and nitrogen content.

Due to the high sensitivity of the NIR/R ratios to soil brightness in comparison to the NIR/MIR ratios, the logarithmic relationship between the dry matter and the index is much stronger than the one obtained with NDVI (Malthus et al., 1993). For assessing of water stress, it was found that the index that is based on the reflectance at 1.53 µm had a better correlation with water content. This match up with previous results (e.g., Sims & Gamon, 2003), but it's still needed to be check in a wider stress range. The better relationship between the TCARI/OSAVI index with nitrogen against the other ones, is due to the fact that background reflectance contribution is reduced and the sensitivity to chlorophyll content is enhanced in this index (Haboudane et al., 2002).

Conclusions

Good results were achieved by analyzing *in–situ* spectral characteristics of wheat in order to determine its biophysical condition with the use of remote sensing techniques. Nevertheless, more effort is needed in order to check these preliminary results on a wider range of data.

References

Gitelson, A.A. and M.N. Merzlyak, 1996. "Signature analysis of leaf reflectance spectra: algorithm development for remote sensing of chlorophyll." Journal of Plant Physiology 148: 494-500.

Haboudane, D., et al., 2002. "Integrated narrow-band vegetation indices for prediction of crop chlorophyll content for application to precision agriculture." Remote Sensing of Environment 81(2-3): 416-426.

Lillesand, T. and Kiefer R., 2000. Remote Sensing and Image Interpretation. New York, John Wiley & Sons, Inc.

Malthus, T., et al., 1993. "Candidate high-spectral-resolution infrared indexes for crop cover." Remote Sensing of Environment 46(2): 204-212.

Peñuelas, J., et al., 1993. "The Reflectance at the 950-970 nm Region as an Indicator of Plant Water Status." International Journal of Remote Sensing 14(10): 1887-1905.

Sims, D.A. and J.A. Gamon, 2003. "Estimation of vegetation water content and photosynthetic tissue area from spectral reflectance: a comparison of indices based on liquid water and chlorophyll absorption features." Remote Sensing of Environment 84(4): 526-537.

Dynamic crop model as an intellectual core of management information system in precision agriculture

Ratmir A. Poluektov[1] and Wilfried Mirschel[2]
[1]*Agrophysical Research Institute, Grazhdansky prospect 14, 195220 St. Petersburg, Russia*
[2]*Institute fuer Landschaftssystemanalyse ZALF, Eberswalder Strasse 84, D-15374 Muencheberg, Germany*
ratmir@mail.dux.ru, wmirschel@.zalf.de

Abstract

A winter wheat crop growth model and the method of its calibration are presented. The model was adapted to different regions in North-West Russia and North-East Germany. It was modified for a purpose of management unit delineation in precision agriculture.

Keywords: Crop model, precision agriculture, model calibration, identification method

Introduction

Crop models seem to be a proper example of application of information technologies for the purposes of agricultural management. They can be considered as an intellectual core of management information systems in precision agriculture. Indeed, the dynamic crop models only adequately reflect the influence of uncontrolled weather conditions and man impacts on the plant production process and the environment. The explanatory model power and its extrapolation character, like intuition, is human distinctive feature, and other formal methods do not possess it.

The model as a software product includes a set of different parameters connected with the conditions of crop cultivation. Some of these parameters reflect physical and chemical soil properties. Another parameter group is related with the climate and weather conditions of the crop cultivation site. Culture and cultivar specific parameters determine the third group. At the final stage of model development all these parameters must obtain their concrete numeric values. The paper is devoted to the parameter estimation problem of the winter wheat crop model developed in the Agrophysical Research Institute St. Petersburg, Russia, [1] and the Institute for Landscape Systems Analysis of the Centre for Agricultural Landscape and Land Use Research Muencheberg, Germany [2]. The model is included into the agro-ecosystem model family AGROTOOL. A new identification method is developed; called "stepwise identification".

Model structure and its identification

The dynamic agro-ecosystem model consists several compartments or units [1]: (1) agro-meteorological unit connected with meteorological database, (2) radiation and photosynthesis unit, (3) soil water dynamics unit, (4) unit of crop growth and plant development and (5) unit for the forecast of the phenological development rate.

The technique of model identification will be described exemplarily for winter wheat cultivated in Leningrad region (Russia) and North-East German region. The results of long-time field research were used for problem solution. The experiments were carried out on experimental plots of the Menkovo Experimental Station ($59^0 21' N$, $30^0 08' E$) in Russia and of the two German Experimental Stations Muencheberg ($52^{00} 1' N$, $14^0 07' E$) and Hohenfinow ($52^0 49' N$, $13^0 56' E$). For the model identification the following characteristics were used: daily meteorological data for respective years; sowing dates and plant development stages onset dates; above ground dry mass at different plant development stages; grain yield; soil water measured in the 0-100 cm layer on these dates; physical properties of the soil such as bulk density, maximum and field water capacity, wilting point.

The model parameter set can be divided into three groups. Physical values, which can be measured directly, compose the first group. These parameters mainly relate to soil characteristics such as soil bulk density, general porosity, field capacity and wilting point. Parameters that can be chosen in the literature compose the second group. Some biological constants, for example, respiration and conversion coefficients, parameters of growth functions and so on form the third group. So the model identification was fulfilled in three steps; the reason to call this procedure "stepwise identification".

Results

In the result an universal model was developed, which describes the plant production process of winter wheat in two different European regions. Some results of the model calibration are presented in the Figure. It includes the comparison of experimental and simulated winter wheat grain yields for eight vegetation periods of different Russian regions and for two vegetation periods of two German regions.

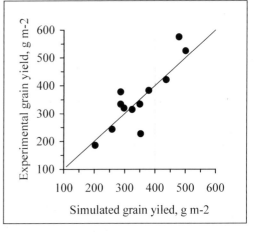

Conclusion

The existing model was transformed in order to use within management information units. First of all, some environment parameters, which were constants in field scale earlier, have been transformed into model input variables. Secondly, the unit of aerial environment was significantly changed in order to take into account the influence of simplest relief peculiarities (field slope and orientation) on incoming radiation and, in turn, on thermal surface balance. Final model version was applied to determination of management units of precision agriculture.

References

Poluektov R.A. et al. AGROTOOL – a system for crop simulation.// Arch. Acker- Pfl. Boden, 2002, Vol.48, pp. 609-637.
Mirschel W., A. Schultz & K.O. Wenkel. Vergleich der Winterweizenmodelle AGROSIM-Wheat und CERES-Wheat. // Berichte der GIL, 1993, Bd.5, pp.29-34.

Acknowledgment

Investigation was fulfilled in the framework of Russian-German program of collaboration in the field of agriculture (Theme 45) and under financial support of RFBR (Project # 01-05-64832).

Adaptation of the FT350 sugarcane fertilizer for precision agriculture

Enrique Ponce[1], Britaldo Hernandez[1], Farnia Fernandez[1], Lazaro Quintana[1], Miguel Esquivel[2] and Jose Julio Rodriguez[3]

[1]Automation and Control Laboratory, [2]Projects Division, [3]Computer Laboratory
National Center for Laboratory Animal Production, CENPALAB, Finca Tirabeque, Bejucal, Havana, Cuba
briti@cenpalab.inf.cu

Abstract

Fertilizer costs represent the highest percentage of the total cost of sugarcane production. One of the main constraints for the quick adoption of precision agriculture systems is the necessary investment in automated machinery ready for such system. Sugarcane fertilization in Cuba is accomplished mainly using the FT350 fertilizer. It has a mechanical system that does not allow variable rate application. It has two fertilizer deposits, but a common rate of application so that only one kind of fertilizer can by applied at once. More that 1200 such machines exist in Cuba and their replacement would be an enormous investment. A study was carried out to test the feasibility of adapting the FT350 for precision agriculture. All mechanical systems were replaced with a hydraulic system composed of hydro-motor, a regulatory hydraulic valve with possibilities of remote electro-hydraulic control and sensor to control the rotation speed of the fertilizer distributor. The system is controlled by a microcontroller (AGROMATIC). When the mechanical system is replaced, there is enough space to include a third fertilizer deposit, all three with independent rate control, making it possible to apply simultaneously three different kinds of fertilizer (e.g. N, P, K) with variable doses. Preliminary field tests showed that the adapted fertilizer machine is able to accurately distribute doses from 10-370 kg/ha of granulated fertilizers.

Introduction

In Cuba, there are more than 1200 FT-350 fertilizer applicators for sugarcane fields. These fertilizers are completely mechanical, imprecise and unreliable. As part of national project for the application of Precision Agriculture (PA) technologies in Cuba, we decided to transform an FT-350, into a hydraulic and automated machine, ready for site-specific variable-rate applications. Needless to say, cost was an important factor in this decision to convert existing machinery, rather than buying PA-ready equipment.

In the figure below, we show the general diagram of the modifications to the FT-350 fertilizer.

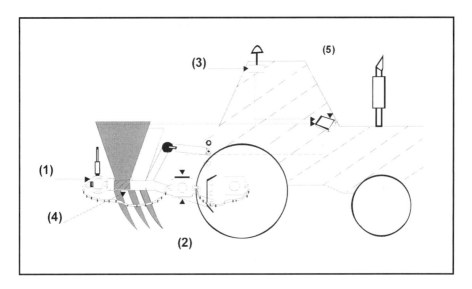

1- Hydro-motor with planetary reduction gearing.
2- Hydraulic flow-control valve.
3- DGPS receiver
4- Worm drive
5- Microcontroller

The worm drive (4) is fixed to the hydromotor (1), worm drive rpm are controlled by a Hydraulic flow-control valve (2) , operated by a microcontroller-based unit (5), which simultaneously receives GPS coordinates and horizontal speed of the tractor from the DGPS receiver (3).
The transformed fertilizer was tested in laboratory and field conditions and managed to deliver from 10 to 370 kg/hectare of Nitrogen fertilizer in granulated form, independently of the tractor speed. For every complete revolution of the worm gear, 200 g of fertilizer are delivered with a precision of +/- 15%.

Online measurement of grain-quality with NIR-spectroscopy on combines

Jens Rademacher
Institute of Agricultural Engineering, CAU Kiel, Max-Eyth-Straße 6 , D-24098 Kiel, Germany
jrademacher@ilv.uni-kiel.de

Abstract

After mapping yield on a combine, now the protein content of wheat as an important characteristic of the quality is measured. Two laboratory instruments, based on the Near-Infrared Spectroscopy were installed on a combine to measure flowing material. After a first year of calibration, online-measurement have been possible in the second year. The SEP(C) of this first calibration have been 0.55 % and decreased in the second year. Parallel data acquisition of protein data and DGPS-data from the yield-mapping system gave the possibility to map the distribution of protein in a wheat field. Zones of different protein contents are visible clearly.

Keywords: Near-Infrared Spectroscopy, Online-Measurement, Quality

Introduction

Different yield-mapping systems for combines are available; on some in series even. Another important factor is the quality of the grain e.g. the protein content. New measuring instruments using the near-infrared-reflection spectroscopy (NIRS) allow the contactless determination of protein content of wheat. This lead to the consideration, to measure the content of protein or other components online on a combine.

Material and methods

Before measurements are possible the instruments must be calibrated. Two laboratory instruments have been installed on a combine: A ZEISS Corona 45 NIR and a PERTEN DA 7000. In a bypass-unit installed after the clean grain elevator both instruments measured the same material of the grain flow. After passing the two instruments the material is carried back into the elevator or into a sample collector (Figure 1). A notebook storages the spectral data as well as the data of the yield-mapping system, include the DGPS-data. During the first season, all data necessarily for calibration were collected. Protein-content from 100 samples of wheat were analysed wet-chemically in a reference laboratory.

Figure 1. Test Combine with mounted NIRS-Instruments.

Results

Figure 2 shows the comparison of reference-data and measured values for the protein content of wheat for the same samples after two years of calibration. On the bisect line the reference values would be equal to the measured protein content. Errors are still existent, but the SEP(C) of 0,4 % is only two times high than the repeatability error of the reference method. Figure 3 shows a map of protein content of wheat. Zones with differentiated protein contents are visible. The stripes are caused by the driving direction of the fertilizer spreader. Different nitrogen application strategies lead to these zones of different protein content.

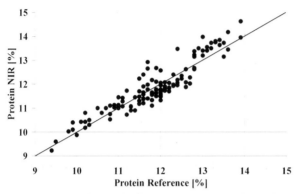

Figure 2. Comparison of reference and measured protein content of wheat.

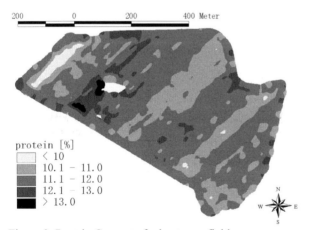

Figure 3. Protein Content of wheat on a field.

Conclusion

The online measurement appears practical feasible, however it is still connected with substantial difficulties. The calibration-files, developed at the Institute of Agricultural Engineering include wheat (protein, starch, moisture), barley (protein, moisture), rapeseed (oil, protein) and corn (starch, fat, protein, moisture). With the help of the continuous NIR measurement level and variability of material contents can be seized and represented.

Measuring quality parameters in combine harvesters online, a new way for precision agriculture

Yves Reckleben
Institute for agricultural engineering technology, University of Kiel, Max-Eyth-Str. 6, 24118 Kiel, Germany

Keywords: precision farming, near infrared spectroscopy, Hydro N-Sensor, online measuring

Precision Farming covers the production of high-quality food optimising sub area-specifically the use of production resources.

The success of sub area-specific management can be evaluated by yield mapping systems economically. This fact forms the basis for the improvement and/or new development of yield measuring systems. Not only the amount of the yield but also its composition (quality) are of increasing importance.

The integration of suitable measuring instruments (e.g. NIRS) into the combine harvester, in order to enable measuring the quality of the thrashing good online, is therefore very important for the farmers, and the processing and agricultural machinery industry. The NIRS method offers generally and sub area-specifically new possibilities to evaluate other influences on the yield (quality).

- In the own project the following variables on protein, fat and starch contents for wheat, barley, yield level
- N-fertilization (amount and exact time)
- soil types and characteristics
- basic nutrients / pH value
- weather condition

oilseed rape and corn are measured and analyzed:

Some of the above mentioned parameters, particularly the N-fertilization and the fertilization time, can be influenced directly and therefore are to be improved pointedly and specifically for each sub area. The aim is to obtain even (homogeneous) portions with optimal yield and quality.

Since 2002 we have been analysing the site specific N-fertilization effect in wheat. For the fertilization the Hydro N-Sensor System was used. This system helps to decide how much nitrogen is needed by the plants. These tests are accomplished by the Institute for Agricultural Process Engineering in cooperation with different agricultural enterprises. These fertilizing tests were made on different fields. The most interesting question is: how large is the influence of the last N-gave to the quality?

To answer this question, tests, using 4 different fertilizing variants, were realized. In all variants the first N-gave was constant, the second and third were fertilized variably by using the Hydro N-Sensor. The fourth gave had the following variants:

v1: no fertilizer
v2: variable fertilization with Hydro N-Sensor
v3: constant fertilization
v4: variable fertilization by using quality calibration of the Hydro N-Sensor

The results are shown in table 1.
Result 1: There is a clear relationship between the N-quantity and the average protein content: more N results in a higher protein content.

Result 2: If the nitrogen employment is reduced, there are clear yield (-10 dt/ha) and quality (-1.5% protein) losses. Therefore the selected fertilization level has been chosen correctly in 2002.
Result 3: A variable fertilization leads to a reduction of the standard deviation in protein values. So the standard deviation could be reduced from 2% (v3) to 1% (v2) resp. 0.76% (v4).Result 4: The quality function of the Hydro N-Sensor produced both the highest yield (+10 dt/ha) and the highest protein content (+0.4 %).
Result 5: The quality function suppresses the dilution effect (high yield means little protein). The dilution effect was not only stopped, but in the test 0.4 % more protein per yield has been produced.

Table1. Results 2002.

	(0 variante)	(N-Sensor)	(constant)	(N-Sensor QC)
-quantity [kg/ha]	147	196	218	211
protein [%]				
maximum	11.68	12.63	13.40	12.70
average	10.06	11.50	11.60	11.97
minimum	8.74	10.68	9.37	11.19
andard deviation	1.48	0.98	2.02	0.76
yield [dt/ha]				
maximum	102	116	117	117
average	78	86	87	98
minimum	59	66	54	63
andard deviation	22	25	31	27

Summary

The tests of 2002 show that site specific differences in yield and quality could be measured with high accuracy. The influence of the last N-gave to quality could be proven clearly. Using the Hydro N-sensor the N-effectiveness (kg N/ dt yield) can be optimized and the product quality can be increased by a higher homogenity. The results of the last year must be corroborated by further tests and are surly not to be generalized.

References

Erekul O. 2000 Einfluß langjährig differenzierter Düngung auf Ertrag und Qualität von Winterweizen und Sommergerste
Ludowicy C., Schwaiberger R., Leithold P. 2002 Precision Farming Handbuch
Rademacher J. Landtechnik 6/2002 Messsysteme für den Proteingehalt während des Mähdrusches
Sturm H., Buchner A., Zerulla W. 1994 Gezielter Düngen

Acknowledgement

The project was funded by the Federal Ministry of Education and Research in Germany.

Proximal sensing of topsoil organic matter and profile features by a hyperspectral scanner and digital camera

G. Reimer and J. Lamp

Institute of Plant Nutrition and Soil Sience, CA-University Kiel, Ohlshausenstr, 40-60, 24118 Kiel, Germany
jlamp@soils.uni-kiel.de, reimer@geries.de

Abstract

An important factor in most PA applications is the organic matter content of topsoils (OM in Ap). A hyper-spectrometer is used to estimate by multiple regressions OM from spectral signatures of topsoils working both, under standard conditions (prepared samples) in the field-laboratory, and ontop of the "SoilRover" terrain vehicle along tramlines of fields. Point and serial measurements of spectral remissions ($\%R_\lambda$) in selected bands resulted in estimates of OM for samples and fields in Northern Germany. Very high multiple correlations were found for standard samples ($r^2 > 0.9$), a high r^2 (0.9) for an areal field estimate at best condition (cultivated bare soil), but medium r^2 (0.5-0.7) for field surfaces interfered by residues, moisture or structure effects. In parallel operation, a digital camera ontop of the "SoilRover" takes laptop-controlled pictures from surfaces, but also from soil auger kernels. A spectral signature and picture data base assists point and areal assessments of surface features or profile diagnostics.

Keywords: organic matter, proximal spectral sensing, soil colour

Introduction

For the mapping approach of Precision Agriculture (PA) it is necessary to have detailed information within fields. In the joint research project *preagro* (WERNER et al. 2002) our part was to develop an efficient soil survey as described by LAMP et al.(2001). Based on a „SoilRover" terrain vehicle which was equipped with dGPS, GIS, EM38 and an auger borer, special results from proximal spectrometer sensings and a digital camera are presented here.

Materials and methods

The hyper-spectral scanner (Zeiss-tec5, about 200 VIS/NIR-bands from 400-900nm) was preared to work under laboratory and field conditions. In the first case, a ceramic plate (white standard), and in the second, a reference-channel ensured that measurements were generated as mostly weather- and skylight-independent relative spectral remissions (R_λ). The exponentially decreasing relationship of R_λ and OM can be substantialy improved by use of logarithms and multiple regression analysis which includes the remissions of several wavelengths and further observations (soil texture and type, landuse). But under field conditions, the relations R_λ to OM are reduced by interference factors, like stones, soil moisture and structures, plant residuals, shoots and weeds which vary locally and seasonally. It was successfully tried to detect and map these factors with a digital camera (Kodak DC 290). All spectral remissions and pictures were geo-referenced by dGPS and stored via a field laptop in PC data bases for further analysis. For user-controlled assessments, the pictures, aside to signature curves or profile descriptions, are displayed per observation point and the viewer is able to enter interpreted informations. So, the areal percentages of plant and soil covers are assessed and digitized for mapping with a high performance. Apart to areal sensings, the camera was used also to take zoomed colour images of the soil kernels in cores of the hydro-powered auger at 4-5 depths of the 1.5m profiles. It turned out to be a valuable tool to assist, control and document soil profile diagnoses, especially of hydromorphic phenomena.

Results

For the estimation of OM by spectral remissions in the field-laboratory >1400 topsoil samples were collected from preagro pilot fields spread over Germany, prepared acc. to standards (sieved, equal moisture, slightly pressed) and analysed. In dependence of regions and OM ranges the multiple r^2-values varied from 0.7 to 0.98. The tramline based field estimates reached under favorable surface conditions (uniform bare soil after sugar beet seeding) a multiple r^2 of up to 0.93. Backed by 47 analyzed soil samples a detailed OM-map for a 40 ha field in Holstein was generated (see fig. 1). Under unfavorable conditions with interfering factors (s.a.) the correlations drop down to r^2-values of 0.5 and less. In order to take care of these factors, maps of plant shoot, stones and crop residual coverages as well as of soil structure classes were generated as mentioned above for a preagro pilot field in Westfalia (21 ha, see fig. 2). The maps show many interactions among the factors which effect the spectral remission. A comparison of >400 surface pictures showed no clear correlation with spectral remission curves under these field conditions.

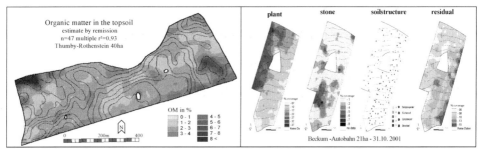

Figure 1. Organic matter map.   Figure 2. Maps of interferencing factors.

Discussion and conclusion

Proximal spectral sensing combined with an efficient soil survey system offers a system for the direct collection, interpretation and validation of surveyed and sensed soil data. In contrast to remote sensing, a main advantage is that the ground truth is performed user-controlled at the same time and locations. For the farmer, who already works with the proximal N-sensor for fertilising, a new tool and application is given without extra costs.

References

Lamp, J., R. Herbst and G. Reimer (2001): Preciser and efficient soil surveys as basis for application maps in Precision Agriculture. In: G. Grenier and S. Blackmore, 3rd Europ. Confer. Precision Agricult., Vol. 1, 49-54, Montpellier

Werner, A., Jarfe, A., Klöpfer, F., Kottenrodt, D. (2002): Forschungskonzept und Projektstruktur von preagro – in KTBL (Hrsg.): Precision Agriculture, preagro Zwischenbericht, Darmstadt, 38, S. 483-500

Acknowledgements

Thy *preagro* project was founded by the German Ministery for Education and Research.

A friendly man-machine visualisation agent for remote control of an autonomous tractor GPS guided

A. Ribeiro, L. García-Pérez, M.C. García-Alegre and D. Guinea
Instituto de Automática Industrial, Consejo Superior de Investigaciones Científicas, 28500 Arganda del Rey, Madrid, Spain

Abstract

This work is framed in a research project that deals with the autonomous navigation of tractors for spraying operations in citric and olive tree fields Present work introduces a friendly man-machine interface, implemented as a universal GIS client. It is a monitoring tool that displays, on a colour digital and georeferenced aerial image of the field, the trajectories described by an automated vehicle from an initial to a final location, GPS guided.

Keywords: autonomous navigation, tractor, GPS, georeferenced maps, agriculture, sensors, aerial image, agent architecture, AMARA.

Introduction

Spraying operation entails risks for both the driver and the environment as they are exposed for long periods to the accumulative effect of toxic products. As a consequence, it would be extremely interesting the automation of tractors for unmanned navigation purposes. In those cases human driver would act as a supervisor that interacts with one or several machines, that communicates through a wireless link. To date, new technological developments are mainly devoted to partially help human driver in tractor guidance, but very few pursue an autonomous navigation to autonomously accomplish an agriculture operation (Garcia-Alegre et al., 2001), (Stentz, 2002).

Sensors and actuators for the vehicle automation

A commercial vehicle, Figure 1, has been fully automated at the IAI-CSIC warehouse and is now able to navigate in two different modes: 1) Teleoperation and 2) Autonomous. It is endowed of two types of sensors. First ones, a bumper and a laser range finder, account for environment recognition and safe navigation and second ones, a dead-reckoning system with odometers, digital compass and a differential GPS receiver, to guide the navigation. The vehicle speed is about 1.8 Km/h.
The vehicle is autonomously steered a multiagent architecture. The architecture is composed of six agents or behaviours: TURN_WHEEL and BRAKE at the first level in direct contact with the sensor signals; STOP, ADVANCE, and AVOID at a intermediate level and finally, GO_TO_POINT agent at the upper abstraction level, to receive/send command/message from the human operator. This architecture has been successfully implemented within a Client/Server paradigm.

The visualisation agent

Within this frame a world visualisation agent has been generated to display the described trajectories on a map. The map is a colour georeferenced aerial image of the navigation scenario, IAI-CSIC campus. It has been digitised and georeferenced by assigning to a set of image pixels their absolute longitude and latitude coordinates, experimentally obtained in the field by means of a RTK DGPS receiver.

Results

The developed application allows a bi-directional communication in such a way that users can define, in the map, the point where the vehicle has to be positioned. From that action, the trajectory followed by the tractor is continually displayed on the aerial digital photography. Present application permits the use of standard operations, such as zoom or pan. A scale bar is always visible, as well as the numerical GPS coordinates of the vehicle. Moreover the cursor movements are also displayed in latitude and longitude coordinates.

The interface window with the trajectory described by the vehicle in autonomous mode is displayed in Figure 2, in an olive tree scenario. Once the operator clicks on the image the final destination, the multiagent control architecture, that mimics human driver actions, is in charge of guiding the tractor to its final destination. The visual interface application has been developed as a generic tool to aid the optimization of the different navigation strategies as well as to supervise the path tracking of one or several vehicles endowed.

Figure 1. Automated vehicle.   Figure 2. Global trajectory (pink).

Conclusions

This work present a visualisation agent based on sensor technologies and wireless (Radio-Ethernet) communication that would help the remote unmanned guidance of tractors in spraying operations. This is an innovative research and development direction to overcome tedious and risky agriculture operations, being an open window for the future of many modern farmers.

References

García-Alegre, M.C., Ribeiro, A., García-Pérez, L., Martínez, R., Guinea, D., Pozo-Ruz, A. 2001, Autonomous Robot in Agriculture Tasks". 3ECPA-3 European Conf on Precision Agriculture, Montpellier.

Stentz, A. 2002, Intelligent unmanned ground vehicles, www.ri.cmu.edu/people/stentz_anthony.html

Acknowledgements

This project was funded by grant AGF1999-1125-C03-03 and AGL2002-04468-C03-01. Authors thanks Eugenio Villanueva for his support in the design and integration of sensors and actuators.

542

A geo-spatial database for precision agriculture information systems

L.C.M. Ribeiro, A.M. Saraiva, E. Murakami and C.E. Cugnasca
University of São Paulo, Polytechnic School, CP 61548, São Paulo, SP, Brazil, 05424-970
luiz.miyadaira@poli.usp.br

Abstract

This paper presents a geo-spatial database for precision agriculture information systems (IS). This database is based on MOSAICo (Object Model for Field Information Systems) model and a Geo-Spatial database conceptual framework. The MOSAICo model represents the business rules for Precision Agriculture domain which has been extended to support geo-spatial features. The main goal is to provide a common way to store and manage geo-spatial data in precision agriculture IS.

Keywords: Geo-Spatial Database, MOSAICo Model, Information Systems

Introduction

Many IS and software tools for Precision Agriculture, PA, are heterogeneous solutions, and generate data with heterogeneous formats. Among other problems it restricts the possibility of integration of software packages that are useful for specific tasks, and also hinders information management, data analysis and interpretation. Heterogeneous data formats make it difficult to retrieve relevant information for decision-making, and are also an obstacle for automated data analysis. An environment for the construction and management of PA information systems is being developed (Murakami et.al, 2002), and its main goal is to provide a way for component-based systems development with reused software. It will also allow systems to evolve according to the needs and possibilities, while maintaining the interoperability among its modules. It is composed of the framework Jakarta Struts, an interoperability bus software, the PA object model MOSAICo (Saraiva et al, 1998), and a geo-spatial database. This paper presents the geo-spatial database, based on MOSAICo and on a geographic information conceptual model (Pires et al, 1993) (Lisboa & Iochpe, 1999).

Materials and methods

The framework MOSAICo is composed by components that encapsulate business rules in a generic PA domain. That framework is based on MOSAICo conceptual model, an object based model previously developed with the main goal of reducing time and cost of PA applications development through the construction of software based on components supplied by the framework. Spatial data vary spatially (position, altitude, soil depth), thematically (geology, soil coverage) or temporally (productivity, soil properties), and are essential for PA systems. In that context, the database will provide a common way for storage, management and representation of those data. It will be based on the MOSAICo model, so becoming a common way for all the applications generated in the MOSAICo environment. Besides that all applications that connect to the environment will be able to generate or use the data available. Figure 1 shows a class diagram that represents a Field Data module of MOSAICo object model. The Polygon and Point classes are geographic classes and the Field, Farm and Sample classes encapusulate the PA business rules and extend geo-spatial features of geographic classes. Additionally, the classes compose two different modules: Geographic Information and MOSAICo Field Data.

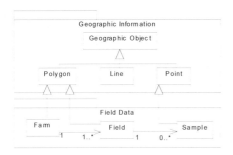

Figure 1. Class diagram for geo-spatial object model.

Results and discussion

From the storage point of view the database contains structures on its schema to support common geometric elements found in Geographical Information Systems, such as lines, points, polygons and others, and relationships among those elements, based on the framework OpenGIS. That framework will be adapted to the characteristics of the MOSAICo model and its spatial data needs. Besides, the model will cover the temporal characteristics of the information, such as data about field productivity obtained every year. Temporal characteristics are fundamental for the model, especially for data analysis. From the management aspect, the database contains functions for information retrieval and manipulation, both the ones defined by the framework OpenGIS, and those of the specific PA domain, such as data analysis, indexing techniques and specific information extraction.

Conclusions

The database will facilitate the application of data retrieval algorithms (i.e. Data mining techniques) to discover non-explicit information, and possibly help to obtain automatically relationships among important variables such as productivity, soil properties, and weather, among others. Besides, that database will contribute to a faithful representation of spatial information and, consequently, it will improve its management.

References

Saraiva, A.M.; Massola, A.M.A.; Cugnasca, C.E. 1998. An Object Model for Field Information Systems. In: Proceedings of International Conference on Precision Agriculture, ASA/CSSA/SSSA. p. 1355-1366. Saint Paul, July 19-22.

Murakami, E.; Ribeiro, L. C. M.;Cugnasca, C. E.; Saraiva, A. M. 2002. An Infrastructure for Development of Information Systems for Precision Agriculture. In Proceedings of 6th International Conference on Precision Agriculture and Other Precision Resources Management. Bloomington, MN, July 14-1.

Pires, F., Medeiros, C. B., Barros, A. 1993. Modelling Geographic Information Systems using an Object Oriented Framework. In Procedddings of XIII International Conference of the Chilean Computer Science Society, pg 217-232.

Lisboa, F.J., Iochpe, C. 1999. Specifying Analysis Patterns for Geographic Databases on the Basis of a Conceptual Framework. In ACM Symposium on Advances in Geographic Information Systems, 7, Kansas City, USA, ACM Press.

Barley yield and yield quality affected by soil properties

A. Ristolainen[1], A. Jaakkola[2], P. Hänninen[3], S. Penttinen[4] and L. Alakukku[1]
[1]MTT Agrifood Research Finland, Soils and Environment, E-House, 31600 Jokioinen, Finland, antti.ristolainen@mtt.fi, laura.alakukku@mtt.fi
[2]University of Helsinki, Dep. of Applied Chemistry and Microbiology, P.O. Box 27, 00014 University of Helsinki, Finland, antti.jaakkola@helsinki.fi
[3]Geological Survey of Finland, P.O. Box 96, 02151 Espoo, Finland, pekka.hanninen@gsf.fi
[4]Geological Survey of Finland, P.O. Box 1237, 70211 Kuopio, Finland, sari.penttinen@gsf.fi

Abstract

A three-year study was begun in Southern Finland where field scale variation of soil physical and electrical properties are mapped. The aim of the study is to establish a relationship between soil physical properties and yield and to test different indirect methods to measure within field variation. According to first year results, soil electrical conductivity and dielectric coefficient seemed to be efficient tools to map soil type differences although no direct relationship between soil properties and yield was observed.

Keywords: Barley, electrical conductivity, nutrient uptake, clay content

Introduction

The determination and control of the variation of soil properties affecting yield and yield quality is essential for precision agriculture. Soil water content and water holding capacity are often the main reasons for yield variation. Soil structure and physical properties affect the availability and movement of soil water and therefore also the nutrient uptake of plants. The physical properties of the soil are seldom determined in field scale and our knowledge of indirect methods to measure soil physical properties is limited.

Materials and methods

Effects of soil physical and chemical properties on spring barley (*Hordeum Vulgare* L) yield and yield quality are studied in Southern Finland on a clay soil (Vertic Cambisol). During the growing season of 2002 spatial and temporal variation of soil electrical conductivity (EC) was measured in different layers across the field. At the same time measurements of soil dielectric coefficient were done in the topsoil. Measurements were done three times during the growing season, after sowing, in the middle of the growing season and after harvest. In addition, measurements of soil mechanical resistance were performed after sowing. Measurements were taken according to 10*10 m grid at an area of 2.5 ha in two experimental fields. One of the fields was mapped for soil texture. The samples were taken from places were clear differences in soil EC existed. Fields were harvested with a combine installed with yieldmapping devices and accurate vegetation and yield samples were collected from 20 places in each field to study the nutrient uptake of plants. From the sample points observations of soil moisture content were done weekly.

Results

Soil dielectric coefficient and soil EC appeared to be almost linearly correlated with soil clay content. Best fit between soil EC and clay content was obtained with EC measurements done in the middle of the growing season.

According to preliminary results, no direct relationship between barley yield and soil EC or soil mechanical resistance was observed during the first experimental year. Still, some evidence was found that changes in soil EC during the growing season are related to nitrogen and nutrient uptake of plants. In some studies soil EC has been used to predict soil nitrate levels. Usually nitrogen uptake of plants is more sensitive to soil physical properties than grain yield.

Discussion

Both soil EC and dielectric coefficent seemed to be efficient tools to map soil type differences in clay soil with relatively hig clay content. The relationship between e.g. nitrogen upptake of plants and soil EC will have to be further studied. More years of soil, yield and EC data is needed to study the complex mechanisms of yield formation in different weather conditions.

Increasing the accuracy of sugar cane yield estimates using NDVI measurements in areas under hydric stress condition

J.V. Rocha, R.A.C. Lamparelli, J. Zullo Jr., H.M. Machado
Unicamp- Cidade Universitária –Campinas - São Paulo – Brasil - CEP:13083-970
jansle@agr.unicamp.br;rubens@cpa.unicamp.br;jurandir@cpa.unicamp.br;machado@agr.unicamp.br

Abstract

In this study the NDVI was calculated with the objective of monitoring sugar cane under hydric stress condition to serve as a support to crop yield estimate. A sequence of 5 Landsat 7/ETM+ images, acquired during the sugar cane vegetative cycle, were used to monitor the plantation, from the end of the raining season (april) to june 2002, which is considered a dry period in Brazil. The results showed good agreement between NDVI data and hydric balance results and it was possible to identify areas where the crop could not recover from the hydric stress.

Keywords: NDVI, yield estimates, biomass variability, Landsat 7

Introduction

The Normalized Difference Vegetation Index (NDVI) is considered a good biomass estimator, is being used to evaluate crop yield (Jackson et al., 1986) and has a good correlation with the plant moisture (Sims & Gamon, 2003). Therefore, it can be used in Precision Agriculture for crop monitoring (Barnes et al., 1996). This study aimed at using NDVI for monitoring sugar cane plantation under hidric stress condition as a support to yield estimates.

Material and methods

The study was carried out using 5 images of the Landsat 7/ETM+, path/row 220/75, acquired during the vegetative cycle of the plantation, from november 2001 to june 2002, in the following periods: november 2001, january 2002, march 2002, april 2002 and june 2002. This period was considered very dry compared to the average standard in the region. NDVI variability maps were generated for each satellite image and, at the same time, the dry period was characterized by the hydric balance for the year 2002, considering 2 soil texture condition, one for clay soils (storage capacity = 70mm) and the other one for sandy soils (storage capacity = 30mm). The NDVI values were monitored during the plantation cycle in order to detect spots of affected areas.

Results

Figure 1 shows the evolution of NDVI in the images of november/2001 (a), january/2002 (b), march/2002 (c), april 2002 (d) and june/2002 (e). There was a correlation between the NDVI values and hydric balance values (Figure 1 f and g). The sequence of NDVI images shows increasing NDVI values up to the april/2002. Figure 1(e) shows a decrease in NDVI values after a hydric stress condition, which indicates areas with less biomass production.

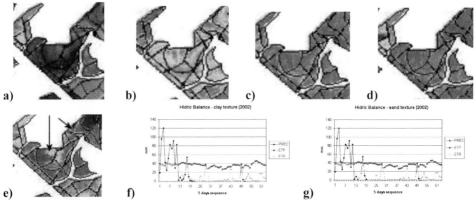

Figura 1. NDVI evolution during part of the sugar cane vegetative cycle (dark grey = high biomass; light grey = low biomass) and hydric stress evolution.

Conclusions

It can be concluded that the NDVI images, associated to hydric balance data can indicate areas were there will be a decrease in biomass production, helping to increase the accuracy of the yield estimates process in sugar mills.

Acknowledgements

The authors would like to acknowledge São Paulo State Research Fund (FAPESP), Brazil, for funding this study (Project 99/07071-1) as well as COSAN S.A., for providing the reference data.

References

Barnes, E.M., Moran, M.S., Pinter, Jr., P.J., Clarke, T.R. 1996. Multispectral Remote Sensing and Site-Specific Agriculture: Examples of Current Tecnology and Future Possibilities. In: Precision Agriculture '96: Proceedings of the 3rd International Conference on Precision Agriculture. 1996. Robert, P.C., R.H. Rust and L.E. Larson (ed.). Minnesota, June 23-26. ASA,CSSA, SSSA, Madison,WI.

Jackson, R.D., P.J. Pinter, R.J. Reginato, S.B. Idso. 1986. Detection and Evaluation of plant stresses for crop management decisions. IEEE Trasactions on Geoscience and Remote Sensing GE-24:99-106.

Sims, D. A.; Gamon, J.A. 2003. Estimation of vegetation water content and photosynthetic tissue area from spectral reflectance: a comparison of indices based on liquid water and chlorophyll absorption features. Remote Sensing of Environment, 84 (4):526-537.

Recommendation of soil nutrients in special care zone based on estimated paddy yield data variability

Swapan Kumar Roy and Sakae Shibusawa
Faculty of Agriculture, Tokyo Univ. of Agriculture and Technology, Tokyo 183-8509, Japan
swapan@cc.tuat.ac.jp

Abstract

Mapping of within field spatial variation in paddy yield was carried out using data obtained from manual harvesting in gird spacing (55 m x 30 m) just before machine harvesting from two plots. The spatial yield maps were produced using point-kriging technique and the yield maps illustrated that the middle portion (>50% of the area) of each plot has lower yield (<3.5 t ha^{-1}) compared to both ends in lengthwise. Special care zones are suggested in the low yielding areas of two plots where soil N and P are needed to improved to moderate (3-6 g kg^{-1}) and low ranges (3-10 mg kg^{-1}), respectively, for obtaining high yield (>5 t ha^{-1}).

Keywords: spatial yield map, management zone, and soil nutrient recommendation

Introduction

Rice monitoring and production estimation has special significance to many Asian countries, as rice is the staple grain. The information about the yield variability in wet paddy cultivation is inadequate. It is important to understand the knowledge of the variability of yield that can be used to enable the farmer to make meaningful management decisions regarding the succeeding crops and it provides a basis for varying inputs. So, the spatial variability of paddy yield was investigated and identified the range of soil N, P and K for maximum production in the management zone based on estimated yield variability.

Methods

The experiment was carried out at two plots (each about 4.0 ha) of a private paddy farm located in Kluang, Johor of West Malaysia. Selected soil properties were measured at different locations in the plow layer and the variations were found to be 0.70-6.00 g kg^{-1} in soil N, 0-13.59 mg kg^{-1} in soil P, 0.24-6.77 cmol(+) kg^{-1} in soil K, 4.25-4.54 in pH, 0.77-0.92 g cc^{-1} in bulk density and the soil texture was clay with 75-83% of clay content (Roy, 2001). Paddy samples were collected manually (using square frame - 1m x 1m) from two adjacent plots B3 and B4. Geostatistical analysis was carried out and the area statistics were obtained while the maps transferred to GIS with GPS (Global Positioning System) co-ordinates.

Results and discussion

Descriptive statistics indicate that the mean yield of B3 and B4 plots (n = 33 for each plot) were 3.99 t ha^{-1} and 2.63 t ha^{-1}, respectively. The random variation of yield (CV) of plot B4 (52%) was more than plot B3 (37%) which indicated the presence of higher heterogeneity in yield variation for plot B4. It was also found that yield variation was significant within each plot (F = 4.58**) and between two plots (F = 40.35**). In Geostatistical analysis, the semivariograms for paddy yields for plots B3 and B4 exhibited spatially dependence with a spherical model (R^2 = 0.71) and a linear model (R^2 = 0.84), respectively. There was no consistent spatial pattern at two adjacent plots though similar management practices (tillage, seeding, fertilizing, weeding, etc.) were followed uniformly where paddy yield was strongly spatially correlated at plot B3 (nugget ratio = 0.21) but moderately spatially dependent at plot B4 (nugget ratio = 0.45). The spatial trend maps (Figure 1), using point-kriging, separated relatively high yielding (>3.5 t ha^{-1}) and relatively low yielding (<3.5 t ha^{-1}) areas. It was observed that low yield was obtained about 46% (1.80 ha) of plot B3 and 64% (2.44 ha) of plot B4. So, the special care zones for improving yield to above 3.5 t ha^{-1} are suggested in the middle of both plots (Figure 1). It was also experienced that the measured values of soil N, soil P and soil K in low yielding areas in both sites were low (<3.00 g kg^{-1}), very low (<3.00 mg kg^{-1}) and moderate (>0.45 cmol(+) kg^{-1}), respectively (Roy, 2001). The existence of soil K over 0.45 cmol(+) kg^{-1} was due to paddy plant residues in soil that left by combine harvester. So, it is necessary to improve soil N and soil P in special care zones.

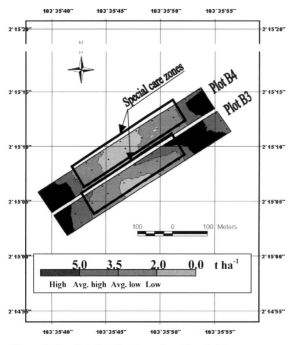

Figure 1. Spatial distribution of paddy yield.

According to the history of the study sites, the field was uneven before starting of paddy cultivation, and the topsoil (with high organic carbon content from the forest) in low yielding area was cut and placed to high yielding area during leveling operation. So, it enhanced the fertility in the high yielding area. However, after comparing the estimated yield data with soil N and soil P, it is suggested that 5 t ha^{-1} yield can be obtained for soil N = 2-3 g kg^{-1} and soil P = 7.17-9.32 mg kg^{-1}. So, the indicated special care zones in plot B3 and B4 can be improved to moderate range for N and low range for P in short and long term implications.

Conclusions

The study demonstrated that significant yield variation was observed in both sites and the soil fertility should be improved in the special care zone for attaining higher yield.

References

Roy, S.K. 2001. Nutrient and yield spatial variability in a commercial Malaysian paddy field. PhD (Precision Agriculture) dissertation (unpublished), Institute of Bioscience, Univ. Putra Malaysia (UPM), Serdang, Selangor, Malaysia.

Using yield maps to predict site-specific phosphorus requirements

A. Sanaei

Farm Machinery Department. College of Agriculture Engineering, Isfahan University of Technology, Iran

Akbars@cc.iut.ac.ir

Abstract

Using the conventional method of fertiliser planning, we can predict the required fertiliser (kg/ha) for each crop based on the previous year's yield mean (t/ha) and soil analysis within each field. In Britain, the soil analysis results of macronutrient such as P, K are classified into 9 indices (MAFF, 1994). Hence, average soil index and previous year's yield map of each field with or without straw removal were used to predict VRP fertiliser recommendations for the wheat or barley.

Keywords: Yield map, Site-specific P requirement, GPS, GIS, Soil index

Introduction

In Precision Farming, reliable historical records of both yield and soil maps based on a synchronised gridding system with identical grid spacing can help to predict the actual fertiliser requirement for each management zone of fields. Therefore two data sets of yield and soil samples for the same sampling area can be compared and evaluated to predict the required nutrient in each cell and management zone based on determination of nutrient removal by the crop and remaining soil nutrient content (Vanschen & De Baerdemaeker, 1991).

Materials and methods

For further processing of the regular normalised yield data to predict variable rates, phosphorus off-take maps of East Hemel field in 1995 and 1996 were prepared and a spreadsheet (Excel) was used to calculate optimum fertiliser (P_2O_5) t^{-1} requirements for each grid square. The phosphorus removed in the cereals is approximately 7.5 kg P_2O_5 /t for wheat and barley with straw ploughed in (MAFF, 1994).

Results and discussion

The average P off-takes for average yield in 1995 and 1996 were 55.6 and 53.2 kg/ha respectively. The subtracted P off-takes of 1995 from 1996's showed a difference of 2.46 kg/ha which is the same as the difference between total average P off-takes. The average of P off-takes: kg/ha in the same classes between 1995 and 1996 (Table 1- last column & row) show relatively similar values. This means that a similar average of P maintenance will be required in same class (as management zone) between 1995 and 1996. Different proportions of total area are covered by each P off-take classes while Z4 (1995) and Z3 (1996) indicate a larger area (31.02%). P off-take maps of 1995/96 at East Hemel field indicated that 26.73%, 35.44%, and 37.62% of total area are covered by AZ and ±1, or > ±2 P off-take classes differed from AZ respectively (Table 2).

Table 1. Calculating agreement zones of P off-takes in 1995 and 1996 at East Hemel field.

Zones: 1996 P kg/ha 1995	Z1 = < 50 m^2	%	Z2 = (50 -55) m^2	%	Z3 = (55 - 60) m^2	%	Z4 = > 60 m^2	%	Total 1995 %	Ave. 1995 P: kg/ha
Z1 = < 50	9870	4.95	10528	5.28	12502	6.27	5922	2.97	19.47	42.35
Z2 =(50 - 55)	8554	4.29	12502	6.27	13160	6.6	9212	4.62	21.78	52.56
Z3 = (55-60)	17108	8.58	13818	6.73	17760	8.91	6580	3.30	27.52	57.61
Z4 = > 60	15792	7.92	14476	7.26	18424	9.24	13160	6.6	31.02	64.39
Total:1996	25.74		25.54		31.02		17.49		99.79	216.91
Ave. 1996 P: kg/ha	41.85		52.48		57.11		63.88		215.32	54.22 53.83

* grid squares: 303 ** Grid area: 658 m^2 /square *** P: Phosphorus

Table 2. Comparing P off-take classes for maps of East Hemel between 1995/96.

Zones	Agreement zone	± 1 P off-take class (%)	> ±2 P off-take classes (%)
1. East Hemel	26.73	35.44	37.62

Conclusion

The potential of yield mapping to predict variable rate P application (VRPA) is that it integrates soil, landscape, crop and climate factors together into an expression of relative productivity (Kitchen et. al., 1995). Therefore, if yield variation patterns within fields changed from year to year and from crop to crop, then yield mapping would offer little guidance for developing VR application strategies. In confirmation of this, our findings showed that the correlation between two sets of regular yield data of 1995 and 1996 was not adequate to develop a VRP application other than to correct for variable off-take by the previous year's crop. Hence, in this case the plan would be to restore fertility to a common base. If correlation was significant (i.e. ~ 0.5) then we could plan greater inputs for high yield areas. Ultimately, in order to develop plans for varying fertiliser rates, defining different management units from yield mapping should be done. For this, accumulation of several years (3-5) of data collection is required to avoid any confusion related to temporal or seasonal yield variability as above which probably requires more soil and climate data too.

References

Kitchen, N. R.; Hughes, D. F.; Sudduth, K. A. and Birrell, S. J. (1995). Comparison of Variable Rate to Single Rate Nitrogen Fertiliser Application. Corn Production and residual Soil NO_3-N Site-Specific Management for Agricultural Systems, American Society of Agronomy, pp. 251-271. USA.

MAFF (1994 & 1997). Fertiliser Recommendations. Reference Book 209, London, HMSO.

Miller, P. C. H. (1993). The Control Of A Spatially Variable Herbicide Sprayer. In SCI (Ed.), Sensing & Control Aspects Of Spatially Variable Field Operations, (pp. 14-15). London: SCI.

Vansichen, R. & De Baerdemaeker, J. (1991). Continuous Wheat Yield Measurement on a Combine. Automated Agriculture for the 21st Century, Michigan, ASAE, pp 346-455, USA.

Acknowledgement

This paper is a partly contribution of the main research project of reliable yield mapping on combine harvesters aided by GPS/GIS that was carried out by the author in University of Newcastle Upon Tyne-UK (1994-98). The project was funded as a part of author's grant paid by both Iranian Ministry of Science, Research and Technology (IMSRT) and The University of Newcastle Upon Tyne-UK.

552

Evaluation of an NIR sensor to measure soil moisture content

L. Sartori[1], M. Bertocco[1] and L. Serva[2]
[1]*Dip. Territorio e Sistemi Agro-Forestali, Viale dell'Università, I-35020 Legnaro, Italy*
[2]*Dip. Scienze Zootecniche, Viale dell'Università, I-35020 Legnaro, Italy*
luigi.sartori@unipd.it

Abstract

Site-specific management of inputs requires accurate knowledge of soil chemical and physical properties. An NIRS sensor was tested on a sandy-silt soil to determine soil water content. Calibrations were done using both row spectra and first derivative mathematical treatments. Partial least square regression, modified partial least square regression, step-up and stepwise multiple linear regression were performed. To limit the study to regions in which there was high NIR prediction accuracy, analysis involved just the field capacity (FC)-wilting point (WP) moisture range: prediction accuracy increased ($R^2 = 0.990$; $1 - VR = 0.986$; SECV = 0.451). To simplify the sensor, prediction was done using just two wavelengths (1876 nm and 2204 nm), with greater prediction accuracy ($R^2 = 0.986$; $1 - VR = 0.985$; SECV = 0.805).

Keywords: field capacity, NIR, soil moisture, wilting point, sensor.

Introduction

Soil moisture determination at sowing can be important for improving conditions for the germination/emergence phase. The accuracy of soil moisture prediction with the NIR method is affected by different factors (Bowers & Hanks, 1965; Dalal & Henry, 1986; Ben Dor & Banin, 1995) such as particle size (Couillard et al., 1996) and clay mineral content. The application of NIR technology to soil properties analysis might consider the data of the complete spectrum, with a high correlation for some regions of the spectrum (Viscarra, 2000). The main objective was to identify a simplified sensor to determine soil moisture content in a field with acceptable accuracy "on the go".

Materials and methods

A sandy-silt soil (sand 57.1%, silt 30.3%, clay 12.6%) with 2.9% organic matter content was analysed. Spectral reflectance was recorded with an NIRs 5000 scanning monochromator (Foss NirSystem), interfaced with WinISI II v1.50 software. Spectra were submitted to first derivative process, with 4 and 5 wavelengths, respectively, compared to the original spectra, the calibration equation was then developed using both scatter detrend correction and no correction; partial least square regression and modified partial least square regression were performed. Analysis was repeated in the FC-WP moisture range. The step-up and step-wise calibration methods were also tested using a wavelength number selected from 1 to 6 (1940 nm, 2204 nm, 1876 nm, 1108 nm, 1380 nm and 1188 nm).

Results and discussion

Results were encouraging and prediction accuracy was high ($R^2 = 0.984$; 1-VR = 0.987; SECV = 0.761), especially on soil samples that were neither too dry nor too wet (Hummel et al., 2001). Analysis was repeated within the WP (6.8 %) and FC (18.5 %) range and prediction accuracy increased ($R^2 = 0.990$; $1 - VR = 0.986$; SECV = 0.451). Accuracy was also high choosing the six more correlated wavelengths ($R^2 = 0.983$; $1 - VR = 0.982$; $SECV = 0.934$ – figure 1), but results were even better ($R^2 = 0.986$; $1 - VR = 0.985$; $SECV = 0.805$) with the prediction run using just two wavelengths (1876 nm and 2204 nm).

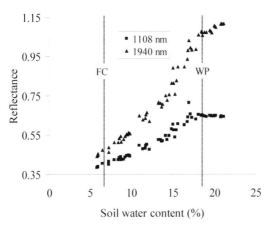

Figure 1. Reflectance at different soil water content with two wavelengths (1108 and 1940 nm). FC = field capacity; WP = wilting point.

Conclusions

Results show that it would be possible to accurately predict soil moisture content using a simplified NIR sensor (just two wavelengths), but it is necessary to use the wavelength more correlated for each soil and different soil properties. Further study is required to find a calibration that can consider the different properties of a soil sample.

References

Ben Dor E. and Banin A. 1995. Near infrared analysis as a rapid method to simultaneously evaluation of several soil properties. Soil Science Society of America Journal 59 364-372.
Bowers S.A. and Hanks R.J.1965. Reflection of radiant energy from soil. Soil Science 100 130-138.
Dalal R.C. and Henry R.J. 1986. Simultaneous determination of moisture, organic carbon, and total nitrogen by near infrared reflectance spectrophotometry. Soil Science Society of America Journal 50 120-123.
Couillard A., Turgeon A.J., Westerhaus M.O. and Shenk J.S. 1996. Determination of soil separates with near infrared reflectance spectroscopy. Journal of Near Infrared Spectroscopy 4 201-212.
Hummel J.W., Sudduth K.A. and Hollinger S.E. 2001. Soil moisture and organic matter prediction of surface and subsurface soils using a NIR soil sensor. Computers and Electronics in Agriculture 32 149-165.
Viscarra Rossel R.A and McBratney A.B. 2000. Laboratory evaluation of a proximal sensing technique for simultaneous measurement of soil clay and water content. Geoderma 85 19-39.

AVIS and GVIS – Hyperspectral imaging spectrometers for economic and ecological optimisation of plant production

Th. Schmidt and W. Mauser

GTCO-Ground Truth Center Oberbayern, Geschwister-Scholl-Ring 3, 82110 Germering, Germany
th.schmidt@gtco.de

Abstract

The growing importance of precision farming requires new data acquisition systems and techniques to obtain detailed information about the vegetation status. It is well-known that the reflection of vegetation depends on a multiplicity of different factors of influence (e.g. LAI, plant components or nutrient supply). Common multispectral sensors are able to supply limited information about these factors. Imaging spectrometry using hyperspectral sensors extends these possibilities substantially. The most important characteristics and specific advantages of this technology can be shown by the hyperspectral imaging spectrometers AVIS (Airborne Visible and Infrared Imaging Spectrometer) and GVIS (Groundoperated Visible/Near-Infrared Spectrometer), which were developed by the Department of Earth and Environmental Sciences of the University of Munich (LMU).

Keywords: remote sensing, hyperspectral sensors, imaging spectroscopy

Introduction

The condition of vegetation can be recognized by analyzing its spectral characteristics by optical sensors. Imaging spectrometers like AVIS or GVIS have more sensitive optical sensors than common multispectral sensors like Landsat ETM, SPOT or IKONOS). The more sensitively the system is, the better spectral characteristics can be recognized and analyzed.

Technical specifications of hyperspectral imaging spectrometers at the example of AVIS and GVIS

The higher performances of this technology can be represented by three substantial criteria:
1. Number of channels and complete cover of a wide range of wavelength
2. Spectral resolution
3. Spectral sensitivity
The number of the channels limits the valuable information. Multispectral sensors have normally up to ten channels (Landsat ETM: 7 channels; SPOT, IKONOS and Quickbird: 4 channels). The hyperspectral imaging spectrometers AVIS and GVIS have up to 128 channels. These sensors cover up a wide range of wavelength (app. 400 - 900 nm) completely. This range contains the complete visible and near infrared region of the spectrum. It includes the so called "red-edge", an area of special interest to derive vegetation parameters. The spectral region between 680 and 800 nm is characterized by a steep increase in the reflectance of vegetation, caused by the chlorophyll content of cells and the cell structure of the leaves. It can be used to determine the conditions of the plants (BACH, 1995; OPPELT, 2002).
One important criteria for analyzing remote sensing data is the spectral resolution. The spectral resolution of AVIS or GVIS is app. 8 nm (resolution of multispectral sensors is about 60 up to more than 100 nm). A high spectral resolution in addition to a large number of channels enables to recognize a precise reflectance curve of vegetation (Figure 1).
AVIS- and GVIS-data have a very high spectral sensitivity by use of 14-bit technology. Each individual channel can acquire more than 16.000 intensity differences (Landsat: 8-bit

technology corresponds to 256 intensity differences). Due to this high radiometric resolution, which exceeds the ability of conventional satellite sensors and human eye as well, small differences within at first sight homogeneous surface are recognizable (Figure 2).

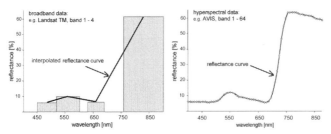

Figure 1. Comparison between broadband and hyperspectral data.

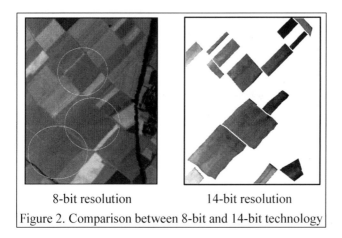

8-bit resolution   14-bit resolution

Figure 2. Comparison between 8-bit and 14-bit technology

Conclusions

In comparison to common optical remote sensing sensors imaging spectrometers enable to acquire much more detailed information of the vegetation status. Therefore AVIS and GVIS can contribute to the economic and ecological optimization of plant production.

References

Bach, H. 1995. Die Bestimmung hydrologischer und landwirtschaftlicher Oberflächenparameter aus hyperspektralen Fernerkundungsdaten. (Determination of hydrological and agricultural surface parameter using hyperspectral remote sensing data) Münchener Geographische Abhandlungen, Reihe B, Band 21.

Oppelt, N. 2002. Monitoring of plant chlorophyll and nitrogen status using the airborne imaging spectrometer AVIS. Ph.D. Thesis Faculty for Geosciences LMU, München, www.geographie.uni-muenchen.de/Internetseiten/Mitarbeiter/Lehrpersonal/ Oppelt/dissertation.pdf

Acknowledgements

This project was funded by the 'High Tech Offensive Zukunft Bayern'.

Multifunctional on-board computers – A key to promote precision agriculture

Steffen Schmieder[1], Beatrix Greifenhagen[1] and Robin Gebbers[2]

[1] Hansenhof Reifland - Vertrieb von WTK-Elektronik, Dorfstr.11, 09514 Reifland, Germany
[2] ZALF e.V., Eberswalder Straße 84, 15374 Müncheberg, Germany
steffen.schmieder@hansenhof.de, rgebbers@zalf.de

Keywords: field-operator, on-board computer, terminal, ISOBUS, multifunctional

Introduction

On-board computers, also called controllers, terminals, com units etc., play an important role in the technical framework of precision agriculture (PA). They are used for implement control, process data recording, and linking of data streams (e.g. GPS and yield monitors).

While PA is evolving by the appearance of new machines, sensors, and concepts, several problems remain unsolved in practice. A serious drawback is the incompatibility and 'mono-functionality' of on-board computers and implements (Kloepfer et al. 2001, Pedersen et al. 2001).

In many cases, every implement needs it's own on-board computer. And despite of the standardisation by ISO 11783, which might apply to new implements, no on-board computer might be available for old, yet good, equipment. The situation is characterized by the incompatibility of data formats and operation features. This causes:

- Higher costs for computer hardware
- Additional learning efforts for each controller
- Handling errors due to dealing with different graphical user interfaces (GUI) and non-ergonomic placement in the cab
- Higher efforts for maintenance of several systems (esp. software updates)
- Limitations in decision making choosing implements with the appropriate functions or the best price while compatibility has to be regarded
- Instability because of workarounds and tying in of many separate systems

Materials and methods

As an OEM manufacturer of on-board computers, job computers, and other agrotechnical equipment for end users and industrial consumers, wtk Elektronik inc. has learned about the problems mentioned above. Therefore, the demands of the users were carefully studied. A list of features was compiled, that has to be offered by a computer instrument for modern agriculture. It turns out to be important, that the computer must be suitable not only for PA but also for standard operations without GPS. On that bases, a new integrated software design called 'field-operator concept' was developed. Technical details of the hardware and the programming are presented by PALLMER 2003.

Results

The hardware platform for the 'field-operator concept' is the 'wtk field-operator 205'.
The field-operator 205 offers the following features:

- Virtual terminal according to ISO 11783
- Task-controller for job processing and processing of raster maps

- Process data recording of implements and as-applied mapping for improved farm management (see DEMMEL et al 2001)
- Tractor electronic control unit (hour meter, area covered, fuel consumption, ground speed)
- Integrated guidance aid for parallel swathing, external light-bar option, and planning of fieldworks by using task templates and field plans.

Additional software modules extend these basic capabilities:

field-probe offers mobile GIS functions. The computer could be used inside and outside the cab for soil mapping, crop scouting, surveying of field boundaries, recording of sensor measurements, and navigation.

field-sense is a universal tool for tying in real-time sensors such as the pendulumeter or the EM38$^{®}$. It enables calibration of the sensors, calculation of target rates and application by real-time sensors, recording of set points and sensor measurements.

Principally, the wtk field-operator is an ISOBUS terminal, but it is downward compatible to non-standard implements through additional software modules. Thus, most of the ISOBUS functions are available for non-ISOBUS implements, including the German LBS standard. Non-standard implements could be controlled via RS232 serial BUS. Even if automatic control is impossible, the field-operator offers GPS-supported acoustic and visual indication of target rates for manual control.

On the host computer, data can be read by most of the commercial PA software (e.g. KemiraLORIS$^{®}$, AGRO-MAP$^{®}$, SSTools$^{®}$). No additional program is necessary.

Conclusion

Meeting the practical demands of the farmers, the 'field-operator concept', will substantially contribute to the promotion of PA. The 'field-operator 205' reduces costs, learning efforts, handling errors and increases stability and flexibility of the PA system. Merging several functions into one system will also induce synergistic effects making field work much more efficient.

References

Demmel, M.; Rothmund, M.; Spangler, A.; Auernhammer, H. 2001. Algorithms for Data Analysis and first Results of Automatic Data Acquisition with GPS and LBS on Tractor-Implement Combinations. In: 3ECPA. G. Grenier, S. Blackmore (eds.), agro Montpellier, France.

Kloepfer, F.; Kottenrodt, D.; Weltzien, C. 2001. Implementation and Acceptance of Precision Agriculture by Farmers and Agricultural Contractors within the Research Project 'preagro'. In: 3ECPA. G. Grenier, S. Blackmore (eds.), agro Montpellier, France.

Pedersen, S.M.; Ferguson, R.B.; Lark, R.M. 2001. A Comparison of Precision Agricultural Practices in Denmark, the United Kingdom and the United States. SJFI-Working Paper no. 2/2001.

Pallmer, M. 2003: ISO-Bus task controller with GSM data transfer. 4ECPA. Poster

Site-specific yield measurement of sugar beet based on single beet mass

O. Schmittmann

Institut für Landtechnik, Rheinische Friedrich-Wilhelms-Universität Bonn, Nußallee 5, 53115 Bonn, Germany

o.schmittmann@uni-bonn.de

Abstract

The site-specific yield of sugar beet is an elementary piece of information for precision agriculture. Harvesting conditions, dirt tare, and the quality of the harvesting process complicate real time measuring. For this reason, two yield measuring systems were developed which are based on the clean, unharvested sugar beet.

Keywords: sugar beet, yield measurement, yield sensor

Introduction

The real time measurement of yield and its spatial assignment are useful for adapting the cultivation process to the requirements of the plant. While yield measuring systems for grain crops already exist, equivalent systems are still missing for the harvest of sugar beets. Harvesting conditions as well as turbulences and impurities in the material flow complicate the development of precise measuring systems.

In order to develop a system suitable for obtaining the clean beet mass on the basis of which site-specific yield can be calculated, two new yield measuring systems based on the single unharvested sugar beet were developed and tested.

System analysis

A systematic analysis of contemporary lifter-based technology for determining crop quantity highlighted the following problems:

1. The material stream in the lifter (mass or volume stream) always contains a mixture of beets, tops, and soil tare. At present, it is impossible to determine exactly the proportions of these components. The variability of their composition results in errors concerning absolute mass. These, in turn, are a source of errors in the determination of site-specific differences in yield.
2. Measuring on the harvester is influenced considerably by exterior factors. For example, real-time measurements are distorted by movements of the harvester and by sloping ground.
3. The measuring of real beet yield depends on lifting and topping quality. Beet tops and severed root ends remain in the field and do not enter into weight calculations.

Solution: Combining yield estimation and online measuring

The discussion of advantages and disadvantages inherent in the different approaches suggested developing a yield measuring system installed in the lifter and drawing on individual beets. The masses of individual beets can be estimated in field on the basis of biotechnical data The addition of these individual beet masses and the integration of cultivation patches makes the site-specific measuring of beet yield in real time possible.

The metrological implementation is made by measuring the profile of decapitated or non-decapitated beets. For measuring the distance between machine and ground surface or beet top, a laser sensor is installed between the topping unit and the lifter; alternatively, a mechanical sensor is used on the row sensing device. For measuring the distance incrementally, a speed sensor is applied. The vectorial combination of both sensor values makes it possible to determine the plant profile. By means of specially developed software based algorithm , the number of beets and their height and maximum diameters can be determined. In a final step, the individual beet mass is estimated by means of a base function and connected with DGPS data. Thus, a site specific documentation of culture and yield data of sugar beet is feasible.

Beet counting

The number of beets is determined by means of counting during the lifting process. The site-specific yield can be derived by multiplication with an average beet weight. The average beet weight can be determined either before lifting by means of random samples, or after delivery when the actual average beet weight can be used to correct estimates.

Beet measuring

The procedure of measuring individual beets was developed as an improvement on beet counting. Instead of a multiplication of assumed average weights by the number of beets, this procedure adds the estimated weights of each individual beet. The estimates of individual beet weights are based on the maximum beet diameter in driving direction. The estimated function delivers a certainty measure of 88%.

Results

Field trials indicate that vibrations of the lifting group, leaves, and beet are disturbing factors which make technical changes necessary. Under optimized conditions, beet detection has an accuracy of about 98%. The overground beet diameter in driving direction can be determined with a standard deviation of about 4 %. On average, this method produces beet diameter values that are 1mm higher than the real beet diameters actually are.

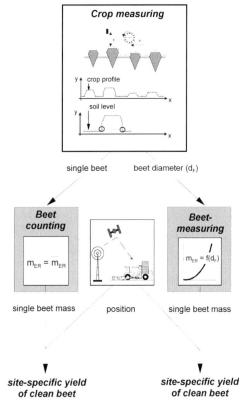

Figure 1. Principles of beet counting and beet measuring.

Conclusion

Two new yield measuring systems based on the clean beet in the soil have been developed. The systems are independent of dirt tare and harvesting quality. While measuring systems so far only produced information about the site specific harvest mass, the new systems additionally produces site-specific information about important factors for plant cultivation and for the interpretation of yield differences (e.g., crop density, beet diameters, and beet height).

Use of remote sensing to identify the spatial distribution of the sugar beet cyst nematode *Heterodera schachtii*

A. Schmitz[1], S. Kiewnick[1], J. Schlang[2], K. Schmidt[1] and R.A. Sikora[1]
[1]*Institute for Plant Diseases, Soil Ecosystem Phytopathology and Nematology, Nussallee 9, 53115 Bonn, Germany*
[2]*Federal Biological Research Centre for Agriculture and Forestry (BBA), Institute for Nematology and Vertebrate Research, Dürener Straße 71, 50189 Elsdorf, Germany*
astrid.schmitz@uni-bonn.de

Abstract

A study was undertaken to map plant stress caused by *Heterodera schachtii* in sugar beets using remote sensing data. Digital colour infrared (CIR) airborne images and ground truth data from a field experiment with increasing nematode densities and sugar beet varieties were used to perform a supervised classification with specific image processing software. The full image was divided into classes by nematode density and used to identify the spectral response patterns for each class. Data from the near infrared and red bands were used to calculate the vegetation index NDVI. The NDVI seems to be a good identificator for characterizing *H. schachtii* damage in sugar beets.

Keywords: *Heterodera schachtii*, remote sensing, sugar beet, cyst nematode.

Introduction

Sugar beet cyst nematode is an important parasite of sugar beets worldwide. Above ground symptoms are not always observed at ground level, yet yields are still affected. Optical remote sensing offers the opportunity to detect plant stress caused by nematodes (Nutter, 2002). An increase of reflectance in the visible and a decrease in the near infrared portion of the electromagnetic spectrum indicate the presence of plant stress that reduces photosynthetic active green leaf area (Nutter, 1987 and 1989). These vegetation characteristics might be detected by spectral vegetation indices such as the Normalised Difference Vegetation Index NDVI (Choudhury, 1988). The objectives of this project were to: 1) geospatially map plant stress caused by the sugar beet nematode by remotely sensed data, 2) investigate whether the nematode can be detected using NDVI and 3) relate spectral reflectance patterns to nematode density to improve practical use by the grower, for example, by optimizing IPM strategies.

Materials and methods

Field experiment: The experiment was located in Niedermerz, NRW, Germany. The area measured 25 x 60 m and was divided into 4 x 5 m quadrats. One part of the field had high and the other part low nematode infestation levels. *H. schachtii* susceptible sugar beet cultivar Penta and the resistent cultivar Nematop were planted in alternate plots.

Nematode densities: The population density of *H. schachtii* was assessed directly before planting and shortly after harvest by the BBA. Soil samples were taken from 12 arbitrarily selected locations in each plot and the number of eggs and juveniles per 100 g of soil was calculated.

Aerial image assessment and analysis: One CIR aerial image obtained in July 1999 from the LIZ (Agricultural Information Service for Sugar beets) was the basis of this study. The image had an spatial resolution of 70 cm. Data from the red and the infrared bands were used to calculate the NDVI. To extract quantitative information from the aerial image we performed a supervised classification. We preselected classes from the image for high and low nematode density and collected a representative sample of pixels which corresponded to these classes. By creating a spectral response pattern for each class the full image could be classified.

Results

The supervised classification showed that infested and healthy sugar beets had different reflectance patterns. Differences in reflectance patterns of the sugar beet varieties were only obvious in highly infested plots. The resistant cultivar in infested plots of the field was classified as healthy. In the low infested area of the experiment the susceptible and the resistant varieties could not be distinguished. Low NDVI values around zero indicated nematode infested plants whereas values close to one indicated healthy plants.

Discussion

Although the NDVI values are based only on one image, the data provide insight into the potential of the index. Experiments on sugar beet with *Rhizoctonia solani* already indicated that the average change in index values assessed at different times in the year might be a possible predictor variable for root rot (Hope et al., 1999). For interpretation of aerial images it is also advisable to distinguish between sugar beet varieties, because resistant cultivars grown on highly infested fields may have the same reflectance patterns as healthy susceptible cultivars.

Conclusions

Future research will compare images obtained at different times of the year to detect characteristic reflectance patterns for healthy and infested sugar beets. Changes in NDVI over the sugar beet growing season might be a potential predictor of nematode infestation. Remote sensing methods for mapping *H. schachtii* would reduce labor costs needed for soil sampling.

References

Choudhury, B. J. 1997. Relationships between vegetation indices, radiation absorption, and net photosynthesis evaluated by sensitivity analysis. Remote sensing of Environment 22: 209-233.
Hope, A., Coulter, L., Stow, D., Peterson, S. and Service, D. 1999. Root rot detection in sugar beet using digital multispectral video. Proceedings of the 20th Asian Conference on Remote Sensing 1999.
Nutter, F. W. Jr. 1987. Detection of plant disease gradients using a hand-held, multispectral radiometer. Phytopathology 77: 643.
Nutter, F. W. Jr. 1989. Detection and management of plant disease gradients in peanut using a multispectral radiometer. Phytopathology 79: 958-963.
Nutter, F. W. Jr. 2002. Use of remote sensing to detect soybean cyst nematode-induced plant stress. Journal of Nematology 34 (3): 222-231.

Acknowledgements

This project is part of a Graduiertenkolleg, a research training program of the University of Bonn, and is funded by the German Research Association (DFG).

On farm experimentation with the help of GPS-based site specific management

H.-G. Schoen and A. Kielhorn
Fachhochschule Osnabrueck, Oldenburger Landstrasse 24, 49090 Osnabrueck, Germany

Abstract

The GPS-based site specific management was used to conduct simple on-farm experiments. Trials were set up easily via application maps. Proper Data aquisition requires careful harvesting. The yield data points still have to be checked and eventually corrected (hl-weight etc.). As a consequence of on-farm experimentation plot size is big. Coefficients of variation were nevertheless surprisingly low.

Keywords: On farm experimentation, GPS

Introduction

On farm experiments are a useful instrument for farmers to optimise the application of inputs. They should be easy to handle. The use of GPS-based application technology and yield mapping enables to conduct simple field experiments easily. One disadvantage is that plot and block size is in general big depending on the working width of the machines. Huge plots normally increase the error term when testing with ANOVA. The investigation was conducted to evaluate the precision of GPS-based on-farm field trials using regular machinery equipment.

Material and methods

Trails were conducted in 2002 on fields near the town of Osnabrueck in North-West Germany using a randomised complete block design with plot sampling. The fields were situated in the mountaineous range of the Wiehengebirge with rolling to hilly land surface. Soils of Field 1 (Wellenkamp) derived from glacial moraines over claystone (Cambisol). Soil texture varied from loamy sand to clay loam. The Trial was placed in the centre of the field with textures varying from loamy sand to sandy loam. The respective crop was winter oats. The first N-rate was given uniformly (70 kg N/ha).In the trial the second and third N-application were varied with three factor levels: site specific according to the farm manager (92 kg N/ha), + 30 % (120 kg N/ha), - 30 % (65 kg N/ha). Fertiliser was applied with the help of application maps generated by the software LORIS and FIELDSTAR (Fig. 1). Application maps were stored as bord computer jobs. The field application was controlled by the bord computer and the job processor of the fertiliser spreader. As-added maps were reread from the job file after application and stored and visualised in a GIS. Plots had the size of around 35 x 35 m each.
Field 2 was a leptic Regosol (Rusterberg). The texture was sandy loam. Plots had the size of around 30 x 60 m. The respective crop was winter barley. N was supplied in a 3-split application. Rates varied with three levels: site specific according to the farm manager (180 kg N/ha), + 20 % (215 kg N/ha), - 30 % (145 kg N/ha). Application was done with the help of GPS and digital application maps as mentioned above.
The fields were harvested with GPS-based yield mapping. The respective yield point data of the core plots was used to calculate ANOVA for a randomised complete block model with plot sampling. ARCVIEW Version 3.1 was used for data visualisation and interpolation and SPSS Version 10 for the statistical analysis.

Results and discussion

The planning and implementation of the trials was very easy and time efficient. Yield data could also be easily identified in the respective plots via GIS. Yield data was first corrected and inconsistent measurements were excluded.

Trial Rusterberg: ANOVA based on point measurements resulted in an extremely low mean square-value of the Block x Treatment interaction. Since this is a hint on violation of the assumptions of ANOVA residues were analysed. H0-hypothesis of normal distribution was not rejected following the K-S test. Nevertheless the distribution seemed to be somehow bimodal. Looking at the yield map gives the impression as if the yield data is partly biased and therefore autocorrelated along the combine tracks due to systematic effects, for instance differing cutting width. Durbin-Watson statistics proved existing autocorrelation. Nevertheless coefficient of variation for the sampling error (innerplot variation=residues) was low with 8.9 %.

Using the mean square of the residues as denominator instead of the block x treatment interaction resulted in an F-value of 0.65 for N-rates. Testing the block effect this way also yielded a non significant F-value. There were no significant differences between the treatments, suggesting that the farmers recommendation seemed to be higher than optimum in this specific year

The ANOVA of the trial Wellenkamp also resulted in non significant differences between the factor levels. Coefficient of variation for sampling error was low with 8.8 %. Assumptions of ANOVA seemed to be fulfilled.

Conclusions

The GPS-based site specific management enabled to conduct simple on-farm experiments. Trials were set up easily via application maps. Proper Data aquisition requires careful harvesting. The yield data points still have to be checked and eventually corrected (hl-weight etc.). As a consequence of on-farm experimentation plot size is big. Trial Rusterberg had mean squares which indicated that assumptions of ANOVA were violated. It seemed as if data was partially biased due to the procedure of data aquisition while harvesting. Trial Wellenkamp did not show these problems. Coefficients of variation were surprisingly low. Because of the huge plot and block sizes the risk of capturing soil heterogenity is high. Therefore the use of latin squares seems to be an interesting alternative if the spatial variability of soil heterogenity is unknown.

References

Dabbert, S.; Kilian, B. (2002): Ökonomie. – In: Precision Agriculture – Herausforderung an Integrative Forschung, Entwicklung und Anwendung in der Praxis, ISBN 3-9808279-0-9, S. 423-446, Darmstadt

Gomez, K.; Gomez, A. (1984): Statistical Procedures for Agricultural Research;ISBN 0-471-87092-7; S. 241-222, New York.

Development of precision agriculture sensing technologies for clam and citrus production

J.K. Schueller, S.M. Baker, W.S. Lee, C.L. Montague, E.J. Phlips, T.F. Burks, J. Jordan, J.W. Mishoe, M. Salyani and A.W. Schumann
University of Florida, Gainesville, FL 32611 U.S.A.
schuejk@ufl.edu

Abstract

Precision agriculture sensing technologies are developed for clam and citrus production.

Keywords: sensing, clam, aquaculture, citrus

Introduction

From the early days (e.g., Schueller, 1992) to the present (e.g., this conference), precision agriculture has been primarily applied to grain crops, especially wheat and corn (maize). Through its Initiative for Future Agriculture and Food Systems (IFAFS) program, the United States Department of Agriculture (USDA) has funded two large multidisciplinary programs to bring precision agriculture technologies to clam and citrus farming.

Clam farming precision agriculture

Florida clam farming has grown from a crop worth US$400,000 in 1987 to US$18,000,000. Clam farmers purchase seed from hatcheries, nurse the young clams in land-based raceways, plant them underwater in large porous bags, and harvest the mature clams after 12-18 months in the ocean. The clams are tumbled and washed, graded by size, and bagged for market. The farmers lease 1 or 2 hectare shallow, off-shore areas from the government.

The results of the project provide farmers "real-time" water quality and weather data gathered from stations in six counties on the Atlantic Ocean and Gulf of Mexico coasts. The parameters measured include air temperature, wind speed and direction, rainfall, barometric pressure, and various water parameters (temperature, salinity, dissolved oxygen, depth, chlorophyll, and turbidity). All stations collect and store data every thirty minutes. Both real-time and archived corrected data can be viewed by selecting "Weather/Water Quality" at http://www.FloridaAquaculture.com. The data helps farmers bid on lease areas, determine management practices, and reduce catastrophic crop loss.

Clam farmers are the pilot program providing federal insurance to aquaculture. The 250 clam growers in the program are protected from hurricanes, tidal waves, storm surge, windstorm, disease, low oxygen, freezes, and salinity. This projects covers the farmer's responsibility to document a covered peril in a loss.

Other components of the project include phytoplankton monitoring, field and lab physiology trials, and simulation model development. The phytoplankton monitoring stations at 28 locations in the Suwanee Sound and 8 locations in the Indian River monitor chlorophyll a, phytoplankton composition, nutrient concentrations, light penetration, turbidity and color, and particulate organic carbon. The simulation model studies practices of seeding date, grow-out density, and seed size to increase growth, farm efficiency, yield and profitability.

Citrus precision agriculture sensing

Previous citrus precision agriculture research (e.g., Whitney, et al., 1999) and commercialization have concentrated on yield mapping and building GIS databases. This project concentrates on tree and soil sensing in co-operation with scientists specializing in other tree fruit (primarily apples and pears) from Washington and Oregon.

One task seeks to sense the nitrogen level in tree leaves to guide fertilization. Over 1000 citrus tree leaves were harvested and measured with a spectrophotometer to determine the reflectance spectra from 400-2500 nm under uniform laboratory conditions. Conversion of the acquired data into absorption spectra and the use of partial least squares resulted in seven and nine factor models which had R^2 values of 0.8234 and 0.8433 between model predicted and experimentally-determined actual nitrogen concentrations for the validation set.

Another aspect involved determining shallow watertables and hydromorphic soils with an EMI profiler. Vertical dipole mode soil conductivity was sensed with an EM38. Correlations with measured soil water tables were significant with R^2 values between 0.48 and 0.93. But they were affected by salinity differences between groves and temporal effects of rainfall.

The tree canopy sensing work builds upon previous work (Tumbo et al., 2002). Fifteen densely foliated and fifteen lightly foliated citrus trees were scanned on both sides by a Durand-Wayland ultrasonic sensor system at three ground speeds and three replications. Overall, there was no significant difference between the sensor data and that obtained from manual canopy measurements, although the differences between those measurements ranged from -17.4 to +28.7%, at the 95 % confidence level.

The tree canopy sensing work builds upon previous work in the area (Tumbo, et al., 2001). Fifteen densely foliated and fifteen lightly foliated citrus trees were scanned on both sides by a Durand-Wayland ultrasonic sensor for three ground speeds and three replications. There was no significant difference between the sensor data and that obtained from manual canopy measurements, although the ultrasonic overcomputed canopy volume by 5.7%.

Remote sensing of citrus tree stress was conducted at four widely-dispersed citrus sites with the cooperation of NASA. Satellite or airborne data was collected using IKONOS, RDACS, ATLAS, and hyperspectral instrumentation. The ATLAS data was corrected for atmospheric and emissivity effects with balloon radiosonde data. Geolocation and reflectance calibration targets were placed in the imagery. This data will be further analyzed to develop methods and algorithms to detect tree stress by airborne and satellite methods.

References

Schueller, J.K. 1992. A review and integrating analysis of spatially-variable control of crop production. Fertilizer Research 33:1-34.

Tumbo, S.D., M. Salyani, J. D. Whitney, T. A. Wheaton, and W. M. Miller. 2002. Investigation of Laser and Ultrasonic Ranging Sensors for Measurements of Citrus Canopy Volume. Applied Engineering in Agriculture 18(3): 367-372.

Whitney, J.D., W.M. Miller, T.A. Wheaton, M. Salyani, and J.K. Schueller. 1999. Precision farming applications in Florida citrus. Applied Engineering in Agriculture 15(5):399-403.

MOSAIC – On-farm-monitoring, geo-spatial-analysis, and deterministic geo-referenced modelling as an approach for spatial crop yield variability and site-specific management decisions

D. Schumann[1], H. Miller[1], P. Jürschik[2], J. Schwarz[3], K.C. Kersebaum[4], H.I. Reuter[5], A. Giebel[4] and O. Wendroth[5]

[1]Suedzucker AG, Agricultural Business Section, 74254 Offenau, Germany
[2]agrocom., Potsdamer Straße 211 33719 Bielefeld, Germany
[3]ATB, Institute for Agricultural Engineering, Department 4, Engineering for Crop Production, Max-Eyth-Allee 100, 14469 Potsdam, Germany
[4]ZALF, Institute for Landscape Systems Analysis, Eberswalder Str. 84, 15374 Müncheberg, Germany
[5]ZALF, Institute for Soil Landscape Research, Eberswalder Str. 84, 15374 Müncheberg, Germany

Abstract

Monitoring soil and crop state variables and their analysis is an important task for deriving on-farm information for better crop management. The aim of the project MOSAIC is, to identify soil variability patterns and their spatio-temporal association with crop development and yield for a better nutrient management that is focused on minimizing ecological risks for the environment and society, and optimizing economical benefits for the farmer.

Keywords: nitrogen dynamics, biomass production, spatial modelling, geo-spatial data analysis

Introduction

Biomass production and underlying processes are known to vary tremendously within arable fields. Patterns of grain yield change between different years and crops, i.e., within the same field, high or low yield regions are not stable in time. Depending on the seasonal weather conditions and crop stress during different development stages, yield patterns are pronounced differently between respective years. Nevertheless, management decisions and adequate variable fertilizer application schemes must be based on some expectation of the local crop yield. In order to enhance understanding of local biomass production, nutrient dynamics, and their spatial and temporal behaviour for prediction and improved management decisions, the research project MOSAIC was founded.
Deterministic modelling of processes related to biomass production on one hand, and geospatial analysis and stochastic modelling of crop yield, soil and remotely sensed state variables on the other hand, both conducted under real farm conditions, are the main conceptual tasks of this project.

Materials and methods

The experimental fields are located on a farm of Suedzucker AG in Luettewitz, south-eastern Germany, Federal State of Saxony. The soils are classified as Stagnic Luvisols with an average soil fertility index (Ackerzahl) of 70. The average annual amount of precipitation is 660 mm, and the mean average temperature is 8.0 °C. So far, cereal crops and rape seed have been investigated.
Two fields are intensively monitored at a 54-m-grid for soil texture (sand, silt and clay fractions, soil organic matter content, soil P- and K-content, and at a 27-m-grid for mineral

soil nitrogen content (0-30, 30-60, and 60-90 cm), crop yield, normalized differential vegetation index (NDVI), surface soil water content in spring, crop-nitrogen-status during growing season, and soil electrical conductivity (EM 38). Furthermore, small scale variability of crop-nitrogen-status and mineral soil-nitrogen content are investigated.

Results and discussion

HERMES (Kersebaum, 1995, Kersebaum et al., 2002) is a deterministic model for simulation of biomass development, and soil water and nitrogen dynamics. Site specific input information is evaluated with respect to adequate process description of biomass development and nitrogen dynamics. Moreover, HERMES was used to calculate site-specific application rates of nitrogen fertilizer.

For the geo-spatial and time series approaches applied, the reader is referred to Nielsen and Wendroth (2003). Description of spatial crop yield distribution and implications for prediction based on soil state variables and on remotely sensed data is given in Wendroth et al. (2003). Reuter et al. (2001, 2003, this issue) analysed the impact of landform on crop yield and yield components, and considered the effect of local shading and associated spatial differences in radiation and its role for biomass production. Spatial description of soil mineral nitrogen and association across four subsequent years was shown by Giebel et al. (2003, this issue). Recent approaches aim at the prediction of crop yield and transferability of state-coefficients for different years (Wendroth et al., 2003, this issue).

References

Giebel, A., Wendroth, O. Reuter, H.I., Kersebaum, K.C. and J.Schwarz. 2003. MOSAIC: Spatial representativity of mineral soil nitrogen monitoring. In: Stafford, J.V. (Ed.). Precision Agriculture '03. Proc. 4th Europ. Conf. Prec. Agric., Berlin, Germany, (submitted).

Kersebaum, K. C. 1995. Application of a simple management model to simulate water and nitrogen dynamics. Ecological Modelling, 81, 145 - 156.

Kersebaum, K. C., Reuter, H. I., Lorenz, K. and Wendroth, O. 2002. Modelling crop growth and nitrogen dynamics for advisory purposes regarding spatial variability. In: Agricultural system models in field research and technology transfer, edited by L. J. Ahuja, L. Ma and T. A. Howell, Lewis Publishers, Boca Raton, USA, pp. 229-252.

Nielsen, D.R., and O. Wendroth. 2003. Spatial and temporal statistics – sampling field soils and their vegetation. Catena, Reiskirchen, Germany, 398 pp.

Reuter, H.I., O. Wendroth, K.C. Kersebaum, and J. Schwarz. 2001. Solar radiation modelling for precision farming – a feasible approach for better understanding variability of crop production. In: Grenier, G., and S. Blackmore (Eds.). ECPA 2001. Proc. 3rd Europ. Conf. Prec. Agric. Montpellier, France. pp. 845- 850.

Reuter, H.I., O.Wendroth, K.C.Kersebaum, and J. Schwarz. MOSAIC: Crop Yield Observation – Can landform stratification improve our understanding of crop yield variability? In: Stafford, J.V. (Ed.). Precision Agriculture '03. Proc. 4th Europ. Conf. Prec. Agric., Berlin, Germany, (submitted).

Wendroth, O., H.I. Reuter, and K.C. Kersebaum. 2003. Predicting Yield of Barley Across a Landscape: a State-Space Modeling Approach. Journal of Hydrology 272:250-263.

The potential for applying precision farming in Iran

F. Shahabazi
Lorestan University Khoram abbad Iran
Shabbazi@yahoo.com

Abstract

Precision farming is an attractive option to promote sustainable agricultural development. It is now possible to base resource utilization and management decision-making on information and knowledge. Many supporting technologies are being developed quickly with possible application in agriculture. Precision farming presents both challenges and opportunities for agricultural engineers. Development of precision agriculture practices will provide good opportunities for renovating conventional farming tools and technology, as well as for disseminating advanced knowledge to farmers. In the forthcoming knowledge-based era, a good vision of what will be happening tomorrow is the key to the further success of modern farming communities. The concept of precision agriculture, based on information technology, is becoming an attractive idea for managing natural resources and realizing modern sustainable agricultural development. It is bringing agriculture into the digital and information age. The practice has smoothly extended into some developing countries. This paper considers the possible adoption of precision agriculture for developing countries and ideas in conducting the practice in Iran.

Keywords: Precision agriculture; Developing countries; Strategic approach in agriculture

Introduction

Toward the new millennium, man is still facing the crucial challenge of ensuring food supplies and sustainability of agricultural development. Ever-increasing popu-lation, resource shortages, and degradation of the ecological environment have added even greater pressure on developing countries. In the past 50 years, world agriculture has experienced enormous changes. Industrialized countries have cre-ated a modernized agricultural system with high productivity and advanced technology nology. In the developing countries, however, the word 'poverty' inextricably remains linked with 'agriculture.' Most of the population in developing countries is still engaged in traditional farming with limited application of modern technologies in rural areas. The key restriction is obviously due to the backwardness of agricultural sciences and technology, and limited input to agriculture. Formulation of the world trade system and the information technology (IT) revolution have changed the external environment of agricultural development for all countries. The information and knowledge-based era will create new opportunities to accelerate the transformation of traditional farming into modern agriculture. Therefore, it is necessary to learn the new trends of modern information technology for agriculture in the developed world and investigate appropriate ways for promotion of new technology application in developing countries. Even within underdeveloped coun-tries, there exists a clear imbalance between less-modern regions and relatively more developed areas. The latter have the potential to accept more advanced technologies and can act as incubators for new ideas and sophisticated technologies based on their domestic conditions. Agricultural engineers are facing new challenges and need to integrate multidisciplinary approaches to solve complex problems in sustainable development of agriculture. They need not only to expand activities for equipment innovation, but also to be experts in production, processing, and management for the entire food production chain. Modern agriculture requires integrated support of agro-biology, engineering technology, and economic manage-ment sciences. Information technology will

569

greatly change the tools for farming with diverse sensors, adaptive actuators, high performance software, and algorithms for data acquisition, processing, conversion, storage and transfer as well as an integrated approach to making management decisions.

Strategies for adopting precision agriculture in developing countries

In the 21st century, man will enter into a knowledge-based era. The poverty of information has become a new problem in restricting social and economic advance-ment for developing countries. However, challenges and opportunities coexist. Many developing countries have paid great attention to speeding up information infrastructure construction in recent years. A favorable development environment to attract new technology and knowledge has smoothly taken shape. There is distinct regional imbalance in socio-economic development even within each devel-oping country. Each country will have to draw its own strategies for using the opportunities provided by the IT era. In the promotion of information high-tech research, the time difference in starting to use high-tech for agriculture among various countries is being shortened. Each country has to identify its strategies and priorities for its development. Agricultural high-tech trial farms are emerging very quickly in many developing countries. The renovation of traditional farming technology through precision farming techniques is a good strategic objective. It integrates a multidisciplinary approach involving agronomists in various fields, engineers, manufacturers, and economists to achieve sustainable development. Such advances could not only integrate a series of innovative tools to support traditional farming reconstruction, but also stimulate a technical revolution among agricultur-ists and farmers. In developing countries, most of the population are still engaged in traditional crop cultivation and are facing the crucial challenge of ensuring grain supply. Improving their crop production systems will be given top priority in their socio-economic development.

The potential for applying precision farming in Iran

Applying precision farming to plantating agriculture requries understanding of difference between farming systems so that potential applycations can be identified. Precision farming involves the adoption of both new technology and new level of management and adoption of both in farms requires new level of knowledge for farmers for applying this new technilogies and development the technology for product the tools for this system of farming in each country. In the other hand the tools and experties required to support a precision farming program can require a substantial investment of time and mony for both farmers and dealers. In iran farmers are lack knowleg about applying the new technology and they have not finacial resources to spend on precision farming and the level of technology is low and there is not enough potential for product the tools for appyling this system of farming in coutary and use of foregion technology is very expensive. The other problem is small-scale farms in iran and small frams budgets will not allow the purchase of much of the new technology designmed for precision farming.

References

Wang, M.H., 1999. The technological advance and equipment innovation in the 'precision agriculture(in Chinese). In: Digital Earth. Publishers of Chinese Environmental Sciences, Beijing, pp. 47–54.

Wang, M.H., 2001. Possible adoption of precision agriculture for developing countries at the threshold of the new millennium. Computers and Electronics in Agriculture .30 (2001) 45–50.

Precision farming Japan model for small farm agriculture

S. Shibusawa
Faculty of Agriculture, Tokyo University of Agriculture and Technology, 3-5-8 Saiwai-Cho, Fuchu, Tokyo 183-8509, Japan
sshibu@cc.tuat.ac.jp

Abstract

A community-based precision farming is an alternative for small farm scale agriculture in Japan. Producing information-oriented fields and information-added products is targeted and it will meet the demand of farmers to create locally branded, highly value-added and traceable products in the market. Wisdom farmers collaborating with technology-innovative companies are organizing a precision farming business model under local constraints.

Keywords: community-based, leaning group, information-oriented, information-added

Introduction

A company-based precision farming is cost-driven and requires big farm scales, which is not available for small scale agriculture in Japan. Adoption of precision agriculture in Japan will be community-based and its story is introduced.

Community-based precision farming

A community-based precision farming is organized by two learning groups of wisdom farmers and technology platform (Figure 1). The wisdom farmers innovate the farming system in order to manage the hierarchical variability: within-field, between-field and farm-style variability. Within-field variability focuses on a single field with a single plant variety being cultivated. Between-field variability implies the variability among fields, where each field can be considered as a unit of maps. Farm-style variability depends on the motivation of farmers who own or manage their respective small fields.

The technology platform organized by innovative companies develops and provides adaptive technologies: mapping techniques, variable-rate techniques and decision support systems available for the rural constraints. Best collaboration of the wisdom farmers and the technology platform can produce "information-oriented" fields and "information-added" products that encourage multi-functions of agriculture creating new value-chains of agro-production-consumption system.

A technology platform "Toyohashi PF-net Society" was founded at Toyohashi city area in May of 2002, in collaboration with a society of expert farmers "Atsumi Farmers Association". Toyohashi PF-net Society provides seminars, extension of information technology, and consulting to collaborate between companies and farmers.

The other is in Honjo city area, and leaders of the farmers founded a learning/practice group "the Honjo Precision Farming Society" in April 2002 (Figure 2). The society strives to produce information-oriented fields and also grow them as new Honjo Branded Agricultural Products through the study of Precision Farming Honjo Model, leading to new value-chains of agro-production-consumption system. They have conducted seminars and experiments as the initial stage of precision farming.

Conclusion

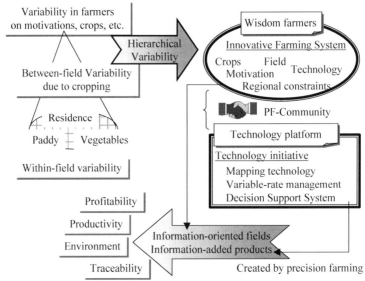

Figure 1. Community-based precision farming Japan model.

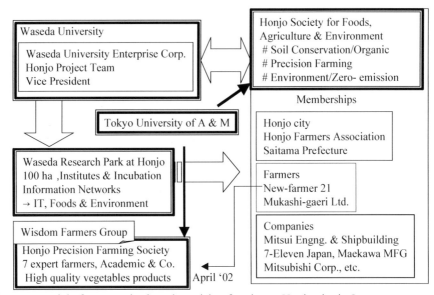

Figure 2. A trial of community-based precision farming at Honjo city in Japan.

A community-based precision farming is an alternative for small scale agriculture and the motivation of adoption is value-oriented and less cost-oriented.

A hand-held terminal for data collection in herb farms

Dror Shriki[1,] Joachim Meyer[1], Yael Edan[1] and Amots Hetzroni[2]
[1]*Department of Industrial Engineering and Management, Ben-Gurion University of the Negev, Be'er Sheva 84105, Israel*
[2]*Institute of Agricultural Engineering, Agricultural Research Organization, Bet Dagan 50250, Israel*
shriki@bgumail.bgu.ac.il

Abstract

This work presents the development of a hand-held terminal for data collection in herb farms where 500 to 1000 data records are generated daily. Data collection by non-professional workers on their various activities in distributed plots is essential to improve management. A prototype of a hand-held terminal for data collection equipped with a bar-code reader was characterized and evaluated. The system was characterized by applying Cognitive Work Analysis (CWA) methods. This included two steps: 1. Data collection tasks were specified by dividing all tasks into sub-tasks using the Sub-Goal Template (SGT) method. 2. The procedures for performing each task either with a bar-code reader or by menu selection were defined and task complexity with each input technique was predicted applying a Cognitive Task Analysis method (GOMS). The simplest input method for each task was selected based on the analysis.

Keywords : Hand-held Terminal, Cognitive Task Analysis, GOMS, data collection.

Introduction

For success in contemporary international agriculture it is necessary to handle the substantial quantity of data generated and to turn it into information. This work is a continuation of previous research that dealt with the development of an information system for a herb farm (Shriki & Cohen, 2001). Evaluation of the information system indicated that automated data acquisition is essential for successful implementation. A crucial component is the hand-held terminal that will be used by workers in the fields. This paper presents use of Cognitive Work Analysis methods (Kieras, 1988) for system analysis and for the determination of interface requirements for such a device.

Materials and methods

Terminal characterization was accomplished in two steps:
1. The tasks for which the terminal is used were specified with the Sub-Goal Template (SGT) method (Ormerod, 1998). In this method each goal starting with the top level is decomposed into sub-goals. Thereby, the smaller tasks required to perform the higher level goals are made explicit.
2. The actions required to perform each task with two different input methods (bar code and menu selection) were analyzed using GOMS modeling (Kieras, 1988). This method guides the development of models that consist of Goals (*what goals can be accomplished*), Operators (*what actions can be performed*), Methods (*what sequences of operators can be used to accomplish each goal*), and Selection rules (*when should each method be used*). The models provided quantitative predictions of the relative complexity of different tasks with different interfaces or input devices.

Results

The herbs' information system contains 23 actions (Shriki & Cohen, 2001). As an outcome of the analysis, it was determined that four of the data collection tasks that are associated with field work will be performed using the hand-held terminal:

- Data collection on field activities (*e.g.*, picking, weeding, pruning).
- Update of working hours of the individual workers.
- Update of workers' yield by activity.
- Update of pesticide application data.

The user interface includes several user profiles corresponding to the different functions on the farm (farmer, manager, workers). Profiles differ in their access privileges to functions (*e.g.*, workers cannot update working hours). The data for each activity can be fed into the terminal by means of a bar-code reader, menu selection or a combination of both (denoted as Combined). For each action and input method, the number of operations was counted. The input method for each action was chosen to minimize the total number of operations (Table 1). Minimum number of operations was required when inserting data using the combined method, yet similar results were achieved using only the barcode reader for two of the operations.

Table 4. Number of operation for each action and input method.

Input method \ action	Field task	Working hour	Workers yield	Dusting
Bar Code Reader	21	10	25	25
Menu Selection	21	11	25	26
Combined	20	10	24	25

Conclusion

Substantial amounts of data records are generated daily in typical herb farm. A prototype of a hand-held terminal was developed to automate the data collection process. The definition of the terminal was achieved using Cognitive Work Analysis methods. Four actions out of twenty-three actions defined by the information system were selected to be performed using the terminal. The data in each action can be fed either by bar-code reader, menu selection or a combination of both. The number of operations for each action and input method was counted. The preferred input method was found to be the combined method, which had only a slight advantage over the bar code method. Empirical evaluations of the different interfaces are currently under way to assess whether the complexity of the combined interface does not impede performance.

References

Kieras, D. E. 1988. Towards a practical GOMS model methodology for user interface design. In Helander M. (Ed.), Handbook of Human-Computer Interaction (pp. 135-158). Amsterdam: North-Holland Elsevier.

Ormerod, T. C. 1998. Using task analysis as a primary design method: the SGT approach. Cognitive Task Analysis, pp.181-199.

Shriki, D., and Cohen, D. 2001. Information system for a herbs farm. Final Project, Dept. of Industrial Engineering and Management, Ben Gurion University of the Negev.

Acknowledgments

The project was partially supported by the Chief Scientist Ministry of Agriculture fund No. 84014118

Spatial application of a time-discrete system to forecast the dynamic of a population of *Heterodera schachtii* with a Geographic Informationsystem (GIS)

G. Spickermann
Institute for Cartography and Geoinformation, University of Bonn, Meckenheimer Allee 172, 53115 Bonn, Germany
spickermann@ikg.uni-bonn.de

Abstract

A system to forecast the density of nematodes (*H. schachtii*) is integrated as a tool into a GIS by a VBA programming. The user gets spatially differentiated results concerning the expected density of varmints on an agricultural lot.

Keywords: *Heterodera schachtii*, nematode forecast, spatial application, geographic informationsystem

Introduction

With a given population of *H. schachtii*, the forecasting system „Nemaplot" will provide the data for the expected infestation of eggs and larva in 100g of soil after three years, depending on crop rotation and temperature change. First, the system treated the agricultural lots only as as a homogeneous area and did not supply site specific prognosis.

Results

With the aid of GIS, however, a heterogeneous dispersion of nematodes can be digitally recorded, too. Through this spatially differentiated registration of the nematode density, the forecast simulation can be made for every partial area of the field; i.e. the program is able to forecast the nematode infestation to be expected within three years for every small part of the lot and the density at every catch crop as well.

In addition, the dynamics of the nematodes' dispersion may also be included in the prognosis. Around each focus of nematodes, a buffer can be placed which exactly corresponds to the average speed of the nematodes' dispersion. Thus, the foci of infestation move or grow depending on the dynamics of the infestation. An important point to be considered here is the working direction of the plough which by transportation of soil significantly contributes to the dispersion. With a declining population, the dispersion can be simulated in the opposite direction, i.e. an inner buffer is generated which documents the decline of the nematodes.

The result of the simulation and the buffer generation will be a map which exactly marks the nematodes' position on the field and their forecasted density. Therefore, it is possible to control the partial areas, i.e. the separate nematode populations specifically; the program also allows to test various possibilities through the variation of crop rotation and to determine which combination of catch crop shows the most negative effect on the nematodes. Just by determining the nematode density at the outset and the temperature pattern, it is thus possible to decrease the nematode density permanently in advance.

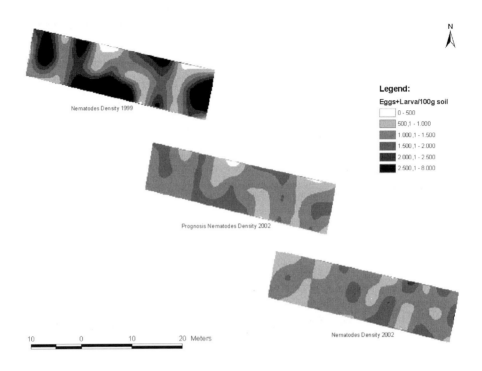

Figure 1. Forecast with „Neamplot".

Discussion

An exact validation of the prognosis by experiments that lead to an improvement of the model is needed, as well as more precise investigations of the spatial dispersion behavior of the nematodes.

Conclusion

This experiment very clearly shows that the combination of a forecast system and a GIS is useful and that by this method a surplus of information for precision farming can be obtained.

References

Razavi, A.H. (2002): ArcGIS Developer´s Guide For VBA. OnWord Press, Delmar, 188 pp.
Schmidt, K. (1992). Zeitdiskrete Modelle zur Vorhersage der Populationsdynamik des Rübenzystennematoden *Heterodera schachtii* (Schmidt) in Abhängigkeit von der Fruchtfolge und des Temperaturmusters (Time-Discrete Systems to Forecast the Dynamic of a Population of *Heterodera schachtii* subject to crop rotation and temperature change).University of Bonn, 142 pp.

Acknowledgements

This project was funded by the DFG (Deutsche Forschungsgemeinschaft).

Examining the EGNOS – reference signal under field conditions for track navigation

J.-M. Stuevel, H.-G. Schoen and B. Lehmann
Fachhochschule Osnabrueck, Oldenburger Landstrasse 24, 49090 Osnabrueck, Germany

Abstract

Computer based track navigation resulted in up to 3.78 m deviation from the expected lane. The EGNOS-correction signal did not improve the precision in all cases. Most probably the used bord-computer was not powerfull enough to process incoming signals. The processor performance of the bord computer must be high in order to keep track.

Keywords: Track navigation, GPS, EGNOS, ESTB

Introduction

Application of inputs with large scale machinery complicates a good alignment of the working tracks. Meanwhile GPS-based solutions are offered for track navigation. Furthermore the European Space Agency offers the GPS-correction signal EGNOS free of charge. In this investigation the precision increase in GPS positioning using the EGNOS reference signal (ESTB) compared to uncorrected satellite signals for field track navigation was observed?

Material and methods

For track navigation a CF-GPS receiver supplied by Satcon Systems was used. This receiver can be enabled to receive the ESTB-correction signal from the AOR-E satellite through a software configuration change. To examine the advantage of this reference signal for field track navigation, a Handera 330 (Palm-OS-Computer) and the respective navigation software PalmNav version 2.14 (Satcon Systems) was used. This software navigates a tractor along a virtual track. The track is calculated according to the working width and the tracks already covered. As alternative a parallel mode can be used in which all requested virtual tracks are calculated an depicted based on a reference track, which has to be covered at the beginning. The software *PalmNav* was used in connection with a *Lightbar*. This lightbar visualizes, if the tractor is leaving the virtual track. In this trial the uncorrected GPS-signal was compared with the ESTB-signal for track navigation. Each of these two factor levels was tested with 3 variations (treatments). The variations differed in the navigation support by flexible rods: treatment 1 had a rod at the starting- and at the endpoint of the track for better orientation, treatment 2 a rod only at the starting point and treatment 3 no additional navigation help at all. In the trial the three respective tracks were passed consecutively. This was repeated three times each in the GPS- and ESTB-mode.

The respective software PalmNav was used in the parallel-mode. The reference track to calculate all following tracks in the parallel mode was established with the help of marking rods. Rods were placed every 10 m along the reference track in order to achive a straight line. The trial distance of the tracks was 100 m, the turning section needed addititional 20 m space (Figure 1). The deviation of the navigated tracks from the expected based on the reference track was measured manualy. The expected tracks had a distance of 10, 20 and 30 m from the base line (reference track). Deviations were measured every 10 m in the track from navigated track middle to reference track middle. Toe width was 180 cm.

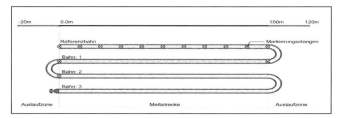

Figure 1. The trial design with the 3 variations (Bahn 1-3), the reference track (Referenzbahn) and the position of the rods (Markierungsstangen).

Results and discussion

The deviation of the navigated tracks from the expected lanes increased with decreasing support of marking rods. Without any additional support the real track deviated up to 3.78 m from the expected when navigating according to the computer calculated track with the help of uncorrected GPS-signals. Figure 2 visualizes the real tracks with and without ESTB-correction. Table 1 contains the measured distances of the real track from the reference track (base line) in the 3 replications. The ESTB-signal (EGNOS) increased precision in treatment 3 (navigation without the help of rods)significantly ($\alpha = 0.003$). Nevertheless absolute deviations were up to 2.17 m. In treatment 1 and 2 the uncorrected GPS signal was not significantly different from EGNOS.

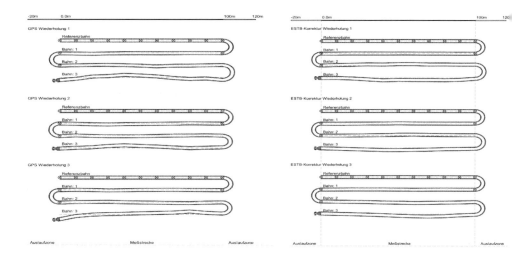

Figure 2. Visualized deviations of the real tracks from the expected lanes in the different treatments with and without EGNOS-signal (Wiederholung=replication).

The comparison of EGNOS with uncorrected NAVSTAR- GPS did not give a clear picture in this field trial. EGNOS-signals partly resulted in higher absolute deviations compared to NAVSTAR-uncorrected tracking. This suggests, that additional factors influence the precision of the track navigation. It was obvious, that the satellite signals were converted in visual position information with a certain time lag. This hindered the performance of the driver. This delay was not due to the data rate of the used 1 HZ-receiver. This time lag could also be observed with a 5 HZ-receiver. The limitation is presumably the processor performance of the Handera 330 processor.

It has to be stated, that the software solution *PalmNav* from *Satcon Systems* helps to navigate along already passed tracks and to optimise the alignment while working with high application widths. Track navigation along virtual lanes calculated by the computer needs a high processor performance.

Table 1. Deviations of the real tracks from the expected lanes in the different treatments with and without EGNOS-signal.

		GPS			EGNOS		
		Rep 1	Rep 2	Rep 3	Rep 1	Rep 2	Rep 3
Treatment 1	Mean	10.31	10.35	10.04	10.12	10.21	10.13
	Std.-deviation	0.19	0.29	0.14	0.07	0.12	0.08
	Min	10.08	9.93	9.87	10.06	10.02	10.04
	Max	10.68	10.89	10.35	10.27	10.41	10.24
	Difference	0.6	0.96	0.48	0.21	0.39	0.20
Treatment 2	Mean	20.16	20.06	20.13	19.14	19.64	20.48
	Std.-deviation	0.34	0.29	0.33	0.35	0.30	0.38
	Min	19.73	19.63	19.63	18.68	19.21	19.88
	Max	20.92	20.48	20.69	19.96	20.08	21.04
	Difference	1.19	0.85	1.06	1.28	0.87	1.16
Treatment 3	Mean	30.06	29.25	29.60	31.23	31.21	31.06
	Std.-deviation	1.44	0.82	1.69	0.7	0.48	0.79
	Min	27.86	27.79	27.53	30.16	30.46	29.81
	Max	32.11	30.43	33.78	32.05	31.87	32.17
	Difference	4.25	2.64	6.25	1.89	1.41	2.36

References

Butzmühlen C., 2002: personal communication, 7. April 2002, EGNOS-Project leader of the Institute of Flight navigation, TU Braunschweig

Gauthier L. et al. 2001: EGNOS: The first European implementation of GNSS-Project status overview
http://esamultimedia.esa.int/docs/egnos/estb/publivations/GNSS%202001/EGNOS%20 Status%20paper%20-%20Seville%20-%20May%202001.pdf

Assessing the EGNOS – correction signal for point positioning

J.-M. Stuevel, H.-G. Schoen and B. Lehmann
Fachhochschule Osnabrueck, Oldenburger Landstrasse 24, 49090 Osnabrueck, Germany

Abstract

A comparison of the standard-GPS-signals and the ESTB-corrected-data with reference to point positioning results in a significantly higher precision of the ESTB-positioning. The ESTB standard deviation values of all position parameters are lower compared to the uncorrected signals.

Keywords: GPS, EGNOS, ESTB

Introduction

Positioning systems are essential for precision farming. The received satellite signals should be corrected in order to get adequate precision. The european EGNOS correction signals are offered free of charge and cover the whole of Western Europe. Which precision increase can be encountered using the EGNOS correction signals in comparison with the uncorrected GPS satellite signals in point measurement?

Material and methods

To examine the precision of positioning with and without the EGNOS System Test Bed (ESTB) investigations were conducted at the Institute of flight navigation at the Technical University of Braunschweig (IFF). The task to be solved was to record the absolute precision of point positioning via satellite signals using a CF-GPS receiver. The standard signals of the NAVSTAR-GPS and the ESTB corrected signals (GPS / SBAS) were compared. The recording of the signals was done for a period of 24 hours each. The used CF-GPS receiver could read the ESTB correction signal (EGNOS) sent by the satellite AOR-E by altering the respective software configuration.

For point measurement at the reference station a laptop was used working with *Microsoft Windows 98*™ operating system. The recording of the satellite information was done by the software Microsoft™ *Hyper Terminal.*

Results and discussion

The results of the point measurements show some remarkable patterns in both modes: the horizontal error is obviously much more influenced by the north position error (ESTB: 95. percentile - 3.84 m) compared to the east position error (ESTB: 95. percentile - 1.36 m). The maximum of the ESTB north position error is remarkably high with 16.16 m and a range of 31.48 m. Surprising is also the huge range of the ESTB altitude position error of 73.30 m (95. percentile: 9.83 m). This error is also strongly influencing the total position error. An association of the precision of the measurement and the number of available satellites could not be verified. During the measurements the ESTB signals could not be received continously. This fact does not influence the shown results since only corrected signals were included in the analysis.

A comparison of standard deviations of the standard-GPS-measurements and the ESTB-corrected-data results in a higher precision of the ESTB-positioning. The ESTB standard deviation values of all position parameters are lower compared to the uncorrected signals. The ESTB-variances of the north-, east- and altitude positioning are significantly smaller than the respective values of the NAVSTAR-signals ($\alpha = 5\%$).

It has to be stated that the recording of the satellite signals in the two different modes was done at different days. Therefore additional factors of the recording conditions could have varied, for instance ionospheric influences and satellite constellations.

North-position error	uncorrected	ESTB-corrected	
North mean	0.135	0.123	
North standard deviation	3.143	2.385	
North 95 percentile	5.508	3.841	
North median	0.137	0.137	
North minimum	-18.568	-16.161North	maximum
	11.990	15.323	
East position error			
East mean	-0.136	-1.048	
East standard deviation	1.694	1.496	
East 95 percentile	2.492	1.360	
East median	0.001	-1.018	
East minimum	-7.358	-7.811	
East maximum	8.946	6.115	
Altitude position error			
Altitude mean	3.739	0.714	
Altitude standard deviation	6.092	5.659	
Altitude 95 percentile	14.326	9.826	
Altitude median	3.426	0.626	
Altitude minimum	-20.574	-34.374	
Altitude maximum	32.526	38.926	
Horizontal position error			
(North error)² + (East error)²			
Horizontal mean	3.001	2.506	
Horizontal standard deviation	1.944	1.661	
Horizontal 95 percentile	6.676	5.576	
Horizontal median	2.584	2.186	
Horizontal minimum	0.048	0.048	
Horizontal maximum	20.61116.415		
Total position error (3D-position)			
(North error)² + (East error)² + (vertical error)²			
Position mean	6.761	5.282	
Position standard deviation	4.262	3.697	
Position 95 percentile	15.259	12.266	
Position median	5.732	4.361	
Position minimum	0.088	0.055	
Position maximum	35.230	41.815	

References

Anonym 2001: EGNOS – European Geostationary Navigation Overlay Service; Programme Overview; Alcatel Space Industries, December 2001.

Butzmühlen et. al., 2000: Pegasus – Prototype Development for EGNOS Data Evaluation – First User Experiences with the EGNOS System Test Bed; Institute of Flight navigation, TU Braunschweig.

Gauthier L. et al. 2001: EGNOS: The first European implementation of GNSS-Project status overview
http://esamultimedia.esa.int/docs/egnos/estb/publivations/GNSS%202001/EGNOS%20Status%20paper%20-%20Seville%20-%20May%202001.pdf

Effect of nitrogen deficiency on laser induced chlorophyll fluorescence from adaxial and abaxial sides of sugar beet and wheat leaves

I. Tartachnyk[1], I. Rademacher[1] and R. Gäbler[2]
[1]Institut für Pflanzenbau, Lehrstuhl für Allgemeinen Pflanzenbau Universität Bonn, Katzenburgweg 5, D-53115 Bonn, Germany
[2]INVIVO GmbH, Gewerbepark 13, 86559 Adelzhausen, Germany
i.tartachnyk@uni-bonn.de

Abstract

The fluorescence ratios (F690/F740) from adaxial and abaxial leaf sides of sugar beet and winter wheat plants grown at sufficient and deficient nitrogen supply were compared. F685/F740 ratios from opposing leaf sides of winter wheat were identical under both nitrogen regimes. Grown under sufficient nitrogen sugar beet, in contrast, showed significantly higher F680/F740 ratios from the abaxial leaf side compared to the adaxial side. This effect persisted tendentiously under nitrogen deficiency.

Keywords: nitrogen deficiency, sugar beet, wheat, laser induced chlorophyll fluorescence.

Introduction

Laser Induced Fluorescence (LIF) is a fast and non-destructive method for the detection of plant stress like nitrogen deficiency. The principle of this method is based on the inverse correlation between leaf chlorophyll content and the ratio of fluorescence intensity at wavelength of 690 nm and 740 nm (F690/F740) (Cerovic, 1999, Lichtenthaler, 2000). The application of LIF for the on-line near–field detection of nitrogen status of vegetation faces the problem, that the sensed fluorescence signals can originate from different leaf sides and organs of the plants. The objective of our study was to compare the F690/F740 ratio from adaxial and abaxial leaf sides on sugar beet (*Beta vulgaris*) and winter wheat (*Triticum aestivum*) grown at sufficient and deficient nitrogen supply.

Materials and methods

The plants were raised from seeds under controlled conditions in growth chambers under optimal nitrogen supply (N+). At the beginning of the experiment nitrogen was withheld in the nitrogen deficiency treatment (N-). The photosystems of dark-adapted leaves were excited by He-Ne laser (633 nm) and LIF-emission spectra were recorded with a radio-spectrometer "FieldSpecTM UV/VNIR" to determine the F690/F740 ratios. Leaf nitrogen and chlorophyll contents were determined on the same leaves.

Results

Significant differences in the F690/F740 ratios of the tested species were observed two weeks after the induction of N-deficiency (Fig.1). At that time nitrogen content in sugar beet leaves declined to 39% and in winter wheat to 48% of the N+ treatment (data are not shown). In both nitrogen treatments the LIF-parameters from the adaxial and abaxial side of wheat leaves were identical. Sugar beet, in contrast, had significantly higher F680/F740 ratios on the abaxial leaf side compared to the adaxial side at sufficient nitrogen supply. Under nitrogen deficiency this effect persisted only tendentiously.

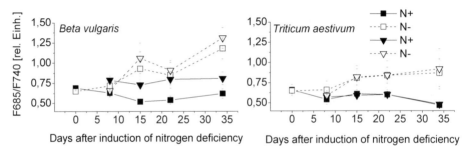

Figure 1. Mean fluorescent ratio (F685/F740) of dark-adapted leaves from sugar beet (*Beta vulgaris*) and wheat (*Triticum aestivum*) grown at sufficient (N+, closed symbols) and deficient (N-, open symbols) nitrogen supply. Measurements were taken from the upper leaf side (squares) and from the lower leaf side (triangles). Bars indicate ± 1 SD (n=10).

Discussion

The higher F685/F740 ratio of the lower leaf side compared to the upper leaf side of sugar beet is a result of the typical bifacial anatomy of these leaves, i. e. the lesser density of chloroplasts and consequently, chlorophyll content in the sponge parenchyma of the abaxial leaf side in compare to the dense palisade parenchyma of adaxial leaf side. The difference in F685/F740 ratios from two leaf sides of bifacial leaves indicate, that a fluorescence spectra can not be representative for the whole leaf, but only for the leaf layer that is faced to the excitation light. A partial penetration of excitation light through the leaf blade, and reabsorption or dissipation of fluorescence, emitted from deeper leaf layers also has to be considered (Cerovic, 1999). Besides a decrease in chlorophyll content, nitrogen deficiency reduces leaf thickness as well (Rademacher, Nelson, 2001). The deeper penetration of excitation light in the tissues of thinner leaf blades, that are formed under the nitrogen shortage, may explain the less pronounced difference in F685/F740 ratios from the upper and lower leaf sides of sugar beet under N- compared to N+ supply.

Conclusions

Filed application of LIF in crops with a typical bifacial leaf structure can complicate the diagnosis of nitrogen deficiency. The heterogeneity in canopy architecture and leaf structure has to be considered for the future applications of LIF-technique in the field

References

Cerovic, Z.G., Samson G., Morales, F., Tremblay, N., Moya, I., 1999. Ultraviolet-induced fluorescence for plant monitoring: present state and prospects. Agronomie 19 543-578.
Lichtenthaler, H.K. 2000. Multicolour fluorescence imaging of sugar beet leaves with different nitrogen status by flash lamp UV-excitation. Photosynthetica 38 (4) 539-551.
Rademacher, I.F., Nelson, C.J, 2001. Nitrogen effects on leaf anatomy within the intercalary meristems of Tall Fescue leaf blades. Annals of Botany 88 893-903.

Acknowledgments

The study was supported by the Deutsche Forschungsgemeinschaft (Graduirtenkolleg 722).

The soil-plant-atmosphere system evaluated by the state-space approach

L.C. Timm[1*], K. Reichardt[1], L. Fante-Júnior[2], E.P. Barbosa[3], J.C.M. Oliveira[4], O.O.S. Bacchi[1] and D. Dourado-Neto[5]
[1]Soil Physics Laboratory, CENA/USP, 13416-000, Piracicaba, SP, Brazil
[2]Methodist University, UNIMEP, 13400-901, Piracicaba, SP, Brazil
[3]Statistics Department, UNICAMP, 13083-970, Campinas, SP, Brazil
[4]Municipal University of Piracicaba, EEP, 13414-040, Piracicaba, SP, Brazil
[5]Crop Science Department, ESALQ/USP, 13418-900, Piracicaba, SP, Brazil
lctimm@esalq.usp.br

Abstract

Using as examples sugarcane and forage oat experiments a contrast is made between classical statistics methodologies and the more recent ones which include position coordinates, and allow a better interpretation of the data. It is concluded that these state-space models improve data analyses of data series related to a given soil-plant-atmosphere system.

Keywords: state-space approach, soil-plant-atmosphere system, spatial variability.

Introduction

State-space models are only recently used in the area of agronomy being potentially valuable tools to study the relations among variables linked to the system under study (Wendroth et al., 2001; Timm et al., 2000, 2003). Two variations of state-space are here presented: one emphasizes the evolution of the state system (Shumway & Stoffer, 2000) while the other, based on the bayesian formulation, emphasizes the evolution of the estimated observations (West & Harrison, 1997).

Material and methods

The state-space model is a combination of two systems of equations:

$$Y_j(x_i) = M_{jj}(x_i)Z_j(x_i) + v_{Yj}(x_i) \quad \text{and} \quad Z_j(x_i) = \phi_{jj}(x_i)Z_j(x_{i-1}) + u_{Zj}(x_i)$$

where the observation vector $Y_j(x_i)$ is generated dynamically as a function of the non observed state vector $Z_j(x_i)$. $Y_j(x_i)$ is related to $Z_j(x_i)$ through the observation matrix $M_{jj}(x_i)$ and by the observation error $v_{Yj}(x_i)$. $Z_j(x_i)$ at position i is related to the same vector at position i-1 through the state coefficient matrix $\phi_{jj}(x_i)$ (transition matrix) and an error associated to the state $u_{Zj}(x_i)$ with the structure of a first order autoregressive model. $v_j(x_i)$ and $u_j(x_i)$ are assumed normally distributed, independent, non correlated among themselves for both lags. For Shumway & Stoffer´s (2000) approach, the matrix M_{jj} is considered as unit (identity), indicating a greater emphasis on the equation of state evolution, the key element Z_j being formed by the filtered version (noise free) of the observable, obtained sequentially by Kalman filter updating. In the bayesian formulation of West & Harrison (1997), M_{jj} is formed by the observed regressors, Z_j by β dynamic (space-varying) regression coefficients, all at position x_i. $\beta_j(x_{i-1})$ is the state-vector formed by the dynamic (space-varying) regression coefficients at position x_{i-1}, and u_j is the system perturbation vector. The presentation of these two different procedures is made for a sugarcane field experiment of Piracicaba (Timm et al., 2003), and an oat field experiment of São Carlos (Timm et al., 2000), both in SP, Brazil.

Results and discussion

Ignoring the locations of the data, we find that not more than 55% of the variance of the soil water content (SWC) data is explained by classical linear and multiple regression using any combination of the soil organic matter (OM), clay content (CC), and aggregate stability (AS) data sets. These classical analyses are based on the assumption that each data set manifests, respectively, a constant mean along the entire transect, and ignore their local spatial crosscorrelations within the transect. From Shumway & Stoffer's approach, considering the locations of the observations, we find that 91% of the variance of SWC data is explained by the state-space equation using CC and AS data. From West & Harrison's approach, the classical static regression was applied and no more than 60% (Figure 1A) of the variance of the root length per unit of soil volume (DRC) data is explained by classical multiple regression analysis using soil total porosity (STP) observations, the only correlated variable with DRC. Considering the state-space model which is local with variable β coefficients, we now find that 99.0 % of the variance of DRC data is explained using STP observations, indicating a significant improvement in relation to the static model (Figure 1B).

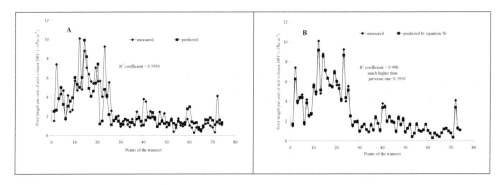

Figure 1. (A) DRC values by classical multiple regression analysis (R^2=0.595); (B) DRC values by state-space model (dynamic model, R^2 = 0.990).

From the results, we conclude that the state-space models improve data analyses and are recommended to analyze space data series related to the soil-plant-atmosphere system.

References

Shumway, R.H. and Stoffer, D. S. 2000. Time Series Analysis and Its Applications. New York: Springer, New York, USA.

Timm, L.C., Fante Junior, L., Barbosa, E.P., Reichardt, K. and Bacchi, O.O.S. 2000. A study of the interaction soil – plant using state-space approach. Scientia Agricola, 57:751-760.

Timm, L.C., Reichardt, K., Oliveira, J.C.M., Cassaro, F.A.M., Tominaga, T.T., Bacchi, O.O.S. and Dourado-Neto, D. 2003. Sugarcane production evaluated by the state-space approach. Journal of Hydrology, 272, 226-237.

Wendroth, O., Jürschik, P., Kersebaum, K.C., Reuter, H., van Kessel, C. and Nielsen, D.R. 2001. Identifying, understanding, and describing spatial processes in agricultural landscapes – four case studies. Soil & Tillage Research, 58: 113-127.

West, M. and Harrison, J. 1997. Bayesian Forecasting and Dynamic Models. Springer-Verlag, London, 2nd ed. 681p.

Sensor system for measuring trafficability on wet grassland sites

R. Tölle[1], A. Prochnow[2], A. Sasse[1] and J. Hahn[1]

[1]Humboldt-University of Berlin, Institute of Crop Science, Department of Agricultural Engineering, Philippstr. 13, 10115 Berlin, Germany
[2]Institute of Agricultural Engineering, Department of Technology Assessment and Substance Flow, Max-Eyth-Allee 100, 14469 Potsdam, Germany
rainer.toelle@agrar.hu-berlin.de

Abstract

A Sensor system is presented as a functional prototype, which will enable to record the trafficability on wet grassland with high spatial density. Therefore the shear strength of the grass mat is determined using a rotatable disc device pulled through the soil. The outcome will be maps of trafficability for site specific management.

Keywords: Sensor, shear strength, trafficability, grassland

Introduction

The results of a project finished in 1999 show that trafficability on wet grassland sites can be predicted by measuring the horizontal shear strength of the ground (Tölle et al., 2000). A shear vane is employed to establish the load-bearing capacity. This is dependent on the type of biotope, vegetation density and soil moisture. These ground properties are taken as a base while calculating current trafficablity situations applying the manually established shear values. The typical spatial distribution of soil and vegetation types on wetlands require at least 25 sample points per hectare to reach sufficient accuracy. In doing this, a reliability of 80% is reached for the entire field, meaning that the grass sod does not break during operation. If knowledge about spatially related load-bearing capacity is aimed to result in site specific management, a different technical form of sensing needs to be developed. Data collection has to be possible during vehicle moving. All measurements have to be taken in one working cycle. This guarantees little effort and the unproblematic spatial allocation of different measurement categories. The shear strength of the topsoil should be taken, but also soil properties like moisture and texture as well as composition and density of vegetation to predict trafficability at a later date. None of these properties can be measured directly. Adequate substitutes have to be found.

Sensor system

Technical realisation:
1. Shear strength of the ground; The shear strength of the grass mat is determined using a rotatable disc device pulled through the soil. A lever transfers the force to a transducer. To determine the actual shear values, further establishment of penetration depth and horizontal speed is needed. This is achieved by recording the angle between the frame carrying the disk and an additional device following the surface of the ground directly. The velocity is measured by a supporting wheel.
2. Soil moisture; The measurement of electrical soil conductivity gives evidence for soil moisture. This is determined by establishing the electrical resistance of the soil between the disk and an electrode following in its track.

3. Soil texture; The evaluation of acoustical noise recorded on the electrode following in the rill made by the disc is used to assess soil texture. This is done by analysing the frequency-related distribution of energy of the recorded oscillations.

4. Type of vegetation and density; A photoelectric sensor fixed on the frame of the device detects the plants passing through. The signal provides information about the height and density of vegetation.

5. Spatial position; The position is determined using a DGPS.

Well-established methods were used for comparison with the measured values resulting from the sensor device. The shear vane was used to establish shear strength, soils moisture was measured using the TDR method and soil cores were taken to determine the amount of organic matter in the soil. The type and density of vegetation was assessed by observation. During test operation a 360-degree prism with automatically following tachymeter was employed for optical positioning to record the exact position and micro-relief. Marks are also positioned at the test sites to be recorded by video.

Results

The efficiency of the sensor as well as its accuracy was demonstrated using a test track.

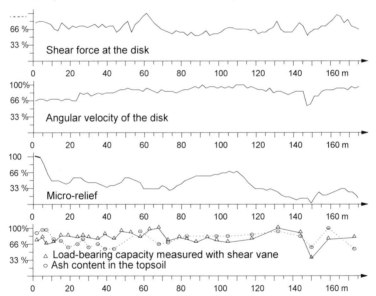

Figure 1. Relationship between records of selected measurement readings.

Conclusions

The principle of the Sensor provides reproducible measurements. The results have to be validated by tests on different sites.

References

Tölle, R., Prochnow, A. and Kraschinski, S. 2000. Measuring Techniques for the Assessment of the Trafficability of Fen Grasslands. Agrartechnische Forschung, Bd. 6, H. 3, S. 54-61

Leaf reflectance changes induced in wheat by cereal powdery mildew (*Erysiphe graminis* f. sp. *tritici*)

S. Turmes, S. Graeff, and W. Claupein
Institute of Crop Production and Grassland Research (340), University of Hohenheim, Fruwirthstr. 23, 70599 Stuttgart, Germany
graeff@uni-hohenheim.de

Abstract

Reflectance changes induced in wheat leaves by powdery mildew were evaluated throughout the visible and the near-infrared spectra to identify important wavelength ranges for the early identification of an infection. Wheat plants were inoculated at the 4th leave stage with spore concentrations of 0 (S0), 2500 (S1), 5000 (S2) and 10000 (S3) spores per leaf. Reflectance of the 4th leaf was recorded during a 26 d period. Reflectance of infected leaves changed significantly 7 d after inoculation in the wavelength ranges 510_{780} nm and 516_{1300} nm. In combination with climate based models reflectance measurements might be a useful tool for the identification of powdery mildew.

Keywords: powdery mildew, reflectance measurements, wheat.

Introduction

The effects of diseases in agricultural crops are of major economic importance in terms of reduced yield. Various attempts using multispectral sensors and satellite images in detection of diseases or plant stress factors have been made. The use of remote sensing techniques, based on measurements of reflectance in the visible and near-infrared region, potentially allows for the early detection of disease outbreaks. However, detection of plant diseases by means of spectral changes may be difficult, because diseases may produce several and similar symptoms. So far, little is known in detail about the effects of plant diseases on the spectral reflectance properties of plants. Thus, the objectives of this study were i) to evaluate reflectance changes of wheat leaves due to an infection by powdery mildew, and ii) to determine important wavelength ranges in the visible and infrared spectra, responsive to a powdery mildew infection.

Material and methods

Summer wheat (*Triticum aestivum* L.) cv. Thasos was selected for inoculation. Wheat plants were inoculated at the 4th leaf stage with four different spore concentrations of *Erysiphe graminis* f. sp. *tritici*. Spore concentrations were 0 (S0 = control), 2500 (S1), 5000 (S2) and 10000 (S3) spores per leaf. All experiments were conducted in a growth chamber. The relative humidity was kept at 70 % ± 5 %, the average temperature was set to 20 °C ± 5 °C. Reflectance measurements were carried out 7, 12, 18, 22, and 26 d after inoculation with a digital LEICA S1 Pro camera in conjunction with a constant light source (HMI 21 W/D   10 W m^{-2}, Sachtler, Germany) of total daylight spectra. Long-pass filters active at wavelengths longer than 380 nm, 490 nm, 510 nm, 516 nm, 540 nm, and 600 nm were used to split the total daylight spectra into various wavelength ranges. Scans were analyzed wtihin the L*a*b*-color system (CIE, 1986) using ADOBE® Photoshop 5.0. Each plant was harvested after scanning and the dry matter was determined. Analysis of variance (ANOVA) was performed on all crop and reflectance data using the general procedures of Sigma Stat 2.0 (Jandel Scientific, San Rafael, CA). Tukey tests were carried out for comparison of means.

Results and discussion

The results showed that the reflectance of uninfected leaves (control) changed very little during the experiment (Figure 1). The reflectance parameter b* of infected leaves changed significantly in the visible and the near-infrared wavelength ranges 7 d after the inoculation. The b*-parameter increased in all treatments S1-S3 over time and spore concentration. Over the duration of the experiment the wavelength ranges 510_{780} nm and 516_{1300} nm were selected for the identification of powdery mildew infection, because leaf reflectance was significantly affected in an early stage of disease.

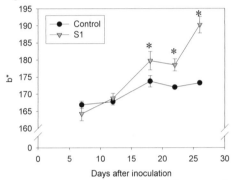

Figure 1. Change of the reflectance parameter b* of the control and the S1 treatment in the wavelength range 510_{780} nm over the time of the experiment.

Lorentzen and Jensen (1989) showed that reflectance of barley leaves infected with powdery mildew changed significantly 6 d after inoculation. They suggested that the degradation of chlorophylls may lead to a change of reflectance in the visible spectra. Further on, morphological, anatomical and physiological changes in leaves may cause reflectance changes in the near-infrared region. Physiological changes in leafs or degradation of chlorophylls might be a reason for the determined reflectance changes in this study. Thus, disease detection by means of reflectance spectra may be possible when changes in pigments or cell constitution create a detectable qualitative or quantitative difference in the spectral reflectance between healthy and diseased plants.

Conclusions

The results indicate that disease detection by means of reflectance spectra may be possible in an early stage of infection by the use of specific wavelength ranges. In combination with climate based models reflectance measurements might be a useful tool for the identification of powdery mildew. Further studies have to be carried out, to discriminate powdery mildew infection from other plant stress factors by means of reflectance measurements.

References

CIE 1986. Colorimetry. 2nd edn., Publication CIE No. 15.2, Central Bureau of the Commission Internationale de L'Eclairage, Vienna.
Lorentzen, B., Jensen, A. 1989. Changes in leaf spectral properties induced in barley by cereal powdery mildew. Remote Sensing of Environment 27 201-209.

The PreciSpray prototype sprayer, first results of spray deposition and distribution in an apple orchard

J.C. van de Zande[1], J.M.G.P. Michielsen[1], V.T.J.M. Achten[1], H.A.J. Porskamp[1], H. Stallinga[1], R.P. van Zuydam[1], M. Wenneker[2] and B. Heijne[2]

[1]*Institute of Agricultural and Environmental Engineering IMAG, P.O. Box 43, 6700AA Wageningen, The Netherlands*
[2]*Applied Plant Research PPO-Fruit, P.O. Box 200, 6670 AE Zetten, The Netherlands*
jan.vandezande@wur.nl

Abstract

Matching spray volume to tree sizes and shapes can reduce chemical application, thus reducing operational costs and environmental pollution. A tree-specific variable volume precision orchard sprayer, guided by foliage shape and volume (canopy density sprayer; CDS) is developed. The spray and air volume is adjusted to tree variation of the foliage shapes and volumes of mapped trees. Tree mapping is done with stereoscopic aerial photography. Position of the sprayer in the orchard is linked to the mapped tree information from a GIS by RTK-DGPS. Spray deposition and distribution measurements with an artificial tree in the laboratory are used to set-up the prototype sprayer in an orchard situation.

Keywords: spray technique, spray deposition, orchards, fruit growing, canopy density sprayer

Introduction

In the PreciSpray project (Meron et al, 2003) a canopy density adjusted segmented cross-flow orchard sprayer equipped with a canopy contour guidance system was developed (Achten et al, 2003). The cross-flow element of the sprayer was equipped with 5 segments of 0.50 m height on top of each other. The segments are capable of moving towards the tree canopy and maintain a constant distance to the outer contour lines of the tree canopy (fig. 1). The sprayer makes use of RTK-DGPS information to know its current position in an orchard. Based on canopy contour information obtained from stereoscopic aerial photography (Shimborsky, 2003), also information in the spray computer is loaded about the place of individual trees and the canopy volume coinciding with the height of the vertical segments of the cross-flow elements next to the sprayer in the tree row. Canopy contour line information is used as input for the sprayer segment movements around the trees on different heights. Spray volume is adjusted according to canopy density information per segment height of individual trees (Van de Zande et al, 2002). Switching the 3 nozzles for each segment on or off changes spray volume. Nozzles were switched on/off depending on the information input from the photographs and the actual position (the view in front of the segment) of the sprayer with RTK-DGPS. Results are presented for initial set-up measurements

Materials and methods

The development of the sprayer started with the testing of the sprayer segment in the laboratory on a spray track spraying an artificial apple tree (Van de Zande et al, 2001) placed inside a conditioned spray chamber (15mx6mx4m,lwh). The effect of sprayer settings: driving speed, air speed, air outlet and spray volume on spray deposition were measured spraying collectors placed in free air at distances of 0.25, 0.50, 0.75 and 1.50 m from the nozzle, behind a single and a double leaf wall and in and behind an artificial apple tree.

591

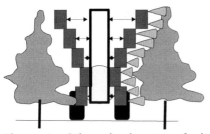

Figure 1. Schematic lay-out of the PreciSpray prototype orchard sprayer.

The optimal settings for different canopy densities were used for the full-sized prototype sprayer to evaluate spray deposition in the orchard. Outdoor measurements were performed on spray deposition with the RTK-DGPS guided segmented cross-flow orchard sprayer. Spray track was set up on a concrete track with additional position information of the place of the artificial tree. In the orchard also measurements on spray distribution were performed in two rows of Elstar and two rows of Jonagold dwarf apple trees planted alternating in an orchard. Spray deposition was measured using a fluorescent tracer (BSF) in individual trees on collectors (chromatography paper and water sensitive cards) placed on standardised places.

Discussion

From the laboratory measurements it was concluded that fan speed and air outlet settings define spray deposition in the tree to the most and can be used to minimise spraying through tree canopy and therefore spray drift potential. Commonly used air speeds of orchard sprayers can be lowered to maximise spray deposition in canopy and to reduce spray passing through canopy. Expressed as percentage of applied volume rate the drift potential can be below the 1% level. A full range of sprayer settings exists to control spray flow depending on canopy structure and density. These data are used to develop the control algorithm used in the Canopy Density Sprayer.

For the orchard situation results are also presented for the spray deposition and distribution of the prototype PreciSpray sprayer in comparison with a standard cross-flow sprayer.

Acknowledgement

This research was sponsored by the Dutch Ministry of Agriculture, Nature Management and Fisheries and the EU project PRECISPRAY (EU-QLK5-1999-1630).

References

Achten, V.T.J.M., R.P. van Zuydam, J.C. van de Zande & P.G. Andersen, 2003. Development of a canopy density adjusted segmented cross-flow orchard sprayer equipped with a canopy contour guidance system (PreciSpray). Proceedings of the 4th European Conference on Precision Agriculture, Berlin.

Meron, M., J.C. van de Zande, R.P. van Zuydam, B. Heijne, M. Shragai, J. Liberman, A. Hetzroni, P.G. Andersen & E. Shimborsky, 2003. Tree shape and foliage volume guided precision orchard sprayer - the PreciSpray FP5 project. Proceedings of the 4th European Conference on Precision Agriculture, Berlin.

Shimborsky, E., 2003. Digital tree shape mapping and its applications. Proceedings of the 4th European Conference on Precision Agriculture, Berlin.

Zande, J.C. van de, A. Barendregt, J.M.G.P. Michielsen & H. Stallinga, 2002. Effect of sprayer settings on spray distribution and drift potential when spraying dwarf apple trees. Paper presented at the 2002 ASAE Annual International meeting/ CIGR XVth World Congress, Chicago, ASAE-Paper No. 2002-1036. 10pp.

Modelling of the impact of variable soil moisture and its composition on the soil properties, energy consumption an quality of tillage

A. Vilde, A. Rucins and S. Cesnieks
Latvia University of Agriculture, Institute of Agricultural Engineering, Ulbroka Research Centre, 1 Institute Street, Ulbroka LV-2130, Latvia
uzc@delfi.lv, arucins@delfi.lv

Abstract

Two of the many factors that influence soil properties and energy consumption are soil moisture and its composition. The correlation derived from our theoretical and experimental research allow to evaluate soil physical and mechanical properties such as density, hardness, friction, adhesion and to assess the draft resistance of soil tillage machines (ploughs, cultivators) depending on the value of the moisture of soil and its composition, as well as on the design parameters and operating speed of the machines, and to determine the optimum range of soil moisture when the energy capacity of tillage is the lowest.

Keywords: soil properties, theoretical correlations, energy capacity of tillage.

Introduction

The specific draft resistance of the tillage machines determines energy consumption for soil tillage. It is known from our previous investigation (Vilde, 1999*)* that the draft resistance of the tillage machines depends on such soil properties as its hardness, density, friction and adhesion. These properties and the tillage quality depend mainly on the soil mechanical composition and moisture (Vilde, 2001). However, there are no correlations that would enable to determine the draft resistance of the tillage machines (ploughs, cultivators), depending on the soil moisture and compositions.

The purpose of the investigation is to estimate the forces acting upon the surfaces of the soil tillage machines as well as their draft resistance and tillage quality in dependence of the soil moisture, mechanical composition and working speed.

Materials and methods

The objects of the research are the draft resistance of the tillage machines, and the tillage quality depending on their design parameters, as well as the soil moisture and composition. On the basis of the previous investigations (Vilde, 1999) a computer algorithm has been worked out for the simulation of the forces exerted by soil upon the operating (lifting and supporting) surfaces of the tillage machines, and the draft resistance caused by these forces. The tillage quality is estimated by testing.

Results

According to our earlier studies (Vilde, 1999), the draft resistance R_x of the tillage machines is determined by the share cutting resistance R_{Px}, the resistance caused by weight R_{Gx} of the strip lifted, by the inertia forces R_{Jx}, by soil adhesion R_{Ax} and by weight R_{Qx} of the machine itself:

$$R_x = \sum R_{ix} = R_{Px} + R_{Gx} + R_{Jx} + R_{Ax} + R_{Qx} \tag{1}$$

The vertical reaction R_z and the lateral reaction R_y of the operating part are defined by corresponding partial reactions:

$$R_z = \Sigma R_{iz}; \qquad Ry = \Sigma R_{iy} \qquad (2, 3)$$

The total draft resistance R_x of the operating part is composed of the resistance of the lifting (share-mouldboard) surface R'_x and the resistance of the supporting (lower and lateral) surfaces R''_x:

$$R_x = R'_x + R''_x = \Sigma R'_{ix} + f_0 \left(\Sigma R_{iz} + \Sigma R_{iy} + p_{Axy}\, S_{xy} + p_{Axz}\, S_{xz} \right), \quad (4)$$

where: f_0 is the coefficient of soil friction along the working and supporting surfaces of the operating part; p_{Axy} and p_{Axz} - specific adhesion forces, respectively, acting upon the lower and the lateral supporting surfaces of the operating part; S_{xy} and S_{xz} - the surface areas, respectively, of the lower and the lateral supporting surfaces of the operating part.

Cutting resistance R'_{Px} is proportional to soil hardness ρ_0 and the share edge surface area ω:

$$R'_{Px} = k_p \rho_0\, \omega = k_p \rho_0\, ib, \qquad (5)$$

where: k_p is a coefficient involving the impact caused by the shape of the share edge frontal surface; i and b - the edge thickness and width.

The resistance caused by the weight and inertia forces of the lifted strip of soil is proportional to soil density and its friction coefficient but *the resistance caused by soil adhesion* is proportional to the specific adhesion force between soil and the surface of the operating part.

The draft resistance caused by the weight of the machine itself is proportional to the friction coefficient.

The total draft resistance of the machine depends on its component resistances. The maximum resistance occurs in caked clay soils, the minimum – in sandy soils. The moisture increase in clay soils to 14...18 %, leads to their decreased resistance, yet at higher moisture it rises again. The best tillage quality is obtained at the optimum moisture and higher speed.

Conclusions

1. The derived analytical correlations allow assessing the draft resistance of soil tillage machines (ploughs, cultivators) depending on the value of soil moisture and composition, as well as on their design parameters and their working speed.
2. The correlations obtained allow determining the optimal soil moisture range when the tillage energy capacity is the lowest. In clay soils it varies from 14 to 18 %. At this moisture level the tillage quality (degree of loosening) is the best too.

References

Vilde A. 1999. Dynamics of the soil tillage machine operating parts and their elements. In: Proceedings of the Latvia University of Agriculture, Vol.1 (295). Jelgava, Latvia, pp. 36-44.

Vilde A. 2001. Physical and mechanical properties of soil affecting energy capacity of its tillage. – Proceedings of the 1st International Conference of BSB of ISTRO "Modern ways of soil tillage and assessment of soil compaction and seedbed quality" – 21-24 August 2001. - EAU, Tartu, Estonia, pp.97-106.

Soil, crop and yield sensing to analyze within-field patterns in a winter wheat field

Els Vrindts[1], Mieke Reyniers[1], Josse De Baerdemaeker[1], Paul Darius[1], Marc Gilot[2], Youssef Sadaoui[2], Marc Frankinet[2,] Bernard Hanquet[3] and Marie-France Destain[3]

[1]*Laboratory of AgroProcessing and Machinery, Katholieke Universiteit Leuven, Kasteelpark Arenberg 30, Leuven, 3001, Belgium*
[2]*Centre de Recherches Agronomiques de Gembloux, Département Production Végétale, Rue du Bordia 4, Gembloux, 5030, Belgium*
[3]*Laboratoire de Géotechnique Environnementale, Unité de Mécanique et Constructions, Faculté Universitaire des Sciences Agronomiques de Gembloux, Passage des déportés 2, Gembloux, 5030, Belgium*

Abstract

In a Belgian precision agriculture project, the data collection on soil and crop and data processing for decision support is investigated. The relations between soil, crop and yield measurements are studied. First results from 2001 and 2002 on one of the fields with winter wheat in Sauveniere are presented. The crop growth and yield are compared to soil variability with maps (spatial patterns) and with correlation analysis. The measured soil parameters were soil texture and chemical components of soil samples of the top 30 cm and apparent soil electrical conductivity over the field. The lower, more sandy soils in the field had a higher crop growth and yield. Possibly, the soils with more clay had reduced crop (root) growth in the very wet conditions in 2001. The soil parameters did not have high correlation to crop growth and yield in 2001, indicating that the measured soil data were not representative of the growing conditions, or that other, non-measured soil parameters had big influence on the crop. In 2002, part of the crop growth and yield pattern was repeated, but an eyespot infection (*Pseudocercosporella herpotrichoides* (Fron) Deighton) greatly influenced the yield results, which were lower than in 2001.

Material and methods

The field of approximately 7 ha, has mainly deep silty soils, with alluvial soils to the northwest side and southern side. Soil samples were taken in spring 2001 from the top 30 cm layer in a 25 by 25 m grid and analysed for texture and chemical components and organic matter content, acidity, total nitrogen content and carbon-nitrogen ratio. Soil, optical crop growth and yield measurements are acquired together with DGPS-coordinates to localise and map all data. Apparent soil electrical conductivity (ECa) was measured with the Geonics EM38 in vertical mode (1.5 m working depth) in April and September 2001, and again in 2002. Soil electrical conductivity describes soil variability and is highly correlated to clay and water content (Hanquet *et al.*, 2002). Crop reflectance in 5 bands (blue, green, red, near infra red (850nm) and middle infra red (1650 nm) was measured with the Cropscan radiometer MSR5 with a field of view of approximately 0.7m diameter, in May 2001 and from end March to June 18 in 2002. The Cropscan was fixed on a tractor-mounted patform for measurement along the tramlines (every 9m) in the field. The normalized difference vegetation index (NDVIg) based on the NIR and green band was chosen to describe crop status and growth in general, since it had better overall correlation to soil parameters and yield than single reflectance bands. The definition of the NDVIg is:

$NDVIg = \dfrac{(R_{NIR} - R_G)}{(R_{NIR} + R_G)}$, with R_{NIR} the reflectance in the NIR band (850 nm) and R_G the reflectance in the green band (560nm).

Grain and straw yield was measured with a New Holland combine harvester type CX820-A1, equipped with sensors for grain and straw yield and grain moisture (Maertens *et al.*, 2000).

All data layers were brought into a GIS (Arview, ESRI Inc.) and mapped to a 6m grid (1599 points) with inverse distance interpolation, using 12 nearest neighbors and power 2. The data that belonged to a wet spot in the field were excluded, because they would distort the analysis result. A correlation matrix was calculated for soil, yield and crop parameters in 2001 and optical crop measurements of 2002.

Results and discussion

Some of the soil texture paremeters are correlated to height, with coarser textures on lower areas (alluvial soils). Clay and fine silt content are positively correlated to calcium and potassium, acidity, ECa, and to a lesser degree to magnesium. Coarse silt and sand are correlated to phosphate and cabon-nitrogen ratio. Grain and straw yield in 2001 have a correlation of +0.45 and are positively correlated to soil parameters soluble phosphate, coarse silt and fine sand, and negatively correlated to ECa, fine silt, clay and potassium (absolute correlation values from 0.34 to 0.47 for grain yield). NDVI in May in both years are correlated to each other (0.30), indicating a reoccuring pattern in crop growth, also visible in the NDVIg maps. NDVIg is correlated to acidity and fine silt in 2001, and to organic matter content and magnesium in 2002, but the correlation values are low (0.2-0.35). The change in NDVI in consecutive measurements is a measure of crop growth and has slightly better correlation to soil texture and chemical components. Grain yield in 2001 is correlated to the topograph related variations in soil texture, and to crop measurements in May, indicating that these parameters could be used in site-specific yield management. The soil measurements of the top 30 cm were not able to explain much of the yield variability. Other soil parameters, like compaction, texture in deeper layers or water condition of the soil profile were probably more important to crop growth and yield. In 2002, part of the yield pattern was repeated, but eyespot disease greatly influenced the yield results, which were overall lower than in 2001

References

Maertens K, De Baerdemaeker J, Reyns P, Missotten B (2000). On-line grain and straw yield measurements during harvest. In: Proc. of 28th International Symposium on Agricultural Engineering, 1-4 February 2000, Opatija, Croatia, pp 25-32

Hanquet B, Frankinet M, Parez V, Destain M-F (2002) Mapping within-field soil varability for precision agriculture using electromagnetic induction. In: AgEng Budapest 2002, EurAgEng, EurAgEngPaper No. 02-PA-004.

Aknowledgement

The financial support of the Belgian Ministry of Small Trade and Agriculture and IWT is gratefully acknowledged.

Development of an in-the-row weeder for horticultural crops

S. Wahlen and H. Ramon
K.U.Leuven, Laboratory for Agricultural Machinery and Processing, Kasteelpark Arenberg 30, B-3001 Heverlee, Belgium
stijn.wahlen@agr.kuleuven.ac.be

Abstract

A new mechanical technique for removing weeds growing *in* the crop rows of transplanted horticultural plants is described. The method benefits of the head start of the crop when it is transplanted in a clean soil. The difference in plant size allows discrimination between crop and weed. The in-the-row weeder consists of a near infrared light source, a module to detect the light reflected by the plant leaves, a computer containing an algorithm for calculation of the exact position of the crop plants and an actuator module performing a weeding action.

Keywords: plant detection, weed control

Introduction

Environmental concerns, increasing herbicide resistance, and a growing demand for organically grown plants urge the revaluation of mechanical weed control. In this context, the mechanical control of weeds growing *in* the crop row is of special interest. The currently used techniques for in-the-row weeding require that the crop plants are more strongly anchored in the soil than the weed plants (finger, brush or torsion weeder), or that the crop plants have a significant larger size compared to the weeds (soil moving hoe). Therefore, these methods are only applicable in a later development stage of the crop. In addition, in leaf crops (*e.g.* lettuce, endive, Chinese cabbage), damaging of the leaves should be avoided. As these horticultural crops are normally sown in peat clods and then transplanted, it takes some weeks before they are strongly anchored into the soil. Therefore, they are easily loosened by the action of the weeding equipment, causing dehydration of the crop plants. In order to find a solution for the above-mentioned problems, a new mechanical technique that removes weeds growing *in* the crop rows of transplanted horticultural plants, has been designed. The construction of this in-the-row weeder and its performance in the weeding of lettuce plants is described.

Construction

The detection is performed in the near infrared (NIR) region of the light spectrum. In comparison to soil, the NIR light is strongly reflected by plant leaves. The detection unit consists of NIR Light Emitting Diodes (LEDs) as light source and a lens system with a photodiode (Siemens) as detector. The LEDs (Siemens, peak wavelength of 880 nm) are pulsed with a 25 kHz square wave, which is also used as reference signal for a lock-in amplifier (Bentham 223). The lens is an aspherical condenser lens with a diameter of 75 mm and an effective focal length of 50 mm. The LEDs are placed symmetrically around the outline of the lens. The photodiode is placed at the focus of the lens. The detector is positioned 35 cm above soil surface. The electrical current of the photodiode is converted to voltage and fed into a lock-in amplifier. Because the light source is modulated and only the modulated part of the photodiode signal is amplified by a frequency and phase dependent amplifier (lock-in), ambient light effects are suppressed. The output of the lock-in amplifier is read into a computer with a data acquisition (DAQ) card (AT-MIO-16E-2, National Instruments).

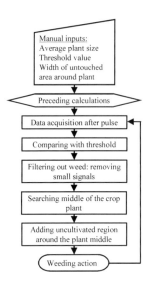

Manual inputs:
Average plant size
Threshold value
Width of untouched
area around plant

↓

Preceding calculations

↓

Data acquisition after pulse

↓

Comparing with threshold

↓

Filtering out weed: removing
small signals

↓

Searching middle of the crop
plant

↓

Adding uncultivated region
around the plant middle

↓

Weeding action

Figure 1. Algorithm

The analog/digital conversions are controlled by a pulse signal delivered by an encoder. This encoder is connected to a wheel and ensures that sampling occurs every time a fixed distance is covered. After sampling, calculations on the data are performed and the output channel is updated. A simplified scheme of the used algorithm is shown in Figure 1. The identification of a plant as "crop" is based on the difference in size of crop and weed. The actuator part consists of a knife that is pulled continuously through the soil at a depth of 2 cm and is rotated back over about 20 degrees when passing a crop plant. This rotation is performed by means of a pneumatic cylinder, acting as a lever. The airflow is delivered by a compressor and directed by a solenoid valve, which is connected to the DAQ card output channel via a transistor. The detection and actuator unit are mounted on a main beam attached to a frame. This frame also carries the measurement equipment and is pushed manually over rails.

Between the rails, lettuce plants were placed on a line right under the detector with an in-row distance of 30 cm. Weed plants were distributed randomly between the lettuce plants. Experiments were conducted on lettuce plants of 6, 7, 8 and 9 weeks age.

Performance tests

After optimization of the manual inputs, the correct detection rates ranged from 90 percent for the plants of 6 weeks to 100 percent for the plants of 8 and 9 weeks. The plants of 6 weeks had not yet many leaves: a bad orientation of the biggest leaves gave a low reflection signal and no or an inaccurate detection of the plant middle. The plants of 9 weeks were well developed with many leaves arranged in a symmetrical way around the plant middle. These plants gave a high reflection signal and consequently a good detection. However, weeds with many leaves were detected as crop plants. Thus, only if there exists a reasonable difference in size between weed and crop plant a good identification is guaranteed. Because of the rotating movement of the knife, some area behind each plant was left uncultivated. The detection algorithm and the knife were responding fast: driving speeds up to 1.4 m/s were possible.

Conclusions

A new weeding equipment consisting of cheap optics has been developed for in-the-row weeding of transplanted horticultural crops. If the difference in size between crop and weed plants is not too small, an almost perfect recognition of the crops was possible. Further experiments will be devoted to the determination of the optimal manual inputs, and the actuator part will be adapted to cover a larger area closer to the crop plant. Moreover, the whole set-up will be tested in real field conditions.

A real-time, embedded, weed-detection and spray-control system

N. Wang[1], N. Zhang[2], J. Wei[2], and Q. Stoll[2]

[1]*Department of Agricultural and Biosystems Engineering, McGill University, 21,111 Lakeshore Road, Ste-Anne-de-Bellevue, Quebec, Canada, H9X 3V9*
[2]*Department of Biological and Agricultural Engineering, Kansas State Unive□sity, Manhattan, Kansas 66502, USA*
wang@macdonald.mcgill.ca

Abstract

Two optical weed sensors and their control modules, a central-control module, a GPS unit, and a spray-control module were successfully integrated into a real-time, embedded system. The system components were networked using a Controller Area Network. The system was tested extensively in two wheat fields. With good training, the system generally reached weed-detection accuracies of over 80%.

Keywords: weed sensor, microcontroller, embedded systems, Controller Area Network, optical sensor.

Introduction

Spot or patch spray has been proven more effective in weed control. Many researches have been conducted on weed control using machine vision (El-Faki, 2000) and remote sensing (Lamb & Brown, 2001). Difficulties encountered in machine vision systems included slow processing speed, large memory requirement, high costs of system hardware, and inaccuracy in classification. Satellite remote sensing is, in general, limited as a tool for weed monitoring due to its insufficient spatial resolution and incapability of detecting weeds in real-time. As a potential solution to fast and low-cost weed detection, optical sensors have advantages on cost, system configuration, and processing speed (Wartenberg & Schmidt, 1999). The objectives of this study were to develop an automated, real-time weed-detection and spray-control system using optical weed sensors (Wang *et al.*, 2001) and multiple microcontrollers networked through Controller Area Network.

Materials and methods

The system block diagram is given in Figure 1. The integrated real-time, embedded weed-detection and spray-control system consisted of two weed sensors, three microcontrollers (C167CR, Infineon Technologies Corporation), a GPS unit (Trimble, *Ag*GPS® 124), a spray unit, an optional PC computer, and a radar velocity sensor. Communications among the microcontrollers and the GPS module were through Controller Area Network (CAN). The system software included programs for data acquisition and preprocessing, weed-classifier calibration using the discrimination analysis algorithm, CAN communication, central control, and PWM spray control. All programs were written in C language, compiled and debugged using the Tasking C166/ST10 Software Development Tools and were running on the C167CR microcontrollers.

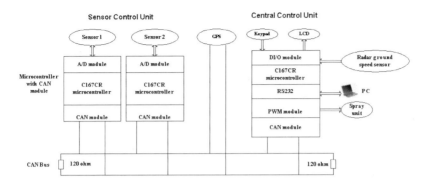

Figure 1. System configuration.

Results

The real-time weed-control system was tested extensively in two wheat fields. Five sets of tests were designed to test the effect of training weed density on weed-detection accuracy, the effect of sensor position relative to crop rows on weed-detection accuracy, the exchangeability of classification models trained during daytime and nighttime, the effect of weed species used in training on weed-detection accuracy, and the performance of the classifier at different weed-density thresholds. In general, test results have demonstrated that the real-time, embedded weed-detection and spray-control system developed in this study was capable of detecting weeds and control the nozzles accurately. With good training, a detection rate of 80-90% can be expected. The system performed well under different field conditions with different weed species, and at different times. Considering the low cost, robust and modular design, and ease of operation, the potential of the designed system in a broad range of applications is great.

Conclusions

The distributed, real-time, embedded weed-detection/spray-control system performed the designed functions. Communications among the system components through CAN were successful. The system was tested in wheat fields and generally achieved classifications accuracies of over 80%. At the current design, the total cost for hardware of the system with two weed sensors is below $2,500.

References

El-Faki, M.S., N. Zhang, and D.E. Peterson, 2000. Factor affecting color-based weed detection. Transactions of the ASAE. 43(4):1001-1009.

Lamb, D.W. and R.N. Brown. 2001. Remote-sensing and mapping of weeds in crops. Journal of Agricultural Engineering Research. 78, Part2:117-126.

Wang, N., N. Zhang, F.E. Dowell, Y. Sun, and D.E. Peterson, 2001. Design of an optical weed sensor using plant spectral characteristics. Transactions of the ASAE. 44(2): 409-419.

Wartenberg G. and H. Schmidt. 1999. Optoelectronic sensors – an alternative for weed detection. Landtechnik. 54(6): 340-341.

Estimation of leaf and canopy variables on EU high-value crops with hyperspectral remote sensing and 3-D models: integration of methods with precision agriculture

P.J. Zarco-Tejada[1], A. Berjón[1], A. Morales[2], V. Cachorro[1], J. Agüera[3], J.R. Miller[4], A.J. Rodríguez[5], P. Martín[1] and A. de Frutos[1]
[1]Escuela Técnica Superior de Ingenierías Agrarias, Universidad de Valladolid, Spain
[2]Escuela Universitaria de Ingeniería Técnica Agrícola, Universidad de Sevilla, Spain
[3]Escuela Técnica Sup.de Ingenieros Agrónomos y Montes, Universidad de Córdoba, Spain
[4]Department of Physics and Astronomy, York University, Toronto, Canada
[5]D.A.P., Consejería de Agricultura y Pesca, Junta de Andalucía, Sevilla, Spain
pzarco@iaf.uva.es

Abstract

Research work conducted under the European Union HySens project aimed to investigate physical methods with ROSIS and DAIS 7915 high-spatial hyperspectral remote sensing imagery to estimate leaf biochemical and canopy biophysical variables in *Olea europaea* L. and *Vitis vinifera* L. canopies. Successful demonstration of accurate estimation of biophysical variables in these crops is critical for the development of effective precision agriculture management practices. Leaf chlorophyll $a+b$ (C_{a+b}) and leaf area index (LAI) are indicators of stress and growth that can be estimated by remote sensing in the 400-2500 nm region.

Keywords: hyperspectral, remote sensing, radiative transfer, olea europaea, chlorophyll

Introduction

Traditional remote sensing methods for vegetation monitoring rely on the calculation of normalized indices such as NDVI as indicators of LAI. However, it is documented that these indices saturate at high LAI values and are affected by canopy structure. The application of physical models has been shown to have particular promise to accurately estimate canopy variables from remote sensing to account for canopy architecture and scene components. However, despite the prevalence of *Olea europaea* L. and *Vitis vinifera* L. landscapes within the EU and worldwide, little research has been conducted examining the reflectance properties of olive tree and vineyards at the leaf and canopy levels. Research that deals with modelling the radiation interception by olive canopies has begun to emerge, with the work by Villalobos *et al.* (1995) for LAI estimation. A potential indicator of vegetation stress is C_{a+b} content because of its direct role in the photosynthetic processes of light harvesting. Differences in remote sensing reflectance between healthy and chlorotic vegetation due to changes in C_{a+b} levels have been detected in the green peak and along the red edge spectral region, thereby enhancing the feasibility of remote detection of crop stress. Chlorosis in olive trees caused by nutrient deficiencies can be treated improving yields and fruit quality (Fernández-Escobar *et al.*, 1993); Fe and N deficiencies developing chlorosis in vineyards cause depression of fruit yield and quality (Tagliavini and Rombolá, 2001).

Materials and methods

Estimation of leaf biochemical and canopy biophysical variables from remote sensing data required appropriate modelling strategies for *Olea europaea* L. and *Vitis vinifera* L. canopies, accounting for structure through its dominant effect on the bi-directional reflectance (BRDF)

signature. 1m pixel-size ROSIS and DAIS 7915 hyperspectral reflectance data in the red edge and NIR region were used to estimate C_{a+b} and LAI (Figure 1). The high spatial resolution ROSIS data allowed targeting of crowns, minimizing structural mixed pixel effects. Linked leaf (PROSPECT) and canopy models (SAILH, SPRINT, FLIM) were assessed for their ability to estimate leaf and canopy parameters by inversion (Figure 1). A field sampling campaign was carried out for biochemical analysis of leaf C_{a+b} content, and leaf sampling was conducted in study areas of *Olea europaea* L. and *Vitis vinifera* L. in Spain. Reflectance (ρ) and transmittance (τ) measurements of olive and vine leaves were carried out using a Li-Cor 1800-12 sphere coupled with a spectrometer. LAI was measured using a PCA LAI-2000, and atmospheric measurements collected at the time of over-flights.

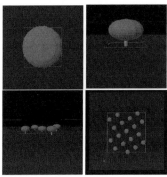
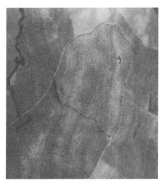

Figure 1. 3-D modelling of olive tree canopies using SPRINT (left). DAIS 7915 imagery (right) collected over olive tree canopies of 5-m spatial resolution and atmospherically corrected (Córdoba, Spain).

Results and conclusions

Results of this research work focused on the application of remote sensing techniques in precision agriculture, studying vegetation indices for leaf biochemistry estimation in *Olea europaea* L. and *Vitis vinifera* L. canopies. The link of PROSPECT and SAILH+FLIM was studied for biochemical estimation in olive canopies, generating relationships between red edge indices and chlorophyll concentration for a set of canopy variables such as LAI and crown density. Validation was conducted using ROSIS and DAIS 7915 for C_{a+b} and LAI variables, and results of the model inversions and *scaling-up* approaches presented in the poster. This methodology enables the estimation of leaf and canopy variables on a tree-by-tree basis, critical for the development of effective precision agriculture management practices.

References

Fernandez-Escobar, R., D. Barranco, and M. Benlloch. 1993. Overcoming iron chlorosis in olive and peach trees using a low-pressure trunk-injection, *Hortic. Sci.* 28:192-194.
Tagliavini, M., A.D. Rombola. 2001. Iron deficiency and chlorosis in orchard and vineyard ecosystems, *European Journal of Agronomy,* 15, 71-92.
Villalobos, F.J., F. Orgaz, and L. Mateos. 1995. Non-destructive measurement of leaf area in olive trees using a gap inversion method, *Agr. and Forest Meteorology,* 73, 29-42.

Suitability of airborne radar-data and aerial color-infrared photographs for the establishment of stem density maps

E. Zillmann[1], H. Lilienthal[1], T. Schrage[2] and E. Schnug[1]

[1]*Institute of Plant Nutrition and Soil Science (PB), Federal Agricultural Research Centre (FAL), D-38106 Braunschweig, Germany*
[2]*Infoterra GmbH, D-88039 Friedrichshafen, Germany*
pb@fal.de, thomas.schrage@infoterra-global.com

Abstract

In this study the properties of radar imagery for predicting stem density of winter wheat were investigated. The results discussed herein are based upon aerial color-infrared photographs and airborne radar-data which were acquired by the German E-SAR (DLR-Oberpfaffenhofen) in northern Germany (13.435 E; 53.893 W) in May 2002. During a ground truth campaign stem density was determined manually in dependence on geomorphology. Single radar bands and several different band combinations were tested for determining stem density. This paper presents an assessment of using radar-data to discriminate in-field variations of stem densities in comparison to aerial photographs.

Keywords: plant density, radar data, remote sensing, TerraSAR

Introduction

The knowledge of the in-field variability of plant density can be used to optimize fertilizer and pesticide application rates (Bjerre, 1999). Recently optical remote sensing has been tested to provide necessary crop information, but little attention has been paid to the suitability of radar data. Radar backscattering from vegetation canopies is mainly influenced by the geometrical canopy structure and the water content. The objective of the study presented here is to analyze wheat canopy backscatter, in order to estimate stem density.

Material and methods

Two test sites are located in Mecklenburg-Western Pomerania near Anklam (13.435 E; 53.893 W) close to the Baltic Sea. In May 2002 (start of stem extension), one field (10-3) was sampled at 50 m intervals along a transect and another (3-2) in a 50 m grid. During the campaign 26 and 18 samples were taken. Stem density was measured by counting stalks in a ¼ square meter. Radar-data was acquired for two different wavelengths X-band (3 cm) and L-band (23 cm) with vertical (VV) and horizontal (HH) co-polarization. Additional cross-polarization (HV, VH) was only available for the L-band. Several different arithmetical band combinations (*polarization ratios*) were generated. Some of them have been used before in other studies (e.g. Pope et al., 1994), others have not been studied yet. Color-infrared photographs were taken and a Vegetation Index (*NDVI*) was computed.
Different statistical analysis were conducted to determine the potential of multi-dimensional radar data to predict wheat stem density. Linear correlation analysis was performed. Three groups of stem density isolated by previous hierarchical cluster analysis were subjected to discriminant analysis. In this model the discriminating variable was the mean radar backscatter value and the NDVI value respectively.

Results and discussion

The field measurements show a high variability of stem density within each field (Tab.1). Field 10-3 is more variable than 3-2. The results show very good correlation with the NDVI.

Table 1. Descriptive statistics of stem density (No. Of stalks) and correlation coefficients.

Test-site	Min	Max	Mean	St.dev.	CV %	r_{NDVI}	r_{sing}	r_{ratio}
3-2	488	860	693	96	14	0.705**	< |0.47|	<|0.55|
10-3	556	1488	780	194	25	0.688**	< |0.39|	<|0.44|

** significant at 0.01 level; CV %: coefficient of variation; st. dev.: standard deviation;
r: correlation coefficient; *sing*: single band; *ratio*: polarization ratio

On the contrary single radar bands and polarization ratio values show only poor correlation with stem density (Tab.: 1). The discriminant analysis obtained better prediction performance for radar-data with percentage accuracy up to 62.5 % and 70.6 % respectively. Tables 2 and 3 contain the summary results of the best discriminant analysis.

Table 2 and 3. Results of discriminant analysis in percent prediction accuracy.

3-2 Group	No. of cases		Predicted Group Membership in %		
			1	2	3
1	8	NDVI	62.5	25	12.5
		Ratio	75.0	25.0	0.0
2	7	NDVI	14.3	57.1	28.6
		Ratio	28.6	57.1	14.3
3	2	NDVI	0.0	50.0	50.0
		Ratio	0.0	0.0	100.0

58.8 % of the grouped cases were correctly classified by NDVI and 70.6 % by L-band radar ratio

10-3 Group	No. of cases		Predicted Group Membership in %		
			1	2	3
1	12	NDVI	66.7	76.9	0.0
		Ratio	33.3	33.3	33.3
2	13	NDVI	23.1	76.9	0.0
		Ratio	7.7	84.6	7.7
3	1	NDVI	0.0	0.0	100.0
		Ratio	0.0	0.0	100.0

73.1 % of the grouped cases were correctly classified by NDVI and 62.5 % by L- and X-band radar ratio

Conclusion

Discriminant analysis revealed that special polarization ratios were useful to discriminate groupings based on stem density differences. The prediction accuracy is similar to that one of NDVI. The relationship between stem density and radar data improved due to generalization. The use of different ratios on each field is due to: (a) different geometric structure of the canopy due to various row direction with respect to the radar view direction, (b) morphology of different wheat cultivars and (c) different homogeneity in each field. Investigations show the feasibility of radar RS to describe agricultural canopies in a similar way as with optical RS (NDVI). Nevertheless, there are some challenges in deriving operational algorithms suitable for different situations on the ground. With the launch of TerraSAR-X in 2005 the temporal gap due to weather uncertainties will be filled and there will exist an operational spaceborne sensor, meeting the time- and resolution-critical requirements of Precision Agriculture.

References

Bjerre, K.D. 1999. Disease maps and site-specific fungicide application in winter wheat. In: *Precision Agriculture '99*, ed. J.V. Stafford, Sheffield, UK: Sheffield Acad. Pr., 495-504.

Pope, K.O., Rey-Benayas, J.M. and Paris, J.F. 1994. Radar Remote Sensing of Forest and Wetland Ecosystems in the Central American Tropics. *Rem. Sens. Environ. 48*, 205-219.

Acknowledgements

The authors of this paper would like to acknowledge BMBF/DLR (grant number 50 EE 0036) and Infoterra GmbH, Germany for its joint funding of this research in the framework of ProSmart IIE project.

605

Retrieving winter wheat LAI and Cab from Xybion airborne data

R. Zurita-Milla, S. Moulin, M. Zago, M. Guérif and V. Houlès

INRA-Unité Climat Sol et Environnement (CSE). Bat. Climat- Site Agroparc- Domaine Saint-Paul, 84914 Cedex 9. France.
moulin@avignon.inra.fr

Abstract

The Xybion airborne multi-spectral video camera was used to retrieve the within-field variability in terms of LAI and leaf and canopy chlorophyll content of a winter wheat canopy. Biophysical variables used in most of the site-specific fertilisation schemes. These variables were estimated with vegetation indices (VIs) calculated with digital numbers because of some calibration problems. Three types of VIs were tested: classical ones, soil-adjusted and atmospherically resistant, in addition to the 6 Xybion bands and to a new proposed index called ANDVI. Results showed that VIs directly calculated with DN performed good, providing satisfactory LAI and canopy chlorophyll content values: R2 up to 0.9 and RMSE around 0.4. The proposed ANDVI index resulted the best of the atmospherically ones.

Keywords: winter wheat, vegetation indices, LAI, Chlorophyll and within-field variability.

Introduction

Most of the site-specific fertilisation schemes need the definition of the within-field variability in terms of LAI and leaf or canopy chlorophyll content. Thus, the goal of this research was to test the feasibility of operating the Xybion airborne multi-spectral video camera for precision agriculture applications.

Materials and methods

The experiment was conducted during the 2001/2002 campaign at a site called Chambry, (1.43° E, 55.07°N), Picardie (France), as part of a long-term precision agriculture project (Guérif et al, 2001), whose objective was to define new strategies for winter wheat nitrogen fertilisation. The study area consisted of a 10 hectare winter wheat field and it was flown over 5 times from March to June with the Xybion multi-spectral video camera MSC-02 (Xybion Inc., CA).
Xybion is an airborne optical sensor that provides coverage in the following 6 bands: 0.433-0.497 (Blue); 0.521-0.598 (Green); 0.564-0.642 (Orange); 0.606-0.694 (Red); 0.711-0.792 (NIR) and 0.812-0.915 (NIR). Flights were performed at 1500 meters high, resulting in a pixel size of 1 meter. In order to calibrate and validate the estimations, LAI and Cab ground data were collected at 81 points throughout the entire field and around the flight dates. Vegetation indices (VIs) were used to estimate the biophysical variables. Among the many different VIs found in literature some of the most common classical ones (RVI, NDVI, WDVI), soil adjusted (SAVI, SAVI2, MSAVI2, ATSAVI), atmospherically resistant (EVI, ARVI, GEMI) and chlorophyll indices (NDVIgreen, R750/R550, R850/R550, TVI, Red/green, SPRI, NPCI) were selected. In addition, the 6 Xybion bands were directly tested and a new index called ANDVI, defined as [(NIR+Blue)-(Green+Red)]/[(NIR+Blue)+(Green+Red)], was proposed for the very first time in this study as an atmospherically resistant index.
VIs regressions were calibrated with the ground truth data: the ancillary dataset was split into a calibration and a validation subset (70 and 30% respectively). Calibration subset was used to fit several functions (logarithmic, power, exponential, 1-logarithmic, 1-power, 1-exponential, and Linear) to the VIs vs biophysical variable empirical relationships.

Results and discussion

Empirical relationships were developed with digital numbers (DNs) and gathering all the dates because of the small VIs variability found for each date. The main results for this study were as follows:

- Most of the relationships (especially for LAI and LAI*Cab) best fit linear functions, although 1-exponential was also found to be a good descriptor.
- VIs performed well to estimate LAI: R^2 around 0.8 and RMSE values of approximately 0.4-0.5 m²/m² for SAVI, WDVI, SAVI2, MSAVI2, ATSAVI.
- Weak relationships were found between the VIs and the Cab in terms of R^2 (<0.55), although the achieved RMSE values seem to be within a good range (0.093 µg/cm²). Top 5 VIs were SAVI2, TVI, ARVI, NDVIgreen and Green Xybion band.
- Chlorophyll estimations improved when working with canopy chlorophyll content: R^2 and RMSE values similar to those found for the LAI estimations. Soil-adjusted VIs did also fine to estimate this variable, together with the R850/R550 ratio.
- Among all the Xybion bands, the blue one had the best R^2 for the LAI and Cab relationships, justifying the inclusion of this channel in the ANDVI index.

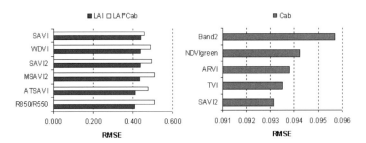

Figure 1. Top 5 VIs, LAI range: 0-6 m²/m², Cab range:0.3-0.8 µg/cm²

Conclusion

Results showed that VIs calculated with the Xybion DNs performed well, providing good estimates of the biophysical variables .

Among all the 24 VIs tested in this study, the SAVI family had the strongest predictive power in terms of RMSE for all the variables, while the R850/R550 ratio yielded the best R^2 value.

Cab was in general poorly retrieved: VIs variability for one date was much smaller than the Cab within-field heterogenity. The proposed ANVI index was the best of the atmospherically resistant ones for the LAI and Cab estimations.

Physically-based inversion might improve these estimation if a good calibration is achieved.

References

Guérif, M., Beaudoin, N., Durr, C., Houlès, V., Machet, J.M., Mary, B., Moulin, S., Richard, G., Bruchou, C., Michot, D. and Nicoullaud, B. 2001. Designing a field experiment for assessing soil and crop spatial variability and defining site-specific management strategies. Third European Conference on Precision Farming, 18-20 June 2001 – Montpellier, France, pp677-682.

Poster abstracts for ECPA

Effect of depth of plowing and soil moisture content reduced expense of secondary tillage

Solhjo A.A. Hojat Ahmadi, M. Loghavi and M. Rozbeh

The tillage equipment should be capable of preparing a suitable seedbed with minimum expense. According to previous research, about 60% of total energy required for preparing the soil is expended for tillage and preparing a good seedbed. It is very important for us to know, which parameters can reduce the cost of tillage and traffic in fields. The soil moisture content is a very important effectiveness parameter for cutting and milling the soil by moldboard plow. With low soil moisture content the cohesion force between particles of soil is very strong and for tillage we need to expend a lot of energy for tillage. Also after tillage we have big clods in the field. With higher soil moisture content, we cannot use tillage equipment in the field. The depth of plowing is also a very important effectiveness parameter. If the depth of plowing is increased the mean weight diameter will go up also. In this research, the effectiveness of depth of plowing (DOP) and soil moisture content (SMC) dry base has been investigated on reduced secondary tillage. Split plots have been used, designed for statistical analysis. The main plot was used for the percentage of soil moisture content and the sub-plot was used in investigation of the effect of depth of plowing, four levels and two levels of the main plot and sub-plot factors were investigated, respectively. Each treatment was replicated three times. For primary tillage and secondary tillage moldboard plow and disk harrow were used, respectively. This research was done in two different land types, one land type had silty clay loam soil (SCLS) and another had loam soil (LS). Based on this research, in silty clay loam soil, the mean weight diameter (MWD) was minimum for two depths of plowing 15-20 and 25-30 cm with 15-18 % of moisture content, and it was maximum for the same depths but with 10-13 % of moisture content. In loam soil (LS), the mean weight diameter was minimum for 15-20 cm depth of plowing with 18-20 % of moisture content and 25-30 cm depth of plowing with 13-18 % of moisture content, and it was maximum for two depths of plowing 15-20, 25-30 cm with 10-13 % moisture content. We can conclude that, if primary tillage was performed optimum soil moisture content, based on type of soil, we will have suitable seedbed with minimum secondary tillage (ST). It can reduce costs of secondary tillage in fields.
Keywords: Soil, Moisture content, Secondary tillage, Tillage, Plowing

Precision farming possibilities in dry onion production in Turkey

Bahattin Akdemir, Simon Blackmore, Can B. Sisman and Korkmaz Belliturk

Aim of this research is to investigate possible precision farming applications in onion production in Turkey. Most processes of the dry onion production have been doing by hand such as planting, hoeing and harvesting. Agricultural machines were used to tillage and to spread fertilizer and pesticides In this research, a field used to measure data for soil properties and yield. Finally, soil map, field slope map and yield map were produced for dry onion. The onion field divided grids and the yield, soil, slope data were determined for each grid. Maps are produced by using a methodology that developed and published by Denmark Royal Veterinary and Agricultural University, Centre for Precision Farming. Data will be use to determine next year strategy for different agricultural applications such as chemical applications, seeding rate, etc. Maps will be used to investigate reason of the yield variability, effect of soil properties and field slope on the yield. According to the results; there is variability in yield, soil properties and field slope maps for the field. Relationship between these factors and yield was discussed according to the maps. All production process of dry onion was also evaluated for application possibilities of precision forming. Yield mapping of onion, variable rate application of fertilizer and pesticide determined possible application process for onion.

Machine guidance with laser and GPS

Andreas Brunnert

On agricultural machines the driver spends about 60% of his attention on steering. For that reason automatic guidance systems take a high work load from the operator. The biggest problem in realizing such systems is to develop suitable sensors which detect the plant rows or the edges of standing crop. Consequently, the first solutions were limited to mechanical scanning of fixed crop rows. The Claas Autopilot for maize harvesting has been representing the state of the art for 25 years and 80 % of the forage harvesters sold by Claas are equipped with this feature. Attempts to detect edges of standing crop or other contours as well remained without success for a long period of time. Three years ago, a breakthrough was achieved in this field by the contactless detection using the Claas Laserpilot. This detection principle has two main advantages. On the one hand very unstable guiding lines can be detected also, on the other hand scanning at a certain distance in front of the vehicle is possible. This allows smooth control of the machine even at high working speed. To use this advantages in other applications apart from the grain harvest, the detection algorithm of the laser scanner was adapted to detect different guiding lines. Especially for the grass harvest with a self propelled harvester an evaluation algorithm has been developed to detect swaths. For the detection of the contour of grain, the travel time of the laser beam is measured and the distance is calculated. At a distance of about 14 meters, the laser beams are reflected by the stubbles whereas reflection of standing crop is at a distance of 10 metres. The position of the crop edge is determined between these areas. The scan profile of a swath looks completely different. Here the centre of the swath has to be determined as a guiding line. This is complicated by different shapes of swaths. Therefore the demands placed on the evaluation software were very high. GPS technology has recently experienced a regular boom in track keeping of agricultural vehicles. The main reason for this were the so-called lightbar systems. These parallel driving aids use LEDs to indicate to the driver if he is on the desired track in parallel to a pre-defined track. Automatic guidance systems, mostly based on real time kinematics technology with centimetre accuracy have gained some importance only in special crops where the system cost can pay off. In the meantime, improved correction signals allow achieving higher degrees of accuracy with D-GPS, broadening the range of applications. On harvesting machines GPS guidance would be a good addition to the Laserpilot. The modular design of the CLAAS Autopilot allows the integration of parallel tracking systems of different suppliers. Future solutions will offer an CAN - interface based on ISO 11783. Apart from track keeping the main advantage of GPS guidance is that computed virtual guiding lines can be scheduled to a work sequence. This will be handled by a path planning module that will be integrated into the machine network.

A self-setting sprayer for vineyards

P. Balsari and M. Tamagnone

A new sprayer prototype, fitted with an electronic equipment able to automatically adjust the spraying parameters according to the features of the vineyard to be treated, has been developed at DEIAFA - University of Turin in collaboration with Unigreen spa company (sprayer manufacturer) and 3B6 srl company (industrial electronics), with the financial support of ENAMA (national board for agricultural machinery). The aim of the project was to establish an automatic communication system (transponder) from the vineyard to the sprayer, in order to provide the vineyard information necessary for the sprayer adjustment. The choice to use the transponder system was made considering that the vineyard lays in the same place for a lot of years, so it is not needed to have an absolute reference system like GPS. The prototype has been developed on the basis of a conventional trailed air-assisted sprayer (Unigreen T10), featured by a 1000 l main tank and a 700 mm axial fan connected with a 2 speed gearbox hydraulic controlled. The fan flow rate could be also changed by varying the diameter of air inlet from 500 to 700 mm using a diaphragm system acting on by an electrical actuator. The direction of the air flow is controlled by means of two deflectors located at the top of the air outlet. Deflector position is controlled by an other electrical actuator. The mixture liquid distribution is made of 12 multiple nozzle bodies, fitted with 3 different sizes of hollow cone nozzle. The acting on of each nozzles are electrically controlled. The operating pressure is setting up by a motorised valve. The sprayer is provided with a receiver, installed under the shaft, able to acquire the information stored in the transponder located in each vineyard to be sprayed in a predetermined position at the edge of the field. In each transponder the vineyard morphological features like row spacing, plant layout, thickness and height of the vegetation in the row and leaf area index are recorded. In order to take in account the canopy evolution during the growing season the software developed is able to receive the trend line that correlate tree dimensions, leaf position and intensity to the date when the treatment is carried out. When the sprayer passes over the transponder, it receives the information related to the parameters of that specific vineyard. These data are elaborated in function of the date of spray application by the central unit present on the sprayer that is able to define the following settings: number of active nozzles on each side of the sprayer, fan flow rate (dimension of air inlet and selection of revolution speed), direction of the air flow and liquid volume rate. During the spray operating pressure is automatically adjusted in function of the forward speed. Thanks to this system it is possible to optimise chemicals distribution, reduce pesticide losses and drift and to avoid mistakes in terms of sprayer setting in each vineyard due also to the change of the tractors driver. The system developed could be very useful especially in large farms with vineyards of different dimensions and leaf intensity.

Development of an environmental information management system for horticulture

H. Bauersachs, H. Mempel and J. Meyer

In the recent years there is an increasing public awareness of environmental and sustainability issues of agricultural and horticultural production systems. As a consequence of this trend the enhancement of the environmental performance of agricultural and horticultural production has become an important policy goal in the agri-sector. For this enhancement approach an increased transparency of the underlying production processes – as a prerequisite for analysis and optimisation - is required. Additionally there are further important benefits of an increased transparency for the grower: proof of environmental quality of his product and therefore increased market value, reduced legal liability as well as the opportunity to improve operational efficiency of his production system. 'Transparency' in this context means the collection and provision of comprehensive information on the production system and the components involved in the production processes. Especially in agriculture and outdoor horticulture these processes and resulting environmental impacts are highly dependent on site-specific factors like climate data and soil properties. Therefore a site-specific approach instead of a global view is essential for an environmental information management approach for the agri-sector. Of course this requirement increases expenditure and costs to operate such an information management system. From an economical viewpoint these costs must not exceed the attainable financial benefits of the system. Within the project 'Environmental Information Management in Horticulture' of the Technische Universität München a conceptual framework for an information management software tool for horticulture is designed and implemented considering the requirements mentioned above. A fundamental task in developing a tool for information management concerns the representation and management of real-world information to abstract entities that can be handled by the computer. For this modelling purpose an object-oriented design approach was selected. In object-oriented programming so called "objects" are used as representations of real-life entities. The representation of a whole real-world system in this modelling approach can be viewed as a population of interacting objects, each of which is an atomic bundle of data and functionality (Yourdon 1979). The aim of the conceptual framework is the analysis of single horticultural companies. Therefore the main object type used in the model represents one company. Each single company object consist of an arbitrary combination of sub-objects. These sub-objects are classified to the following object types: Area objects: represent the fields that are available for crop cultivation within the company Crop objects: represent a plant species cultivated on a specific area object at a specific date Device objects: basically represent machine objects like tractors, tillage machinery, irrigation equipment etc. Process objects: an abstract object type defining the usage of one or more device objects on a crop or area object (e.g. a "ploughing process" could consist of a tractor object and a plough object which are used on a specific date for tillage of a specific area object) Each of these basic objects may interact with other objects, e.g. each crop object is associated with exactly one area object, a process object is associated with zero to many device objects etc. By choosing specific components from a database of predefined object templates, customisation of these objects and definition of associations between single objects it is possible to represent an individual horticultural company. The resulting custom and site-specific object structure can then be used to assess the environmental impacts related to the production system. For the assessment of the impacts current scientific models and methods will be used. The input data required by these models are provided by the objects of the object model: the objects state, the objects behaviour and the objects associations. In the paper the basic concepts of this framework model are presented as well as the first results of the ongoing implementation process.

Recording of georeferenced data about yield and cereal quality at the harvest on combine harvester

Daniel Boffety, Frédéric Vigier, Erwan Ollivier, Bruno Viallis and David Spano

Recording of georeferenced data about yield and cereal quality at the harvest on combine harvester Daniel Boffety, Frédéric Vigier, Cemagref Clermont-Ferrand Erwan Ollivier, Bruno Viallis, David Spano, ULICE Today the evolution of the requirements expressed about cereal quality in the agricultural markets leads all the cereal corporation and the farmers to a more detailed knowledge of the production quality for a better valuation. Now various recent technological headaways in GPS location and in yield measurement on a combine harvester allow to record the yield of cereal field at the harvest. In most part of the cases, the mapping realization is a first step in a more global step of precision farming, where the necessity of recording additional multiannual informations takes on all its interest. As regards the evaluation of the cereal quality at the harvest directly on combine harvester and mainly the evaluation of protein rate, the difficulty lives in the sensors reliable availability, easy to implement and of limited costs. Already several attempts, with more or less success, allowed experimentally to embed of laboratory sensors on combine harvesters. During the 2002 harvest season, experiments were realized in France by the Cemagref near Clermont Ferrand - Montoldre (Research institute for the agriculture engineering and the environment). A combine harvester equipped with a Agrocom yield mapping system connected with a DGPS Omnistar was used to harvest a wheat field. A protein sensor was also fitted on the grain elevator near the grain flow sensor. This Near Infrared Spectrometer (NIRS) is associated with a grain sampling device connected to a central unit by an optical fiber to realize deported measures. A specific computer application was developed under C++ Builder to allow the simultaneous recording of the positions DGPS and the protein rate values of cereal harvested. In the same time of protein rate measurement by this embedded sensor, a sampling of grain was realized. After in laboratory, the real protein rates of harvested cereal is checked with Infratec method. Graphic representations of cereal yield and proteins rates measured with the both methods were realized with a Geographic Information System (GIS). In this field, the wheat yield evolved from 2t / ha to a little more of 8t / ha. The proteins rates measured by the embedded sensor or obtained with Infratec method are included between values slightly lower than 11 % and slightly higher to 15 %.The proteins rates measured by the embedded sensor or obtained with Infratec method are included between values slightly lower than 11 % and slightly higher to 15 %.The implemented of such protein sensor on a combine harvester allowed to approach several technicals aspects about to its functioning such as, principle of measure, calibration, dynamic precision with a variable grain flow and data quality recorded. The implemented of such protein sensor on a combine harvester allowed to approach several technicals aspects about to its functioning such as, principle of measure, calibration, dynamic precision with a variable grain flow and data quality recorded.

Sensing wheat for end use harvest and late nitrogen fertilization requirement decisions

D.J. Bonfil, M. Raz, Z. Schmilovitch, I. Mufradi, S. Asido, H. Egozi, A. Hoffman and A. Karnieli

Yield of dryland wheat in semi-arid and arid zones is limited by precipitation. Nitrogen application and precipitation distribution determine biomass production, soil water depletion and grain quality. Precise base nitrogen is applying according the annual precipitation quantity. In case of more rain, higher biomass production will dilute the nitrogen and low quality of grain would be harvested. Under drought condition, harvest for hay or silage increase income more than living the crop for grain production. Our objective was to establish a quick sensing method based on the leaf spectra that would show out the plant status according water and nitrogen content without a within field reference requirement. This study tested near infrared transmittance (NIT) and near infrared reflectance (NIR) chemometric use for monitoring water and N content in flag leaves at heading of spring wheat grown in semi-arid region during the three growing seasons (2000 to 2002). Measurements from four optical devices were compared. These are: the Minolta SPAD-502 that measure transmittance of two waves (650 & 940 nm); transmittance spectra (530-1100 nm) by Ocean Optics S2000; reflectance within the range of 400-1100 nm by the Licor Li-1800 spectroradiometer equipped with an external integrating sphere; and reflectance within the SWIR region (1100-2498 nm) by Foss NIR System TR-3950. Leaves water and nitrogen content were calibrated with the whole spectra and did not limit to specific vegetation indices. It was found that the SPAD values were dependent on water supply and were not correlated to water content. Reflectance values were found to be an accurate indicator of water and nitrogen content in the flag leaf. Hence, these data take place in a agronomic decision support system develop that would help decision making of early hay harvest or enhance grain protein concentration, which are important for increasing income and quality of wheat yields in semi-arid regions. Further experiments would be established for fine-tuning and method validation in commercial fields.

Understanding GPS guidance before you buy

Roz Buick and Thomas Muhr

GPS (Global Positioning System) for guidance is a rapidly growing area in precision agriculture. GPS guidance technologies can be either manual (lightbar) or fully automated steering systems for farm vehicles. Guidance benefits include improved in-field productivity by avoiding overlaps and skips, reduced input costs (fuel, fertilizer, chemicals), reduced labor costs, improved crop yields, extended hours of operation at critical times (e.g., planting), improved utilization of capital equipment and reduced operator fatigue. All these benefits save the farmer and custom applicator time and money.

In the agricultural industry, there is considerable debate about GPS accuracy and performance. Before purchasing a GPS guidance system it is important to become informed and to understand GPS accuracy, what type of GPS you require for various agricultural applications, what affects GPS performance in the field and how you can determine payback on your GPS guidance system. It is also critical that you understand how to get an optimal GPS guidance system for the specific agricultural operation that you run or serve. GPS receivers vary based on their internal design and method of calculating positions. Typically GPS accuracy in agriculture ranges from 5 meters down to centimeter-level. GPS performance in the field is influenced by many variables including position and geometry of the GPS satellites, quality of the GPS receiver, atmospheric conditions, interference and noise, terrain, GPS and guidance systems settings, and the capabilities of the user. It is important to know what types of accuracy estimates a GPS guidance manufacturer is providing – absolute, static, dynamic or relative GPS performance. For example, absolute static GPS accuracy estimates are more useful for applications such as mapping and guidance back to recorded features (such as the first swath or irrigation drip tape). Relative accuracy is the more useful estimate in pass-to-pass field guidance where you do not need to return to the same vehicle tracks each time. GPS receivers can be configured to optimize absolute or relative accuracy as needed for each application.

In addition to understanding the performance of the GPS guidance system and how to optimize it, a prospective customer needs a way to evaluate how long it will take to pay back the system. How long will it be to see a return on their investment. This is very specific to each farming operation. However, a number of key variables can be used to derive a relatively robust payback model for each agricultural operation. Factors to consider include reduced overlap which equates to reduced fuel use, lower use of fertilizer and crop protection products, reduced labor, improved crop yields based on more precise placement of seed relative to irrigation drip tape or strips of fertilizer (e.g., strip till) and reduced probability of not getting a crop planted in time during critical windows, no need to purchase and maintain traditional methods of guidance (e.g., row markers, foam markers, using bag techniques).

This presentation/paper highlights the list of key features, benefits and other buying decisions that a prospective user should consider before purchasing a GPS-based guidance system.

Development and testing of a simple instrument to be fitted on tractors to aid farmers and contractors measure the size and shape of farmlands

A. Calcante, S. Landonio, M. Lazzari and F. Mazzetto

In Italy, it is difficult to adopt precision agriculture based on Yield mapping and VRT except in some very special areas and certain conditions. In practical terms, the precision problems faced are quite different. Indeed, it is often found, for various reasons, that both the farmers and the contractors have a vary vague notion of the dimensions of the areas to be cultivated. Trials carried out in other countries have shown how GPS, without differential correction and possibly built-in to devices fitted on tractors, can be used for the measurement of farmland to facilitate the current management of farm workers (who generally give little importance to topographical precision). For this reason, the authors decided to set up an experiment in Italy to investigate the practical advantages of using these devices. To do this, a series of field surveys were made on the Milan university experimental farm to assess the performance of the 3 following devices (2 commercial models and 1 innovative prototype, devised by a private company in cooperation with a university research unit - DPVTA; Univ.Udine, and IIA, Univ.Milano): Sys1. ARVAnavCE: GPS on-board guidance tool for rice paddies, produced and marketed by ARVAtec (list price: 5700 Euro); Sys2. eTrex VISTA: Portable GPS with inexpensive receiver on a dedicated palm-top, produced and marketed by GARMIN (list price: 600 Euro); Sys3. ARVAareaCE: inexpensive experimental prototype, based on the use of electronic components and a simple GPS (estimated cost: 2000 Euro); The surveys were carried out in 3 plots (A: 39,268 m2; B: 11,387 m2; C: 44,155 m2) selected to compare very different shapes and sizes. The real data (area and perimeter) for each plot were measured manually with topographical instruments to ensure a one centimetre accuracy. The measurements were performed by fitting, simultaneously, the 3 GPS antenna for the test devices on the tractor cabin roof and then driving the tractor, as close as possible to the border, around the edge of each plot. At the end of each trial, the following controls were made: · the paths were observed to check for any obvious interference with reception; · the area was calculated, taking into account the width of the vehicle; · the perimeter was calculated. Finally, to obtain a better understanding of the reasons for the different performance of the three systems, the most significant raw data were processed again with GIS to georeference the targeted paths on aerial photographs of the farm. The results showed that all the devices gave markedly lower values, for the measurement of the areas and the perimeters, than the real references for each plot. As regards the areas, the greatest differences were seen in the smallest field (B:1.11 ha). The worst performance was obtained with Sys2 (6.79%), followed by Sys3 and Sys1, with deviations of 2.75% and 1.66%. The results were more reassuring with the other plots, with the Sys2 showing deviations of between 2 and 4% , even better for Sys1 and Sys3 with mean deviations always less than 0.5%. Similar results were obtained with the measurement of the perimeters, again with greater differences for Sys2 (2.5-4.5%) and comparable results for Sys1 and Sys3 (1-1.5% in the larger fields). It can therefore be concluded that the measurement of areas using GPS devices fitted on-board agricultural vehicles is certainly feasible provided that the information obtained is used exclusively for technical and agronomic management purposes. Indeed most of the errors found in these trials can generally be considered as acceptable, even though the experimenters did not follow any particular procedure when making the measurements. These experiments did however provide some practical indications which should be observed to ensure acceptable results.

Survival of soil-borne pathogens for decision support and modelling of soil solarization on protected crops in southern Italy

G. Cascone, G. Polizzi, A. D'Emilio and I. Castello

Soil plant pathogens and pests frequently cause heavy losses in major crops, by affecting both yield and quality. In intensive agriculture, the same land is continuously cultivated, thus leading to a rapid increase of parasitic populations in the soil, especially those attacking the roots. The aim of soil disinfestations is to eradicate populations of soil-borne agents to the desired soil depth, with minimal disturbance of biological activity. Soil solarization is a hydrothermal process that occurs in moist soil which is covered by plastic film and exposed to sunlight during the warmest summer months. The success of soil solarization is based on the fact that most plant pathogens and pests are mesophylic and they are killed directly or indirectly by the temperatures achieved during the soil heating. Soil-borne organisms if not directly inactivated by heat, may be weakened and became vulnerable to changes in the gas environment in solarizing soil or occasionally to changes in the populations of other organisms which may exert a form of biological control. The decline of soil-borne organisms during solarization depends on both the soil temperatures and exposure time, which are inversely related. The effectiveness of solarization for disinfesting soil depends on soil colour and structure, air temperature, soil moisture, length of day, intensity of sunlight, and the thickness and light transmittance of the plastic film.

In this paper the effectiveness of soil solarization treatments during summer and autumn was evaluated in Sicily in two multi-span tunnel greenhouses through the survival of two soilborne pathogens Fusarium oxysporum f.sp. radicis-lycopersici (FORL) and Cylindrocladium pauciramosum. The greenhouses were covered with a 130-µm thick ethylene vynilacetate film (EVA) and with a 60 µm-thick ethylene tetrafluoroethlylene film (ETFE). Nylon mesh bags containing carnation leaves and basal tomato stems infested respectively with C. pauciramosum and FORL were buried 15 and 30 cm deep. Soil solarization was performed during a 3 to 30-day period using a 30 µm-thick clear EVA film and a 35 µm-thick green coextruded film. The viability of the two pathogens was constantly monitored after different days of soil solarization treatments using a recently modified semi-selective medium (CRM) amended with 3 µg a.i./ml of benomyl (Benlate 50 WP, Dupont) for the detection of FORL isolate and with 100 µg a.i./ml of benomyl for C. pauciramosum isolate. Air temperature, air relative humidity and total solar radiation flux were measured inside and outside the greenhouse. Soil temperatures were measured with thermocouples buried at 1 cm, 5 cm, 15 cm, and 30 cm. All the measured parameters were sampled every fifteen seconds and averaged for half-hour readings by a data-logger. The mortality of the two tested pathogens resulted strictly related to the thermal regimes measured in the soil under the different treatments.

Laboratory trials were also conducted by means of a specific equipment able to reproduce the thermal regime recorded during the trial under greenhouse. The obtained results showed that laboratory techniques can represent a valid alternative to the field trials and can help the definition of mathematical models aimed to the determination of the optimal treatment duration in relation to the physical conditions.

A methodology to define soil units in support to precision agriculture

A. Castrignanò, M. Maiorana, M. Pisante, A.V. Vonella, F. Fornaro and G. Fecondo

In the precision agriculture soil tillage and crop management within an agricultural field are adjusted to the local soil properties and crop requirements. That requires a detailed knowledge of the spatial variability of the soil properties in the field. Research suggests to divide a field into a series of homogeneous zones on the basis of soil properties, which can be assumed as a basis of varying agronomic inputs.

This paper presents a methodology to derive basic units for precision agriculture, referred to as soil units. Their main purpose is to reduce the actual soil variability in the field to a limited set of clusters, needed to enable the incorporation of mechanistic models in precision agriculture. Geostatistical methods provide a means to study the heterogeneous nature of spatial distribution of soil properties.

The research is carried out in the experimental farm of the Institute, located at Foggia (Apulian Tavoliere), the main durum wheat cropped area of southern Italy.

The climate is "accentuated thermomediterranean" (UNESCO-FAO classification), with winter temperatures which can fall below 0 °C, summer temperatures which can rise above 40 °C and rains unevenly distributed during the year, being concentrated mainly in the winter months.

The soil is a silty-clay vertisol of alluvial origin (classified as Typic, chromoxerert, fine, Mesic by Soil Taxonomy-USDA), the physico-chemical mean characteristics of which are the following: total N = 0.151%; total P (as P_2O_5) = 0.168%; available P (as P_2O_5) = 70 ppm; total K (as K_2O) = 1.675%; exchangeable K (as K_2O) = 1556 ppm; $CaCO_3$ = 4.70%; organic matter = 2.48%; C/N = 7; pH = 8.0; coarse sand = 5.3%; fine sand = 14.5%; clay = 26.3%; silt = 53.9.

The experiment was set up on a field of 2 ha cropped with durum wheat (Triticum durum Desf.) under continuous cropping, with zero-tillage.

From four plots of 0.5 ha each, the most important soil characteristics (N, P, K, pH, limestone content, organic carbon, sand, clay and silt) were determined on a grid 10 x 20 m at two depths.

The spatial structure was investigated by variography and by the spatial interpolation techniques of kriging and cokriging. Most of variables, with the exception of P and N, looked well spatially structured and dependent up to distances of 40-50 m.

To divide the fields into a number of soil clusters or classes, without any previous information about the existence and the number of the previous groups, we selected a clustering method which makes use of density estimation. The basic idea of this approach is that the clusters in the set of the estimated values of the soil parameters correspond to modes or peaks in the density estimate constructed from these points.

The results of this study pointed out that the field can be divided into 5 distinct classes from the physical and fertility properties. In particular, the wide central cluster showed the highest content of clay and N.

The proposed approach offers a quick method to identify potential soil units as a basis for the variable-rate application of inputs.

Spatial reflectance at sub-leaf scale discriminating NPK stress in barley

L.K. Christensen and R.N. Joergensen

Discriminating between nitrogen, phosphorous and potassium stresses, non-destructively, in a canopy is an essential element in the process of protecting our natural resource base. Non-destructive stress symptom diagnostics is essential optimising variable nutrient application with a minimal environmental load. The identifying characteristics when spectrally determining nitrogen, phosphorous and potassium stress in a canopy are dominantly due to the varying levels of chlorophyll in the leaves. The challenges arise when aiming for spectrally discrimination of N, P and K stress. Due to the pigment, chlorophyll is predominantly influencing the characteristics of each of the three target symptoms; it is essential to include spatial information in order to reach discrimination abilities of N, P and K stress symptoms. This paper introduces a methodology potentially able to discriminate between N, P and K stress symptoms by identifying the most significant characteristics on the plant separating each of the three stress symptoms. The methodology was tested in a spring barley crop. Equipment The equipment used was a Zeiss monolithic miniature spectrometer (MMS 1) NIR enhanced (Fig 1). It is an OEM (Original Equipment Manufacture) device. It is a tec5 company multi operating spectrometer system. The spectrometer system consists of a light source (12V/100W tungsten halogen lamp), which is controlled by a photodiode sensor, and optical components of the spectrometer. The optical components consisted of a body made of UBK 7 glass with aberration corrected grating. The Zeiss MMS 1 NIR was the sensor. The claimed detection range of the sensor was 306-1132nm in 2nm broad bands (Zeiss company, 1999), however, the wavelength range used was 450-1000 nm. Fig 1: The tec 5 spectrometer system (Abu-Khalaf, 2001) N, P and K Stress Symptoms Establishment Clear visible stress symptoms of N, P and K respectively were established under glasshouse conditions in an inactive media (perlite) applied with sufficient amount of essential nutrients (micro and macro) with the exception of the respectively target nutrient stress components (N, P and K) Three replicates in time were carried out and ten sub-samples for each nutrient stress symptom (N, P and K stress) were established. Control plants, applied with all essential nutrients and in sufficient amounts, were further established. Data Acquisition and Analysis Directed sampling technique (DST) was carried out on each plant at three different growth stages. The principle of the DST is to reach very specific, spatially information and relate the spectral information according to the spatial information on the plant. Each leaf on the barley plant was measured at the top, mid and bottom. This acquisition of data was carried out at three early growth stages. The trial and measuring design generated 4-dimensional data consisting of the specific plant, the actual leaf number, the leaf position, and the spectral dimension. N-PLS is used as prediction model since it uses latent structures for making predictions of dependant variables (stress symptom N, P, K) within the empirical four-way data set. The strength of N-PLS is that it summarizes all latent information from a large N-way dataset of object variables (X) and relates it to dependent variables (Y) using a relatively low number of parameters, which makes the prediction more robust. Conclusion The preliminary results indicate that the N-PLS methodology can reveal the location of the significant characteristics determining and discriminating the three nutrient stresses.

Using indicator kriging for evaluating spatial soil quality chemical indicators

J.E. Corá, A.V. Araujo and G.T. Pereira

Awareness of the importance of soil quality for agricultural sustainability has increased in the last years. The terms surrounding the concept of a high-quality soil relate to its ability to supply crop nutrients, resist erosion, provide habitat to microorganisms and microfauna, and ameliorate environmental pollution. To quantify soil quality, specific soil indicator need to be measured spatially. These indicators are mainly soil attributes related directly to the crop productivity but also include policy, economic, or environmental considerations. Because assessing soil quality is complex, the individual soil quality indicator need to be integrated to form a soil quality index. The objective of this work was to identify site-specific management zones throughout chemical soil quality indicators on Latosols (oxisols), which have been managed with sugarcane, using indicator kriging procedure, aiming a better use of the precision agricultural techniques. The study was conducted in a 90 ha section of a larger field (1900 ha) located at Jaboticabal, São Paulo State; Brazil (21o15' S e 48o 18' W). A 50 m regular grid was imposed on the experimental area (300 by 3000 m). Soil samples were obtained at each 50 m grid point (421 points) from the top 0.2 m. Organic matter, P, K and V% values were determined for the soil samples. The soil attributes data were coded as 0 or 1 if they lie above or below the threshold value chosen for each variable, respectively. Using the results of geostatistical analysis for the coded data set and the indicator kriging procedure, counter maps of probability of meeting the critical threshold value for each of the individual soil attribute and combinations of them were produced. These maps indicated areas in the landscape that have different probability of having good soil quality according to predetermined criteria. The procedure used allowed the identification of the indicator parameters responsible for zones of relativity low soil quality, thus allowing specific management plans or land use to be developed.

Response of some crops to short-term flooding under field conditions in the Indo-Gangetic plains of India

I.S. Dahiya, S.B. Phogat, R.S. Hooda and Y.P.S. Solanki

The water logging (saturated and/ or temporary flooding conditions) is an alarming problem threatening the sustainable crop production of those crops which are not adapted to wetland conditions in many parts of the Indo-Gangetic Plains. Approximately 6 million ha land in India are under water logging of which about 3.4 million ha are affected by flooding for different durations in several states. Field experiments were conducted to quantify the effect of short-term flooding on pigeon pea, pearl millet, sorghum and wheat from 1992 to 2000. In pigeon pea, flooding treatments were applied 30 days after sowing for 0 (drained), 12, 24, 48 and 72hr. Flooding up to 48 and 72hr decreased the survival of plants significantly with survival percentage of 54 to 85 at the four durations of flooding. Flooding decreased plant height and dry shoot weight significantly except under 12hr flooding. Percent decrease in the two parameters was 18 to 36 and 17 to 67, respectively, under the four duration of flooding. Flooding for all the four durations significantly decreased branching, pod formation and grain yield by 37 to 52, 33 to 63 and 21 to 60%, respectively. The 1000-grain weight was not affected significantly. In the case of pearl millet, flooding treatments were applied for 0, 45, 90, 135 and 180 hr after 30 days of sowing. Flooding for 135 and 180 hr significantly increased plant morality and reduced tillering, dry shoot weight and dry root weight. Plant height, ear head length and grain yield were reduced significantly by 90, 135 and 180 hr of flooding. The 1000-grain weight and ear head thickness were not affected significantly. The yield reduction was 23 to 67% under the four flooding durations. The corresponding figures for decrease in tillering, dry shoot weight, dry root weight, plant height and grain yield were 16 to 36, 30 to 58, 6 to 61, 6 to 20 and 8 to 60%, respectively. The Post-flooding N and Zn spray significantly increased dry shoot weight, dry root weight, plant height by 15 to 105, 11 to 261, 3 to 27 and 32 to 132%, respectively, at all levels of flooding. In the case of sorghum, flooding treatments were given for 0, 90, 180 and 270 hr after 30 days of sowing with and without post-flooding additional N and Zn application through spraying. In general, flooding for 180 and 270 hr significantly decreased dry shoot weight, dry root weight, plant height and green fodder yield by 5 to 28, 5 to 28, 18 to 37 and 6 to 29%, respectively, at all the levels of post-flooding N and Zn spray. Post flooding N and Zn spray, however, resulted in a significant increase in all the four crop parameters studied at all durations of flooding. The increase was 2 to 17, 4 to 17, 3 to 17 and 2 to 19% for dry shoot weight, dry root weight, plant height and green fodder yield, respectively. The results established the fact that amongst the three crops studied here, pigeon pea is the most sensitive to flooding than sorghum. The wheat crop was subject to flooding for 0, 24, 48, 96, 144 and 192 hr. Dry shoot weight, dry root weight, plant height and grain yield decreased significantly under 96, 144 and 192 hr of flooding. The decrease was 1 to 60% in dry shoot weight, 0 to 58% in dry root weight, 2 to 34% in plant height and 3 to 62% in grain yield under 24, 48, 96, 144 and 192 hr of flooding.

Dynamic behaviour of combine yield measurement systems investigated on a test stand

M. Demmel, T. Steinmayr and H. Auernhammer

Local yield data and yield maps have become a key element of precision farming. They deliver input information for the determination of management zones and are a tool to evaluate the results of site specific farming. Today all manufacturers of combine harvesters are producing and selling yield measurement systems for their products and also first systems are available for choppers and root crops. Because yield measurement systems for combines are on the market now since nearly 10 years, a number of investigations on their accuracy has been made and published (Auernhammer et al. 1993, Macy et al. 1994, Grisson et al. 1999, Demmel 2001). Most of these investigations have tried to get information on the measurement accuracy of the systems in the field under real work. To gain this information they have counter weighed tank loads of the combines (Auernhammer et al.- 1994, AL-Mahasneh et Colvin 2000). Other groups have investigated the influence of isolated different errors sources on test rigs or test stands (Steinmayr et al. 2001, Demmel 2001). A small number of engineers has tried to compare the yield readings of the measurement systems with the yields of reference plots directly beside the track of the combines (Searcy 1998). Following this activities the American Society of Agricultural Engineers therefore is defined an ASAE Standard for testing yield measurement systems. But all of this published investigations only give information on the overall accuracy of the systems based on different types of aggregating or accumulating procedures . To get an information on the quality of yield data from a yield measurement systems also their dynamic behaviour is very important because yield data and yield maps should deliver a picture of the changes in the field. Slow or fast, smooth or sharp changes in yield while the combine is driving through the field in a first and major step are smoothed by the combine and its threshing mechanism. The important question is how are the measurement systems reacting on the remaining changes? How are their dynamic behaviour? This important question cannot be answered in the field because the changes in the yield are not defined or know. Therefore the only way to get information is to run dynamic tests on the test stand with different pattern or shapes of mass flow changes. Three different shapes of mass flow changes for yield measurement systems have been proposed: "Ramp flow", "step flow" and "alternating flow". On the test stand for combine yield measurement systems at the Technical University of Munich three different yield measurement systems for combine harvesters have been investigated with these three proposed procedures and different tilt (side hill) angles. As expected they have shown different behaviour. They are reacting with different delays and in different speeds on the changes in the mass flow. In the proposed presentation the test stand, the testing procedures and the results of the tests on the dynamic behaviour of yield measurement systems will be described. The consequences on the possibility to create yield maps will be discussed. Based on the results the authors will try to define requirements for improvements of the dynamic behaviour of yield measurement systems for combine harvesters.

Prospects for site-specific management of corn rootworms

M.M. Ellsbury, D.E. Clay, S.A. Clay, C.G. Carlson and D.D. Malo

Distributions of pest populations commonly are characterized as field means expressing density of insects per unit area or per sample unit for a given field. Variogram analysis following field-scale grid sampling of corn rootworm life stages and injury to corn have shown that infestations of corn rootworms vary spatially, that is, they are not constant over the entire expanse of a field. Spatial variability in rootworm incidence suggests that a site-specific approach to the management of corn rootworms may be feasible. This concept is attractive because of the environmental benefits from possible reduction in pesticide usage and lower input costs for the grower. Long-term grid sampling of spatial dispersion suggests that corn rootworm distributions change over multiple growing seasons as the population develops. Populations of extended egg-stage diapause northern corn rootworm more than doubled over a 6-year observation period in a corn-soybean agroecosystem. Spatial distributions of the egg, larval, and adult stages do not necessarily correlate. Interaction between temporal and spatial variability thus makes precision corn rootworm management difficult to undertake. Nonetheless intensive grid sampling of rootworm life stages has provided potentially valuable knowledge in the form of GIS-managed data layers relating corn rootworm dispersion and distribution to soil properties, fertility, weeds, landscape, and yields in soybean and corn fields. But, the usefulness of this information is limited by the high cost of sampling. The necessary GIS/GPS capabilities are available but have not yet been effectively incorporated into decision support systems for site-specific corn rootworm management. Better integration of economical scouting methods or real-time monitoring and mapping of corn rootworm variability with map-driven application technology must be achieved to enable site-specific management of corn rootworms.

Semi-automatic VRT-based fertilization system utilizing GPS guidance

Moustafa A. Fadel, A. El-Mowafy and A. Jomaa

Modern technologies gave the agricultural engineers a historical chance to improve the agricultural production environment. Precision farming is one of the most growing arenas in the agricultural engineering field. Employing real-time GPS positioning methods, made it easy to build accurately guided machines. On the other hand, environmental concerns made it necessary to apply Variable Rate Treatment (VRT) concept in fertilization. Precise fertilizers application is used widely where fully automatic systems are extensively presented commercially. In this research, s new semi-automatic system is designed and developed locally. The proposed system employs a real time GPS positioning system to guide a tractor mounted rotary spreader to apply phosphate to the field. In order to estimate soil requirements, the field was gridded in 50-meter mesh and soil samples were collected from eight GPS defined locations. Soil analysis results were mapped and studied to find out the needed amount of Calcium Super Phosphate to be added per hectare and prescription map was developed. In order to evaluate system performance and its feasibility, soil chemical analysis of the same GPS defined locations, were carried out after applying the fertilizers using the designed system. It was noticed that, assessment of VRT-based systems is clearly stated anywhere in the literature. Different methods of performance assessment are presented in this research. The developed system was tested and evaluated in Al-Oha experimental farm, UAE University.Field tests were carried out in different methodologies to evaluate the overall system performance. I t was found that system performed satisfactorly under field working conditions.

Evaluation of a crop - denitrification/ decomposition model for nitrate leaching based upon a U.S. Midwestern row-crop field

N. Farahbakhshazad, C. Li, D.B. Jaynes, D.L. Dinnes and D. McLaughlin

The application of fertilizers in the U. S. corn-belt region has been increased substantially during the last decades. One of the environmental consequences of these agricultural practices is the contamination of surface and groundwater resources by the leached nitrate from soil profiles. According to improve nitrogen management practices, the Denitirificaiton-Decomposition (DNDC) model has been applied for a corn-soybean farm field in Iowa that has had intensive nitrate concentration and flow rate measurements from field drainage tiles. Four years of field data (1996-1999) from nine drainage tiles with three different N-fertilizer treatment rates (low, 67 kg ha -1 in 1996 and 57 kg ha -1 in 1998; medium, 135 kg ha -1 in 1996 and 114 kg ha -1 in 1998 and high, 202 kg ha -1 in 1996 and 172 kg ha -1 in 1998, applied to corn) were used for calibration and validation of the model. Simulated water fluxes at 1.0 m depth correlated well with the measured flow in tile drains with an underestimation of -9%. The model responded to the increase in N fertilizer application rates. The comparison between the simulated and field measured nitrate concentration for four years shows that the model underpredicted leaching losses for the low N-application, and is in good agreement with the medium and high N-applications. On average, the simulated annual nitrate leaching fluxes from all the N-treatments differed from the measured values by -8%. The results demonstrate that the DNDC model is capable of estimating N leaching losses from cropland. However, the model may need further improvements in accounting for soil moisture content under differing conditions and in its nitrogen cycling algorithm.

Development of site-specific nitrogen fertilization strategies for sites with high and low plant available soil water capacity

D. Geesing, R.Gutser and U.Schmidhalter

Rationales for site-specific nitrogen fertilization on sites differing in water availability are not well defined. Field studies were conducted to determine the interacting effects of N fertilization (N treatment), irrigation and drought stress (water supply treatment) on wheat yields, and yield components on two sites within a heterogeneous field. N rates were 120 kg ha-1 and 180 kg ha-1. Irrigations were applied or water stress was induced by withholding rain-water by sheltering plots during rain events between two-node stage and flag leaf stage in 2000, and between three-node stage and appearance of awns in 2001. Soil water content was monitored during the spring- summer growing season. Two-years data suggest soil water depletion down to 50 cm of soil depth on the loam-sandy sites (low plant available soil water capacity) and down to more than 90 cm on the loamy and silt-loamy sites (high plant available soil water capacity). The total amount of soil water consumed down to 50 cm was similar on both sites of each field. Soon after the water supply treatment was terminated, the level of soil water depletion was similar for all water supply treatments of each site. The effect of the site on grain yield interacted with water supply treatment but not with N treatment. Within the sites, nitrogen was the yield- limiting factor on the sites of high plant available soil water capacity, but on the sites of low plant available soil water capacity, water prevailed over nitrogen as limiting factor. The factor site was, however, the dominant cause of variation in grain and straw yield, seed weight, N, P and K uptake (higher at the sites of high plant available soil water than at the sites of low plant available soil water) and grain P concentration (lower at the sites of high plant available soil water than at the sites of low plant available soil water) in a field. Irrigation reduced substantially the differences between the sites. Water use efficiency was higher at the site of high plant availability of soil water than at the site of low plant available soil water and much higher in the second experimental year than in the first. Based on the consumption of soil water only, water use efficiency was considerably higher on the site of low plant available soil water than on the site of high plant available soil water site in the second experimental year. Soil mineral N, monitored only in 2001, was higher on the site of low plant available soil water than on the site of high plant available soil water before, during and after the trial. The findings of the present study show the importance of the site and the distribution of precipitation during the growing season and suggest that distinct fertilizer recommendations for each site of a field are necessary when soil texture is variable. Yield is strongly influenced by precipitation during heading until grain-filling, but stress during jointing limits yield potential that is hardly regained when stress is relieved. On sites of high plant available soil water, plants are less susceptible to stress during jointing and thus, an increase in applied N if required for higher yield is in most cases appropriate for these sites. On sites of low plant available soil water, only if rainfall during jointing is above average, an increase in the amount of applied N to obtain higher yield on sandy sites may be indicated, but also augments the risk of N leaching

Evaluation of yield mapping data analysis procedures in Italy

Rino Gubiani and Massimo Lazzari

Introduction Increasing attention is being paid in yield mapping systems to evaluating both the quality and precision of the gathered data and their subsequent processing for decision-making purposes. Errors in the gathering of yield data have been described by many Authors. To utilise the data from yield mapping in decision-making, Juerschik et al. (1999) propose setting up successive attempts to correct the raw data collected based on excluding values that diverge from determined thresholds and processing them on the basis of initial areas of different sizes. This contribution reports a study case used as an example of the application of a procedure deriving from the one proposed by Juerschik et al. (1999). Materials and methods The study used yield mapping data gathered over 4 successive years on a rectangular plot of 8.4 ha. The data obtained from the "smart card" were: · plotted on geo-referenced maps; · used to calculate average yields; · transformed into percentages for statistical comparison. This was done calculating the correlations between the different time periods using both all the data and excluding deviant values from the mean by: ± 3s, ± 2s, ± 1.64s. The plot was also divided into header and central areas to evaluate possible concentrations of errors. The correlations were then calculated for the points of the nodes of the square side grids equal to: 10 m, 20 m, 30 m, 40 m, 50 m. The values assigned to each node of the grids were calculated using Nearest Neighbor Search with 4 points per node weighted on the basis of the reciprocal of the distance to the square of the node (1/d2). Lastly, the Kriging method of geostatistical analysis was applied to obtain further information on the trend of the yield data. The yield data were also correlated with those of the soil analysis. Results An initial examination of the maps of raw data leads to the hypothesis that a yield deformity exists on the plot because the correlations between the single points on a temporal basis are extremely low (0.15; 0.33; 0.08 for the three comparisons). Even with using the different filtering systems the correlations do not improve. Analyses of the values by grids using dimensions of 30 m, 40 m, and 50 m lead to a clear rise in the correlation values (from a minimum of 0.28 to a maximum of 0.60). Lastly, the variogram obtained by Kriging analysis demonstrates that there is a trend of a slow drift of the data from one end of the field to the other. Conclusions In the studied case, processing the data for small areas leads to unsatisfactory results because the measurement errors in the field are big enough to render the use of the point data of little significance. Also the use of filtering systems does not give appreciable improvements in the analysis. Only by working on large grid dimensions (the best correlation is found at 2500 m2) leads to good results. Yield mapping, which at first sight would appear to be a panacea for decision-making problems at management level as it provides the user with a large amount of easily interpreted data, is instead a process that requires a lot of knowledge on the part of the technicians who handle such large amounts of data, otherwise there is a risk that they are unable to interpret the real phenomena. References Juerschik P., Giebel A., Wendroth O. (1999) - Processing of Point Data from combine harvester for precision farming – 2nd European Conference on Precision Agricolture Odense Congress Centre, Denmark, 11-15 July.

Yield and soil variation for precision farming in semi-arid region: A case study in Turkey, Ankara

A. Gucdemir, U. Turker, A. Karabulut, C. Arcak and K. Karuc

Yield and Soil Variation For Precision Farming in Semi-Arid Region The average annual gain in the yield of wheat in Turkey has stagnated during the past 15 years particularly in dry land areas. Soil fertility, soil erosion, weed and root diseases made cereal production increasingly less profitable. For the past three decades farmers agronomic practices have not changed and often resulted in increased soil deterioration and decreased soil fertility and organic matter content. The practice and management of dryland areas urgently are needed to be revitalize by testing and implementing precision farming techniques in the dry land region. 84% of the cereal fields of the central Anatolia is smaller than 50 acre. So, a small-scale precision farming project which was supported by the scientific and technical research council of Turkey and soil and fertilizer research was commenced in 1999 in the region. The aim of the study was to measure and analyse variations in soil properties and winter wheat yields in semi-arid region. The study was carried out in 2 different field respectively 15, 18 acre and about 30 km away from each other. The annual perticipation rate is less than 500 mm in the region. The common practice in the region is that the farmer leave the land fallow for a year. Commercially available grain yield monitor and a satellite computer (DGPS) were installed on a John Deere combine. Table switch, speed sensor, cut out switch, hillside sensor were also included in yield monitoring system. The data string was recorded as latitute and longitute in ceres monitoring system and transferred to micotron satellite computer via RS232 connection. This data string was matched and recorded with yield data in satellite computer. Field 1 was monitored and yield mapped in 2000 and 2002 and Field 2 was monitored and yield mapped 1999 and 2001. Soil samples and various soil and plant measurements were collected from each field on 30-40 meter-grid from entire field. Measured soil characteristics were pH, organic matter, nitrogen, phosphorus, potassium, lime (CaCo3). Plant leaf nitrogen, phosphate, potassium were also measured and mapped. Other measurements at each sampling point were micronutrients like copper, iron, zinc, mangan. Weed density was also mapped by ground scouting and compared with yield maps. Soil and plant data were imported into ArcView and AMCS software. Contour maps were produced for field 1 and field 2 using ArcView software. Geostatistical, descriptive statistics and multivariate regression analysis of data were also prepared. It was found that the spatial variability of yield and some soil properties differed substantially in two fields. There was a relationship between variability of yield maps and some soil properties. Fe, Mn, Cu and potassium have not shown any correlation with yield. Weed density has shown a high correlation with the yield. The level of the zinc and organic matter content across the field and phosporus and soil EC were significant in explaining the yield and yield variability. From the analysis results, it is apparent that an application to decrease the pH level is necessary. The zinc level was under the critical level (less than 0.5) and limiting the potential benefit of nitrogen, phosporus and other nutrients. As a result, there is a variation in soil properties and relation to crop performance for the semi-arid region fields. Precision farming has the potential to make crop production more profitable and sustainable in semi-arid fields. But practical implementation are necessary to the commercial fields in the region.

On-farm research by grower learning groups in Minnesota, USA

J.A. Hernandez, C.J. Iremonger and P.C. Robert

The use of differentially corrected global positioning systems (DGPS) and yield monitors in precision agriculture allows farmers to obtain rapid and accurate geo-referenced crop productivity across their fields. These technologies also give investigators the opportunity to perform large-scale experimentation. One of the advantages of on-farm experimentation is the availability of many locations and crop-years that can be assessed, and moreover research is performed at the same scale that farmers use. The Precision Agriculture Center at the University of Minnesota has been testing and adapting an outreach model based on Grower Learning Groups (GLG). The emphasis of this GLS is on empowering growers and their advisors to conduct their own experiments, build local databases, and create local recommendations. The PAC is sponsoring GLG dedicated to sharing knowledge of precision agriculture practices and conducting on-farm experiments. The GLG represent distinct regions defined by soils/landscape, climate, and cropping system (agroecoregions). In collaboration with PAC outreach specialists, growers identify limiting factors for crop yield, quality or water quality management and then run experiments to test new management approaches. Several (GLG) have been established in Minnesota dedicated to sharing knowledge about precision agriculture practices and conducting on-farm scale experimentation. Sixteen farmer members across the three GLG agreed to conduct a total of 26 trials. During 2001-2002 members of the GLG were interested in three different studies: (a) planting speed trials (b) plant population trials and (c) a nitrogen rate and timing trial. Each farmer laid out randomized complete block experiments with treatments depending upon the study. Farmers recorded grain yield with their own combine yield monitors. This poster will present results from several on-farm research trials, analysis will be based on classical analysis of variance (ANOVA), pair wise treatment comparisons and will include the effect of soil type in each agroecoregion.

Application of reflectance measurements of maize under field scale conditions for the attainment of site-specific nitrogen uptake targets

G. Huber, F.-X. Maidl and A. Schmid

Heterogeneous Fields are characterised by a spatial variability in water supply and nitrogen mineralisation. Site specific cropping takes care of these parameters. According to Hatfield et al. (2001) a variation of nitrogen fertilisation can change the water use efficiency. A vital approach for variable cultivation of maize can be the adaptation of nitrogen application and plant density to the characteristics of the soil in different field parts. Variable fertilisation regimes could be realised according to the current biomass production and nitrogen uptake. For this purpose the non-destructive methods, successfully used for detecting the biomass and nutrition status of cereals, have to be adapted to the specific demands of row crops such as maize. Test plants were conducted in the upper Bavarian Tertiär-Hügelland on a field that is characterised of a broad heterogeneity in yielding capacity. Between the three leaf stage and silking, intermediate harvests of the whole plant were made at six dates. The dry matter yield and the nitrogen concentration of every biomass sample was detected. The seed density and the nitrogen level that leaded to the highest corn yields, and the best nitrogen fertiliser utilisation respectively, were identified. This data can serve as a basis for the suggestion of site specific values for optimal seed densities and nitrogen fertiliser amounts. The biomass information attained by intermediate harvests allow the construction of particular nitrogen uptake progression optima for different yield expectation levels. An additional field trial with the factors nitrogen amount, application date and cultivar took place on a uniform site. Four varieties that differ visibly in terms of colour and leaf orientation were grown and detected periodically using a reflection spectrometer. The reflectance bands were recorded in the range of visible and the near-infrared radiation (360 nm - 1050 nm). The spectral resolution was 3 nm. The recorded reflection signature was analysed in regard to an evaluation of well-known vegetation indices such as e. g. REIP (Guyot et al., 1988). The suitability for assessing the status of maize canopies was examined. This occurred on the basis of the correlation between the wavelength-related multi-spectral reflectance and the ground truth parameters. The appreciation was done by drawing correlograms resembling those described by Yoder et al. (1995). The reflectance bands were classified in regard to the specific application potential for an estimation of maize biomass. Particular attention was paid to sensibility towards the disturbing signal from soil background. The difference between the high-yielding site and the low-yielding area was 12.5% for maximum total dry matter yield and 15.6% for maximum grain yield. The advisable values of plant density were positively correlated with the site-specific yielding potential, but they were the same whether an improvement of the yield or an optimisation of the nitrogen efficiency was aspired. This was in contrast to the nitrogen fertilisation. Application amounts that led to the highest yields exceeded the range of optimised utilisation rates. The results of the reflectance measurements show that spectroscopic methods are basically suitable for the non-destructive detection of the nutrition status of maize. An optimisation of the assessment at early growth stages, when the soil background is still covered inadequately, is essential for agronomic purpose but also seems to be feasible. A practical approach for a site-specific fertilisation can be constructed on the basis of nitrogen uptake targets for every field part. After a uniform start application of a small nitrogen rate a supplementary fertilisation would be added. The difference between the target nitrogen content and the actual N uptake, estimated by spectrometry, would determine the supplementary rate.

Variable nitrogen fertilisation for maize on heterogeneous sites

G. Huber, F.-X. Maidl and A. Schmid

Introduction Heterogeneous Fields are characterised by a spatial variability in water supply and nitrogen mineralisation. Site specific cropping takes care of these parameters. A vital approach for a variable cultivation can be the adaptation of nitrogen application and plant density to the characteristics of the soil in different field parts. Variable fertilisation regimes could be realised according to the current biomass production and nitrogen uptake. Non destructive methods for detecting the biomass and nutrition status of crop stands have to be developed. Methods Similar test plots were placed on two sites that differ clearly in cropping capacity and water supply. In the period between early June and the middle of August, intermediate harvests of the whole plant were made at six dates, every two weeks. The dry matter yield and the nitrogen concentration of every biomass sample was detected. At dough stage and at the time of physiological maturity the cob and the rest of the plant were harvested and analysed separately. An additional field test with the factors nitrogen amount, application date and cultivar was conducted on a uniform site. Four varieties that differ visibly in terms of colour and leaf orientation were grown and detected periodically using a reflection spectrometer. The recorded reflection signature was transformed into the vegetation index "Red Edge Inflection Point (REIP)" (Reusch, 1997). Results Four weeks after the emergence of the plants the spectrometer detection was still affected by the soil background. After another week of continuing growth, the plant canopy covered the ground much better and the coefficients of determination of the linear regression between applied fertiliser quantity and REIP reading got clearly higher. When the development of the stem was fully completed, the REIP showed the fertilisation very well for all varieties. The r^2 values of the regression, pooled across all varieties, are lower than those calculated separately for every single cultivar. Thus it appears that a specific calibration for every variety is required. The maximum corn yield was achieved by the highest fertilisation (170 kg N/ha). The best nitrogen utilisation was attained with 120 kg N/ha. Under every fertilisation level and on both locations a raised plant density caused a higher nitrogen efficiency. Discussion High amounts of applied nitrogen result in high corn yields but however exceed the range of optimised nitrogen efficiency. Under constant fertilisation doses, a raised plant density results in higher corn yields and in better fertiliser utilisation. Moderate nitrogen doses can lead to very high utilisation rates. The results show that the used spectroscopic method is suitable for the non-destructive detection of the nutrition status of maize. There are restrictions at early stages of growth, when the soil background is not yet covered enough. A practical approach for a site specific fertilisation can be constructed on the basis of nitrogen uptake targets for every field part. After a uniform start application of a small nitrogen rate a supplementary fertilisation is added. The difference between the actual N uptake and the target determines the supplementary rate. References Reusch, S., 1997, Entwicklung eines reflexionsoptischen Sensors zur Erfassung der Stickstoffversorgung landwirtschaftlicher Kulturpflanzen, Diss. Uni Kiel

Combine modifications for precision agriculture research

J. Hummel, K. Sudduth and S. Drummond

Use of the Global Positioning System (GPS) and grain yield monitors has become more and more commonplace in agriculture. Commercial grain producers are in the forefront in embracing the new technology, and farm machinery manufacturers are equipping new combines with the appropriate wiring harness to accommodate the appropriate sensors and monitor, either as a factory installation or by the dealer subsequent to delivery. Research scientists and engineers are asked to provide the scientific knowledge base so that producers can relate yield variability to spatial variability of other plant and soil parameters. Simultaneous collection of multiple datasets, including yield data, creates a need for a data collection workstation within the combine cab. Therefore, we undertook the development and installation of such a workstation on a full-size combine harvester. The objective was to develop a combine cab extension that would accommodate a 2nd person for data collection during plot and field harvesting operations. The combine being modified, a John Deere Model 4435, was manufactured in Germany. This combine model was marketed in both Europe and the United States. The door was removed from the combine cab, and the combine?s cab was extended to the left of the combine operator?s station. The cab extension provided a workstation for the data collection technician alongside the combine operator. This arrangement facilitated communication between the combine operator and the data technician, gave access to data collection monitors to both persons, and allowed cab heating and cooling equipment to service both work workstations. The yield monitor, other data acquisition equipment, a portable computer, an auxiliary power supply, and GPS receiver were added to the data collection workstation. The modified combine has been used to collect yield data and, simultaneously, to collect data on plant diameter, plant population and spacing in maize. Both mechanical and photoelectric plant sensors were implemented on all rows of a 4-row corn head. Real-time kinematic (RTK) GPS was used to obtain distance data, for use in calculating plant stalk diameter from the photoelectric sensor light and dark periods. A major problem with photoelectric maize plant detection and measurement at harvest is interference by plant leaves, weeds, etc. Filtering methods were developed and used to improve the prediction capability of the data set. A Windows®-based post-processing program was developed to convert raw sensor data into useful information about the corn crop. The program allows supervised comparison of sensor data with manually collected data. Data from both the mechanical and photoelectric sensors were compared with manually collected plant diameter and spacing data. Multi-season analyses show that filtering techniques can significantly improve population estimation. The yield data set and plant population data set were integrated and mapped to show spatial variation in individual plant grain yield. The implementation of other sensors being developed to assess additional plant and soil parameters at harvest will make further use of the cab extension. Plant disease, weed infestation, and insect infestation and/or damage are additional candidates for sensor development that may be sensed and measured simultaneously with yield. Thus, the addition of a combine cab extension to provide a safe workstation for a data collection technician will facilitate future precision agriculture research.

Evaluation of mineralisation zones in winter wheat on field with high variation in OM-content

Anders Jonsson, Thomas Börjesson, Bo Stenberg and Mats Söderström

The objective was to validate mineralization zones calculated based on three years mineralizations studies using "zero-N-plots". Zones of expected N-mineralization was calculated based on three years data from zero N-plots in winter wheat on a field in south-west Sweden. The mineralization, measured as N-uptake in straw and grain, varied between 16–130 kg N/ha. Seven zones were constructed and experimental plots in winter wheat were placed in 3 of them representing "expected N-mineralisation" of 30, 60 and 110 kg N/ha for site 1,4 and 7, respectively. The N-demand for each plot was calculated based on earlier years yield data and the expected N-mineralization. Three N-treatments were applied at each site; +30 kg N and –30 kg N compared to expected N demand based on yield. The treatments were fertilized three times (60 kg N, DC 30 + adjusted according to treatment, DC 32 + 30 kg N, DC 49). The plots were harvested and the N-content determined in grain and straw. The yield in the experiment varied between 6458 kg to 8361, protein content of 10,6% to 12,4% and the N-yield between 119 to 166 kg N/ha. The same N-yield, 140 kg N/ha was observed in all three sites and a similiar yield of 8000 kg was reached in each of the experiments after addition of 160, 135 and 70 kg N/ha at site 1, 4 and 7, respectively, ie reflecting the estimated N-differences between zones of 30 kg N between 1 and 4 and 50 Kg N/ha between 4 and 7. The results indicate that the adjustment was rather successful at the high OM part of the field and that unsufficient supply of N is easily detected in reduced protein content. In conclusion the zones based on earlier years mineralisations reappeared even year 4 and could be confirmed using simple N-treatment experiments. For the future a combination of expected N-mineralization and estimates of expected yields could be important PA-tools to determine correct fertilization level.

The DGPS device to investigating chosen physical soil properties and taking soil samples of undisturbed structure

Jan Jurga

In the Institute of Agricultural Engineering (Szczecin/Poland) the device for investigating some chosen soil properties meeting the needs of precision farming have been designed and manufactured. The installation is pre-programmed and needs no specialised service when making measurements. To limit costs of production it does not posses its own drive and is hanged to the three-point suspension system of the tractor. The tractor allows the device to be transported and supplies execute elements with power from its own electrical system or from external hydraulic one. The device make possible to investigate the physical soil properties on the depth up to 60 cm. Soil samples may have volume even to 1.5 dm3. Basically the replaceable measuring cylinders with diameter from 40 to 100 mm were used. The general principle of operation is as follows: on the site where measurements are executed, the device fastened to the three-point suspension system is lowered until it rests on the support foot that is also the base for measuring the sinking depth of measuring sensors. Measuring starts when the sensor of the position of the support foot comes into contact with the ground. Then, depending on the program of measurements, there has a place direct measuring of soil compaction, volume moisture or taking samples of soil with undisturbed structure. Simultaneously the site is localised with the help of GPS. All the data are logged in the deck computer. Next, these data are converted to standards meeting the protocols used in other systems employed in precision farming. In the case of undisturbed soil samples, after they are taken, laboratory analyses regarding mainly nutrient contents or physical properties such as volume moisture, porosity, weight moisture, air and water permeability are needed. Having finished the single measurement the installation is raised and transferred to another measurement point. When the cycle is ended the data are immediately ready to further use in the electronic form or as the maps of distribution of these measurement data.

Managing spatial variability of N-mineralisation within a field

Thord Karlsson, Sofia Delin and Thomas Kätterer

The goal of the project is to investigate the possibility to improve fertiliser recommendations in precision agriculture. In precision agriculture, the aim is to manage the existing variability within fields and thereby reduce the risk of N leakage and improve yield quantity and quality. A key-process for determining the amount of N to be supplied as a side dressing is the amount and timing of N that can be mineralised during the growing season. The intention was to calculate the mineralisation, using only a limited number of commonly available field data. The organic matter turnover model ICBM/N was used to calculate N mineralisation for 34 locations within a field in south-western Sweden and the results were compared with measured mineralisation during a four-year period. Data about soil organic matter content, texture and incorporated crop residues were collected from the different locations. Clay content was found to be between 7 and 27% and soil organic matter content between 2.7 and 4.6%. Weather data were collected in the field during the growing seasons and data from a meteorological station in the region were used for the remaining part of the year. To adapt the model to site-specific conditions, available site-specific data and climate data were converted by pedo-transfer functions to model parameters such as carbon and nitrogen input and decomposition rate. We discuss the potential of our approach to predict the probability distribution of N mineralisation within a field at the time for a supplementary N application for the rest of the growing season using historical yield maps, site-specific soil information and a stochastic weather generator.

Decision support technique for precision agriculture: intelligent geoinformation system approach

Dmitry Kurtener and Vladimir Badenko

The applications of intelligent technology for decision making and policy development related to agricultural management and sustainable growth have increased over the last few years. In connection with this elaboration of spatial intelligent technology for precision agriculture holds the greatest interest today. It is estimated that 80% of data used by managers and decision makers is related geographically. It is believed that the majority of agronomic information is referred to unique location based on geographical co-ordinate system. Geographical Information Systems (GIS) are important tools for spatial planning that encompasses three key activities: location choice, land suitability assessment, and collaborative decision-making Location choice is defined as the identification of the best location for a particular activity or investment. In the sense that it is searching for an optimum, land suitability assessment is similar to choosing a location, but its goal is to map a suitability index for an entire study territory. Collaborative decision support systems are user-friendly computer systems combining GIS, decision-making techniques and environmental modelling. Spatial planning involves decision-making techniques that are associated with techniques such as Multi Decision Criteria Analysis and Multi Criteria Evaluation . Another category of decision-making techniques utilised in spatial planning is based on the application of fuzzy set theory. It is well known that many elements of precision agriculture include levels of uncertainty. However, these uncertainties may not always be expressed with probabilistic methodologies. Therefore fuzzy set theory and fuzzy logic could be an alternative method to characterise these uncertainties. In this paper we present decision support technique for precision agriculture. In particular it is considered a prototype of intelligent geoinformation system for precision agriculture (PAIGS). The structure of PAIGS contains two main components: commercial geographical information system (GIS) and decision support subsystem (DSS). DSS includes algorithms of fuzzy set theory and the combination of empirical and deterministic models. In particular, it was studied the combination of MapInfo GIS with existing deterministic soil water regime model. Also it was considered interfacing MapInfo GIS with existing deterministic model of soil heat flow. Algorithms of fuzzy set theory utilised in DSS are based on methodology multi-criteria assessment and choice of alternatives, the use of triangular fuzzy numbers and a fuzzy reliability indices. An intelligent geo-information system approach (IGIS) in the agriculture domain can be used for a variety of purposes. This approach was applied by authors of this paper to assess polluted agricultural fields in order to design a strategy for territorial prophylactic actions, for land suitability evaluation in the process of agricultural experimentation, for the assessment of agricultural lands to plan site-specific residue management, and for the multi-dimensional evaluation of areas on the land market. Information in details is given in publications: Victor Yakushev, Dmitry Kurtener, Vladimir Badenko. 2002. Monitoring frost injury to winter crops: an intelligent geo-information system approach. In Physical Methods in Agriculture – Approach to Precision and Quality (editors: Juri Blahovec and Miroslav Kutelek), Kluwer Publishers, Dordrecht, The Netherlands
(http://kluwer.m0.net/m/s.asp?HB6690050063X1488931X121017Xjo%40plenum.co.uk).
Kurtener, Dmitry, Vladimir Badenko. 2002. Fuzzy Algorithms to Support Spatial Planning. In Planning Support Systems in Practice (editors: Stan Geertman & John Stillwell), Springer Publishers, Berlin (http://www.springer.de).

Forward and inverse modeling of gpr signal for estimating soil water content

S. Lambot, I. van den Bosch, E.C. Slob, M. Vanclooster, B. Stockbroeckx and B. Scheers

For a large variety of environmental and agricultural applications, the dielectric characterization of the shallow subsurface using ground penetrating radar (GPR) is a matter of concern. Actually, the dielectric properties can be related to different important functional characteristics such as soil water content and soil water composition or salinity. The main interest of the technology arises from the fact that subsurface properties can be identified by nearby remote sensing, thereby allowing non destructive mapping with high spatial and temporal resolutions. Notwithstanding considerable research and applications have been devoted to the use of GPR since the 1960s, particularly in archaeology and civil engineering for detecting buried objects, the use of GPR for assessing the shallow subsurface dielectric properties is presently still limited. For practical applications, the current state of technology still needs improvements and new developments. Research has focused on the development of an integrated approach to estimate simultaneously the depth dependent dielectric constant and electrical conductivity of the shallow subsurface from the GPR signal. This implies the design of a radar system being appropriate at the same time for practical applications and for the use of advanced GPR signal analysis methods. Indeed, it is proposed in this study to identify the dielectric parameters by inverse modeling. An appropriate forward model describing the radar-antenna-soil system is consequently required. We propose to use a stepped frequency continuous wave radar with an ultrawide band dielectric filled TEM horn antenna used in monostatic mode. This radar configuration presents many advantages compared to the traditionally used pulse radars with respect to both practical concerns and modeling concerns. Mainly, the monostatic mode allows for a more realistic forward modeling of the radar-antenna-soil system. Forward modeling was based on the exact solution of Maxwell's equations assuming the soil to be represented by a horizontally multilayered medium. The source of the antenna was reduced to a infinitesimal x-directed electric dipole and the x-directed component of the upward polarization of the electric field, denominated as the Green function, was computed at the same point. The antenna was considered as a time-invariant linear system and the Green function was accordingly related to the radar measurements using the frequency response of the antenna. Inversion was formulated by the classical least square problem. Given the inherent complex topography of the objective functions to optimize in electromagnetic inversion problems, a global optimization approach is required. We used the recently developed global multilevel coordinate search algorithm that we combine sequentially with the local Nelder-Mead simplex algorithm . Radar measurements were conducted in laboratory conditions at the Royal Military Academy (Belgium). The forward model was tested for different configurations: the antenna being above a metal sheet and above a tank filled with a homogeneous sand subject to different water content levels. Measured and simulated data for the antenna being above the metal sheet agreed very well. The inverse estimation of the soil dielectric constant was remarkably well in accordance with each water content level and the corresponding theoretical values of the dielectric constant for the sand. Comparison of GPR measurements with estimations from time domain reflectometry (TDR) were also well in close agreement. However, the inverse identification of the electrical conductivity of the sand led to less satisfactory results. In this study, we have presented and validated an integrated approach to identify the dielectric properties of the soil from off-ground radar measurements. The method allows to estimate soil water content with a high accuracy in laboratory conditions. The overall methodology is currently improved and tested for conditions closer to reality.

The basic study for yield monitoring system development of head-feeding combine (part 1)

Choung-Keun Lee, In-Gyu Jung, Sang-Chul Kim, Je-Hoon Sung, Woo-Pung Park, Won-Kyu Park and Jae-Un Park

Yield sensing is one of the basic components for precision agriculture, where investigations have recently been started in paddy fields. Yield maps have become an important tool for precision agriculture. Investigation of spatial yield variability is important both for input data for management decisions and evaluation of decision processes. However, generating yield maps by manual grid sampling is laborious work, is time consuming, and may result in inappropriate information. Therefore, a yield monitoring system is undoubtedly necessary as a fundamental tool to measure yield in real-time in paddy field. A number of methods have been developed to sensor grain flow such as impact, gamma radiation, volumetric means, optical sensors and weighing conveyors. And also a number of methods have been developed to sensor grain moisture content as dielectric constant and electric resistant methods. In upland crop areas like European countries and the U.S.A, for such crops as wheat, corn, soybeans and cotton, a yield monitoring system with yield sensor, grain moisture content sensor and GPS equipment is installed in the combine. Up to now, the various yield monitoring systems, however, are not used in paddy field areas, because the system is not appropriate for head-feeding combines. For example, the mentioned system is structurally too large and has low sensitivity at less grain flow rate than 1kg/s. Generally, the flow rate of a head-feeding combines is about 0.2~1kg/s. And also harvesting season in paddy field grain moisture content is so high compare to upland crops. Grain moisture content sensor with high moisture content and high sensitivity is need in head-feeding combines. The objectives of this study, therefore, is to develop a grain flow sensor, grain moisture content sensor which have high sensitivity for the head-feeding combines. The proposed grain flow sensor, grain moisture content sensor has been in development since 2001 in Korea. The grain flow sensor was an impact type using loadcell, impact plate, transducer, signal treatment, rubber plate and power supply. The grain moisture content sensor was an electric resistance method using two crushing roller, log amp, inlet of grain, temperature monitoring sensor, cleaning brushes and power supply. At the same time, combine speed sensor, global positioning system(GPS) and touch sensor were installed in the combine in order to generate yield map. Data acquisition system and output monitor were installed in the combine. Tested data was recorded the system. Recorded data information was grain flow rate, grain moisture content, harvesting velocity, touch sensor signal, positioning information, calculated yield information. This system was developed and tested in indoor test. The grain flow sensor was installed near the inlet of the grain tank to receive the impact of the grain emitted from the blade of the vertical auger conveyer. The grain inlet of moisture content sensor was installed the internal wall of the grain tank to receive the grain emitted from the blade of the vertical auger conveyer. Speed sensor was installed the right wall of front the head-feeding combine. Touch sensor was installed the wall of front the head-feeding combine. GPS antenna was installed the upper of the combine cabin. The data acquisition system was installed inner cabin of the combine. Each sensor and system was checked indoor test in National Agricultural Mechanization Research Institute(NAMRI). The results of indoor test, the grain flow sensor and grain moisture content sensor showed high sensitivity. The coefficient of determination was 0.99 and 0.98, respectively.

Site-specific accuracy of grain yield maps

K. Maertens, M. Reyniers and J. De Baerdemaeker

Research on grain yield mapping tools has been carried out for many years at both industrial and academic research institutes. Although different commercial yield mapping packages have proven their ability to exhibit within field grain yield variations, it is still not clear how accurate those yield maps correspond with the real crop variability. The creation of accurate yield maps is a process made difficult by a number of possible error sources. Yield measurement are subject to many random and systematic errors. To determine the quality and applicability of grain yield maps, an accurate estimation of the yield map errors should be acquired. The classical way to estimate the accuracy of yield sensors is to compare an integrated grain yield signal with the total weight of the harvested crop. Latter evaluation procedure only assesses the static error of the flow sensor and not the total accuracy of the grain yield map. To determine the effective accuracy of grain yield maps, both original and reconstructed yield maps should be compared and analyzed geo-statistically. The limited accuracy of reference measurements on local grain yield makes it difficult to estimate the real yield distribution on the field (Reyniers et al., 2002a).

Since small-scale information is lacking in case of real field experiments, virtual experiments are carried out on a drawn field with known yield distribution and accompanying geo-statistical yield properties. This virtual approach has already proven its use for the design of yield mapping systems (Maertens et al., 2002). It is possible to harvest the same virtual field several times, but with different machine settings, ground speed strategies and measurement devices. The accuracy of each individual measurement device can be quantified. The accuracies as calculated by Reyniers et al. (2002b) are used in the virtual harvesting experiments. They also suggested a method to process the data: while many of the researchers used filtering or error removing/smoothing in the spatial domain, a time domain algorithm has bee introduced. The same method is used to process the yield data of the virtual field.

The main advantage of this virtual harvest approach is that both the original yield distribution on the field and the reconstructed yield map is known. Subsequently, a grain yield error map can be constructed. The application of geostatistical techniques in agriculture has multiple and different purposes, such as descriptive studies of the spatial variability of certain crop properties. Geo-statistics uses the understanding of statistical variation as an important source of information. With the site-specific analysis of harvested yield data from a virtual field experiment, the error on yield in function of distance can be determined. In this way, it is possible to determine the ability of yield mapping systems to reconstructed short range yield variations.

Setting up and testing of an on-board guidance tool for agricultural vehicles

F. Mazzetto, R. Bonera, A. Calcante and S. Landonio

Driving a tractor during agricultural operations requires concentration and skill on the part of the driver, especially for operations which involve: a) the use of tools with a particularly large working span, and b) the impossibility of using the classical mechanical marking systems (row markers, foams, guide-wheels). A new on-board guidance tool for agricultural vehicles was designed and produced by the Authors with the dual purpose of: a) reducing the cost by creating a modular architecture based on the use of inexpensive commercial GPS receivers and palm-top computers to obtain on-board guidance tools, able to deal with data processing as well as display the path to be followed; and b) aid the driver with specific utilities to control and facilitate his work (guided return to a point, control of missed/repeated passages, corner trajectory control, recording of the paths executed for the company database). The system can function both with and without DGPS correction (supplied by the public satellite WAAS-EGNOS, completely free). In choosing the GPS receiver, 2 commercial models were considered which were already programmed for the reception of the WAAS differential correction service: 1) Canadian Marconi Smart Antenna, with the service already available, although often working only in testing mode; 2) Garmin GPS 17N, with a service that will only be available as of 2004. Their performance was compared in a series of field trials with a high precision control receiver (Novatel Millenium), and another inexpensive commercial vehicle model (Canadian Marconi Navistar) without DGPS correction. For each trial, the same tractor was fitted with both the control receivers and the test model. The real position was determined with post-processing correction of all the fixings recorded by the control receiver. The tractor simulated the path of the vehicle along parallel paths, assuming a working width of 12 and 18 m, on 2 plots of different sizes (4 and 10 ha). The results showed how the CMC-Navistar receiver, although economical, was not suitable - without DGPS - for this kind of application as it was unable to guarantee an adequate parallelism between successive paths (with errors of even >1 m as compared to the targeted path). The other two receivers, again without DGPS, showed a similar performance with acceptable errors of around the same degree of accidental deviation from a straight-line (0.3-0.6 m). The results obtained with the CMC Smart Antenna, working with WAAS DGPS correction, cannot be considered as definitive as during these trials the service was functioning in "mode 2" (testing) as opposed to the "mode 0" which should be its working standard. Even under these conditions, however, the WAAS signal markedly reduced the path errors (always <0.5 m). In this case, the greatest problems were caused by the interruptions in the signal reception due to repeated interruptions of the service. In the end it was this model which was chosen as the most suitable GPS, for the following reasons: technical (the rugged base version, with good prospects for the final standard performance; more reliable firmware); economic (it costs approximately 500 € more than potential competitor models, but in return offers a saving of 800 €/year from the elimination of the annual fee for correction). In conclusion, it should be emphasized that: 1. for operations which do not require centimetre precision (e.g. spreading fertilizer, seed scattering, weed-killers) and which do not allow the use of traditional mechanical row markers (as in flooded rice-paddies), the proposed system offers considerable production savings as it drastically reduces overlapping and distribution errors; 2. the positioning error with GPS is however comparable to the driving errors on the part of the driver; therefore there is obviously no point in investing in high precision GPS receivers if it is too complicated to drive the tractor, with centimetre precision, along the targeted path.

Definition of an algorithm for automatically detecting and recording farm field activities

Fabrizio Mazzetto, Bonera Roberto, Calcante Aldo, Landonio Savio and Massimo Lazzari

A research to develop an automatic identification and recording system able to provide the most relevant information related to all the farm field mechanised operations was undertaken in Italian farms. The information recorded is used to build a database representing the farm 'historical memory' to be retrieved by Decision Support Systems constituting the core of a new and advanced form of farm Information Management system. The data supplied by such a system will aid farmers in their management and strategic decision-making processes and will be an essential requisite to any application of precision farming techniques. The detection system works on the basic assumption that every field activity is always carried out using specific machinery on specific locations. Each activity is an event in which the following specific objects act: tractor, implement, worker and location. These objects are provided with GPS and radio frequency transmitters sending code-signals that are automatically identified by a tractor-mounted monitor device connected to two receiving antennas. Data are then recorded into an on-tractor memory device (a simple flash card) and periodically transferred into a farm PC hard disk, where a software carries out their elaboration producing the actual timetable of scheduled operations. A specific algorithm was defined and implemented for constructing such a timetable from a sequence of elementary data, each fixing a 'photograph' of the particular operation that is currently performed. Each 'photograph' includes several details used to better explain what type of 'photograph' is. This is a sort of 'reduction approach' that could be explained even throughout a literary metaphor: a single farm 'operation' is analogous to a period (i.e., a complete sentence) in a novel. An operation is composed by a set of farm main and subordinate 'activities' (analogously to the main and subordinate clauses composing a period), on their hands determined by a sequence of elementary 'working states' (the words). Finally, each state is composed of a set of 'elementary working conditions' (the letters in a word), each of them providing simple but essential information on the type of work currently carried out. The full comprehension of a complete sentence is reached only when the grammatical and syntactic rules of the language used to write the novel are well known. Such rules drive the reader with a constructive, down-top approach throughout the steps (letter)-(word)-(clause)-(period). In the same way, we defined specific 'grammatical' and 'syntactic' rules to build the farm history database (the novel) starting from the sequence of the elementary data recorded by the system throughout the steps (elementary working condition)-(state)-(activity)-(operation). In the algorithm here described, an 'elementary working condition' is a Boolean value calculated based upon the sequence of the data recorded into the flash card. A working state, on its hands, is a set of 8 different types of elementary working condition (i.e., our vocabulary is composed only of 8-letter words) having a fixed position order with a different Boolean 'meaning' (e.g.: 'an implement is coupled to the tractor', or 'the tractor is moving', or ' the tractor is inside the perimeter of a field'). The resulting 'value' of a working state is firstly used to distinguish between actual working and auxiliary states, where the meaning of 'actual work' state strictly depends on the type of the implement coupled to the tractor. The sequence of several auxiliary and working states leads then to the construction of activities and operation lists. Some application examples and experimental results will be presented and discussed in the paper. They refer to operations carried out in a 60-ha arable farm where the prototype of the detection system is currently used and tested.

Teaching principles of precision technologies with a variable-rate seeding simulator

G.E. Miles, D.L. Dux and D.R. Ess

This presentation will describe a Variable-Rate Seeding Simulator (VRSS) used in formal academic courses and workshops offered by Purdue University. The Variable-Rate Seeding Simulator is a mobile apparatus which simulates all the functions of variable-rate seeding in a laboratory environment. A CNH series 1200 planter row unit is mounted on a frame above a treadmill which collects and conveys seeds to where students can catch them. The speed of the electric motor which drives the treadmill is controlled by manually depressing the up or down button on the control panel. The planter metering device is turned by a chain drive connected to a Rawson Accu-rate controller. The power required to turn the stepper motor in the Rawson Accu-rate controller is provided by a Parker-Hannifin hydraulic power supply which is connected to an electrical outlet. The number of seeds metered by the row unit per rotation of the Rawson stepper motor must be calibrated once and recorded in the Accu-rate controller. The Rawson controller adjusts the speed of the seed metering device on the row unit in accordance with the desired plant population and the treadmill speed which is sensed by a Dickey-john radar. The plant population can be set manually on the Accu-rate controller, or electronically by an Ag Leader PF3000 monitor. A map of seeding rates must be transferred to the monitor by a PCMCIA data card. A program running on a laptop Personal Computer (PC) is used to send GPS position data in the form of NMEA strings to the PF3000. Although the GPS data must be pre-recorded, this provides a convenient method of simulating movement without the hazards of being in close proximity to a machine operating in the field. The PF3000 accepts the simulated data just as if they were sent from a real-time DGPS receiver. Once the field position is determined from the GPS NMEA string, the PF3000 uses the seeding rate map to determine the desired plant population and sends the appropriate signal to the Rawson controller. The VRSS can be used to provide experiential learning experiences related to calibrating planter seed metering devices, variable-rate seeding, and automatic controls. In each case, students can interact with the apparatus by programming it to deliver a desired seeding rate, then measuring the results. For planter calibration, the Accu-rate controller is set to a desired population and row width, and students collect seeds for a specified time period (usually 30 seconds or 1 minute). Then the students must calculate the seeds per land area (seeds per acre, or seeds per hectare) using the row spacing and the seeds per minute data. For variable-rate seeding, students begin with building a map using a GIS. Usually SSToolbox is used, but other mapping software packages are available. Unless students have collected their own data, they must use a field for which a GPS path has been recorded previously. The seeding rate can follow any formula which the student chooses, but they are given references to agronomic extension publications. Once maps are constructed, they are transferred to the PF3000 via a PCMCIA card. Students follow instructions for operating the VRSS and collect seeding rate data at prescribed locations in the field, and compare the results to the planned seeding map. For studies in control, students collect seed drop data for a range of seeding populations and forward speeds. The actual data are compared to the desired data and the potential sources of error are discussed.

An integrated GPS, and internet based GIS application for the support of farm level agricultural advisory systems

Tamás Németh; József Szabó and László Pásztor

Precision agriculture requires three main components: accurate positioning, real-time technology for controlling agro techniques and appropriate spatial databases that provide information necessary to obtain the promised benefits. In this context soil represents one of the most important factors to be involved, and the larger the applied scale the more accurate is the achieved result. For site-specific management the relevant map-equivalent scale of spatial information is at least 1:10,000. Larger scale base data could provide better results but can seldom obtained. It is partly large-scale spatial soil information systems that made the evolution of precision agriculture possible. Agricultural advisory and recommendation systems summarize current knowledge on soil fertility, nutrient supply and limiting factors with respect to the expected yield. In Hungary in the 70's a systematic program was introduced for controlling the plant nutrient status of cultivated soils. In the 80's a new computerised fertiliser system was elaborated, which was used on more than 500,000 ha. After the political and economical changes in the late 80's - early 90's, which had drastic influence on the Hungarian agriculture, a demand rose for a new, cost saving, environmentally friendly fertiliser recommendation system. The next step in the improvement of these advisory and recommendation systems should be the extension of their spatial features that is integrating them with GIS. GIS adaptation and digital reambulation of large-scale information on land resources originating from various surveys and sources has become a key issue in Hungary due the recent challenges (like requirements raised by Hungary's EU-accession, introduction of national agri-environmental program, operational practice of precision agriculture etc.). At the same time, access and interpretation of existing data and information are problem areas. A complex GI system was elaborated by the GIS Laboratory of RISSAC in close co-operation with the Plant and Soil Conservation Service of Fejér County for the whole territory (about 3,500 ha) of a co-operative farm in Central Hungary. In addition to carrying out the digital reambulation and GIS adaptation of large-scale soil surveys further information was integrated into the system: (i) topographic data, ensuring the delineation of the geographical environment and helping information processing and positioning; including a DEM; (ii) cadastre information: property register data determining the ownership and registration conditions of parcels; (iii) multitemporal data of nutrient control based fertilization advisory system monitored yearly. As a next step an Intranet/Internet based realization of the system was worked out using Autodesk MapGuide Program. This type of user interface can be learnt and used by anyone and could provide fast access to the spatial data and is operating system independent. Internet also provides other useful advantages since Internet browsers are available almost everywhere and this technology provides access to the server for unlimited number of users. The digital processing and treatment of spatial databases (GI systems) has been so far dominated by immovable, geographically bound technology. A great breakthrough is provided by the handheld computers i.e. Personal Digital Assistants (PDAs). Our intranet GIS server can also be reached from these mobile clients, which (supplied with suitable GPS-cards) make GIS supported, real-time, in-situ survey possible. The intranet based GI service together with the mobile technology jointly can be efficiently used during the preprocessing works and the elaboration of sampling design, on the field as well as during the postprocessing period. Consequently, the elaborated system can prove to be very useful in the whole process of the agricultural counselling practice thus it is not suggested to be used only on a single demonstration area, but can serve as a model for the preparation of similar applications all around the country.

Airborne remote sensing and plant disease management : A case study with *Gaeumannomyces graminis* pv. *Tritici* on winter wheat in west of France

H. Nicolas, N. Garcelon, Z. Thomas, Y. Fouad, P. Lucas and M. Lennon

The objective of this study is to reduce pesticides applications on crops by site specific management The purpose is to analyse some potentialities and limitations of airborne remote sensing for plant diseases management in field conditions of Western Europe.. A spatial description of the rapid changing pathological status of the crop is therefore essential. We analyse the optical range (visible, near infrared) and the thermal range (thermal infrared) aerial images. The agronomic support was focused on winter wheat infested by Gaeumannomyces graminis pv. Tritici, a telluric pathogen. The different infestation levels were created by appropriate rotations of crops. This gravity of a pathogen attack is hard to quantify and to map by only ground observations with an adequate accuracy for site specific management. During the crop growth at specific phenological stages, we reported on each spot observation, the percentage of nodal roots infested in the whole root system and calculated locally an average gravity index of crop disease. The experimental site was located in the west of France The remote sensing data in the optical range were acquired from an aircraft scanner (CASI radiometer) with 9 and 5 spectral bands. Radiometric distortion was corrected with ITRES software. Geometric distortion was corrected fist from the aircraft attitude data and then, completed by mapping polynomials and ground control points for a better accuracy. In the thermal range, the images were acquired from a thermal infrared camera. The geometric distortion of thermal images was corrected by mapping polynomials and the crop radiative temperature was determined from ground measurement with handheld radiothermometers at the same time than aircraft scanner measurements. In the optical range, we evaluated several methods of image processing for their possible use in an agricultural context, when only a small amount of information is a priori available about crop disease. An appropriate methods give access to information that permitted classes identification. We compared three supervised classification methods after selecting training data from the spectral space. In addition, we obtained a class image by applying a threshold method to a vegetation index (NDVI) image. Results were different from one method to another. The best results when compared to ground observations, were obtained with the maximum likelihood method and the class image obtained from the NDVI. The data available for classes identification are the spectral characteristics of the training data and the mean vegetation index value of each class. A knowledge of plant-pathogen relations and spectral characteristics of crop is therefore essential to identify the classes on thematic images. Finally, the intersection of the two previous thematic images provide the best accuracy for plant disease mapping. In the thermal range, radiative temperature of infested plants was less than 1K higher than radiative temperature of healthy ones and the relation between radiative temperature and plant disease gravity is scattered. The radiative temperature difference changes under different climatic conditions and the quantification of the attack is not easily accessible. The use of thermal images only is consequently difficult for plant disease application at field level. But, when used simultaneously with optical remote sensing, thermal images bring new relevant information for class identification. The extraction of mean radiative temperature from each class obtained from optical range images is a pertinent parameter with high physical significance. These results help to characterize the place of the remote sensing techniques in plant disease management at field scale. In addition to remote sensing, data fusion methodologies could be consequently a relevant and complementary tool for mapping pathological status of a crop.

Exposure and animal induced soil compaction in a grazed silvopastoral system

L.C. Nwaigbo

Topographic exposure (topex) effects and soil resistance to a probing instrument in a predominantly rye grass (Lolium perenne)pasture planted with sycamore (Acer pseudoplatanus L), and Hybrid larch (Larix eurolepis Henry)) at different spacings and grazed by sheep were studied. There were two spacing treatments for sycamore -SYC100 (10m x 10m) and SYC 400 (5m x 5m). The Hybrid larch had three spacing treatments - HL 100 (10m x 10m), HL200 (7m x 7m) and HL400 (5m x 5m). All the treatments were replicated over 3 blocks of land in a randomized complete block design. Each plot was 0.8ha. The plots were grazed by sheep yearly from April to October and received an annual application of 160kg Nha-1. The Glensaugh research station site of the Macaulay Land Use Research Institute is 38km South of Aberdeen. Located in the Grampian region, it is about 180m above sea level. Exposure is considered in relation to the topographic features surrounding the site which integrates a number of site characteristics such as aspect and geomorphic shelter. This is relevant as the effects of increasing elevation on the wind exposure of site are modified by the influence of surrounding topography. On exposed sites there is always the need for the animals to use trees as shelter from wind. The presence of surrounding high grounds can be particularly important in reducing and accentuating wind speed which may influence animal behaviour and movements. These plots were therefore assessed for degree of exposure, soil compaction around the trees caused by animal foot traffic with recourse to compass direction. The influence of local topography on site was assessed in degrees and were classified based on exposure ratings. This topographic exposure was determined as the total skyline elevation angle in eight compass directions - N, NE, E, SE, S, SW, W and NW. All the six plots of SYC100 and SYC200 were found to be moderately exposed. The plots of HL400 were moderately sheltered while the six plots of HL200 and the HL100 were moderately exposed. The resistance of a soil to the penetration of a probing instrument (cone penetration) is an index of soil compaction. The soil compaction as measured by a penetrometer is indicative of resistance to root penetration. Two trees 60-75m apart were randomly selected in each plot of SYC100, SYC400, HL100, HL200 and HL400 for soil compaction study. Assessment was made around each tree in 8 compass directions using the hand - held Bush recording cone penetrometer. In each direction assessment started at 0.5m from the base of the tree and at 1m intervals. Expected penetration depth was 52.5cm and readings were taken at 3.5cm dept intervals. These were recorded in kilogram force (kgf) and converted to megapascals (Mpa). Soil compaction was observed in each plot at a horizontal distance and depth from the trees. Highest values were observed closest to the base of the trees in each plot. As exposed site, soil compaction due to animal traffic as influenced by direction of wind on the site was explored under (i) North-axis and (ii) East-axis using 5 code levels. For the North-axis, the compass directions taking a deviation from the north with the code values in parenthesis were N(1), NE(2), E(3), SE(4), S(5), SW(4), W(3) and NW(2). For the East-axis, taking a deviation from the West, they were W(1), NW(2), N(3), NE(4), E(5), SE(4), S(3) and SW(2). Soil compaction as evidence by the search for shelter by animals in the Hybrid Larch plots was unidirectional in the North-axis while in Sycamore plots it was in the East-axis, a reflection of the exposure levels of the plots.

An automated method to locate optimal soil sampling sites using ancillary data

David Olsson and Mats Söderström

Production of variable-rate application maps for fertilisation and liming in precision agriculture has made evident that the common grid soil sampling strategy sometimes is inadequate. It has become increasingly common to use other information (for instance electrical conductivity), with higher spatial resolution, in an effort to make better predictions of various soil properties. One way in which such information may be used is to modify the soil sampling scheme, in order to obtain the samples from locations which are likely to produce maps which capture the within-field variability of the field. The goal here was to develop a tool for simplifying the process of interpreting secondary information. A software was developed which produced suggestions of soil sampling sites based on such information. The strategy was based on backward interpolation, which means that interpolation of the secondary variable, measured at the sites suggested automatically by the program, would create a similar surface as a map created from all samples of the same variable. The problem is to find the sites which minimise the differences between these maps, and at the same time fulfil some other criteria, such as fairly even spatial coverage and not sampling too close to field borders. Three methods were implemented: i) Directed sampling: the points suggested will be selected without restrictions concerning distances between samples and is selected one at a time at the location which deviates most from the original map ii) Stratified extreme sampling: the field is grided and in each cell is selected 0-3 sites depending on the field area in the cell. Local min and max are sampled. iii) Stratified directed sampling: this is a combination of method i and ii which successively selects samples in grid cells (until a maximum number is reached) at locations which deviate most from the original map. Method i and iii performed best in a comparison with grid sampling. The latter method should be more appropriate since we also may analyse soil properties that are not directly correlated to the ancillary data.

Radar reflection to analyse soil and crop variability

Wolfgang Paul and Hermann Speckmann

In recent years there has been remarkable progress in radar technology. Components have become smaller, the radiated energy had been reduced to a very safe level and the working frequency had been increased. For process engineering there are many distance measuring radars for tank gauging at a price of just over 1000 € on the market. For safety aspects in cars a new generation of radar equipment has been announced, probably leading to new possibilities of measurement at an even better price / performance ratio. So regarding the behaviour of soil and crop at radar frequencies in the GHz range is evident. The basic principles to analyse the backscattering of a radar beam are understood. But the intrinsic irregularities and variability's of agricultural matter can not be modelled in all aspects. The purpose of this presentation is to examine the possibilities of these emerging technologies for non contact measurements in precision agriculture. To do this, first an overview is given on the radar principles and the theory of radar backscattering. With the newly developed pulsed radar as well as with the FMCW (frequency modulated continuous wave) -radar the power of a reflected signal over the distance can be recorded. This is done by measuring the signals time of flight and by recording the relative weakening of the echo compared to the transmitted signal. In agricultural applications the echo is determined mainly by the water content, because of the very high permittivity of water compared to soil or organic matter. Then the basic principles of radar backscattering, attenuation, dispersion and depth of entrance will be discussed. It can be shown that the reflected power is under agricultural circumstances equivalent to the change in the apparent dielectric number at the interface between air and crop or air and soil. But disturbances due to an uneven surface or a stochastic distribution of leaves or stems must be considered. Nevertheless, it is shown that as practical results the detection of a crop edge, the measurement of the soil humidity, the assessment of the crop density and an alarm on soil compaction can be given. In detail the assessment of crop density for exact fertiliser or agent inputs and the measurement of soil humidity for precise irrigation are presented. Regarding crop density, as examples the echoes of wheat with different densities are given. The reflections at the crop edge are mapping the density. The discrimination is even greater when measuring in horizontal mode with a metal reflector. Here the reflected power is inverse to the crop density, because the wheat is radiated in a pseudo-transmission mode. Similar results are obtained with other crops (corn, barley, sugar beets, grass). Regarding soil humidity, reflections when looking vertical into soils are given. Dry and moist soil can clearly be distinguished. With the given radar device (working frequencies 5.8 or 26,1 GHz) one can reach only a shallow depth of penetration. So at these frequencies the moisture in the upper layer of the soil can be detected. At lower frequencies the available water in deeper layers will also become measurable. It is known from airborne geo-radar investigations, that the reflected signal depends on the free, plant available water between wilting point and field capacity, so giving an excellent signal for irrigation. As conclusion, with the emerging radar technology new possibilities for non contact measurements in precision plant production will arise. The developments in car electronics will lead to further improvements.

Which role can play precision techniques in the sustainability of mountain agriculture?

R. Raffaelli, G. Gios and O. Clauser

The object of this paper is to contribute to substantiating which role Precision Agriculture (PA) can have within a strategy which aims at pursuing sustainability in mountain agricultural production. The starting point is the belief that mountain development does not solely depend upon the possibility of utilising the available local natural resources but, above all, on the profitable adoption of new production technologies and new organisational models, even if, in many cases, these were designed for completely different contexts. This could be the case of Precision agriculture meant as "the management of part of fields for actual needs rather than whole fields for average needs "(Mangold 1995). The environmental potential of this new technology has been in evidence since the beginning of the scientific debate but empirical studies of its effects in mountain areas do not exist. Nevertheless it is here argued that environmental sustainability is to be inherently found in this new technology whereas it is also important to verify if this technology is economically and socially sustainable. The first experiments in precision viticulture in Italian highlands seem to confirm that these techniques may be adaptable to permanent arboriculture in the mountains. But there are difficulties to overcome regarding the use of both mechanised farming methods and information technology. Given these difficulties, PA appears as non economically sustainable if considered only from the point of view of costs. But if positive externalities, product quality and traceability are properly assessed, conclusions can be completely different. From the point of view of social sustainability, the diffusion of precision agriculture technology can bring some substitution of non-specialised forms of work with more technological advanced ones and will surely influence the sector's structure. Therefore, a government of the technological transition is necessary. But there are other relevant conditions in order to avoid mountain zones being excluded from this process of innovation and evolution. First of all there is the necessity "to adapt" this technology to the specific environmental context of the mountain territory. The adaptation process regards not only the technology in itself but also its objectives -in connection with farmers' overall strategy- and the related performance monitoring indicators. To be successful this process requires a strong involvement of farmers together with the scientific research organisations operating in the territory. In order to render this technology adoptable, it seems necessary to intervene by using economic incentives (grants or subsidies) on one hand and a taxation system of variable inputs on the other. On the institutional side, "the development of institutions that will enable us to take advantage of the technological capabilities we possess" (Zilberman et al. 1997) may constitute a big challenge. This could be the case of community or co-operative management centres for PA services to be created in those contexts where the tradition of associated service management is solid. On the other hand, public institutions are necessary in order to facilitate the development of "measurement methods, new approaches to basic agronomic research, unbiased evaluation of pros and cons of PA technology and education and training (Heimlich 1998). Finally, all interested parties (farmers, technology providers, policy makers) must become conscious and aware of PA implications. Given all these conditions, a coherent plan of political and economic measures appears as indispensable to "promote, adapt and adopt" PA techniques. Only in this case, notwithstanding some limitations and negative aspects, precision agriculture can become an interesting opportunity -from the point of view of sustainability- for Alpine areas. In fact, precision agriculture may reinforce the pursuit of overall quality and product traceability, contribute to recover the "social legitimacy to produce", represent an opportunity of human capital development and enhance a new "connection to the rest of the world", helping to reduce the risk of marginalization of some Alpine areas.

Investigations of positioning systems of shuttle gantry aggregates

Bareisis Ramunas, Viselga Gintas and Sniauka Povilas

All kinds of vehicles with mobile technique damage the soil. It is very important to diminish the damage but traditional means are quite limited. The diminution of driving number and track area is still rarely used means. These indicators can be improved while using field gantry systems. The classification and analysis of gantry modules revealed significantly wider possibilities for energy saving and determination of working parts moving accuracy, while moving shuttle straight-line, when positioning shuttle modules were used. Here the implements are directed along the beam and the chassis of their carts are moving along the hard beam with the least resistance. The energy demand for beam chassis and the movement of implements does not coincide in time. This diminishes the installed power. Expressions to define the undriven field area were determined with an attempt to compare shuttle modules with cantilever and uncantilever beam taking into consideration the indicators of driven field area. The area of module with uncantilever beam ruts is insignificantly smaller in the limits of 1,0-2,5 % than the area of module with cantilever beam. Module with cantilever beam makes an additional rut in contiguous plot. When the beam is longer than 18 m and the width of wheels is 0,4 m, the level of field area usage approaches to 93-94 %. Experimental researches, rolling wheel and stepping chassis imitative modules through the field, showed that the biggest influence of field relief and micro-roughness is made on wheeled chassis curvature, while the smallest is on the curvature of stepping chassis, moving on the tops of field roughness. When the imitative model cross was stepping on the random contact points of field roughness, the data of track curvature are intermediate. The constant use of rectifiers for both kind tracks brings their curvature and divergency of inter-rows nearer the dependence on relief. These researches grounded the expediency of stepping chassis with track rectifiers in the technological scheme of positioning shuttle modules. While using the set automatic positioning and straight-line movement maintenance system in the field laboratory conditions, the chassis directed along the laser ray, when the distance between photodiodes is 2 cm, deviates from the straight-line ±1 cm. The deviations of the second chassis did not exceed 5 cm and depended on the roughness of track and relief. For the beam moving between positions with the speed of 0,5 m it was enough to have the 4,4 kW power installed with reserve. When the engines of stepping chassis and other modules operating in 'start-stop' conditions are replaced by hydraulic automatic drive it is possible to cut down the power and energetic consumption. Experimental researches allowed to draw up the alignment chart of efficiency dependence on the main parameters of module and operation conditions. As the efficiency and energetic consumption mainly depend on the length of beam, lengthening this parameter the analysis of possibilities to avoid the more significant bend of beam, while projecting lintel beams and shroud stiffened suspension bridge construction, was carried out. Profile 'T' for implement cart movement is recommended to be used in the lower beam construction in all the cases. Practical analysis gives a chance to lengthen the lintel beam up to 30 m, and the shroud stiffened suspension bridge construction one even more. The, if no other components of efficiency are changed the energy consumption would decrease about 1,17 time if compared with 18 m beam energetic module and about 2,15 times if compared with 6 m beam energetic module.

Development of a pistachio yield monitor

U.A. Rosa, B.C. Heidman, P.H. Brown and S.K. Upadhyaya

An electro-mechanic weighing system based on a load cell was designed and installed on a commercial pistachio catch-frame harvester. Yield data obtained with the weighing system during the first year of this study indicated the transport lag in the conventional catch-frame corresponded to approximately the time required to harvest five trees. However, the objective of this study was to develop technology to measure individual tree yield based on the catch-frame harvester. Another catch-frame harvester was completely re-designed for use with the weighing system, in a subsequent year, to reduce the transport lag. When the harvester works at commercial speeds all the nuts harvested from the current tree need to be transported through the harvester in about 15 seconds, before the next tree is harvested in order to avoid mixing of nuts coming from different trees. Experiments were conducted with the re-designed harvester in a 32 ha area selected in a pistachio orchard at Paramount Farming Company, California. Over 6000 trees were successfully harvested in this area by using the pistachio yield monitoring system. The results showed that the transport lag problem was successfully resolved. The combined weighing system and the re-designed catch-frame harvester accurately measured individual yield of trees in the pistachio orchard when the harvester worked at commercial speeds. Tree identification and positioning were obtained for production of yield maps. Tree yield measuring and mapping were successfully accomplished by using data obtained from the yield monitoring system. Aspects of the system design, results showing extent of yield variability and mapping will be presented in this poster.

Draft force mapping as a tool to map spatial variability of soil properties

B. Schutte and H.D. Kutzbach

Site specific farming in precision agriculture requires knowledge about spatial differences and properties of soil in order to make management decisions for sowing, fertilizing etc.. Current practice is analysing soil samples, taken at several points in a field, to win the knowledge about the variable soil physical properties. Due to the large amount of work in collecting and analysing the soil samples in some cases mapping of electrical conductivity is taken to describe the variability of soil within a field. However, these data has to be analysed subsequently and statistical calculations are necessary in order to get area wide soil property maps. Maps off electrical conductivity have to be critical analysed because the measured parameters depend on a lot of soil properties like soil type, water content, salinity or soil compactions. The recent development in precision agriculture has shown the benefit of automatically techniques during anyway necessary working processes such as yield mapping on combine harvesters.

A promising way to collect data about local soil properties is the position-related measurement of draft forces during conventional tillage operations as cultivating or ploughing. Especially in view of modern tractors which are already equipped with sensors to determine draft forces, mapping of draft forces in connection with a (D)GPS-system might be a low-cost, time-saving and exact way to win information on spatial variability of soil properties.

Former experimental investigations have shown the potential of mapping draft and/or PTO–forces values during the conventional tillage operations. The measured data is influenced by lots of parameters. Once there is the tillage tool itself, than different soil parameters like moisture, compaction, soil type and humus matter and process determined influences like speed, working depth and width. At the Hohenheim Institute of Agricultural Engineering these measurements were carried out by the use of an 6 component measurement frame between tractor an implement. Six load cells measure all three components of forces and torques between tractor and implement. In combination with high accuracy position data of an RTK GPS unit it is possible to map the spatial variation of draft force and PTO - force over a plot.

A new far-reaching measurement system allows logging of the engine and ground speed, working depth/width, implement draft forces (draft measurement frame), implement draft forces (lower link draft sensors), fuel consumption and positioning data during normal field work. Measurements can be done with various implements like ploughs, cultivators or rotor harrows. In addition the influence of soil moisture and soil compaction on the draft values is being examined. The test fields are selected to determine the practical value of draft maps for site-specific managements systems. A comparison between other ways (mapping electrical conductivity, soil samples) to get soil information in site specific management is accomplished in the test fields. With regard to the implementation of draft measurements on standard tractors and implements in practice the suitability of the lower link draft sensors compared to the measurement frame is tested. To enable the application of mapping draft forces in practical use under various conditions and with different implements/tractors algorithms on base of the experimental data are drawn up.

Possible adoption of precision agriculture for developing countries at the threshold of the new millennium

F. Shahbazi

Possible adoption of precision agriculture for developing countries at the threshold of the new millennium feizollah shahbazi scientific member of lorstan university faculty of agirculture department of farm machinery khoram abbad iran email: shabbazi @ yahoo. com fax: + 98 661 430289 tel: + 98 661 430012 Abstract Precision farming is an attractive option to promote sustainable agricultural development. It is now pos sible to base re source utilization and manage ment decision-making on inf ormation and knowl edge. Many supporting technologies are being developed quickl y with possible ap pli cation in agriculture. Precision farming presents both challenges and opportuniti es for agricultural engineers. The 21st century is a century of biological and information technology revolutions. The fundamental forces for ensur ing worldwide food security and human welfare will lie in scientific and technological progress. Development of precision agriculture practices will provide good opportunities for renovating conventional farming tools and technology, as well as for disseminating advanced knowledge to farmers. In the forthcoming knowledge-based era, a good vision of what will be happening tomorrow is the key to the further success of modern farming communities. The concept of precision agriculture, based on information technology, is becoming an attractive idea for managing natural resources and realizing modern sustainable agricultural development. It is bringing agriculture into the digital and information age. The practice has smoothly extended into some developing countries. This paper considers the possible adoption of precision agriculture for developing countries and ideas in conducting the practice in Iran. Keywords: Precision agriculture; Developing countries; Strategic approach in agriculture

Developing sustainable cropping systems - characterization of a clay soil

Maria Stenberg, Mats Söderström, Carl-Anders Helander and Karl Delin

An efficient use of added nitrogen and of site-depending available nitrogen is one of several aims of great concern for developing sustainable arable farming systems. On sandy soils excess nitrogen is easily leached through the soil profile as nitrate. On heavier soils with higher clay content, denitrification and other processes for nitrogen losses can be more crucial. Nitrogen use efficiency is of importance for production economy as well as environmental concern, product quality and bio-diversity. For this, there are guidelines and laws for measures in agriculture. This study is carried out on a farm in south-west Sweden having soils with an average clay content of 40 %. Organic matter content varies between 2-3 %. Since 1991, the farm has been managed divided into three different arable farming systems without manure: conventional farming (A), organic farming (B), and integrated farming (C). Each system has an eight year crop rotation. In both B and C, green manure leys with forage legumes are used. In B, nitrogen fixing crops are grown every second year. Yields of all crops are recorded every year, from 1999 by yield mapping. All crops and all measures are documented since 1991. Crop and soil nitrogen dynamics have been followed in reference areas. In 2003 the tile drain system at the farm will be replaced by a new system. Changes in soil conditions after tile draining will be followed. For this, the soil is being characterised during 2002 and 2003. Soil physical, biological and chemical parameters will be determined. Soil sampling points were chosen by using the software FuzMe to create zones from soil electrical conductivity determinations (EM 38), topography data and data from soil mapping in 1991. Discharge from each field will be measured and drain-water analysed with the new system. Losses of N2O from the system will be quantified as well.

Potential applications of wireless internet technology in agriculture

M.D. Steven, D.L. Swain, K. Raatikainen and S.C. Tan

Advances in mobile communications offer opportunities for applications of Wireless Internet Technology (WIT). A desk study addressed potential applications in agriculture, with particular emphasis on Malaysia. Generic issues include the need for spatially distributed information services, i.e. in the field, and whether the application is time critical. The evaluation of different agricultural sectors also considered the size distribution of the enterprises and their spatial concentration. While farmers are clearly the largest target market, the group most likely to see immediate benefits in WIT are managers and advisers in a diverse range of support services. Such users could benefit from mobile access to a regional GIS containing background information on land ownership, service provision, administrative details, crops planted, environmental sensitivity, satellite image data etc. Data and analyses could be downloaded to the field and the results of field checks transferred rapidly for processing. In precision agriculture, a similar approach could be used to link tactical decision-making in the field to strategic information held on a more detailed farm-scale GIS. In livestock agriculture, animal health scares such as foot and mouth and BSE have initiated a more detailed procedure for recording animal movements. Greater automation in this process could be achieved by M2M wireless interrogation of animal transponders or ear tag bar-codes, with data automatically downloaded to a central database. Automated systems could similarly monitor animal health, nutrition and fertility. Timely intervention with artificial insemination is particularly valuable in maintaining the production cycle. In Malaysia, the ministry provides a free service to identify insect pests, fungi, viruses and other plant diseases and weeds. Many of the symptoms are visual, suggesting the use of an integrated phone and camera such as the Nokia 7650. As well as speeding up the identification process to allow more timely action, it would cut costs and increase diagnostic capacity by reducing the need for expert visits to field sites. It would also eliminate any risk of disease transmission by contaminated samples. Alternatively, two-way image transfer via a multimode radio card such as the Nokia D211 to a portable computer would allow the farmer to refer to a reference key to make a preliminary diagnosis. Once the problem is diagnosed, appropriate treatment could be guided by interaction with an online Decision Support System, taking into account the stage of crop development, the cost of remedy, market prices and weather information. Paddy rice is mostly produced in large irrigated regions, requiring complex management of water. Published plans include telemetric monitoring of rainfall and of water levels in rivers and dams, as well as remote control of water gates, offering scope for M2M applications of the kind supported by the Nokia 30 connectivity terminal. Malaysian plantation crops would benefit from WIT for labour management. Whereas rubber is mainly grown on smallholdings and is in long-term decline, oil palm is 8.7% of GDP and increasing. It is mainly concentrated in large commercial estates and major national schemes, suggesting a greater likelihood of resources to invest in new techniques. The application of wireless internet technologies in agriculture also requires consideration of data transfer rates. 3G technology can transfer data at up to 2 Mbps and would be acceptable for most applications, with the possible exception of online decision-making on moving vehicles. This rate would be reduced by competition for signal, but given the likely density of users in rural areas and the infrequency of high intensity demand for data transfers, it seems unlikely that competition will be a major problem.

Development of coverage and gray scale measuring system for field information acquisition of paddy rice

J.H. Sung, C.K. Lee, W.P. Park, I.G. Jung and S.C. Kim

Fertilizer application is very difficult, delicate and laborious. The farmer should decide on the amount of fertilizer to be applied by looking at the leaf's color, by considering the field's history and variety as well as the area's climatic conditions, among other factors, and ultimately, from experience. Furthermore, for optimum fertilizer application, site-specific management is needed that considers variances in the growth status of crops on the field. Manual application also becomes more difficult as the size of the field grows. Environmental issues in agriculture likewise have to be considered, such as the impact on the environment and ground water contamination. To address these issues, precision agriculture (PA), also called variable management, has been developed. The key concept behind PA is the automation of fertilizer application and its optimization based on the crop's health status, which result in reduced agricultural inputs and environmental impact. The key feature of fertilizer application automation under the PA method is the development of the sensor, which assumes the role of the human eye. To increase the effect of fertilizer application, the sensor should be non-destructive, should not make contact with moving agricultural vehicles and acquired with other growth information. The PA method also effectively senses light reflectance. The reason for this is that sunlight is reflected, absorbed or transmitted on the crop's canopy in an open field. The canopy's health status or natural disposition also transforms this energy from the sun. This study was carried out to develop a coverage and gray scale measuring system for a paddy field. The system thus developed grab the image of rice canopy on the paddy field using color VGA image sensor, and calculates the coverage and gray scale of rice canopy at the position data by way of a GPS receiver. This system was stands alone system and a part of field information acquisition system. The results of this system were compared to other field information. The gray scale of rice canopy was compared to the value of SPAD, 25 points was averaged at unit mesh, measured by SPAD-502, height measured by manually using scale, 9 points was averaged at unit mesh, and the value of chlorophyll meter measured by CM100. All of these data were acquired 5 by 5 meter square (unit mesh) in 30 by 100m field and compensated by light intensity (quamtum) that was measured by light meter LI-250. And compared to the intensity of satellite image (KOMPSAT-1, Korea). The satellite image was acquired 6.6 by 6.6 meter square and re-sampled by 5 by 5 meter square at 8 bit resolution. The coverage was compared to the amount of quantum, value of phyranometer and value of photometer these were acquired by light meter (LI-250) for measured to grow luxuriantly of paddy rice. These data were acquired upper side and down side of paddy rice. And compared to the tiller no that was acquired by manually at 5 by 5 meter square in 30 by 100m field. The system could group the coverage and gray scale by health status for five stages over 70% accuracy, and for three stages over 80% accuracy.

Optical sensortechnique for site-specific application of farm chemicals

Eiko Thiessen and Hermann J. Heege

A lot of information about site-specific characteristics of a field can be obtained with optical sensors. This presentation focusses on the nitrogen supply of the patches as well as on the plant stress, in particular fungal infestation. While the technique for the former already exists and is in use, the latter still requires further investigations before plant protection by means of optical sensors can be tested in practice. The reflectance of the crop, normally caused by solar irradiance, serves as a measure of the nitrogen supply. This is due to the strong influence of the chlorophyll concentration and leaf area index to the reflectence spectra in the visible and near infrared. The stress can be detected much earlier by means of fluorescence analysis than by means of reflectance which only changes with chlorosis or necrosis. The spectra and kinetics of fluorescence are influenced by plant substance, such as phenolic compounds, and physiological state. These parameters can indicate an infestation before visual symptoms occured. The red edge inflection point of the reflectance spectra is taken as a sensorvalue for the sensor-controlled nitrogen fertilisation. There is a broad variability of the sensorvalues within a field. This is due to the nitrogen supply as well as the soils' characteristics. The sensorvalues alone are not a measure for the nitrogen requirement of the patches; they serve only for the plants' characterisation. Neverless, the sensor-controlled nitrogen fertilisation requires a calibration, i.e. each sensorvalue is related to a nitrogen dose. The tests on sensor-controlled nitrogen fertilisation in cereals show that sensible calibration increases average yields while at the same time saving nitrogen input compared to uniform application. These calibrations have to be carried out with regard to locally applied crop science aspects. Using only the calculation of the sensorvalues to create a theoretically optimal, automatic adjustment of the nitrogen dose, the site-specific calibration was compiled. This method uses a reference strip that had been created to that purpose. However, it is too complicated for practice. It was shown that the plot-specific calibration usually suffices in practice: a nitrogen dose is simply assigned linear to the sensorvalue. For this type of sensor-controlled nitrogen fertilisation, correspondence with crop science nitrogen recommendations, a more even crop and less lodging referring to uniform application are documented. A weak point of this calibration is that it considers only the sensorvalues. It should be eliminated in the future by using a map overlay. The basic studies on the identification of fungi infected leaves with the help of fluorescence indicate a possibility for the site-specific controlling of a plant protection sprayer with fungicides. Variable chlorophyll fluorescence suggests a large potential for the infestation diagnosis through its causal linkage to the plants' physiological status - the photosynthesis. The constant blue fluorescence serves as an indicator for phenolic compounds in the leave, such as phytoalexines, and hence reveals a defence mechanism of the plant against fungal infection. Infected and healthy leaves were measured with fluorometers during the timeperiod from infection until the appearance of symptoms. Modifications show up both in blue fluorescence and in induction kinetics with infected leaves compared to healthy leaves even before visual symptoms occurred. Particularly, newly compiled characteristics, like the normalized slope of fluorescence in a time interval 5 s after actinic illumination, were usually more significant than the classical characteristics Fm and Fo. The minimisation of other stress factors as disturbances and the earliest possible infestation recognition are still aims of research.

Mississippi cotton yield monitor: research, development, and commercialization

J. Alex Thomasson and Ruixiu Sui

The spatially variable information that is most important to the practice of precision agriculture is crop yield, which affects all management decisions involving profitability. Knowledge of the variation of yield within a field requires a system to monitor the harvesting process. To build a yield monitor for any field crop, one must combine a position sensor with a real-time crop-flow sensor. Positioning technologies, largely GPS, have become readily available and increasingly affordable, and fairly reliable crop-flow sensors have been developed for grains. However, sensors have been slower to develop for certain crops, notably cotton, which flows sporadically in inhomogeneous clumps of mixed fiber and seed. Several attempts at developing cotton-flow sensors have used optical transmission in the harvester air-flow stream. Some of these attempts have recently acheived commercial status. Scientists at Mississippi State University developed a novel optical cotton-flow sensor in 1999, and they used that technology in the Mississippi Cotton Yield Monitor, which has been under development since then. One prototype of the yield monitor was field tested in Mississippi in 1999, and three prototypes were constructed and field tested in Texas, Georgia, and Mississippi in 2000. About 530 ha of cotton with different varieties and large yield variations were harvested. In the Texas test, average absolute load-by-load errors for the two fields were 5.9% and 5.4%. Results from the Georgia test showed an average absolute error of 5.7%. System reliability was tested in Mississippi by harvesting more than 445 ha of cotton, and it was found to be quite good. Based on these results, a new version of the Mississippi Cotton Yield Monitor was designed to include anti-stray-light and temperature-compensation features. Five prototypes of the new version were fabricated and field-tested in 2001 on three cotton pickers and two cotton strippers at five locations in Georgia, Texas, and Mississippi where 1230 ha of cotton were harvested from September to December. Results from the tests showed that the average absolute errors of the system on a cotton picker were 3.7% and 4.9% in two fields. Results with a cotton stripper also indicated average absolute errors consistently below 5% over eight days of harvesting. Cotton yield maps were created with the data, and the maps realistically exhibited yield variations within the fields, based on the expectations of experienced producers and consultants. The tests also indicated that the system was reliable, and easy to install, operate, and maintain. The technology on which the Mississippi cotton yield monitor was based has been issued a provisional patent and also submitted for US and international patent protections, which are currently pending. Mississippi State University has contracted with a private company in order to give the company an option to manufacture and market the Mississippi Cotton Yield Monitor. A new data acquisition system was developed as part of the commercialization efforts, and 10 full prototype systems were tested on commercial and research farms in the 2002 season.

Algorithms for sensor based N-fertilisation

Anton Thomsen, Ole Møller Hansen, Jørgen Berntsen, Henning Hougaard and Rita Hørfarter

During 2001 and 2002 plot experiments were placed on 6 privately owned farms. Three farms were included each year. The major objective of the experiments was to develop algorithms for sensor based nitrogen fertilisation of winter wheat. Sensor measurements included the spectral canopy index, RVI, and soil electrical conductivity. RVI was measured using a set of two band (red and near infrared) spectral sensors. Soil electrical conductivity was measured using the EM38 instrument. The selected fields were all highly variable with respect to soil texture and the experiments were placed in order to include as much of the soil variability as possible. Some of the fields also included significant topographical differences. At all sites the general layout of the experiment was the same including typically 140 plots. The individual plots were fertilised at 4 nitrogen levels: 60, 120, 180 and 240 kg N/ha. All plots were fertilised with 60 kg N during early spring. The remainder was applied during early may. All plots were harvested at ripeness using specialised equipment. Based on the considerable amount of data three algorithms for optimising the grain yield with respect to N-fertilisation were developed: (1) algorithm based on spectral (RVI) measurements, (2) algorithm based on measurements of electrical conductivity (EM38) and (3) algorithm combining both types of measurement. The algorithms were developed from statistically significant models describing the relationship between N-fertilisation, sensor measurements and yield levels. The algorithms were limited to redistributing a specified amount of N-fertiliser within the field. By applying the algorithms the expected yield was calculated and compared to a uniform N-application. For all the included sites the yield increase was small and in all cases less than 1 hkg/ha. The small yield increase was found for both low (100 hkg/ha) and medium (160 hkg/ha) N-levels.

Mobile TDR for geo-referenced measurement of soil water content and soil electrical conductivity

Anton Thomsen, Per Drøscher and Flemming Steffensen

Water availability and soil texture largely govern the yield potential of agricultural fields. Soil properties often vary significantly between and within agricultural fields. This variability is both the major rationale and challenge for developing precision farming methods. The development of site specific crop management is constrained by the availability of sensors for monitoring important soil and crop related conditions. Time domain reflectometry (TDR) instrumentation has become the preferred technique for measuring both soil water content and soil electrical conductivity. Electrical conductivity and water content are measured on identical soil volumes. To exploit the potential of the TDR technique for rapid geo-referenced measurements a farm tractor mounted frame, supporting hydraulic subsystems for the insertion and retrieval of a robust TDR probe, has been developed. The design has proven to be sufficiently robust for making a high number of measurements under varying soil conditions. The TDR probe is vibrated and not simply forced into the ground. This allows the instrument to be used on dense and stony soils where the probe would otherwise be damaged and require frequent repair. A compact electronics module including a PC104 format PC, display, TDR instrument and DGPS receiver is placed inside the tractor cab. The mobile TDR system has been used for the measurement of water content and electrical conductivity under spring and summer conditions. The early measurements are related to the amount of plant available water. Summer measurements have been related to crop water use. Combined measurements of water content and electrical conductivity are closely related to the clay and silt fractions of a variable field.

Visualisation methods in precision agriculture

Jonas Tornberg, Mats Söderström, Susanne Van Raalte and Johan Martinsson

Background and objectives The intensive spatial and temporal data collection in precision agriculture makes it interesting to use non-traditional mapping and visualisation methods. Geographic information systems (GIS) are in general used for the management and analyses of precision agriculture data in two-dimensions, but videos, animations, 2.5 and 3D GIS models as well as true virtual reality models are possible options when we want to present and facilitate interpretation of complex data that vary in space and time. This poster exemplifies how such methods can be used to visualise and model data in a precision agricultural context. Applications 1. Virtual reality (VR) may be defined as a computer-based representation of a space in which users can move their viewpoint freely in real time. The level of detail is accomplished through a combination of photographs and spatial objects. Simple VR-models may be created directly from vector data in some geographic information systems and used on the Internet without expensive software. In this study we have created three examples: a simple model of a test farm, a model for presenting soil maps to farmers, and a more technically demanding test of the use of a 3D cube. 2. Precision farming data are typically spatio-temporal and animation is an excellent technique to introduce the temporal component of spatial data. To exemplify this, we produced a video clip from repeated Hydro N-Sensor scannings, an animation of the effect on soil pH of precision liming and a fly through/fly-over model from aerial photographs, digital elevation models and yield maps. 3. Even if it is not common in precision farming, soil-sampling data may be available from several depths in some fields. In general, it is possible to generate perspective views in most GIS, but it is more difficult to visualise true three-dimensional GIS-data. Here is visualised three-dimensional soil data for an agricultural field by combining different types of simpler presentation techniques such as various diagrams, animations and VR-models.

Geostatistical analysis of precision agriculture yield maps

S.R. Vieira, P.R.R. Martinho, S.C.F.Dechen and I.C. De Maria

The field work involved in this study was conducted in two private farms. One of them is located at Campos Novos Paulista, SP, Brazil, has 35ha of summer soybean, autumn corn or milet and black oats as a winter green manure crop, all of them under no tillage. The soil at this location is a Red Latosol, with high aluminium saturation and medium texture (LVd1 - CNPta). The other field is located near Angatuba, SP, Brazil, in an area of 77ha, under center pivot irrigation system, in which no tillage is used for the rotation of beans, corn, and potatoes, where the potatoes are planted using conventional tillage. The soil at this location is a Red Latosol, with high aluminium saturation and clay texture (LVd2 - Ang). Field sampling was done according to: 1. For chemical analysis, the LVd1 - CNPta was sampled on 50x50m grid, while the LVd2 - Ang was on 100x100m grid. At each sampling point, one sample composed of 9 sub samples, collected on 5x5m grid, represented that point for 0-10cm, 10-20cm, and 0-20cm depths; 2. At the LVd2 - Ang, and at the 0-20cm depth, additional samples were collected as follows: a) The sample collected was divided in two and sent to two different laboratories in order to assess the variation between them and its effect on precision agriculture recommendation; b) In 5 locations within the field, the 9 sub samples were sent to the laboratory for analysis without bulking, in order to evaluate the sampling scheme; 3. Samples were collected for particle size distribution analysis on a 100x100m grid on both locations, at 0-20cm depth; 4. Permeability measurements were made using constant head field permeameter at 10 and 20cm depth, on 50x50m grid on both locations. Crop harvesting was done on this same grid. All sampling points were georeferenced using a GPS with differential correction, in order to apply geostatistics and construct and manipulate the resulting maps. The main objective of this work was to correlate the soil variables among themselves and with crop yield in order to identify homogeneous management zones. In general, all variables were well represented by the sampling at 50x50m grid, except for the permeability parameters which indicated the need for further sampling at finer spacing. The data referring to soil fertility, sampled at 100x100m grid, presented a larger number of variables which showed pure nugget effect semivariograms or larger values for the nugget effects, indicating weak spatial dependence. The particle size distribution parameters were well represented by this sampling spacing. The LVd1 - CNPta presented medium spatial dependence for the soil fertility variables at 0-10cm depth, and showed large coefficients of variation indicating the effect of the fertilization and the no tillage mulch. The use of the potato crop on the LVd2 - Ang causes changes on the 0-20cm depth leading to a more homogeneous layer. None of the parameters evaluated showed strong correlation with the crop yield. Crop yield at both sites seemed to have been affected by unidentified causes. At the LVd1 - CNPta, even though there was not a strong correlation between any of the variables sampled and crop yield, potassium content had the highest correlation. For the LVd2 - Ang, the bad distribution of water by the center pivot system seemed to be the major cause for the variability in yield.

Weed - crop discrimination using Neural Network computing : application upon multispectral images

J.B. Vioix, J.P. Douzals, J.W. Lu, F. Truchetet and J.P. Guillemin

In order to minimize economic and environmental consequences of agriculture, site-specific applications of pesticides and fertilizers are studied. Prior to do theses process a characterisation of agronomic variability must be done. For example, site-specific applications of herbicides involve first to detect and to localize weeds. As a research level, this information would also be useful to understand the mechanisms and consequences of weed-crop competition during their growth. Plants detection and characterization are widely studied through spectrometric measurements. This method generally gives reflectance spectrum (in visible and near infrared wavelenghts) that are specific of each plant. To give quantitative results, light conditions must be strictly controlled. That's why this method is rather restricted to laboratory measurements. Moreover, as spectrometer are generally composed of a linear CCD, spatial information cannot be integrated. An alternative study is to collect radiometric values of few spectral wavelengths using pass-band filters. When several bands are taken into account it is called multispectral imagery. In this case, the use of a matrix CCD allows two dimensions images analysis, then the spatial information can then be integrated. Multispectral images can be obtained either with number of CCD sensors coupled with specific band-pass filters or a single CCD and various band-pass filters that can be rapidly shifted in front of the CCD. Due to a lower cost, the second way was chosen in our case. Nevertheless a non-neglictable time was spent for electronically controlled synchronizations (CDD and carousel, specific exposure time of each band-pass filter). The objective is to integrate this acquisition device into a remote control aircraft but first trials related in this paper were conducted at ground level. Our study concerned onion plants growing with several weeds into micro-containers. For each scene, 4 images (768*288 pixels) are acquired representing the reflectance information for the red, green, blue and near-infrared bands. As we weren't able to formalise the spectrometric relationships between reflectance and plant type (weed or crop) a priori, a simultaneous analyzis of theses four images has been done. In these conditions, the most appropriate method is a Neural Network The network entries are the values of each image component (R,G,B and nIR) for each pixel. The learning rule is based as follow : 10 pixels owning to weeds are labelled with the value of -0.95 and 10 pixels owning to crops with the value of +0.95 as a basis of the learning rule. A better result is obtained when integrate soil pixels (as value 0) . An improvement of the discrimation among crop or weed leaves borders is obtained. This method gives valuable results and allows real-time processing. No specific information about either filters band-pass or CCD gain is necessary to allow images computing. Moreover the learning rule is 'capitalized' for further analyzes and lowers the results of subsequent shifts of light conditions. Further developments will concern the integration and combination with spatial information. Previous works showed that weed / crop discrimination was possible when crops appears in row. The results mentionned in this paper would allow to improve the global methodology.

Development and evaluation of decision rules for site-specific nitrogen fertilization of wheat

Peter Wagner and Georg Weigert

The potential benefits, both economic and environmental, of variable rate nitrogen fertilizer application have been widely reported during the development of precision farming. The benefits depend largely upon the quality of appropriate site-specific N-management recommendations within partfields. In spite of the fact, that variable rate technology is already in use, there is a lack of valid management guidelines for fertilizer application. Furthermore, one of the major constraints in adoption of precision farming techniques by farmers is this lack of readily available, definitive guidelines on variable rate nitrogen fertilizer management. Three major nitrogen fertilization techniques can be identified. These techniques can be used for both, variable or constant N-application: 1) applying N-fertilizer in one, but more common in three to four rates (intensive N-strategy). 2) fertilization in one or two rates with or without the use of nitrogen inhibitors. 3) NH4-injection in one or two rates with injection spreaders (CULTAN). Each of the techniques needs a specific set of decision rules. The decision rules presented in the paper deal with the first strategy throughout. Furthermore three specific approaches of site-specific N-fertilization exist: 1) Mapping approach (fertilizer decisions based upon thematic maps). 2) Sensor approach (fertilizer decisions just based upon the reflection behavior of plants). 3) Sensor approach with map overlay. The decision rules presented in the paper mainly focus on the third, most sophisticated approach. Two different decision strategies will be presented. The strategies have been developed within the project "Information System for Site-Specific Management & quot; (http://ikb.weihenstephan.de) founded by the German Research Foundation. 1) "rule of thumb" strategy: site-specific fertilizer decisions made by human decision makers. The site-specific fertilizer recommendations depended on three different "management-zones". The mapping approach with sensor overlay has been tested against the mapping approach and against two reference strategies. The experiment has been conducted in 2000 on a 10 ha field. The results presented show ecological effects by means of nitrogen efficiency and economic effects by means of gross margins of each of the four strategies. 2) "automatized strategy", site-specific fertilizer decision made by algorithms. This approach is based on the finding, that farmers will not accept any strategy, which is time consuming and expensive. Almost all of the traditional "whole field approaches" do not fulfill this condition if it comes to site specific applications. But, due to precision-farming-techniques new possibilities arise. New kinds of data can be gathered, not known before. There are, for example, EM 38 data (electrical conductivity) which can lead to site specific water capacity statements or draught force resistance data as an indicator for the texture of soils. So far it is not very clear, which of these "new" data are meaningful to apply for decision rules. The paper wants to show which possibilities exist. Furthermore an experiment has been conducted in 2002 in which a new algorithm based on automatized gathered data (yield maps and REIP (red edge inflection point) data) has been tested against uniform fertilization. The results of this experiment will be presented, again by means of nitrogen efficiency and gross margins. Although these very early results do not lead to euphoria, the potential of those highly automatized decision rules can be seen. Additionally the paper wants to emphasize the role of different production functions in the process of creating decision rules by means of an example.

Development of a precision aerial imaging and orthorectification system

N. Wells, S. Pocknee and C. Kvien

Aerial images have been shown to be a valuable tool for highlighting and monitoring within-field variability. Despite the existence of much evidence supporting the use of aerial imagery in farm management, farmer adoption has been limited. Cost, spatial resolution, timeliness, format, and availability issues are all factors in this lack of adoption. Manual rectification of photographs is a tedious and time consuming process which involves the matching of known coordinates with features that can be unambiguously identified in the image. In an attempt to address the inefficiencies of current methodologies a project was initiated to develop a relatively low cost imaging system to generate high resolution, fully rectified digital images with minimal post processing for rapid turnaround. To provide a cost effective solution only commercial ?of-the-shelf? (COTS) components were used. The resultant system includes a six megapixel digital camera, an inertial navigation system (INS) consisting of a Differential Global Positioning System (DGPS) and a Fibre Optic Gyroscope (FOG) Inertial Measurement Unit (IMU), on board computer, and post processing software. Mission planning and flight guidance is achieved using the TrackAir aerial survey system. This software provides comprehensive planning options, ensuring optimal coverage of the target image area by triggering the camera through timed or coordinate based firing. The system measures the position of the camera perspective center in three axis of space (x,y,z) as well as the three dimensional orientation (omega, phi, kappa) of each photograph. Combining data from the INS and Camera requires precise time synchronization. This is achieved by using the flash synchronization port of the camera to send a pulse to the INS each time an image is captured. Data from the camera system is then input directly into aerial triangulation software (ERDAS Orthobase) where block adjustments are performed and individual images are mosaiced into a seamless coverage of the area of interest. The entire system occupies a volume of less than one cubic meter, making it suitable for use in most small single engine aircraft that have a ?belly mount? camera port. It is portable and can be moved from aircraft to aircraft in minimal time. This paper describes the economic and practical rationale for building this system. The system design and component hardware is discussed. Additionally, results from several aerial calibration tests are presented. At several stages in it?s development the system was flown over farmland containing geo-referenced aerial targets. Currently, average absolute errors of 3 to 4 meters in the resulting imagery are being observed. These errors are similar to what was previously seen using manually digitized and geo-referenced data. The system has been successful in drastically reducing the man hours involved in capturing and processing aerial photography. The higher initial investment in technology should be quickly offset by the reduction in processing costs and the greater volume of data that can be accumulated and marketed. Currently, steps are being taken to further streamline the processing of the data. We are also looking to implement an efficient methodology for making the aerial imagery available and accessible for farmers. The appropriateness of an online distribution system is being investigated.

Mitigation of environmental impacts and precision agriculture

Elith Wittrock, Tania Runge, S. Bauer and H. Strasser

Large-scale projects like road construction or urban development require extensive compensatory measures to balance the impact of these projects on nature and landscape. In Germany the implementation of compensatory measures is regulated by law. Farming is usually affected twice through an impact. Areas are not only withdrawn from agricultural production through the projects but are also needed for compensatory measures. Farming can now become involved in a new service sector such as taking over the implementation of compensatory measures on agricultural areas where cultivation and ecological improvement go hand in hand and are honoured with an appropriate payment. Within the framework of a project supported by the 'Deutsche Bundesstiftung Umwelt' concerned with 'Impact Mitigation Regulation and Agriculture ' (Eingriffsregelung und Landwirtschaft) a method has been developed to determine the requisite payment for the farmers. This happens on the basis of agricultural production systems and under consideration of necessary compensatory measures A computer-based economical and ecological management model (MODAM) for the integration of nature conversational objectives in farm production has been further developed. This model allows an economic and ecological evaluation of compensatory measures. The model computes the low cost solution for ecological improvement for any agricultural business/ farm taking into account the internal company correlations as well as the requested amount of ecological improvements. The ecological requirements can be varied step by step in different model variants in order to estimate the spectrum of possible compensatory measures in the respective farm. On the one hand precision agriculture can provide methodical help to control the implementation of compensatory measures by means of a small-scale definition of cultivated areas. On the other hand the presented method is a step towards small-scale ecological and economical optimisation. It helps to determine specific cultivation measures on the farm in view of economy and operating procedures and their effects on the environment.

The analysis of the spatial variability of field environmental factors for precision agriculture

Gaodi Xie, Wenhui Chen and Qingqing Zhuo

China is a country with large population. Precision Agricultural production is of great strategic significance to her. It's known to all that the inherent soil nutrient factors always have an effect on the yield. This paper is mainly about the processing and analyzing of the data collected in the experimental section. Based on the experimental data, we have worked out the variability of farmland basic environment information and their relationship with the variability of paddy yield, and tried to make an explanation for it; by which, can we provide scientific basis for the fertilizing of precision planting. To obtain basic information for variable rate fertilizer for rice production, spatial variability of soil chemical properties were evaluated in a slight alkaline paddy field approximate 130 hm2. Two hundred and eighty-one surface soil samples (0-20cm) were collected from each of the 57m×57m plots after harvest to investigate the spatial variability of their chemical properties: organic matter, hydrolysis nitrogen, available phosphorus, and available potassium. Their spatial distribution has been mapped by using different kinds of interpolation methods. These maps are very useful references for variability fertilizing in precision paddy field management. The densely soil sampling, that explores the spatial variation of soil biochemistry characteristic, is the starting point of precision agriculture concept and development. Through the practice of precision agriculture in recent 20 years, soil sampling on the contrary has become the bottleneck for its extension and deepening. For reasons of costs of both of economy and time for soil sampling and experimental analysis, the spatial intervals between samples have to be at a level of several decade meters or even more than a hundred meters. The spatial resolution of produced soil maps are hardly matching with yield map, and nor with the control accuracy of variable application machinery. It is significant to improve the accuracy of practicing precision agriculture and constructing mathematical models for revealing the relation between crop yield and its environment for precision agriculture, through indirect methods to raise the resolution of soil maps and evaluate its reliability. Based on the evaluation of interpolation methods, we adopted Kriging interpolator to produce soil maps of different factors in Shanghai Base, using 280 soil samples of systematic sampling of 3 samples per hectare. The comparison between Kriging and inverse distance function interpolating was carried out using the same data set, the result showed the Kriging interpolator is better to show spatial variation than the later in this situation. We use the dispersion index to indicate the spatial variability of the environmental factors in the scale of the field. Comparing with the yield Variability, Total P, Available P , Cu, Zn, Salt, SO42-, Cl-, Ca2+, K+, Na+ show a Great Variation, Available K, B, Fe and HCO3- show a middle Variation, SiO2, Al2O3, Fe2O3, CaO, K2O, MgO, Na2O, TiO2, MmO, Organic Matter, Total N, Available N, Mn and pH show a little variation in the scale of field. That means the factors with great variability in the field play a more importent role than other factors. With field precision farming or management should be implied according to the spatial variation of these factors.

Basic study of a GPS-operated boom-sprayer

Chung-Kee Yeh and Ying-Tar Liao

In the past, the field operation of the farmers was identified: same spraying rate of pesticide and herbicide in field for the plant protection. In order to increase the crop productivity of field, variable spraying rates will be applied to the specified fields recently and these are called precision spraying. The first purpose of this study is to establish a precision spraying system of the boom sprayer that can be controlled by a GPS. The sprayer can be operated at a destined field by a fixed flow rate. The necessary apparatus and instruments include: 1) Trimble GPS Pathfinder Pro XR: includes a receiver, an antenna and a data collector/controller; 2) Control and analysis software: includes Asset Surveyor, Pathfinder Office, ArcView and ARC/INFO; 3) Boom sprayer: manufactured by Zetor Co., Model: ZTSL500; 4) Sprayer control kit: produced by Spraying System Co., Model: 744A; 5) GPS controller: made by a domestic Company. The methods are: 1) Actual measurements by GPS: Positioning by an actual dynamic path. 2) Actual operation of the boom sprayer in field: a) Coordinates saved in a CF card by means of a computer. b) CF card inserted into the GPS controller. c) Boom sprayer operated in field, observed whether the spraying valves are actuated. d) A signal lamp mounted for the purpose of monitoring in a long distance. Results and Discussion: 1) The positioning experiments of GPS show that all the errors can be kept within three meters, either the error from the static positioning point or the error from an actual measurement by various dynamic paths. 2) When the boom sprayer is operated in a destined field, a lamp will flash and rotate. The control valves will be actuated simultaneous for spraying. If the sprayer leaves the preset field, the lamp will not flash, the valves will also be closed and the pesticide will not be sprayed any more. The second purpose of this study is to develop a variable-rate spraying equipment for a boom sprayer operated under a precision agriculture system. This variable-rate spraying equipment is used to achieve the pesticide and herbicide applications according to different predetermined spraying rates that are judged in classification of insect pests and spread of weeds in the field at the right location by the right spraying quantity. The experiment methods are described as followed. First, a variable-rate spraying system was designed and assembled. In the spraying system, system backflow rate is used to control the spraying rate. Then this system was tested dynamically in order to study the influence of various opening values of the proportional control valve vs. backflow rate under the different pressure setting of the spraying pump. Finally, the data was analyzed by means of the regression method to get the characteristics curves of the system performance. These experimental results show that for the backflow rate the proportional valve controllable range is within 11%~64%. The range is then divided into 3 stages: 20%, 40% and 60%, with 5 various pump spraying pressures, so 15 different variable-rate spraying effects could be got. For the purpose of variable environment, the spraying capacity is from 7 LPM to 34 LPM.
Key words: GPS, GIS, Boom sprayer, Variable-rate spraying, Precision agriculture.

Measuring soil complex permittiivity within a low frequency range

Naiqian Zhang and Kyeong-Hwan Lee

Complex dielectric permittivity of soils includes both conductive and capacitive properties. These properties are closely related to several important soil physical and chemical properties, such as water content, salt content, organic matter, texture, pH value, and density, that affect crop yields. Fast, in-field measurement of soil complex permittivity can provide valuable information to assist decision makings in precision-agriculture practices. Commercial soil-conductivity sensors only allow fast measurement of soil conductive properties. It is believed that simultaneous measurement of both conductive and capacitive properties of soils is possible only when the measurement is conducted at high frequencies. This presentation reports the progress made by the authors towards a fast, simultaneous measurement of soil conductive and capacitive properties made within a relatively low frequency range. Laboratory tests have been conducted to simultaneously measure soil water and salt contents based on a four-electrode method. Soil bulk density and electrode penetration depth were strictly controlled during the experiment. A Partial Least Squeres method was used to establish calibration models for predicting water and salt contents within the frequency range of 1Hz to 15 MHz. The R-square values for predicting water and salt contents at the 30-mm penetration depth were 0.88 and 0.83, respectively. Hardware and software to realize real-time measurement using this technology is under development. Based on the progress made during the laboratory tests, the authors believe that the sensor, the experimental procedure, and the statistical models developed in this study have demonstrated a great potential for a practical technology that would allow fast, real-time, simultaneous measurement of multiple soil properties.

Full non reviewed papers for ECPLF

Australian Precision Livestock Farming (PLF) workshops

Thomas Banhazi[1], John Black[2] and Matthew Durack[3]

[1]*Livestock Systems Alliance, South Australian Research and Development Institute, Roseworthy Campus, Adelaide University, Roseworthy, SA 5371, Australia*
[2]*John Black Consulting, Warrimoo, NSW 2774, Australia*
[3]*National Centre for Engineering in Agriculture (NCEA), The University of Southern Queensland Campus, Toowoomba, QLD 4350, Australia*

Abstract

Recent technology adoption and R&D planning workshops conducted in Australia significantly contributed to the development of current livestock management practices. Producers, scientists, consultants and other industry representatives attending the workshops were leading experts in different areas of pig production, including Information Management, Intelligent Bio-system Engineering and Livestock Management. It was generally agreed during the workshops, that information technology has to become part of modern livestock management systems and improved data acquisition, data analysis and production control systems have to be utilised by producers, if livestock industries are to remain competitive.

Keywords: technology adoption, information technology, precision livestock farming.

Introduction

"Precision Livestock Farming" (PLF) principles and techniques are already widely utilised within the broad-acre and row-crop industries of Australia and overseas. The principal component of precision farming in these industries is the development of accurate real time yield or performance mapping systems. The development of new sensing and data management systems enabled the development of analogous systems within the Intensive Animal Industries. In other industries the key benefit of these technologies has been in allowing producers to target specific areas in their production more efficiently for improvement (Lemin *et al.* 1991). The potential benefits of PLF systems are believed to include:

- Improved production efficiency and optimised level of inputs, which will limit waste and decrease the costs of production.
- Better marketing and health control of animals as animals are surveyed throughout the growth period.
- Improved scientific understanding of nutritional, health and environmental effects on the animals and the opportunity to undertake detailed on farm production R&D at limited cost to the producer.
- Improved IT management on farms, including the increased adoption of management softwares within the Industry, such as AUSPIG and PIGPULSE.
- Greater level of management control without human intervention, which in turn can reduce the potential for human error and allow faster response to identified problems.
- Improved traceability and product consistency and through these improvements, ultimately improve product safety for consumers.

Most of the technological components of Precision Livestock farming (PLF) systems such as climate control equipment, automated feeding systems, computer models and decision support softwares are well developed and readily available commercially. However, the integration and on farm implementation of PLF components need further research and development work (Frost *et al.* 1997). Therefore, the aim of the first workshop was to bring a body of experts together and generally examine the opportunities and barriers of implementing a real time "Precision Livestock Farming" (PLF) system in the Australian pig industry. Specifically, the second workshop aimed at listing all important actions needed to enable implementation of electronic measurement, analysis and control systems on Australian pig farms that will result in improved efficiency of production, enterprise profitability and long-term sustainability.

Materials and methods

The methodology used during the implementation of the workshops had the following main components:
Pre-workshop review of PLF technologies – All participants (prior to attending the workshops) were provided with an extensive report to achieve a level of common understanding regarding terms, status and potential opportunities of the technology.
Organisation of National workshops - A representative cross-section of the industry was invited to identify the level of interest and potential expertise in the precision livestock farming area.
Summary of workshop outcomes - A final report/workshop summary was produced, with specific advice to the main funding body representing the interest of the Australian pig industry in relation to potential research opportunities.

Main presentations

The main aim of the first workshop was to review the current status of PLF technologies in Australia and overseas and to identify the opportunities in the Australian pig industry for use of these PLF systems. The following presentations were given during that workshop:
- Introduction and review of principles of PLF – T. Banhazi (Australia); Review of International developments in the PLF area – Prof. C. Wathes (UK, SRI); Electronic aids including AUSPIG and continuous improvement strategies – Prof. J. Black (Australia); Technical review of Precision Farming – Dr M. Durack (Australia, NCEA); IMS related projects at Silsoe, UK – Dr T. Demmers (UK, SRI); Bio-system Engineering in the Livestock Industries – Prof. R. Gates (USA, University of Kentucky); Practical experiences with Precision Farming – H. Crabtree (Farmex, UK); Individual identification of farm animals – Prof. I. Naas (Brazil, University of Campinas); Data analysis, interpretation and its value – D. Rutley (Australia); From Data to Informed Decision Making in the Pork Industry – M. Schuster (Australia, QLD DPI); How consultants can assist producers – Dr C. Cargill (Australia); Summary – A. King (Australia)

Group discussions

The main aims of the group discussion components of the workshops were to (1) evaluate the capacity of the Australian industry to adopt these technologies, to (2) develop a plan for the appropriate implementation and (3) to generate a team of experts capable of driving forward the suggested projects.

Results-issues discussed at the workshops

Both workshops were successful in attracting a group of high quality participants. During the first National PLF workshop, delegates were presented with an extensive review of available PLF technologies, while during the second workshop, priority research areas were identified.

Need for improvements

The presumption in promoting the PLF approach is that the utilisation of modern production management principles in livestock industries will create a framework, which will ensure the identification of inefficiencies and allow for continuous improvement of animal production, which in turn will improve profitability (Black 2001). Data acquisition, data management and process control were the main areas identified during the workshops, which need urgent improvements. Based on the workshop participants understanding of research in the area to date, It was agreed that the essential components of a well-designed and controlled production process should be:

1. Incorporating data acquisition systems into the production chain (Frost 2001)
2. Establishing protocols for data-integration and automated data analysis to identify inefficiencies and facilitate analytical decision making (Schofield *et al.* 1994)
3. Transferring the results of data analysis as inputs into automated decision making processes and activating certain management actions (Banhazi *et al.* 2002)
4. Activating control systems, which could be either automated or appropriately documented manual control functions (Gates and Banhazi 2002; Gates *et al.* 2001)
5. Using different procedures (feedback-loops) to monitor the outcome of control actions (Black 2001)

Suggested production process diagram for livestock industries can be seen in Figure 1.

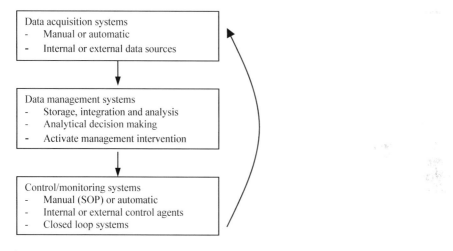

Figure 2. Data management and control diagram for a livestock enterprise (Banhazi et al. 2002).

Data collection systems

Electronic capture of the information is essential to reduce the labour cost and improve reliability of the information that must be collected if pig growth models are to be readily accepted by commercial pig enterprises. In the past, information collection on such wide ranging parameters has been difficult and costly and was therefore often neglected (Black *et al.* 2001). However, recent advancement in sensor, computer technology and modern analytical tools have made the collection and analysis of large amounts of data highly feasible (Black *et al.* 2001). During the second workshop participants specifically agreed that the following measurements should be made electronically on farms:

Pig weight and variation, Pen and sow feed intake, Pen and sow water intake, Shed temperature, Shed humidity, Shed dust concentration, Outside temperature, humidity, wind speed and wind direction, Shed air speed, Proportion and duration of wet skin on pigs in pens, Shed CO_2 concentration, Shed ammonia concentration, Audio capture and analysis for disease diagnosis, Video capture for identifying behavioural changes, disease and mortality, Oestrus detection, Shed and outdoor odour levels

Data integration, analysis and modelling

Pig simulation models should in the future become an integral part of integrated pig management systems. The models would be capable in real time of identifying individual pigs that are of the correct weight and body specifications to maximise payment form a wide range of buyers. The pig growth models would be capable of assessing whether individual pigs or groups of animals have an excess or deficiency in amino acid supply for every diet presented, whether ambient temperature is above or below the zone of thermal comfort and whether the pigs are under- or over-stocked (Black 2002). In addition, the models could assess whether the pig is being fed an energy intake that is coincident with the minimum needed for maximum protein deposition, where the efficiency of feed utilisation is maximised. The factors limiting feed intake could be predicted and excessive feed wastage identified. Diets could be reformulated automatically at specified times to most economically meet the nutrient requirements of each group of pigs and feed intake modified to achieve greatest economic efficiency of feed utilisation (Wathes et al. 2001). Shed temperatures, air movement and pig wetness could be controlled to ensure that the animals were in the zone of thermal comfort for sufficient time each day to maximise profit in relation to constraints on growth and costs of environmental control (Gates and Banhazi 2002). Similarly, stocking rates and diet composition could be adjusted through the automatic movement of pigs and modifications to diet energy density to ensure maximization of profit.

A variety of approaches can be used for analysing recorded information from simple graphing and spreadsheet analysis to sophisticated statistical analyses, such as data-mining and computer simulation modelling (Aerts et al. 2001; Bird et al. 2001; Durack 2002; Schmoldt 2001; Stafford 2000). Although the simpler techniques can provide useful insights into production inefficiencies, they are often limited and do not account for important interactions between factors that affect pig performance and enterprise profitability (Black 2001). Consequently, combining data (data-fusion) obtained from different areas of the livestock enterprise is essential and best achieved by using computerised and automated information management systems (Thysen 2000). Computerised models that account for total enterprise resource use such as AUSPIG (Black et al. 2001) could be used to identify likely inefficiencies in production methods and profit generation.

Process control

It would be ideal to either automatically modify certain aspects of the livestock production process and/or prompt the farm manager to execute certain control functions in order to achieve the optimal overall outcome for the production process (Wathes et al. 2001). If manual control is used, it is important to have proper Standard Operating Procedures (SOP) in place, so the control function is standardised and documented in advance.

Other aspects

Several other important reasons for collecting farm information were identified, including benchmarking between and within production units, long-term planning based on various economic and production models and development of capital investment strategies.

Results - R&D opportunities identified during the workshops

It was agreed during the workshops, that a formal organisational structure would greatly assist the implementation of Precision Livestock Farming practices in the Australian livestock industries. Many of the Precision Farming technologies would apply across the livestock industries, including pigs, poultry, feedlot cattle, dairy cattle and sheep. Several of these industries are already investigating the role of Precision Farming technologies for enhancing their international competitiveness. A formal organisational structure encompassing the Australian livestock industries would enhance the capabilities of each industry and the general competitiveness of Australia. The proposed Australian Centre for Precision Livestock Farming (ACPLF) would play a major role in coordination of activities across Australia, accessing funds from Government and industry, integrating equipment R&D with implementation of systems onto farms, software research and student training.

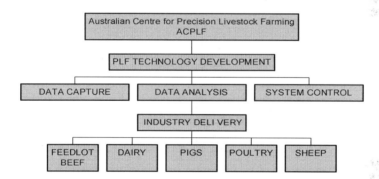

Figure 3. Suggested organisational structure and main Precision Livestock Farming R&D areas (Durack 2002).

Measurements, data analysis and control systems required

The measurements, specifications for the measurements, frequency of measurement, data analysis & interpretation and actions & control systems were identified as far as possible for each important decision affecting pig enterprise profitability during the second workshop. The results from the discussions are summarised in the Table 1-6.

Table 1. Proposed measurement, analysis and control systems for decision 1 - Herd feed conversion efficiency (feed intake, wastage and cost).

Parameters	Optimal specification for measurement	Optimal frequency of measurement	Data analysis & interpretation required to:	Actions & control systems needed
Feed intake	Feed intake for pens of growing pigs and individual sows	Continuous and real-time measurement	Plot feed intake; Identify blockages and significant variation; Set maximum & minimum limits; Design an analytical system to identify what is the likely problem and where it may occur	When limits to measured values are reached; Site of blockages flagged; When reason for changes in feed intake are identified through data analysis & interpretation systems such as AUSPIG
Feed waste	Accurate measurement of feed waste, location and causes	Continuous and real-time, but values over extended periods would be also useful	Identify proportion of feed disappearance wasted Identify location & cause of feed waste	Action to eliminate or reduce feed waste
Feed Cost	Available nutrient specifications and cost of all ingredients as they are being mixed into a diet	Accurate knowledge of ingredient costs and nutrient specifications prior to diet formulation.	Least-cost feed formulation packages are needed that can utilise batch or in-line information on ingredient costs & nutrient specifications and animal nutrient requirements	An automated mixing system that can change ingredient mixes on the basis of available nutrients, ingredient costs and animal nutrient specifications

Table 2. Proposed measurement, analysis and control systems for decision 1 - Herd feed conversion efficiency (growth rate).

Parameters	Optimal specification for measurement	Optimal frequency of measurement	Data analysis & interpretation required to:	Actions & control systems needed
Body weight	Mean weight and variation of pigs; capacity to automatically draft pigs on pre-determined specifications of weight, size and/or shape	Continuous and real-time	Plot weight and variation; Identify significant changes using tools such as PigPulse and predict likely causes of significant variation, including changes in feed intake, climatic environment, air quality, disease etc.; Set maximum and minimum limits for drafting of pigs in relation to uniformity of requirements or buyer price grid specifications	When limits to measured values are reached; When reason for changes in weight are identified through data analysis & interpretation systems such as AUSPIG
Diet nutrient specifications	Continuous identification of available nutrients and costs of ingredients and nutrient requirements of groups of pigs and the variation within a pen	Accurate knowledge of animal genotype & sex on growth characteristics; Accurate knowledge of ingredient available nutrient composition	Least-cost feed formulation packages needed that can utilise batch or in-line information on nutrient availability for ingredients and nutrient requirements for specified animals; Ability to predict when a new formulation will be most economical in relation to changing animal specifications, herd variation and ingredient costs and nutrient contents	An automated mixing system that can change ingredient mixes on the basis of batch or in-line measurements of available nutrients, ingredient costs and animal nutrient specifications
Climatic environment	Continuous measurement throughout the shed of temperature, humidity, air speed and proportion of wet skin	Continuous and real-time at various locations within each shed for temperature, humidity and air speed; Continuous and real-time estimation of skin	Plot climatic variables continuously; Continuous prediction of critical temperatures to identify when pigs are outside thermal neutral conditions and consequences on daily feed intake, growth rate and efficiency of feed use with software such as AUSPIG	Actions and automatic climate control through altering heating, ventilation rates, spray cooling and other means

Parameters	Optimal specification for measurement	Optimal frequency of measurement	Data analysis & interpretation required to:	Actions & control systems needed
		wetness would be best, but predictions would be useful		
Air quality	Continuous measurement in several locations throughout the shed of respirable dust, ammonia, carbon dioxide and total bacteria	Continuous and real-time at various locations within each shed for respirable dust, ammonia and carbon dioxide; Frequent measurements of total bacteria over periods of a day would provide valuable information	Plot air quality variables continuously; Set maximum values for each variable; Continuous prediction of the effects of air quality on daily feed intake, growth rate and efficiency of feed use would be valuable.	Actions and automatic control of air quality through various means
Disease	Continuous measurement of health status	Continuous and real-time, perhaps using sentinel pigs; Video imaging and analysis may be of great assistance for documenting welfare status and health of pigs	Plot water consumption, deep body temperature and breath-salivary variables continuously; Maximum and/or minimum values set for each variable; Systems need to be developed for analysing audio and video data as indicators of disease status	Actions need to be determined for response when measured variables are outside the limits set

Table 3. Proposed measurement, analysis and control systems for decision 1 - Herd feed conversion efficiency (reproductive performance).

Parameters	Optimal specification for measurement	Optimal frequency of measurement	Data analysis & interpretation required to:	Actions & control systems needed
Body condition of sows	Still needs some clarification; affected by parity	Weekly	A data analysis system needs to be developed to determine the effects of condition score as measured and reproductive performance with maximum and minimum limits set for the measured variables	Alter feeding strategies as determined
Heat detection	Continuous monitoring of LH concentrations	Continuous and real time when approaching ovulation	Data analysis systems, maximum and minimum limits to values need to be determined for optimal insemination timings	Insemination at appropriate times to maximise pregnancy rates

Table 4. Proposed measurement, analysis and control systems for decision 2 - Marketing.

Parameters	Optimal specification for measurement	Optimal frequency of measurement	Data analysis & interpretation required to:	Actions & control systems needed
Body weight	Daily measurement of individual pigs with automatic drafting systems	Daily measurement of mean weight and variation; capacity to draft pigs of specified weight	Understand and document variation within herd	Capacity for automatic drafting of pigs at predetermined weights
Body composition	Daily measurement of P2, body fat & lean and body fat distribution. Measurements to coincide with components of	Daily measurement of mean P2 or body composition and variation; capacity to draft pigs of specified maximum or minimum values for body composition	Analyse distribution and ratio of fat/lean within herd	Capacity to draft animals with specified body composition values

	Optimal specification for measurement	Optimal frequency of measurement	Data analysis & interpretation required to:	Actions & control systems needed
Buyer price grids	buyer payment grids; Electronic transfer of all buyer price grids to a central database: Individual enterprises down-load price grids from web based database	variables used by specific buyers; Continuously available on the Web prices and grid specifications for all buyers of pigs in Australia	Software is needed to up-load all buyer price grids on a daily basis and to allow down-loading by individual enterprises; Software such as AUSPIG to identify the sale weight and buyer that will maximise income for an enterprise	To mach the pigs available on farms with buyer specification

Table 5. Proposed measurement, analysis and control systems for decision 3 - Effluent management.

Parameters	Optimal specification for measurement	Optimal frequency of measurement	Data analysis & interpretation required to:	Actions & control systems needed
Odour	A system that measures likely odour levels continuously and that alerts management to possible off-site odour impact	Continuous measurement of odour levels in specific places within sheds and outside odour sources such as lagoons and deep litter stockpiles	Plot likely odour values at specific sites; Set maximum values for each site; Predict likely impact on receptors outside the enterprise boundaries using software such as AUSPLUME	Actions need to be devised to reduce odour when it reaches predetermined maximum values
Nutrient output	A system that predicts or measures the output of N, P, Na and K on a regular basis	One monthly or for each batch of pigs	Plot N, P, Na and K output from each building and whole enterprise on a monthly or batch basis; Set maximum values for each nutrient	Actions to change diet formulation to reduce nutrient output
Site application of effluent	A measurement system and/or model that predicts nutrient application, uptake by plant biomass, stored in soil and losses to the air and ground water	Monthly to yearly (?)	Plot likely nutrient application, use by plants, stores in soil and loss to the air and ground on a frequent basis; Set maximum for nutrient application and loss	Appropriate actions need to be determined for each enterprise

Table 6. Proposed measurement, analysis and control systems for decision 4 - Staff training.

Parameters	Optimal specification for measurement	Optimal frequency of measurement	Data analysis & interpretation required to:	Actions & control systems needed
Staff training	Automatic identification of short comings in staff performance on farms	Daily	Data analysis systems are required to identify whether increased variation in measured values is due to errors in SOP's or staff compliance with SOP's. Systems need to be developed for using measured values as a basis for staff training and reward	When shortcomings are identified; Perhaps automated downloading of relevant information (training package) via the Internet from training providers to personal computers or palmtops (to be used by staff while performing the task)

Discussion - workshop outcomes

The Workshop participants agreed that the Australian pig industry needs to adopt Precision Farming systems if it is to retain its international competitiveness and continually improve efficiency, productivity and profitability. The two most important electronic measurement systems to introduce into Australian pig enterprises were identified as live weight with automatic drafting and accurate measurement of feed intake on a pen or sow basis. Some R&D is probably required before these systems can be implemented on farms. Other measurement systems for internal shed temperature, humidity, air speed, CO_2 and ammonia are already well developed and can be introduced into piggeries relatively easy. These measurement systems are already available for poultry sheds. However, development of common software interface systems will be required if all the measuring devices are to be linked through a single system. Other measurements requiring further development were given high priority, including respirable dust and bacteria levels in sheds, P2 and body composition, skin wetness, oestrus detection and audio and video surveillance systems for disease and welfare diagnosis. It appeared, that the area of data analysis, interpretation and economical modelling are the potentially viable research areas to be developed in Australia. The establishment of centralised data-warehouses was also identified during the workshops as essential components of a well functioning PLF system.

Summary and conclusions

Although, the market place demands safe, uniform, cheap, environmental and welfare friendly products, the current experience based, non-integrated, and poorly controlled production processes are undermining the long-term sustainability of the Australian pig industry and perhaps other livestock industries as well. Better integrated, information-based and better controlled production systems are needed in the Australian livestock industries. Applying modern methods of production management involves establishing data acquisition systems, analysing the recorded information, triggering management actions and activating either automatic control systems. Utilising more advanced IT techniques as well as monitoring the outcomes of control action is also important part of advanced production management. New technologies, such as Precision Livestock Farming methods, have to be adopted by the Australian livestock industries. These technologies have the potential to improve production efficiency as well as welfare and health of animals, and could also potentially reduce the environmental impact of livestock production (Banhazi et al. 2002).

Acknowledgments

These workshops would not been possible without the generous support of many individuals and organisations. We wish to particularly acknowledge the support and assistance of the following colleagues and organisations - Dr Ian Johnson, Mr Geogy Philip and the financial support of the Australian Pork Limited (APL); Karl Hillyard, Tim Murphy, Dr Colin Cargill and Dr Paul Hughes, Pig and Poultry Production Institute - South Australian Research and Development Institute; Prof. Richard Gates, University of Kentucky; Prof. Christopher Wathes and Dr Theo Demmers, Silsoe Research Institute, UK; Peter Finlayson, Finlayson Consultancy; Hugh Crabtree, Farmex Technologies, UK; Greg Ludvigsen, SOCOM Piggeries, SA.

References

Aerts J.M., Wathes C.M., Berckmans D. (2001) Applications of process control techniques in poultry production. In 'Integrated Management Systems for Livestock'. Selwyn College, Cambridge, UK. (Ed. C Wathes) pp. 147-154. (BSAS, Edinburgh).

Banhazi T., Black J.L., Durack M., Cargill C., King A., Hughes P. (2002) Precision Livestock Farming. In 'Australian Association of Pig Veterinarians Adelaide Proceedings 2002'. Adelaide. (Australian Association of Pig Veterinarians).

Bird N., Crabtree H.G., Schofield C.P. (2001) Engineering Technologies Enable Real Time Information Monitoring In Pig Production. In 'Integrated management systems for livestock- Occasional Publication No 28'. (Eds CM Wathes, AR Frost, F Gordon and JD Wood) pp. 105-112. (British Society of Animal Science and Institution of Agricultural Engineers: Edinburgh).

Black J.L. (2001) Swine Production - Past, Present and Future. In 'Palestras XXXVII Reuniao annual Sociedade Brasileira de Zootecnia'. Brazil. (Sociedade Brasileira do Zootecnia).

Black J.L. (2002) Experience in the successful adoption of AUSPIG By Industry. In 'Animal Production in Australia'. Adelaide, South Australia. (Eds D Taplin and DK Revell) pp. 442-448.

Black J.L., Giles L.R., Wynn P.C., Knowles A.G., Kerr C.A., Jone M.R., Gallagher N.L., Eamens G.J. (2001) A Review - Factors Limiting the Performance of Growing Pigs in Commercial Environments. In 'Manipulating Pig Production VIII'. Adelaide, Australia. (Ed. PD Cranwell) pp. 9-36. (Australasian Pig Science Association, Victorian Institute of Animal Science, Werribee, Victoria, Australia).

Durack M. (2002) 'Precision Pig Farming- Where Are You Pigs And What Are They Up To?' National Centre for Engineering in Agriculture, Toowoomba.

Frost A.R. (2001) An overview of integrated management systems for sustainable livestock production. In 'Integrated Management Systems for Livestock'. Selwyn College, Cambridge, UK. (Eds C.M. Wathes, A.R. Frost, F. Gordon and J.D. Wood) pp. 45-50. (BSAS, Edinburgh).

Frost A.R., Schofield C.P., Beaulah S.A., Mottram T.T., Lines J.A., Wathes C.M. (1997) A review of livestock monitoring and the need for integrated systems. *Computers and Electronics in Agriculture* 17, 139-159.

Gates R.S., Banhazi T. (2002) Applicable Technologies for Controlled Environment Systems (CES) in Livestock Production. In 'Animal Production in Australia'. Adelaide, South Australia. (Eds DK Revell and D Taplin) pp. 486-489.

Gates R.S., Chao K., Sigrimis N. (2001) Identifying design parameters for fuzzy control of staged ventilation control systems. *Computers and Electronics in Agriculture* 31, 61-74.

Lemin C.D., Casey K.D., Foster M.P. (1991) Environmental monitoring in pig housing. *Paper American Society of Agricultural Engineers* 91, 1-9.

Schmoldt D.L. (2001) Precision agriculture and information technology. *Computers and Electronics in Agriculture* 30, 5-7.

Schofield C.P., Beaulah S.A., Mottram T.T., Lines J.A., Frost A.R., Wathes C.M. (1994) 'Integrated Systems for Monitoring Livestock.' MAFF.

Stafford J.V. (2000) Implementing Precision Agriculture in the 21st Century. *Journal of Agricultural Engineering Research* 76, 267-275.

Thysen I. (2000) Agriculture in the Information Society. *Journal of Agricultural Engineering Research* 76, 297-303.

Wathes C.M., Abeyesinghe SM, Frost AR (2001) Environmental Design and Management for Livestock in the 21st Century: Resolving Conflicts by Integrated Solutions. In 'Livestock Environment VI. Proceedings of the Sixth International Symposium'. Louisville, Kentucky pp. 5-14. (The Society for Engineering in Agricultural, Food and Biological Systems).

Portable ultrasound diagnostic system for pregnancy diagnose in swine

M. Cegarra, J.J. Anaya, M. Duque and M.G. Hernández
Instituto de Automática Industrial, CSIC. Ctra Campo Real, km. 0,200. La Poveda. 28500. Arganda del Rey. Madrid. España.
macp@iai.csic.es

Abstract

Real time ultrasound B-scan systems have been used for the swine pregnancy monitoring during last years because they improve the pregnancy detection, thus reducing the non-productive sow days and increasing sows per year in a livestock farm.
The present work deals with a portable hand-carried ultrasound diagnostic system (B-scan in real-time) applied to pregnancy diagnose in swine. The diagnosis system is formed by a small ultrasound system which operates as a simple peripheral device, connected to a mobile computer (laptop, PDA, etc.) through a USB standard communication interface. Consequently, the system can exploit the possibilities of computers and peripheral standard interfaces, and incorporate the technological advances with low development cost. Moreover, it is a low cost system that can use software applications installed in a laptop as databases, expert systems or image processing software.

Keywords: Real-Time Ultrasound B-Scan, USB, Laptop PC, swine breeding.

Introduction

The ultrasonic technology in A-Scan mode and Doppler has been applied for swine breeding during the last 20 to 25 years. The inherent limitations of these techniques guided to the research and development of real time ultrasound in B-Scan mode (RTU B-Scan), whose use has become acceptable for the swine pregnancy monitoring during last 5 years. The advantages of using B-scan ultrasonic systems for pregnancy detection in swine have been described in references [1-3]. For instance, RTU B-Scan systems for pregnancy detection, takes 23 days from the pregnancy beginning instead the 28 to 35 days of other ultrasound systems.
The acquisition of RTU B-Scan equipment requires higher initial inversion than other ultrasound equipment, making this cost an important factor for the small and medium scale swine livestock reproduction farms. The equipment cost mainly depends on the electronic and visualisation system, which are critical factors as they limit the image resolution. Moreover, some equipments offer added features such as, exchanging probes with better flexibility, capture and processing of ultrasonic images, access to external displays, computers and printers, etc. However, these added features involve more complexity and cost. An additional drawback is that the new technology associated to information and communication areas can hardly be incorporated in the present RTU B-scan equipments.
In this work, a portable ultrasonic diagnostic system for swine breeding is presented. The principal guide for its design is: to exploit the possibilities of common computing tools and electronic resources as PC laptops and peripheral standard interfaces. This way, we have developed a prototype, which could be the basis for an inexpensive and competitive product in the market of portable RTU B-scan systems.

Portable ultrasound diagnostic system

As said above, the main goal has been to exploit the possibilities of common computing tools and electronic resources as PC laptops and peripheral standard interfaces. Then the ultrasonic diagnostic system can be divided into two parts (figure 2): a small ultrasonic system and the user interface based in a laptop PC, linked by a standard communication pipe.

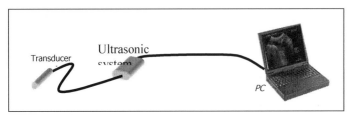

Figure 2. The diagnostic system.

The ultrasonic system is the part of the instrument with electronic circuits to accomplish the tasks of transducer control, signal conditioning and acquisition, digital signal processing and acoustic image formation in real time. A high level of integration issues in a small system eases its operation.

The use of a laptop PC as the user interface brings us some advantages: it is a common device that can incorporate support software to aid the diagnostic system, such as data bases or image storage, etc. It can also incorporate artificial intelligence techniques for diagnosis help or tele-diagnostic. It presents a high quality and large digital colour display. PC computers can incorporate micro-display glasses devices where image is projected on the lenses; by this way the mobility and comfort is improved. Moreover, technological advances can be incorporated to the diagnosis system with low development cost. For instance, it is not difficult to replace the PC laptop by other smaller electronic portable devices, such as PDA's or last generation mobiles.

For communication, any standard system satisfying the bandwidth required by the application can be used. In this sense, the USB interface reaches the features for ultrasonic diagnosis, being at the same time very simple, economical and widely spread in the computing systems.

Ultrasonic system

The ultrasonic system is formed by the following parts: an ultrasonic scanning probe and electronic circuits for probe control, signal processing and communication (fig. 3)

Our prototype is based on a microcontroller, which includes an USB interface and Programmable Logic Devices (PLD). The microcontroller is used for controlling the motor movement, synchronism of emission and reception, and communication management through the USB interface. PLD devices perform digital processing of the ultrasonic signals. The USB interface has the mission of coding/decoding serial data, errors checking, and transferring data between the host computer and the microcontroller memory, in a transparent way, so the microcontroller can perform other tasks.

The motor control permits several sweep velocities. The motor moves following an easily programmable curve of velocity, causing a soft movement and avoiding strokes in the motor borders, which produces a disturbing sensation to the person using the probe.

Reception uses a linear amplifier with variable gain up to 80 dB, including selectable filtering circuits and signal envelop extraction. Controlling the gain amplifier, curves of distance-amplitude correction can be used.

Conclusions

This work presents an echographic system based on laptop PC, which is linked to a small ultrasonic system through a USB standard interface. Systems using this architecture are light, ergonomic, cheap and of easy maintenance. Moreover, they can exploit the possibilities of computers and peripheral standard interfaces, and incorporate technological advances with low development cost. For instance, it is not difficult replacing the laptop PC by other smaller electronic portable devices, such as PDA's or next generation mobiles, which make the pregnancy diagnosis in swine more comfortable and reliable in relation to other traditional ultrasonic diagnostic systems. Using the high processing capabilities of PC's, additional tasks such as image improvement, data base, case-histories, statistics, etc can also be incorporated.

Acknowledgements

This work has been supported by the European Union FEDER 2FD1997-1581 and MAGAPOR company (Zaragoza, Spain).

References

[1] Belstra, Brad. "Reproductive Uses Of Ultrasonography In Swine Breeding Herds: Present And Future Applications". Proceedings of the North Carolina Healthy Hogs Seminar.
[2] Rozeboom, Kevin J. "Current and Potencial Applications of Emerging Reproductive Technologies". North Carolina Porks Conference 2001.
[3] Flowers, W.L. "Real-Time Ultrasonography and diagnosis of pseudopregnancy in swine reproduction". Annual Swine Report 2001.
[4] Hyde, John. "USB Design By Example". John Wiley & Sons 1999.
[5] C. Fritsch, M. Parrilla, O. Martínez, D. Jiménez, "A multirate scan conversion method", Ultrasonics, n. 38, 2000, pp. 179-182

Operation of electronic identification of cattle in Finland

H.E.S. Haapala
MTT Agrifood Research Finland, Agricultural Engineering Research (Vakola), Vakolantie 55, FIN-03400 Vihti, Finland

Abstract

The aim of the project was to evaluate electronic identification (EI) in northern winter climate that is characterized by extremely low temperatures (down to -50 °C) and rapid and frequent fluctuations (+5 … -25 °C). The work was based on the I.D.E.A. project (IDentification Electronique des Animaux: basis for the future implementation of the Electronic Identification System in livestock of the European Union) of JRC (EU Joint Research Centre, Ispra, Italy). I.D.E.A. suggests requirements that might be inadequate in northern climate. The project gave information on the actual operating conditions especially as regards animals kept in uninsulated loose housing. The results enhance JRC´s work in defining test procedures and standards. They also give authorities, slaughterhouses and farmers indications of the possible new requirements for production conditions and buildings.

Keywords: electronic identification, cattle, cold climate

Introduction

Several research projects on livestock electronic identification have been performed during last years (i.e.: FEOGA, AIR2304, I.D.E.A.) which included several EU countries. These research projects have come to a general conclusion that the electronic identification system is an efficient and reliable method for the traceability of the animals from birth to the slaughterhouse.

The I.D.E.A. project that ended on December 2001 was a large-scale project lasting four years and including approximately one million animals in six EU member states. The I.D.E.A. was started because of new legislation proposals regarding livestock identification in EU. However, none of the Northern European countries participated in the IDEA Project mainly because, at that time, they had recently reached the Union.

In the Northern European countries livestock is kept under cold conditions, characterized with low temperatures (down to -50 °C) and rapid and frequent temperature fluctuations between +5 … -25 °C. It was anticipated that the results of the ongoing I.D.E.A. project could suggest requirements that might be inadequate in northern cold climate.

The main objective of this Finnish project was performance evaluation of the electronic identification devices in northern winter climate. The project was also set up to give information on the actual operating conditions especially as regards to animals kept in insulated loose housing. Finally, the results could enhance JRC's work in defining appropriate test procedures and standards for the RFID devices.

Materials and methods

EI devices complying with the ISO 11784 (code structure) and 11785 (technical concept) standards were tested in the field on 3 farms with ca. 270 ruminants. Reference tests were done in climatic chambers. Electronic ear tags and ruminal boluses were used as transponders. Both static and dynamic reading methods were tested. To achieve this, hand-held readers and portable stationery readers were tested.

The reference tests were first started in the climatic chambers at the TEMPEST Laboratory of the JRC Ispra, Italy. Selected pieces of equipment were tested in climate chambers mainly according to JRC´s standard test procedures. In addition to standard tests some additional ones were performed according to the Finnish environmental conditions. In reference tests no ruminal boluses were tested, as ambient conditions do not affect their performance. The tested equipment in Ispra were:

-Allflex standard electronic eartag (JRC IDEA certificate No. 11/1997)
-Allflex portable reader (JRC IDEA certificate No. 050/1997)
-Gesimpex Hokofarm Portoreader (JRC IDEA certificate No. 08/1997)
-Datamars Isomax III (JRC IDEA certificate No. 035/1997)
-Gesimpex F100 stationary reader (JRC IDEA certificate No. 12/1997)
-Gesimpex F200 stationary reader (JRC IDEA certificate No. 46/1998)
-Gesimpex GO3C stationary antenna (JRC IDEA certificate No. 13/1997)

In Finland, all animals on the selected farms were equipped with transponders. The researchers visited the farms regularly and read the tags with both systems. The information was stored in a database, which followed the approved data dictionary standard. The success in identifications was recorded together with environmental conditions. Operation environment was characterized with temperature and relative humidity. Environmental measurements were done with recording instruments to save the temperature history of the devices. Necessary data transfers were arranged to communicate with JRC. The animals to be slaughtered during the project period were followed and identified in the slaughterhouse as far as it was technically possible. The used ear tags and boluses were collected for later inspection.

Results

For the first reference test in JRC selected eartags (10 Allflex standard electronic eartags, JRC IDEA Certificate no 011) were tested at - 40 °C for 96 hours (according to IEC 68.2.1 standard). During the test, each eartag was read inside the climatic chamber using a Datamars Isomax II portable reader. The results were positive: no mechanical or electronic problems occurred.

In the next test three portable readers were introduced from ambient temperature inside a climatic chamber at –25 ° C with their batteries fully loaded. At this low temperature, a Rumitag bolus was continuously read. The number of readings and the reading autonomy were checked. Reading autonomy was considered as the time after introducing the reader inside the chamber when it was impossible to read the bolus code. (Table 1)

Table 1. Reading autonomy of portable readers at –25 °C.

Reader	Number of readings	Autonomy (min)
Allflex Portable Reader (JRC Cert. 050)	790	25[1]
Datamars Isomax III (JRC Cert. 035)	230	20[2]
Gesimpex Hokofarm Portoreader (JRC Cert. 008)	112	11[3]

[1]Frost on the display, [2]after 6 minutes, display response slow and not as clear as at ambient temperature; [3]after 8 minutes, display response slow and not as clear as at ambient temperature.

For the third reference test the selected readers were placed, batteries fully loaded, at –10 °C for 96 hours (according to IEC 68.2.1 standard). During the test the readers were checked every day by reading the ID code of a Rumitag bolus (Table 2).

Table 2. Reading efficiency of portable readers at –10°C.

Date	Hour	Allflex	Isomax III	Hokofarm
11-29-99	13:00	1	2	1
11-29-99	14:55	1	2	1
11-30-99	08:10	OK	3	OK
11-30-99	09:45	OK	3	OK
11-30-99	11:15	OK	2	OK
11-30-99	13:20	OK	2	OK
11-30-99	15:50	OK	2	OK
12-01-99	07:50	OK	2	OK
12-01-99	12:00	OK	2	OK
12-01-99	14:15	OK	2	OK
12-01-99	15:50	OK	2	OK
12-02-99	07:30	OK	2	OK
12-02-99	10:39	OK	2	OK
12-02-99	13:45	OK	2	OK
12-02-99	15:55	OK	2	OK
12-03-99	07:45	OK	2	OK
12-03-99	09:39	OK	2	OK
12-03-99	12:05	OK	2	OK

1: Limit of reading on the display; 2: Quite impossible to read the bolus ID number on the reader's display; 3: Battery charger connected to the reader

In order to study the recovery time from –25 °C to ambient temperature (i.e. the ability of the portable readers when transferred from –25 °C to ambient temperature to read again a transponder), the following special test was conducted: the readers were placed for 48 hours at –25 °C then quickly placed at room temperature. Pressing continuously their reading switch, it was recorded after how much time they were able to read again a Rumitag bolus and to display clearly the ID number. (Table 3)

Table 3. Recovery time of readers from –25 °C to ambient temperature.

Reader	Recovery time (min)	Ambient temperature (°C)
Allflex	9	24.9
Isomax III	24	20.6
Hokofarm	11	25.3

According to IEC 68.2.1 standard, the two stationary readers with their antennae were placed at -40 °C for 96 hours. The temperature decreasing speed (from ambient to -40 °C) and increasing time (from –40 °C to ambient) was set to 2 hours. During the test, the readers were continuously powered and a Rumitag bolus was read twice a day. Both tested readers, Gesimpex F 100 and Gesimpex F 200, performed perfectly.

One of the important aspects to be considered on the readers' performance is the temperature variation with a short time period. Due to the important temperature variations occurring in Finland, the following test was conducted on the two stationary readers: two readers (with their antennae) were submitted during 10 days to a thermal cycle (Fig. 1).

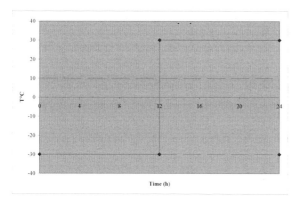

Figure 1. Thermal cycle –30 °C/+30 °C for stationary readers.

During the test, the readers were continuously powered and a Rumitag bolus was read regularly. The results were perfect so that both of the tested readers were operational throughout the test time.

In Finnish farms a total of 273 beef and dairy cattle were electronically identified with ruminal boluses (n=165) and electronic eartags (n=108). Animals were located in four farms with different breeding aptitude and temperature conditions. (Table 6).

Table 6. Number of animals identified on each farm and temperature conditions. Very Cold Conditions: outside without any insulation (*VCC*), Cold Conditions: inside without heating system (*CC*) and Warm Conditions: inside with heating system (*WC*).

Farm	Breeding	Tag	Temperature conditions			Total
			VCC	CC		
Suckler cow unit, Tohmajärvi Experimental Farm	Beef cattle	Eartags	36	65	7	108
		Bolus	5	8		13
Sampo Rauma	Beef calves	Bolus		56		
Kitee Learning Centre		Bolus			14	
Suitia Research Farm		Bolus			42	
Total			41	169	63	273

Attaching of the both tag types was quite easy if only the workers were trained enough. Two persons were needed to complete both the attachment and bookkeeping. Extra persons moving and treating the animals helped a lot in the practical work. According to the work time measurements the ear tags were much faster to attach than the boluses. Aggressive animals were difficult to handle when boluses were attached. The bottleneck in time use was the electronic bookkeeping with the reader device.

It took sometimes a long time (up to 0,5 hour) for the boluses to settle down. This caused problems in workflow since the boluses could not be checked for operation right after the attachment.

It was easier to attach the tags in dairy cattle than in beef cattle. Dairy animals were much more used to handling than the beef animals were.

All the transponders worked satisfactory. The readers, on the other hand, were quite different from each other. (Table 7)

Table 7. Summary of the readings with the reading date, type, the reading efficiency and T/HR conditions.

Farm	Reading type / date				
	1 day	1 month	7 months	Every 7 months	Every 7 months
Tohmajärvi Exp. Farm	21.12.1999 Dynamic 100% (48/48) -16 °C, 75%	22.3.2000 Dynamic 100% (95/95) +3 °C, 70%	13.7.2000 Dynamic (n=101) Static (n=7) 100% (108/108) +23 °C, 80%	12.12.2000 Dynamic 100% (101/101) +8 °C, 98%	29.3.2001[1] Dynamic 100% (99/99) -1 °C, 83%
Kitee Learning Centre	21.12.1999 Static 100% (14/14) -6 °C	22.03.2000 Static 100% (10/10) +4 °C	All animals slaughtered		
Sampo Rauma	22.3.2000 Dynamic 91% (41/45) +4 °C	13.7.2000 Dynamic 100% (19/19) +23 °C	7.3.2001 Static 100% (2/2) -6 °C	All animals slaughtered	
Suitia Research Farm	30.3.2000 Static 100% (82/82) Dynamic 0 % +13 °C, 57%	18.5.2000 Static 100% (82/82) Dynamic 0 % +18 °C, 53%	19.12.00 Dynamic (n=17) Static (n=65) 100% (82/82) +13 °C, 80%	8.3.2001 Dynamic (n=57) Static (n=4) 100% (61/61) +13 °C, 75%	25.4.2001 Dynamic 93.4% (57/61)[2]

[1]An additional reading was performed with 100% or reading efficiency (101/101)
[2]Not checked with Portable Reader

Differences in reading time and physical dimensions and field shape of the antenna caused great variation in reading work. When using handheld readers with short antennas and short reading distance the work postures were quite difficult and the work was also potentially dangerous. The casing material of some readers was not suitable for cold environment since it turned slippery in low temperatures. There were also no carrying straps in the readers to hold them firmly during reading so that they could easily be dropped. It was also observed that in almost every reader there were too low-level audible and/or visible signals indicating a successful reading.

In static readers there were some annoying faults in the antennas. The antennas were very sensitive to e.g. electromagnetic disturbances and required correct installation to avoid malfunction. There was also a lot of trouble with the connection software to external computers.

In slaughterhouses it was quite easy to retrieve the boluses and ear tags. Data handling and integration to existing registers is quite easy.

Discussion and conclusions

Most of the problems encountered were due to other reasons than the cold environment. It can be concluded that the operation and functioning of the electronic identification devices used is not affected when used in different types of cattle kept at extreme temperature and humidity conditions. However, other aspects such as training of the operators and usability of the devices in cold conditions need further research to be solved before the full implementation of the electronic identification system in northern countries starts. Management issues such as arrangements in animal traffic seemed to be the most tedious challenges to meet. Finnish farms lack good solutions for catching the animals for tagging so new solutions have to be developed and tested in practical environment.

It would be rational to use the same tag for all the appliances where animal identification is needed so that e.g. the automatic feeding systems would use the system as well. This would also reduce the possibilities of electromagnetic disturbances between the two systems.

The EI system, if implemented, has wide effects on the whole production system in EU. There is great variation in local circumstances. Thus, before implementation, the readiness of the whole production chain needs to be analysed in order to minimize disorder.

Acknowledgements

This project was funded by Joint Research Centre of European Union and Finnish Ministry of Agriculture and Forestry.

References

Haapala, H. 2001. Eläinten elektroninen tunnistus – apuväline tulevaisuuteen. Lihatalous 6.

Haapala, H. & Havento, J. 2002. Electronic identification of cattle in Finland. In: Russian Academy of Angricultural Sciences. Proceedings of the 3rd Scientific and Practical Conference. June 5-6 2002. Saint-Petersburg. Vol 2.

Haapala, H., Havento, J., Kangasniemi, R. & Peltonen, M. 2002. Eläinten elektroninen merkintä elintarvikeketjussa. Maataloustieteen päivät 2002. Electronic publication: http://www.maataloustieteellinenseura.fi/mtp2002

Ribó, O., Havento, J., Haapala, H., Kangasniemi, R., Mainetti, S. & Korn, C. 2002. Draft version 2. EUROPEAN COMMISSION. Institute for the Protection and the Security of the Citizen (IPSC) Non-Proliferation and Nuclear Safeguards Unit. Technical Note. Electronic Identification of Cattle in Finland: Operation and Functioning of Electronic Identification Devices in Northern Cold Climate.

JRC 2002. Monitoring of Livestock (IDEA). CODE ISIS-11. Electronic publication: http://www.jrc.cec.eu.int/download/workprogram2001-en.pdf. p. 141.

Telemetric measurements of heart rate variability in dairy cows: assessment of stress in automatic milking systems

I. Neuffer, H. Kuendig, C. Kaufmann, R. Hauser and B. Wechsler
Swiss Federal Veterinary Office, Centre for proper housing of ruminants and pigs, FAT, 8356 Taenikon, Switzerland

Abstract

In previous studies, researchers came to no consistent response whether milking dairy cows in automatic milking systems (AMS) is more stressful for the animals than milking in conventional milking parlours. To address this question, different parameters (successful attachment rate, behaviour during milking, heart rate variability, milk cortisol), were collected on four farms each with two different AMS-systems and with auto-tandem parlours. On each farm, 10 cows were included in the data collection.
Data on heart rate variability of the cows during milking are presented.

Keywords: Automatic Milking System, AMS, stress, dairy cow, heart rate variability

Introduction

In Switzerland, an authorisation procedure regarding animal welfare is established since 1981 for mass-produced housing systems and equipment for farm animals. Contrary to conventional milking systems, automatic milking systems are included in the test procedure, as the functioning of such systems is not constantly supervised by the farmer.
Previous studies came to different conclusions concerning the welfare of dairy cows milked by automatic milking systems (AMS). Wenzel (1999) found that cows milked by AMS experience greater stress than cows in milking parlours, while Hopster et al. (2000) concluded that cows milked by AMS were subjected to less emotional stress than cows in tandem parlours. As these studies provided only information for specific AMS models, comparative studies with different models are needed. Many of the studies concerning AMS and animal welfare were carried out on experimental farms with only small herds of dairy cows; the results thereof are difficult to apply on practical farms.
Besides other parameters such as video observations and milk cortisol levels, heart rate variability was chosen to evaluate the stress of the animals in automatic milking systems. The physiological variability in heart rate is considered a better indicator of the individual capacity to cope with the environment than simple heart rate (Porges, 1995). By quantifying the heart rate oscillations the animal's reactivity and vulnerability to stress can be assessed (Porges, 1985). Analysis of interbeat intervals may reveal HRV to be a potential tool for the assessment of autonomic function and therefore a useful indicator of emotional state in farmed animals (Forde and Marchant, 1999).
The data gained in this research will be used as a data basis by the Swiss Federal Veterinary Office in the context of the testing procedure for mass-produced housing systems.

Animals, materials and methods

The investigations were carried out on practical farms in Switzerland. For each of the two AMS-systems, four practical farms were included in the study. To ensure a sufficient habituation of the farmer and the herd to the new milking system, each AMS had be in use since at least six months at the time of the investigation. Only farms with one-box-systems were examined. As a control, four practical farms with recent auto-tandem-parlours were investigated. Of all conventional milking systems, auto-tandem-parlours resemble AMS the most (for example automatic entry and exit gates, individual entry of the cows, tandem stalls). To avoid any influence on the behaviour of the animals due to high temperatures and insects, the investigations were carried out in autumn / winter 2001/2002, spring 2002 and autumn / winter 2002/2003. Each farm was examined for five successive days; the first was used for preparing the equipment and the last for dismantling the material. The goal was to record data for three successive days, thus gaining data of at least six milkings per cow assessing a minimum milking frequency of two milkings per day.

Non-invasive, commercially available systems for heart rate measurements in dairy cattle are not able to transmit the signal for a period longer than a few hours. For our investigations, a system was needed that transmits the signal for a few days without having to approach the animal. An elastic belt, mounted behind the forelegs of the cows, was equipped with three electrodes, an amplifier and a radio transmitter, which was able to transmit the signal continuously for three to four days. The corresponding receiver unit was connected to a demodulator, which transformed the analog signal first into the original and afterwards into a digital signal. Finally, the peaks in the R-R interval curve were detected and saved as interbeat intervals for later analysis.

On each farm, ten lactating animals were equipped with a belt. These cows did not show lameness or signs of acute illness and were of normal behaviour in the milking parlour respectively in the AMS. To represent a cross-section of the herd, the selected cows were of different age and stages of lactation.

Due to the great interindividual variability in the level of HRV, a direct comparison between animals is difficult. Instead, data of a given cow obtained during the visits to the automatic milking systems and the milking parlours were compared to data of the same cow while lying quiet.

Results

The results of the farms equipped with the different AMS-systems are compared with the results of the farms with milking parlours to assess the stress the animals are exposed to during milking in AMS. Data collection was finished in January 2003, results are available in spring 2003.

References

Hopster, H., van der Werf J.T.N., Korte-Bouws, G., Macuhova, J., van Reenen, C.G., Bruckmaier, R.M. and Korte, S.M., 2000. Automatic milking in dairy cows: welfare indicators of Astronaut effectiveness. In: Robotic Milking: Proceedings of the international symposium held in Lelystad, The Netherlands, 17-19 August 2000, Wageningen Pers, Wageningen, The Netherlands, pp. 259-266.

Forde, R.M. und Marchant, J.N., 1999. Heart rate variability: a novel non-invasive means of assessing fear responses in animals. In: ISAE '99 - Proceedings of the 33rd International Congress of the International Society for Applied Ethology, Lillehammer, Norway, p 121.

Porges, S.W., 1985. Spontaneous oscillations in heart rate: Potential index of stress. In: Moberg, G.P., Animal Stress, American Physiological Society, Waverly Press, Inc., Baltimore, Maryland, pp. 97-111.

Porges, S.W., 1995. Cardiac vagal tone: a physiological index of stress. Neurosciences and Biobehavioral Reviews 19, 225-233.

Wenzel, C., 1999. Untersuchungen zum Verhalten und zur Belastung von Milchrindern beim Melken in einem Automatischen Melksystem. PhD-Thesis. Ludwig-Maximilians-Universität München.

Continuous measurement of the ensiling process in pressed sugar beet pulp

H.G.J. Snell[1], H.F.A. Van den Weghe[2] and W. Lücke[3]
[1]*Research Centre for Animal Production and Technology, Universitätsstraße 7, 49377 Vechta, Germany*
[2]*Research Centre for Animal Production and Technology*
[3]*Institute of Agricultural Engineering, Gutenbergstraße 33, 37075 Göttingen, Germany*
hsnell@gwdg.de

Abstract

In five experiments, studies as to whether modern measuring technology can make a contribution towards the continuous measurement of the ensiling process of pressed sugar beet pulp were carried out. The results showed that a suitable pH-electrode can provide reliable, continuous measurement of pH although over a limited period of time. The use of a fibre-optic oxygen sensor leads to a plausible indication of the variations in oxygen concentration during ensilage. In combination with temperature measurement, information about significant parameters of the ensiling process can be gathered.

Keywords: silage measurement techniques, pressed sugar beet pulp, pH-value, oxygen concentration

Introduction

Continuous measurement of the ensiling process offers perspectives both for scientific research and for use in practice. Reports on continuous measurement of this process are rare. Since on the one hand temperature affects the biochemistry of silage, e.g. enzyme activity, and on the other hand silage temperature can be used as an indicator for microbial activity, temperature measurements are often carried out during ensiling experiments. The continuous measurement of the pH-value was impossible for a long time, because electrodes, which provide stable measurement values in a relatively dry substrate (i.e. 200 – 600 g DM kg[-1] WM; DM, dry matter; WM, wet matter) over several weeks, were not available (R. Muck, US Dairy Forage Research Center, personal communication). An initial report on measurements of the pH-value in compressed silage, which lasted 14 days, can be found in Snell et al. (2000). Gas concentrations can be measured even in compressed silage by gaining gas samples (Williams et al., 1997; O'Kiely et al., 2002). However, gaining gas samples takes gas out of the silage, enforces gas flow in the silage and allows only discontinuous measurements. Snell et al. (2001) measured the oxygen concentration during ensiling with the aid of fibre-optic sensors.

On this basis, experiments were carried out to determine whether modern measuring technology can contribute towards the continuous measurement of the ensiling process.

Materials and methods

Experiments

Five experiments were carried out using pressed sugar beet pulp. Experiments P_1 (03 Dec. 99 – 16 Dec. 99) and P_2 (17 Nov. 00 – 01 Dec. 00) focused on the continuous measurement of the pH-value. In both experiments, four ensiling treatments were set up, which differed with regard to silage compression, i.e. DM density, and storage temperature. The density of the strongly

697

compacted treatments A, B amounted to 217 and 240 kg DM m^{-3} in the experiments P_1 and P_2 respectively. The corresponding densities of the slightly compressed treatments C, D were 136 and 150 kg DM m^{-3}. The silage was stored at room temperature (A, C) or in a cooling cell (B, D). For each treatment, eight mini-silos were set up, which were opened at regular intervals. Subsequently, the pH-value of the mini-silo content gained was measured in the laboratory and compared with the results of continuous measurement.

In experiments P_3 (28 March 01 – 04 April 01), P_4 (05 July 01 – 19 July 01), and P_5 (25 July 01 – 08 Aug. 01), interest primarily focused on the measurement of oxygen concentration in the silage. In these experiments, only four mini-silos each were set up, which were stored at room temperature and opened at the same time. Two ensiling treatments were compared in each experiment, which differed with regard to the thickness and, hence, the gas permeability of the film used to cover the mini-silos (A, 90 µm; B, 200 µm).

Set-up of the mini-silos

For the first two experiments, pressed pulp was used which was picked up at the sugar factory Wabern, Germany, and ensiled on the same day. The last three experiments were carried out using the same material as in P_2. After having been picked up, however, this material was first frozen in single portions, and thawed before the beginning of the individual experiment.

For all experiments, vertical plastic tubes (h = 400 mm, d_i = 100 mm, V = 3142 cm^3) were used as mini-silos. All mini-silos of the experiments P_3, P_4, and P_5, and the mini-silos of the slightly compressed treatments C, and D in the experiments P_1, and P_2 were filled with 2.0 kg WM each through manual stuffing. In the case of the highly compressed treatments A, and B in the experiments P_1, and P_2 the mini-silos were filled with 3.2 kg WM with the aid of a pneumatic press.

During the trials P_1, and P_2, both ends of each mini-silo were sealed with rubber caps. The upper caps featured one valve each, which was intended to avoid excess pressure in the mini-silos. During experiments P_3, P_4, and P_5, the lower end of the mini-silo was closed using stable, sealed plastic caps. The upper end had a cap consisting of several layers. Given proper sealing, this cap permitted gas exchange only in a round area (50 cm^2) covered with commercial silo film out of low-density polyethylene. Design details can be found in Snell et al. (2001).

During experiments P_4, and P_5, a vacuum (900 mbar, hysteresis 50 mbar) was established in a bore hole (d = 15 mm) at the bottom of the mini-silos after ca. 7 days. The objective of this measure was to cause a gas flow through the silage. This allowed the response of the measuring signals to altered conditions in the silage to be examined.

Laboratory analyses

Samples of the ensiling material were first frozen and later analysed in the laboratory of the Department of Animal Nutrition and Animal Health of Kassel University (P_1) and in the laboratory of the Research Centre for Animal Production and Technology, FOSVWE, (P_2, P_3).

In experiments P_1 and P_2, the mini-silos were opened at given intervals. Subsequently, the content of the mini-silos was homogenized, and the pH-value was measured in the lab. Due to experiences with previous experiments, the measurement of the pH-value with the aid of different measuring methods seemed advisable:

a) 100 g of material + 1,000 cm^3 of water$_{\text{deionized}}$ were left for 12 h, then shaken thoroughly, and filtered. The measurement was then carried out in the filtrate.

b) like a). However, 200 mg of streptomycin sulphate were added to the infusion water.

c) like a). However, the infusion was stored in a refrigerator.

The measuring methods b) and c) were intended to reduce microbial activity in the infusion. Each measurement was carried out twice.

In experiments P_3, P_4, and P_5, the mini-silos were opened as simultaneously as possible. The content of each mini-silo was homogenized, deep frozen, and later analysed in the laboratory of the FOSVWE.

Continuous measurements

The continuous measurements were gradually extended as the experiments progressed.

Temperature measurement in the silage was carried out using cubic (56 x 51 x 32 mm) data loggers with an integrated sensor (Tinytag, Gemini Data Loggers, UK; accuracy, ± 0.2°C). While the mini-silos were filled, the data loggers were put between the material to be ensiled. In experiment P_2, one such data logger was installed in one mini-silo per ensiling treatment, whereas in P_3, P_4, and P_5 each mini-silo featured a data logger. In experiment P_1, temperature measurement remained limited to discontinuous checks with the aid of a glass insertion thermometer.

In each experiment, the measurement of the pH-value during the ensiling process was carried out in 2 mini-silos of each experimental treatment. For this purpose, pH-electrodes (HA 405-DXK-S8/120, Mettler-Toledo, Steinbach, Germany) were vertically inserted into the silage. The electrodes were specified by the manufacturer as follows: slope pH 4...7, % of theoretical value, > 98%; zero point, pH 7.00 ± 0.25; response time ≤ 20 s. The measurement values were stored at 15 min intervals using data loggers (Micromec, Technetics, Freiburg, Germany). The mean value of the past measuring period was stored. In experiment P_1, the measurement values could only be registered discontinuously.

Oxygen concentration in the silage (P_3, P_4, and P_5) was measured directly under the film and in the centre of the containers with the aid of a fibre-optic system (Mops-8, Comte GmbH, Hanover, Germany; accuracy, 2% of value). The measuring principle of the sensor is based on the effect of the fluorescence quenching of special colourants caused by molecular oxygen (Comte, 2002).

Results

Ensiling material and silage composition

The DM content of the ensiling material in the experiments P_1, P_2, P_3 amounted to 213, 236, 238 g kg^{-1} WM respectively. The corresponding values for crude ash; crude protein; crude fibre and sugar amounted to 56, 81, 82; 85, 92, 89; 183, 236, 237 and 139, -, - g kg^{-1} DM.

With respect to the composition of the silage at the end of the experiments, there were few remarkable differences between the experiments and between the treatments within the experiments.

Measurements during experiments P_1 and P_2

The courses of the pH-value and the temperature were strongly influenced by storage temperature in the experiments P_1, and P_2 (Figure 4). The degree of compression did not exert any influence on silage temperature. Its influence on the pH-value was small and could not be detected in all cases.

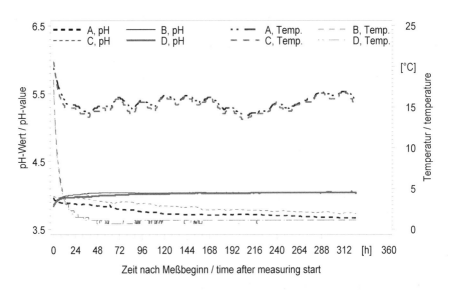

Figure 4. Experiment P_2, course of the pH-value and course of the temperature depending on the treatment. A, B, C, D, treatments.

Each of the graphs, which show the course of the pH-value, represents the mean value of the measurements with two electrodes in two different mini-silos of the same treatment. During both experiments the correspondence between the electrodes within one treatment was close (Figure 5).

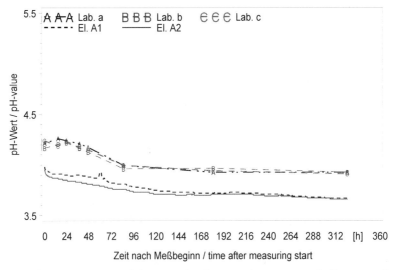

Figure 5. Experiment P_2, course of the pH-value in strongly compressed silage stored at room temperature (treatment A). Results of different laboratory analyses and the continuous measurement with two electrodes. Lab. a, b, c, laboratory method a, b, c; El., electrode.

In addition, Figure 5 shows that the three laboratory methods applied parallel led to virtually identical results. In all cases, however, the analyses in the laboratory yielded slightly higher measurement results. After the completion of the experiments, the electrodes were removed from the mini-silos and examined in a buffer solution with an indicated pH-value of 4.0 (20°C). On average, the measurement result after the end of experiments P_1 and P_2 amounted to 4.04 ± 0.02 and 3.99 ± 0.05.

Measurements during experiments P_3, P_4, and P_5

In experiment P_3, the fibre-optic measurement of oxygen concentration showed a plausible development of this parameter. Within the first 36 h after the beginning of ensiling, virtually all the oxygen in the mini-silos was consumed.
In experiment P_4 the pH-value dropped gradually at first. After a vacuum was established in the mini-silos, the pH-value began to rise quickly after a certain delay. The speed of this increase diminished towards the end of the experiment. In connection with this rise in the pH-value, an increased silage temperature was measured for several days.
In experiment P_4, the oxygen in the mini-silo was consumed after ca. 24 h. In all three mini-silos for which usable measurement results are available oxygen concentration did not increase again at first. Immediately after the establishment of the vacuum, the oxygen content rose sharply again. At virtually all measuring points, the following phase was characterized by long-term fluctuations in the measurement results, which described a cycle of ca. 24 h.
In experiment P_5, the development of the pH-value and the course of the temperature were virtually identical in all mini-silos. The measurements of oxygen concentration also led to comparable results. If air flowed through the silage, the fluctuations in the measured oxygen concentrations also occurred in experiment P_5, though in a less pronounced cycle. Figure 6 visualizes different phases of the ensiling process.

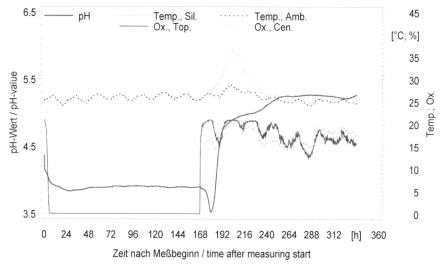

Figure 6. Experiment P_5, courses of the oxygen-concentration under the film and in the centre of the mini-silos, courses of the pH-value and the temperature. With exception of the ambient temperature the graphs represent average values of several mini-silos. pH, pH-value; Temp., temperature; Sil., silage; Amb., ambient temperature; Ox., oxygen concentration; Top, under the film; Cen., centre of the mini-silos.

Discussion

Ensiling material and silage composition

The ensiling material of experiments P_1 and P_2 exhibited some remarkable differences, which may be attributed to the fact that the pressed pulp was gained during two different harvest campaigns. The material for P_3 was gained from the same lot as the material for P_2. Therefore there were no significant differences between the ensiling material for these experiments. If one assumes that the distribution of the material among the three lots, which were used after deep freeze storage, had the same degree of uniformity, this result also applies to experiments P_4 and P_5.

Continuous measurements

In ensiling experiments, the measurement of silage temperature is a well-known standard procedure. It provides initial indications that microbial processes occur in the silage, as has been reported repeatedly (e.g. McDonald et al., 1991). In the present study, these processes were reflected by a quick increase in silage temperatures after the establishment of the vacuum in experiments P_4 and P_5. However, the relevance of silage temperature is limited.
According to Pahlow and Weißbach (1996), not only the final degree of acidification, but also its temporal course is significant for the evaluation of the ensiling process. As mentioned above, continuous measurements of the pH-value have not been carried out thus far because no electrodes were available which allowed stable measurement values to be obtained in a relatively dry, compressed substrate over several weeks. The electrode type used in the experiments described here fulfils the high requirements of such a measurement, which is reflected by the excellent consistency of the electrodes within the same ensiling treatment and the high long-term stability of the measuring signal.
With regard to the course, the consistency of the continuous measurement with the three laboratory methods applied parallel was good in experiment P_2. The fact that in all cases the laboratory analyses led to slightly higher measurement results cannot be explained using the available data material. However, this is of little importance for the assessment of the ensiling process. The significant deviation of the laboratory results from the continuously registered data, which was mainly observed in the second half of experiment P_1 (Snell et al., 2000), must likely be attributed to problems regarding the carrying out of the chemical analyses in the lab.
In experiment P_1 as well as during ensilage at room temperature in experiment P_2, stronger compression led to slightly lower pH-values during continuous measurement. If one assumes that this observation can be repeated, the question arises whether the differences have a biological or a measurement-technological reason. In the present study, this aspect cannot be given due consideration. While the results of the laboratory measurements in P_1 might also indicate an influence of compression on the pH-value of the silage, the results of the laboratory measurements in P_2 were not affected by silage compression.
The measurement of oxygen concentration in the silage provides a significant contribution towards the description of the living conditions of microorganisms. As compared with conventional methods of oxygen measurement, the fibre-optic measuring method employed here is interesting because the sensors are small, metal-free, and can be sterilized. The measurement does not consume any oxygen.
In all experiments where oxygen concentration was measured, its course was plausibly registered through fibre-optic measurement. An explanation for the heavy oscillation of the measurement values after the establishment of the vacuum cannot be given here. Very generally speaking, however, one must take into consideration that alterations in the ensiling

conditions always led to a change in the measuring conditions. This has already been discussed in connection with pH-measurement at different degrees of compression. If air was forced to flow through the silage by means of a vacuum, this not only led to altered gas concentrations, but, among other effects, also to an increased, inhomogeneous DM content of the silage.

Conclusion

Suitable pH-electrodes allow for the reliable continuous measurement of the pH-value of pressed pulp silage for at least two weeks. The use of fibre-optic oxygen sensors leads to a plausible description of the alterations in oxygen concentration during ensiling. The combination of the mentioned measurable variables with temperature provides information about significant parameters of the ensiling process.

References

Comte GmbH. 2002. Fiber Optic Oxygen Sensor Technology MOPS. http://www.comte-online.com/index.htm (01.02.2002).

McDonald, P., Henderson, A. R., Heron, S. J. E. 1991. The biochemistry of silage. Chalcombe Publications, Marlow, Bucks, UK.

O'Kiely, P. O., Forristal, D. P., Brady, K., McNamara, K., Lenehan, J.J., Fuller, H., Whelan, J. 2002. Improved technologies for baled silage. Beef production report series no. 50. Grange Research Centre, Dunsany, Co. Meath, Ireland.

Pahlow, G., Weißbach, F. 1996. Bedeutung sowie Möglichkeiten einer Beeinflussung der Säuerungsgeschwindigkeit in Grassilagen [*Possibility to influence the speed of the fermentation in grass silage and its importance*]. 37. Tagung des DLG Ausschusses für Futterkonservierung am 19. - 21.3.1996, Aulendorf.

Snell, H. G. J., Oberndorfer, C., Van den Weghe, H. F. A., Lücke, W. 2000. Continuous measurement of the pH-value during ensiling - potential, requirements and reliability of the procedure. ASAE Annual International Meeting. Paper No. 001033.

Snell, H. G. J., Van den Weghe, H. F. A., Lücke, W. 2001. Continuous measurement of the oxygen concentration during ensiling - technology and preliminary results. ASAE Annual International Meeting. Paper No. 011092.

Williams, A. G., Hoxey, R. P., Lowe, J. F. 1997. Changes in temperature and silo gas composition during ensiling, storage and feeding-out grass silage. Grass and Forage Science 52 176-189.

Abstracts from oral presentations for ECPLF

Controlling growth of broiler chickens on-line based on a compact predictive growth model

J.-M. Aerts, S. Van Buggenhout and D. Berckmans

Selection programs for broilers, combined with improvements in nutrition, have led to a reduction of the age at which the animals reach the commercially desired slaughter weight. This evolution has also resulted in several negative growth responses (increased body fat deposition, a decrease of reproduction capacity, metabolic diseases and a high incidence of skeletal diseases). A possible way to correct for these negative growth responses is to alter the growth trajectory in such a way that the initial growth is lowered followed by an accelerated growth, i.e. compensatory growth. The objective of the reported research was to control the growth trajectory of broiler chickens during the production process based on an adaptive compact dynamic process model. More specifically, the daily food supply was calculated, based on a model-based control algorithm, with the aim to follow a previously (compensatory) growth trajectory as close as possible. For modelling of the dynamic growth response of broiler chickens to the control input food supply, an on-line parameter estimation procedure was used. It was assumed that the relation between cumulative food intake and weight, which is non-linear, could be described by recursive linear regression relations between both variables. A recursive moving rectangular window algorithm was applied. This algorithm estimated the parameter values on-line (every time new process information becomes available) based on a small window of actual and past measured data of process input(s) and output(s). In order to develop the control algorithm, Model Predictive Control theory (MPC) was used. In general, it is tried to predict and to control the future process output (y) in such a way that it follows a predefined reference trajectory (r) and at the same time to minimize the control effort (Du), necessary to follow this predefined trajectory. This was achieved by minimising an objective function J. The MPC method uses the receding horizon principle. This means that after calculation of the optimal sequence of the control input over the control horizon (4 days in this case), only the first value of the calculated control inlet is applied; then the window is moved with one measuring unit (1 day) and subsequently the procedure of optimisation is repeated with new measured information. Three experiments were conducted on a small scale (50 animals) and three experiments were carried out on a large scale (2900 animals). In the small-scale experiments, Ross 308 broilers, mixed sex, were used. Birds were kept in two climate-controlled compartments with 4 pens, each with a stocking density of 16 birds per m². Average weight and food consumed was recorded daily. In the large-scale experiments Ross 308 broilers, mixed sex were used. Birds were kept in two climate-controlled compartments with 2900 chickens each (stocking density of 16 birds per m²). Both compartments were in the same stable. Average weight and food consumed was recorded daily. All experiments lasted 42 days. To test the usefulness of the model-based algorithm, different target trajectories were defined ranging from severe restricted growth trajectories to pronounced compensatory growth trajectories. The control algorithm managed to follow the previously defined target trajectory with a mean relative error that varied between 3.7 % and 5.3 %. Controlling the growth trajectory had also a positive effect on the welfare of the broilers. The controlled groups had lower mortality than the reference (ad libitum fed) groups in 4 out of 6 cases. It can be stated that the integration of dynamic data-based modelling approaches with new hardware techniques and sensors and/or sensing techniques, should make it possible to model animal responses (the most important process part) to variations of process control inputs. This is a basis for advanced model predictive control of bioprocesses.

Development of a standard for the networking of computer systems of the interior economy

Rudolf Artmann

The computer systems available for the automation of animal husbandry offer in the best case Island-like solutions for one animal species. In the most cases neither environmental conditions nor the overall monitoring of the total production processes can be realized. The most serious deficiencies are the individual manufacturers working styles and the lack of a standardized solution for the data exchange between different computer systems. In addition to internal concerns of complete control of animal production, all links with industry, commerce, administration and advising take more and more place via intra-farm management. It is therefore necessary that the dependency of the flow of goods can be monitored from the perspective of incomming/outgoing, quantity, quality as well as the time period in question and the financial effects. A standard for intra-farm networking must be capable of carrying all required data transactions between all electronic devices, which make data available and/or those which require data. The standard should not only enable networking in animal husbandry, but rather also in storage, ventilation, sources, sales, information gathering and dissemination (external farm communication) as well as to external agricultural electronic systems (LBS or ISO 11783, weather or weighing systems). Only a thorough and complete network system providing overall observation of the reciprocal relationships, and an optimal control of the production, permits the meeting of the requirements of both animal and environment. Additionally, a reliable and credible record of the production methods is also possible. Such credibility is becoming even more important for consumer acceptance of agricultural products. It is therefore a necessity to develop an international standard from economical, ecological and animal protection aspects. As a consequence of the current situation a working group under the roof of Bauförderung Landwirtschaft; was created with the aim to force and develop a standard on ISO level for the networking of computers on farms. Until now a New Work Item; was specified and distributed for voting. During the voting process the need for an International Standard is confirmed but not enough national bodies will participate as active members. Meantime the established working group has decide to start the definitions of the necessary parts of the standard. At the same time activities are started to win international members which support this work and gives the opportunity to shift it to an international level (ISO). The preliminary specifications for the standard are: Use of Ethernet with TCP/IP and UDP/IP. Connections with hub or switches and industrial plugs RJ45 or M12. The minumum data rate shall be 10 MBit/s. The whole task is now transfered to three under groups. Group 1 working on the physical layer. Group 2 search for a middleware to made applications as easy as possible. Group 3 evaluate possibilities for a common data description on the basic of ADED and XML with easy access and update as well as long run availability and maintenance. Contents of the presentation on the Conference will be the actual status of the tasks in the working groups. The aim is to get international support and cooperations as well as to give a first insight into a coming standard.

Results of continuous measurements of ear temperature in boars

J.H. Bekkering and St. Hoy Brandt

In the future, there will be more regulations in the EU member states concerning monitoring and documentation of the health status in animal production. The aim of our investigations includes the following subjects: - test of a computer supported system of continuous measuring of ear temperature in boars in an AI (artificial insemination) station - automatically registration of ear temperature considering different environmental factors and the clinical situation (health status - fever) of boars. At the end, the question has to be answered if health control of boars kept in AI centers is possible by automatic measurements of temperature as early warning system (as tool for veterinarians to survey the health status). Measurements started with 51 boars single kept in one house of an AI station. The boars have got an ear tag with the sensor covered in plastic material. The sensor measures every 3 min skin temperature at the ear and creates a mean temperature on the basis of five 3-min-values. These means are transfered via aerial to a computer with a special software programme (QSS 2000 = quality control system). The software generates a continuous curve of dynamics of temperature. Additional informations about the individual animals can be added to the programm. On the PC monitor all boars measured are configured with individual numbers. Upper and lower bounds of warning temperatur can be set. If the temperature curve of one animal crosses the upper or lower bound a warning signal is given by programme. So, the figure of this individual on PC monitor is colored red. The datas are stored in an Access file and can be exported to other programmes to make statistical analysis. Results The measurements showed differences in ear temperature measured by sensor and by digital thermometer. The sensor measured lower temperature by approximately 4 Kelvin. x s Min Max ear temperature left 29,3 5,06 17,6 34,9 (digitalthermometer) ear temperature left 25,3 4,39 16,4 32,5 (sensor) On the basis of 29 measurements at 6 days with 6 boars a coefficient of correlation between ear temperature measured by sensor and ear temperature measured by digital thermometer of 0.84 was calculated. There was in all boars observed during 134 days a typical daily rhythm of ear temperature with a drop in temperature in the early morning (between 3 and 5 am) and a peak in the afternoon (between 3 and 5 pm). During the night the standard deviation of the ear temperature was much higher (up to +/- 5.8 K) than during the afternoon (between +/- 2.0 and 2.8 K). So, it seems to be impossible to define a fixed upper bound of warning temperature. The mean temperatures of the days before, at and after semen collection were compared because there were no sick boars in the first six months of investigations. There were no differences in the mean ear temperature between the days around semen collection (days -2, -1, 0, +1, +2). There seems to be only a very short rise in ear temperature during semen collection with large individual differences between boars. The investigations will be continued waiting for sick boars to check whether these configuration can be efficiently used as a warning system.

Is there a possibility of clinical application of infrared-thermography for diagnostics in oestrus detection in dairy cows ?

U. Brehme, D. Ahlers, H. Beuche, W. Hasseler and U. Stollberg

Introduction: Infrared-thermography is a non-invasive diagnostic method. The body surface temperature of an animal is measured from a distance with an infrared camera, and the photographed part of the animal's body is displayed as a thermal profile picture. No direct animal contact or immobilisation is necessary. In our proposal a thermal profile is defined as an infrared picture of a detail of an animal body with its surroundings. Infrared thermography in zoo and wildlife medicine has been applied very successfully in recent years for diagnosing injuries, as well as for oestrus and pregnancy detection. The question considered was whether it is possible to use this non invasive method for monitoring oestrus detection of dairy cows in farms too. An investigation was started this summer in cooperation between ATB Bornim and the Veterinarian University in Hanover.

Material / Methods: Six dry dairy cows from the Clinic for Cattle Obstetrics and Gynaecology at the Veterinarian University in Hanover were included in an 14 day investigation. The animal body surface temperature was measured from a distance (1.5m) with an infrared camera. The aim of our investigation was to establish whether there was a possibility of identifying the oestrus climax, considered as the optimal point for successful insemination of dairy cows, from a thermal profile in an infrared picture when the subject of the picture was only the external vaginal area of the cow. All cows were oestrus-synchronised before the investigation started. At the beginning of the test all six cows were equipped with a data logger inside the vaginal tract to measure body core temperature and activity. The data logger was configured with a measuring interval of 5 min for both parameters for a continue measuring. Main problem was to establish whether we could identify the very slight change in body core temperature during oestrus of only $0.3...0.6°C$ in an infrared thermal picture of the external vaginal area of the cows. The time interval for infrared pictures of the external vaginal area was two hours. Five days after starting the investigation all the cows received a Prostaglandin F^2 application to stimulate artificial oestrus three to four days after the application. We ensured precise animal housing conditions -no wind or draught, heat was constant and indoor temperature between $22.0....27.0°C$.

Results: All cows show excellent course of body core temperature and activity with oestrus cycle two to four days after Prostaglandin F^2 application. The average body core temperature of the cows are between $38.0°C$ to $38.5°C$. The oestrus climax is also evident for all cows.

The oestrus of the test cows runs from :
- start of oestrus : 29th July – around 02:00 a. m. – $37.9°C$
- oestrus climax : 30th July - around 08:00 p. m. – $39,0°C$
- end of oestrus : 31st July - around 17:00 a. m. - $38.1°C$

For all cows we have exact data of an oestrus cycle. This was possible by measuring important physiological data such as body core temperature and activity with data logger. The difference between the average body core temperature before and after oestrus is $1.1°K$.

Infrared photos shows the development of body surface temperature of the external vaginal area. We can identify a temperature increase on measured points of the labia from $0.2°C$ to $0.4°C$. This trend indicates oestrus or illness of the animals.

Conclusions:
- The oestrus climax of this test animal was noted by a data logger.
- We also found precise indications of oestrus via infrared thermography of the external vaginal area.
- A temperature increase of $0.2°C$ to $0.4°C$ at the labia was discovered.

A new concept to control the indoor climate of broiler houses

R. Brunsch

The rapidly increase of the growth rate in young broiler chicken within the last twenty years has changed the heat and carbon dioxide production. This has an influence to the climate control because former calculation data do not work in a successful way. The shares of sensible and latent heat in the broiler house depend from a few factors, biological as well as physical. Behaviour studies have shown that especially heavy animals under warm conditions have a problem to get free from the heat, produced by the body. Extra heat production in the litter and high animal densities in the stables increase the problems. In this way it is necessary to improve the climate control algorithm. The idea behind the new control concept is to integrate the thermophysiological response of the broiler chickens on changes of thermal conditions in the broiler house. In the thermophysiology there is a definition about the biological optimised temperature (BOT). This physiological model means that the minimum thermal stress to the animal is when the share of latent heat is at the minimum too. In this case the ambient temperature is the BOT (this is only a very short definition). With the help of basic equations from thermodynamics it is possible to describe the amount of the latent heat on the total energy content of the air as the ratio between enthalpy and water content in the air. The ratio of the differences between indoor and outdoor enthalpy and water content gives an information about the total heat production and the share of latent heat. This ratio is sensitive to the animal response and the building. Because of these facts it is necessary to find out the special "building factors" and the total heat production of the animals. As a result of measurements in some broiler houses at different times of the year it is known that the "building factors" have an importance to the efficiency of climate control. Modern management programs measure the averaged feed intake per animal and day. Based on this it is possible to calculate the total heat production. It changes in broiler production at least daily and has a typical diurnal rhythm under special management conditions. In the first step the control concept works on the base of daily average data for heat production, but has a continuously response on an increase in the share of latent heat. That means there is a reaction to higher evaporation from building surfaces, litter or mainly from animals. Model calculations, based on measurement data from a practical farm with some houses, are shown, that in addition to the common used temperature control, the calculation with the ratio between the differences of indoor and outdoor enthalpy and water content in the air, gives the better understanding of the animals response and improves the possibilities to react on changing climate conditions. The full paper gives the thermophysiological basic understanding, equations and some calculation results for a discussion to improve the living conditions for the animals. As a result of such a advanced control strategy it is possible to reduce heat stress in broiler houses especially at the end of the fattening period. In this way the control concept is a step to increase the animal welfare an lower the losses.

Portable ultrasound diagnostic system for pregnancy diagnose in swine

M. Cegarra, J.J. Anaya, M. Duque and M.G. Hernández

Ultrasound technology, A-scan and doppler modes, have been applied to swine research and production for the last 20 to 25 years. However, its regular use in pregnancy diagnose has grown during the last 5 years with the use of B-scan in real-time ultrasound. The timely identification and removal of non-pregnant females from the breeding herd is important because it reduces both non-productive sow days and production costs. With an ultrasound diagnostic system we can shorten the time of detection of pregnancy which leads to a more profitable situation in the swine production. Many studies have revealed that the use of ultrasound diagnostic systems is more effective than traditional methods. In conclusion, ultrasound technology can be used to reduce the non-productive sow days, and increase sows per year in a livestock farm. There are several ultrasound diagnostic systems in the market, which are used in livestock farms. The present work deals with a portable hand-carried ultrasound diagnostic system (B-scan in real-time) applied to pregnancy diagnose in swine. In this kind of application, portability is an essential stuff, because of the problems involved in moving the animals. Other important requests of this ultrasound application, reached by this system, are the robustness, lightness and ease of use. A feature of this system is that it is a simple peripheral device, which runs connected to wide spread existing mobile computers (laptop, PDA, etc.) and a standard communication interface between the ultrasound system and the mobile computer, so it is a low cost system and can use software applications installed in a laptop as data bases, expert systems or image processing software. The communication interface used is USB 1.1, that reaches the requirement of transmission velocity, and it is a standard interface for laptops and most of PDA, making use of the features of USB as hot-pluggable connection, up to 127 peripheral devices running simultaneously and power supplied by USB cable. The visualization screen depends on the mobile computer used, and the display resolution can be modified, so it is possible to use a version of a ¼ VGA size resolution, considering the possibility of visualization with special glasses, with the ultrasound image in one lens. We can visualize the image on a bigger format in a laptop display. The image control includes a four-depth levels zoom and it is possible to storage images in a bitmap format at any time. The electronic architecture consists of a general purpose microcontroller, the Cypress EZ-USB model, together with High Density PLD (Programmable Logic Device), allowing digital processing of the ultrasonic signal. The probe has a built-in three keys set, which allows modify all the system parameters: gain, brightness, contrast, etc. Another feature of the system is that it is extendable. The prototype designed verifies the requirements to be used in basic pregnancy diagnose, as detection of embryonic vesicles, but it can be complemented with accessories to improve its features and communication capabilities with the end user, as connection of other peripheral devices and software applications running simultaneously. Improving its features and capabilities in the future, the system could be used for more specific applications related to swine reproduction, as examination of the ovaries or uterus to diagnosis several diseases. Acknowledgement This work has been supported by the European Union FEDER 2FD1997-1581 project and MAGAPOR company. References 1. B. A. Belstra. Reproductive Uses of Ultrasonography in Swine Breeding Herds: Present and Future Applications. Proceedings of the North Carolina Healthy Hogs Seminar, 2000. 2. W.L. Flowers. Real-Time Ultrasonography and Diagnosis of Pseudopregnancy in Swine. Annual Swine Report 2001, NC State University.

A proposal of a new empirical model for lactation curve of cow and its use with the milking-database for the farm management

Giovanni Carlo Di Renzo, Giuseppe Altieri and Antonio Colangelo

A successful dairy farm management normally involve the strict observation of standard healthful rules and careful genetic selection of cows, these are considered the first steps to increase the farm production level in terms of quality and quantity amount levels and possibly can help to increase the production of high quality milk. In a cattle farm located in Basilicata (Southern Italy) over a period of three years, for dairy cows milked twice per day, milking data were collected using an automatic system based on electronic individual identification of cows (using a transponder ruminal bolus) and the related milk yelds (measured using an electronic milk flowmeter). Using collected data a new empirical model of cow's lactation curve has been tested, with the aim to verify its forecast capacity in comparison with others existing models. The model proposed uses the cubic b-splines function as approximating function (via the least squares criterion) on 6 variable knots sequence; such function, and more generally piecewise polynomial functions, it is used to successful approximate experimental data or design curve measurements; due to its polinomial nature it is easily handled by computer programs; moreover it don't require trascendental functions. The algorithm with variable knots sequence was preferred because it greatly increase the flexibility of the b-spline approximation. The parameters considered to estimate of good approximation between the models and the physical data were: the squared correlation coefficient (R2), the standard error (SE) for each lactation curve (R2c and SEc) and for the averaged lactation curve (R2a and SEa) versus the model results. Only for the averaged lactation curve, the global error (SSR) as the sum of the absolute values of the local error between averaged lactation curve and lactation model curve, were calculated. These parameters contribute together to the score (S) of good approximation with the physical data obtaining the score S for each models. The models considered were: spline3-6 (cubic b-spine with 6 variable knots), spline3-5 (cubic b-splines with 5 variable knots), lognormal4 (a general peak approximant function with 4 parameters), Guo95, Wilmink87, Ali87, Emmans86, Wood67, Nelder66, Brody45. The scores obtained for cows with one, two, three and greater parities, were respectively: spline3-6 0.93-0.80-0.97, spline3-5 0.83-0.53-0.96, lognormal4 0.61-0.60-0.92, Guo95 0.62-0.61.0.90, Wilmink87 0.49-0.44-0.90, Ali87 0.74-0.71-0.94, Emmans86 0.42-0.41-0.89, Wood67 0.65-0.11-0.91, Nelder66 0.00-0.10-0.73, Brody45 0.45-0.44-0.00. The cubic b-spine with 6 variable knots has reached the best score (S) than the other models even with respect to cow's parities. A possible use of the model respect to the lactation curve database involves the definition of the mean productivity index after 305 days from calving (IPM305): this is defined as the average value of the lactation curve onto the specified 305 days period. Two useful guideline applications are proposed and discussed. The first involve the positioning of the daily milk yield onto the herd lactation curve model: a milk yield below the alarm bands (e.g. +/- 10%) of the herd lactation curve model can predict a stressed cow if this bands are well tuned in cooperation with the veterinarian. The second one involve the comparison of IPM305 of each cow with respect to the the IPM305 of the herd averaged over all the cows (HIPM305). This comparison can help the farm manager in genetic analysis and traits selection and these characteristics make it more suitable for using in farm management software.

The development of integrated management systems for livestock production

A.R. Frost, D.J. Parsons, A.P. Robertson, C.P. Schofield and K.F. Stacey

The objective of most livestock production systems is to achieve an acceptable economic performance, subject to constraints of animal welfare and environmental impact. This is achieved by regulating the system inputs such as nutrition and housing conditions. Livestock systems are comprised of sets of complex interconnected processes, each with their own inputs and outputs, and the interconnections between the various processes make simultaneous, independent control impossible. For example growth, health, welfare and environmental emissions all depend on the animals supply of nutrients. A change of diet may increase growth rate, but will the increased feed cost be justified, and will the change of diet make the animal too fat or lead to increased emissions or welfare problems and how should the environmental temperature be adjusted. Such dilemmas arise because each of the individual processes is controlled separately. The weak connections between the management systems for the various processes need to be strengthened through the development of integrated management systems, designed to control simultaneously more than one, and ideally all, of the interrelated processes. It will not be possible to develop a fully integrated management system in one step because of the number and complexity of the processes involved. It is necessary to target those processes which are the most important according to economic, political or some other criteria. At Silsoe we have begun by developing a real-time, model-based nutrition and growth control system for broiler chickens. Broilers are traditionally fed according to a schedule which dictates how much energy and protein should be supplied as the birds reach a given age. If they are not growing as planned the manager can change the diet. The difficulty is that the manager has to identify the problem and then decide how to change the diet. One of our aims was to develop a computer-based controller that will do this daily diet calculation automatically. This was achieved through the development of a real-time, model-based, nutrition control system. The system has performed well in full-scale trials and is now being tested commercially. The project has shown that broiler growth can be regulated by automatically varying the birds diet. This should result in improved profitability. We have also shown that we can quantify many of the factors involved in pollutant emissions. The growth model can predict the amount of nitrogen that is excreted - a necessary link to integrating the effect of diet on growth as well as emissions. Further work is required to investigate more closely the relationships between diet and emissions. When this has been done it will be possible to design a system which will enable a poultry producer to control bird growth and pollutant emissions at the same time. The principles of integrated management systems can be applied to all livestock production processes. As another example, at Silsoe we are also working towards a system in which the management of pig growth and environmental emissions are integrated. The approach being applied involves measuring process outputs including pig growth rate derived from video images of the pigs. This is fed into a process controller which calculates the difference between actual (measured) and optimal (calculated from a model) growth, enabling the feed input to be altered to minimise error. If further integrated management systems for animals are to be developed it will be necessary to devise sensors to measure the important states of the animal and its environment. Assuming that relevant information is available from sensors the next requirement is for the development of mathematical models which will enable actions to be taken. The final stage is to assemble the various sensors, models and actuators into working systems which are practical and robust, and will be welcomed by potential users.

Information monitoring system for surveillance of animal welfare during transport

G. Gebresenbet, D. Ljungberg, G. Van de Water and R. Geers

Transport of animals is increasing both on national and international levels in relation to marketing system. Though transport plays an important role in stimulating economy and increase quality of life, and at the same time transport and handling plays negative role in several aspects. It is reported by many researchers that transport and handling activities are compromising animals health and welfare, and meat quality, increases the risk of spread of infectious diseases over large distances, and also degrades environment in terms of pollution emanated from vehicles. Currently, drivers make route planning manually, and therefore, un-necessary transport distance, time, and roads couldn't be avoided. More over, no continuous information is available on (a) the climatic conditions, i.e., temperature, relative humidity, and level gases in the loading compartment, (b) vehicle's performances (particularly vibration), (c) driving routes and performances, and (d) animal behaviour when subjected to uncomforted conditions during transport. Besides the improvement of vehicles design and handling methods, continuous and accurate measurement and report of stress inducing factors and stress response parameters, and continuous observation of animals are necessary and essential to improve animals welfare and the quality of meat, the final product. New welfare regulations will impose surveillance systems so that information on the quality of transport conditions is available. Moreover a route description is useful for optimisation of transport logistics, but also in relation to estimating of sanitary risk and food safety, including trace ability of individual animals. In line with the above objectives, instrumentation of different categories has been developed to study air quality, vibration, animal responses in terms of physiological and behavioural alteration during handling and transport from farms to abattoirs. More over, logistical and control system has been developed and tested. The components of the transport surveillance system are: (a) integrated sensors and data transferring component in the animal loading compartment of the vehicle (information on individual identification of animals, air quality (temperature, relative humidity, emissions), vibration and behaviour of the animals, loading and un-loading place and time, and route tracing); (b) Information storing and data transfer unit in the cabin. Component (b) is also integrated with GPS and GSM module. Data transfers from component (a) to component (b) wireless, and the from component (b) transfers to server using GSM. Hence, information is available on-line and on disk, so that the driver can be informed and corrected at the spot. It may noted that sufficient data (that was not transferred to the control centre) are stored in the component to be used if necessary. Another instrumentation package that comprises sensors of heart rate and vibration on the animal has been integrated. These sensors can be mounted on animals and the data is transferred to a database through a wireless network. Dynamic route optimization of cattle collection from farms and logistical activities of abattoirs are considered in relation to animal welfare Comprehensive field measurement has been made to evaluate the system and found that the package performs well. It can be noted that using the developed system accurate measurement can be made and enable optimisation of the handling and transport system to improve animal welfare. Hence, advice will be generated for vehicle manufacturers, haulers, farmers, slaughterhouses and retailers.

Operation of electronic identification of cattle in Finland

Hannu Haapala

Introduction

The aim of the project was to evaluate electronic identification (EI) in northern winter climate that is characterized by extremely low temperatures (down to -50 °C) and rapid and frequent fluctuations (+5 … -25 °C). The work was based on the I.D.E.A. project (IDentification Electronique des Animaux: basis for the future implementation of the Electronic Identification System in livestock of the European Union) of JRC (EU Joint Research Centre, Ispra, Italy). I.D.E.A. suggests requirements that might be inadequate in northern climate. The project gave information on the actual operating conditions especially as regards animals kept in uninsulated loose housing. The results can enhance JRC´s work in defining test procedures and standards. They also give Finnish authorities, slaughterhouses and farmers indications of the possible new requirements for production conditions and buildings.

Material and methods

EI devices complying with the ISO 11784 (code structure) and 11785 (technical concept) standards were tested in the field on 3 farms with ca. 270 ruminants. Reference tests were done in climatic chambers. Electronic ear tags and ruminal boluses were used as transponders. Both static and dynamic reading methods were tested. To achieve this, hand-held readers and portable stationery readers were tested.

All animals on the selected farms were equipped with transponders. The researchers visited the farms regularly and read the tags with both systems. The information was stored in a database, which followed the approved data dictionary standard. The success in identifications was recorded together with environmental conditions. Operation environment was characterized with temperature and relative humidity. Environmental measurements were done with recording instruments to save the temperature history of the devices. Necessary data transfers were arranged to communicate with JRC. The animals to be slaughtered during the project period were followed and identified in the slaughterhouse as far as it was technically possible. The used ear tags and boluses were collected for later inspection.

Results

Attaching of the both tag types was quite easy if only the workers were trained enough. Two persons were needed to complete both the attachment and bookkeeping. Extra persons moving and treating the animals helped a lot in the practical work.

According to the work time measurements the ear tags were much faster to attach than the ruminal boluses. Aggressive animals were difficult to handle when boluses were attached. The bottleneck in time use was the electronic bookkeeping with the reader device.

It took sometimes a long time (up to 0,5 hour) for the boluses to settle down. This caused problems in workflow since the boluses could not be checked for operation right after the attachment.

It was easier to attach the tags in dairy cattle than in beef cattle. Dairy animals were much more used to handling than the beef animals were.

All the transponders worked satisfactory. The readers, on the other hand, were quite different from each other. Differences in reading time and physical dimensions and field shape of the antenna caused great variation in reading work. When using handheld readers with short antennas and short reading distance the work postures were quite difficult and the work was also potentially dangerous. The casing material of some readers was not suitable for cold environment since it turned slippery in low temperatures. There were also no carrying straps in the readers to hold them firmly during reading so that they could easily be dropped. It was also observed that in almost every reader there were too low-level audible and/or visible signals indicating a successful reading.

In static readers there were some annoying faults in the antennas. The antennas were very sensitive to e.g. electromagnetic disturbances and required correct installation to avoid malfunction. There was also a lot of trouble with the connection software to external computers.

In slaughterhouses it was quite easy to retrieve the boluses and ear tags. Data handling and integration to existing registers is quite easy.

Discussion and conclusions

Most of the problems encountered were due to other reasons than the cold environment. However, ergonomics of the readers have to be developed. Management issues such as arrangements in animal traffic seemed to be the most tedious challenges to meet. In Finnish farms there are few good solutions for catching the animals for tagging so new solutions have to be developed and tested in practical environment.

It would be rational to use the same tag for all the appliances where animal identification is needed so that e.g. the automatic feeding systems would use the system as well.

The EI system, if implemented, has wide effects on the whole production system in EU. Before implementation the whole production chain needs to be analyzed in order to minimize disorder. There is great variation in local circumstances. Extensive user training is also needed.

Modeling individual feed intake of the dairy cow

I. Halachmi, Y.Edan, U. Moallem and E. Maltz

Voluntary feed intake of the dairy cow is an important variable in the modern dairy herd management. Existing formulas that calculate dry matter intake (see Mertens, 1985; NRC 2001) are not applicable for estimating dry matter intake (DMI) of an individual cow. New formula is necessary to adjust individual cow ration density by a computerized concentrate self-feeder and by a milking robot combined with best possible milking frequency for the individual cow in real time. Unfortunately the individual forage feed intake of each cow can not be explicitly measured in a for-profit farm since the cows are kept in a group, fed total-mixed-ratio (TMR) in a feeding lane or grazing. In our study, we developed a model predicting the individual DMI. The input parameters are body weight and milk yield, both are measurable on-line in an industrial dairy farm. The model result is voluntary dry matter intake (kg) per day for each unique cow, can be embedded in the nutrition calculation of the individual cow. Thirty-nine high-yielding multiparous Israeli-Holstein cows of the Bet Dagan experimental farm calving between November February were housed in loose covered pens with adjacent yards. Cows were fed from an individual electronic feeding system that recorded the intake of each cow. All cows were adapted to the feeding system for 2 wk during the dry period. Body condition was determined 1 wk before the expected parturition. Cows were selected according to parity, date of parturition, BCS, and FCM yield during the first 150 d of the previous lactation. The net energy (NE) of feeds was calculated with NRC values; all feeds were mixed and fed once daily from a mixing wagon. Each cow was fed from an individual trough in a specific position reached via an automatic yoke gate. The system enabled free and easy entry to the trough and ad libitum feed consumption for each individual cow; measurement of the duration and feed intake of every visit was recorded as previously described (Halachmi,1998). Orts were weighed automatically and removed daily at 9 a.m. The intake of the previous day was determined and 107 to 110%of DMI was offered in every individual box at 10 a.m. The daily intake of each cow was determined from parturition to 260 days in milking. Samples of feeds and orts were analyzed for each group of cows every week. Cows were milked three times daily, and milk yield was recorded electronically. Milk composition from a composite of three consecutive milkings was determined every 2 wk. From parturition, cows were automatically weighed three times daily after every milking. BCS on a 1 to 5 scale was determined weekly by an experienced technician. Energy balance was calculated for each cow based on daily food intake, daily milk production, daily BW, and milk fat concentration. In early lactation, four phases in DMI were observed in all diets treatments: I) a rapid increase in feed consumption during the first 25 to 30 days; II) a more moderate increase in DMI between 30 and 55 days; III) a period with little change between 55 to approximately 105 days (this differed somewhat among treatments); and IV) a slow decline in feed consumption. More quantified results are given below: cow #1: mean error (ME)=-0.69 kg, error standard deviation, std(e)=1.78 kg; mean absolute value of the errors (MAE)=1.43kg. cow #2: ME=-0.76 ; std(e)=2.09 ; MAE=1.66, while error is the actual feed minus model's output. 82% of cows passed the statistical T-test, has p-value < 0.05. Under the conditions of our trial, daily milk yield and body weight proved to be the only variable required for estimating DMI of the individual cow. However, this has yet to be tested for different cows and conditions.

Improvement of teat cup attachment in automatic milking systems by using teat coordinates

J. Harms and G. Wendl

Introduction When using automatic milking systems quick and sure attachment of the teat cups is very important for the capacity but also for animal welfare. Attaching teat cups has been improved very much during the last years, but still in some cases the attach-ment fails. Normally, the average teat positions of the last successful milkings are used for locating. When the teat position is located exactly, the cup is attached. After that the system has to wait for a feedback like milk flow, constant vacuum or sound to decide whether the attachment was successful or not. Aim of this investigation was, to check the possibility of using the teat coordinates to make a decision whether the found teat position is plausible, before the teat cup is attached. Method and material The investigation based on 59012 milkings of 94 different Simmental cows which were milked by an automatic milking system Fullwood-Merlin. For every milking the coordinates of each teat, the milking time of each teat and a system message de-scribing the success of the milking were available. The vertical teat coordinates (dis-tance of the teats to the floor) of all successful milkings (no negative system mes-sage) were compared to those where the robot attached the cups, but no milk flow occurred at one teat at least (1136 milkings = 1.9%). The differences between the actual height of each teat and the average of the last ten successful milkings were analysed for all milkings in order to find a rule to detect incorrect attachments. Two limits for a supposed wrong detection of udder height were used (The absolute deviation to the last ten successful milkings and the deviation between the teats). Different combinations of the limits (AND/OR) were tested. Results For the individual animal only small variation coefficients of the vertical teat positions were found. The averaged values of all successful milkings were: Right front 5.3%, left front 5.3%, right rear 5.8% and left rear 5.8%. Moreover no significant correlation between the day of lactation and the udder height was found. Using the rule: The actual attaching procedure is supposed to be unsuccessful, if the absolute deviation is bigger than 55 mm or the deviation between the teats is bigger than 65 mm, 52.5 % of all failed attachments could be detected (sensitivity). In these cases, a wrong teat cup position was supposed because no milk flow was observed. An error rate of 1% was found (wrong teat cup position supposed, but milk flow ob-served). To analyse the impact of different limits on sensitivity and error rate, the limits were varied +/- 10 mm. The error rate then ranged between 0.3% and 2.6%, the sensitivity between 42.1% and 60.1%. Conclusions Udder height is changing only very little during lactation and from milking to milking. Based on this, a plausibility check for the teat position before attaching the teat cup is possible. A sensitivity of more than 50% combined with an error rate below 1% could be reached. The low error rate would offer the possibility to restart searching the teat while taking only low risk of making a wrong decision. So the time could be saved, which is normally needed to wait for a feedback of the system like milk flow, constant vacuum or sound. This would extend the capacity of the system and have positive effects on animal welfare.

Individual identification of milking cow by their voice

Yoshio Ikeda, Gerhard Jahns, Takahisa Nishizu, Kunio Sato and Yoshinari Morio

Generally speaking, the animals have the central nervous system and take the autonomous behavior in response to the environmental conditions surrounding them and their physical states, such as sickness or oestrus. Therefore, it is a very attractive research to understand the animal behavior through the computer from the academic point of view. From the practical point of view, and of course, for automatization of animal breeding for the precision livestock farming, it is necessary to understand the animal behavior by some means or other. The behavior of the animals can be generally understood through the auditory and visual senses from the viewpoint of non-contact and non-invasive detection of conditions. In this research, we will try to analyze the cow vocalization and to identify the individual animal by the characteristics of the voices. As the merits of the voice signal, the data acquisition system is comparatively simple and low cost. On the other hand, as the demerits, since the animals move in the yard or pen, the distance and attitude between the signal source and the receiver cannot be controlled easily and freely, and then the signal becomes unstable on collecting the data. Therefore, in order to compensate the amplitude variation due to these causes, it is desired to normalize the signal amplitude by some means. Or, it is necessary to extract the features unrelated with the signal amplitudes. In this research, we will propose the new method to compensate the variation of the sound source. The spectra of the sequence of short time variance for 5, 10 and 15 ms were determined by the Maximum Entropy Method (MEM) for estimating the parameters contained in the linear prediction filter because of the smoothness of the estimated spectra. The selection of the number of the terms of the linear prediction model is of importance, but is very difficult. In this research, the number of terms of the linear filter was so determined as for the MEM spectra to trace the FFT spectra as smooth as possible. The number of terms in this research was 7. The straight line fitting of the spectra on the log-log axes of the frequency and power was used to characterize the voice signal. And the amplitude of the signal may not affect the slope of this straight line. Hence, the straight line fitting of the log-log spectra compensate the variation of the signal amplitude due to the relative distance and attitude between the sound source and the receiver. The slope may have the values between zero and negative infinite and expresses physically the degree of the fluctuation of the loudness of the voice during vocalization. That is, the extreme slope of zero, the horizontal line, means that the loudness of the voice changes randomly during vocalization. And the other extreme aspect of the slope of the negative infinite, the vertical line, means that the loudness does not change and remain constant during vocalization. The wide range of the value may be promising for characterizing the voce signal. The coefficients of the linear filter reflect the characteristics of the voice signal and may be used as the feature parameter. The feature space is composed with the slope of log-log spectra and the prediction coefficient to discriminate the voice. Results Three cow A, B and C, the cow A and B could be identified on the feature plane of the slope of the log-log spectra and the linear prediction coefficient

Infrared positioning system to identify the location of animals

R. Kaufmann, H. Bollhalder and M. Gysi

Animal behavior occupies an essential place in farm animal research. Animal behavior is one of the most important criteria when assessing the animal friendliness of a housing system. Another factor of interest is the frequency the animal group stays in different housing zones. Until now behavioral research has been based primarily on direct observations or video observations. The drawbacks of both methods are high labor input and problems with nocturnal observation. Direct observation can influence the natural behavior of the animals, particularly when a minimum amount of light is required in nocturnal situations. The objective of a new development at the Swiss Federal Research Station for Agricultural Economics and Engineering, Tänikon (FAT) was to create a system which could be flexibly used for the observation of animals both indoors and out, and which yielded reliable time and space-related information on individual animals. The newly developed FAT- For Animal Tracing 2002â positioning system comprises the following elements: · Video cameras fitted with infrared filters · Animal collars for the transmission of infrared pulses · Image analysis and evaluation software. A camera mounted 4 meters above the test area covers a range of approx. 8 x 6 meters. In the present layout, a computer station can process the signals from four cameras. This is the equivalent of a test area of around 200 m2, or the area occupied by 20 - 30 cows in a loose dairy housing system. Specially equipped collars emit infrared impulses from individual animals. The signals received by the cameras are evaluated by analysis software according to time and location. The individual sections of floor space to be separately recorded and analyzed in a particular experiment can be configured on the PC in any graphic form desired. With the aid of variable cameras, the system is able to record the position of up to 100 different animals and to relate the signal to a maximum of four groups. The observation area can be divided into altogether 100 subfields. The lower time resolution is a one minute positioning interval and the spatial resolution is ten centimeters. It is designed on a basis of 98 % system reliability. Experience in the field and outlook Initial experience in the field has been positive and has proved that the system functions. Equipped with the appropriate analysis software, control functions can be transmitted to the system as well as recording the desired behavioral parameters. For example, correct herd allocation in the housing area can be monitored with electronic access gates. The system is being further developed in the following directions: 1. Mobile version: designed to support on-farm research projects. 2. Outdoor version: can conceivably be used outdoors if suitably suspended, protected from the weather, and the recording technology adapted appropriately. 3. Recording movement in the third dimension: this will make it possible to distinguish between a standing animal and a resting one. Results of several tests are going to be presented and explained. Conclusion The FAT- For Animal Tracing 2002â positioning system can record and evaluate the locational characteristics of individual animals in space and time. This permits the rational observation of behavior in housing systems. Strengths of the system: · The animal is not influenced by the observation procedure · Nocturnal observation is possible without artificial light · Animal numbers, groups, and the shape and size of observation areas can be flexibly defined and adjusted to suit the experiment. · Flexibility of research location. The system supports on-farm research, but use is even possible outdoors. Further stages of development are planned, the most important being the simultaneous recording of resting behavior and the outdoor application. However applications in the industrial field are also possible, e.g. the automatic tracking of material flows or the provision of logistical information on production procedures.

Pastures from space - improve Australian sheep production

R. Kelly, C. Oldham, A. Edirisinghe, G. Donald and D. Henry

The capacities of Australian agricultural industries to continue to survive and improve are increasingly affected by their ability to make timely decisions for the expanding area of their business. Improved utilisation of pastures by sheep and cattle across southern Australia provides such an opportunity, as it is generally very low. Studies have shown that in many seasons the amount of feed grown that is consumed by sheep and beef cattle varies between 20 and 30%. Economic analyses estimate that better utilisation of pastures could double farm profit - for every 10% increase in utilisation, profit increases by $20/hectare.year. There has been considerable investment by many Australian organisations into providing the farmer with skills to improve pasture utilisation. Over the past decade more than 10,000 farmers have attended courses on how to manage their pastures to improve animal performance. In many respects these courses have failed to achieve their potential as they rely on the farmer to measure the biomass of their pastures; a high cost, low accuracy input. For a 1500 ha property, fortnightly pasture biomass assessments alone would take ~2 days each, and still miss the essential measure of pasture growth rate unless combined with meteorological information. The paper reports how the provision of information over the Internet on pasture biomass (www.pgr.csiro.au; www.spatial.agric.wa.gov.au/foo) generated from NOAA AVHRR Landsat TM and SPOT satellite imagery is being used in South Western Australia to improve sheep and wool production. Weekly, reliable, accurate, low cost information on biomass and growth rate of pastures (Henry et al. 2002), combined with animal measurements (Oldham et al. 2002; C.M. Oldham and S.G. Gherardi pers. comm.), were used in the period 2000 to 2002 to investigate their potential to help manage increases in stock carrying capacity to manipulate wool and meat production on 50 farming properties. In 2001, predicted pasture growth rate (PGR) explained up to 80% of the variance in ground measurements. Pasture biomass or feed on offer (FOO) was predicted with a standard error of +/- 10%. Feedback from farmers indicated that most do not consider themselves accurate enough to visually estimate either PGR or FOO, and hence are enthusiastic about what the satellite based technology can offer. Over the three-year evaluation period, the farmers were generally able to achieve substantial improvement of performance. In general higher stocking rates were associated with decreases in wool per head but more wool per hectare that was finer and stronger and therefore more valuable ($/clean kg). In addition, wool producers believed that the remote sensed estimates of FOO and PGR would enable them to repeat this outcome in most years. Therefore they were encouraged to experiment with increases in their stocking rate over the whole-farm in future years when the technology was commercially available. The remote monitoring of pasture supply in paddocks, and its predicted growth rate, provided in a cheap, accurate and timely manner, offers substantial potential to raise the productivity of the farm business and improve its environmentally sustainability. The future farm may well combine these pasture measurements with electronic systems to monitor animal performance and control access to feed. The PGR service has recently been expanded through a commercial e-commerce portal: The Farmshed (www.thefarmshed.com.au). References: Henry, D.A., Edirisinghe, A., Donald, G. and Hill, M. (2002). Monitoring pastures using satellite remote sensing. Proc. Aust. Soc. Anim. Prod. 24: 81. Oldham, C.M., Gherardi, S.G., Paganoni, B. and Yelland, M. (2002). A new approach to managing wool production; Measure as you grow; Proc. Aust. Soc. Anim. Prod. 24: 161.

Detection of udder diseases in automatic milking systems by using fuzzy logic

Stefan D. Koehler and Otto Kaufmann

The development of improved sensor technology leaded to a constantly growing amount of available data relevant to the control of single animals, environment and production. On the other hand, the benefit which can be reached by this abundant data is to be maximized only by the application of integrated monitoring systems able to concatenate information from sensors and data bases with mathematical models and expert knowledge. Methods of Fuzzy logic have been used in agricultural research and in the development of decision support systems only for the last ten years. Some publications state that besides graphic programming languages and neural networks, Fuzzy logic might become a future trend for automation and Precision livestock farming in dairy farms. Nevertheless, there is only a small number of publications related to topics in animal science. Practical use of Fuzzy logic methods in dairy production is as yet unknown. The early recognition of udder diseases represents a crucial challenge to the further use of automatic milking systems (AMS). Existing recognition models are to be supplemented by intelligent methods which enable the system to combine several process parameters for early and reliable conclusions on the udder health. The present research labor was centered on the evaluation of different milking and milk parameters regarding their suitability as input variable of recognition models for udder diseases as well as on the development and optimization of Fuzzy Logic models. In the context of a preliminary study, it was examined which variables should form the input of the recognition model. For this purpose, besides the collection of the process data of an AMS Astronaut® 754 measurements of single udder quarters were accomplished with measuring devices LactoCorder® low flow and individually processed. The evaluation of this data resulted in certain differences between inconspicuous (healthy) and conspicuous (more ore less ill) udder quarters regarding the parameters milk production rate, milk flow and milking interval which are usable for model building. Also the electrical conductivity should be included into further procedures. The functionality of simple Fuzzy logic models was tested successfully on the basis of 74 generated data records. The main investigation took place in a dairy farm with two parallel working AMS Astronaut®; in a herd of 103 milking cows. 4,282 measurements with LactoCorder® low flow -devices at 359 udder quarters were analyzed individually. After removal of possibly incorrect data records or those which can not clearly be assigned remained a usable database of 2,826 records measured at 195 inconspicuous and on 41 conspicuous udder quarters. The statistic analysis of the absolute parameter values resulted in highly significant differences between inconspicuous and conspicuous quarters for the examined input variables milk production rate, milk flow, milking interval and electrical conductivity. For the further modeling, 527 standardized data records computed from the raw data were used because these are largely uninfluenced by disturbing individual and environmental influences. Three different approaches of model building for the standardized parameters were pursued, analyzed and compared among themselves as well as with the alerts of the AMS: a) One-parameter-models (threshold models) b) Index-models c) Fuzzy logic While the threshold models pointed out the highest specificity values, the best sensitivity values were determined for the linkage of the four input variables by index-models. However, the optimal result of modeling was obtained by Fuzzy logic. It resulted in the lowest probability of false classifications (wrong diagnoses) with 5,9 per cent.

Improving oestrus detection in dairy cows by combination of different traits using fuzzy logic

Joachim Krieter, Regina Frik, Eckhard Stamer and Wolfgang Junge

In modern dairy farming the request for reliable systems for automatic oestrus detection increases due to downsized staff per herd and due to a reduced reproductive efficiency in dairy cows. Undetected and falsely detected oestrus result in missed and untimely inseminations with high losses of income. In order to improve the accuracy in systems for automatic oestrus detection, oestrus detection should be performed by multivariate analyses of different traits. In this study a fuzzy model was developed for the multivariate analysis of the traits activity, milk yield, milk flow, conductivity and period since last oestrus. The collection of data was performed on a commercial dairy farm in 1998. During this period 862 inseminations were verified as oestrus cases by a following calving. For oestrus detection time series consisting of 15 days before oestrus, the day of oestrus and 15 days after oestrus were analysed. Over this period serial information were available for activity, milk yield, milk flow and conductivity on a daily basis. The parameter period since last oestrus included information about previous inseminations and previous oestrus cases. For each day considered in the analyses the period since last oestrus was calculated from the difference of the actual day and the day of previous information. Oestrus detection was performed with univariate methods (day to day comparison, moving average, exponential smoothing) with different threshold values and multivariate models (fuzzy logic). Fuzzy model was implemented with the elements fuzzification (membership function), inference and defuzzification. Input values for fuzzification were the relative deviations between estimated and observed values. These values were transformed into fuzzy values by linguistic interpretation (membership functions with grades between 0 and 1). The linguistic combination of the traits was performed in the fuzzy inference with the form if condition; then conclusion. By defuzzification fuzzy values were transformed back into crisp results. Values higher than 0.5 were treated as oestrus alerts. The evaluation of the different models based on the parameters sensitivity and error. Univariate analysis of the trait activity for a threshold of 40% (deviation from individual mean) indicated a sensitivity of 94.2% and an error rate of 53.2%. Increasing of the threshold value to 120% reduced the sensitivity (71.0%) and enhanced the error rate (21.5%) as expected. Single analysis of milk yield, flow rate and conductivity showed senitvities in the range from 31 to 99% (depending on the threshold), but error rates always exceeded 90%. Therefore the additional gain by using different combination of activity with milk performance and conductivity was only small. Oestrus detection was significantly improved by using activity and the period since last oestrus in a multivariate fuzzy model. Sensitivity ranged between 87.9 and 90.1% and error rate decreased to 23.5 and 12.5%. The results of multivariate analysis with information about previous oestrus cases indicates that the automatic oestrus detection could be improved by considering the trait period since last oestrus. Especially the error rate was enhanced in comparison with univariate analysis of the traits activity. It is therefore preferable that oestrus detection is performed before the aspired breeding period is reached. The results of the study also demonstrate that fuzzy logic is a suitable method to support decision making in dairy farms.

Telemetric measurement of heart rate and heart rate variability for evaluating psychological stress induced by visual discrimination tasks in dwarf goats

J. Langbein, G. Nürnberg and G. Manteuffel

Since the publishing of the Brambell report, applied animal science has accumulated important knowledge about various behavioural needs of farm animals which has mainly been used to configure housing systems more in line with animal welfare. Surprisingly, little research has been conducted during this period concerning the cognitive abilities of farm animals. So we need to further investigate the sensory and cognitive properties of farm animals as well as how they learn to cope with challenging cognitive situations that test their behavioural adaptation capabilities. In order to examine visual learning in dwarf goats we employed three visual discrimination tasks. Each task lasted for thirteen days. We tested groups of 12 naive young males using a computer based learning device integrated into the home pen of the animals. In each task four visual stimuli were presented on a computer screen in a multiple choice design. To get a water reward goats had to choose the correct stimulus by pushing one of four buttons in front of the screen. Each individual trial was followed by an inter-trial interval lasting six seconds (black screen) before the same stimuli were presented in a distinct pattern following a quasi random series. All manipulations of the buttons by an animal were recorded by a computer program. On the basis of this data we analysed individual learning success on a daily basis. To indicate changes in the sympatho-vagal balance related to psychological stress during learning we telemetrically measured heart rate (HR) and heart rate variability (HRV) on the first two days and the last two days of each single task. At the same time the behaviour of the animals was recorded on video. HR and HRV were analysed separately for periods when the animals were inactive and active in the pen, as well as when they were active at the learning device. HRV was analysed in the time domain and the frequency domain following the rules of the Task Force of the European Society of Cardiology. For the analysis of non-linear HRV phenomena we applied Recurrence Quantification Analysis. Learning success started at levels below 25% correct choices (random level) on day one of the first discrimination task. The animals reached a stable level of 70% of correct choices not until day 12. The mean learning curve followed a sigmoidal type. In tasks two and three, animals started from 45% correct choices at day 1 and reached a stable level of 70% of correct choices at day 8 or day 7, respectively. The mean learning curves were similar in both these tasks and followed a logarithmic type. HR and HRV results indicate a changing level of psychological stress with rising learning success in each task as well as differences in the stress level between the three consecutive tasks. Whereas HR increased significantly with a rise in learning success in task one, this relationship was inverted in tasks two and three. As concerns the coping predictability concept, loss of control over drinking water at the beginning of task one seems to provoked a certain level of helplessness in the animals. Sings of activation of a coping reaction in the animals were not seen before they had reached a certain level of learning success at the end of task one. When the animals were familiar with the learning procedure in tasks two and three, the coping reaction was activated right from the start of the tasks and led to a reduced level of stress with increased learning success. Results of HRV suggest that changes in HR related to psycholgical stress are primarily caused by withdrawal of vagal control of the heart.

Variable milking frequency in large dairies: performance and economic analysis - models and experiments

E. Maltz, Y. Edan, S. Matza, N. Livshin, A. Antler and A. Antman

Increasing or decreasing milking frequency (MF) can be a tool to influence milk yield (MY) to meet temporary constraints without changing the number of cows or applying nutritional means. Large dairies in which the milking parlor is operated most of the diurnal hours, maintain the capacity to milk different groups of cows in different MFs. Manipulating MF is expected also to effect food consumption that might impose additional operational costs, while not all cows will respond as expected. Therefore, an optimal MF policy should include a physiological and economic analysis of each individual cow in the herd to evaluate suitability and ability to meet performance expectations when changing MF while considering the economical benefits. In this paper we will present a preliminary model that evaluates the economic value of changing the MF of the individual cows in accordance to their physiological capacity and economical goals of the dairy. The model was designed using initial evaluation in the first step and then applying results from changing MF experiments. The experiment included reducing MF from thrice to twice daily to produce below the quota (winter), followed by resuming thrice daily milking and then increasing to 4 milkings a day (summer). The rational behind this strategy is that the farmer in Israel is allowed to shift up to 20% of the monthly quota from winter months (January-May) to summer months (July-November), and get a bonus of 26.5% for each kg "shifted". The model uses daily milk yield and body weight data and monthly-analyzed milk composition as inputs. Food consumption of the individual cow is estimated by applying a model that uses these daily records. Prediction of the influence of changing the milking frequency and consequently the feed consumption as well as parlor operations and human labor are considered in a full economic analysis. The initial model (using preliminary evaluation of milking frequency policy results) resulted about 10% prediction errors. The experiments of changing MF from 3 to 2 milkings a day were conducted only on cows that their milk economical value (ECM) was lower than their MY because of its "poor" composition. In addition, no physiological risks were taken when MF increased from 3 to 4. This change was done only for cows that were pregnant, had over 200 day in milk, weighing at least 500 kg with an increasing body weight curve when the MF was increased. MF reduction from 3 to 2 times daily yielded a higher reduction in MY than in ECM compared to the control as a result of milk quality improvement after MF was reduced. When thrice daily milking was resumed, both MY and ECM increased above (by about 5.5%) the production of the control cows. When MF was increased to 4 times daily production (both MY and ECM) increased by about 4% above the control cows. The response of the cows to the changing MF was analyzed in relation to days in milk, lactation number and initial MY and ECM, and the results were incorporated into the model, which improved its prediction capacity. It was concluded that in large dairies production can be controlled by changing MF by using physiological-economical prediction models.

Telemetric measurement of heart rate variability in dairy cows: assessment of stress in automatic milking systems

I. Neuffer, H. Kündig, C. Kaufmann, R. Hauser and B. Wechsler

In Switzerland, an authorisation procedure regarding animal welfare is established since 1981 for mass-produced housing systems and equipment for farm animals. Contrary to conventional milking systems, automatic milking systems are included in the test procedure, as the functioning of such systems is not constantly supervised by the farmer. Previous studies came to different conclusions concerning the welfare of dairy cows milked by automatic milking systems (AMS). Wenzel (1999) found that cows milked by AMS experience greater stress than cows in milking parlours, while Hopster et al. (2000) concluded that cows milked by AMS were subjected to less emotional stress than cows in tandem parlours. As these studies provided only information for specific AMS models, comparative studies with different AMS models are needed. Many of the studies concerning AMS and animal welfare were carried out on experimental farms with only small herds of dairy cows; the results thereof are difficult to apply on practical farms. Three different AMS models are tested and compared to conventional milking systems, a modern tandem milking parlour. Investigations are carried out on three farm each for the different AMS models as well as for the control systems. On each farm, twenty dairy cows showing no clinical signs of illness and representing all ages and stages of lactation are chosen for the investigation. As a consequence, the parameters chosen to assess stress in the cows had to be applicable on practical farms and non-invasive. Besides other parameters such as video observations and milk cortisol levels, heart rate variability was chosen to evaluate the stress of the animals in automatic milking systems. The physiological variability in heart rate is considered a better indicator of the individual capacity to cope with the environment than simple heart rate (Porges, 1995). By quantifying the heart rate oscillations the animal's reactivity and vulnerability to stress can be assessed (Porges, 1985). Analysis of interbeat intervals may reveal HRV to be a potential tool for the assessment of autonomic function and therefore a useful indicator of emotional state in farmed animals (Forde and Marchant, 1999). Non-invasive, commercially available systems for heart rate measurements in dairy cattle are not able to transmit the signal for a period longer than a few hours. For our investigations, a system was needed that transmits signals for a few days without having to approach the animal. An elastic girth belt was equipped with three electrodes, an amplifier and a radio transmitter, which is able to transmit the signal continuously for three to four days. The corresponding receiver unit is connected to a demodulator, which transforms the analog signal first into the original and afterwards into a digital signal. Finally, the peaks in the R-R interval curve are detected and saved as interbeat intervals for later analysis. On each farm, ten animals of the selected group mentioned above are included in this investigation. Due to the great interindividual variability in the level of HRV, a direct comparison between animals is difficult. Instead, the data obtained during the visits to the automatic milking systems will be compared to data of the same cow during quiet standing and lying. Data collection will be finished in December 2002, results will be available in spring 2003.

Evaluation of udder cleanliness by spectroscopy

Dieter Ordolff

Actually automatic milking systems are unable to clean udder and teats before milking according to the demand for doing so, given in directives at EU and national level, and with respect to the success of the operation. No detection of damaged teats is possible at the moment. To overcome that problem up to now some basic research has been done on optical evaluation of the surface of udder and teats, using experimental sensors and video systems. Although results were mainly positive no standard procedure for evaluation of udder cleanliness was to be deducted. Therefore the efficiency of a sensor system using the industrial standard CIE-lab according to DIN 6174 was investigated over three milking sessions with 76 milked cows at the experimental station of the Federal Agricultural Research Centre (FAL) at Braunschweig (D). The optical condition of surfaces was described by three parameters indicating the proportion of reflected light (parameter l), the shift of surface colour to red or green (parameter a) and to yellow or blue (parameter b). In a second experiment including 13 cows, six of them with black teats, it was investigated whether it was possible to detect bloodstained surfaces. For data recording the measuring device (SPECTROPEN, Dr. Lange GmbH & Co. KG, Düsseldorf, D) was to be brought into mechanical contact with the surface to be evaluated. In both experiments data were recorded before and after manual cleaning of udder and teats. Results were evaluated applying ANOVA-procedures. It was found that the proportion of reflected light after cleaning was higher at white surfaces of udder and teats. The generally lower level of this parameter at black teats was farther reduced after the cleaning procedure. All differences due to cleaning of that parameter were significant. The parameter red/green, which in general was dominated by red (positive values) was practically not affected by cleaning. The variation due to cleaning of yellow/blue was not uniform. While there was no clear change at white surfaces the proportion of yellow was clearly reduced at surfaces of black teats. The proportion of reflected light was best suited for evaluation of cleanliness of udder and teats, corresponding to visual inspection by milking staff. The structure of data included a neutral area, corresponding to dirty surfaces. Clean surfaces were to be found above and below that range of data. The detection of bloodstained surfaces was possible in the most reliable way by using the parameter red/green which also was efficient at black surfaces. An obvious limit was to be found for undamaged surfaces. If it was exceeded blood was present with a high probability. Similar to results of earlier experiments cited in literature the investigations indicated that parameters defined in industrial standards can be useful to evaluate cleanliness of udder and teats and to detect bloodstained surfaces in that area. For practical application sensing devices requiring no mechanical contact with surfaces to be evaluated may be more useful, e.g. industrial machine vision systems. Summary Evaluation of cleaning efficiency of udder and teats by spectroscopic parameters indicated an obvious change of intensity of reflection from surfaces due to manual cleaning. The parameters red/green and yellow/blue were not useful to indicate cleaning efficiency at white surfaces. A significant reduction of the parameter yellow after cleaning was observed at teats with black surface only. Bloodstained surfaces were indicated by the parameter red/green in a most complete way. For this parameter an indication for a threshold value for bloodstained surfaces was found, independent from pigments at teats evaluated. A more practical evaluation of visual cleanliness of critical areas can be expected by application of image processing systems, using various optical aspects for scoring structures of surfaces.

Designing new Individual Ration Systems for dairy cows with the IDEF0 method

J.C.A.M. Pompe, S.J.M. Wantia, C. Lokhorst and A.H. Ipema

Cow recognition systems, automatic concentrate feeders, milking robots and the accompanying management software are now well-accepted technologies for dairy farmers. These technologies have provided the opportunity to optimize the milk production and the concentrate intake of individual dairy cows. Precise and automatic supply of the complete ration - including roughage - to individual cows offers the potential to further optimize the feed intake by dairy cows, reduce the labor requirement and increase the flexibility of the dairy farmer to plan his daily activities. Other researchers described Individual Ration Systems (IRS), but these were all designed for experimental purposes. The IRS in our work is intended for practical farms. In our presentation we would like to focus not so much on the design of the feeding station itself, but on the logistics of an IRS which will result in a practical solution for various farms. An IRS needs to fulfill the functions of a feeding station, an assembly and a mixing unit, one or more transportation devices, one or more unloading systems, and one or more storage systems. All these functions are subject to various requirements, such as determined by the feeding management, the feeding strategy, the productivity of the farmland and the possibility to purchase various ingredients. The feeding process itself is affected by the components of the ration (roughage, concentrates, by-products, water and vitamins, minerals and spore-elements), the behavior of cows with respect to feed- and water-intake and the mixability of the feedstuffs. Possibilities to plan an IRS are affected by the layout of the farm buildings and rules and regulations by the government. An IRS itself has an influence on the labor organization, the mineral balance, the utilization of the farm machinery and the financial budget. From the review of the aspects that influence an IRS it became clear that different disciplines are involved; the relations between the various functions are so complex that it proved necessary to use a methodological approach for designing an IRS. For this purpose we applied the IDEF0 functional modeling method. IDEF0 is a standard method to produce a structured representation of the functions within a system and the data and objects that inter-relate those functions. It results in a hierarchical series of diagrams, text, and glossary cross-referenced to each other. IDEF0 is a member of the IDEF (Integrated DEFinition) family and is widely used in the government, industrial and commercial sectors. This was the first application of this method to an agricultural system. During our presentation we would like to demonstrate how we applied IDEF0 to construct our model for the composition of the ration, the acquisition of the various feedstuffs, the storage of the feedstuffs, the assembly of the various feedstuffs to a ration, the distribution of the ration, and the registration of the feed-intake. By combining this model with a process of eliminating, combining or changing the sequence of (sub)functions alternatives we can now analyze the options for the design of an IRS. We developed two options for a fully automatic IRS for a high tech farm with a given layout, herd composition and milk production. We will also discuss future plans for work in this area.

Development status of a BUS system for livestock farming

J.-P. Ratschow and R. Artmann

Electronic components are being used increasingly in livestock farming to monitor, control and regulate: - liquid feeding systems have a process computer; - animal housing climate computers have replaced simple transformers; - electronic identification of individual animals is possible; - milk quantities are measured electronically; - the State Monitoring Association (LKV) makes the data available via E-mail or the Internet; - animal data are transmitted via the Internet to the HIT database; - sow and cow reproductive management systems facilitate the organisation of labour and the managerial analysis. The relevant firms and institutions offer technical solutions for all the above mentioned examples. However, it is very difficult to transfer the data from one stage, from one company-specific solution, to another stage or the solution presented by a different company. For instance the data from feeding computer A and the data from climate computer B cannot be introduced expediently into the sow management system C, and most definitely not into programs for plant production, such as arable field files. A working party has been set up in the BFL to simplify communication between equipment and exchange of data between the equipment and the various stages. It is preparing standardisation at ISO or DIN level in the VDMA. This working party "BUS System Livestock Farming" has formed three international work groups. Work group 1 is dealing with the physical layout of networking. The special conditions of use in animal housing such as temperature, humidity, or aggressive gases, as well as costs, will be taken into account. The concept should make it possible to integrate mobile systems and wireless radio transmission too in future. Extensions or limitations in relation to existing standards that may become necessary for this will have to be worked out. Work group 2 is preparing a concept for the "Middleware";. - For this it is necessary to check whether e.g. the NDDS can be used as middleware, - i.e. whether it suffices as regards performance capability and transferability to agricultural applications and whether the licence conditions and costs are supportable, - or whether it is necessary to fall back on alternative processes with appropriate performance capability. Work group 3 is devising a data model and data management. For this purpose it is necessary to investigate - whether ADIS/ADED and/or XML/ADED can be used, - how the necessary icons can be generated, - how the data can be managed and how the access to data can be controlled, - how the process control data can be integrated, - how data updating and keeping can be secured in the long term. International specialists have been nominated for the three work groups. The members communicate with each other via a mailing list or a chatroom, which the BFL has set up. The initial work results will be discussed within the scope of a joint meeting scheduled for 2 0 September 2 0 0 2. The results of the deliberations can be presented at the conference via the status paper that will then be available.

Quantitative analysis of ultradian, circadian and seasonal time patterns of behavioural rhythms for the evaluation of well-being of animals in captivity and wildlife conditions

K.M. Scheibe, A. Berger, W.J. Streich, J. Langbein and K. Eichhorn

A chronobiological procedure has been developed for evaluation of living conditions, behavioural and physiological state especially for animals on pasture or free-ranging animals. It is based on recordings of activity and feeding with subsequent comparison of levels, daily patterns as well as daily and ultradian rhythms. For continuous recording of these behaviours from free ranging animals, a storage telemetry system (ETHOSYS) has been developed. The system is able to record behaviour patterns as activity and grazing over complete years and results can be downloaded on intervals like days or weeks on distances of up to 200m. Telemetric observations by means of this system were carried out on alpaca, sheep, cattle, domestic horses, Przewalski horses, roe deer, red deer and mouflon under various conditions. The time patterns of the different species were analysed macroscopically and by autocorrelation function and power spectral analysis. The different species-specific ultradian rhythms and their annual variation correlate to the nutritional strategy of each species. Especially for ruminants in stress-free conditions we found a stable ultradian structure of behaviour which was synchronised with the 24-h component of the spectrum. A more unstable adaptive time pattern was found in Przewalski horses. But also these animals showed a clear tendency towards a stable rhythmic pattern. Activity as a multiple motivation behaviour was generally more variable than feeding which in most cases was of clear rhythmic and harmonic structure. Degrees of Functional Couplings (DFC) and harmonic parts (HP) were used for quantitative comparison of rhythmic structures in activity and feeding. DFCs express the percentage of the circadian component and harmonic ultradian components in relation to all significant rhythmic components of a spectrum, while HPs express the percentage of harmonic power in relation to the total power of a spectrum. Accordingly, they can be regarded as a measure of internal and external synchronisation. They were found to be high in well adapted, healthy and undisturbed individuals but were lowered during periods of adaptation and disturbance in all species and situations. Such situations were disturbance by shots (Przewalski horses), several human activities (red deer), wounds (alpacas), high parasite load (sheep), transport and introduction into a new herd (Przewalski horses), or social interactions (red deer). Changing feeding conditions in farmed red deer caused a reduction in DFCs for feeding but not for motor activity. Cutting antlers reduced DFCs for activity but not for feeding. This demonstrates, how disturbances can be detected within the functional behavioural circles affected by a special external disturbance. In sheep with parasitic infection, about one month before death structural changes could be found in activity and grazing. They consisted of reduced general activity and grazing but especially nocturnal activity. In parallel, the noise - rhythm relation became gradually reduced as indicated by the HP, while the DFCs were reduced only during the transition period from high to low activity level. The time of lambing in free ranging mouflon could be determined by an automatic identification procedure, based on quantitative and pattern analysis of activity. We offer the chronobiological approach to compare and evaluate the quality of living conditions of animals quantitatively and to follow normal changes of general state such as birth or adaptation of annual rhythms. The approach can be applied to identify stress, disturbance, disease and endangering situations. It is especially convenient for free ranging animals but can also be applied to other conditions of animal keeping.

High resolution electronic weighing for experimentation and farm practise

Jürgen Schellberg, Arno Lellmann and Walter Kühbauch

The precise determination of animal live weight with high temporal resolution is a justified demand, because live weight is a key parameter in animal, grassland and forage science. Traditional cattle weighers do not allow frequent weighing. It has influence on animal behaviour and thus on the result of the experiment. Moreover, long weighing intervals confine available information about the effects of age, pregnancy, forage allowance, feed intake, environment and livestock management on live weight evolution. We demonstrate a prototype of an electronic weighing system (Texas Trading®), which overcomes the most striking disadvantages of a mechanical cattle weigher. It has been tested under various field conditions on the Research Farm Rengen (Eifel) throughout 2 summers (field) and winters (cow house). The system consists of 2 electronic balances recording data in intervals of 1 second (www.texas-trading.de). The balances are installed underneath a platform surrounded by a steel fence, which also encloses a water dispenser. To satisfy water uptake, animals are forced to enter the grating several times a day for some minutes, during which electronic weighing is conducted. The animal identity is detected through belt carried transponders. Animal weight is recorded by a computer connected with the transponder interface. Entering and leaving the platform induces variation of the animal weight records and is eliminated (Auerswald et al., 2001). Only constant readings of the animal weight is stored and analysed. During water uptake, the animal stands sufficiently still to obtain at least about 200 uniform records. However, increasing water fill of the rumen influences the recorded actual live weight. The electronic weighing system was applied in a large scale grassland experiment with 2 suckler herds containing each 20 cows with their calves on 40 hectares. In this experiment we investigated the overall nutrient circulation under intensive and extensive grassland management. Irrespective of occasional variations of live weight records due to water uptake, the electronic weighing sufficiently documented the daily live weight increase of calves. Additionally, the live weight measurements provided strong relationships with forage intake (pasture and silage), which has been simultaneously estimated from sampling in and alongside exclosure cages (Frame, 1993). Live weight increase of pregnant cows was documented precisely as well. The paper demonstrates functional relations between live weight measurements, sward yield and forage quality relevant for intake and animal production of the suckler herds. It includes the evaluation of the measurement procedure and data processing as well as the handling in animal and grassland experiments and on dairy farms. The major advantages of the automatic weighing system are (1) labour extensive and unperilous handling of the animals without stress, (2) precise and frequent measurements, (3) easy handling of digital data, and (4) accurate prediction of weight evolution. Once installed and settled on dairy farms, the system offers a wide range of application in animal husbandry. It provides valuable information about animal welfare, health, productivity and hence improves production and quality of beef and milk. Animal selection within a dairy herd can be improved through precise data on heifer growth as well as weight control of the dairy cows. The water dispenser inside the corral can be replaced by a licking block with salt, from which ruminants commonly consume 1 to 2 times a day. This would avoid the influence of water uptake during recording, but the frequency of weighing per day would be lower. Another problem still is, that under field conditions measurements are influenced by rainfall, as the wet animal fur influences correct live weight determination. Rainfall information - commonly available on most farms - may help to identify and eliminate records during such critical weighing periods.

Modelling the nutrient cycle on grassland dairy farms

Jürgen Schellberg and Ingo F. Rademacher

The understanding of nutrient circulation on dairy farms is a key issue of agronomic and environmental research. Precise calculations of quantity, linkage and loss of nutrients is a prerequisite to prevent environment pollution. Additionally, dairy farmers are required to evaluate nutrient input (e.g. fertiliser, feed concentrates, N2-fixation of legumes, deposition) carefully to improve nutrient use efficiency wherever possible. The economic and ecological benefit of precise nutrient management is beyond dispute. However, we need time consuming and costly experiments to understand the effects and interactions of management (e.g. animal diet, feed concentrate import, lactation period, grazing practise) and grassland characteristics (e.g. yield, forage quality) on nutrient circulation. Computer modelling can overcome present difficulties and avoid immense experimental effort. Although some models on N circulation on dairy farms exist (e.g. Rotz et al., 1989), they all include forage production on arable land (maize, red clover, alfalfa), and none of it refers to grassland exclusively. However, for climatic and topographic reasons, arable crop production is mostly impossible in mountainous grassland. Hence, a different structure of simulation model is needed, which considers grassland farming based entirely on pastures and meadows. This paper demonstrates the GRASFARM model (Schellberg and Rademacher, 2001). We proved and quantified the effects of changes in management with respect to the internal flow, import, export and overall balance of nitrogen. GRASFARM is a semi-empirical model that includes the following four sub-routines, (i) intake of forage, energy and protein, (ii) milk production based on grassland forage, (iii) intake of feed concentrate, and (iv) nitrogen excretion and balance calculation. External files contain daily data of forage yield, protein, energy, milk yield and actual live weight of the dairy cows. Additional parameters can be selected by the user and can be varied step-wise in repeated simulation runs to perform sensitivity analysis. The model was edited using MODELMAKER Software, Cherwell Scientific Ltd., UK and is currently converted into MATLAB SIMULINK (Mathworks Coop., US). Simulation runs with GRASFARM clearly indicate that the overall N budget increases with N import from feed concentrates, N2-fixation of pasture legumes and atmospheric deposition. Compared to import, the simulated N export with milk out of the farm is low. Increasing annual milk yield per cow, simulated through increasing concentrate feeding, additionally intensifies N accumulation on grassland. However, it can be shown that a reduction of N concentration (protein) in applied concentrates can lower the surplus. Grassland forage quality generally increases the intake by ruminants and hence the calculated amount of net energy intake provided for lactation. As a consequence, the simulated milk yield that is produced on the basis of grassland forage increases also with forage quality. The N use efficiency of the whole farming system is high under these conditions, due to the fact that the demand for feed concentrate import remains low. However, forage conversion efficiency of the ration gets low, when feed concentrate consumption decreases. A series of sensitivity tests with GRASFARM shows, that the balance between N input through feed concentrates and the output with milk is reached at an annual milk yield of about 8500 [kg cow^{-1}]. Under future EU regulations, grassland farms have to calculate their N balances precisely. This holds especially when they increase their milk production on the basis of feed concentrates and when the area of available grasslands for N reflux is scarce. In this study we found a significant exponential discharge of the N balance under conditions, when grassland area was increased and stocking rate was low. Referring to a more precise handling of nutrients on grassland farms we recommend to further develop and improve applied models, which are robust and easy to handle. Furthermore, there is the option to couple such models with geographic information systems where ever available.

Risk Analysis as a decision support tool within a quality information system to establish a chain oriented information and communication system

Thomas Schmitz and Brigitte Petersen

Failures occurring during the establishment of a new process cause high costs of correction. They may often lead to losses of market share and image of the supply chain. It is therefore reasonable to pay attention to potential failures in the early stage of process planning. The Failure Mode and Effect Analysis (FMEA) as a risk analysis is an appropriate method for a systematic and early identification of potential failures. Due to the early search for potential failure sources, a failure prevention strategy can be implemented to avoid a costly correction of failure. Potential causes and effects of failures are listed and assessed. This risk assessment supports a FMEA team to develop control actions during the process planning. The establishment of new chain oriented identification and traceability systems may be an operational field of the FMEA in the agriculture and food industry. This team oriented and preventive quality technique is particularly helpful for single companies and supply chains that plan to implement targeted check and preventive actions within a comprehensive quality system in the sense of the EN ISO 9000 standard series beyond legal claims. At present a number of amendments which must be planned chain oriented queues in the meat branch. Examples are the implementation of hygiene management systems, the translation into action of hygiene decrees like HACCP-concepts, the improvement of the data acquisition, the implementation of a tracking and tracing system, of a report system and of a early warning system. Once a mistake is made with the help of the FMEA the cooperation in the chain may be improved. Through risk assessment in the planning phase weak points can be identified early and priorities for an action catalogue for failure prevention can be set. Improvement of quality in processes resulting from the implementation of the FMEA reduces the risk that failures occur when the process is running. In a pilot project the strategy of a computer-aided FMEA is being tested at present at the implementation of a chain oriented information and communication system within the monitoring of salmonella. It is the prerequisite for this that experts from the different steps of the meat production chain and of different service providers work together on the establishment of a FMEA. The knowledge concerning the complete process from the animal identification, the barcode identification of the sample to the control of actions within the monitoring of salmonella is distributed to many persons. Therefore a FMEA team is formed in which these persons transmit there experience as experts. A computer-aided FMEA system can be exceptionally helpful to a risk analysis of complex systems like the chain oriented information and communication system. Because this quality management technique permits a team to develop the solution of a chain oriented problem systematically. Furthermore the analysed system is documented in a knowledge database completely. As a result it is possible to retrieve the gathered expert knowledge in future. A future aim should be to use and also to apply the advantages of the computer-aided FMEA in the meat producing chain to other corporate divisions.

A neural network for the analysis and monitoring of stress calls of pigs

P.-C. Schön, B. Puppe and G. Manteuffel

In the last years sound analysis has become an increasingly important tool to interpret the behaviour, health condition, and well-being of animals. This applies especially to stress situations. Vocalizations may deliver a useful tool for evaluating the emotional state of animals under captive and natural conditions. The important advantage of this approach is a relatively objective, non-invasive and real-time monitoring method of stress caused by environmental conditions. In pigs, the stress vocalization is a rather sustained cry with high-frequency bands that may be highly dynamical. The analysis and classification of pigs screams may deliver the species and the individuals phonetic characteristics that can be attributed to a particular stressor. If information on this interdependence is given it will be possible to judge the individual stress perception of an animal and, thus, its state of welfare or suffering. The presentation presents a procedure that allows to discriminate stress calls of domestic pigs from other calls and environmental noise. Based on the acoustic model of sound production the extraction of features from calls was performed with linear prediction coding (LPC). As the result an extremely compact short time representation of the call with a relatively low effort of calculation and a low number of features (LPC- coefficients) was reached. The LPC-coefficients represent the resonance frequencies of the vocal tract and deliver a model of the produced call. For practical purposes 12 LPC-coefficients, equivalent to the first 6 resonance frequencies, are calculated using time windows of 100 ms duration. The classification of the calls was obtained with an artificial neural network. Artificial neuronal networks are proven tools for many classification tasks. They consist of one- to multi- dimensional groups of nodes (neurons), each calculating a scalar output value from input vectors of arbitrary dimensionality. The inputs excite the neurons by an mostly simple transfer function (weight). Each neuron sums its inputs and outputs the result by another transfer function. Since the input space of all input vectors usually is transferred to many or all neurons a single neuron does not represent a specific feature of the input signal but rather the sum of arbitrary components (sub-symbolic distributed signal processing). The output of the network is observed and can be adapted by changing the input weights as long as it has not reached the desired value supervised learning). Alternatively other network-types may self organize the input space according to specific rules (unsupervised learning). Basically both types of networks are suitable for classification and have their own advantages and disadvantages. We obtained good results using the supervised Perceptron type or Self-Organizing Feature Maps taking the LPC-coefficients of single time windows as 12-dimensional input vectors. The neural network was used in two phases. In the first phase (training) the network was trained with known calls (stress vs. non-stress) and noise. In the second phase (application) the assignment of unknown calls to the corresponding class took place. The procedure allows the development of online monitoring of stress calls in farming environments. It can be employed in environments of breeding, transportation and slaughtering. We were able to detect stress vocalizations of pigs in noisy stables with only small recognition errors (< 5 %) The programmed system (STREMODO: stress monitor and documentation unit) running on a commercial laptop is insensitive to environmental noise, human speech and pig vocalizations other than stress calls. As a stand alone device it can be routinely used for objective measurements of acute stress occurring in various farming environments and conditions. The procedures to obtain the vectors and for network programming were developed using the graphical language LabVIEW with the supplementary software DataEngine V.i.

Imaging spectroscopy for grassland management

A.G.T. Schut, J.J.M.H. Ketelaars and C. Lokhorst

The potential of an imaging spectroscopy system was explored as tool for grassland management. The system detects reflection in image-lines in the wavelength range from 405-1650 nm with three sensors, from 1.3 m above the soil surface in a regular sampling pattern. The spectral resolution varied between 5-13 nm, and spatial resolution between 0.28-1.45 mm2 per pixel. Aspects of grassland management considered were growth monitoring and assessments of clover content, feeding value and sward production capacity and nitrogen and drought stress detection. Experiments were conducted with mini-swards grown in containers. Under a rain shelter in 1999 and 2000, an experiment was conducted with 36 mini-swards were sward damage varied, from undamaged to severely damaged due to drought or removal of circular patches from the sward. In two experiments (2000 and 2001), nitrogen application varied (0, 30, 60, 90 and 120 kg N ha-1 harvest-1). The drought stress experiment was conducted in a climate chamber and there was a control with high N supply and two drought stress treatments with low and high N supply. Drought stress treatments received no additional water during growth. During the experiments, 42 image-lines per mini-sward were recorded 2-4 times per week. Image-lines were classified to separate pixels containing soil, dead material and green leaves. These classes were subdivided into reflection intensity classes. Ground cover (GC), reflection intensity, image-line texture, spatial heterogeneity and patterns, and spectral characteristics of green leaves were quantified. The index of reflection intensity (IRI) quantified the distribution of green pixels over intensity classes. Spatial heterogeneity was quantified with the spatial standard deviation of ground cover (GC-SSD), and image-line texture and spatial patterns with wavelet entropy (WE). Spectral characteristics were quantified with shifts of spectra at various edges. Partial least squares (PLS) models combining spectral and spatial information were calibrated and validation on two separate data-sets from the sward damage and N experiment in 2000, to predict DM yield, feeding quality and nutrient content. As result of the design of the experimental system, reflection intensity varied with sward height and leaf angle. Therefore, IRI quantified vertical canopy geometry. GC was linearly related to light interception under a cloudy sky and logistically under a clear sky (R2adj = 0.87-0.95). The relations differed significantly between dense and open swards. Growth was accurately monitored with GC and IRI. Seasonal means of GC and IRI were strongly (R2adj = 0.77-0.93) related to seasonal dry matter yield (SDM) and light interception capacity. There was a clear relation between seasonal means of GC-SSD and spatial variability of tiller density (R2adj=0.69). Seasonal means of GC-SSD differentiated dense from damaged swards. GC-SSD and WE of GC were related to SDM. The WE of image-line texture differentiated clover from grass swards, where mixtures had intermediate values. Drought stress was detected in an early stage, when DM content of leaves was below 20%, with shifts at the edges of water absorption features. The combination of inhibited or reversed shifts at the green and red edge was strongly related (R2=0.95) to yield reduction relative to high N supply. The PLS models accurately predicted DM yield, nutritive value and nutrient contents with mean prediction errors of 235-268 kg DM ha-1, 0.96-1.68 % DM, 5.8-6.5 g kg-1 DM ash, 8.4-10.4 g kg-1 DM crude fiber and 0.24-0.34 % N. The predictions of P, K, S, Mg, Na and Fe were robust in both experiments, in contrast Zn, Mn and Ca. It is concluded that imaging spectroscopy can provide valuable information for grassland management on dairy farms.

Comparison of grass sward dry matter yield assessment with imaging spectroscopy, disk plate meter and Cropscan

A.G.T. Schut, J.J.M.H. Ketelaars, M.M.W.B. Hendriks, J.G. Kornet and C. Lokhorst

The accuracy of an imaging spectroscopy system was compared with a disk plate meter and Cropscan for non-destructive assessment of grass-sward dry matter (DM) yield. The imaging spectroscopy system detects reflection in image-lines in the wavelength range from 405-1650 nm with three sensors, from 1.3 m above the soil surface in a regular sampling pattern. The spectral resolution varied between 5-13 nm, and spatial resolution between 0.28-1.45 mm2 per pixel. Two experiments were conducted in 2000 with Lolium Perenne L. mini-swards grown in containers in were sward damage (exp. 1) and nitrogen (N) application (exp. 2) varied. There were 36 mini-swards in exp. 1, from undamaged to severely damaged due to drought or removal of circular patches from the sward. In exp. 2, there were 15 mini-swards and N application varied from 0, 30, 60, 90 up to 120 kg N ha-1 harvest-1. There were data from 9 (exp. 1) and 6 (exp. 2) harvests available. At all harvests, crop height was recorded with a disk plate meter (DPM). For three harvests, NDVI and WDVI were calculated from crop reflection recorded with a Cropscan. Just before harvest, 42 image-lines per mini-sward were recorded. Samples of harvested material were analyzed on feeding quality and nutrient content. Image-lines were classified to separate pixels containing soil, dead material and green leaves. These classes were subdivided into reflection intensity classes. Per mini-sward, ground cover (GC), index of reflection intensity (IRI), quantifying the distribution of green pixels over intensity classes, and a mean spectral curve was calculated from all grass pixels. Partial least squares models including GC, IRI and spectral information were calibrated and validation per experiment to predict DM yield, feeding quality and nutrient content. Replicate observations may reduce the non-biased part of the model error. Therefore, the potential reduction in prediction error was explored for 10, 25 and 50 observations per field with low (0.25) and high (0.5) fractions of model bias. The crop height was very sensitive for lodging of the swards, what occurred in the N experiment under high N supply. Crop height was linearly related to DM yield (R2adj=0.55-0.66), after logarithmic transformations of crop height and DM yield. The regressions yielded for NDVI R2adj values of 0.59-0.84 (with a power-fit) and R2adj values of 0.47-0.65 (linear) for WDVI. The NDVI and WDVI were severely limited in their applicability at higher DM yields, as sensitivity strongly decreased above 2000 kg DM ha-1. Combinations of GC and IRI were strongly linearly related to DM yield (R2adj of 0.82-0.95). With spectral information only, DM yield predictability resulted in prediction errors of 286-344 kg DM ha-1. Combining GC and IRI with mean sward spectra further reduced the prediction error to 235-268 kg DM ha-1. Multiple observations may reduce the mean prediction error of DM yield with 2-10%, 27-40% and 37-54% for 10, 25 and 50 observations respectively for low and high model bias fractions. It is concluded that the accuracy of DM yield assessment with imaging spectroscopy is better than the disk plate meter or Cropscan.

Secure identification, source verfication and traceback of livestock - the value of retinal images and GPS

John A. Shadduck and Bruce L. Golden

Secure source verification of livestock is needed to assure food safety and high quality retail meat products, prevent fraud in animal subsidy programs, and enable efficient, effective traceback systems for control of animal diseases. The purpose of this presentation is to describe the use of biometric markers, particularly retinal vascular patterns combined with the use of the Geopositioning Satellite System (GPS), as a system for secure identification, source verification and traceback of livestock. Source verification and traceback of animals is now a worldwide issue, with concerns over BSE and FMD being especially important. Traceback of animals to farms of origin, tracking of movement of animals from farms to markets and abattoirs, and monitoring of animal movement within and between countries are all critical to animal disease control and verification of food safety. Any solution that will meet these diverse needs must have several major features. · First, it must resist fraud and must include information on location as well as identity. · Second, it must be based on a robust biometric marker. · Third, the acquisition of the needed biometric information must be rapid, inex pensive and accurate. · Fourth, the data must be easily and rapidly transmitted, stored and retrieved. · Finally, the method must be humane and ideally, non-invasive. Only a biometric method has the potential to completely satisfy these requirements of producers, consumers and regulatory agencies. A biometric identifier is any measurable, robust, distinctive physical characteristic ... that can be used to identify or verify the claimed identity of an individual. Measurable means that the characteristic can be easily presented to a sensor and converted into a quantifiable, digital format. The robustness of a biometric is a measure of the extent to which the characteristic is subject to significant changes over time. Distinctiveness is a measure of the variations or differences in the biometric pattern among the general population. The higher the degree of distinctiveness, the more unique the identifier. [Woodward John D., Webb Katherine W., Newton, Elaine M., Bradley Melissa, and Rubenson David (2001). Army Biometric Applications - Identifying and Addressing Sociocultural Concerns. Santa Monica: RAND.] Only iris imaging, retinal scanning, and retinal imaging are potentially suitable as a secure biometric identification method for animals (other than humans). We believe retinal imaging is the preferred method because an image of the retinal vessels is the most powerful and flexible biometric identification method for livestock. The patterns created by the retinal blood vessels are highly unique, not only between animals, but also between eyes within individual animals. The image of the retinal vessels is collected via a digital camera linked to a hand-held computer with an internal GPS receiver and clock. Software in the computer selects and encrypts the image and the date, time and location of the animal at the time the image is collected. Other data can be added via a keyboard and/or via a USB port. Image storage can be either local or transmitted via the Internet to a central archive. The image, GPS data and other data are stored and retrieved via a number of different queries, including a search for a match of a reference image in the database. Storage, retrieval and database searches take fractions of a second. Because the identification (ID) is based on a unique marker, the match between the animal and its ID in the database is essentially perfect. The retinal patterns are highly variable, so the probability of images from different animals being identical is essentially zero. The animal's ID can be linked to other information such as specific owner identification and animal health and/or performance records. Searches of the archives can be done in fractions of a second, even when millions of images are stored.

Continuous measurement of the ensiling process

H. Snell, H. Van den Weghe and W. Lücke

The possibility to preserve fodder by ensiling is known for a long time. Still only the enormous scientific and technical progress since the 1960s made the rapid expansion in the quantities of forages conserved by ensilage possible. However, under practical production conditions the quality of silages often does not meet the requirements of ruminant feeding and of the performances aspired. A representative study of WEISSBACH (1997) in Northern Germany revealed that about one third of the grass silages was unsuitable for producing high quality raw milk. Continuous measurement of the ensiling process offers perspectives both for scientific research and for the use in practice. In the future, this measurement could be used as a means of process monitoring under practical conditions. In five consecutive, completed experiments it was investigated, if modern sensors can contribute to a continuous measurement of the ensiling process. The experiments P1 and P2 focussed on the measurement of the pH-value. 4 treatments (2 densities; 2 storage temperatures) were compared. Several tubes per treatment were filled with pressed sugar beet pulp. The tubes were opened in defined intervals and the pH-value was measured in the laboratory. Finally the results of the continuous measurement and of the analysis in the lab were compared. The experiments P3, P4 and P5 focussed on the fiberoptic measurement of the oxygen concentration in the silage. 4 tubes were filled. 2 tubes were covered by a thin LD-PE film, 2 tubes were covered by a thick LD-PE film. The fiberoptic measurement of the oxygen concentration is based on the fluorescence quenching of specific fluorophors by molecular oxygen. The sensor, called optode, is manufactured by immobilizing an oxygen sensitive fluorescence dye, dissolved in pure silicon, on top of one end of a single optical fiber. The other end of this fiber is connected to the measuring device which houses a blue LED for excitation of the dye. The resulting fluorescence at the sensor-tip is traveling back through the fiber into the instrument where it is measured by a photodetector. Both beams, excitation and fluorescence light, are split by a mirror. Excitation and detection is controlled by a microcomputer. The oxygen concentration is calculated by the measuring device. The results of the investigation point to the fact that a reliable measurement of the pH-value is possible over a period of at least several weeks with the aid of appropriate pH-electrodes. The fiberoptic measurement of oxygen is an important step towards the completest possible online recording of the ensiling process. Several characteristics (very small sensor, no consumption of oxygen during measurement, no disturbance due to electric or magnetic fields, suitable for very low oxygen concentrations, sensors can be sterilized chemically or thermally, optode may be up to 100 m long) of this technology make it extremely interesting for ensiling experiments and as a means of process monitoring. The use of fiberoptic sensors leads to a plausible documentation of the oxygen concentration in the silage. Combining these two variables with the temperature leads to valuable information about the ensiling process. However, further suitable variables have to be looked for, to allow also the documentation of micro-biological processes.

Research on sensor- supported cleaning and disinfections of the surface of Automatic Milking Systems (AMS)

Jens Unrath and Otto Kaufmann

One of the major goals of dairy farming is to produce milk under conditions characterized by a high level of hygiene. This level of hygiene is required by the milk industry as well as by the consumer. The reputation of the product milk, which is considered to be pure due to its white colour, is influenced by the assurance of the quality of raw milk. Today questions about how and where the milk is produced become more important and this is summarized in the term process quality.

In the near future questions about the stabilization of all functions of a milking unit will arise. This applies in particular to the hygiene of the periphery of the milking unit. An insufficient hygiene management leads to a lower milk quality and has a negative impact on the udder health of the entire herd.

While AMS parts in contact with milk are automatically cleaned in regular intervals, the periphery such as the bionic arm must be cleaned by hand. The higher frequentation of an AMS results in a higher contamination compared to conventional milking systems. This requires a manual cleaning of the AMS periphery several times a day.

This is opposed to the expectations of the farmer, who does not want to depend on a fixed schedule. In practice, that leads to the fact that between the cleaning times the systems get considerably dirty. This problem has to be solved, because it is not only an ethical problem regarding a vitreous production.

Our research has shown, that after cleaning in the course of time the germ load, which is dangerous for the milk and the udder health of the entire herd, quickly rises again.

The aim of our research is to analyse the defilement with the help of image processing and then to calculate the correlation between defilement and germ load.

The matrix for the automatic cleaning and disinfections of the AMS periphery will be derived from the results.

Long time experience with electronic identification in pig husbandry

Jan van de Veen

Long time experience with electronic identification in pig husbandry. In May 2002 we provided pig number 35.000 with an electronic ear tag. Ever since 1995 all pigs on the Institute De Schothorst carry an electronic device in their left ear from weaning to the end of the fattening period. The Institute for Research on Animal Nutrition De Schothorst acts as the joined R & D organisation for the Dutch compound feed co-operatives. Its research includes physiological and feeding research with dairy cattle, swine and poultry. The Institute is situated in one of the new polders, 12 kilometres from Lelystad. The swine department has a capacity of 230 breeding sows and 1800 fatteners From all pigs older then 4 weeks, all data is recorded by means of electronic identification and automatic data logging. Software for automatic data transfer to the existing herd-management program of the Institute is available and being used. Five hand-held dataloggers are used in the different units on the farm, all equipped with a tailor made database for their specific task. In the slaughterhouse an antenna was installed, for automated identification and recording of the classification record of each individual animal. From all slaughtered pigs a complete record is available, including pedigree information, sex, weight (at partus, weaning, start and end of the finishing period), feed intake, health record, classification record and abnormalities on carcass and organs. We have used this database intensively for statistical analyses. There is also the possibility to calculate the effect of a disorder from individual pigs in for example the suckling period on the performance in the finishing period. The application of the ear tags does not need much skill. Inflammation does occur occasionally, but never forced the removal of the ear tag. Ear tag loss from weaning to finishing is less than 1 %. The ear tags are protected against heat and flames during the slaughter process by means of a heat shield. The loss of ear tags in the slaughter process is 2 %, making a total loss of 3 %, which is an acceptable level. After removal at the slaughterhouse, the tags are returned to the Institute and prepared for a new cycle. The estimated lifetime of 8 cycles has long been passed as the majority of the responders are now used for the 15th consecutive time. After 7 years of experience with electronic identification some firm conclusions can be made. The project has proven that the use of electronic identification on pig farms can be done successfully. No special knowledge or skills are needed and the paperless data collection is superior to the methods that use pen and paper. Ear tag loss is about 3% and the recycling of the responders keeps the costs to an acceptable level. The ID-loggers are reliable and light weighted so they can be carried around without any trouble. The worldwide production of pork has become a real bulk industry. At the same time the consumers are increasingly demanding specific quality with respect to animal welfare, leanness feed additives, etc. In order to survive in a competitive and changing market the pork industry has to find new ways to improve and assure the quality of their products. It is hard to believe that they will be able to do so without the use of electronic identification as a livestock precision tool. My presentation will challenge the pork industry in a positive way and is of interest for all that are involved in animal production as the use of electronic identification is not limited to the swine business only. Lelystad, the 10th of September Jan van de Veen

The importance of sampling-time for on-line mastitis detection by using the Electrical Conductivity or Na$^+$ and Cl$^-$ content in milk

M. Wiedemann, D. Weiss, G. Wendl and R. Bruckmaier

Introduction Electrical conductivity (EC) as an indicator for mastitis in bovine milk has been discussed since decades. The influences of the milking interval and milk fat constuents on the EC had been evaluated. On-line measuring systems for EC patterns during milking were established, but the detection rate of mastitis is not satisfying. Despite of these shortcomings EC is currently the main parameter for detecting mastitis in automatic milking systems. The EC is mainly determined by the concentration of the ionic active elements e.g. Na, Cl or K. The concentrations of these elements in milk is lower compared to their concentrations in blood serum. In case of mastitis the mammary epithelial tight junctions get leaky to allow diapedesis of leukocytes to the infection site. The aim of the study was to analyse the patterns of the EC and the Na+ and Cl- content in relation to the time of taking the sample in comparison to the somatic cell count (SCC) in foremilk (exp. 1). In additional foremilk samples the significance of these parameters for detecting mastitis was proved (exp. 2). Method and material In experiment 1, 20 cows without indications for clinical mastitis but SCC between 104 and 5*106 cells/ml in the total milk were used to take different milk fractions per quarter. Samples were taken every 20 s from start of stimulation (t=0 s) until t=120 s in the right front quarter at three consecutive milkings. In experiment 2, foremilk samples of the two right quarters were taken in six cows over a period of two weeks to evaluate if the occurence of spontaneous mastitis can be detected by the methods, described above. EC was measured at 25 °C using the LDM electrode from WTW (Germany). For Na+ and Cl- determination the ion selective electrodes model 9811 and 9617BN (Orion, USA) were used in raw milk. For statistical analysis the data were clustered for SCC in the milk sample at t=0 s (I: <100,000/ml; II:100,000 to 500,000/ml and III: >500,000/ml). Results were analysed using the repeated measures analysis of the MIXED procedure (SAS) and tested for significance (p<0.05) using the Least Significance-Difference Test (LSD). Results In exp. 1, milk with less than 100,000 SCC/ml at t=0 s had no big changes in EC, Na+ and Cl- between t=0 s and t=120 s. In milk with more than 500,000 SCC/ml at t=0 s the EC is 1.20 times higher than the EC of milk with less than 100,000 SCC/ml. While there is still a difference at t=80 s between SCC-group I and III it is no more possible to distinguish between these SCC-groups at t=120 s. Because of a higher percentual difference between the foremilks of SCC-group I and III by using the Na+ content (t=0 s: 180% and t=120 s: 90%) a better detection of quarters with increased immunological activity is possible compared to measurement of Cl- and EC. The analysis of exp. 2 showed the following ranking for detecting quarters with increased immunological activity regarding to specifity, sensitivity and error rate: 1. Na+, 2. Cl- and 3. EC. Conclusion By using the EC or ionic content for detection of udder health status the time of taking the sample is crucial. In connection with automatic milking systems it should be possible to collect the first milk strips for analysing before the occurrence of milk ejection. The results indicated that the direct measurement of ionic content, especially Na+, is advantageous in detecting the status of udder health as compared to EC.

Full poster papers for ECPLF

Precision livestock farming in Turkey

H. Bilgen[1] and H.Öz[2]
[1]*Department of Agricultural Machinery, Faculty of Agriculture, Ege University, 35100 Bornova-İzmir, Turkey*
[2]*Department of Farm Machinery, Ege Vocational Training School, Ege University, 35100 Bornova-İzmir, Turkey*
hbilgen@ziraat.ege.edu.tr

Abstract

The number of big farms that have 200-500, cows herd-size is getting increased quite rapidly in recent years. From the beginning of year 2000, the application of the precision livestock farming by using electronic data collection and evaluation systems has been still increasing rapidly as well. Therefore, there is a big potential for the precision livestock farming applications in the future in Turkey. Hence, the objective of this study was to determine the improvement and to evaluate the expectations of the precision livestock farming for the future in Turkey.

Keywords: Turkish livestock farming

Introduction

The cattle population of Turkey is about 10,5 million heads (FAO,2002), and 50% of this consists of cows. In other words, the total milk production of Turkey for 2002 is 9,5 million tons (FAO) and 89% of this production was harvested from cows. It is so obvious that dairy farming is the most common farming type in the Turkish animal husbandry. The small-scale family farms having less than 10 cows each are on majority and form the main the structure of the Turkish animal husbandry. In all of these farms, animal monitoring has been done by the farmers themselves without using any electronic system.

On the other hand, an increasing trend of the number of the animals with high milk yield in total population and the average of herd size indicate that the use of the precision farming applications in the animal husbandry will be much more popular in the near future in Turkey.

Evaluation of precision livestock farming in Turkish animal husbandry

The number of the farms employing precision livestock farming (PLF) applications increases both in Turkey and around the world. From the standpoint of Turkey, the reasons for this improvement depend on several factors. These include,

1. Increasing of herd size of dairy farms,
2. Decreasing of rural population,
3. Increasing percentage of high milk yield cows in total cattle population,
4. Recently introduced premium system based on milk quality and quantity,
5. Naive employees working in milking parlours,
6. Owners' increasing trend to use their farm technology as a prestige,
7. Increasing number of high-educated farm owners,
8. Dealing with animal husbandry as an additional enterprise area besides industrial ones by private entrepreneur,
9. The reliability on electronic systems for accurate evaluations of both, herd and employees.

PLF applications are implemented usually in dairy farms in the size of ranging between 35 and 1200 cows. DeLaval, Westfalia and SAC are the main companies that sell milking and farm management systems. The first application of PLF in Turkey was established in 1992. In the last decade, these applications, especially 2000s, increased quite rapidly as seen from Figure 1. The data obtained that include March 2003, show that PLF is applied 48 dairy and beef cattle farms in Turkey.

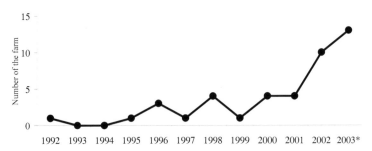

Figure 1. Number of farms applying precision livestock farming by years.
* End of March

Some of the characteristics of these farms can be summarized as follows:
1. 60% of the farms change their system to the PLF during the technological upgrade of their milking parlour, and rest of the farms use PLF at the beginning of the installation of them.
2. 52% of these farms are private owned farms, the others belong to public.
3. Individual milk yield measurements of the cows are made by using milk meters in all of the farms, excluding one.
4. Herringbone milking parlour is the most common type with a ratio of 70%, side-by-side and rotary milking parlours follow this type.
5. In 94% of these farms, identification of the each animal is achieved by using neck tags. Ear tag is ranked as the second one.
6. Animal activity is monitored by using activity meters in 89.5% of the farms.
7. 21% of these farms have feeding stations to determine the concentrate feed consumption of each animal.
8. Especially in all of the slaughter farms, alive weight of the animals are measured by electronic scales and data obtained are stored for comparison purposes.
9. Obligation of herd-book recording system forced by government.

Conclusion

The most important reason to start using PLF applications in the Turkish husbandry is to control milk yield and to determine the insemination time of the cows individually in the herd. Due to the reasons mentioned above, it is expected that the precision livestock farming will be much more popular in parallel to the development of the Turkish agriculture.

References

Personal contact with the agencies of DeLaval, Westaflia and SAC companies in Turkey. Interviews with some farmers or managers.

Application of lumped-parameter and fuzzy model of cow movement in modern dairy farms

D. Cveticanin[1], G. Wendl[1] and M. Klindtworth[2]
[1]*Bayerische Landesanstalt für Landwirtschaft, Am Staudengarten 2, 85354 Freising, Germany*
[2]*Department für Biogene Rohstoffe und Technologie der Landnutzung, Technische Universität München, 85354 Freising, Germany*
dragan.cveticanin@lfl.bayern.de

Abstract

In this paper a procedure for "in-motion" weighing of dairy cows is investigated. The scale for automatic weight measurement is constructed and tested. Two models are created to simulate cow motion. The first is the lumped parameter model, while the other is a fuzzy logic based model. The developed models are used in body weight recognition on the automatic weighing scale. Both the single and crowded crossing cases are investigated. The accuracy of 2 % is achieved for single crossing case with lumped parameter method. The first method does not work for the case of crowded crossing. The fuzzy method works with accuracy of 2 % for single and 3 % for crowded crossing the weighing scale.

Keywords: dynamic, weighing, cow, modelling.

Introduction

In nowadays of modern dairy farming the objective is to automate every process from feeding to milking in order to reduce time and costs of production. To create a fully automated high productive dairy farm the important parameters need to be measured and recorded. Among these data are: quality and quantity of produced milk, quality and quantity of food intake, activity, body weight, body temperature, barn temperature, etc. As it is discussed by Maltz *et al.*, 1992, the information on milk yield, body weight and food intake are dependent on each other. Averaging the signals as done by Ren *et al.*, 1992 and Peiper *et al.,* 1993, the problem of dynamic weighing was solved, but inaccuracies still occurred when crowding of animals on the weighing scale and fast crossing of the scale.

Materials and methods

The experimental weighing scale was positioned in the exit corridor of the milking parlour. The dimensions of the weighing platform were 2.5 meters in length and 1 meter in width. The dimensions guarantee that there is a moment where a cow is with its full body mass on the scale. In crowding the cows cannot fit side by side nor ride on each other hindquarters but follow each other in quick succession over the scale.

The lumped parameter model is an inverted pendulum with attached point mass and free bar (pendulum) at its end. The assumption is that four legged being can be represented as two legged since the diagonal legs are moved simultaneously. The equations of motion and force of the model are denoted. The model has three parameters: body weight, walking velocity and vertical force to the ground. The model is used in body weight recognition. The force in time is measured for a crossing situation. The body mass and walking velocity are optimised in the model in order to get the best match to the measured force. The calculated force matches partially to the measured values, since the stochastic influences during walking are neglected in the model and the simplifications assumed.

The fuzzy model has two Gaussian membership functions, product inference engine, normalised part and two Takagi-Sugeno fuzzy rules. The time of crossing is the input and weighing force is the output of the model. A database of model parameters for various crossing velocities and masses is created. It is used in body weight recognition. For the new measured crossing over the weighing scale, the force in time is approximated using fuzzy method creating a curve. The parameters of the model from the database are chosen according to the measured time of crossing. The new created curve and the database curve are compared. The difference between the integrals of the two curves denotes the new body mass of the crossing weight.

Results

Using the lumped parameter method in body mass recognition for single crossing case a maximum error of 2 % was achieved for twenty tested situations with five cows weighing between 550 and 860 kilograms. The method did not work for crowded crossing case since the briefness of the measured signal when the cow is alone on the scale. The fuzzy method showed accuracy of 2 % for single crossing case for the same measured situations as in previous method. The maximum error of 3 % was reached in crowded case for twelve measurements with the combination of the five mentioned cows.

Conclusion

Comparing the value of the calculated mass to the statically measured mass the maximum error is 2 % for single case crossing for both methods and 3 % for crowded case for fuzzy method. The advantage of the methods developed in this paper compared to the known procedures of dynamic weighing is that this method does not require extra helpers like slow down stepper nor has constrained error margin for the new measurement compared to the previous one and shows better preciseness in dynamic weighing. The body mass recognition methods give results after each day of measurements, while the known methods discard the results when bigger than allowed error and keep the previously measured mass as valid. The disadvantage of the suggested fuzzy weighing method of livestock is that a number of "in-motion" measurements for a variety of animals have to be done in order to create a proper database of models. The fuzzy method is better than lumped parameter since it does not work for crowded case. However, lumped parameter method is useful in predicting the amount of force created during walking for a certain body weight and speed, and so the endurance of barn floor can be inspected.

References

Maltz, E., Devir, S., Kroll, O., Zur, B., Spahr, S. L. and Shanks, R. D. 1992. Comparative Responses of Lactating Cows to Total Mixed Rations or Computerized Individual Concentrates Feeding. Journal of Dairy Science 75 1588-1603.
Peiper, U. M., Edan, Y., Barak, M. and Maltz, E. 1993. Automatic weighing of dairy cows. Journal of Agricultural Engineering Research 56 13-24.
Ren, J., Buck, N. L. and Spahr, S. L. 1992. A dynamic weight logging system for dairy cows. Transactions of the ASAE 35 719-725.

Evaluation of a beef-cow pregnancy model

Cristian R. Feldkamp and Horst J. Schwartz
Humboldt-Universität zu Berlin, Institute of Animal Science, Livestock Ecology Section, Phillipstr. 13, 10115 Berlin, Germany
cristian.feldkamp@agrar.hu-berlin.de

Abstract

Prediction of the pregnancy distribution through the breeding season is a crucial part of cow-calf models. This paper evaluated the suitability of a logistic reproduction model developed by Feldkamp and Schwartz (2002) to estimate pregnancy distribution. A data set containing 125 cases of adult beef cows was analysed. Breeding season was divided in three periods. ANOVA showed that there were not significant differences between observed and estimated pregnancy rates in each period (P=0.36). Correct predictions were 63.8, 57.8 and 68.1 % for the first, second and third period. Model predictions were close to the theoretically maximum accuracy. Therefore, the model it is suitable for the objectives of its development.

Keywords: model evaluation, pregnancy distribution, modelling of pregnancy.

Introduction

Modelling of cow-calf enterprise may give insight of the functioning of the agro-ecosystem and also assist in the decision-making process. Reproduction is the main factor affecting production efficiency of beef cattle (Short et al., 1990). Both calving rate and calving distribution pattern play a crucial role in herd functioning. These variables heavily depend on nutritional and other aspects related to management. Accurate estimations of the moment of pregnancy would allow for instance evaluating different mating season length and different nutritional management on the production efficiency. Therefore, considerable interest exists in estimating the pregnancy probability for periods shorter than a whole mating season.

In a previous work (Feldkamp and Schwartz, 2002) we have proposed a model for the estimation of pregnancy probabilities and presented an overall evaluation. It is also necessary to evaluate the ability of this model to predict monthly pregnancy rates within a breeding season. In this paper we present the evaluation of the model for predicting the distribution of the pregnancy through the breeding season.

Materials and methods

A data set containing 160 cases of adult beef cows was analysed. Data are from the Mercedes Agricultural Experimental Station, Argentina. Details about the model and materials used are explained in Feldkamp and Schwartz (2002). The breeding season was divided in three periods of the same duration. Both observed and estimated pregnancies were assigned to one of the three periods. Two approaches were used to evaluate the model. First, differences between observed and estimated pregnancy rates (PR) for each period were analysed using Statistica 6.1 (2002), with year as replication. PR for each period was the dependent variable and Period, Method (observed or estimated) and the interaction Month-Method were the categorical variables included. Year was used as replicate because there were not data available to perform ANOVA for each year (i.e. there were only one value of observed PR for each period and year). Besides ANOVA, percentages of correct predictions and other indicators of binary modelling suitability were estimated for each period pooling the data of the ten years.

Results and discussion

Statistically significant differences were not detected between observed and estimated pregnancies when using year as repetition (P=0.36). PR was significantly different between for different months (P<0.01). However, the interaction Month-Method was not statistically significant different (P=0.43). This lack of significance between observed and estimated PR is partly due to the high variability within replications (years) for PR.

Maximum attainable correct predictions percentage for this sort of models is approximately 70 % (Feldkamp and Schwartz, 2002). Model predictions are very close to the potential accuracy of the pregnancy distribution (Table 1). Sensitivity and specificity values indicate the accuracy of the model to classify pregnant and non-pregnant cows respectively. Values calculated are within the acceptable margins for this sort of models. However, low sensitivity and odds ratio in period 3 is related to the low number of cases.

Table 1. Suitability indicators of binary modelling.

Indicator	Period 1	Period 2	Period 3
Correct predictions [%]	63.8	57.8	68.1
Sensitivity [%]	66.2	41.7	11.1
Specificity [%]	61.4	70.2	81.6
Odds ratio	3.1262	1.6837	0.5536
N	160	83	47

Conclusions

This work shows the suitability of the model to assess pregnancy rate and pregnancy distribution pattern. Model predictions were close to the theoretically maximum accuracy; thus it is adequate for the objectives of its development. Nevertheless, assessment of the effect of suckling on pregnancy estimation is required before it can be used in a whole-farm model.

References

Feldkamp, C.R. and H.J. Schwartz. 2002. Logistic modelling of beef-cow pregnancy probability. Vortragstagung der Deutschen Gesellschaft für Züchtungskunde e.V. und der Gesellschaft für Tierzuchtwissenschaft Halle (Saale). Martin-Luther-Universität, Halle-Wittenberg.

Short, R.E., R.A. Bellows, R.B. Staigmiller, J.G. Berardinelli and E.E. Custer. 1990. Physiological mechanisms controlling anestrus and infertility in postpartum beef cattle. Journal of Animal Science 68: 799-816.

Acknowledgments

The authors are grateful to Mr. Daniel Sampedro and Mr. Oscar Vogel from Mercedes Agricultural Experimental Station, INTA (Corrientes, Argentina) for permission to use their data.

Food authentication of beef – The use of stable isotope signatures for tracing dietary composition

T. Gebbing and W. Kühbauch
University of Bonn, Institute for Plant Production, Katzenburgweg 5, 53115 Bonn, Germany
t.gebbing@uni-bonn.de

Abstract

Meat samples from a slaughter house were analysed for C- and N-isotope composition. Stable isotope signatures showed significant variation between farms, and indicated different amounts of maize in the steer's diet.

Introduction

In the past, milk and beef production in Europe was mainly based on fodder production from grassland. More recently, economic motives led to an increase of daily live mass gain, which required a high nutrient concentration in the ruminants diet. This was achieved by an increased use of concentrates and/or the introduction of silage from maize.

In human nutrition the consumption of saturated fatty acids has been assumed as a risk factor for coronary heart disease. Recent studies showed that fresh grass in the ruminants diets increases the content of polyunsaturated fatty acids in meat, which is recommended by nutritional scientist (Enser *et al.*, 1998). Beside this advantage in human nutrition, the meat production in grassland systems is associated with relatively low environmental impacts, and the grazed ruminants are allowed to realise their species specific behaviour.

During photosynthesis C isotope discrimination is much less expressed in C_4 plants than in C_3 plants (Farquhar *et al.*, 1989) and thus ^{13}C content in organic matter is significantly higher in maize ($\delta^{13}C$: ~-12‰) than in the majority of C_3-plants abundant in European grassland ($\delta^{13}C$: ~-24 - -28‰). Given the dependence of the stable isotope signatures in animal tissues on the isotope composition of the diet (DeNiro & Epstein, 1981; Hobson, 1999), stable isotope analysis may be used successfully to trace the amount of maize in the animals' diet. In the present study, variability of stable isotope signatures (C and N) of meat produced from different farms was analysed to investigate this relationship.

Material and methods

Meat samples were obtained from a slaughter house. Lipids were removed from freeze dried samples by washing the sample three times with methanol and dichlormethane. At the experimental station Rengen, (Daun, Germany), blood samples were taken from previously grazed steers, which had been finished on a grass silage based diet. Blood samples were freeze dried and lipids were removed. The residuals were analysed for C- and N-isotope composition. Isotope composition of samples were converted to δ values by analysis of reference gauged [PeeDee Belemnite C (Farquhar *et al.* 1989) and Air for N] standards between samples.

Results and discussion

C isotope composition of meat samples indicated the expected large contribution of maize to regional beef production (Figure 1). The blood samples (without maize) and meat produced on the 'organic' farm 'a' showed similar δ values. In most cases meat samples of different animals from the same farm were similar, but there were significant exceptions (farm l and o).

The reason for this deviation is not known, but may be attributable to recently purchased animals. The time required to establish a new equilibrium between isotope composition of animal tissues and isotope composition of the diet is subject to further investigation.

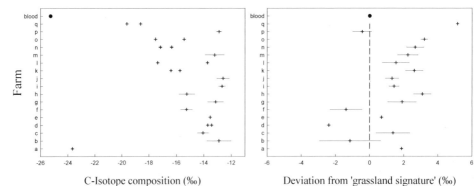

Figure 1. The mean C-isotope composition of defatted meat from steers fed at different farms. Horizontal lines indicate 2SE. For less than three animals per farm single measurements are shown. Blood samples taken at the experimental 'grassland' station Rengen are also indicated.

Figure 2. Mean deviation of N-isotope composition of defatted meat samples from N-isotope composition of blood taken from steers exclusively fed on grassland (termed 'grassland signature', experimental station Rengen). Horizontal lines indicate 2SE.

N-isotope composition of blood samples from grass fed steers showed only small variation (Figure 2). Because we were sure that these steers were fed on grassland exclusively, we termed the N isotope composition of the blood a typical 'grassland signature'. Though, one must be cautious to compare different tissues (i.e. blood *vs.* meat), N isotope signature of meat samples is expressed as deviation from this 'grassland signature' (Fig. 2). N isotope composition in meat samples showed a strong variability, but still, significant differences between farms could be found. In conclusion, a multi element stable isotope analysis may serve as a promising tool in authentication of agricultural products in regard to diet composition. Such signatures may help to establish a market for meat produced from grassland exclusively. A prerequisite for such an application of isotope analysis is the knowledge of the natural variation of stable isotope ratios in the different compartments of the 'grassland production system' and the isotope fractionation factors from soil to forage to food. Thus, stable isotope ratios may give also insight into the fluxes of these elements in an agricultural ecosystem, without disturbing the system by experimental treatments.

References

De Niro, M.J. and Epstein, S. 1981. Influence of diet on the distribution of nitrogen isotopes in animals. Geochimica and Cosmochimica Acta 45 341-351.

Enser, M., Hallett, K.G., Hewett, B., Fursey, G.A.J., Wood, J.D. and Harrington, G. 1998. Fatty acid content and composition of UK beef and lamb muscle in relation to production system and implications for human nutrition. Meat Science 49 329-341.

Farquhar, G.D., Ehleringer, J.R. and Hubick, K.T. 1989. Carbon isotope discrimination and photosynthesis. Annual Review of Plant Physiology and Plant Molecular Biology 40 503-537.

Hobson, K.A. 1999. Tracing origins and migration of wildlife using stable isotopes: a review. Oecologia 120 314-326.

Evaluation of different ruminal boluses with passive transponder for animal identification

G. Genovese[1], G. Altieri[1] and A. Calbi[2]

[1]University of Basilicata, Dept. DITEC, c.da Macchia Romana, I-85100, Potenza, Italy
[2]Associazione Provinciale Allevatori di Matera e di Potenza, Via Trabaci G. M., I-75100, Matera – Via della Edilizia, I-85100 Potenza – Italy
Altieri@unibas.it

Abstract

The ruminal bolus enclosing passive transponder is a safe method of electronic animal identification. It could be used indistinctly for cow, sheep and goat and it should be removed only after animal died. In this work were compared the boluses by different firm: Gesimpex, DeLaval, O.PI.VI. (transponder encapsulated in ceramic jacket) and Westfalia (transponder encapsulated in plastic jacket). The readability distance doesn't change significantly when the transponder is inside or outside the bolus; moreover the bolus readability is due to the sensitivity of transceiver antenna. The maximum readability distance is due to bolus position.

Keywords: animal identification, ruminal bolus, transponder, hand-held transceiver.

Introduction

Animals individual identification using passive transponder represents an important step in the organisation of herd management system based on the on-line control of cow milking parameters (Eradus & Jansen, 1999).
In order to evaluate the transponder reliability in terms of reading accuracy in the present work are reported the results of experimental trials carried on different ruminal boluses equipped with passive transponders for the permanent electronic identification of animals. Passive electronic identification trough transponder is a good method to identify the animals (Artmann, 1999; Jansen & Eradus, 1997; Kampers et al., 1999.). Furthermore the bolus with passive transponder is designed to be permanently retained in reticulum of ruminants without any trouble for the animals, it presents high specific weight and it should be removed only after animal died (Caja et al., 1999; Fallon, 2001). Considering that the same bolus could be used for cow, sheep or goat at different age (Caja et al., 1997; Caja et al., 1999), it has been verified the bolus stability in terms of material and readability error referred to distance from the reader and to relative position.

Materials and methods

In this work boluses produced by different firm were compared: Gesimpex, DeLaval, O.PI.VI. and Westfalia. DeLaval, O.PI.VI. and Gesimpex transponders are encapsulated in an atoxic ceramic jacket (minimum 90% of alumina), while Westfalia transponder is encapsulated in a plastic jacket and it presents a steel cartridge enclosure. In laboratory were used different hand-held transceivers: Allflex, Gesreader with stick antennas and Daisy Reader Module DHP 112 coupled with a PSION Workabout computer with stick antennas. In order to carry out the tests, each transceiver have been fixed to a graduated bar with the aim to exactly measure the transceiver distance from its target (bolus), then the bolus has been positioned on the orthogonal axis (reference axis) of a horizontal plan that was constituted by concentrical circles distributed every 5 cm.

Moreover, in order to verify the low porosity of the ceramic material, the boluses have been placed in an acid water solution under weak pressure.

Results

The reading test using different hand-held transceivers gave the following results:
1. the compatibility between bolus and transponder available on the market is not complete;
2. the readability distance is near 20 cm and it is similar for all the boluses, except for one, Westfalia bolus coupled with transceiver Allflex, where it is near 50 cm (Table 1);
3. the readability distance when the transponder is inside or outside the ceramic bolus doesn't vary significantly (Table 1);
4. the maximum reliable reading is due to position of bolus.

The ceramic material of bolus enclosing the transponder is a material with near zero porosity (Table 2).

Table 1. Maximum distance of bolus identification in horizontal and vertical position.

Position Bolus: horizontal	TRANSCIEVER			Position Bolus: vertical	TRANSCIEVER		
	GESREADER	ALLFLEX	DAISY Reader Module DHP112		GESREADER	ALLFLEX	DAISY Reader Module DHP112
BOLUS FIRM	cm	cm	cm	BOLUS FIRM	cm	cm	cm
GESIMPEX	14.5	19.5	13.5	GESIMPEX	19.5	25.5	19.5
O.PI.VI.	14.5	19.5	13.5	O.PI.VI.	21.5	27.5	21.5
ALFA LAVAL	13.5	19.5	13.5	ALFA LAVAL	21.5	27.5	21.5
Only Transponder	19.5	19.5	11.5	Only Transponder	27.5	27.5	19.5
WESTFALIA	-	-	-	WESTFALIA	no readable	50.5	33.5

Table 2. Percentage of humidity retained in boluses used in the tests.

Bolus firm	Wet weight (gr)	Dry weight (gr)	Retained humidity (gr)	% humidity on wet basis
GESIMPEX	75.0413	75.0296	0.0117	0.016
O.PI.VI.	67.0711	67.0682	0.0029	0.004
DE LAVAL	63.7668	63.7661	0.008	0.001

Discussion

The material of bolus doesn't significantly affect the readability distance of transponder.
The maximum reliable readability is due to position of bolus. In fact the reading range is influenced by the orientation of bolus toward the transceiver's antenna. The bolus is identified when it is positioned parallel to the reference axis but it doesn't identified when it is orthogonal to the reference axis. The pressure and temperature are irrelevant on the humidity absorption of the bolus material, so that after 12 months in the water under 3 bar of pressure the water content retained in the bolus material resulted less than 0.1% without significant variation with the control.

Conclusions

The ruminal bolus represents an efficient solution for permanent electronic animal identification in the farm management and it can be used indistinctly in sheep, goat and cattle at different ages.

The identification reliability trough the passive ruminal bolus it doesn't depend on the kind of bolus but it depends on the sensitivity and power of transceiver antenna. Moreover, assessed the not excessive distance of reading of the single hand-held transceiver, it is opportune that such transceivers are equipped with extendible stick antenna so as to facilitate the operation of animal identification and to allow the achievement of greater identification distances.

References

Artmann R., 1999. "Electronic identification system: state of the art and their further development". Computers and Electronics in Agriculture, 24: 5-26.

Caja G., Barillet F., Nehring R., Marie C., Conill C., Ricard E., Ribò O., Lagriffoul G., Peris S., Aurel M.R., Solanes D., Jacquin M., 1997. "State of the art on electronic identification of sheep and goat using passive transponders". Option Mediterranèennes, Série A: Séminaires Méditeranéens, 33 : 43-57.

Caja G., Conill G., Nehring R., Ribò O., 1999. "Development of a ceramic bolus for the permanent electronic identification of sheep, goat and cattle". Computers and Electronics in Agriculture, 24: 45-63.

Eradus W. J., Jansen M. B., 1999. "Animal identification and monitoring". Computers and Electronics in Agriculture, 24: 91-98.

Fallon R.J., 2001. "The development and use of electronic ruminal boluses as a vehicle for bovine identification". Rev. sci. tech. Off. Int. Epiz., 20 (2): 480-490.

Jansen M. B., Eradus W., 1999. "Future developments on devices for animal radiofrequency identification". Computers and Electronics in Agriculture, 24: 109-117.

Kampers F.W.H., 1999. "The ISO standard for radiofrequency identification of animals". Computers and Electronics in Agriculture, 24: 27-43.

Acknowledgements

This project was funded on the "Programma Operativo Multiregionale Misura 2 -Progetto B15 - Informatizzazione dei servizi veterinari e di assistenza tecnica alle aziende zootecniche".

Study on the effect of fan-to-pad evaporative cooling on broiler production during dry and wet summer season in a hot and arid climatic areas

Osama Abdel Hafiz[1] and Abdien Hassan Abdoun[2]
[1]Omdurman Islamic University, P.O.box 382, Omdurman, Sudan
[2]P.O.Box 1333, Khartoum, Sudan
elhafiz999@yahoo.com

Abstract

Poultry housing with negative pressure fan-and- pad evaporative cooling system was studied. The experimental data and the production performance were made and analyzed as a comparison between two old identical broiler test houses during Dry and Wet Summer seasons. One of these houses was completely rehabilitated and made more airtight and the other house was used as is. Test results shows higher dry summer average temperature inside the non-rehabilitated house (31.3 °C) whereas the complete rehabilitated house was found to be 30 °C. Pad -to- Fan temperature gradient was found to be (3.3 °C & 6.5 °C) and (3.4°C & 7.1°C) for the complete rehabilitated house and non-rehabilitated house during Dry and Wet Summer seasons respectively. The saturation efficiency was found to be (76.5% & 67%) and (75.5% & 71.0%) for the complete rehabilitated house and non-rehabilitated house during Dry and Wet Summer season respectively. The effect of cooling on broiler production performance was then evaluated.

Keywords: Cooling, Temperature, Environment, Broiler

Introduction

In hot and arid climatic conditions, as the case in Sudan, a negative pressure fan-and-pad cooled broiler houses have an inherent problem that drastically affects their cooling effectiveness. Due to the negative pressure created in the house by the exhaust fans, hot ambient air usually leaks, in through cracks and other openings that develop; by time, in the walls and roof of the house. This tends to greatly reduce the effectiveness of the cooling system resulting in low production rates and efficiency.

Materials and methods

Two identical broiler houses, 13 years operational age (108.5m long and 12.5m wide), were used for the experimental work. One of these houses was made more airtight, known as a completely rehabilitated house (H1). The other house was left as is (non-rehabilitated house, H2).Two identical cooling pad banks (28m long and 0.91m high), made of specially impregnated cellulose paper of high wettability were located along the middle of both side walls of each of the house. The flute height of the pad was 6.3 mm and water rate of 1 liter/m^2 was used for pad wetting. The cooling pads were protected against direct sunlight and pollution. Sixteen exhaust fan (10,050 m^3/hr) were distributed over the four sides at each and of the house. The records of the environmental conditions during Dry and Wet Summer season were taken at mid-day house at four location along the house. The performance of birds were recorded and evaluated accordingly. The other factors which concern the management, veterinary and feeding systems were the same during the experimental period, in both houses.

Results and discussion

The results of the field measurements of the two broiler test houses, obtained for the two cycles of production, are presented in the Table below:

Cycle	House	Outside Conditions				Inside Conditions					Saturation Efficiency (%)
		Dry Bulb Temp. (^{o}C)	Wet Bulb Temp. (^{o}C)	RH (%)	House Temp. (^{o}C)	House Wet Bulb Temp.	RH (%)	Cooling Pad Temp. (^{o}C)	Exhaust Fan Temp. (^{o}C)	Pad-to-Fan Temp. (^{o}C)	
Dry Summer	H1	40.5	24.8	28.0	30.0	25.2	68.2	28.8	32.1	3.3	76.5%
	H2	40.5	24.8	28.0	31.3	25.7	63.6	30.5	37.0	6.5	67%
Wet Summer	H1	39.1	27.9	43.4	29.8	25.4	69.2	28.8	32.3	3.4	75.2%
	H2	39.1	27.9	43.4	30.9	25.3	64.5	29.3	35.4	7.1	71.0%

The design parameters were made as 80% pad efficiency based on 45°C & 23°C for dry and wet bulb temperatures as ambient conditions respectively, in order to maintain a cooled air temperature equal to 27.4^{o}C and average pad-to-fan temperature not exceeding 3°C. The Table above shows that, although the saturation efficiency of H1 has shown better performance than that of H2 in both Dry and Wet Summer cycle but all seemed to be lower when compared with the designed saturation efficiency. Also, it has been clearly shown that, during dry summer, the saturation efficiency dramatically decreased in the H2 even when it was compared with Wet Summer of the same house. Pad-to-Fan temperature gradient has indicated greater value than the design one (3°C). This was mainly due to the effect of air infiltration as it possibly leads to higher temperatures inside the house and tends to reduce cooling effectiveness. The cooling effects during Dry and Wet summer makes a considerable difference on bird performance as shown in the figure below:

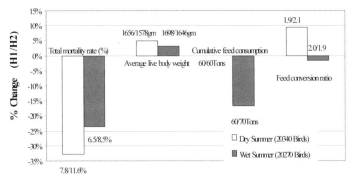

The figure indicates that, H1 has shown better production performance than that of the H2 during both Dry and Wet summer season.

Conclusion

The effect of ambient air infiltration is passively affected the performance of broiler production during the two seasonal variations (Dry Summer and Wet Summer season) although one of the houses is environmentally controlled.

References

1. Abdalla A.M. 1994 "Design of Poultry houses and improvement of thermal environmental conditions to maximize the production efficiency under hot climatic conditions of Sudan" Publications of Middle East and North Africa Poultry, No. 188.
2. Albright L.D. 1990 "Environmental control for animals and plants" International Standard Book; an ASAE Textbook, No. 4, USA.

Acknowledgment

Arab Authority for Agricultural Investment & Development (AAAID).

Application of thermal imaging for cattle management

H. J. Hellebrand[1], U. Brehme[2], H. Beuche[1], U. Stollberg[2] and H. Jacobs[1]

[1]*Department of Technology-Assessment and Substance Cycles, Institute of Agricultural Engineering (ATB), Max-Eyth-Allee 100, D-14469 Potsdam, Germany*
[2]*Department of Engineering for Livestock Management, Institute of Agricultural Engineering (ATB), Max-Eyth-Allee 100, D-14469 Potsdam, Germany*
jhellebrand@atb-potsdam.de

Abstract

To study the possible application potential of thermal imaging for health and fertility diagnostics, series of thermal images of grazing suckler cows were taken over a period of several months. The initial results indicate that cows with hot udders can be monitored. On the other hand, it was not possible to determine pregnancy by means of changed skin temperature distribution. The skin temperature is dominated by the ambient conditions to which the cows are exposed and not by the changes due to gravidity. The oestrus cycle affects the core body temperature of cows and could produce partial changes in skin temperature as well. The measurement of core body temperature, activity and pudendum temperature by thermography from oestrus-synchronised cows proved that the temperature maximum during oestrus climax could be shown by thermal imaging.

Keywords: thermal imaging, suckler cows, hot udder, gravidity, oestrus

Introduction

Thermal imaging can be utilised in animal husbandry for the control of buildings (e.g. heat insulation), processes (e.g. air conditioning, eating behaviour), and state of animals (e.g. stress susceptibility or health state of animals). Fast and cost effective measuring methods are looked for to improve cattle management particularly in the fields of health and fertility diagnostics. Sickness, local inflammations, injuries and other changes of the state of animals may produce an increase of the core body temperature and/or a change of the skin temperature distribution. As the skin temperature is influenced by the intensity of blood circulation, gravidity of cattle might be visible in thermal images. The state of animals can be sensed by thermography, if increased skin temperature or a change of its distribution reflects the change of the state of animals. Therefore, grazing suckler cows, heifers at different gravidity states, and oestrus-synchronised cows were studied by thermal imaging. The aim of this study was to evaluate the potential of thermography for cattle management.

Materials and methods

Remote temperature measurement techniques utilise infrared radiation in the range of about 8 to 12 µm. Whereas for the measurement of static scenes like analysis of buildings scanning cameras with one image per second are suitable, FPA-cameras (Focal Plane Array) with a capability of 50 images per second must be used for dynamic scenes with animals. In the measurements here, a camera with a non-cooled micro-bolometer array was utilised, operating in the 10-micrometer waveband at a temperature resolution of about 0.1 K at 300 K. The suckler cows were on the pasture for the entire time. The heifers were in a free stall barn connected to a feedlot. Images were taken in the open stable and at the feedlot. For the oestrus study, six dairy cows were investigated at the Clinic for Bovine Obstetrics and Gynaecology,

School of Veterinary Medicine Hannover (TiHo). Prostaglandin $F_{2\alpha}$ was applied to stimulate the artificial oestrus. The oestrus climax was evaluated by traditional measurements of the core body temperature and monitoring of animal activity.

Results and discussion

Thermal images and photos of grazing suckler cows on a pasture were collected for several months to study the practicability of monitoring cows at field conditions. It was found that cows with hot udder could be detected (figure 1). However, the as the skin is so well insulated, it makes it difficult to determine pregnancy by means of changed skin temperature distribution. There is no easily recognisable change in the skin temperature, which could be assigned to gravidity. Additionally, the black-white pattern in the visible range (photo below in figure 2 and figure 3) shows a similar pattern in the thermal image. This is valid for shadow (figure 2) as well as for basking heifers in the sun (figure 3). The differences in spectral emissivity (and thus the reflectivity; "black-white pattern") of the skin seem to range at first glance from the visible range (0.5 µm) to the thermal range (10 µm). Usually, the thermal emissivity of biological objects ranges between 0.90 and 0.99. Since the radiation interaction with the surrounding environment in connection with heat transfer from the body and cooling by transpiration and convection produces the skin temperature distribution, more detailed studies are necessary to determine the heat and radiation balance on the skin of animals.

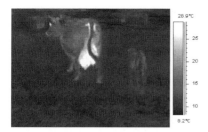

Figure 1. Suckler cow at pasture with hot udder.

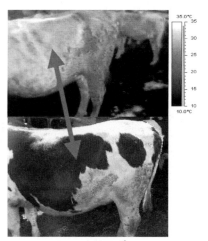

Figure 2. Heifer 516 (8th month pregnant) in the shadow (open stable; April 22, 2002).

During oestrus climax, the core body temperature of cows increases between 0.5 K and 1.0 K for a period of about 6 to 12 hours. An additional indicator of the oestrus climax is a maximum in activity of the cows. In an experiment at the School of Veterinary Medicine Hannover, the core body temperature, the activity and the pudendum temperature by thermography from six oestrus-synchronised cows was measured over a fortnight. The result was that thermal imaging could show the temperature maximum during oestrus climax.

Figure 3. Heifer 531 (4th month pregnant) in the sun (feed lot; May 22, 2002).

Conclusions

Cows with hot udder can be easily detected by thermal imaging on pastures as well as in barns. Gravidity of heifers in natural ambience (pasture or barn) cannot be determined by simply monitoring with a thermal imager. The temperature distribution of the skin is mainly influenced by the black-white pattern of the cows (visible range) and by the heat exchange between skin and surrounding. Since the skin is a good thermal insulator, ambient conditions and not the minor changes in body temperature usually determine the skin temperature. In contrast, the external pudendum temperature follows the core body temperature, and infrared thermometry can be utilised for oestrus climax determination.

Modeling and sensors application for air control in rabbits houses

Pavel Kic
Technical Faculty Czech University of Agriculture Prague, 165 21 Praha 6 – Suchdol, Czech Republic

Abstract

The aim of this paper is to describe the use of methods for air control in buildings for rabbits using the simple sensors and instruments for internal microclimate measurement. Selecting the right type of sensors depended on measuring tasks. Thermocouples and resistor-based sensors were used for temperature measurement; psychrometers and capacitive humidity sensors for the humidity measurement. The measurement of CO_2 was provided by the sensors based on infrared optics, etc. Several different farms with intensive rabbits breading and different technological situation were analysed. Interaction between the housing technology and air-conditioning system should be included in the designing procedure. The general strategy and structure of mathematical model for calculation of main designing capacity parameters is described.

Keywords: sensors; air; rabbits; technology; microclimate

Introduction

The rabbits farms for intensive breeding and fattening can be equipped by technique of different quality and properties, which influences the performance of production results. Interaction between the housing technology, building, animals and ventilation or air-conditioning system should be included in the designing procedure. The final production success in the farm performance is based partly on the design quality. The intensive breeding and fattening farms of rabbits with large capacity, provided in technologies with one, two or three tiers cages in batteries were analysed from different point of view, especially the quality of microclimate. Different sensors, detectors and instruments were used for the measurement of microclimatic parameters. The results show that the quality of indoor microclimate is rather strongly dependent on the interaction between the technology, building and ventilation or air-conditioning system.

Material and methods

The simple sensors and instruments were adapted and used for internal microclimate measurement. Data acquisition system 55-2s-T65, data loggers ZTH 65, data loggers Almemo 2290-8 with sensors thermocouples and resistor-based sensors (Pt 100 and Ntc); psychrometers FP A846, FP A846-2 and capacitive humidity sensors FH A646 were used for temperature and humidity measurement. The measurement of carbon dioxide gas was provided by the sensor module FY A600-CO2 based on infrared optics, the oxygen measurement by the sensor cells FY 9600-O2, ammonia and hydrogen sulphide by the special instrument Oldham, etc. Four farms of different capacity, equipped by different technological equipment for housing of rabbits, situated in the central part of the Czech Republic were tested. The basic data of the experimental farms are included in the table 1.

Table 1 Basic data of the experimental farms.

Parameter	Farm 1	Farm 2	Farm 3	Farm 4
Capacity	2 200	7 500	900	4 500
Technology	MBD	Dervaco	Dervaco	Kovobel
No of Tiers	2; 3	2	2	1; 2
Ventilation System	Ventilation + Heat Recuper.	Ventilation	Air-conditioning	Ventilation

The general strategy and structure of mathematical model for calculation of main designing capacity parameters is described in some other publications, e.g. [1]. The values of biological productions used in this program are based on some practical values recommended in [2]. The values used for calculations in Czech Republic are dependent on internal temperature t_i and mass of animal m_z [3], e.g. total heat production q_s and production of CO_2 per one animal are presented in the equations 1 and 2.

$$q_s = (6,8 - 0,05.t_i) . m_z^{0,64} \qquad (W.animal^{-1}) \qquad (1)$$

$$m_u = 1,5 . m_z^{0,64} \qquad (mg.s^{-1}.animal^{-1}) \qquad (2)$$

Results and discussion

The principal parameters during the fattening period were collected and together with other technical data analysed and generalized. The measured parameters of indoor thermal state and air quality (oxygen, noxious gases etc.) were collected and statistically analysed. The main final results of some trials are presented in the table 2.

Table 2 Final results of noxious gases measurement.

Parameter	Farm 1	Farm 2	Farm 3	Farm 4
CO_2 [% vol.]	0,27 – 038	0,1 – 0,245	0,047 – 0,183	0,06 – 0,086
NH_3 [ppm]	≈ 26	1,27 – 4,3	≈ 5	2,5 - 13

Conclusion

The results of measurements of various microclimatic parameters verified that the quality of internal microclimatic conditions in the buildings for rabbits is created by the housing technology, building properties, ventilation or air-conditioning system and strongly influenced by the farm management. Interaction between the housing technology and air-conditioning system should be included in the designing procedure.

References

1. Kic, P.-Chiumenti, R.-Bortolussi, S.-Da Borso, F.: Simulation of ventilation and indoors air conditions of agricultural buildings. In: Building Simulation Fifth International IBPSA Conference. Proceedings II. Prague, 1997, s. 261-268
2. CIGR WG: Report of working group on climatization of animal houses. Aberdeen, 1984, 72 s.
3. ČSN 730543-2 Vnitřní prostředí stájových objektů.Část 2: Větrání a vytápění. 1998

Automatically controlled preparation of cow's udder for milking

A. Laurs[1] and M. Lusis[2]

[1]Latvia University of Agriculture, Faculty of Engineering, Liela 2, Jelgava, LV-3001, Latvia
[2]Ltd. A.M.L., Valdlauči, Riga, LV-1076, Latvia
office@aml-ramava.lv

Abstract

This research aims at finding a solution for mechanical cleaning and drying of udder prior to milking. The cleaning of teats by jets of water and rotating brushes was tested. For cleaning of the udder the best results were obtained by using wet rotating brushes which were placed on each side of the udder and which were moved along the udder forwards and backwards, alternating their direction of rotation. A high quality cleaning was effected in 69% of cases. For drying of the udder, the best results were obtained by using dry rotating brushes simultaneously blowing water drops off the teats.

Keywords: cleaning of udder, drying of udder, washbox, rotating brushes, jets of air

Introduction

Due to widely spread automation of milking processes and introduction of AMS, a mechanized preparation of cows' udders for milking has become topical. Three modes have proven to be more suitable: cleaning of the udder by jets of water, by brushes and by a special shell for teat cleaning. We were looking for a rational solution, by applying cleaning of the udder by jets of water and brushes.

Material and methods

Research was effected in a special washbox where it was possible to change efficiently both elements of work and also their performance. Cleaning and drying efficiency was assessed by using napkins and merely visually. Research of udder cleaning was effected with a group of 12 cows in 6 (spray of water) and 7 (brushes) repetitions, research of drying – with a group of 15 cows in 4 repetitions.

Results and discussion

In the first series of tests for cleaning of the udder in a washbox there were applied two rotating brushes placed on each side of the udder. Water with intensity 6 l/min was supplied onto the brushes through nozzles. In previous tests there was set an optimal duration of udder cleaning – 20 s. Taking into account results obtained by measuring the cows' udders, an optimal kinematics of the brushes operation was set. Brushes are to be moved towards the udder with a radius of 400 mm, and its surface operation diameter should be 160 mm. The rotation axis of the brush in horizontal position should be in 570 mm height from the floor, which corresponds to a medium udder height; the lifting mechanism should ensure a possibility for the brushes to be lifted in 750 mm height.
In the second series of tests udders were cleaned by using shower roses. We used 4 shower roses, the jets of which were directed towards the udder. Also in this case washing lasted 20 s. Results of research are shown in Table 1.

Table 1. Quality of udder cleaning by means of various elements of work.

Element of work	The number of repetitions	Udder cleaned completely (%)	Not cleaned teats (%)	Not cleaned base of udder (%)
Brushes	84	69	20	12
Shower	74	43	46	24

For drying of the udder there were three types tested: by warm air jets, by dry cleaning brushes and by dry cleaning brushes by simultaneously blowing air to the udder under pressure. Duration of drying was 1.8-3.2 min (on average 2.3 min). Technological calculations proved that the duration of drying should not exceed 30 s. Otherwise throughput of the milking parlor becomes so low that such technology proves inefficient. Therefore drying with dry brushes was tested for 10; 20 and 30 s. By treating the udder for 10 and 20 s, in all cases there remained drops of water on the teats. By increasing the time for drying up to 30 s, only in 20% of cases udders were completely dry. Drying of the udder by dry brushes and compressed air was checked in the mode of 10, 20 and 30 s.
Results of tests are shown in Table 2.

Table 2. Quality of udder drying by means of dry brushes and blowing off moisture by air jets.

Duration of udder drying (s)	The number of repetitions	Udder cleaned completely	
		repetitions	%
10	80	42	53
20	80	68	85
30	80	72	91

In Table 1 one can see that the udder can be most efficiently cleaned by brushes, however in 20% of cases the surface of teats remains unclean. The reason for that is the rotation of brushes only in one direction. The teats bend down; one side of them remains unclean. It could be avoided by changing the direction of rotation of the brush. To increase the efficiency of cleaning, it would be reasonable to increase also the diameter of brushes.
In Table 2 one can see that the most efficient result can be obtained by drying the udder by dry brushes and air jets. By treating the udder for 10 s, only 53% of udders become completely dry. By increasing the treatment time up to 20 s and 30 s, 85% and 90% of udders are dried completely. The reason for the fact that udders of separate cows remain non-dried are morphological features of some cows' udders.

Conclusions

Mechanized cleaning of the udder by using wet rotating brushes placed on each side of the udder when the direction of rotation during operation is changed moving along the udder forwards and backwards, performs its function fully in 69% of cases.
Drying of the udder by using rotating and by simultaneously blowing off drops of water by pressurized air is efficient in 85-95% of cases.

Trace metal contamination of pasture grasses in Dar es Salaam

G.B. Luilo and O.C. Othman
C/o Department of Chemistry, University of Dar es Salaam, PO Box 35061, Dar es Salaam, Tanzania, East Africa.
gebhardk@hotmail.com

Abstract

Animal husbandry constitutes the most important urban agricultural activities in Dar es Salaam City. Lack of pastures force the city farmers to resort to the grasses in road reserves to graze their animals. However, the vegetation is subject to pollution from motor traffic which threatens the quality status of the pastures. This study has ascertained that pasture grasses studied are contaminated with Cd, Cr, Pb and Zn. In such a situation human beings are at high exposure risks to these toxic metals through food chain. Therefore urban agriculture in growing cities should be practiced with public health concerns.

Keywords: *Cynodon* grasses, toxic metals, traffic pollution, urban agriculture

Introduction

Dar es Salaam City is under high pollution pressure from anthropogenic activities and one of which is from operating motor vehicles in city. Motor vehicles emit gaseous and trace metal contaminants into the roadside environment. Urban agriculture in the city is growing faster and farmers make use of roadside pasture grasses animal feed (Sawio, 1998). These grasses may be contaminated with trace metals from point and non point sources. Nevertheless, little has been done with regard to levels of trace metals pasture grasses and product quality management Thus, this paper aimed at assessing levels of cadmium, chromium, lead and zinc which are trace metals of health concerns, in grasses along roadsides in Dar es Salaam.

Materials and methods

Plant samples (*Cynodon dactylon and Cynodon plectostachyus*) were collected at distances of 1, 5, 15, 35 and 50 m from the road edge at. Lugalo-Area E, Makongo, Cafe Latino, Ubungo and Victoria areas in Dar es Salaam. These sites differed in traffic densities. The plant samples were oven dried at room temperature for a week and then charred at 450 $^{\circ}$C in muffle furnace for 30 minutes followed by grinding in porcelain motor to pass through a 250µm sieve. Then 5.000 g of the ground samples were in dissolved in a 250 ml flask using 4 M HCl followed by filtration through a 60-Wattman filter paper. The filtrate was made to the mark by distilled water and the resultant solution was used for analysis of Cd, Cr, Pb and Zn by Perkin Elmer AAS (model Aanalyst 300) at wavelengths of 228.8 nm, 357.9 nm, 283.3 nm and 213.9 nm using air-acetylene flame respectively.

Results and discussion

The results showed that the pasture grasses collected with 50 m distance off the road edge were contaminated with Cd, Cr, Pb and Zn and their concentration were generally above the natural trace metal levels in non-edible plants (Ward, 1995). The concentrations (ppm) of Cd, Cr, Pb and Zn in *C. plectostachyus* ranged 0.85 – 2.95, 0.05 – 3.25, 0.1 – 31.25 and 17.20 – 80.55 respectively. Whereas, the levels (ppm) of Cd, Cr, Pb and Zn in *C. dactylon* ranged from 0.05 – 4.60, 0.1 – 4.55, 0.03 – 35.55 and 20.20 – 162.50 respectively. The mean levels of the trace metals in pasture grasses for each site are presented in Table 1.

Table 1. Mean trace metal concentration (ppm) in pasture grass species at different sites.

Sampling site	Pasture species	Cadmium	Chromium	Lead	Zinc
Lugalo Area-E	C. dactylon	0.26	1.18	1.39	27.24
	C. Plectostachyus	1.96	1.62	7.9	32.63
Makongo	C. dactylon	0.58	0.56	5.66	46.44
	C. Plectostachyus	2.62	0.70	3.94	47.01
Café Latino	C. dactylon	2.09	0.80	5.80	41.21
	C. Plectostachyus	2.67	1.45	26.09	43.80
Ubungo	C. dactylon	0.36	2.33	13.84	64.10
	C. Plectostachyus	0.99	2.06	20.55	53.03
Victoria	C. dactylon	0.74	2.45	17.75	95.36
	C. Plectostachyus	1.15	1.65	15.69	61.33
Ward (1995)		0.1	0.2	1	30

It is found that the levels in grasses were in the order of Zn > Pb > Cr ~ Cd and among these metals only lead in C dactylon showed a distinct increase concentration with change in sites. Among the four metals studied Cd and Pb are the most toxic ones and therefore have to be monitored in animal feeds. The levels of Cd in particular were above the German Feed Crop Value Limit Regulations (ZEBS, 1990) while that of Cr, Pb and Zn were above natural trace metal level in plant as per Ward (1995). Due to scarcity of arable land in the city the urban dwellers resort to the open spaces for cultivating vegetable and where there are pasture grasses animal keepers graze their cattle or goats on. These grasses, as shown above, are subject to contamination and therefore regular grazing of dairy cattle in grasses along highways is potentially risk. The animals are fed with grasses with toxic metals. Through food chain these metal may reach human body whose consequences have never been studies yet here in Dar es Salaam. But it well known that these metals are toxic particularly cadmium and lead. The city dwellers make use of these grasses as pasture out ignorance and also lack of data to substantiate unsuitability of these grasses as animal feed. The concept of precision facing and the technology are very new to Tanzania and none can afford to practice it in the near future as majority of the population rely on peasantry farming. Nevertheless, it high time now for developed countries to help poor countries to adapt this new farming technology for the public and environmental health.

Conclusion

Pasture grasses in road vicinity in Dar es Salaam are contamination with Cd, Cr, Pb and Zn which in turn threaten the quality of animal food products. Based on the toxicity of metals such as Cd and Pb, human being are at risk of getting poisoned by ingestion milk and meat products from cattle or goats fed by grasses from these areas. It is therefore recommended that further studies have to be carried out to study levels of these metals in milk and meat products from domesticated dairy cattle and goats.

References

Sawio, C.J. 1998. Managing Urban Agriculture in Dar es Salaam, Cities Feeding People Report Series Report 20. International Development Research Centre: Ottawa, 40 pp.

Ward, N.I. 1995. Trace metals. In: Environmental Analytical Chemistry '95, edited by F.W. Fiffield and P.J. Haines, Blackie Academic and Professionals, London, pp. 321–351.

Zentrale Erfassungs- und Bewertungsstelle für Umweltchemikalien des Bundesgesundheits-amtes (ZEBS).1990. Richtwerte für Schadstoffe in Lebensmitte. Bundesgesundhbl 5 224-226.

Site-specific yield sensor for forages

K. Nishizaki and T. Tsukamoto
Obihiro University of Agriculture and Veterinary Medicine, Engineering in Agricultural and biological Systems, Inada Obihiro Hokkaido 80-8555 Japan
nisizaki@obihiro.ac.jp

Abstract

Two yield sensors for forages was constructed and investigated.
One type is the dynamic weighing system in a round baler. A round baler was equipped with 3 load cells (the drawbar coupling and right and left-side wheel axle of the baler) for measuring the bale weight in the baler. Three sensors were installed for measuring the round bale mass of the baler.
Another type is a pendulum system that is measuring the angle of a suspended pendulum through the resistance of grass stem. A pendulum-meter was developed to measure fresh plant mass by V. Hammen, D. Ehlert (Germany). Field attempts were carried out in using new materials and structure.

Keywords: PF, forages, yield sensor, yield map

Introduction

The cost reduction and the productivity growth of forages by technological advance is a fundamental source of solution for the improvement of self-sufficiency and food safety. Precision agriculture can increase yield, reduce cost, conserve resources, and minimize environmental impacts. Sensor-technology applications include information management system, capable of integrating precision agricultural devices and decision making for operation.

Materials and methods

Both systems were tested for measuring on the go. DGPS was equipped to detect the track of the tractor and the data sample locations. Finally, yield maps of timothy were created from the field weight data by using the dynamic weighing system (Figure 1, 2).

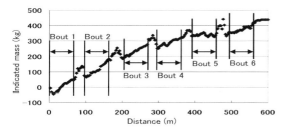

Figure 1. Registered weight signal from the round baler.

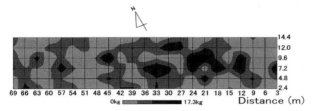

Figure 2. Yield map (weighing system).

And yield maps of alfalfa were also created from the angle data of the pendulum sensor (Figure 3,4).

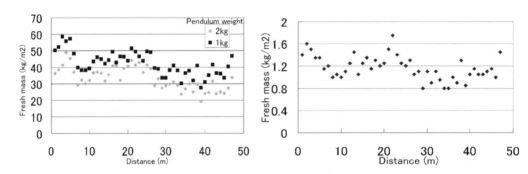

Figure 3. Angle of deviation.　　　　　　　　Figure 4. Fresh mass.

Results

Measurements of forage yield are important for the decisions of site-specific management. The information for spatially difference of forage yield contributes to improve fertilization system. It is possible to make site-specific fertilizer input and consequently reducing economic cost and environmental protection. Further research is planned to measure plant mass including above two methods and to detect the moisture content of forage in harvesting on the go.

How does feeding frequency influence performance and prevalence of gastric lesions in fattening pigs?

E. Persson[1], M. Wülbers-Mindermann[2], C. Berg[2], and B. Algers[2]
[1]*Swedish University of Agricultural Sciences, Department of Agricultural Research Skara*
[2]*Swedish University of Agricultural Sciences, Department of Animal Environment and Health*
Eva.Persson@jvsk.slu.se

Abstract

In the same stable, 360 fattening pigs were divided into two groups; control group and treatment group, and fed three or nine times daily to evaluate how feeding frequency affects performance and prevalence of gastric lesions. The pigs were weighed individually every second week and at slaughter the stomachs were examined for lesions in pars oesophagea and scored on a scale from 1 (normal) to 6 (ulcer).
More frequent feeding occasions influenced both performance and gastric status of the pigs negatively. The results indicate a reduced welfare in the more frequently fed pigs. Higher incidence of gastric lesions can be assumed to be associated with a higher level of stress, since the feed content and ration was identical.

Key words: Slaughter pigs, health, animal welfare, liquid feeding, feeding frequency

Introduction

In Sweden the animal welfare legislation states that farm animals should be able to express their natural behaviours. According to the Swedish animal welfare legislation the pen area for a fattening pig should be at least 0.86 m^2 (SJVFS, 1993:129). With such housing density in combination with a relatively barren environment the possibility of fulfilment of the pigs' natural behavioural repertoire such as foraging and social behaviour is limited. Feedings therefore becomes the "happening of the day".
This study is linked to the so-called "Anti Stress Theory" developed by Uvnäs-Moberg (1998). The theory is based on the calming effect of the hormone oxytocin, which is released when the parasympaticus (calm status) is activated in combination to positive social or physical contacts. Feeding behaviour also activates the parasympaticus and enables a higher oxytocin level, which leads to improved digestion and therefore a better growth rate.
The aim of this study was to evaluate how feeding frequency affects performance and the gastric lesions of pigs raised under conventional conditions.

Materials and methods

360 fattening pigs (26-115 kg) were kept in the same environment in the same house. The pigs were weighed and assigned to pens (9 pigs/pen) according to weight and sex at the beginning of the trial. Until the first pig was sent for slaughter they were weighed every second week and thereafter weighing was carried out once weekly. A compound feed mixed with whey and water to a liquid feed were given and the feed ration was correlated to the live weight of the pigs, according to the Swedish standards. The 180 pigs in the control group were fed three times a day at 7 am, 13 and 19 pm. The 180 pigs in the treatment group were fed nine times a day (three times at every standard feeding time with 45 minutes interval). Individual recordings were made on daily weight gain, live weight at slaughter in relation to

carcass weight and meat percentage. At slaughter the stomach of each pig was examined for lesions in pars oesophagea and scored on a scale from 1 (normal) to 6 (ulcer).

Results and discussion

Pigs in the treatment group grew slower compared to pigs in the control group, 723 g/day versus 834 g/day ($P<0.001$). Pigs in the control group had a lower prevalence of gastric lesions measured in pars oesophagea according to the scale applied; 2,4 compared to 3,0 ($P<0.01$). There was no correlation between daily weight gain and gastric lesion score.

These results indicate a reduced welfare in the more frequently fed pigs. Higher incidence of gastric lesions can in this case be assumed to be associated with a higher level of stress, since the feed content and ration was identical. Stress effects the digestion negative and can lead to higher feed conversion rate and development of gastric lesions. However, it could not be shown that individual pigs with gastric lesions grow slower than without lesions (Dobson *et al.*, 1978). The hypothesis of the "Anti Stress Theory" regarding feeding did not hold in this situation. A possible explanation is that a stressful situation occurred for the treatment pigs at the feeding through to fight more often for small servings of feed, which in turn reduced weight gain and gave a higher incidence of gastric lesions.

Conclusion

In this study we could not find that more feedings daily helped the fattening pigs to a better performance because of a feeding routine more close to their natural behaviour.

References

Djurskyddsförordningen 1988:539. Statens Jordbruksverks föreskrifter, SJVFS 1993:129, SJVFS 2. (Swedish Board of Agriculture; animal welfare legislation).

Dobson, D. J., Davies, R. L., Cargill, C. F., 1978. Ulceration of pars oesophagea in pigs. Aus. Vet. J. 54: 601-602.

Uvnäs-Moberg, K., 1998. Antistress Pattern Induced by Oxytocin. News Physiol. Sci., Vol. 13, pp 22-26.

Acknowledgements

This project is a part of FOOD 21 - research project for a sustainable Food Production, funded by MISTRA.

Decision-tree induction for the classification of services per conception in dairy cattle

D. Pietersma[1], W. Grzesiak[2], P. Blaszczyk[2], P. Sablik[2], R. Lacroix[1] and K.M. Wade[1]
[1]McGill University, Department of Animal Science, Macdonald Campus, 21111 Lakeshore Road, Ste. Anne de Bellevue, QC, H9X 3V9, Canada
[2]Agricultural University of Szczecin, Department of Ruminant Animal Science, ul. Dr Judyma 10, 71-466 Szczecin, Poland
diederik.pietersma@mail.mcgill.ca

Abstract

Decision-tree induction was used to generate classifiers to predict the number of artificial insemination services per conception in dairy cattle. Data comprised 1054 cows and the "services per conception" attribute was categorized into three classes. A decision tree induced from the entire dataset had six nodes with the attributes "body condition score" and "calving difficulty" being most important. Cross-validation tests yielded an accuracy of 88%. A second tree was induced to focus on classifying cows with many services per conception correctly. These trees may help identify cows with potential reproductive problems.

Keywords: decision-tree induction, data mining, insemination services, dairy cattle.

Introduction

Decision-tree induction is a data-mining technique that can be used to discover interesting patterns among routinely collected data in livestock farming. Data such as milk yield per test and lactation, calving details, and body condition may help to identify cows with reproduction problems (Stevenson, 1997). Unsuccessful artificial insemination services can lead to substantial reductions in profit due to increased reproduction costs, more culling, and reduced milk yield (Britt, 1985). In this study, decision-tree induction was used to generate classification models to predict the number of services per conception in dairy cattle.

Materials and methods

Data comprised 1054 cows from six dairy herds. The number of services per conception was categorized into three classes: class A for 1 or 2 services (809 cases); class B for 3 or 4 (168 cases); and class C for 5 and 6 (77 cases). The predictive attributes consisted of percentage Holstein-Friesian genes, lactation number, calving month, calving difficulty (levels 1 through 4), 305-day milk yield of preceding lactation, milk yield for the lactation's first two months, average 305-day herd production for milk, fat, and protein, body condition score (BCS) at early lactation (ranging from 1 to 5), and quality of artificial insemination service (evaluated using a 1 to 5 range – only values 4 and 4.5 in dataset). Data preprocessing, decision-tree induction, and performance estimation were performed with Kensington Discovery Edition 1.8 (Inforsense Ltd, London, UK). The decision-tree algorithm parameter for minimum number of cases per child node was set at 15. Ten-fold cross validation was used to estimate the classification performance (Pietersma et al., 2003).

Results and discussion

The decision tree induced from the entire dataset is shown in Figure 1. Decision node 1 shows the number of cases per class for the entire data set and the condition to split the data (BCS $\leq$ 1.5). Cases that comply with the condition are assigned to leaf node 1 predicting class C, while the other cases go to decision node 2. The two most important attributes in this tree are BCS and calving difficulty, with a low value for calving difficulty and a high BCS leading to few services per conception (class A). Ten-fold cross validation revealed an average accuracy of 88.2 (0.9) percent (standard error). Correctly classified cases for classes A, B, and C were, on average, 96.7 (0.8), 64.9 (3.6), and 50.9 (6.3) percent, respectively. In order to improve the relatively poor performance of class C, the dataset was balanced to 77 training cases for each class resulting in a slightly different tree of five decision nodes: calving-difficulty ≤ 2: class A, calving-difficulty > 3: C, BCS > 2.25: B, milk-first-two-months ≤ 28.1: B, 305-day-milk $\leq$ 7489: C, and remaining cases: class B. For this tree, the estimated accuracy was 86.2 (0.6) percent and correctly classified cases for classes A, B, and C were 94.8 (1.0), 54.2 (4.9), and 66.4 (4.4) percent, respectively. With the balanced data, the classification performance substantially improved for class C at the expense of reduced performance for classes A and B.

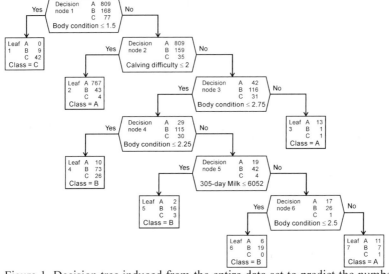

Figure 1. Decision tree induced from the entire data set to predict the number of services per conception using classes A (1 or 2 services), B (3 or 4), and C (5 or 6).

Conclusions

These results suggest that body condition score and calving difficulty are the most important attributes in predicting services per conception. The induced decision trees allow dairy producers to identify cows at risk for reproduction problems. Data from a larger number of herds will be required to confirm these results.

References

Britt, J.H. 1985. Enhanced reproduction and its economic implications. Journal of Dairy Science 68 1585-1592.

Pietersma, D., Lacroix, R., Lefebvre, D. and Wade, K.M. 2003. Performance analysis for machine-learning experiments using small data sets. Computers and Electronics in Agriculture 38 1-17.

Stevenson, J.S. 1997. Management of dairy cows to minimize reproductive problems after parturition. In: Proc. of Western Canadian Dairy Seminar, Volume 9, Department of Agriculture, Food and Nutrition Science, Univ. of Alberta, Edmonton, Alberta, Canada.

Benefits from pig welfare

Arnolds Skele, Imants Ziemelis, Aldis Putans and Uldis Iljins
Ulbroka Research centre, Ulbroka 1 Instituta str., LV-2130, Latvia

Abstract

One of the most power-consuming factors in the swine-breeding energy balance is local warming of piglets . Optimum temperature for young piglets, first days after birth is 32 - 34^0C. Indoor temperature in swine houses in winter time is 8 - 16^0C and piglets local heating in Latvia is practiced. Different ways of local heating have been tested. The most effective was automatically controlled infra-red lamps. The automatic system provides heating intensity control adjusting ambient temperature and age of pigs. The investigations show, that automatic control of pig heating systems are more favourable than ones of on unregulated heating systems.

Keywords: energy saving, piglets, automatic system.

Introduction

High piglet productivity depends mainly on breed, adequate feeds, and keeping conditions. In special investigations was found, that during the first days of life the optimum temperature in the piglet lairs is 32 ... 34 ^{0}C, what depends mainly on mass of piglet. The optimum air temperature in the pigsty for sows is 16 ... 20 ^{0}C. A research was conducted in order to determine the best type of local heating to ensure the welfare of young pigs. The investigation showed, that more effective are infra-red warming devices.

The procedure of the research

To be sure, that the automatic piglet warming system developed is better, than that used before, it was envisaged to do the following investigations:
1. To measure the consumption of electric energy of one electric infra-red heater working with an automatic voltage regulator in accordance with the zootechnical recommendations and to compare it with the identical heater operating in the previous manner;
2. To investigate how the piglets use the warming systems with automatic regulation and without it by carrying out detailed timing of the behaviour of animals in the experimental and the reference groups.
3. To control the growth of piglet's mass in 5 experimental and 5 reference groups.

The results of the research and evaluation

Since the room temperature (9 ...16^0C) during the experiments changed, the consumption of energy was determined by the counters of electric energy throughout the whole growing period. The consumption of electric energy in farm measured for one heater without automatic regulation was 1447 kWh, with automatic regulation - 630 kWh. The measurements of energy consumption for warming piglets show, that despite fact that the experiments were carried out in the period of cold weather, when the temperatures in the pigsty were low, automatic power regulation of the heaters decreases the consumption of electric energy more than 2 times.

The daily research of the behaviour of fifteen days old piglets, who had already developed physical thermal regulation, have shown, that under unregulated heating conditions (control group) the piglets stayed under the heater approximately half (48.7%) of their sleeping time and the rest it they were sleeping in other places (51.3%), whereas in the case of regulated heating they stayed 81.4% of the whole sleeping time under the heater and 18.6% in other places of the trough. Much less the heater with unregulated warming mode was used by piglets of 55 days - only 21.9 per cent of their total sleeping time, whereas under regulated heating piglets lay 3 times longer (78.1%). Under regulated and cyclic heating conditions (experimental group) on the contrary - they slept 77.6% under the heater and 22.4% from sleeping time - in other places. The walking (10.6 and 18.6%) and sucking (23.8 and 7.8%) time for 15 and 55 days old piglets in non regulated conditions was almost the some as in conditions of regulated warming system (9.7 - 14.9% and 22.5 - 8.8% accordingly), it means, that movements for physiological needs was not influenced by different conditions of warming. The investigation show that the unregulated heating is not used effectively by the 15th nor 55th days old piglets as they were sleeping mainly out of the zone of heaters action. In the case of regulated and cyclic heating the piglets used the heated zone more effectively during action periods of the lamps. So, 15 days old piglets in control group had bigger motion activity, what we can see by twice bigger number of position changes (25.5 and 16.6) in comparison with piglets in experimental group (11.0 and 12.5). At the end of the research the life weight growth of the experimental piglets was on average for 2.04 kg bigger for every piglet, than for piglets grown under unregulated heating (P<0.005). Our opinion is that the increased growth and 70 g bigger daily life weight of piglets under controlled heating conditions is the result of additional feed consumption. In additional to this - the 100% piglets saving was observed just in this group at the end of suckling period. It testifies, that regulated warming is more agreeable (better) for the piglets than unregulated warming.

All piglets of these experiments had been under the same feeding and growing conditions with remarkable changes

Conclusions

- System of automatically regulated warming of piglets, in comparison with usually used infrared lamps, allows saving electric energy more than 50%.
- The amount of the saved electric energy by automatically regulated warming system depends on the outside and inside temperature and the age of the piglets.
- Automatically regulated piglet warming does not show any harmful effect on the physiological functions of organism.
- Automatically regulated piglet warming was favourable on the behaviour of the piglets and has stimulated about 15% greater growth of their mass and therefore it can be recommended to piglet breeders.

Continuous measurement of the ensiling process in chopped maize

H.G.J. Snell[1], H.F.A. Van den Weghe[1] and W. Lücke[2]
[1]*Research Centre for Animal Production and Technology, Universitätsstraße 7, 49377 Vechta, Germany*
[2]*Institute of Agricultural Engineering, Gutenbergstraße 33, 37075 Göttingen, Germany*
hsnell@gwdg.de

Abstract

In four consecutive experiments it was investigated, whether modern sensors can contribute to a continuous measurement of pH-value and oxygen concentration during ensiling of chopped maize. Even with different degrees of silage compaction the pH-values in different mini-silos correlated strongly. There also was a strong correlation with the results of discontinuous laboratory analyses. Combining the measurement of temperature, pH-value and oxygen concentration is a valuable tool for process monitoring.

Keywords: silage, measurement techniques, maize, pH-value, oxygen concentration

Introduction

Continuous measurement of the preservation of silage maize offers perspectives both for scientific research and for the use on farms. In the future, farmers may conduct those measurements as a means of process monitoring.

Materials and methods

Overview
The experiments M_1 (09.10.99-25.10.99) and M_2 (22.09.00-09.10.00) focused on the measurement of the pH-value. There were no different treatments in M_1. In M_2 there were 3 treatments, which differ with respect to silage density. In both experiments several tubes were filled with chopped maize. They were opened in defined intervals and the pH-value was then measured in the laboratory. Finally, the results of the continuous measurement and of the analysis in the lab were compared. M_3 (10.10.01-29.10.01) and M_4 (15.11.01-06.12.01) focused on the measurement of the oxygen concentration in the silage. There were 2 treatments, differing with respect to the thickness of the PE film covering the tubes.
Set-up of the mini-silos
Plastic tubes (V = 3142 cm^3) were filled with fresh maize (M_1, 2.5 kg wet matter, WM, per tube; M_2, treatment a, 1.5; b, 2.5; c, 3.0 kg WM per tube) or maize which was frozen and thawed immediately before the beginning of the experiments (M_3, M_4, 2.0 kg WM per tube). 13 mini-silos were filled in M_1. In M_2, M_3, and M_4 8, 2, and 2 mini-silos were filled per treatment respectively.
During M_1 and M_2, both ends of each mini-silo were sealed with rubber caps. In case of M_3 and M_4, the lower end of each mini-silo was closed using plastic caps. The upper end had a cap, which only permitted gas exchange in a small area (50 cm^2) covered with silo film (film thickness, treatment a, 90 µm; b, 200 µm). During M_3 and M_4, a vacuum (M_3, 950 mbar; M_4, 990 mbar) was established at the bottom of the mini-silos after about 7 days. In M_4 the vacuum was increased to 950 mbar 335 h after ensiling.
The analyses and measurements were carried out according to Snell et al. (2003).

Results and discussion

In Figure 1 the high correlation between the pH-values in different tubes becomes evident. Even with different degrees of compaction (M_2) this correlation could be observed.

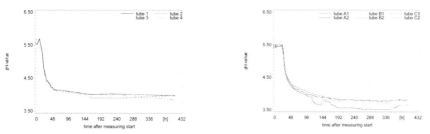

Figure 7. Course of the pH-value in the experiments M_1 (left) and M_2. Correlation of the measurements within different mini-silos (A, treatment a; B, treatment B; C, treatment c).

In Figure 2 the correlation between the continuous measurements and the analyses in the lab are visualized. The course of the pH-value is represented by both techniques in similar way. The initial values were lower with the laboratory analyses.

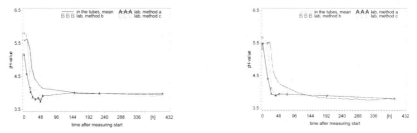

Figure 2. Course of the pH-value in the experiments M_1 (left) and M_2. Results of different laboratory analyses (Snell et al., 2003) and the continuous measurement in the mini-silos.

Figure 3 demonstrates the potential of a combination of different measurements. The oxygen trapped in the silo was consumed rapidly. A vacuum in the silo caused an increase in oxygen concentration, pH-value and temperature.

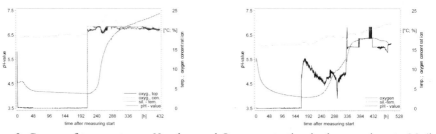

Figure 3. Course of temperature, pH-value, and O_2-concentration in the experiments M_3 (left) and M_4 (oxyg, oxygen; cen., centre; sil., silage; tem., temperature; mean values of 4 tubes).

References

Snell, H.G.J., Van den Weghe, H.F.A. and Lücke, W. 2003. Continuous measurement of the ensiling process in pressed sugar beet pulp. Proceedings of the 1st European Conference on Precision Livestock Farming, 15.06.-18.06.2003, Berlin.

Modelling, flow and use of data from automatic milking systems

J. Spilke and R. Fahr

University of Halle-Wittenberg, Agricultural Faculty, D-06099 Halle (Saale), Germany
spilke@landw.uni-halle.de

Abstract

In dairy farming the amount of data per individual animal will increase in the future. Therefore, an appropriate data management is indispensable in order to meet the danger of bottle-necks in time-critical case-related provision of data or to avoid mistakes in long-term consistent data storage. Furthermore, in this context the data management has to serve as a basis of any kind of decision support. The given example of data utilization is the application of mixed linear models to generate predicted milk yields and test day effects.

Keywords: relational data model, operational data, analytical data, decision support

Introduction

Today's milk production is more than ever before faced with consumer- and animal welfare concerns, quality assurance, ecological and economic sustainability as well as crisis prophylaxis. In addition to rising herd sizes and economical reasons, the transition to group-housing systems with respect to animal welfare, must not necessarily lead to a decrease in individual animal care. Above all, the current frequent occurrence of mastitis creating welfare problems for the cows along with a dramatically reduced longevity force us to a stronger monitoring on the level of the individual animal. Therefore, a stronger continuous monitoring on the individual level is necessary, than this is the case at present. However, the monitoring conditions need to be improved like developments in process-controlling, information- and communication systems. In addition, the complexity of these demands require an integrated approach of all involved scientific disciplines in order to guarantee the demanded quality of the total process. A promising approach for dairy production is Precision Dairy Farming (PDF). PDF is a special interdisciplinary approach of different scientific disciplines (among others Informatics, Biostatistics, Ethology, Economics, Animal Breeding, Animal Husbandry, Animal Nutrition and Process Engineering). Within these disciplines, Informatics has a special integration and interface function. For that reason Informatics must ensure for example that (grounded on a data base concept) relevant data are kept consistently serviced for any upcoming tasks. A particular meaning has the use of this data through decision support systems.

Data flow and storage

A Data Base Management System (DBMS) which is based on the relational data model forms the core of the data administration. The data produced by the "process computing technology" were written continuously into the tables of the concerned DBMS. A permanent consistency test is connected with the data transfer. Accordingly, a consistent database is always available for any kinds of evaluations. By using the Structured Query Language (SQL), arbitrary subsets of the data base are held available for these evaluations and handed over to respective analysis programs. In the present case the statistic package SAS (Procedure Mixed) is used according to the special tasks. Equally, the results of the evaluation are entered in the data base and thus they are permanently available thereby on a long-term basis. For example, the data sets resulting from Automatic Milking System (AMS) are neither manageable on a long-term basis without the use of a DBMS nor is it possible to hold partial data sets available for specific evaluations.

This kind of data is part of the so called "operational data". Derived from the data amount as well as the necessity to provide different levels of data aggregation it is indispensable to build up so called "analytical data", stored in a Data Warehouse. For example, parts of the results of the method, described in the next section, may be content of a Data Warehouse.

Data use

An example of data utilization is the generation of predictions of individual animals, for example for the milk yield. The predicted milk yields are based on a mixed linear model with the effects lactation number (fixed), test day (fixed), animal (random) as well as 4 coefficients of regression (fixed) for modelling the lactation curve (X_1 = day of lactation/400; $X_2 = (X_1)^2$; $X_3 = \ln(400/\text{day of lactation})$; $X_4 = (X_3)^2$).

This model allows the generation of predicted milk yields for the time t+1 from the true milk yields of the animal under consideration at time t. The forecast functions, however, help identifying the yield deviations and thereupon derived individual controls. Thus, the arising decrease of individual animal control, when using AMS, can be at least partly prevented.

Furthermore, the test day effects included in the mixed model represent fluctuations caused by the specific environment of the herd under consideration. In this sense they can be used as an management assistance instrument. The practical background is the aim of an evenly good environmental design. If this is perfectly reached, test day effects are very slight.

Conclusions

The effectiveness of using an increasing amount of individual animal data is only warranted if they are used for defined meaningful decisions on the individual dairy cow. Thus, a large range of available data is not beneficial, if strategies are missing to its continuous use for a decision support. A consistent data storage is the basis for any decision-supporting model. Therefore, the utilization of data models is indispensable. The relational data model used in our investigation proved to be sufficiently flexible and efficient for operational data. The results of a mixed linear model provide the basis for the use of management assistance tools, which support the individual-referred control in particular. However, the use of these options require an integration of the data supply and model solution into the operational information system on the farm level as well as the development of Data Warehouses to storage analytical data, for example parts of the results of statistical models as described. That is a research task to be solved with priority in the future.

Acknowledgements

The authors gratefully acknowledge the cooperation and help provided by Lely Germany and the Agrargenossenschaft Polleben.

Profitability of precision technology: introduction on reconstructed dairy farms in Latvia

V. Zujs and J. Priekulis

Institute of Agricultural Engineering, Latvia University of Agriculture, LV-3001, Jelgava

Abstract

In the research different reconstruction versions for large dairy farms with 400 and more cows, but in the 70 – is and 80 – is, have been compared. It has been stated that the most economically efficient reconstruction version is transition to loose handling of cows in boxes. This version requires higher capital investments but the prognosticated milk production cost is the lowest.

Introducing precision technology the capital investments are even higher. Therefore this technology will become profitable mainly after Latvia enters the European Union when the salaries of workers and other production costs will become higher.

Key words: milk cows, technology, cowshed reconstruction, costs.

Introduction

In the 70-ties and 80-ties in Latvia 240 dairy farms were built for 400 to 600 cows. According to the number of cows they correspond to the present requirements. Nevertheless, their technology is out-dated and these farms need reconstruction.

Materials and methods

Seven cowshed reconstruction versions have been compared in the research. In the firsthand second versions tied handling of cows is maintained but technological equipment is replaced. In the third version handling of cows in combi – boxes is introduced, but in the fourth – on slope floor. In the last three versions loose handling of cows in boxes is introduced and in the sixth and seventh versions with applications of precision technology.

The reconstruction versions are compared according to the necessary capital investments, consumption of work and planned milk production cost. Two production cost levels are included in the comparison: the level existing in Latvia and the level in the European Union countries. These levels differ in salaries for workers, electroenergy and fuel costs, animal feed costs etc. The calculations are done according to the known production technology economic evolution methods developing computer programs for this reason.

Results

Milk production cost is essentially influenced by the production cost level. Introducing the EU cost level, milk production cost increases 2,1 – 2,7 times. The highest production cost is with tied handling of cows as a large number of workers is required. Therefore, considering the possibility that Latvia enters the EU it is necessary to center on loose handling of cows in boxes (reconstruction versions 5 – 7).

With the existing cost level in Latvia, version 5 is the most economically efficient. In this case the planned milk production cost is 0,124 €/kg but repayment of capital investments does not exceed 6 years. With introduction of the EU cost level the sixth reconstruction version becomes economically more efficient with additional application of precision technology.

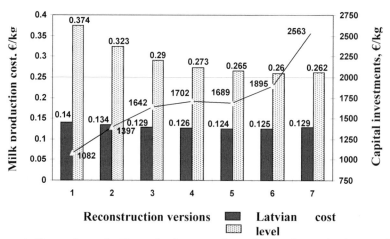

Figure 1. Comparison of milk production costs depending on the production cost level and large farm reconstruction version (including replacement of technological equipment): version 1 - tied handling of cows is maintained and machinery produced mainly in Latvia is used; version 2 - tied handling of cows is maintained but machinery produced abroad is introduced, including milking of cows with automated machines; version 3 – handling cows in combi-boxes is introduced; version 4 - handling cows on slope floor is introduced; version 5 - handling cows in boxes is introduced but comparatively simple machinery is used for mechanization of work; version 6 - handling cows in boxes and up – dated technology are introduced including application of precision technology; version 7 – in distinction to the version 6 liquid manure is collected through grated floor.

Conclusion

The most efficient large farm reconstruction version is loose handling of cows in boxes (version 5). In this case the milk production cost is the lowest and the planned capital investment repayment time does not exceed six years. Introduction of precision technology becomes economically efficient after Latvia enters the EU when salaries of workers and other production cost will increase.

References

Gartung J. 1992. Baukosten. Milchviehhaltung. BauBriefe Landwirtschaft No.33. Landwirtschaftsverlag Münster-Hiltrup,136 S.
Priekulis J. 2000. Efficient Technology and Mechanisation in Dairy Farming. Jelgava: LLU, pp. 101-107.

Saving operating time by automatic milking

Peter Zube and Jürgen Trilk
Department of consumer protection and agriculture, Institute for animal breeding and husbandry, state Brandenburg, Germany

Introduction

In the experimental farm of our institute nearly 140 dairy cows are held in a loose box with outside climate since May 1999. Two robot milkers (automatic milking system, AMS) of a single box type ("Astronaut") are used for milking. There are additional four places on a pipe milking station (PMS) for cows which are not suited for AMS. In general PMS is used for first milking after calving.

Determination of operating time in practice

There are considerable differences in published results dealing with saving operating time by use of AMS. We registered operating time by using robot milkers in a quadruple repetition. The objective of our investigation was to detect if it is possible to save working time. The results (Tab. 1) were compared with recommended standards for milking in herringbone milking parlour (HMP).

Table 1. Operating time for milking in AMS, in PMS and total.

date of registration	Parameter	AMS [1]	PMS	whole flock
November 2000	number of cows	118	15	133
	min in total	200	199	399
	min/cow and day	**1,69**	**13,27**	**3,00**
February 2001	number of cows	118	11	129
	min in total	214	84	298
	min/cow and day	**1,81**	**7,64**	**2,31**
April 2001	number of cows	100	11	111
	min in total	138	121	259
	min/cow and day	**1,38**	**11,00**	**2,33**
August 2001	number of cows	108	17	125
	min in total	173	109	282
	min/cow and day	**1,60**	**6,41**	**2,26**
$\bar{x}$	number of cows	**111** (100-118)	**14** (11-17)	**125** (111-133)
	min/cow and day	**1,62** (1,38-1,81)	**9,58** (6,41-13,27)	**2,48** (2,26-3,00)

[1] On average every cow was milked 2,7 times a day. Per day 21 times cows must be fetched to the milking box, because the not entered voluntary the robot milker. Measured operating time also includes 17 minutes per day, which were spend for service.

The working expenditure per cow and year was calculated on the basis of measured operating time per cow per day. The comparison to conventional milking systems included both twice and triple milking per day in a HMP (see figure 1).

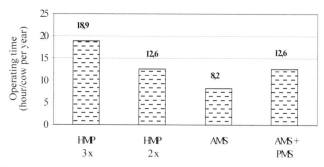

Figure 1. Operating time for milking per cow per year.

Operating time in AMS is reduced by 56% and 35% respectively compared to triple and twice milking in HMP. It is necessary to include the operating time for milking cows not suited for AMS in comparison. This cows were milked in the pipe milking station. In this case time for milking in AMS is only reduced by 33 % compared to three times milking in HMP. In comparison to two times milking operating time was not saved.

Total operating time in AMS has been split in different categories. The special shares of these categories are shown in figure 2.

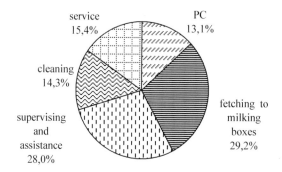

Figure 2. Share of special activities on the working time for milking by using AMS

It was necessary to fetch about 10% of cows to milking boxes, manual assistance and supervising. This seems to be a small proportion. But regarding operating time, this corresponds to 57,2 % of operating time for milking of the whole flock in AMS.

Conclusions

1. The operating time for milking can be reduced considerable with automatic milking
2. systems .
3. A high proportion of working expenditure in AMS is used for fetching cows to milking
4. boxes, supervision and manual assistance in unsuited cows.
5. Cows are not suited for milking in robot milkers should be selected if there is not conventional milking state in addition. The necessary time for assistance is so high as saved time for all others.

786

Poster abstracts for ECPLF

A dynamic model to control production processes in pig farming

H. Andree, T. Hügle and P. Leuschner

Improvements in information technology and novel sensors facilitate new approaches to control pig production processes. Several issues and alternatives such as quality of the end products, welfare of the animal, environmental impact and economic viability have to be considered in the selection of a housing/ feeding system and production management. Basically, the fattening pig can be regarded as an evolving dynamic system. Interactions with the environment force the animal to physiologically and behaviourally regulate internal parameters to maintain life-functions and growth. The regulatory capacity is limited, which, when exceeded, results in depressed performance and finally in death. An economic viable pig fattening production system has to provide the fattening pig with an environment that enforces as little regulatory effort as possible, thus to shift the utilization of nutrients and feed energy from maintenance to growth. It also has to be assured, that the animal receives a feed that precisely matches the nutrient demand in amount and nutrient content at any time, not more and not less. Accurate decision-making in pig fattening processes can only be achieved by considering simultaneously the effects of many interacting factors that influence the outcome of a production system. At present, only little available information is used to control pig production. Pig feeding is usually predetermined and follows assumed growth curves derived from the average animal age. Changes of the thermal and infectious environment, as well as individual reactions to these challenges, affect nutritional demand and feed utilization. In dependence of housing type, feeding technique and group size the pig farmer relies in his management and feeding decisions to more or less accurate visual control of the animals. He can only vaguely estimate average growth and has barely any information on utilization of the expended feed in terms of lean tissue growth. Visual based observations of the stockpersonel also lead to delayed reactions that bare a high risk not to meet the situation any more that caused the changes in animal growth and/or behaviour. Another important aspect of livestock keeping, but especially in pig fattening, is environmental pollution. Usually finisher feedstuffs are in abundance of protein. To not limit protein accretion owing to a lack of individual essential amino acids, most diets are formulated to meet the higher protein demand of the small animals at the beginning of the fattening period. In practice, only little effort is made to adapt the feedstuffs to the changing demand of the energy-protein-ratio in the course of a fattening period. This can lead to excessive nitrogen deposition into the environment, especially to high ammonia emissions. In some regions and countries this might be a limiting factor to pig production. In future, different approaches to regulate and control pig production are essential. Within the bounds of "precision livestock farming" new concepts have to be developed. Efforts have to be done to create dynamic real-time-control-systems, based on sensor data and mathematical models. Interesting parameters to monitor either by sensors or observation could be feed-intake, feed-intake-behaviour, social behaviour, water consumption, growth, protein accretion, health-status, slurry composition and changes in indoor-climate. Aim of this presentation is to point out an approach of a dynamic model to control production processes in pig fattening. Several sensors and methods to monitor the effects of changes in the nutritional, thermal and social environment already exist or have to be adapted to specific requests. Feeding techniques which anable short-term reactions to changing nutritional demands already exist - for example sensor contolled liquid feeding with BIO-FEEDER. With such a feeding system it is possible to adapt feed allotment and nutrition content with each meal. Experimental work to identify further essential parameters and interactions will start in January 2003, followed by model validation in several fattening trials.

An ammonia-controlled ventilation system for fattening pigs

Walter Grotz, Stefan Neser and Andreas Gronauer

Though German legislation has been setting limits for the concentration of certain gases, e. g. 20 ppm ammonia (NH3) in calve- and pig housings, there is still a gap between the demands on stable air condition and the situation on the farms as different studies proof. Therefore an ammonia-controlled ventilation system with NH3 as an additional control variable besides the temperature was developed at the Bavarian Research Center for Agricultural Engineering, in cooperation with ventilation companies. The study was focussed on three targets: The first was to select and test low cost ammonia sensors, the second was the design of the controlling process and the third was to estimate the environmental impact (emissions of gases and odours) compared to a temperature controlled ventilation system. Two opto-chemical, two electro-chemical and two metal-oxide sensors were selected in a first step as low cost ammonia sensors and, in addition, a photo-acoustic Infrared-Spectrometer (PAS-IR) with integrated multiplexing unit. The sensors were tested under controlled conditions in the laboratory at a gas dilution system and later, under practical conditions, in pig houses. In the laboratory tests the opto-chemical sensors and one metal oxide sensor did not meet the selection criteria. In practical operation one metal oxide sensor and one opto-chemical sensor did not stay within the limits of the permissible measuring inaccuracy (5 ppm at concentration values 0 to 20 ppm) either. At low or high concentrations, the electro-chemical sensors deviate far from the reference value, or they show non-uniform deviations from the reference value over longer periods and thus do not provide the user with reliable values. One opto-chemical sensor has been improved in a second version. However, its values fluctuate heavily over time. Especially at high NH3 concentrations, one metal oxide senor shows large deviations from the reference value. At least during the test phases, the tested PAS-IR did not prove to be very reliable because it had to be repaired quite often. Otherwise, it would be an alternative to the low-cost sensors. The ammonia-controlled ventilation system consists of a process controller which compares the command variables of the control variables temperature and NH3 with the current values measured by the specific sensors. This serves as the basis for the calculation of the manipulated variables for the actuators ventilator and heating system. To gain reliable NH3-concentrations, a PAS-IR Spectrometer was used as sensor in the controlling process. In general, the control system proved functional. For the calculation of the gaseous emissions multigas (H2O, CO2, NH3, CH4, N2O) analysis has been realised by a Fourier-Transform-Infra-Red-Spectrometer (FTIR-S). Odour concentrations were measured with a Multi-Sensor-Array (MSA). By calibration with a vane wheel anemometer it was possible to specify the air flow by associating it to the control voltage of the motors of the ventilators. Average concentrations of ammonia and methane nearly have been halved and carbon dioxide concentrations have been reduced by a third in the ammonia controlled compartment compared to the temperature controlled compartment. There was only little difference in odour concentration. However, there were only remarkable differences in the emission rates of methane, which was about 20 % higher, and nitrous oxide, which was about 50 % higher in the finishing fattening period in the temperature controlled compartment.

Artificial neural network in classification of the number of dairy cow insemination services

W. Grzesiak, R. Lacroix, P. Sablik and P. Blaszczyk

The study was aimed to determine, using an artificial neural network (ANN), the number of services which are necessary for successful insemination of a cow. The analyses covered 1054 Black-and-White (BW) cows with varied percentage of Holstein-Friesian (HF) genes (85 % on average), which were kept in 6 barns. The predictive, or input variables were the following: month of calving, lactation number (from 2 to 10, with majority of cows in lactation 2 to 6), preceding 305-d lactation milk yield (average of the studied period was 6201 kg of milk), average daily yield of two initial months of the current lactation (26.39 kg), average indices (average 305-d lactation yield of milk, fat and protein content) for the barn productivity during the studied periods, degree of calving difficulty in previous parturition normal delivery; (40.9 %), assisted delivery; (36.3 %), ;difficult, veterinarian-assisted delivery; (19.3 %), or ;very difficult delivery, abortion, or stillbirth; (3.5 %), current cow's condition (evaluated in 5-point scale; average for all cows - 2,77 point), professionalism of artificial insemination service (evaluated in 5-point scale - only 4 and 4.5 point), percentage of HF genes in the genotype. Three neurons represented the hidden layer. The output layer was a single neuron with three possible classes of the number of insemination services needed for the cow to conceive (class A meant 1 or 2 services, class B - 3 to 4 services, and class C - more than 4 services). The dataset was divided into three subsets: training (620 records), verifying (217 records), and testing (217 records). The ANN was trained with the quick propagation method though 10,000 epochs; learning rate = 0.1, acceleration = 2, and noise = 0.001. Next, the network was additionally trained by means of backpropagation. The proportion of accurate classifications was 90.99% for the whole network, 92.58% for the training set, 88.02% for the verification set, and 89.40% for the testing set. The network showed the highest accuracy while assigning the observations to class A (1.89% of erroneous classifications were found in the training set only, 1.18% in the verification set, and 0.61% in the testing set). The assignment to class B was less accurate (31.25% wrong classifications in the training set, 54.28% in the verification set, and 40.54% in the testing set), while slightly better in the case of class C (14.89% in the training set, 38.46% in the verification set, and 41.17% in the testing set). Analysis of sensitivity demonstrated that the strongest influence on the classification belong to, in turn: type of parturition, cow's condition, cow's productivity in the previous lactation, average yield of the barn where the cow had been managed, percentage of HF genes in the genotype, and the number of lactation. Basing on the studies one may conclude that an artificial neural network may constitute a good tool for initial diagnosis of the herd in relation to conception rate. The producer is informed by the network about the issues that may affect the cow's conception. Thus, reproduction performance in the analysed herds of dairy cows can be improved, which may indirectly lead to production costs reduction.

Artifical neural networ in classification of the number of dairy cow insemination services

W. Grzesiak, R. Lacroix, P. Sablik and P. Blaszczyk

The study was aimed to determine, using an artificial neural network (ANN), the number of services which are necessary for successful insemination of a cow. The analyses covered 1054 Black-and-White (BW) cows with varied percentage of Holstein-Friesian (HF) genes which were kept in 6 barns. The predictive, or input variables were the following: month of calving, lactation number, preceding 305-d lactation milk yield, average daily yield of two initial months of the current lactation, average indices (average 305-d lactation yield of milk, fat and protein content) for the barn productivity during the studied periods, degree of calving difficulty in previous parturition (normal delivery, assisted delivery, difficult, veterinarian-assisted delivery, or very difficult delivery, abortion, or stillbirth), current cows condition, professionalism of artificial insemination service, percentage of HF genes in the genotype. The output layer was a single neuron with three possible classes of the number of insemination services needed for the cow to conceive (class A meant 1 or 2 services, class B, 3 to 4 services, and class C, more than 4 services). The proportion of accurate classifications was 90.99 percent for the whole network, 92.58 percent for the training set, 88.02 percent for the verification set, and 89.40 percent for the testing set. The network showed the highest accuracy while assigning the observations to class A The assignment to class B was less accurate, while slightly better in the case of class C. Analysis of sensitivity demonstrated that the strongest influence on the classification belong to, in turn: type of parturition, cows condition, cows productivity in the previous lactation, average yield of the barn where the cow had been managed, percentage of HF genes in the genotype, and the number of lactation.

Automatic milking and heat stress

Ilan Halachmi

In a conventional farm a cow is cooled in the milking-parlor waiting yard or in the feeding lane, to which all the cows rush after milking, to consume forage feed. In a robotic milking-oriented design these options are not available, since only a few cows are waiting at the robot entrance at any moment, and they move toward the forage lane one-by-one, 24 hours a day round the clock, according to their biological needs. Therefore, the aim of the study was to design and validate robotic milking farm, maintaining a cow-cooling system with high-pressure sprinklers and mechanical ventilation. The design criteria were: up to five milkings per day; 65-75 cows in milking; free-traffic routine; no cubicle housing; more than 20 m2 floor space per cow; desired robot utilization of 85%; up to four cows waiting in front of a robot. Two feeding options were: TMR (total mixed ratio); and individual concentrate feeding with 3-6 kg per cow per day according to her individual milking frequency and milk yield. The design process involved simulation of four alternative layouts. Simulation runs were performed in the presence of the farm managers, the veterinarian and the nutritionist. Then, after choosing the conceptual layout, we optimized the facility allocation by fine-tuning the chosen layout until we reached the so-called optimal layout for the local conditions. The last design phase involved integrating the cooling system, which was evaluated in conjunction with the management practices, cow behavior and the proposed physical layout of the barn. Locating a cooling system at the robot entrance (as in a milking-parlor farm) is not effective, because it provide only 17.3 min of cooling per visit. Location at the forage lane is somewhat more effective; it provides 244.6 min of cooling per cow per day but, because of the long length of the forage lane, we have divided it into sections, each automatically activated by a photocell. An additional option, that was simulated, involved varying the area at the robot exit and installation of a one-way gate after the self-feeder, thus providing a control tool that delayed specific cows and kept them longer in that cooling area, which was eventually chosen for cooling. The layout was designed accordingly. We have validated the design at the first summer after installation (July 2002). Sixty cows were kept in the cowshed, TMR, Holstein-frizzy, 10,000–11,000 litter per cow per year, At 0600 forage feed is distributed by a tractor-driven mixer wagon, at 0600 and 1200-1300 the farmer brings in a few cows who have not visited the robot for more than 8 h. Slowdown from 0400-05:30 is a well known cow behavior phenomenon in a robotic milking barn. However there was also a problem of cows gathering in one corner, toward the wind direction, while 80% of the cowshed is vacant. Because of planned future expansion, we could not locate the robot on the windward side, but on the other side of the barn; the feeding routine was supposed to attract the cows to the robot even during the hot hours by the use of sophisticated time windows and location of the self-feeder at the robot exit did not generate the expected performances. During the hot hours, around 1100 to 1400, all the cows gather at the far corner of the cowshed, the side remote from the robot, and almost none of them reach the robot at these hours, at that time robot utilization drop down from 75% utilization to less than 30% utilization and consequently decline up to 2 litters per cow per day. The solution, a combination of a cow cooling system and concentrate feeding management improved the situation: the paper presents the performance of the 60 cows during 15 days before installing a cooling system and 15 days after installing it.

Technical and financial feasibility of electronic identification in the mandatory I&R system for cattle in the Netherlands

A.H. Ipema, A.C. Smits, P.H. Hogewerf, K. van der Walle and A.G.J. Velthuis

The requirements for an Identification and Registration system for cattle are mainly based on European and national legislation. The implementation of the I&R regulation is part of the Animal Health and Welfare regulation. The practical consequences are that identification means have to be acquired and applied and that a number of administrative duties have to be fulfilled. In an evaluation it was concluded that the existing I&R system needed revision. The new I&R system should be based on three pillars: 1) individual identification, 2) central registration and 3) use of electronic identification means. The use of electronic identifiers is in line with the conclusion of the EU project named: Large-Scale Project on Livestock Electronic Identification - The IDEA project. An important finding in the Dutch part of the IDEA project was that electronic ear tags could be applied to 100% of the animals at a young age and that the identification with electronic ear tags fits in with the present practise. The decision to use electronic ear tags in a new I&R system was the main reason for the Dutch Ministry of Agriculture and Fisheries to order for a study after the technical and financial feasibility of electronic identification. The scope of the research concerns the complete production chain from primary cattle farms, slaughterhouses, sale yards, cattle dealers and carriers. The research is divided into three phases, in which successively inventories are made from: 1) the cattle sector, 2) available technical and administrative systems, 3) models for the future technical infrastructure with their economic consequences. Phase 1 comprises an inventory of the types, sizes and numbers of primary farms with the numbers of cattle. Data about the movements of animals between different properties were analysed from the existing I&R database. This information is needed for the calculations of the economic consequences of certain future technical infrastructure models. In phase 2 the technical feasibility of the electronic ear tags as well as all other devices for receiving and handling the electronic signals from the ear tags are studied. Particular attention is paid to available readers (from hand readers till advanced panel readers), handling of animals without or with inferior identification means, reading of animals at sale yards, slaughterhouses and in combination with carriers, possibilities for automatic transfer of reports to and from the I&R database. An additional aspect is the use of electronic ID for process control and farm management purposes. In phase 3 the technical options for the use of electronic ID in the I&R system are summarised in models. These models will show which hardware (identification means, readers, communication means and other devices) and software is needed to identify an animal from birth or import, and at all transfers until slaughterhouse, export or destruction. The level of automation can vary from the use of a handreader with manual input into the I&R database till fully automated reading and data transfer into the database with an immediate confirmation. Attention is also given on possibilities to prevent and solve failures like not identified animals. The effect of the level of automation in electronic identification and registration financial requirements will be quantified for all links in the chain from farm until slaughterhouse, including transport. Finally potential profits in labour requirements for the identification and registration procedure, in investments for identification systems used for process control and farm management and an improved control of infectious diseases will be discussed.

Problems of mastitis detection in automated monitoring systems and possibilities of implementation

Paolo Liberati

The study is focused on the reliability of mastitis detection and its possible implementation using the following milk parameters measured for each quarter: electrical conductivity, opacity, flow and total production. Data measured for several months in two farms adopting automated milking systems (De Laval) have been analysed in order to verify the time and individual variation and check the correlation with the occurrence of mastitis. For what's concerning the firs aspects is has been observed that: a) the electrical conductivity is strongly related to the hour of milking, decreasing from morning to night; b) the conductivity values are also correlated to the air temperature, with a clear seasonal variation; c) the conductivity increases with the interval between milkings; d) the opacity values show a lower variability, more related to the individual differences than to external factors; e) the production values appear very related to the previous parameters, being also depending on the temperature. The automated mastitis detection revealed not very efficient, even though the number of cases found in a period of about 7 months were very few, thanks to the good hygienic conditions (in average 80000 scc/ml). The conductivity values did non present a significant increase with mastitis occurrence; on the contrary the opacity appeared more efficient for this purpose. A correlation can be found also between mastitis and production, with some delay with the beginning of the pathology. A way for implementation of efficiency in detection can be found in a more accurate consideration of the variability shown by each parameter as indicated before. In general better results can be obtained by means of model combining more parameters together in individual way. This can be done by a neuro-fuzzy procedure for data treatment which is finally presented in a preliminary version.

Accurate prediction of egg mass production and grows curves for poultry

V. Narushin and C. Takma

The contemporary egg production is highly mechanized and automated industry. The feed consumption is calculated by several factors, e.g. hen's age and weight, average daily temperature and the weight of an egg being produced. Then the data is processed by a special computer program and transferred to the feeder dosing system. There are no problems to indicate the average temperature and to register the hen's age. But there is a problem to measure daily eggs weights being laid in a poultry house. A lot of additional equipment is need which leads to increase of production costs. It much more easier to predict indirectly the daily egg mass production by the hen housed. For this purpose the predictive model for calculation the daily egg mass production is need. The objective of this study was to choose the best predictive model of the daily egg mass and body weight being produced by different poultry species during the productive period. It was proposed a model, represented as the ratio of polynomials of the 3rd and the 2nd powers, for description of growth and production curves in different poultry species, such as laying hens, ducks, ostiches. The deduced Narushin-Takma model was tested for the accuracy of the results prediction in comparison with the following models: the logistic, Gompertz, von Bertalanffy, Richards, Weibull and Morgan-Mercer-Flodin functions, as the growth models; and the Adams-Bell, logistic-curvilinear, compartmental and Lokhorst functions, as the egg production models. It was found that the proposed Narushin-Takma model was the best in description both body weight growth and egg production curves.

Portable and mobile instrument for continuos stable climate measurement

Gottfried J. Pessl and Heinrich Denzer

The stable climate has a decisive influence as you can utilize the genetic potential of your mast or breeding animals. High relative humidity, high ammonia values in the air or a cold main body of the building influence negatively the health of pigs and chickens in stables. High temperatures in stables or on the feedlot will decrease the efficiency of milk or mast cattle. μMetos zt is an instrument for continuos recording of all relevant data in a stable or on the feedlot. Measuring of: Temperature Relative Humidity Restlessness Noise Brightness Infrared main body building temperature Ammonia content of the air Easy mounting on various vehicles with immediate display of data. All operations are carried out easily and simple by a rotary switch and two push buttons. μMetos® zt informs the user through a two line display on all actual or instantaneous as well as all stored values. μMetos® zt measures every 5 minutes and stores the values for at least one full year. Wireless data transfer to the PC – μMetos® zt has a inbuilt infrared interface and communicates wireless to the Handheld or PC. Recorded data can be downloaded to any commercial Handheld`s (Pocket PC Series) without dismantling the device from the site. Interfaces to various stable climate computers are available. Software μLink - μMetos® zt comes with the comprehensive and easy to operate software platform μLink which allows all kinds of useful tables as well as graphic presentations. With one software license an unlimited number of stations can be managed. μLink is an open software platform with various data export features as well as an e-mail function which allows data to be sent automatically via Internet. Simple and fast desinfection of the device – μMetos® zt is build in a way that fast and secure disinfection of the main unit and the sensors are possible. This feature allows the user to move the unit whenever needed from one stable to another without long quarantine (delay). Temperature and relative humidity - The convection cap allows natural ventilation and gives protection to dust or other problem factors. The sensors for temperature and relative humidity have a special teflon filter cap over the sensor element and are mounted inside the convection cap. Sensors used are: The main body building temperature is measured accurately by a wireless IR (infrared) temperature sensor. Special interest for these values are to find reasons why animals tend to pollute the stable. In the firm holding of animals a too low main building body temperature can promote clinical diseases. Especially in pig holdings the critical (not damaging) temperatures should not be below 18°C. Ammonia will come predominantly from faeces excreta of the animals which is lighter than air and comes from the slurry channels and rises to the boxes of the animals. Toxic level of ammonia will create lung disease only with concentrations over 50 ppm but concentrations of higher than 10 ppm will systematically boost existing infections. The sensor used for ammonia is the VAISALA AMT100 which delivers very accurate data in the important operating range of 0-100 ppm. Noise and restlessness of animals: Noise creates stress – stressed animals are making noise. Pigs and chickens are reacting with stress and disturbances (restlessness and noise) to factors such as too little or too much feed. These two parameters are monitored with infrared motion sensor and noise is measured in a wide frequency spectrum (20Hz to 20kHz) which evaluated by the software. Thresholds are selected by the user and therefor all important information is available in a time sequence (i.e. noise during feeding hours is normal for some time and not relevant) over 24 hours and stored for many weeks for information and later analysis.

Development and testing of a practical measuring method for air velocity in the animal occupied zone of pig facilities

A.V. van Wagenberg and M.T.J. de Leeuw

Air velocity in the Animal Occupied Zone (AOZ) of a pig facility influences the thermal comfort of the pigs, and thereby affects animal production, animal health and animal welfare. The air velocity in the AOZ in livestock buildings is, similar to ventilation effectiveness, one aspect of the building performance, and is importantly determined by the ventilation system design. The paper describes the development and a practical test of a method for measuring air velocity in the AOZ and gives insight in the occurring air velocities and some other climatic factors in the AOZ of a door-ventilated room for weaned piglets. In a door- ventilated room air enters the room through an opening in the door and flows over the operator walkway. From the operator walkway it flows over the solid pen partitions into the pens. The development and test op de measuring method is divided in several separate parts. These parts are to determine: (1) the effect of a wire protection cage around the sensor (ultrasonic anemometer) on the measured air velocity; (2) the optimal aggregation interval for time-averaged air velocity; (3) the effect of interaction between animals and the sensor on the measured air velocity; (4) the effect of the location of the sensor in a pen on the measured air velocity and on the airflow direction and (5) the effect of presence of a sensor in a wire protection cage on the lying behaviour of the piglets. With the measuring system developed air velocity measurements are done in three pens during three batches in the experimental door-ventilated room. During the three batches air temperature and CO_2 concentration in the pens were also measured. The preliminary results show that the method developed is suitable for measuring air velocity in the AOZ, and thereby quantifies one important aspect of the building performance. Between pens and even within the relatively small pens (3 * 1 m) used in the research, there are differences in air velocity. Therefore measurement location needs to be well considered and chosen near the lying area of the animals. Animal lying behaviour was not importantly affected by the presence of the sensor, and animal activity did not importantly affect the measured air velocity. In the door-ventilated room air velocity in pens further from the door is in general higher than in pens closer to the door. In one batch in a pen farthest from the door the air velocity exceeded the advised maximum (0.15 m/s) in 21% of the time. In another pen closer to the door this was in 2% of the time. Furthermore, temperature measurements show that pens further from the door have lower temperatures and lower CO_2 concentrations than pens closer to the door. These differences in AOZ climatic conditions between pens within one room could result in differences in animal production, animal health and animal welfare per pen. The measuring method developed, in combination with a method to quantify ventilation effectiveness or fresh air supply to the AOZ, can in future be used to measure climatic conditions in the AOZ and evaluate the design of ventilation systems under practical conditions.

Preagro Section

Precision agriculture as an example of inter- and transdiciplinary research in agriculture: Results[3] of the German joint research project: preagro

Armin Werner[1] and Andreas Jarfe[2]

[1]Head of the joint research project preagro, Head of the Dept. of Land Use Systems and Landscape Ecology

[2]Chief coordinator of the joint research project preagro, Centre for Agricultural and Land Use Research (ZALF) in Müncheberg, Eberswalder Str. 84; D-15374 Müncheberg, Germany
AWerner@ZALF.de

Introduction

A large joint research R+D-project for relevant scientific and practical aspects in arable precision agriculture was conducted in Germany during the time from January 1999 until end of June 2003. With the project-title *"Management system for site-specific crop production to increase the economic viability of agriculture and to promote its environmental performance"* the main intention of the project is summarized.

The German Federal Ministry for Education and Research (Bundesminsterium für Bildung und Forschung = BMBF, Bonn) has supported a nation-wide interdisciplinary research and development project with the objective of developing site-differentiated measures and support tools for arable plant production based on satellite navigation or sensors (site specific farming, precision agriculture..). The final goal is a management system and to help precision agriculture to become practice-ready (Werner and Jarfe 2002). The necessary site information, the rules for defining proper cropping measures as well as the economic and ecological impacts of this technology had to be analysed by doing research on practical farms and with conventional cropping systems. Using this innovative technology, farms should achieve to work in a more economic and environmentally friendly way.

Structure of the joint R+D-project *preagro*   (see also figure 2)

Subprojects:
    22 subprojects in 5 project areas:
    Practical testing **(I)**
    Site and canopy analysis **(II)**
    Management of information processing **(III)**
    Impact analysis on economy and ecology **(IV)**
    Project-coordination **(V)**

Project partners:
    17 project partners from science, research, industry and service companies in Germany

Agriculture:
    At 8 locations in Germany with:
    16 agricultural companies
    3 agricultural subcontracting companies
    1 cooperative in farm machinery

Timeline:
    01.01.1999 - 31.12.2002 main project, until 30.06.2003 final works

Project coordination:
    Centre for Agricultural and Land Use Research (ZALF) in Müncheberg

---

[3] The organizing committee of the 4th European Precision Agriculture Conference in Berlin 2003 invited the German joint research project *preagro* to present their activities and results in a special seminar on the conference and in this booklet. The project *preagro* covers the costs for this publishing and those for the seminar rooms by funds from the German Federal Ministry of Education and Research (BMBF, Bonn).

Scientific background and problem-setting

The research and development project *preagro*: "*Management system for site-specific crop production to increase the economic viability of agriculture and to promote its environmental performance*" was established to support the development of the technology GPS- or sensor driven crop production for practical use in agriculture. Conventional aspects of precision farming played therefore a prominent role. It was intended, that for this type of site-specific crop production, the management principles and rules will be elaborated. This should be done with a targeted control of agricultural measures on the fields, based on site characteristics and plant-growth as well as on economic and ecological criteria.

As a technical aid for such oriented soil- and crop management, geographic information systems (GIS) as well as GPS satellite navigation will be used. The agricultural bus system (*LBS*, or *ISO-Bus*) will also support this type of crop production by offering the appropriate electronic interface for tractors and equipment which is necessary to control the management measures. The information basis (methods, software) necessary for an economic use of this technology as well as the necessary cultivation principles for plant production must be developed and tested in agricultural practice. Only in this way the preconditions soon will be available in order to create, with this technology, new and economic production processes for agriculture. Besides this, it will offer through innovative products (appliances, software, tools) and thus chances for the agricultural engineering industry or the agro industry in general. In addition it will allow additional functions in a new agricultural service sector. This form of management is an additional important step towards the improvement of agricultural "good technical practice" and a keystone for sustainable development in land use (Werner 2003).

Many fields (i.e. areas used for farming) have very large differences in their preconditions for plant development and growth (local site characteristics, i.e. the interacting effects of soil, relief, local climate and their neighbouring influences onto growth conditions). These differences can be related to soil or relief as well as being a result from cultivation measures (soil tillage, straw management, erosion etc.). They cause uneven plant stands and thus differences in natural yields. The actual cultivation practices of farmers can only take into account these location differences only to a limited extent. The farmer adapts his practices (tillage, sowing, fertiliser application, plant protection etc.) to an average site quality for the field). The location or plant stand differences within the field have, to date, remained largely unexplored. Therefore, where field parts with high yield potentials are concerned, the potential is not exploited. On the other hand, areas with low fertility are, for example, oversupplied with fertiliser. This is an inefficient resource use and leads to economic losses but also to potential ecological problems. In addition, long term efficiency of resource-use in production is a major goal in the sustainability track (de Wit 1992).

Subdividing the fields into uniform sub-fields according to local site differences is mostly not feasible because of farm-organisation and economic constraints (e.g. field size and form). Furthermore, the differences in site quality often appear within a few metres and are therefore small-scale. Relevant site quality differences can be found within short distances in moraine terrains, lowland areas with random waterway impacts and low mountain areas. These regions have a high share of agricultural land use in Central Europe. To these are added agricultural-related differences in site and crop stand. These are resulting from differences in the activities and quality of cropping measures. But also ditches for pipes, tilling of old ways, filling of trenches etc. cause small-scale site differences in the fields. These activities often cause effects for many years on plant stand and yield development of the crops.

Technical development has had ready for some years a satellite navigation instrument which creates important preconditions so that these differences in fields can be taken into account during tillage in a targeted way. But there are still no generally valid, transferable rules and principles according to which sowing, fertiliser and pesticide use can be adapted to site differences with proper cropping measures. In addition, each cropping measure must especially be consistent with the other agronomic activities in the set of cropping measures for one crop and the crop rotation (fig. 1). Only such consistent or integrated measures can ensure high economic efficiency and broad ecological performance (Auernhammer et al. 1995).

Sub field-specific cultivation has already the focus of some agricultural pilot-projects with mostly narrow defined contents and research aspects. Thus the lack of an integrated system which could support the application of plausible rules in crop production, as well as the lack of data on the economic viability of this technology leads to actually still low acceptance by farmers (Jürgens 2002). Also the technical equipment for applying site specific measures is often only available as a prototype.

Goals of the joint R+D-project *preagro*

The primary intention of the joint project is to design a decision support system for site specific crop management in the form of rules or algorithms for the appropriate crop management. These should support the use of site specific cropping measures in plant production as a software. The economic viability of this technology as well as the possibility for consideration of ecological aspects should be assessed. In essence the following steps are necessary in this respect:

i. Identification, description and spatial defining of the different site characteristics of small areas (sub-units), almost homogenous in their growth conditions within fields.

ii. Derivation of the specific crop growth potentials and the ecological sensitivity corresponding to the different site characteristics of such sub-units.

iii. Development of algorithms or rules for sound organisation of plant production according to the site characteristics, actual and expected plant-growth, economic and ecological criteria of such sub-units ('Management-units').

iv. Design and develop efficient and cheap management of gathering data or information and their interpretation.

v. Analyse the effects of site-specific crop management on agricultural business in regard to economic and organisational processes.

vi. Analysis of site-specific crop management in regard to the potential effects onto the environment and towards nature conservation goals.

vii. Development of practical suggestions for integration of environmental and nature conservation goals by spatially organised and site specific farming.

viii. Integration of algorithms and data processing into computer-based management systems (Software)

ix. Support in the development of practice-ready cropping equipment for site specific applications.

Basics in spatially differentiated, site specific crop production

With the new position technology and additional information systems by maps or sensors at the traveling device on the field (board-computer, control-units) the cropping measures can be selected and done according the site potentials within fields and in dependence of the farmer's decisions (Earl et al. 1996). This 'site specific crop production' includes all steps of crop production, from soil cultivation until harvest. Because this new technology allows a spatially more precise action in crop management and because some farmers who use it are more precise than before, this new technology is currently named precision farming or precision agriculture.

More and more equipment for crop production will be supplied with the necessary technology for precision agriculture (controller, GPS-receiver, flow through scales etc.). In table 1 the usefulness of spatially differentiated measures for the crop management as well as for the environment are identified. For most of these cropping practices a differentiation in measures is feasible and sensible. To differentiate the cropping measures across a site-heterogeneous field it is necessary to have principles or rules, how these measures should be differentiated in quantity and quality (Werner 1990). The necessary knowledge can be derived from the current agronomic knowledge. With this it will be possible to differentiate most of the cropping measures. With the interdisciplinary project *preagro* some of these rules and algorithms should be developed.

In the near future the current research activities will provide also complete procedures for automatic data acquisition of crop and soil variability and support the agronomic decision making (Bootlink and Verhagen 1997). New sensors will probably allow to differentiate cropping measures according to actual soil or plant conditions ´on the go´ (online-sensors). That means, the decisions are automatically done during crossing the field (Viscara Rossel and MacBratney 1997)

Table 1. Possible site specific variation of crop production measures and the relevance of such site specific measures for crop production and environmental protection (from Werner et al. 1999).

Cropping measure	Varying effects on yield formation	Relevance for	
		Crop production	Environment/ nature
Primary soil tillage	Depth	++	+
	Intensity	+++	++
Sowing	seed bed preparation	++	+
	sowing rate	+++	++
	Distribution of plants over the field area	++	0
	sowing depth	+++	0
Fertilisation			
	nutrient type	++	+++
	nutrient amount	+++	+++
Weed control	type of weed control	0	++
	Selection of herbicide and additions	++	+++
	Spraying rate	+++	+++
Application of	type of action	+	+++
Insecticides	Selection of pesticide and additions	+	+++
Fungicides	Spraying rate	+++	+++
Plant growth regulators	Application rate	+++	++
Work flow	action control (trafficability due to soil moisture etc.)	+++	+
	Production related information flows	+++	0
	Supervision of the farm	+++	+++

0 = no relevance; + = weak influence, ++ = moderate influence, +++ = strong influence

As an example for elaborating rules for site specific crop management should serve the differentiation of the sowing density in small grains. With this example also the general principles of deriving the necessary information how crops should be managed in site specific crop production are described.

The general criteria of the actual best management practices in crop production is to adopt the growth and yield formation of a crop (figure 1) onto the site specific growth potentials and actual conditions (Heyland 1991, Werner et al. 2000b). The main goal is to build up a crop stand that (i) is best suited to the site (and thus the specific growth conditions) and (ii) still has the ability to be controlled with usual cropping measures (nitrogen fertilisation, growth regulators etc.). The latter (ii) is necessary, when during crop growth unexpected situations cause changes in the desired crop development. A crucial point for this ability is the right plant and tiller density (figure 1).

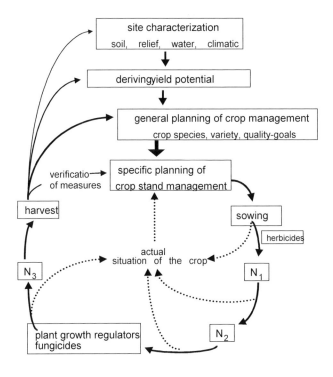

Figure 1. Crop management principles in site specific farming as applied by the joint research project preagro (example: small grain production, e.g. winter-wheat) (from: Werner et al. 2000a).

The plant and tiller density is a function of sowing density, growth conditions for the seedlings during emergence or winter and the strong regulation power of the nitrogen supply (Kochs 1986). Table 2 shows the principal way to find the correct sowing rate for a cereal crop on the different sites within a field that is heterogeneous due to varying soil and relief properties. The main differentiating forces are the expected yields of the sub-units (Werner et al. 2000b). From these estimates, the optimal ear-density as well as expected tillering rate can be derived. The respective ratio defines the necessary number of plants in spring. With estimates form the survival of plants over winter and the emergence rate, the sowing rate can be calculated consecutively. Due to differentiating influences of site (soil and position in the relief) it is necessary to adopt the losses and emergence rates according the soil conditions of the different sites within a heterogeneous field (table 2).

Table 2. Example of the calculation steps in determining sowing rate of winter barley for a heterogeneous field in the southern Uckermark (state of Brandenburg, Germany). The calculations are being made differently for the site related yield potentials (according the site quality, derived from index of land quality) and position of the sowing activity within the undulating relief of the field (from Werner et al. 1999).

Index of land quality:	< 30				30 – 39				40 – 45				> 45			
Position in the relief	normal	crest	lower	slope	normal	crest	lower	slope	normal	crest	lower	slope	normal	crest	lower	slope
Estimated yield (dt/ha)	30				45 / 50				55 / 60				65 / 70			
Tillers with ears (per m²)	350				475				575				650			
Ears / plant	1,2				1,9				2,9				2,9			
Plants in spring (per m²)	300				250				200				225			
Loss of plant in winter (%)	10	30	30	20	10	25	25	15	10	20	25	10	10	15	25	10
Plants in fall (per m²)	330	390	390	360	275	310	310	290	220	240	250	220	245	260	280	245
Seedlings (per m²)	335	400	400	370	280	315	315	295	225	245	255	225	250	265	285	250
Emergence rate (%)	95	75	90	95	90	85	90	95	90	80	90	95	90	75	90	95
Sowing rate (seeds per m²)	355	500	440	375	310	360	345	310	250	295	280	250	275	325	310	260
Sowing rate (kg/ha)[1]	182	260	230	190	159	180	170	160	130	149	150	125	139	170	160	140

[1] at 45g thousand kernel weight and 89% germination capacity

Conclusions

Due to the benefits for the farm economy, for the environment and on the long run for the region and its people, the technology of precision agriculture can bring a substantial step for land use towards a more sustainable development in land use.

The activities concerning research and development related to precision agriculture have to be done jointly with several scientific disciplines. The scientific activities in the field of precision agriculture will force scientists of different disciplines and methodological approaches to work together (Buchleiter et al. 1997). This will be a great chance for the agricultural sector to be in forefront when developing new strategies for the concept of sustainable development.

Thus we state, that an interdisciplinary, integrative development and the use of a proper and well designed technology of precision agriculture will be one of the major steps in the next decades to put agricultural land use onto the track of sustainable development.

The definition of the regional and local environmental goals (Plachter and Werner 1998) as well as the definition of the proper actions in the crop production have to be a joint and iterative process of farmers, environmentalists and conservationists. They should be accompanied by representatives for the regional development. Thus all aspects of the sustainability concept: economy, ecology, social responsibility could be handled at the same time. Solutions that do not sufficiently fit to all or one of theses criteria would lead to a recurrent review of the pre defined goals and the existing or planned frame conditions. All these steps have to be done by the relevant groups and their representatives on the regional as well as on the federal or state level. This participial and iterative approach (Werner et al. 1997) is currently the only possible way, how the concept of sustainable development for land use could be put into action.

Project design and knowledge-transfer

The scientific core of the joint R+D-project *preagro* is the development and implementation of crop management rules (algorithms, models) for site-specific management in agricultural practice as well as the software modules which will derive from this development. Therefore scientific-technical institutions as well as 16 farms, 3 contractors and 1 farm machinery cooperative are partners in the project and work closely in an integrative way. The linking brackets that will support joint activities and an interactive research are the decision support modules (like defining sowing density or amount of N-dressing for management-units). To analyse, design and test the algorithms or rules for single cropping measures, to define their necessary and available input-data, their sensitivities and integrating the agronomic fundamentals of such site specific measures is only possible when working together with different scientific disciplines. The project *preagro* was set up to develop practical solutions for precision agriculture, that are based on former experiences and research but also on own studies and experiments. The research in *preagro* was done in a constructive, intensive and integrated manner.

Figure 2. Locations of the project farms of preagro in Germany.

The site analysis, method development, the experiments and the practical applications all were set up on practical farms. In this conducting research for precision farming under practical conditions, the chance was high, that the results of the project are of high practical value and can be implemented for practical farming soon.

The locations of the project farms have been selected so that they correspond to the different climatic and geographical relationships of agricultural land in Germany. Consequently the locations of the farms range from north (Schleswig-Holstein) to south (Bavaria), from east (Saxonia-Anhalt) to west (Westphalia). The farms also represent typical farm-types and sizes.

Together with the farms, the project-related measures of the subprojects have been performed, scientifically supervised and evaluated. The results of the application attempts will be used to monitor and further develop the developed methods for site analysis and the crop management rules.

In a second step, at least three additional farms (validating farms) were integrated in the project. The first introduction of site specific farming should be monitored on these farms. These farms are, furthermore, intended to represent typical agricultural locations in Germany so as to be able to monitor the algorithms and rules developed for their transferability.

Linking the subprojects

17 institutions from the sciences, services and industry were participating to work on the problems that have been identified in the associated research projects (see box below). They were working on the scientific, technical and practical problems in a highly interlinked way concerning a total of 22 subprojects. The results of the individual subprojects are linked to each other in a logical and consistent manner. The partial aspects to be worked out were therefore divided into five project domains (see figure 3):

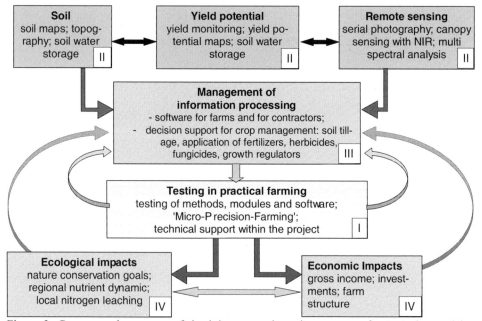

Figure 3. Content and structure of the joint research project preagro for R+D in precision agriculture.

I Testing in practical farming

The practical testing represented the general interface for the other subprojects at the data assessing and application level. This project domain thus organised the technical work on the farms and carried out experiments with the farmers or the industry.

II Site and canopy analysis

The project domain for site and canopy analysis develops methods to identify the site characteristics and their spatial distributions on the project fields. In addition, this domain supplied all the data that were necessary as local site information for field management from the different subprojects. In addition to the newly collected data, official and private data will also be made available. The subprojects also carried out methodical work scientific problems characterising the sites. Evaluations of available soil maps, relief analysis, remote sensing data from air images, model simulations, soil tests and soil conductivity measurements are also used among other methods.

III Management of information processing

The project domain ´management of information processing´ defined and developed the necessary plant growth algorithms and rules according to the plant stands and site data. Newly developed software modules were used to derive recommendations for plant growth measures in sub-units.

IV Impact analysis

This project domain studied the economic and the ecological effects caused by precision farming. Together with the other subprojects a further development of the precision agriculture technology was intended additionally orientated towards the goals of economy and ecology.

V Project coordination

The coordination project domain organised the internal project links, the information flows, the practical applications, the knowledge transfer into practical farming and extension service. In addition the external representation of the project and the public relationship was organised.

Subproject partners

The institutions and businesses listed below have been integrated in the project as scientific-technical partners. Through their own preliminary work in precision agriculture, almost all of the project partners had already been working in the area of precision agriculture as specialists. Within the project, the partners are participating in the formulation of project goals and activities on an equal basis.

The composition of the joint R+D-project was of high relevance to the agro-technical industry and practical agricultural business. The scientific or technical results and the technical development were worked out with the future users. From the project's beginning, co-operation with farmers, future marketers and commercial users of this technology was a major intention of the project.

Project partners:

1. Agri Con GmbH, Jahna
2. Agrocom GmbH + Co Agrarsystem KG, Bielefeld
3. Agro-Sat Consulting GmbH, Baasdorf
4. Federal Research Institute for Agriculture, Braunschweig
   ▶ Institute for Farming Techniques
5. German Agricultural Society (DLG) e.V., Groß-Umstadt
   ▶ Agricultural Technology Ministry, Test Center for Agricultural Machinery
6. geo-konzept GmbH, Adelschlag
7. Institute for Agricultural Engineering, Potsdam Bornim e.V. (ATB)
   ▶ Dept. for Engineering in Crop Production
8. Association for Technology and Building-Construction in Agriculture e.V., Darmstadt (KTBL)
9. Technical University of Munich (TUM), Centre of Life and Food Sciences
   ▶ Dept. of Technology in Plant Production, Section Biological Raw Materials and Technology in Land Use
   ▶ Chair of Plant Nutrition
10. University of Hohenheim, Stuttgart
    ▶ Dept. of Production Theory and Resource Economics
11. Christian-Albrechts-University of Kiel
    ▶ Institute of Plant Nutrition and Soil Science, Workgroup Soil Informatics
    ▶ Ecology-Centre of the University of Kiel
12. Philipps-University of Marburg
    ▶ Dept. of Nature Conservation
13. University of Rostock
    ▶ Institute for Geodesy and Geo-Informatics
14. Centre for Agricultural Landscape and Land Use Research (ZALF) e. V., Müncheberg
    ▶ Institute of Land Use Systems and Landscape Ecology
    ▶ Institute of Landscape Systems Analysis

Cited literature

Auernhammer, H., T. Muhr & M. Demmel (1995): GPS and DPGS as a Challenge for Environmentally Friendly Agriculture. Journal of Navigation 48(2): 268-278

Bootlink, H.W.G. and J. Verhagen (1997): Using decision support systems to optimise barley management on spatial variable soil.- In: Kropff, M.J.,Teng, P.S., Aggarwal, P.K., Bouma, J., Bouman, B.A.M., Jones, J.W., van Laar, H.H. [eds.]: Applications of System Approaches at the field level. Vol 2. Kluwer, Dordrecht, The Netherlands: 219-233.

Buchleiter, G.W., W.C. Bausch, H.R. Duke and D.F. Heerman (1997): Multidisciplinary approach for precision farming research.- In: Stafford, J.B. [ed.]: Precision Agriculture '97. Papers presented at the First European Conference on Precision Agriculture, Warwik. UK 7.-10. September 1997. BIOS Scientific Publishers Ltd.

de Wit, C.T. (1992): Ressource use efficiency in agriculture. Agricultural Systems 40: 125-151.

Earl, R., P.N. Wheeler, B.S. Blackmore and R.J. Godwin (1996): Precision farming - the Management of Variability.- Landwards 51(4): 18-23.

Jürgens C, 2002: Akzeptanz von Precision Agriculture. Ergebnisse einer repräsentativen Umfrage. Neue Landwirtschaft 6/2002, S. 35

Heyland, K.-U. (1991): Integrierte Pflanzenproduktion. System und Organisation.- 296 S., Stuttgart (Eugen Ulmer).

Kochs, H. J. (1986): Saatstärke und Düngung: Computerempfehlung über das Telefon. DLG-Mitteilungen, 101: 26-29.

Plachter, H., Werner A. (1998): Integrierende Methoden zu Leitbildern und Qualitätszielen für eine naturschonende Landwirtschaft. Zeitschrift für Kulturtechnik und Landentwicklung. Zeitschr. für Kulturtechnik und Landentwicklung, 39, 121-129

Viscara Rossel, R.A. and A.B. MacBratney (1997): Preliminary experiments towards the evaluation of a suitable soil sensor for continuous ΄on-the-go΄ field pH measurements.- In: Stafford, J.B. [ed.]: Precision Agriculture ΄97. Papers presented at the First European Conference on Precision Agriculture, Warwik. UK 7.-10. September 1997. BIOS Scientific Publish. Ltd.

Werner, A. (1990): Anbauverfahren auf heterogenen Flächen: Werkzeuge in der Entscheidungsfindung für durch Fernerkundung charakterisierte Teilstücke. Gesellschaft für Pflanzenbauwissenschaften; 34. Jahrestagung vom 27.-29. September 1990 in Bonn.

Werner, A. (2003): Precision Farming als Schlüsseltechnologie zur nachhaltigen Entwicklung der Landnutzung. In: „Bewertung von Umweltschutzleistungen in der Pflanzenproduktion". KTBL-Schrift 415. Kuratorium für Landtechnik und Bauwesen in der Landwirtschaft (KTBL), Darmstadt, p 116-134

Werner, A., Jarfe, A. (2002): Precision Agriculture - Herausforderung an integrative Forschung, Entwicklung und Anwendung in der Praxis. KTBL-Sonderveröffentlichung 038. Kuratorium für Landtechnik und Bauwesen in der Landwirtschaft, Darmstadt, 522 S., ISBN 3-9808279-0-9

Werner, A.; Müller, K.; Wenkel, K.-O.; Bork, H.-R. (1997): Partizipative und iterative Planung als Voraussetzung für die Integration ökologischer Ziele in die Landschaftsplanung des ländlichen Raumes. Zeitschrift für Kulturtechnik und Landentwicklung 38: 209-217

Werner A., Jarfe A, Roth R, Pauly J. (1999): Precision Agriculture, a new Technology in Crop Production - will it enhance Sustainable Development in Land Use? -. In: Natural and technological Problems of Protection and Development of Agricultural and Forest Environment. Rocz. AR Pozn. CCCX, Melior-Inz. Srod. 20, ISSN: 1230-7394, Band I: 327-342

Werner, A., Bachinger, J., Sodtke, R., Roth, R., Jarfe, A., Zander, P., Schuler, J. (2000a): Decision support systems in crop production on the farm and field level. In: Kukula, S., M. Fotyma, T. Gorski, M. Jurzysta, J. Kus, S. Martyniuk, J. Mazurek, H. Terelak [eds.]: Development of sustainable agriculture on the verge of the 21th century: 511-535.

Werner, A; Dölling, S., Jarfe, A., Kühn, J. , Pauly, J., Roth, R. (2000b): Deriving Maps of Yield-Potentials with Crop Models, Site Information and Remote Sensing. In: P.C. Robert, R.H. Rust, and W.E. Larson, [ed.] 2000: Proceedings of the Fourth International Conference on Precision Agriculture, July, Minneapolis, Minnesota, ASA-Press, St. Paul

Acknowledgements

The joint research project *preagro* was funded through the German Federal Ministry of Education and Research (BMBF, Bonn; ID: 0339740). The research activities of the Centre for Agricultural Landscape and Landscape Research (ZALF) are made possible through basic funding by the German Federal Ministry of consumer protection, food and agriculture (BMVEL, Bonn) as well as by the state ministry of environment and regional planning of Brandenburg (MLUR, Potsdam).

Precise site assessment by data of German soil ratings, geoelectric conductivity, terrain modelling and digital farm soil mapping[4]

extended summary of the respective subprojects of preagro, dealing with methods for site characterizations

J. Lamp, R. Herbst, G. Reimer, F. Schmidt, E.W. Reiche, U. Schmidhalter and J. Bobert

Knowledge about soil and relief heterogeneity within fields is a prerequisite for Precision Agriculture (PA) and a key factor to site specific variable rate technology (VRT) of various applications from tillage to sowing and fertilising. In water limited regions, available soil water capacity (AWC) of the rooting zone determines mainly the yield, whereas in more humid areas complex interactions of soil, weather and relief become more important. With increasing depth soil surveys require higher expenditures, but a data amelioration by all static site features will be profitable on a longer run. In preagro several subprojects aim for alternatives and a most efficient site inventory by using simplified support information either of existing or new spatial data sources.

Inventories on preagro pilot fields: On 64 fields of pilot farms (in total 1.262ha), covering main soil(land)scapes of Germany, site assessments by various approaches were performed. Initially, the old German soil rating data ("Reichsbodenschätzung") were digitised and used mainly for VRT maps of the first trial year. In addition, ECa maps by an electric conductivity sensor (EM38) and digital terrain models (DTM) either from State Geodetic Agencies (LVA) or by very precise Global Positioning Systems (RTKGPS) covered increasingly all fields. From the second trial year, an AWCoptimized version of digital Farm Soil Maps was used instead which tries to integrate all support information and soil boring data of ground truth surveys. On selected project fields special investigations were performed about quality (calibration, validation) and functional aspects (relations to plant and yield parameter) as well as on statistical measures of heterogeneity.

Geostatistical Results: Measures of local variabilities of soil and nutrient attributes are important prerequisites for assessing the profitability of PA and for planning efficient surveys. From random grid surveys on selected fields, spread statistics of sum frequency and variogram curves were evaluated for soil and available nutrient attributes. In few cases of the fields and attributes (ca. 1015%) the geostatistical measures enhanced non applicability or non profitability of PA by extrem narrow or very wide autocorrelations, respectively. Some fields show a very distinct separation of field management zones and a high attribute concordance which can be used for method testing and demonstration purposes of PA. But in most cases soil correlations tend to be mediumlow. The semivariances approach the critical interpolation sills at ca. 70m. Therefore, site mapping by interpolating data from dense raster samples (sample density > 2/ha) will often not be acceptable by farmers from an economic point of view. But which areal support information can be used instead?

Old German soil ratings: These data source, based on a standard 50m grid survey all over the farmer German territories, was made available from local fiscal offices. Rating class maps (1:2000) and soil pit field descriptions designating the polygons were digitised, georeferenced and translated into a modern nomenclature. For PA purposes, the spatial density of pits is rather low with one sampling point per 24 ha and the quality of the data assessment was not uniform. Soil texture data seem to be most reliable, though silt fractions have not been classified. But the old geology and pedogenetical diagnostics, either due to soil and land use changes or to scientific progression, often need careful checks and validations.

---

[4] Reprint from: Werner, Jarfe (2002): Precision Agriculture – Herausforderung an Integrative Forschung, Entwicklung und Anwendung in der Praxis, KTBL Sonderveröffentlichung 038, Darmstadt, pages 30 - 33

Soil electrical conductivity (ECa): Diffusely induced by the EM38 probe into soil depths of about 2m, the electrical current is mainly influenced by the soil texture or clay content as well as by the amount and the salt content of the soil moisture. A high survey performance of up to and more than 100 ha per day is achievable and EC maps can be generated with often very low costs per ha. EC maps from different instruments or sampling times reproduce relative soil texture patterns rather stable, especially for sandy soils and dry regions. But loamy to clayey soils and humid conditions may show time and location dependant moisture effects more strongly which overlay and cover the texture information. These complex interactions and an increased depth resolution need further research emphasis.

Proximal spectral sensing of soil surfaces: A hyperspectral scanner (Tec5Zeiss) and a Trimble dGPS, both mounted on the "SoilRover" and connected to a laptop data logger, were used to capture spectral signatures. These were interpolated to remission maps (RI). The spectral signatures were analysed by bi and multivariate methods to determine the optimal wave length and a best combination of spectral and soil field parameters for estimating the humus content (Corg) of topsoils, both under laboratory and field conditions. Colour images from many observation points along the tramlines of fields were also taken by a digital camera (Kodak DC290). A semiquantitative visual evaluation of the images, assisted by a picture database, helped to code and eliminate interferences by varying coverages of plants, residues and stones or by soil structures. But these factors, as mapped by interpolation, are of high interest for plant management, too.

Digital Terrain Modelling (DTM): Instead of data from the geodetic surveys, very precise elevations were measured by the Real Time Kinematic GPS-technology and used in the preagro project. By GIS interpolation techniques detailed elevation data on raster basis were generated which can be used to compute slope, exposition and relief parameters for assessing local climate conditions. Especially, near surface water flow patterns have been computed and mapped by means of the Topographic Wetness Index (TWI, a transformed ratio of the local catchment and the slope for each raster element). The results show the potential wetness at convergent zones, stream lines at slope rills or potentially dry zones at concave areas. These morphometric zones can be used with profit by several modules of management algorithms constructing application maps. A data resolution of 25m in grid size and ± 10cm in elevation is sufficient to produce high quality results of the water fluxes and moisture predictions. For this purposes the quality of data from the geodetic surveys often proved to be unsufficient. An ArcView extension is being developed which predicts by lowpassfiltering and a modified TWI more applicable soil moisture potentials. Optionally, the overlay with ECa maps will be integrated in order to allow the production of moisture maps by two sensor data sources from one tramline drive only.

Soil water storage: The project aimes to assess the variability in time and space of the water regime of plants and soils by an optimal combination of capacity and gravimetric as well as geoelectrical methods. In 2001 the calibration of the capacitive sensors the year was enforced which predict the relative soil moisture and thus the dynamics very well, but which have an offset of ca. +30% in the absolute measurement of volumetric water contents (WCv). In order to calibrate the sensors for the soil conditions of the Wulfen pilot farm, six representative soil samples (from sand to silty loam) were analyzed in the laboratory. Calibration functions of "scaled frequencies" of the sensors were established with respect to the moisture status, the texture class, mineralogy and humus contents of the soils (r^2 =0.86, without humus rich samples: r^2=0.91). The water contents were measured on two fields at 16 sites of the farm with the Diviner system and at four sites with the system EnviroScan. Soil densities were assessed from texture data acc. to Saxton (1986). The transferability was proved by applying the method on another field of Wulfen (for a depth of 0130cm r^2 is now 0.67). Texture specific calibration functions for single soil profile depthes shall increase the estimations.

Combination of terrain and moisture regime models: By example of field 641 of the Wulfen farm the concordance of TWI based moisture predictions and soil water measurements have been investigated and the transferability of the approach been studied. The results show high correlations (r^2 up to 0.81) between predictions and measurements for specific, agronomic relevant seasons and soil depthes. Seasonal dynamics have to be considered: with the dryness r^2 increases at the topsoils in spring time, too, and after progressed soil evaporation until sommer higher correlations can be expected at greater depthes, only. The areal moistures of wet or very dry soils are better predictable by texture than by relief indicators. The assessments of clay contents by geoelectrical (EC) signals could be enhanced by taking the TWI values in account, too. Therefore, the combination of both approaches is rational for improving the predictions of the moisture regimes on heterogeneous fields. High EC values correspond usually with high clay and humus contents, but occur also in depressions and waterways which show relations to high TWI values and in normal years with higher yields. Sandy to gravelic hilltops show contraversal indicator patterns. But if TWI and EC values do not correspond, a local study soil is neccessary. The transferability of the complex relationships is critical. In the rolling landscapes of young moraines of the Kassow farm, high TWI values at erosion valleys correspond with low EC signals and low humus amounts, but at lower hillfoot positions both signals have a high level.

Indicator quality, efficient boring technique and soil inventory: The geostatistical results show on one side that soil mapping by interpolating raster survey data are not acceptable by farmers due to cost reasons. On the other side, quality assessments of the areal support information like German soil ratings, geoelectrical and proximal (or remote) spectral sensing and of terrain modelling remind repeatedly not to forget the ground truth checks by direct soil borings, especially if modells have to be transferred to different soilscapes. This demand is higly supported by the summary evaluation of existing preinformations while performing on all 65 pilot fields of preagro soil inventories based on totally 1680 soil borings. The indicator quality, as assessed by criteria of data availability, transferability and prediction power for single target parameter, varies by landscapes and conditions very much. This hinders the setup of generic prediction models. Therefore, emphasis was laid on the development of an efficient soil boring technology by the "SoilRover": by a dGPS and GISaided drive along all tramlines of the fields, geoelectrical and spectral signals as well as colour images are being collected, georeferenced and interpolated afterwards. Together with other important support informations (German soil ratings and multiannual yield maps) conceptual soil maps are constructed as basis for an optimal sampling design. Soil profile data from a Concord hydraulic borer help to identify correlating map zones, delineate the borders and to setup an additional boring survey if discrepancies occur or models have to be extended.

PESWoptimized and special farm soil maps: On all pilot fields a total of 798 soil polygons were delineated and characterised in a corresponding GIS table by several soil attributes (soil and substrate type, humus and stones in topsoils). Following the main agronomic target, prediction of yield potentials, and predominantly by help of EC values the soil polygons were constructed with respect to include similar AWC values. But for practical reasons, too intricate patterns and too small units were avoided. From the second year, these management units with fixed and sharp boundaries have become the basis for all targetyield dependant application maps of the preagro - experiments. But many application algorithms which are being developed by preagro also demand special soil parameters, especially from topsoil's (eg. for VRT tillage depths, seeding and base fertiliser rates). Therefore, a set of special soil maps (esp. humus and soil texture/clay content in Ap) has to be generated, each based on the best possible prediction model and data sources. This approach will be followed also by 3Dfuzzyset methods.

Remote sensing[5]

extended summary of the respective subprojects of preagro, dealing with "Remote Sensing"

R. Bill, G. Grenzdörffer, T. Foy, T. Selige, U. Schmidhalter, B. Dohmen and A. Reh

Remote sensing data for variable rate treatments may be separated into two categories: spatial base information of soil properties or yield potential (= site (potential) map) and spatially dynamic information of the canopy development, the soil water dynamics or the quality of recent crop management decisions (= status map).

The contribution of remote sensing for "static" applications e.g. seeding, base nutrient fertilisation is to deliver reliable base information based upon the spectral reflectance such as quantitative information of the top soil, the available field capacity or the yield potential. For the long term site information the specification for the sensor platform, processing, ancillary data etc. have to meet certain standard that allow a multi temporal analysis.

The contribution of remote sensing for "dynamic" applications, e.g. fungicides, Nfertilisation, is to provide current crop status information, e.g. biomass, nitrogen content which is of great importance for several treatments. With the online approach the current crop stands or weeds are identified and analysed by a terrestrial sensor for a near realtime or realtime treatment. Airborne or satellite remote sensing may also deliver current and large scale crop status information, e.g. for Nfertilisation or for certain plant protection issues. Due to importance of the turnaround time (the time between image acquisition and delivery) and in accordance to a specific size of a management unit in relation to the treatment a variety of remote sensing sensors are suitable for different treatments, figure 1.

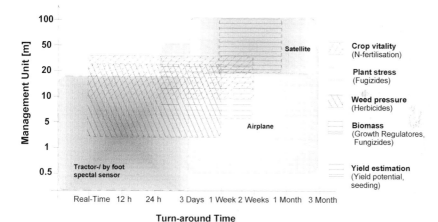

Figure 1. Suitability of different sensors for various treatments in terms of turnaround time and ground resolution (» 1/3 of management unit size).

Beside the use of remote sensing for application support remote sensing data is also a valuable source of information for site specific farm management, which is especially important for large farms to obtain a current overview. From the remote sensing data the farmer may benefit in different ways, e.g.:

---

[5] Reprint from: Werner, Jarfe (2002): Precision Agriculture – Herausforderung an Integrative Forschung, Entwicklung und Anwendung in der Praxis, KTBL Sonderveröffentlichung 038, Darmstadt, pages 93 - 96

1. Obtain a representative and current overview of the whole field / farm through a targeted addressing of certain spots of interest.
2. Receiving the site properties of the farm quickly after a change in the management personnel or the purchase of new fields
3. Objective documentation for legal issues, e.g. (wildlife damage, hail damage, street planning, environmental protection, ...)

The interpretation of such images lies in the hands of the farmers and or consultants in contrary to the application support, thus setting particular demands for a sensor system and the necessary high ground resolution of < 1 m.

The core of the image acquisition system FIFF developed by the TP II2 ‚aerial images' is a highresolution digital colour Rollei DSP 104 camera with a resolution of 2010 * 2018 pixel, which is used on board a Cessna 172. At an altitude of 2.500 m 3.000 m a ground resolution of 0.75 – 0.9 m is reached. With the digital workflow and image processing procedures it is possible to preprocess and geocode the images within a few, thus making them quickly available to the project partners. The imagery which is taken at different pheonological stages is also used as a management tool to support soil and inventory management and for near real time support of sitespecific applications.

The transformation of remote sensing information to specific treatments or for combined utilisation in a GIS requires an precise and quick geocoding of the imagery. Due to this reason a GPSAHRS system was developed and different photogrammetric procedures which require none or very few ground control points were investigated and used for automatic aerotriangulation.

With regard to crop development, remote sensing imagery may deliver information is quickly outdated. For applications such as the 2nd and 3rd Ndressings, relevant derivations must therefore be extensively automated and generated objectively. Due these reasons digital image analysis procedures such as the Visible Atmospherically Resistant Index (VARI) were tested successfully. This index relies only on the spectral bands in the visible spectrum and is linearly correlated to the vegetation fraction (= crop density). Furthermore object oriented classification strategies were also evaluated. However, for absolute quantitative measurements and time series analysis there is still demand for further research, especially in the field of preprocessing and atmospheric correction of the image data.

Beside this, the focus of the research activities has been widened significantly by investigating additional sensors in the second half of the project. On one hand satellite sensors, like Landsat TM and on the other hand terrestrial sensors like the HydroN sensor in the spectral mode and field spectroradiometer measurements. The research focused on time series analysis at different spatial and temporal scales including a comparison of different sensors. At the field Kiesberg, farm Wulfen it could be shown that an upscaling from a hand survey to terrestrial, airborne and satellite borne sensors is possible. At the field 1114 in Kassow the development of the senescence was investigated by several overpasses with the HydroN sensor in the spectral mode. The calculated vegetation indices were well correlated with the biomass and less correlated to the grain yield, due to a complex yield structure 2001. The spatial pattern of the biomass clusters was done by a supervised multi temporal classification of the NDVIdata sets.

The „crop stand information' subproject TP II4 deals with the analysis of spectral differentiation's between vegetation and soil in the near infra red (NIR) range using digital video technology. The system used consists of 2 video cameras builtin one behind the other in the ground hole of a Cessna 172 so that one camera records in NIR and the other as a colour camera senses the visible light (RGB). Flight altitude is adapted to the size of the fields and provides a ground resolution of 3 4 m at altitudes of 2.000 3.000 m above ground. The GPS signal recorded on the video's soundtrack enables the targeted search for the fields to be evaluated. The necessary geocoding of the imagery is done one field at a time by means of ground control points which are either obtained from digitally available field boundaries or from a largescale topographical map. In this way it is possible to precisely overlay imagery acquired at different times and perform change detection calculations. The formation of 16 reflection classes of a video image enhances significant soil and crop differences.

To record different development stages of the plant canopy repeated aerial surveys are necessary. With the flight campaigns of the year 2001 the existing data base for multitemporal NIRanalysis was extended. The time of senescence is among others a phenologically important flight date. Plants react to stress syndromes such as food or water shortage with premature ageing (senescence) thereby changing the cell structure. This leads to a reduced NIR reflection in these areas of the field.

For the examination of coincidences between the senescence patterns of a field over the years, data of the field Finkenherd (141) in Wulfen was analysed. Because difference images are preferable used for the comparison of two different dates, a new approach was first explored in 2001 that enables multi temporal analysis of different type of data. A variety of different data sources including soil information, yield maps and remote sensing data was incorporated into a supervised classification procedure (maximum likelihood) of the field 141. As a result management zones of different yield potential could be determined. These management zones could be used in the decision making for variable rate treatments such as seeding, fertilisation and plant protection. The best results with this approach was realised in areas in which the plantavailable water storage capacity becomes a limiting parameter factor for crop production. Areas with less than 500 mm of rainfall and a unfavourable distribution of the precipitation are considered to be "dry sites" and are often found in eastern Germany.

The use of NIR-imagery for the determination of management zones and decision support for other sites with more precipitation requires further research. The image data of 2002 will surely deliver more valuable information for this and other questions.

The "soil water storage" subproject TP II6a develops techniques for the derivation of soil and location maps from multispectral remote sensing data. The goal is to develop remote sensing supported methods for the derivation of spatially differentiated maps. The plantavailable soil water storage capacity is of mayor importance, because the it is the most important parameter for variable yields within a field.

Remote sensing technologies are not able to view into the soil profile but are recording the spectral characteristics of soilcrop surfaces. For the differentiation of soil related zones a model has to use the canopy reflection as an indicator of the subsurface situation. The development of a soilplantsensor model was thoroughly validated with the data from the year 2000. The investigations to forecast grain yield demonstrated the opportunities to determine the relative yield variations early and accurately with the multi spectral data of the Deadalus ATM scanner of the DLR. The derivation of the soil bound yield potential was validated with selected hand measurements.

In regions with a negative water balance during the vegetation period the plantavailable water storage capacity becomes the most limiting parameter of soil productivity. Beside local influences due to ground water and lateral water the crops only have access to the plant available water storage capacity of the root zone (AWCrz). This soil parameter has been identified as central factor for soil productivity. Therefore the AWCrz correlates strongly with the biomass and the crop stand condition. To determine the site specific productivity the influence of the ground water and the lateral water has to be considered. In the reaction of the crops to the local water resources this effect is already included. This allows for a more precise determination of the soil productivity than conventional soil taxation based on drill holes. Research work has shown that very good differentiation can already be achieved in the area of low stress effects by the use of the thermal emission response. Classes of 50 mm plantavailable water storage capacity could be separated significantly.

Based on the map of the plantavailable soil water storage capacity the site specific efficiency of the N-fertilisation was investigated. The results underlined the central importance of the plantavailable soil water storage capacity for the site specific efficiency of the N-fertilisation. Considering the findings the N-fertilisation should be coupled to the water availability, to increase the efficiency of the fertilisation and to guarantee an ecologically responsible use of the fertiliser. Because at a site with poor water capacity a standard fertilisation with uniform high dressings will lead to a high amount of residuals and thereby increase the risk of N-leaching.

Technology – quality[6]
extended summary of the respective suprojects of preagro dealing technology – quality

P. Noack and C. Weltzien

Introduction: The work of the subprojects TP II5 'Processing of Yield Data' and TP I1 'Technology Support and Evaluation' can be addressed under the keywords technology and quality. Assurance of working quality is the common aspect of these two subprojects.

The management system Precision Agriculture stands at the edge of broad introduction into practice. It still requires various developments for the system of PA to reach the quality standard needed for broad application. Some agricultural businesses already work with PA technology. Certainly those experiences made by the first users e.g. in the preagro project are watched closely and critically by interested potential users. Hence the assurance of the working quality especially at the pilot farms is an important step in promoting the broad introduction and acceptance of PA. In order to promote the acceptance of the management system PA it is necessary to control the working quality this means reliability, accuracy, comparability and repeatability of the applied technology as well as the data. It is necessary to assure and improve the working quality in order to describe the effects of PA in detail and more than that to prove the benefits.

In order to identify error sources and weakpoints of the technology applied in PA working quality, reliability, functionality/compatibility and handling are analysed and necessary developments of PA technology pointed out. This knowledge is made accessible to the manufacturers to speed up further developments.

The data raised through and for PA is being checked very carefully. E.g. filter tools are developed to assure the quality of the data. Furthermore it's being promoted that uniform measurands (measured quantity) and data formats are defined in order to make results from different sensors interchangeable and comparable. To achieve this considerable amounts of data are analysed.

In order to assure the working quality for the On-farm-research it is necessary to determine the reasons for e.g. the lacking success of a measure. Defects could be introduced through corrupt data or else insufficiencies of the technology, a faulty operation, a misplanning or an erroneous algorithm. A lot of attention needs to be paid to quality control during the application. The influencing parameters and their origin have to be determined one by one in order to be able to neutralise the error sources.

Processing of yield data: In the context of the subproject II5 the quality of raw yield data from different Yield mapping systems is being examined.

---

[6] Reprint from: Werner, Jarfe (2002): Precision Agriculture – Herausforderung an Integrative Forschung, Entwicklung und Anwendung in der Praxis, KTBL Sonderveröffentlichung 038, Darmstadt, pages 150 - 152

Combine harvesters present a harsh environment for online sensors. Therefore yield data and position information collected with grain flow sensors and DGPS receivers are more or less likely to be defective. Data collected with different yield monitors differ caused by different builds of sensors and varying steps of internal pre-processing. Variations show also in terms of information depth and information quality.

The need for accurate information on local yields has increased with the deployment of fertiliser spreaders and seeding machines being capable of interfacing with on board controllers and terminals enhancing variable rate application. In order to identify management zones spatial information is needed. Local yield data collected with combine harvesters is often the only information with a high spatial resolution available for site specific management.

In real life the yield data available to produce yield maps is often data from different combines. These combines are eventually equipped with different yield mapping systems. For the consistency of yield maps from data of several yield mapping systems this data has to be preprocessed and filtered. Last but not least It should be possible to compare yield data which was collected on the same field in subsequent years.

In order to understand the differences between data collected with different yield monitors yield data collected in eight different regions of Germany using three different types of yield mapping systems will be monitored and analysed by the subproject I15 over 4 years (1999 to 2002). According to the results of these analyses an approach is going to be developed which is aiming at:

1. optimising data in terms of comparability of yield maps generated from yield data collected with different yield monitors in subsequent years and
2. the generation of consistent yield maps from yield data collected with different yield monitors on one field.

Analysing yield data trackbytrack and comparing yield values from neighbouring tracks can help to detect and correct erroneous yield values before applying geostatistical methods. This will result in yield maps where low residuals in gridcell values can be maintained when decreasing gridcell sizes.

This method will allow to label the yield values of a grid cell with a quality criterion in the sense of quality assurance. This quality criterion can be used as a limiting quality level to reject corrupt data for the further use e.g. to determine management zones.

Precision Agriculture Technology at use: Technical support at the farm sites as well as evaluation of functionality and working quality focused on fertiliser spreaders and GPS receivers are the main tasks of the subproject I1.

The partner farms are equipped with a lot of different (individual) technical solutions for PA. This represents the state of the art of PA-technology. Compatibility and reliability of the systems were not fully satisfactory. The multilayer structure of PA-systems makes them versatile but operation is complex and requires very knowledgeable users. Technical support for the users is essential. Most farmers will rely on support by PA experts after the project running time. The functionality and working quality of fertiliser spreaders and GPS receivers is evaluated in cooperation with the DIAS/RCB. Results from the tests on working quality of fertiliser spreaders show that application rate variations do influence the working quality. Main influencing parameters are the working width, the physical properties of the fertiliser, the type of machinery, the setting of the spreader, and the flow rate. Exact calibration and frequent control by the user is critical for the working quality during PA applications. A conflict of interests arises where the development of spreader technology is producing ever larger working width (up to 48m) whereas PA has a need for small scale resolution as narrow as 10m x 10m.

D/GPS receivers are used for all PA applications but precision requirements may vary. As result of the evaluation classes of precision (<10m, <5m, <1m, <5cm) are defined for common types of D/GPS receivers. There haven't been a lot of evaluations of dynamic measurements of position during field operations, thus the test conditions have to be developed. The use of a differential signal is recommended for most agricultural applications. Developments in Research and Market: Current electronic development focuses on logistics, documentation and Teleservice. The standardisation process follows the direction of ISO 11873. Standardisation and harmonisation is a basic requirement for the broad introduction of PA. Further necessary developments are sensors to control for the working quality during the PA application and documentation of process data both for process optimisation and proof of quality. Possible future (political) requirements will urge even more quality control.

Yield potentials of sub-units within fields as a key input for crop management in precision agriculture[7]

extended summary of the respective subprojects of *pre agro*, dealing with "yield potential"

A. Werner, E. Kettner, J. Pauly, E. Reining, R. Roth, J. Kühn, T. Selige, J. Bobert, U. Schmidhalter and J. Hufnagel

Introduction

Spatial distributions in the conditions of plant variables or spatial differences in crop yield within fields are caused by (i) small scale site effects (soil, relief, lateral impacts, depth of water table etc.) or (ii) by anthropogenic impacts (historic land use, previous management, variance of crop management techniques, mishaps in technical measures etc.). The total result of all these impacts and effects is a heterogeneous crop with spatial differences in plant density, canopy structure, biomass, yields and qualities of the harvestable products. The *crop management technology of precision agriculture* tries to regard such differences and controls the cropping measures[8] accordingly. There are several ways, on what terms such a control of the cropping measures is planned and can be operated. These different approaches relate either the degree of the measure to (I) an average or specific previous site conditions of the sub-unit (mapping approach), to (II) actual conditions of the crop and/or that of the site of the sub-unit (online approach) or (III) both (online approach with map overlay). Especially the third possibility promises to be the main approach in the future of precision agriculture.

A process of planning the measures of precision agriculture is only possible when the average site conditions (approach I) or average site conditions plus the actual conditions of crop and/or site (approach III) are known in advance. From such information the feasible cropping measures can be derived within several crop management strategies. In most of these cases the cropping measures are related to the goals of the very crop production. The absolute levels of these goals[9] or at least their range are an input in the decision making process when selecting the quantity or quality for a cropping measure or for a set of cropping measures. Knowing the goal that should and could be reached, is a general prerequisite when planning or steering a crop stand with cropping measures. Therefore it is necessary to have an understanding of the possible development of the crop stand in its future. At least it is necessary to have an estimate of the final condition of the crop stand, the height of the yield.

---

[7] Reprint from: Werner, Jarfe (2002): Precision Agriculture – Herausforderung an Integrative Forschung, Entwicklung und Anwendung in der Praxis, KTBL Sonderveröffentlichung 038, Darmstadt, pages 197 - 200

[8] Important cropping measures are: soil tillage, seedbed preparation, sowing, Nitrogen-fertilisation (N), weed control, crop protection and fertilisation of Phosphorous (P) and Potassium (K).

[9] Yield in level and stability, concentration of specific compounds in the harvested products etc..

The yield levels or the absolute values of expectable yields are necessary inputs for crop driven decision support systems in precision agriculture. In the practical application of such decision support systems for precision agriculture, the farmers will be able to select from a set of different methods to estimate the yield characteristics of its sub-units of the field. The choice for a specific method will depend on available data necessary for the method, the reliability and robustness of the method and its support by the developer over time.

Several methods to determine the possible yield and its characteristics for a single subpart of fields are scientifically developed and compared in this book.

Terms

Yield potentials are defined as long term averages of yields for a specific site, a selected crop species and a range of similar varieties growing under a predefined type of homologous management strategies. Such yield potentials are necessary input information within crop management planning to look ahead into a still unknown future. The exact future cannot be determined in advance. But it is obvious, that the pattern of crop development, crop growth and yield formation of a (f. e. not yet sown) crop will most likely be very similar to one of those that can be estimated by experience or will be derived from available data or their appropriate inter- and extrapolations. Thus yield potentials are theoretical values for sub-units of heterogeneous fields for a general situation. The yield potentials can be absolute values as averages with their distribution, caused by the weather-variance of that site. Yield potentials can also be classified values as levels of yield height[10].

Yield expectations are defined as values or levels of yields that can be achieved with a specific crop, a specific variety, a specific history of crop management and with average or distributed weather conditions. In most cases of operative decision making for cropping measures, it is necessary to look into the future of a crop from a time-point within the crop development. From that point on the crop development and the yield are determined by the previous growth conditions, the applied cropping measures and the pattern of possible weather conditions. These can be characterized by the long-term weather data for that site. The crop stand itself has only a restricted number of possibilities for its future development. This number is higher before sowing[11] and reduces gradually until the end of the growing season. Thus yield expectations are theoretical values for sub-units of heterogeneous fields for a specific situation. The yield expectations can be absolute values as averages with their distribution, caused by the weather-variance of that site. Yield expectations can also be classified values[12].

Methods

A simple but pragmatic approach (Roth, 1995) was applied in the joint research project *preagro* to determine the yield potentials of sub-units when preparing the site specific sowing rates and nitrogen-dressing. By using a table function, empirically developed from measured yield data and expert knowledge, the site specific yield of small grains (winter wheat as a reference) is derived in dependence of the German soil classification number and annual average precipitation. The model works on most sites relevant for precision agriculture in Germany. The estimated yields correspond well with the experiences of the farmers working in the joint research project *preagro*. These site-specific yield estimations are supplied as options in the

---

[10] Yield classes: high, medium, low etc.; and adding an absolute reference value for reason of comparability.
[11] *Yield potential and yield expectation of a sub-unit can be equal at the evaluation point of pre sowing, but because yield potential is derived for a set of varieties and yield expectations can derived for a specific variety, there can be differences at that specific point.*
[12] Yield classes: high, medium, low etc.; and adding an absolute reference value for reason of comparability.

preagro modules for supporting the decision-making in the management of sub-units with precision agriculture. The farmers still can override these estimations with their own experiences. The estimated values correspond well with simulations.

In many regions of Central Europe the yield patterns within fields are caused by differences in local water supply of the very site of a sub-unit. In most cases this will be a lack of water, not fulfilling the demands of the crop for a high yield. In other, fewer cases this might be a too high water supply caused by shallow groundwater tables, collected water in depressions or water accumulation due to impermeable layers in the soil. Because of this, it is reasonable to develop methods for determining yield potentials and yield expectation of sub-units with respect to the water supply of the very sites.

With a combined approach of remote sensing and soil surveying, a regression-based method is developed that allows estimating the *site potentials* and *yield potentials*. Based on the reflection characteristics of crop canopies in the thermal and the near infrared-range of light, the water status of the crop-plant complex is determined. Relating the different values of biomass and yield to the water status of sub-parts in the field the method estimates the water holding capacity of the soil profile. From that maps of potentially plant available water of a field can be derived. These data can be a base for yield estimations with models.

When looking for very detailed aspects of *yield potentials* or *yield expectations*, then methods are necessary, which evaluate the appropriate yield values for the interactions of *species* x *variety* x *site* x *management* for all sub-parts within an arable field. A feasible method is the application of validated crop growth models. It is described, that with the use of such models (CERES-family), the yield potentials can be estimated without long-term data sampling. Only data are used that are available for farm fields on an overall base. With additional calibration experiments the data input for soil information, evaluated with the soil water balance are determined. The yield potentials are determined as averages of multi-year simulations, using long term weather records of the very site. The first results show a good fit of the estimated yields compared to those being measured on the fields, especially when delineating the observed zones in the fields with remote sensing.

Management units

When determining the yield potentials of sub-parts of fields, it is important to know, how far in distance these values are valid. The spatial range of the yield potentials or yield expectations is a crucial point for managing the sub-parts. Areas, that show homologous characteristics of yield or of crop stand variables can be delineated as management zones or management units. Methods to determine such units are described in the book.

Discussion

Yield potentials or *yield expectations* are theoretical values. It is therefore obvious, that it will not be possible to determine exact values of yield potentials by yield mapping or through correlations from soil and relief related site information.

Yield mapping will predominantly represent values of a single or of few years for a specific variety and the specific management of that year or those few. These data hardly can be reliable information for a view onto yields on a long-term base. Only in a few regions and with few fields it will be possible to determine persistent spatial yield patterns with a few years of yield mapping. Changing in varieties, crop rotations and crop management over time will restrict this. Site information does not properly consider the dynamics of a crop stand over time and the autocorrelation of certain later situations with earlier conditions in crop stand or that of the site.

Integrated management of soil and crop in precision agriculture[13]

extended summary of the respective subprojects of *pre agro*, dealing with integrated management of soil and crop

A. Werner, R. Roth, J. Kühn, H.-H. Voßhenrich, C. Sommer, K.O. Wenkel, S. Brozio, R. Gebbers, K.-H. Dammer and D. Ehlert

Introduction

Crop management of integrated farming procedures is adapted to the site specific conditions for growth and yield formation of the very crop. In addition the different measures[14] are created compatible to the other measures within the cropping system to ensure a development and yield that is most suitable to the expected results of the crop production. Such an approach is rather complex already for single, uniformly cultivated field. All measures are planned according the site conditions and linked logically to their specific impacts onto soil and crop. Those measures that aim directly onto the soil are important additions to the crop management measures, because they prepare physical conditions as important prerequisites for the crop development and yield formation.

In precision agriculture these principles for soil and crop management are applied for single sub-units in the heterogeneous fields. According to this, each step in planning and managing the crop is related to the site conditions of the sub-unit, its crop management history and linked with the other measures on that very sub-unit. To ensure the proper use of the site data and deducting feasible activities for cropping measures for each sub-unit in the fields, it is necessary to support the decision making of the farmer. The joint-research project *pre agro* develops such decision support tools as software modules for farmers and extension service. These modules aggregate the appropriate existing knowledge on crop management. The knowledge is stored in mathematical algorithms and logical rules. The algorithms and rules are developed in a general approach and are still site specific. Thus these rules can be applied on most locations in Germany that are suitable for precision agriculture. The modules are programmed in standard software.

Soil cultivation

A prototype of a soil cultivator was developed in *pre agro* that is capable of tiling the soil in different depths. The depth of cultivation can be changed during the soil cultivation according to a set of rules. These rules refer to the site conditions of sub-units, especially the soil aeration due to more or less dense and compacted soil as well as water flow in the soil. Reducing the soil tillage depth can lower the fuel consumption for the tractors up to 60%.

Sowing (module: *pre agro*-sow)

For sowing winter wheat differentially according the site specific conditions of sub-units within field, the prototype of a decision-support module was developed. It was programmed as an ArcView-extension. Sowing winter wheat differentially according site conditions was done on all *pre agro* fields in the last three years. The planned plant densities as well as the

---

[13] Reprint from: Werner, Jarfe (2002): Precision Agriculture – Herausforderung an Integrative Forschung, Entwicklung und Anwendung in der Praxis, KTBL Sonderveröffentlichung 038, Darmstadt, pages 222 - 224
[14] Important cropping measures are: soil tillage, seedbed preparation, sowing, Nitrogen-fertilisation (N), weed control, crop protection and fertilisation of Phosphorous (P) and Potassium (K)

planned ear densities were achieved on the fields in some cases rather well. Due to bad weather conditions (drought) the emergence rate was reduced and a much lower plant density was achieved in autumn of 2000. Thus, due to favorable weather conditions later on during the relevant growth period, almost similar yields were harvested in many sub-parts of some of the different fields in the project. On some locations the planned ear densities were far to low compared to the counted results in the fields in the year 2001.

N-fertilization (*pre agro*-N)

A module for N-fertilization in precision agriculture was developed that helps to make decisions in different nitrogen fertilization strategies. The module was programmed as an ArcView-extension. On-Line as well as Off-Line approaches (mapping approach) and combinations (on-line with mapping overlay) were developed. Some of the algorithms were tested with plot experiments as well as with on-farm experiments. The analyzed N-fertilization strategies resulted in similar yields as the usual N-fertilization strategy of the farmers.

P-/K-/Mg-fertilization, liming (*pre agro*-basic fertilization)

Algorithms for site-specific fertilization of P, K, Mg and for site specific liming were developed. New approaches were designed to calculate the demand and to typify the proper soil-nutrient level. The algorithms are specific for whole farms and less for single crops. These rules and balancing algorithms were programmed as prototypes.

Weed control

With a prototype of an optical weed detector the local weed density in row crops can be analyzed. According to the detected weed density the proper amount of herbicide can be adjusted on the sprayer accordingly. With such an approach the necessary herbicide amount could be reduced about 12,7 % in comparison to the standard application rate.

Crop protection

With the use of a mechanical biomass sensor (pendulum-meter) the leaf area of sub-units in heterogeneous crop stands can be determined. With this information the necessary amount of fungicide spray to cover to canopy completely can be adjusted. With this approach the amount of sprayed fungicides could be reduced about 14 % in comparison to the standard application rate. The pendulum-approach was compared with other methods, especially those from non-destructive LAI-measurements as well as from remote sensing.

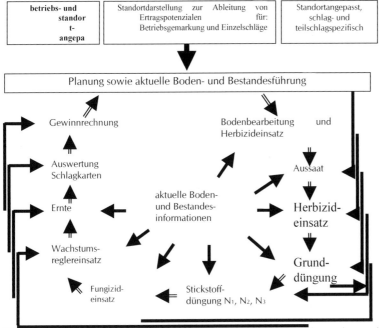

| betriebs- und standor t- angepa | Standortdarstellung zur Ableitung von Ertragspotenzialen für: Betriebsgemarkung und Einzelschläge | Standortangepasst, schlag- und teilschlagspezifisch |

Planung sowie aktuelle Boden- und Bestandesführung

Gewinnrechnung

Bodenbearbeitung und Herbizideinsatz

Auswertung Schlagkarten

Aussaat

Ernte

aktuelle Boden- und Bestandes- informationen

Herbizid- einsatz

Wachstums- reglereinsatz

Grund- düngung

Fungizid- einsatz

Stickstoff- düngung N_1, N_2, N_3

Figure 2. Illustration of the integrated management of soil and crop in precision agriculture.

Software for a precision agriculture management system[15]

extended summary of the respective subprojects of *pre agro*, dealing with "data management"

S. Böttinger, K. Oetzel, R. Schwaiberger and P. Leithold

Also in agriculture software will be the backbone of upcoming business decisions. The central role of software in a management system especially for the use in agriculture will make high demands concerning the functionality and the user interface. The corresponding tasks for the software are derived from the necessity to collect data of different qualities and to calculate with these data. Especially with geographic data the processing of the results has to be in corresponding graphics and maps. Only summaries can be outputted as text in a table.

Software has to support decision processes by enabling the planning in different variations and by the comparison of the results. There are special demands in agriculture due to the complexity of all parameters in plant production. Therefore the local knowledge of the farmer or of his advisor and also expert knowledge has to be integrated. In the preagro project the necessity for the integration of software modules via common interfaces has be derived. These modules has to be developed further on by the experts and the researchers.

The broad spectrum of the skills of the users is generating high demands towards the graphical user interface. This spectrum ranges between beginners in Precision Agriculture and with PCs, who are using such a software only sometimes, via advisors, who are working daily with such a software, and service providers, who has high demands to the automation of the

[15] Reprint from: Werner, Jarfe (2002): Precision Agriculture – Herausforderung an Integrative Forschung, Entwicklung und Anwendung in der Praxis, KTBL Sonderveröffentlichung 038, Darmstadt, page 312

data processing. The needs and the demands of all these users can not be covered with one single software solution. Therefore two different software has to be developed with regard to common expert modules and common data interfaces to cover this spectrum.

Local nitrate leaching[16]

K.C. Kersebaum and K. Lorenz

To study effects of various measures of uniform and site-specific nitrogen fertilization on crop yield and nitrate leaching model simulations were performed based on grid oriented soil sampling. At the grid points time invariable soil characteristics were measured as input for a spatially distributed simulation as well as temporal dynamic state variables like soil moisture and mineral nitrogen content for model validation. Scenarios were calculated to study the effect of different soil data bases and different fertilization strategies. A reduction of nitrate leaching or an increase of grain yield by the applied site specific fertilization could not be achieved so far. Model based optimization of N-fertilization show, that the spatial pattern as well as the quantity of the nitrogen fertilizer recommendation is dependent on the spatial resolution and accuracy of the available soil data.

Quantitative description of matter balances on the landscape level[17]

W. Windhorst, E.-W. Reiche and A. Rinker

The task of the subproject "Quantitative description of matter balances on the landscape level" is to evaluate the effects of precision agriculture concerning nutrient and water balances covering the field and the regional level too. This is modelled with a GIS-based model system, which could be validated on different spatial scales. The simulations are based on highly resolved spatial site data collected in the *pre agro*-project. The model tools are used to develop alternative land use scenarios estimating losses of matter caused by recent agricultural management (a) on the one hand and by site specific methods of precision agriculture (b). In both scenarios two optimisation strategies are distinguished, one aiming to maximise the economic yield and another aiming to avoid negative environmental effects. In order to take site specific differences into account as far as possible not only on the site level but on the regional level too, easy accessible digital landscape information and easy transformable analog information are used to assemble the data needed to run the model system. The following data sources are exploited: soil maps, digital terrain models, land use data, hydrological maps and meteorological data sets. If feasible, official data are used preferably and enriched by local information. Based on this, it is possible to elaborate an integrative site and regional evaluation covering following steps:

- regional separation of catchments in order to define the spatial area of matter balances,

---

[16] Reprint from: Werner, Jarfe (2002): Precision Agriculture – Herausforderung an Integrative Forschung, Entwicklung und Anwendung in der Praxis, KTBL Sonderveröffentlichung 038, Darmstadt, page 369

[17] Reprint from: Werner, Jarfe (2002): Precision Agriculture – Herausforderung an Integrative Forschung, Entwicklung und Anwendung in der Praxis, KTBL Sonderveröffentlichung 038, Darmstadt, page 386

- Harmonisation of all available site information in order to set up the data sets needed by the model (preprocessing),
- Simulation and estimation of matter losses caused by conventional agriculture,
- Integration of model features in order to take site specific differences of plant growth into account, which are caused by the precision agriculture (site specific plant numbers, site specific tillage depth, site specific nitrogen fertilisation),
- Calibration and optimisation of the model based on measured data about yield and soil nitrogen content,
- Simulation and estimation the effects of precision agriculture concerning regional nutrient and water balances,
- Model based development of rules to minimise matterlosses by site specific management,
- Development of strategies to allow the integration of landscape based environmental quality targets in precision agriculture.

Environmental protection targets[18]

H. Plachter and B. Janßen

The transfer of general nature conservation issues to the level of single agricultural measures is a prerequisite for the realization of these targets by means of precision agriculture. Therefore, the identification of target areas must be possible without comprehensive investigations in that specific place, and the relations between agricultural measures and their effects on targets must be known. Our investigations show with the example of amphibians that the target area identification by models of orientation and habitat preference is not yet possible with the necessary degree of certainty. The efficiancy analysis shows – with the example of weeds and carabid beetles – effects even in small spatial units with reduced pesticide use, supporting the concept of habitat patch networks inside of arable fields. Nevertheless distinct modifications of agricultural measures are necessary to support nature conservation targets in a comprehensive manner.

Economy[19]

S. Dabbert and B. Kilian

Currently no established methods for an economic assessment of site-specific field treatments exist. Thus, this paper presents two approaches to assess site-specific nitrogen fertilization on project fields.

The first approach derive site-specific yield-response functions from standardised yield-response functions to conduct economic analysis. The results of this method show an increase of gross margin due to a site-specific nitrogen fertilization of 25 €/ha on average for the

---

[18] Reprint from: Werner, Jarfe (2002): Precision Agriculture – Herausforderung an Integrative Forschung, Entwicklung und Anwendung in der Praxis, KTBL Sonderveröffentlichung 038, Darmstadt, page 399

[19] Reprint from: Werner, Jarfe (2002): Precision Agriculture – Herausforderung an Integrative Forschung, Entwicklung und Anwendung in der Praxis, KTBL Sonderveröffentlichung 038, Darmstadt, page 424

experimental fields. Furthermore it was demonstrated that the increase of gross margin is determined by yield variation and yield level.

The second approach use a plant-growth simulation model (HERMES), in cooperation with the project partner TP IV Kersebaum, to identify the optimal site-specific nitrogen application. Due to the complexity of the model only one experimental field could be analysed. The simulation results demonstrate a potential gross margin increase up to 50 €/ha for the investigated field. The results also show that the kind of information used to derive the site-specific fertilization recommendations has an impact on the economic success. The highest gross margin increase was achieved by using soil information of the "Reichsbodenschätzung", while information based on intensive soil sampling and the advanced information method of "Hof-Boden-Karten" lead to lower gross margin increases.

These potential increases of gross margin due to a site-specific fertilization were opposed to the average investment for precision farming of farms participating in "preagro". This leads to the conclusion that a profitable use of precision farming technology for a site-specific fertilization seems to be possible in farms exceeding 300 ha of cereals.

Transborder farming[20]

H. Auernhammer and M. Rothmund

The idea of transborder farming is to cultivate the area of a transborder field over the property borders as if it were one plot, thus using the declining effects of labour and costs of increasing size of plots without implementing a land consolidation. Due to the plots arisen this way it is possible also in small structured areas to use variable rate technologies in a precision farming system in a reasonable way. Property orientated, uniform and site specific cultivation are possible enterprise strategies in transborder farming. By automated collection of data by means of a GPS based determination of position during field work a property orientated assignment and evaluation of yield, input of production facilities and working times is possible. In Zeilitzheim in lower Franconia (Bavaria) transborder farming systems are being tested in practice trials.

---

[20] Reprint from: Werner, Jarfe (2002): Precision Agriculture – Herausforderung an Integrative Forschung, Entwicklung und Anwendung in der Praxis, KTBL Sonderveröffentlichung 038, Darmstadt, page 447

Soil and site inventory for application maps in precison agriculture

J. Lamp, R. Herbst and G. Reimer
Institute of Plant Nutrition and Soil Science, WG Soil Informatics, C.-A.-University Kiel, Olshausenstr.40, D24118 Kiel, Germany
jlamp@soils.uni-kiel.de

Abstract

Efficient inventories of minimum sets of soil and relief parameters are the basis for various VRT applications in the PA mapping approach. By joint projects, variogram analyses, areal pre-informations and tramline sensor surveys (German soil ratings, DTM, EM-38 sonde, remote/proximal spectral sensing) were investigated. Special parameter and integrated soil farm maps were generated on > 60 preagro pilot fields in Germany.

Keywords: precision agriculture, soil survey, variogram, electric soil conductivity, proximal spectral sensing, digital soil farm maps

Introduction and targets

Precision Agriculture based on Maps (PAM) aims to till, seed and apply agrochemicals, on each site or soil of a field specifically, with an economic and ecological optimal rate (or depth). Thus, in the annual course for all Variable Rate Technologies (VRT) - from tillage and seeding in autumn, via base fertiliser (P, K, lime) and nitrogen to pesticide applications until summer - digital maps have to be generated in time for each field which specify local rates with respect to management, weather, relief and soil conditions. Figure 1 gives an overview of different PAM applications in rainfed crop production and a minimum set of soil parameters which are needed to predict the optimal rates, locally. An analysis of algorithms for application maps which were generated and approved within the joint German PAM research project *preagro* (Werner & Jarfe, 2002) showed these soil parameters to be most important

- *in topsoils* (Ap, ~0-30cm): texture (or clay content), humus content and structure.
- *in subsoils* (~30-70 cm): texture, bulk density, colluvia (M hor.) and hydromorphy.
- *in rooting plus capillary rise zone* (soil profile ~0-150cm): texture and density dependent available water capacity (AWC) incl. capillary rise (from groundwater).

The lines in figure 1 connecting the various applications and soil parameters indicate attributes which are most important for VRT application maps, only.

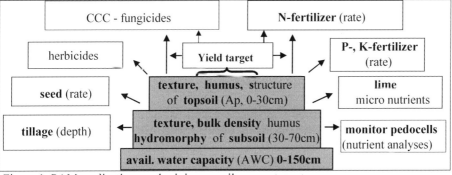

Figure 1. PAM applications and minimum soil parameter set.

Preagro acted on 11 farms spread over Germany with different management systems and climates (Werner & Jarfe, 2002). Between and within preagro fields various associations of very different soil taxa and land conditions occur (details: Herbst, 2002). Studies on site factors were performed in *preagro* directed to terrain analyses (Bill & Schmidt, 2002), ECa and water availability (Schmidthalter et al., 2002) while we had to develop an efficient soil survey technique and to apply it to all *preagro* fields (Lamp et al., 2002).

Soil heterogeneity measured by geostatistical variogram analyses

A PAM-essential question is how to survey and handle site and soil heterogeneities, field and application specifically. Good criteria to serve this question are evaluated by *median distances* MD of variograms which relate semi-variances ($\gamma(h)=1/2n\Sigma(x_i-x_{i+h})^2$) with distance classes (h) of geo-structured variables (x) at locations i. From fitted variogram curves a summary criteria MD is extracted at 50% of $\gamma(h)$ between PAM-relevant heterogeneities (maximal sill minus nugget, fig.2, inlay). The latter comprises „noise" within smallest mappable and steerable management units: *pedocells* of ca. 20*20m.

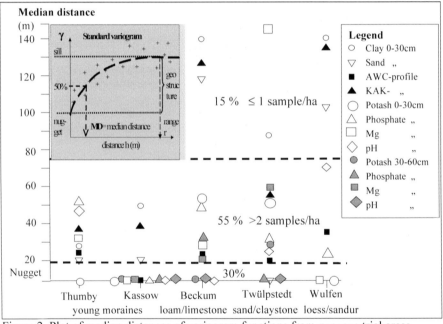

Figure 2. Plot of median distances of variogram functions from *preagro* trial areas.

Selected trial areas on five farms have been intensively surveyed by profile borings, topsoil samplings and analyses based on regular grids varying from >100m down to 25m distances (Herbst 2000, 2002). Figure 2 summarizes results as evaluated for various soil parameters, nutrient analyses and soil depths (total ~200 variograms). About 30% of MDs occured in the range of nuggets: this very high (short distanced) heterogeneity will be very difficult to map and manage by PAM. Many nutrient parameters of Kassow farm belong to this class. About 15% showed large distanced heterogeneities which can be subjected to soil analyses based on regular grids in affordable ranges ≤ 1 sample/ha. But in most cases (55%) heterogeneities are short distanced, but manageable by VRT. Cost problems of mapping will arise in this class, because often >2 samples/ha are neccessary to capture only 50% of geo-structured variances by regular grids. As a consequence, efficient and cost-acceptable survey techniques based more on areal pre-information and sensor maps than regular grid samplings should be applied.

The integrated „SoilRover" survey technique

In preagro, digitization and use of pre-information and sensor maps in form of remote sensing data, digital terrain models (Bill & Schmidt, 2002), pre-war soil ratings and electric conductivity sensing with EM-38 (Schmidthalter et al., 2002) have been investigated. These sources were checked and used by our group for their ability to predict soil conditions or parameters and thus reduce the number of soil augerings which are rather expensive, but neccessary to take (Lamp et al., 2002). The availability in time, the precision and cost of external sources showed often to be a practical and logistical problem. Therefore, an integrative soil survey approach was developed and approved by the „SoilRover" technique (Herbst & Lamp, 2002). A hydraulic auger (Concord 9300) extracting soil profile kernels quickly down to 1,5m was mounted on a Landrover (Defender 110). On this terrain vehicle, which is able to drive in the tramline spurs, an EM-38 sonde, a hyperspectral scanner (tec5, Zeiss spectrometer) and a digital camera were installed. All sensors are linked to a laptop logger with common GIS (ArcView) and data base software (MS-Access). Data are geoferenced by dGPS (Trimble). Without impacts on crops, this integrated technique - optionally to be extended by further sensors - collects by a single drive along all tramlines the most important areal predictors. Point- or cellwise measurements are at once interpolated geostatistically to produce areal maps of apparent soil conductivity (ECa), spectral remissions at selected visible or near infrared wavelengths (R_λ) or of interpreted picture elements. These sensor maps are the basis for an efficient survey plan and - in a second field drive – for representative soil borings (standard profile 1.5 m) and pedocell samplings for topsoil nutrient analyses. Usually, ≤ 1 boring or sample per ha is enough to calibrate and interprete all areal sensor data and to check existing or delineate new soil maps. Overview results can be presented here, only.

Checking existing German soil rating data („Bodenschätzung")

For taxation all agricultural land of Germany has since 1936 been and is still surveyed and rated with respect to the productivity of soils. The rating is based on intensive soil surveys with borings, first at 50m grids (mini-pits, 0-~80cm), and afterwards by „designating pits" (DP), to characterize the contents of class polygons (CP). CPs are attributed by class symbols and rating points from 0-100 (RP, Bodenzahlen) and are delineated on detailed topomaps. Increasingly, this consistant soil data source is digitized and can be used for PAM purposes, too. The conclusions about the usability are for

- *soil (layer) attributes of DP:* often features of textures are confident and relevant, of humus rather unprecise, and of hydromorphy unreliable or out of date.
- *areal extendability of DP:* as checked by geostatistical MDs (s.a.) most soil parameters can by far not be extended to the border lines of rating classes.
- *productivity rating of CP:* Often, RPs do not correlate well with relative yields on preagro and other fields (Reimer 2001, 2003). Only at well-structured fields under rain-limited conditions sufficient correlations(r~0,6-0,8) can be observed. Thus, climatic conditions must be more included in soil productivity ratings.

Calibrating and interpreting electric conductivities (ECa) of soil profiles

The Geonics EM-38 sonde induces currents into the soil until ~2m, electromagnetically. The conductivity of these electric currents depends on the clay and humus content, but also on the moisture, temperature and density of soils (Schmidhalter et al., 2002). The responding signals of apparent soil conductivity (ECa) are thus only very relative ones, especially if only the clay content averaged over the profile (0-1,5m) is the target of sensing. This parameter can be used in PAM as an important predictor for AWC and the rain-limited yield potential. Unfortunately, also high humus (organic soils) and moisture contents (in depressions) will show also high conductivities (see figure 3).

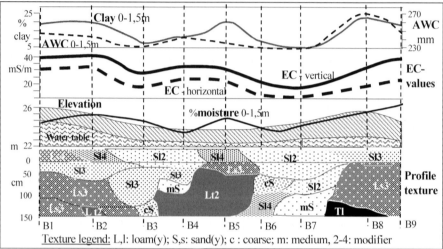

Figure 3. Transect-correlations of ECa- and soil observations at Thumby-Rothenstein.

Conclusions drawn from these studies (Herbst, 2002; Schmidhalter et al., 2002) are for:

- *practical use:* ECa mapping by EM-38 is generally a robust and efficient technique and a valuable pre-information source for precision soil surveys.
- *reproducebility and specifity:* Absolute ECs depend not on profile clay contents only, but ECs are reproduceable, relatively. They need calibration by soil augering.
- *soil depth resolution:* ECs merge conductivities over soil profiles which is unsufficient with respect to the dominating horizontation and texture layering of most soils. The direct multipol conductivity technique is more promising in this respect.

Capture, calibration and use of proximal spectral sensings of topsoils

Spectral remote sensing was investigated in *preagro* mainly as tool for crop monitoring. If cloudfree weather conditions are given, indirect indication of soil water limitations including near infrared bands is possible and profitable, but the crop stress signatures are phenologically very variable at seasons and need „ground truth" by parallel soil surveys. Direct sensing of topsoil conditions by airplane or satellite platforms is rather limited because the signatures from a main target parameter, the humus content, will be superposed by effects of the atmosphere, reflection geometry, topsoil structure, texture and moisture as well as by crop stands and residuals. These „noise" factors will also hamper the proximal sensing by the

832

hyperspectral scanner. But in contrast to remote sensing, these factors are under visual control by the surveyor and specific ground truth actions can be located and performed on the spot. Figure 4 gives a map and regression plot by example of field Thumby-Rothenstein. Topsoil humus contents were estimated with a fairly high correlation by one (of ~200) spectral band, only. Conclusions are that

- *practically,* the tec5 spectrometer is a robust scanner and very efficient, if applied in one integrated tramline survey (or with the N-sensor).
- *multivariate analysis of spectral signatures* will increase the predictability of organic matter under favorable topsoil conditions from about 50 to 90% (r^2).
- *the phenotype conditions of topsoils* are very important for success of humus predictions. Best times are from plowed/prepared seedbeds until < 25% crop coverage.
- *pattern and element analyses of colour images* taken by digital camera can aid the recognition and quantification of topsoil features (stones, structure, residues, crop).

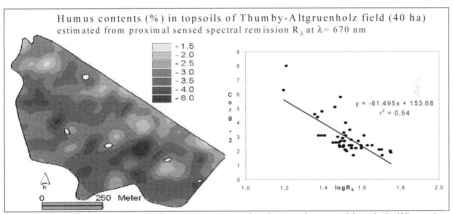

Figure 4. Humus contents predicted by one proximal sensed spectral band (SoilRover).

Yield maps as indicator for soil conditions

Yields maps have been produced on *preagro* fields for up to four years (1999-2002). They can be valuable sources for predicting soil conditions, but need careful succesive evaluation. First, technical malfunctions (a), calibration errors and local noise variation must be removed (Muhr & Noack, 2002) which may accumulate even >50% of total field variances. Second, linear patterns caused by tillage and managements along tramlines (b) are often detectable. Finally, three soil complexes, often interacting with each other and weather conditions, are

- *limited AWC features* (coarse textures, shallow soils on hard rocks), especially at elevated relief and under rain-limited conditions,
- *moisture and nutrient conditions at slopes* where an optimal water supply often will occur at mid- to footslopes (figure 5, c), and yields decrease either by water excess in depressions (d, esp. in wet years) or deficiency at hilltops (e, dry years).
- *soil structure (compaction)* limiting yields. This occurs predominately at headlands of fields (f) or at eroded hilltops with raw subsoil material inverted to topsoils.

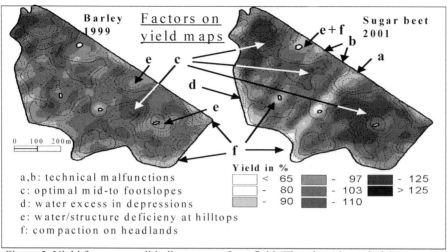

Figure 5. Yield factors as soil indicator on a farm field (Thumby-Altgrünholz).

References

Herbst, R. & J. Lamp (2000): Geostatistische Analyse für effiziente Probenahmen im Präzisen Landbau. Wasser & Boden 52 (6): 7-10.

Herbst, R., J. Lamp & G. Reimer (2001): Inventory and spatial modelling of soils on PA pilot fields in various landscapes of Germany. Grenier & Blackmore: 3rd Eur. Conf. Prec. Agric. 1: 395-400, Montpellier.

Herbst, R. (2002): Bodenschätzung, geoelektrische Sondierung und geostatistische Modellierungen als Basis digitaler Hof-Bodenkarten im Präzisen Landbau. Dissertation, Pflanzenernähr.&Bodenkunde, University, Kiel

Lamp, J., R.Herbst & G. Reimer (2001): Preciser and efficient soil surveys as basis for application maps in Precision Agriculture. Grenier & Blackmore: 3rd Eur.Conf. Prec. Agric. 1: 49-54, Montpellier

Reimer, G. & J. Lamp (2001): Ertragskarten als Indikator für Standort- und Bodeneigenschaften. Mitt. Dt. Bodenk. Ges. 96 (2): 443-444.

Reimer,G. (2003, in prep.): Spektrale Naherkundung und Ertragskarten als Basis digitaler Hof-Boden-karten im Präzisen Landbau. Dissertation, Pflanzenernähr.& Bodenkunde, University, Kiel

Werner, A. & A. Jarfe (2002, Eds.): Precision Agriculture, with site contributions of

Bill,R. & F.Schmidt: Relief, p. 65-76

Lamp,J., R. Herbst & G. Reimer: Digitale Hof-Bodenkarten, p.35-52

Muhr,T. & P. Noack: Aufbereitung von Ertragskarten, p.169-178

Schmidthalter,U.,J.Raupenstrauch &.T.Selige: Geophysikalische Erfassung von Standorteigenschaften, p. 53-63;

Tagung, März 2002, KTBL-Sonderveröff. 038, KTBL, Darmstadt

Acknowledgement

We thank the German Ministry for Education and Research (BMBF, Bonn) for generous funding.

Authors index

A

Abdoun, A.H., 757
Achten, V.T.J.M., 345; 591
Adamchuk, V., 153; 193
Adams, M., 180
Adams, M.L., 287
Adams, N., 267
Aerts, J.-M., 707
Agüera, J., 347; 601
Ahlers, D., 710
Ahmadi, S.A.A.H., 611
Akdemir, B., 612
Alakukku, L., 421; 545
Albertin, I., 383
Aldo, C., 644
Alexandre, C., 245
Algerbo, P.A., 277
Algers, B., 773
Ali, M., 184
Al-Karadsheh, E., 154
Altieri, G., 713; 753
Alvarez, R., 351
Amy, K., 216
Anaya, J.J., 685; 712
Andersen, P.G., 252; 345
Anderson, J., 209
Andree, H., 789
Antler, A., 726
Antman, A., 726
Apan, A., 226
Araujo, A.V., 623
Araújo, J.C., 503
Arcak, C., 631
Ardoin, N., 317
Artmann, R., 708; 730
Asadi, M.E., 155
Asido, S., 617
Assémat, L., 156; 353
Auernhammer, H., 77; 111; 188; 199; 273; 297; 625; 828
Autrey, L.J.C., 449

B

Bacchi, O.O.S., 585
Bach, H., 157
Backes, M., 158; 355
Badenko, V., 639
Bailey, J.S., 159; 247
Baker, S.M., 21; 565
Balasundram, S.K., 357
Balsari, P., 614
Banhazi, T., 675
Barbosa, E.P., 585
Barbottin, A., 377
Baret, F., 509
Bareth, G., 473
Barkusky, D., 172

Barroso, J., 71; 160; 189
Basso, B., 161; 162; 365
Batchelor, W.D., 163
Bauer, S., 668
Bauersachs, H., 250; 615
Baxter, S.J., 164
Beesley, B.J., 359
Bekkering, J.H., 709
Belliturk, K., 612
Benet, B., 361
Berckmans, D., 707
Berg, C., 773
Berger, A., 731
Bergerou, J., 89; 233
Berjón, A., 601
Berntsen, J., 661
Bertocco, M., 162; 553
Betteridge, K., 237
Beuche, H., 710; 761
Bilgen, H., 745
Bill, R., 815
Black, J., 675
Blackmore, S., 165; 202; 278; 612
Blanco, G., 347
Blanke, M., 479
Blaszczyk, P., 775; 791; 792
Blesse, D., 315
Blumenthal, J., 201
Bobert, J., 812; 820
Boess, J., 166
Boffety, D., 616
Boisgontier, D., 205; 283
Bollen, A.F., 286
Bollhalder, H., 721
Bonera, R., 643
Bonfil, D.J., 529; 617
Bongiovanni, R., 240
Booker, J., 266
Börjesson, T., 167; 636
Böttcher, U., 363
Böttger, H., 65; 185
Böttinger, S., 168; 825
Bourennane, H., 169
Braga, R., 365
Braga, R.P., 389
Brandhuber, R., 196
Brandt, S.H., 709
Bravo, C., 256; 367; 369
Bredemeier, C., 170; 301; 497
Brehme, U., 710; 761
Brodahl, M., 219
Broge, N.H., 171; 371; 373
Brown, P.H., 653
Brozio, S., 172; 823
Bruckmaier, R., 742
Brunnert, A., 613
Brunsch, R., 711
Bryson, R.J., 173

Kaiser, T., 194
Kalk, W.-D., 453
Karabulut, A., 631
Karlsson, T., 638
Karnieli, A., 529; 617
Karuc, K., 631
Kato, Y., 455
Kätterer, T., 638
Kaufmann, C., 695; 727
Kaufmann, O., 723; 740
Kaufmann, R., 721
Kelly, R., 175; 722
Kerry, R., 227
Kersebaum, K.C., 228; 239; 292; 305; 331; 409; 567; 826
Keska, W., 457
Ketelaars, J.J.M.H., 303; 499; 736; 737
Kettner, E., 820
Khatry, R.S., 459
Khosla, R., 141; 204; 223; 332
Kic, P., 765
Kielhorn, A., 229; 563
Kiewnick, S., 561
Kilian, B., 828
Kim, S.C., 641; 658
Kim, Y., 461
King, D., 169
Klindtworth, M., 747
Kock, B., 332
Koehler, S.D., 723
Kogan, M., 271
Köller, K., 334
Kondo, N., 513
Köppen, D., 463
Korduan, P., 231
Kornet, J.G., 499; 737
Korsaeth, A., 465
Koskinen, W.C., 209
Kovacs, L., 397; 399
Kovaříček, P., 18; 443
Kraatz, S., 198
Kramer, E., 143; 328
Kravchenko, A., 232
Krejčová, J., 415
Krieter, J., 724
Krohmann, P., 190
Kroulík, M., 415
Kuendig, H., 695
Kues, J., 166
Kühbauch, W., 225; 732; 751
Kühn, J., 290; 820; 823
Kumhála, F., 467
Kündig, H., 727
Kurtener, D., 639
Kutzbach, H.D., 304; 654
Kvien, C., 285; 321; 667

L

Lacroix, R., 775; 791; 792
Ladányi, M., 469

Lahoche, F., 89; 233
Lakota, M., 234
Lambert, D., 240
Lambot, S., 640
Lamp, J., 431; 539; 812; 829
Lamparelli, R.A.C., 471; 547
Landonio, S., 643
Langbein, J., 725; 731
Lantmännen, S., 167
Lapen, D.R., 249; 320
Larcher, J.-M., 298
Lark, R.M., 235
Larsen, R., 171
Larson, J.A., 295
Lascano, R.J., 266
Laudien, R., 473
Launay, M., 214
Laurs, A., 767
Lavialle, O., 182
Layrol, L., 89; 233
Lazzari, M., 246; 630; 644
Lee, B.-W., 515; 517
Lee, C.-K., 515; 517; 641; 658
Lee, H.J., 475
Lee, K.-H., 671
Lee, S.H., 475
Lee, W.S., 21; 565
Lehmann, B., 577; 581
Lei, T., 216
Leithold, P., 825
Lelarge, C., 377
Lellmann, A., 732
Lemaire, D., 369
Lemoine, G., 236
Lennon, M., 647
Lenthe, J.-H., 477
Lepoutre, D., 89; 233
Leuschner, P., 789
Leuthold, D., 523
Li, C., 628
Liao, Y.-T., 670
Liberati, P., 795
Liberman, J., 252
Lilienthal, H., 237; 603
Lindenthal, M., 479
Link, A., 238
Link, J., 481
Linseisen, H., 327
Linz, A., 224
Liu, G., 483
Livshin, N., 726
Ljungberg, D., 715
Loghavi, M., 611
Lokhorst, C., 259; 303; 729; 736; 737
Lorenz, K., 228; 239; 826
Lowenberg-DeBoer, J., 202; 240
Lu, J.W., 665
Lucas, P., 647
Lück, E., 302; 485
Lücke, W., 697; 739; 779

Stallinga, H., 591
Stamer, E., 724
Steffensen, F., 662
Stein, A., 333
Steiner, U., 479
Steinmayr, T., 625
Stempfhuber, W., 77; 199
Stenberg, B., 636
Stephens, P.R., 340
Steven, M.D., 312; 657
Sticksel, E., 300; 445
Stockbroeckx, B., 640
Stoll, A., 313
Stoll, Q., 599
Stollberg, U., 710; 761
Stombaugh, T., 314
Strasser, H., 668
Strauss, O., 275
Streich, W.J., 731
Strickland, R.M., 255
Strong, W., 175
Stuevel, J.-M., 577; 581
Sudduth, K., 635
Suguri, M., 505
Sui, R., 660
Sullivan, M., 200
Sung, J.H., 641; 658
Svoboda, P., 415
Swain, D.L., 657
Szabó, J., 646

T

Tabbagh, A., 183
Takenaka, H., 427
Takma, C., 796
Tamagnone, M., 614
Tan, S.C., 657
Tarkalson, D., 201
Tartachnyk, I., 583
Thiessen, E., 659
Thomas, D.L., 321
Thomas, Z., 647
Thomasson, J.A., 660
Thomsen, A., 661; 662
Thoren, P., 315
Tillett, N., 316
Timm, L.C., 585
Tisseyre, B., 275; 317
Tölle, R., 587
Topp, G.C., 320
Tornberg, J., 319, 663
Toulios, L., 489
Trautz, D., 229
Trilk, J., 785
Trouche, G., 195
Truchetet, F., 665
Tsipris, J., 251
Tsukamoto, T., 771
Turker, U., 631
Turmes, S., 589

Turner, S.B., 248
Turpin, K., 320

U

Umeda, M., 53; 505
Unrath, J., 740
Upadhyaya, S.K., 653

V

Valdhuber, J., 234
Van Buggenhout, S., 707
Van de Veen, J., 741
Van de Water, G., 715
Van de Zande, J.C., 252; 345; 591
Van den Bosch, I., 640
Van den Weghe, H.F.A., 697; 739; 779
Van der Voet, P., 236
Van der Walle, K., 794
Van Raalte, S., 319; 663
Van Wagenberg, A.V., 798
Van Zuydam, R.P., 252; 345; 591
Vanclooster, M., 640
Vangeyte, J., 322
Vellidis, G., 321
Velthuis, G.J., 794
Verboom, W., 181
Verch, G., 172
Vergil, G., 282
Vergil, M., 282
Verschoore, R., 322
Viallis, B., 616
Vieira, S.R., 664
Vigier, F., 616
Vilde, A., 593
Vioix, J.B., 665
Viscarra Rossel, R.A., 323
Völker, U., 453
Vonella, A.V., 381; 621
Voßhenrich, H.-H., 324; 823
Vrindts, E., 595
Vrsic, S., 234

W

Wade, K.M., 775
Wagner, P., 327; 666
Wahlen, S., 597
Walter, C., 323
Walters, D., 201
Wang, M., 325
Wang, N., 599
Wantia, S.J M., 729
Wartenberg, G., 65; 185
Watts, P., 316
Weber, G., 326
Wechsler, B., 695; 727
Wei, J., 599
Weigert, G., 327; 666
Weiss, D., 742
Weisskopf, P., 143; 328

Printed in the United States
by Baker & Taylor Publisher Services